T0310259

DRIED BLOOD SPOTS

WILEY SERIES ON PHARMACEUTICAL SCIENCE AND BIOTECHNOLOGY: PRACTICES, APPLICATIONS AND METHODS

Series Editors:

Mike S. Lee
Milestone Development Services

Mike S. Lee • *Integrated Strategies for Drug Discovery Using Mass Spectrometry*

Birendra Pramanik, Mike S. Lee, and Guodong Chen • *Characterization of Impurities and Degradants Using Mass Spectrometry*

Mike S. Lee and Mingshe Zhu • *Mass Spectrometry in Drug Metabolism and Disposition: Basic Principles and Applications*

Mike S. Lee (editor) • *Mass Spectrometry Handbook*

Wenkui Li and Mike S. Lee • *Dried Blood Spots—Applications and Techniques*

DRIED BLOOD SPOTS

Applications and Techniques

Edited by

WENKUI LI
Novartis Institutes for BioMedical Research
East Hanover, NJ, USA

MIKE S. LEE
Milestone Development Service
Newtown, PA, USA

Published by John Wiley & Sons, Inc., Hoboken, New Jersey.
Published simultaneously in Canada.

Library of Congress Cataloging-in-Publication Data

Dried blood spots : applications and techniques / edited by Wenkui Li, Novartis Institutes for BioMedical Research, East Hanover, NJ, USA, Mike S. Lee, Milestone Development Services.
pages cm
Includes index.
ISBN 978-1-118-05469-7 (cloth)
1. Blood–Collection and preservation. 2. Phlebotomy. 3. Forensic serology. 4. Forensic hematology. I. Li, Wenkui, 1964- editor of compilation. II. Lee, Mike S., 1960- editor of compilation.
RB45.15.D75 2014
616.07′561–dc23

2013047841

Printed in the United States of America

ISBN: 9781118054697

10 9 8 7 6 5 4 3 2 1

CONTENTS

PREFACE viii

CONTRIBUTORS x

PART I HISTORY, APPLICATIONS, AND HEALTHCARE

1 Overview of the History and Applications of Dried Blood Samples 3
W. Harry Hannon and Bradford L. Therrell, Jr.

2 Dried Blood Spot Cards 16
Brad Davin and W. Harry Hannon

3 Dried Blood Spot Sample Collection, Storage, and Transportation 21
Joanne Mei

4 Dried Blood Spot Specimens for Polymerase Chain Reaction in Molecular Diagnostics and Public Health Surveillance 32
Chunfu Yang

5 Application of Enzyme Immunoassay Methods Using Dried Blood Spot Specimens 40
Mireille B. Kalou

6 Applications of Dried Blood Spots in Newborn and Metabolic Screening 53
Donald H. Chace, Alan R. Spitzer, and Víctor R. De Jesús

7 Dried Blood Spots for Use in HIV-Related Epidemiological Studies in Resource-Limited Settings 76
Sridhar V. Basavaraju and John P. Pitman

8 **Use of Dried Blood Spot Samples in HCV-, HBV-, and Influenza-Related Epidemiological Studies** 95
Harleen Gakhar and Mark Holodniy

9 **Applications of Dried Blood Spots in General Human Health Studies** 114
Eleanor Brindle, Kathleen A. O'Connor, and Dean A. Garrett

10 **Applications of Dried Blood Spots in Environmental Population Studies** 130
Antonia M. Calafat and Kayoko Kato

11 **The Use of Dried Blood Spots and Stains in Forensic Science** 140
Donald H. Chace and Nicholas T. Lappas

PART II PHARMACEUTICAL APPLICATIONS

12 **Pharmaceutical Perspectives of Use of Dried Blood Spots** 153
Christopher Evans and Neil Spooner

13 **Punching and Extraction Techniques for Dried Blood Spot Sample Analysis** 160
Philip Wong and Christopher A. James

14 **Considerations in Development and Validation of LC-MS/MS Method for Quantitative Analysis of Small Molecules in Dried Blood Spot Samples** 168
Wenkui Li

15 **Challenges and Experiences with Dried Blood Spot Technology for Method Development and Validation** 179
Chester L. Bowen and Christopher A. Evans

16 **Clinical Implications of Dried Blood Spot Assays for Biotherapeutics** 188
Matthew E. Szapacs and Jonathan R. Kehler

17 **Potential Role for Dried Blood Spot Sampling and Bioanalysis in Preclinical Studies** 195
Qin C. Ji and Laura Patrone

18 **Clinical and Bioanalytical Evaluation of Dried Blood Spot Sampling for Genotyping and Phenotyping of Cytochrome p450 Enzymes in Healthy Volunteers** 202
Theo de Boer, Izaak den Daas, Jaap Wieling, Johan Wemer, and LingSing Chen

19 **Application of Dried Blood Spot Sampling in Clinical Pharmacology Trials and Therapeutic Drug Monitoring** 216
Kenneth Kulmatycki, Wenkui Li, Xiaoying (Lucy) Xu, and Venkateswar Jarugula

20 **Automation in Dried Blood Spot Sample Collection, Processing, and Analysis for Quantitative Bioanalysis in Pharmaceutical Industry** 229
Leimin Fan, Katty Wan, Olga Kavetskaia, and Huaiqin Wu

21 **Beyond Dried Blood Spots—Application of Dried Matrix Spots** 235
Shane R. Needham

PART III NEW TECHNOLOGIES AND EMERGING APPLICATIONS

22 Direct Analysis of Dried Blood Spot Samples 245
Paul Abu-Rabie

23 Paper Spray Ionization for Direct Analysis of Dried Blood Spots 298
Jiangjiang Liu, Nicholas E. Manicke, R. Graham Cooks, and Zheng Ouyang

24 Direct Solvent Extraction and Analysis of Biomarkers in Dried Blood Spots Using a Flow-Through Autosampler 314
David S. Millington, Haoyue Zhang, M. Arthur Moseley, J. Will Thompson, and Peter Smith

25 Development of Biomarker Assays for Clinical Diagnostics Using a Digital Microfluidics Platform 325
David S. Millington, Ramakrishna Sista, Deeksha Bali, Allen E. Eckhardt, and Vamsee Pamula

26 Applications and Chemistry of Cellulose Papers for Dried Blood Spots 332
Jacquelynn Luckwell, Åke Danielsson, Barry Johnson, Sarah Clegg, Mark Green, and Alan Pierce

27 Derivatization Techniques in Dried Blood Spot Analysis 344
Ann-Sofie M.E. Ingels, Nele Sadones, Pieter M.M. De Kesel, Willy E. Lambert, and Christophe P. Stove

INDEX 355

PREFACE

By definition, dried blood spot (DBS) sampling involves the collection and storage of a small volume of blood obtained via a simple prick (heel, finger, toe, or tail) or other means from a human or study animal. The blood sample is typically collected on a paper card made of cellulose or polymer materials. The use of filter paper for blood specimen collection can be traced back to a century ago when Bang (1913) introduced a method of absorbing blood onto filter paper for the analysis of glucose. However, it was not until 1960s when Dr. Robert Guthrie reported a method for the collection of newborn blood obtained from a heel stick onto a special filter paper for phenylalanine measurement that the use of DBS sampling started to play an important role in human life sciences.

For the past half century, the novel and low-cost approach of blood collection introduced by Dr. Robert Guthrie has led to the nationwide or worldwide population screening of inherited metabolic diseases in newborns, and the application has been expanded to many other important areas of human life, from epidemiological (e.g., HIV, HBV, HCV), general human health (e.g., DNA, RNA), environmental and forensic studies, to therapeutic drug monitoring. Lately, the technique has been heavily evaluated and/or implemented in the pharmaceutical industry, where the technique is deeply appreciated for the potential of cost-effectively facilitating the drug discovery and development process to bring new therapies to patients.

Given the exponential growth of the knowledge and experience of using DBS sampling in various studies in human healthcare, it is timely to conduct a broad overview of the principles and applications using the technique. This book is the first comprehensive book for DBS sampling and provides an update on all important aspects of the technique. The book provides a unique resource on the history and fundamentals of the technique as well as the emerging experimental procedures and applications of the technique for newborn screening and diagnosis, epidemiological studies, general human health studies, drug discovery and development, and therapeutic drug monitoring.

The book is divided into three parts. Part I provides a comprehensive overview on the history and applications of dried blood spots (Chapter 1), dried blood spot cards (Chapter 2), dried blood spot sample collection, storage, and transportation (Chapter 3). This is followed by detailed procedures of polymerase chain reaction in molecular diagnosis and public health surveillance testing using DBS (Chapter 4) and application of enzyme immunoassay methods using DBS (Chapter 5). Various applications of DBS in newborn and metabolic screening are captured in Chapter 6. The use of DBS in epidemiological testing is highlighted in Chapter 7 for HIV-related studies in resource-limited settings and Chapter 8 for HCV-, HBV-, and influenza-related studies. Chapter 9 highlights some major applications of DBS sampling in general human health studies. This is followed by some unique applications of the techniques in environmental population testing (Chapter 10) and forensic studies (Chapter 11).

Part II of the book is focused on pharmaceutical applications. Chapter 12 highlights the pharmaceutical perspectives of using DBS. This is followed by description of various punching and extraction techniques that are used for DBS sample collection and analysis in support of pharmaceutical research and development (Pharma R&D) (Chapter 13). An in-depth review on how to develop and validate a robust DBS-LC-MS/MS assay for quantitative analysis of small molecules in DBS samples is captured in Chapter 14. This is followed by various practical approaches in addressing the issues that renders DBS assay method development and validation more difficulties than the conventional aqueous samples matrices (Chapter 15). Chapter 16 presents the current clinical implications of DBS assay for biopharmaceutics. Of special interest are the applications of DBS sampling in preclinical studies (Chapter 17), genotyping and phenotyping in drug metabolism studies (Chapter 18), and clinical pharmacology trials and therapeutic drug monitoring (Chapter 19). The automation in DBS sample collection, processing, and analysis for quantitative bioanalysis in pharmaceutical industry is captured in Chapter 20. Chapter 21 concludes Part II by extending dried blood spots to other dried form of biological specimen, such as dried plasma spots and dried urine spots, and their applications.

New technologies and emerging applications are captured in Part III, which begins with a comprehensive review of various technologies for direct analysis of DBS samples (Chapter 22). This chapter is followed by a focused overview on the utility of paper spray ionization for direct analysis of DBS samples (Chapter 23). Direct solvent extraction and analysis of biomarkers in DBS samples using flow-through autosampler and digital microfluidics platform were, respectively, highlighted in Chapters 24 and 25. Applications and chemical treatment of cellulose papers for DBS is featured in Chapter 26. To conclude Part III, derivatization techniques are elucidated in Chapter 27 for enhancing DBS assay sensitivity.

Our hope is to provide human healthcare professionals involved with newborn screening, clinical diagnostics, epidemiological studies, public health studies, disease prevention, forensic studies, environmental protection, drug discovery and development, and therapeutic drug monitoring with the "important points to consider" as well as with the "procedures and technologies that are available" for collection of DBS samples for qualitative and quantitative analysis/diagnosis for the intended studies. We are delighted with the contributions from all of our distinguished authors and we are grateful for their significant contributions to the field. Indeed, the book represents a major undertaking which would not have been possible without the contributions of all the authors, and of course, the support of their families. We also wish to thank the terrific editorial staff at John Wiley & Sons and give a special acknowledgment to Michael Leventhal, Managing Editor, and Robert Esposito, Associate Publisher, at John Wiley & Sons, for their premier support of this project.

Wenkui Li, PhD
Mike S. Lee, PhD

CONTRIBUTORS

Paul Abu-Rabie, Bioanalytical Sciences and Toxicokinetics (BST), Drug Metabolism and Pharmacokinetics (DMPK), Platform Technology and Science (PTS), GlaxoSmithKline, Ware, UK.

Deeksha Bali, Duke Biochemical Genetics Laboratory, Duke University Health System, Durham, NC, USA.

Sridhar V. Basavaraju, HIV Prevention Branch, Division of Global HIV/AIDS, Center for Global Health, Centers for Disease Control and Prevention, Atlanta, GA, USA.

Chester L. Bowen, Bioanalytical Science and Toxicokientics (BST), Platform Technology and Science (PTS), Drug Metabolism and Pharmacokinetics (DMPK), GlaxoSmithKline, King of Prussia, PA, USA.

Eleanor Brindle, Center for Studies in Demography and Ecology, University of Washington, Seattle, WA, USA.

Antonia M. Calafat, Division of Laboratory Sciences, National Center for Environmental Health, Centers for Disease Control and Prevention, Atlanta, GA, USA.

Donald H. Chace, The Pediatrix Analytical Center for Research, Education and Quality, MEDNAX Services/Pediatrix Medical Group/American Anesthesiology, Sunrise, FL, USA.

LingSing Chen, Translational Medicine, QPS Delaware, Newark, DE, USA.

Sarah Clegg, GE Healthcare Life Sciences, Cardiff, UK.

R. Graham Cooks, Department of Chemistry, Purdue University, West Lafayette, IN, USA.

Åke Danielsson, GE Healthcare Life Sciences, Uppsala, Sweden.

Brad Davin, Emeritus (Retired President), ID Biological Systems Inc., Greenville, SC, USA.

Theo de Boer, Analytical Bioechemical Laboratory, Assen, The Netherlands.

Víctor R. De Jesús, Newborn Screening and Molecular Biology Branch, Centers for Disease Control and Prevention, Atlanta, GA, USA.

Pieter M.M. De Kesel, Laboratory of Toxicology, Faculty of Pharmaceutical Sciences, Ghent University, Ghent, Belgium.

Izaak den Daas, QPS Netherlands BV, Groningen, The Netherlands.

Allen E. Eckhardt, Advanced Liquid Logic, Morrisville, NC, USA.

Christopher Evans, Platform Technology & Science, DMPK Bioanalytical Science & Toxicokinetics (BST), GlaxoSmithKline Research and Development, King of Prussia, PA, USA.

Leimin Fan, Department of Drug Analysis, AbbVie Inc., North Chicago, IL, USA.

Harleen Gakhar, Division of Infectious Diseases and Geographic Medicine, Stanford University School of Medicine, Stanford, CA, USA; VA Palo Alto Health Care System, Palo Alto, CA, USA.

Dean A. Garrett, ICF International, Rockville, MD, USA.

Mark Green, GE Healthcare Life Sciences, Cardiff, UK.

W. Harry Hannon, Emeritus (Retired Chief), Newborn Screening Branch, Centers for Disease Control and Prevention, Buford, GA, USA.

Mark Holodniy, Division of Infectious Diseases and Geographic Medicine, Stanford University School of Medicine, Stanford, CA, USA; VA Palo Alto Health Care System, Palo Alto, CA, USA.

Ann-Sofie M.E. Ingels, Laboratory of Toxicology, Faculty of Pharmaceutical Sciences, Ghent University, Ghent, Belgium.

Christopher A. James, Bioanalytical Sciences, Pharmacokinetics & Drug Metabolism, Amgen Inc., Thousand Oaks, CA, USA.

Venkateswar Jarugula, Drug Metabolism & Pharmacokinetics, Novartis Institutes for BioMedical Research, East Hanover, NJ, USA.

Qin C. Ji, Bioanalytical Sciences, Bristol-Myers Squibb Co., Princeton, NJ, USA.

Barry Johnson, GE Healthcare Life Sciences, Cardiff, UK.

Mireille B. Kalou, International Laboratory Branch, Division of Global HIV/AIDS, Center for Global Health, Centers for Disease Control and Prevention, Atlanta, GA, USA.

Kayoko Kato, Division of Laboratory Sciences, National Center for Environmental Health, Centers for Disease Control and Prevention, Atlanta, GA, USA.

Olga Kavetskaia, Department of Drug Analysis, AbbVie Inc., North Chicago, IL, USA.

Jonathan R. Kehler, Bioanalytical Science and Toxicokinetics (BST), Drug Metabolism and Pharmacokinetics (DMPK), Platform Technology and Science (PTS), GlaxoSmithKline, King of Prussia, PA, USA.

Kenneth Kulmatycki, Drug Metabolism & Pharmacokinetics, Novartis Institutes for BioMedical Research, Cambridge, MA, USA.

Willy E. Lambert, Laboratory of Toxicology, Faculty of Pharmaceutical Sciences, Ghent University, Ghent, Belgium.

Nicholas T. Lappas, Department of Forensic Sciences, George Washington University, Washington, DC, USA.

Wenkui Li, Drug Metabolism & Pharmacokinetics, Novartis Institutes for BioMedical Research, East Hanover, NJ, USA.

Jiangjiang Liu, Weldon School of Biomedical Engineering, Purdue University, West Lafayette, IN, USA.

Jacquelynn Luckwell, GE Healthcare Life Sciences, Cardiff, UK.

Nicholas E. Manicke, Department of Chemistry, Purdue University, West Lafayette, IN, USA.

Joanne Mei, Newborn Screening Quality Assurance Program, Newborn Screening and Molecular Biology Branch, Centers for Disease Control and Prevention, Atlanta, GA, USA.

David S. Millington, Division of Medical Genetics, Department of Pediatrics, Duke University Health System, Durham, NC, USA.

M. Arthur Moseley, Proteomics Core Facility, Duke University School of Medicine, Durham, NC, USA.

Shane R. Needham, Alturas Analytics Inc., Moscow, ID, USA.

Kathleen A. O'Connor, Center for Studies in Demography and Ecology, Department of Anthropology, University of Washington, Seattle, WA, USA.

Zheng Ouyang, Weldon School of Biomedical Engineering, Purdue University, West Lafayette, IN, USA.

Vamsee Pamula, Advanced Liquid Logic Inc., Morrisville, NC, USA.

Laura Patrone, Drug Safety Evaluation, Bristol-Myers Squibb Co., New Brunswick, NJ, USA.

Alan Pierce, GE Healthcare Life Sciences, Cardiff, UK.

John P. Pitman, HIV Prevention Branch, Division of Global HIV/AIDS, Center for Global Health, Centers for Disease Control and Prevention, Atlanta, GA, USA.

Nele Sadones, Laboratory of Toxicology, Ghent University, Faculty of Pharmaceutical Sciences, Ghent, Belgium.

Ramakrishna Sista, Advanced Liquid Logic Inc., Morrisville, NC, USA.

Peter Smith, LEAP Technologies Inc., Carrboro, NC, USA.

Alan R. Spitzer, The Center for Research, Education and Quality, MEDNAX Services/Pediatrix Medical Group/American Anesthesiology, Sunrise, FL, USA.

Neil Spooner, Platform Technologies & Science, Drug Metabolism & Pharmacokinetics, GlaxoSmithKline Research & Development, Ware, Hertfordshire, UK.

Christophe P. Stove, Laboratory of Toxicology, Faculty of Pharmaceutical Sciences, Ghent University, Ghent, Belgium.

Matthew E. Szapacs, Bioanalytical Science and Toxicokinetics (BST), Drug Metabolism and Pharmacokinetics (DMPK), Platform Technology and Science (PTS), GlaxoSmithKline, King of Prussia, PA, USA.

Bradford L. Therrell Jr., National Newborn Screening and Genetics Resource Center, Austin, TX, USA; Department of Pediatrics, University of Texas Health Science Center at San Antonio, San Antonio, TX, USA.

J. Will Thompson, Proteomics Core Facility, Duke University School of Medicine, Durham, NC, USA.

Katty Wan, Department of Drug Analysis, AbbVie Inc., North Chicago, IL, USA.

Johan Wemer, QPS Netherlands BV, Groningen, The Netherlands.

Jaap Wieling, QPS Netherlands BV, Groningen, The Netherland.

Philip Wong, Bioanalytical Sciences, Pharmacokinetics & Drug Metabolism, Amgen Inc., Thousand Oaks, CA, USA.

Huaiqin Wu, Department of Drug Analysis, AbbVie Inc., North Chicago, IL, USA.

Xiaoying (Lucy) Xu, Drug Metabolism & Pharmacokinetics, Novartis Institutes for BioMedical Research, East Hanover, NJ, USA.

Chunfu Yang, HIV Drug Resistance and Molecular Bioinformatics, International Laboratory Branch, Division of Global HIV/AIDS, CGH, Centers for Disease Control and Prevention, Atlanta, GA, USA.

Haoyue Zhang, Duke Biochemical Genetics Laboratory, Duke University Health System, Durham, NC, USA.

PART I

HISTORY, APPLICATIONS, AND HEALTHCARE

1

OVERVIEW OF THE HISTORY AND APPLICATIONS OF DRIED BLOOD SAMPLES

W. Harry Hannon and Bradford L. Therrell, Jr

1.1 HISTORY

"Everything should be made as simple as possible, but not simpler."

Albert Einstein (1879–1955)

For almost 100 years, the simple concept of applying biological fluids to filter paper, drying, transporting to a laboratory, and taking aliquots has been adapted into a variety of innovative methods to generate a suitable specimen and sampling matrix for analytical testing. These methods for collecting blood and other biological fluids created a novel way to sample biological fluids and are responsible for initiating the first interest in analytical micromethods (Schmidt, 1986).

Filter paper sampling greatly simplified blood sample collection, handling, and storage over other methods in use at the time. Over the years, the availability of analytically acceptable dried blood spots (DBSs) has significantly impacted a variety of fields of study including newborn screening (NBS); epidemiology (field testing); infectious diseases; environmental research; forensics; therapeutic drug monitoring; illicit drug analysis; toxicology; and toxico- and pharmacokinetic studies of drugs and candidate drugs. Filter paper samples, although conceptually simple with few anticipated matrix effects, have resulted in many unexpected analytical complexities. The utility and acceptance of DBSs have increased over the last 30 years primarily because of concerted efforts to control, minimize, and eliminate these analytical variations.

Perhaps the earliest reference to blood collected on paper can be seen in the well-preserved Mayan murals of Bonampak, Mexico, dated 780AD. One part of the mural shows women pricking their tongues, fingers, and lips while letting the blood drops collect on paper that is then placed in a container and burned to summon the gods. Another section of the illustration portrays a child spreading his fingers with one finger being pricked for blood collection on paper. A cup on one end on the table appears to contain smoldering blood spots. A row of blood-spot images is depicted on the table's edge (Miller, 1995).

With his modification of the Wassermann test in 1911, Noguchi reported using hemolytic amboceptor (old term for hemolysin/antibody) serum absorbed onto filter paper to increase the stability of this reagent for his complement fixation method for syphilis (Noguchi, 1911). But the first credit for using filter paper for specimen collection is attributed to Bang, who reported its use in this manner for an analytical method in 1913 (Bang, 1913). His ingenious approach introduced a method for absorbing blood onto filter paper, drying the spots, and then determining glucose concentrations from the eluate. He first mentioned his specimen collection method in 1907, although it was not published until 1913 (Schmidt, 1986). His method for blood sugar analysis was both practical and reliable. Bang introduced the use of micro-samples of blood (about 100 mg) absorbed into prewashed and dried filter paper and weighed on a balance to measure the sample aliquot size. This sampling and testing technique was designated "Bang's Method" (Van Slyke, 1957; Schmidt, 1986). Bang also performed Kjeldahl nitrogen/protein determinations with a filter-paper specimen-based micromethod. Ivar Christian Bang has been declared the "founder of modern clinical microchemistry" (Schmidt, 1986).

In 1924, Chapman reported using several varieties of absorptive papers in his studies and noted that "any good

Dried Blood Spots: Applications and Techniques, First Edition. Edited by Wenkui Li and Mike S. Lee.

grade of a nonalkaline filter paper of low ash content and good absorption power may be used" (Chapman, 1924). He typically used Schleicher and Schuell No. 595 paper but reported that "Whatman No. 3 paper was found to be most satisfactory for the Wassermann test." Chapman conducted a series of investigations to determine whether patient's blood could be collected on filter paper, dried, and then used in the complement fixation test for syphilis. He used DBSs for the following reasons: (1) less blood was necessary (especially important for children); (2) blood collection supplies were simple and inexpensive; (3) risks of specimen spoilage (by bacterial contamination) or hemolysis could likely be eliminated; and (4) specimen preservation for long periods of time with little deterioration would be possible. Blood was absorbed onto filter paper strips (about 4 in.), dried, and then cut into smaller pieces for elution and analysis. Chapman also noted that blood collected on filter paper and dried could be kept for at least 1 month with little deterioration of its complement fixing power (Chapman, 1924).

By 1939, filter paper was more commonly used as a transport medium for blood samples for serological examination (Zimmermann, 1939). Different types of filter paper were available and recommended or preferred by different investigators. For example, one investigator declared that only Canson 435 blotter paper was usable for his studies, while electrophoresis paper and membrane filters were not suitable (Vaisman et al., 1963). Interestingly, serum collected on filter paper was thought to be less satisfactory than blood. Additionally, efficient extraction techniques were considered to be problematic, so elution methods were recommended. Zimmerman evaluated different blotter and filter papers to determine the best product for his test. These papers included Canson 435 blotting paper and several filter papers: Delta No. 310, Schleicher and Schuell No. 589, and Whatman No. 1 (Zimmermann, 1939). He used a sharp hollow pipe to punch 15 mm paper circles for sampling. The amount of serum and whole blood in a sample was determined by weighing. Samples were declared dry after 2 hours of air drying at ambient temperature. Zimmerman also investigated the impact of different solutions, temperatures, and elution times on DBS elution efficiency (Zimmermann, 1939). The best elution was reported to be obtained from phosphate-buffered saline at pH 7.2 when eluted for 2 hours at 37°C. All papers studied were suitable for transporting dried blood and serum. Consequently, the paper with the smallest variation in the absorbed quantity of blood and serum (determined by weighing dried paper punches before and after soaking in blood or serum) was preferred (Zimmermann, 1939).

In 1950, Hogan noted that "A simpler method of collecting blood would be useful for diagnostic activities in general and would be particularly desirable for the diagnosis and treatment of congenital syphilis" (Hogan, 1950). "Indeed the nonavailability of such a simple test has hampered congenital syphilis control programs" (Hogan, 1950). He described a syphilis test that used whole blood from a finger prick collected on filter paper and dried (Eaton and Dikeman No. 613). Strips of paper were saturated with blood from a finger stick (heel, toe, or ear lobe puncture in infants and children) and the strips were air dried on a clean flat surface. Small squares were cut from the dried filter paper strips and microscopically analyzed. This procedure avoided jugular punctures in infants and children—the standard blood specimen collection procedure at the time. Although the new DBS test was not considered as "good" (sensitive) as the standard liquid serum serologic test, it fulfilled an apparent need (Hogan, 1950). The filter paper method was particularly desirable when screening children for congenital syphilis, especially for mass testing programs and home collections. The finger prick procedure quickly overcame parents' objections to jugular punctures on infants and small children. Nevertheless, some investigators continued to downplay the use of DBSs because of the lower test sensitivity when compared with liquid samples in standard laboratory tests and suggested using these DBSs only when no other specimen collection method was possible (Freeble and Orsburn, 1952).

Filter paper techniques for collecting whole blood and serum for studies on eastern equine encephalomyelitis were reported in 1957. However, the reporting investigators found that hemolytic eluate from the dried blood samples impeded readings on samples with low titers for complement fixation reactions (Karstad et al., 1957). They determined that paper discs made of highly absorbent, commercially available white paper were equivalent to discs cut from white blotting paper. Schleicher and Schuell filter paper was reported to be superior in uniformity and tensile strength when wet. Blood samples were collected from birds and horses, and 12.5 mm discs of the type used for antibiotic sensitivity tests were used. Serum was dropped onto the discs until they appeared saturated. Discs of 12.5 mm and 15 mm diameter were found to absorb 0.16 mL and 0.20 mL serum, respectively, and overnight elution was reportedly most effective (Karstad et al., 1957). Adams and Hanson found that exposure of paper discs to 56°C for 1 hour or exposure to 37°C for 7 days resulted in no appreciable titer loss for neutralizing antibodies; and therefore, the technique was declared valuable for epizootiological studies (Adams and Hanson, 1956).

Early methods to obtain DBS aliquots relied on scissors, copper tubes, or precut paper discs. In 1957, the first use of a paper-hole punch device for collecting aliquots was reported. "The filter paper strip is dipped into the blood and allowed to saturate half of the filter paper strip, dried at ambient temperature and then one-quarter inch punches are taken from the dried-blood tip of the paper strip with an ordinary ticket hole puncher to produce standardized dried-blood aliquots for testing" (Pellegrino and Brener, 1958).

Advantages of the filter paper method for specimen collection were summarized by Wolff in 1958 (Wolff, 1958).

In particular, he noted several advantages: transmittal speed (air mail was quick and economical); minimal chance of contamination; and elimination of breakage. He also noted that in complement fixation tests, it was preferable to dry serum on blotting paper rather than to collect whole blood (Wolff, 1958). The next year, Farrel and Reid (1959) conducted assays for poliovirus antibody using paper discs moistened with serum and applied to the surface of poliovirus-injected agar (with overlaid tissue culture to show zones of inhibition for the poliovirus).

Also in 1958, a method for collection of urine from infants onto squares of Whatman filter paper was described as part of a program for early detection of phenylketonuria (PKU) (Berry et al., 1958). The method involved placing a 2 in. square absorbent filter paper into an infant's diaper, allowing it to become wet (or dipping it in urine), and then allowing it to dry. If the urine appeared very pale and dilute, then the square was dipped a second time and dried again. Lead pencil identification marks on the filter paper did not interfere with any of the tests described. A 1×0.5 in. rectangle was cut from the filter paper that contained approximately 50 μL urine (Berry et al., 1958). This technique of using dried urine papers was proven important in testing for phenylpyruvic acid and a number of other excreted substances caused by metabolic defects (Berry et al., 1958; Berry, 1959).

A figure published as part of a January 1961 report (Anderson et al., 1961) was the first to illustrate inked rings (circles) printed on the filter paper as targets for blood collection. Earlier publications had only indicated techniques for dipping paper strips into blood or saturating paper squares or discs before allowing them to dry. In the inked target procedure, small quantities of blood were to be placed inside the printed circles. For convenience, four circles were printed on a small sheet of filter paper and the volume of blood needed to fill a circle was determined to be 65 μL. For field use, sheets of filter paper were placed in booklets separated with cellulose sheets. Dried specimens were placed in envelopes and mailed to a central laboratory where batches of tests could be performed with standardized reagents (Berry et al., 1958).

In November 1961, Dr Robert Guthrie reported a unique bacterial inhibition assay (BIA) for measuring phenylalanine to detect PKU. He reported collecting specimens by absorbing a few blood drops from a heel stick into special filter paper (Guthrie, 1961). The combination of an easily transportable specimen and an inexpensive and accurate screening test made large-scale screening for PKU feasible. Subsequently, NBS for PKU using DBSs became widespread as a critical prevention strategy in public health practices. Despite the fact that various techniques using filter paper as a collection medium for biological samples had been reported for at least 50 years (several times as a newly developed technique), many investigators at the time attributed the first use of DBSs to Guthrie, often referring to them as "Guthrie spots." This trend even continues today.

For Guthrie's PKU test, blood from a skin puncture was spotted onto a piece of filter paper (Whatman No. 3), dried, and mailed to his testing laboratory where 200 specimens could be tested in a batch by a single technician. The DBSs were steamed to coagulate blood proteins, after which discs were punched with an ordinary paper-hole puncher from each spot (Guthrie and Susi, 1963). The method provided a simple way of sample collection for screening babies at the time of discharge from a birthing hospital. The first case of PKU detected by the Guthrie procedure occurred in a pilot study at the Niagara Falls City Health Department Laboratory (New York) after only 800 newborns had been screened (Guthrie and Susi, 1963). For a period of time, Guthrie's laboratory also confirmed PKU from dried urine-impregnated filter paper collected by the mother 2–3 weeks after discharge and mailed to his laboratory. This procedure was later discontinued because it lacked sufficient sensitivity (Guthrie and Susi, 1963).

Interestingly, Guthrie's 1961 article was initially rejected because another recently published study using the Guthrie BIA had experienced a high number of false-positive results and had recommended not using Guthrie's test for PKU screening (Scheel and Berry, 1962). Guthrie later noted that a different filter paper source used in that study might explain their high false-positive rate (Guthrie and Susi, 1963). He felt that the filter paper used by other investigators (Scheel and Berry, 1962) might not be absorbent enough for uniform spotting. Guthrie noted that Schleicher and Schuell Grade 903 filter paper was best suited for collecting uniform DBSs and developed some minimal criteria for uniform DBS assay performance. In particular, the DBS should appear similar on both sides of the paper, should be at least 3/8 in. in diameter (but not more than 1/2 in.), and should be located close enough to the edge of the paper to allow easy punching with a paper-hole puncher (Guthrie and Susi, 1963).

In the 1960s, Schleicher and Schuell began to manufacture a special specimen collection paper called "Grade 903" and referred to as "filter paper." This paper was highly controlled to yield a consistent cotton linter paper that contained no hardeners or additives (Harvey, n.d.). Today, blood collection papers are expected to meet the ASTM International (formerly the American Society for Testing Materials) consensus standards (Hannon et al., 2013) and are certified to meet the performance standards for sample absorption and lot-to-lot consistency as set by the approved standard, NBS-01-A6 (Blood collection on filter paper for NBS programs) of the Clinical and Laboratory Standards Institute (CLSI) (Hannon et al., 2013). The introduction of this CLSI approved standard in 1982 contributed to the quantitative consistency of filter paper within and among lots of paper produced by the manufacturer and guided the extensive applications of samples dried in this matrix across a variety of fields during

TABLE 1.1 Historical and Present Sources of Filter Paper Used for Sample Collection

Filter Paper	Year in Use[a]
Schleicher & Schuell No. 595	1924
Schleicher & Schuell No. 589	1939
Delta 310	1939
Canson 435	1939
Whatman No. 1	1939
Eaton-Dikeman (ED No. 613)	1948
Munktell No. 3	1955
Schleicher & Schuell No. 740	1956
Schleicher & Schuell Grade 903	1961
Whatman No. 3	1961
Whatman No. 4	1978
Toyo Roshi (Advantec)	1979
Schleicher & Schuell Grade 2992	1983
Macherey-Nagel	1989
Whatman BFC 180 and Grade 903[b]	2005
Ahlstrom 226[b]	2007
Munktell TPN	2009

[a]Earliest year of documented use in our literature search. The indicated year may not be the first actual usage.
[b]Bolded paper sources are approved/cleared sources by the Food and Drug Administration and are monitored routinely by the Centers for Disease Control and Prevention's Newborn Screening Quality Assurance Program for sustained adherence to established performance criteria (Centers for Disease Control and Prevention, n.d.).

the last two decades. Table 1.1 contains a list of the historical sources of "blotter" and filter paper that have been or are presently used by investigators for preparing dried biological samples.

A sample aliquot obtained from a DBS is expected to achieve the same level of accuracy and precision as that from an equal pipette volume of a liquid blood sample. Since most clinical decisions are based on measured concentrations of a biomarker, it is important that the concentration in the eluate from the extracted DBS punch replicates the level in the original blood sample. Studies have demonstrated that different lots of filter paper from sources cleared/approved by the Food and Drug Administration (FDA) for blood collection are consistent with lot-to-lot variance within established performance limits (Hannon et al., 2013). The criteria for quality filter paper are established for a specific set of parameters: (1) volume of serum contained in a 3.2 mm punch from a DBS prepared with a standard blood volume and hematocrit; (2) time required to absorb a standard size blood volume; (3) appearance of the DBS; and (4) size of the spot created by a standard blood volume. Each criterion is evaluated for new lots of manufactured paper by the Newborn Screening Quality Assurance Program (Centers for Disease Control and Prevention, n.d.) at the Centers for Disease Control and Prevention (CDC) before its release to the user community.

1.2 HISTORICAL APPLICATIONS

A 2001 table contained a list of over 175 analytes that have been measured in DBSs. The analytes varied widely and included acylcarnitines, C-reactive protein, cyclosporine A, cytokines, hepatitis B virus, gentamicin, glucose, lipoproteins, prolactin, selenium, specific antibodies for over 30 viruses and microorganisms, trace elements, vitamin A, zinc protoporphyrin, among others (Mei et al., 2001). Because of the large number of biomarkers that have been analyzed from blood and other biological specimens on filter paper, it is not feasible to describe all of the applications. Thus, a representative historical review is provided with examples selected because of their impact at the time, their overall importance, and their degree of innovation.

Early in the twentieth century, several investigators reported using blood and serum collected and dried on filter paper for serologic testing for syphilis (Chapman, 1924; Chediak, 1932; Zimmermann, 1939; Hogan, 1950). Both field and home sample collections were described. The use of blood samples collected on either blotting paper or filter paper for viral serologic studies was also reported periodically. In 1956, Adams and Hanson absorbed neutralizing antibodies of vesicular stomatitis virus on blotting paper discs (Adams and Hanson, 1956). Karstad (Karstad et al., 1957) used serum and whole blood absorbed onto filter paper discs in his studies on eastern equine encephalitis. Green and Opton in 1960 (Green and Opton, 1960) and Kalter in 1963 (Kalter, 1963) reported using whole blood on filter paper for poliovirus antibody measurements in population surveys. Anderson et al. reported filter paper blood collection for fluorescent antibody studies of schistosomes (Anderson et al., 1961). Filter paper disc methodology facilitated seroepidemiology field studies, which measured antibody titers in infants and children for many viruses including measles, mumps, poliovirus, parainfluenza virus, respiratory syncytial virus, and others (Klas-Bertil and Heimburger, 1953).

The first report of blood absorbed onto filter paper for enzyme measurements was published in 1953 (Chin et al., 1966). To simplify cholinesterase measurements, a method was developed to determine two activities from one drop of blood. This method used the same principle as "Bang's Method" (Bang, 1913), discussed earlier. With two separate blood spots, the enzymes were analyzed separately with two distinct substrates to measure two types of blood enzymes: plasma butyrylcholinesterase and erythrocyte acetylcholinesterase (Klas-Bertil and Heimburger, 1953).

In 1962, Berry considered filter paper urine samples for population-based screening programs (Berry et al., 1962). She planned to screen for conditions in which a relatively large increase in certain amino acids was expected. Two possible conditions included "Hartnup disease" and "maple sugar [sic] urine disease" in which valine, leucine, and isoleucine were excreted in large amounts. The *Lactobacillus arabinosus* microbiological assay commonly used for these amino acids in urine was modified for filter paper urine specimens. Small pieces of filter paper saturated with urine (approximately 25 μL of urine in 0.5 $in.^2$) were dried and incubated in each of three special media inoculated with the test organism. Since the organism required valine, leucine, and isoleucine for growth, absence of one or more of these amino acids was indicated by minimal growth. Each of the special growth media lacked one of the amino acids so that no bacterial growth was observed unless the dried urine contained increased quantities of the specific amino acid (Berry et al., 1962).

The use of blood spot screening for PKU became a widespread public health program during the 1960s. The Massachusetts Department of Health decided to offer a population-based screening program for PKU in July 1962. The Guthrie BIA was used on DBSs obtained from heel sticks on newborns and screening was intended to occur in all maternity hospitals in the State. When three PKU cases were found in 4 months, participation grew rapidly and by midwinter all maternity hospitals in the state were participating. This achievement made Massachusetts the first state to have state-wide NBS (MacCready, 1963).

The first report of a unique procedure for recovering blood from DBSs and matching it with liquid serum was published in 1966 (Chin et al., 1966). Using a lancet to cut the skin and induce blood flow with forceps to hold the disc, capillary blood from a finger or heel stick was saturated onto a precut disc until no white spots were showing. Discs were then placed in small petri dishes without airtight covers to dry. Each disc was assumed to contain 10 μL of serum. Blood was eluted from the spots by loading them one at a time into a syringe barrel without a plunger and forcing 25 μL of saline through each disc. There was a good correlation between the matched liquid serum and dried blood discs. A slightly lower titer was observed from the blood discs (Adams and Hanson, 1956; Chin et al., 1966).

In 1973, NBS for congenital hypothyroidism (CH) began in Quebec, Canada, expanding the testing beyond amino acids. Dussault reported success in determining thyroxin concentrations in blood eluted from filter paper discs collected from 3- to 5-day-old newborns (Dussault and Laberge, 1973). Detecting CH within the first few days of life has been recognized by some as the single most important screening test in newborns since it allows prevention of mental retardation if lifelong treatment is initiated by 2–3 weeks of life. NBS programs began to add this testing to their ongoing PKU screening programs; and by 1978, almost one million newborns had been screened in North America. To avoid a high recall rate, a certain percentage of newborns with low thyroxin were selected for secondary screening with thyrotropin (thyroid stimulating hormone [TSH]) using the original DBS sample. The combined results of the two tests were used to better estimate the likelihood of disease (Walfish, 1976). Screening for TSH alone was used in some programs, primarily in settings where screening occurred several days after birth.

Filter paper discs have also been used to estimate lysozyme activity in tear fluid. In 1976, Mackie and Seal collected tear samples by placing a 6 mm filter paper disc in the lower conjunctival fornix and allowing it to become entirely wet (Mackie and Seal, 1976). The quantity of tears collected was determined by first weighing a bottle containing the dry 6 mm filter paper disc and then weighing it after the disc was saturated with tears. The filter paper was transferred to an agar plate for lysozyme estimation. The quantity of tears held by the disc varied up to fourfold, but the lysozyme concentration could be reported in units of activity per microliter of tears when the weight was known (Mackie and Seal, 1976).

An immunochemical test for colon cancer screening using fecal occult blood smears on specially treated filter paper was reported in 1980 (Songster et al., 1980). The test employed high titer monospecific antisera to intact human hemoglobin in a radial immunodiffusion (RID) assay. A 1/8 in. disc was punched from a uniformly dry area of the fecal-smeared filter paper using a commercial conductor's ticket hole puncher. The disc was placed on the surface of an RID plate, rehydrated, and incubated at room temperature. Perpendicular diameters of precipitin rings were measured at equivalence using indirect illumination. Punched discs soaked in standard hemoglobin hemolysate were used to validate constant sensitivity of each RID plate (Songster et al., 1980).

In 1987, McCabe first reported successfully extracting DNA from blood collected on "blotter" paper and dried (McCabe et al., 1987). His results, combined with the refinement of nucleic acid amplification techniques, made genetic testing from DBSs possible. With these advances, neonatal screening laboratories began to add genetic tests for hemoglobinopathies, Duchenne muscular dystrophy, and cystic fibrosis as second-tier tests to reduce recall rates and unnecessary confirmatory testing. However, good reproducibility from DNA extractions was technically difficult compared with routine NBS methods. While direct genotyping secondary to screening (i.e., from the same DBS) was shown to facilitate diagnostic confirmation and improve the time to diagnosis and management (McCabe et al., 1987), widespread testing as part of routine NBS protocols has developed slowly and is not yet a reality. Progress in genetic analysis with samples collected on filter paper has also been achieved using buccal, saliva, and urine samples (Harvey, n.d.).

In the late 1980s, screening for lead poisoning was widely performed by determining lead concentration in blood or the concentration of another hematological indicator such as erythrocyte protoporphyrin. While routine screening usually involved liquid blood collected in small containers, blood was also collected by finger pricks onto filter paper cards and mailed to Guthrie's Central Laboratory for testing (Shaltout et al., 1989). Recent publications largely conclude that filter paper sampling is a viable alternative for blood-lead screening, and a few laboratories routinely perform filter paper blood-lead testing on pediatric specimens (Verebey et al., 1991; Yee et al., 1995).

Polymerase chain reaction (PCR) based methods for DBS samples have proven effective and convenient for the early diagnosis of neonatal HIV-1 infection, especially in developing countries. Samples collected and dried on filter paper have facilitated large field studies by simplifying all aspects of sample collection, storage, and transport and minimizing biohazard risks. DBSs have also been found to be satisfactory for virology studies. They have been used for monitoring the emergence of drug-resistant mutations, characterizing the genotype of transmitted virus, determining viral load, and tracking global spread (Cassol et al., 1997). Filter paper–based methods have also been used for measuring viral RNA in dried plasma spots (DPS). This DPS method is cost-effective and has broad implications for population-based research and surveillance (Hamers et al., 2009).

The introduction of tandem mass spectrometry (MS/MS) as a multi-analyte platform for NBS permitted the analysis of a wide array of analytes from a single DBS punch (Rashed et al., 1999; Chace et al., 2003). The MS/MS platform was a significant factor in introducing a recommended uniform screening panel (RUSP) for newborns using DBSs in the United States (Watson et al., 2006). The RUSP currently includes over 50 conditions (including secondary targets) using a variety of testing methods (e.g., immunoassays, high performance liquid chromatography (HPLC), isoelectrofocusing, MS/MS, enzyme assays).

The feasibility of detecting patients with several inherited lysosomal storage diseases (LSDs) using DBSs as the sample matrix in appropriate enzyme assays has been reported. Detection of LSDs is based on specific enzymatic assays using plasma, leukocytes, fibroblasts, and, more recently, DBSs. Despite the low individual incidences of these conditions, enzyme assays for their detection are being used in some NBS programs internationally and are implemented presently in a couple of US state programs, while a few other states are in start-up phases. (Civallero et al., 2006).

In 2001, Severe Combined Immunodeficiency (SCID) was identified as a possible goal for public health screening with DBS at a workshop convened by CDC in Atlanta, GA (Lindegren et al., 2004). The lack of functional T or B lymphocytes in SCID can serve as a screening tool prior to the onset of symptoms. Shortly after the workshop, Chan and Puck published a method using quantitative real-time (qRT) PCR to detect SCID in newborn by measuring in DBSs the T-cell receptor excision circles (TREC), an extra-chromosomal DNA fragment uniquely created during T-cell formation (Chan and Puck, 2005). The initial results suggested that the TREC assay could be a sensitive and specific method for SCID screening. A qRT PCR TREC assay with high throughput capacity now exists (Baker et al., 2009).

The DBS samples remaining after completion of NBS (residual DBSs) are used effectively for analytical method development, method comparison, and result validation (Therrell et al., 2011). Whole blood absorbed into filter paper and dried also offers an efficient mechanism for creating a repository or bank of samples with potential for research using DNA or other stable biomarkers. DBS banking systems already exist in the US military and in a few states and countries (e.g., California, Michigan, Denmark, Thailand) (Therrell et al., 1996; Therrell et al., 2011). The use of residual DBS samples for quality assurance and public health epidemiology are generally considered important public health applications; however, their potential for research use has been controversial, primarily because of the lack of consent processes in many NBS programs. A highly successful historical example of public health research using residual DBSs was the CDC's HIV-seroprevalence survey among childbearing women that provided critical public health data on the spread of HIV (Gwinn et al., 1991).

1.3 OVERALL ADVANTAGES AND DISADVANTAGES OF DRIED-BLOOD SPOT SAMPLING

1.3.1 Advantages

DBSs can be obtained with relatively little training, require minimal manipulation at the collection site, are generally considered nonhazardous, and can be transported from remote sites easily for analysis. Filter paper collection devices are low cost, stored and handled easily, offer a stable analyte matrix, and are easily used in resource-restricted settings (Hamers et al., 2009; Harvey, n.d.). DBS samples have proven to be inexpensive and reliable for testing, particularly for large-scale testing in remote populations (Oppelaar, 1966). They offer several advantages over conventional liquid whole blood, plasma, or serum sample collection. For example, the DBS sampling method is less invasive (simple heel, finger, or ear lobe prick, rather than venous needle insertion), which helps in recruiting subjects for blood studies. Sample storage is simpler and transfers are easier because the specimen is nonbreakable and not required to be frozen or shipped on dry ice for most applications. DBSs can be conveniently collected by parents or other adults with minimal training, and they can be cheaply and easily conveyed to the testing laboratory. The collection process reduces infection risk, and the volume of blood required is smaller than the liquid blood requirements (Green and Opton, 1960; Cassol et al., 1997; Harvey, n.d.).

Over time and with experience, guidance has evolved for efficiently transporting DBS specimens. DBS samples should not be packaged in airtight, leak-proof sealed containers (e.g., plastic or foil bags) because the lack of air exchange in the inner environment of a sealed container causes heat buildup and moisture accumulation, which causes specimen degradation. Heat, direct sunlight, humidity, and moisture are considered to be detrimental to the stability of most analytes contained in DBS samples, leading to poor analyte recoveries. Desiccant packs included in shipping containers can aid in preventing moisture accumulation; however, shipping conditions are generally not controllable, resulting in limited desiccant effectiveness (Hannon et al., 2013). DBS samples can be shipped or transported by mail or courier without expectations of occupational exposure to blood or other potentially infectious material. Nevertheless, "standard precautions" and compliance with local regulations and institutional policies are required in preparing DBS samples for shipment. Despite the fact that few pathogens are known to survive for extended periods of time in dried blood on filter paper (Hannon et al., 2013), the packaging should adhere to basic triple-packaging recommendations: (1) blood is absorbed into the paper; (2) an inner envelope or other protective cover (flap) should be present containing a biohazard label; and (3) the sample should be enclosed in an outer (gas permeable) envelope of high quality paper (or tear-resistant material) (Hannon et al., 2013).[1]

For many years, investigators have studied the relationship between temperature, humidity, and stability of biomarkers in DBSs (Cassol et al., 1997). Most analytes appear to be stable in dried matrices stored at low temperature with low humidity (~30%), similar to a lyophilized environment (Therrell et al., 1996). An extensive search of the literature regarding stability of analytes in DBSs was of minimal value in assessing the impact of long-term storage. For stability studies, differences in markers studied, study duration, storage conditions, and data analyses made comparisons difficult. Reported studies used a variety of procedures and conditions, and most did not provide meaningful conclusions about long-term storage outcomes. Analytical reference points for assessing stability were often weak. Studies covered relatively short time periods and provided data relevant only to testing environments for identifying disorders among newborns. Interpretations of stability data were inconsistent and evaluated only analyte recoveries or concentrations for disease classifications. DBS elution studies were usually limited to fixed time intervals; therefore, data based on samples that eluted slowly could be misinterpreted as resulting from sample instability (Therrell et al., 1996).

[1] Note: The US transport standards (DMM Reference 601.10.17.8) described here are harmonized with the World Health Organization's Guidance on Regulations for the Transport of Infectious Substances (World Health Organization, 2011) and the International Civil Aviation Organization's Technical Instructions for Safe Transport of Dangerous Goods by Air (International Civil Aviation Organization, 2006).

Available data on long-term storage of DBSs suggest that, for maximum recovery of most analytes, DBSs should be stored at low temperature with controlled low humidity (Therrell et al., 1996). Stability of DBS samples has been evaluated for most analytes currently included in NBS with varied outcomes. For the majority, sample degradation is minimal when stored at room temperature for several weeks (Therrell et al., 1996). Cooler, drier conditions can extend sample stability. Hot and humid storage conditions for up to 20 weeks have been found to cause a progressive decline in HIV-1 antibody titers, especially with low titered samples. The use of gas-impermeable bags containing a desiccant improved DBS sample stability for HIV-1 antibody detection over time and is recommended for the storage of whole-blood spots on filter paper in harsh tropical field settings (Behets et al., 1992).

Several stability studies have demonstrated the extractability and stability of DNA over time in residual NBS DBSs. While genomic DNA appears to be stable under tropical conditions for at least 11 years at ambient temperature, the quality of DNA for amplifying larger DNA fragments appears to decrease in specimens stored beyond 10 years (Chaisomchit et al., 2005). A study of 70 residual NBS DBSs stored for 19 months at ambient temperature gave adequate DNA for forensic studies (Kline et al., 2002). Likewise, whole genomic amplified DNA from residual DBSs that were archived for 15–25 years were acceptable for genome-wide scans and it was a cost-effective alternative to collecting new specimens (Hollegaard et al., 2009). Conversely, the stability of non-DNA biomarkers commonly used in NBS varies by analyte and appears highly dependent on storage conditions (Strnadova et al., 2007; Therrell et al., 2011; Adam et al., 2011).

1.3.2 Disadvantages

The disadvantages and assay variables potentially affected when using filter paper samples are complex. Concerns include sample volume, humidity, hematocrit, chromatographic effects (homogeneity), analyte recovery, specimen source, anticoagulant, and filter paper characteristics. For both DBSs and other biological fluids collected on filter paper, there are reported variations in quantitative analyte recovery between dried samples and liquid whole blood or other samples. Questions regarding dried samples often concern sample homogeneity and how best to minimize this variability. Issues include the best location for obtaining the optimal specimen punch (variability across and within the specimen related to absorption and chromatographic effects), the size of sample applied to the filter paper (variations in analyte concentration dependent on sample volume), and the consistency of filter paper performance (variations in performance related to punch size). Random sampling errors related to small testing aliquots from paper punches are also a concern. Some of these issues are discussed below.

Filter Paper Effects: Throughout the history of biological fluids collected on filter paper there have been concerns, both perceived and real, about the sources and types of paper selected for a particular analytic system. As a prime example, in an early publication Guthrie successfully used Whatman No. 3 filter paper (Guthrie, 1961). Shortly afterwards, he stressed the improvement in his test results when he used Schleicher and Schuell No. 903 filter paper, which he observed to be more absorbent (Guthrie and Susi, 1963). Commenting later on reported poor results with his method by another user (Scheel and Berry, 1962), Guthrie (Guthrie and Susi, 1963) noted that "While it is not possible to account for the discrepancy between Scheel and Berry's (Scheel and Berry, 1962) results and ours, one likely source is the filter paper used in their study, Whatman No. 3." Interestingly, Partington and Sinnott (1964) were unable to find an advantage of one filter paper over the other; and they reported that there was no evidence that the Schleicher and Schuell paper facilitated any better collection of blood from patients. Direct comparisons of BIA growth zones for the two papers using the same blood sample yielded similar results with no particular advantage to either paper (Partington and Sinnot, 1964).

While less invasive, collection of quality DBS samples requires practice and plays a major role in the quality of test results. Conversely, the quality of a liquid blood specimen is seldom identified as a factor in the quality of analytical results. As one example, when blood does not completely and evenly penetrate through the filter paper to the other side, it is generally labeled as a "poor" sample. In early comparisons with the BIA for PKU measurements, growth zones obtained from "poor" samples were the ones that were significantly smaller than those from "good" samples. A correlation factor used with poor samples assumed that they were at least half soaked, so a high correction factor was applied to guard against underestimating the phenylalanine content (Partington and Sinnott, 1964). Correction factors were eventually replaced by a request for a properly collected replacement specimen, which continues as the practice today.

As test sensitivities increased, the quality of DBS specimens became more important. Paper quality was increasingly scrutinized, particularly when radioimmunoassay techniques for detecting CH became widespread in the1970s (Dussault and Laberge, 1973). Responding to continuing complaints about filter papers' reliability and reproducibility, the CDC initiated a filter paper evaluation project in 1981 (Centers for Disease Control and Prevention, n.d.). Filter paper quality was assessed using a DBS-based quantitative isotopic test for measuring and monitoring an established set of performance characteristics. Each filter paper production lot was measured against a defined set of performance criteria. Filter paper manufacturers voluntarily collaborated in CDC's evaluation of their products as an integral part of their quality assurance protocol. This project ultimately resulted in a national standard for blood collection on filter paper developed by the CLSI, currently in its sixth edition. The standard (LA04/NBS01) defines proper specimen collection protocols and addresses related issues including paper quality (Hannon et al., 2013). The FDA defined filter paper manufactured for blood collection as a medical device and requires FDA clearance. The combination of activities above has resulted in minimizing lot-to-lot and source-to-source variations in the quality of filter paper currently used (Slazyk et al., 1988). Data from a recent CDC study (Mei et al., 2010) indicated 4–5% variability in analytical results between approved filter paper sources, and these results were similar to that observed for the lot-to-lot variability (Mei et al., 2010). It is generally agreed that for the best performance, the same filter paper lot and source are used throughout a study. A research study examining results for 28 elements (e.g., lead, iron, zinc, nickel) simultaneously measured in DBS samples produced more variation than expected. The authors concluded that random contributions from the background elements in the filter paper were the primary obstacle to measuring elements in DBSs (Langer et al., 2011).

Sample Volume: The process of collecting blood on filter paper from newborns is considered difficult by some, but result imprecision and variability can (and should) be minimized by using standardized collection procedures. Using preprinted target rings on the filter paper can help control the volume of blood collected, particularly when the target circle is filled correctly to its edge during blood collection. While not the preferred collection technique for NBS, careful blood collection using capillary tubes of predetermined fill volume may be helpful for some investigations. (Mei et al., 2001; Hannon et al., 2013). A study of the potential recovery variability from differing volumes of blood on filter paper has demonstrated the importance of controlling blood spot–volume for improved quantitative measurements. Five different volumes of blood ranging from 25 mL to 125 mL were spiked with 125I-thyroxin at a constant hematocrit and applied to paper from a single production lot from each of two filter paper manufacturers. The samples were air dried overnight and a single 6.4 mm punch was taken from the center of 50 DBSs from each blood volume applied on each filter paper lot. Isotopic count–measurements were used to compare the serum volume of the center punches. A 13% increase (approximate) in serum volume, sufficient to potentially affect analyte measurements, was observed between center punches taken from the lowest blood volume to the highest volume; however, essentially no differences were noted between the two paper sources in this study (Mei et al., 2001).

Dried whole blood, plasma, and serum may provide suitable matrices for HIV type-1 viral load determinations and for drug resistance genotyping (particularly useful in antiretroviral therapy for HIV-1-infected individuals living in low to middle income countries). Limitations of the DBSs include reduced sensitivity resulting from small sample volume, nucleic acid degradation under extreme environmental conditions, impaired efficiency of extraction, and interferences with proviral DNA. Improved analytical sensitivity is necessary for routine applications to monitor therapeutic treatments, especially at the onset of treatment failure (Hamers et al., 2009).

Humidity Effects: Humidity is a factor in the way blood spreads in filter paper, how the sample dries, and analyte stability in the dried blood matrix. Each of these factors has been closely examined (Cernik and Sayers, 1971; Hannon et al., 2013; Adam et al., 2011). The separate contributions of elevated temperature and elevated humidity changes in levels of 34 analytes in DBS samples have been reported (Adam et al., 2011). Results from these accelerated degradation studies showed that degradation of 27 of the analytes was primarily caused by high humidity in the DBS storage environment; whereas, the degradation of 4 analytes was primarily caused by the 37°C storage temperature. Markers varied widely in the degree and rapidity of degradation during storage at both low and high humidity and elevated temperature (Adam et al., 2011). Other investigators found that IgG deteriorated faster in samples stored on filter paper relative to serum samples. As a result, precise evaluation of the serum dilution represented by the DBS eluate and its loss of activity relative to temperature variations and humidity were suggested (Guimaraes et al., 1985)

Hematocrit Effects: The newborn's hematocrit (proportion of red blood cells) can significantly affect analytical results. The blood volume within a given 3.2 mm disc may increase significantly when hematocrit levels fluctuate from 30% to 70% (Adam et al., 2000). Hematocrit variability and its effect on quantitative results represent significant analytical challenges in method validation (Carter, 1978). A 6.0 mm punch from the center of a 100 μL DBS made from 30% hematocrit blood contained 47% more serum volume than a similar punch from 70% hematocrit blood (Adam et al., 2000). Blood samples with low hemoglobin concentrations (and hence low hematocrits) spread more widely than those with normal and high hemoglobin concentrations. This was confirmed in a study of blood lead values in laboratory-prepared samples with low hemoglobin values. The results were found to be unreliable unless adjustments were made relative to the area over which the blood had spread (Carter, 1978).

Analysis of DBSs by MS/MS has vastly expanded the detection of inborn errors of metabolism in NBS. Despite its proven sensitivity, many issues related to DBS sample preparation remain unresolved (Holub et al., 2006). The hematocrit profoundly affects blood viscosity and may influence blood diffusion properties. Newborns show a considerable inter-individual variability in hematocrits. A recent study showed that levels of most amino acids and guanidinoacetate increase significantly with increasing hematocrit, while hematocrit has a less pronounced effect on some amino acids. Total acylcarnitines correlated positively with hematocrit levels. In low hematocrit samples, levels of most amino acids and free carnitine were higher in the peripheral spot punches than in the center spot punch. Both hematocrit and position of the punch within the DBS can significantly affect analyte values (Holub et al., 2006).

Chromatographic Effects: Chromatographic effects can affect test results when fibers in the filter paper matrix influence the spread of analytes within the sample as blood is absorbed into the paper. In such cases, analyte concentrations can vary from the center of the specimen to the edges depending on the magnitude of the effect. A CDC study statistically assessed evaluation study data collected over 3 years for different lots of quality control materials (Adam et al., 2000). Within a manufactured lot of filter paper, the average serum volumes of peripheral punches (north, east, south, and west positions) were compared with the average of the corresponding center punch. Center punches were found to have slightly higher serum volumes (diluting the analyte concentration) than peripheral punches. The volumes differed by 1–2% for each filter paper lot, indicating that chromatographic effects may account for up to 2% of analytic variations, assuming the punched discs stayed within the confines of the recommended printed circle size. Metabolites may be distributed as a result of chromatographic effects within a spot; and therefore, punch location may have a significant effect on the metabolite concentration detected (Holub et al., 2006). Therefore, it is theoritically possible to miss an out-of-range analyte depending on the hematocrit and position of the punched disc.

Anticoagulants: Some filter paper specimen collection techniques, particularly the use of capillary tubes, may involve anticoagulants, which may affect assay results. As an example, an enzyme-linked affinity assay was used to quantify PCR products from whole blood, plasma, and separated mononuclear cells collected in the presence of four common anticoagulants: acid citrate dextrose, ethylenediaminetetraacetic acid (EDTA), potassium oxalate, and sodium heparin. In the case of sodium heparin, reduction of the product signal was observed after amplification of nucleic acid extraction from whole blood, washed mononuclear cells, and plasma. The inhibitory effects on gene amplification could be reversed with heparinase. The addition of as little as 0.05 U of heparin completely inhibited amplification of the HLA-DQ sequence from placental DNA. No adverse effects were detected with acid citrate dextrose and EDTA (Holodniy et al., 1991).

Trace amounts of EDTA have been shown to interfere with DBS NBS lanthanide fluorescence/time-resolved immunofluorometric assays (Holtkamp et al., 2008). EDTA may be distributed unevenly throughout the filter paper card, even throughout a single blood spot, especially when

EDTA-coated capillary devices are used. It is possible to use MS/MS to simultaneously detect the presence of EDTA while routinely analyzing DBSs for amino acids and acylcarnitines. This information can then be used to identify samples containing EDTA, which may potentially interfere in the analysis of thyroxine, thyrotropin, immunoreactive trypsinogen, and 17-hydroxyprogesterone (Holtkamp et al., 2008).

Analyte Extraction: Extraction solutions and buffers used with DBSs may limit the sensitivity of some tests (Zimmermann, 1939). Samples are usually diluted during extraction processes, and clean sample reconstitution may be difficult. The choice of extraction solvents and elution conditions may affect testing and extraction efficiency and can lead to assay problems (Kayiran et al., 2003). Elution efficiency with DBSs depends on the sample volume required and may vary by analyte. Depending on assay sensitivity, it may be difficult or impossible to extract sufficient amounts of an analyte from small volume DBSs (Kayiran et al., 2003). It is important for the eluate from a DBS to quantitatively reflect the original blood sample content. At least one study has demonstrated that different production lots of filter paper from a manufacturer consistently and accurately reflect the original sample concentrations (Mei et al., 2010).

Extracts from DBSs used to monitor recovery amounts of specific antibody activity have exhibited a spectrum of results related to drying times. Antibody activity of IgM immunoglobulin declines rapidly on storage, regardless of temperature. As an example, a study of yellow fever IgM immunoglobulin antibodies found that they were not detectable in eluates stored for 1 month, even at −20°C. In contrast, antibody activity residing in the IgG immunoglobulin was shown to be relatively stable on filter paper discs over a 4-month period, especially when stored at lower temperatures. Stability of IgA antibodies requires further study. Therefore, interpretation of serologic data from studies using DBS samples must account for the differences in storage stability of the various immunoglobulins classes (Cohen et al., 1969).

Blood Source: Questions often arise concerning potential differences between capillary specimens collected by skin puncture versus venous specimens. DBSs generally are not considered "standard" diagnostic samples and their results may not be directly comparable with those from liquid serum or plasma. Analytical decision levels established for diagnostic clinical specimens may not correlate well with DBS results, so different decision levels should be established for samples in a DBS matrix. In serologic studies of children, the correlation of titers of paired sera and DBSs was excellent with the exception that titers of the capillary DBS samples tended to be slightly lower than the matched serum samples (Chin et al., 1966). On the other hand, differences in analyte concentration have been reported between capillary and venous blood sample sources for cholesterol studies (Greenland et al., 1990). A study of various parameters in the complete blood count versus differential counts found differences depending on the type of blood sample used, venous or capillary (Kayiran et al., 2003). In term neonates capillary blood samples had higher hemoglobin concentrations, hematocrits, red blood cell counts, white blood cell counts, and lymphocyte counts than venous blood. With a higher hematocrit, the serum content is lower in the DBS punched disc. The terms "capillary blood" or "venous blood" should be used rather than the term "peripheral blood" when results are reported so that sample accuracy will be better reflected (Kayiran et al., 2003).

1.4 CONCLUSION

For nearly 100 years, the process of collecting biological fluids on filter paper has waxed and waned as a useful scientific application for laboratory testing. This sampling technique has been periodically rediscovered and considered to be "new." The dried sample matrix has presented unique challenges, but these have been overcome with perseverance and innovation. As with most scientific discoveries, the collection devices used and the techniques for using dried samples have improved and expanded over time, particularly in the past 25 years. Samples collected on filter paper have been useful for many different technologies and in a variety of fields of study. Today, there is widespread acceptance of DBSs and other dried biological fluids as valid samples useful for many different purposes. The future of this simple idea for collecting and transporting a valid biological sample seems unlimited. Experiences from using filter paper sample matrices are likely to evolve into similar uses with some of the newly developed non-cellulose products currently available. This evolution will likely provide both improved analytical precision and accuracy. In turn, skepticism about the use of dried samples should be minimized or eliminated when considering their efficacy in high quality research.

REFERENCES

Adam BW, Alexander RJ, Smith SJ, Chace DH, Loeder JG, Elvers LH, Hannon WH. Recoveries of phenylalanine from two sets of dried-blood spot reference materials: prediction from hematocrit, spot volume, and paper matrix. *Clinical Chemistry* 2000;46:126–128.

Adam BW, Hall EM, Sternberg M, Lim TH, Flores SR, O'Brien S, Simms D, Li LX, De Jesus VR, Hannon WH. The stability of markers in dried-blood spots for recommended newborn screening disorders in the United States. *Clinical Biochemistry* 2011;44:1445–1450.

Adams E, Hanson RP. A procedure for adsorbing virus neutralizing antibodies on paper disks. *Journal of Bacteriology* 1956;72:572.

Anderson RI, Sadun EH, Williams JS. A technique for use of minute amounts of dried blood in the fluorescent antibody test for schistosomiasis. *Experimental Parasitology* 1961;11:111–116.

Baker MW, Grossman WJ, Laessig RH, Hoffman GL, Brokopp CD, Kurtycz DF, Cogley MF, Litsheim TJ, Katcher ML, Routes JM. Development of a routine newborn screening protocol for severe combined immunodeficiency. *Journal of Allergy and Clinical Immunology* 2009;124(3):522–527.

Bang IC. *Der Blutzucker*. Wiesbaden, Germany: JF Bergmann; 1913.

Behets F, Kashamuka M, Pappaioanou M, Green TA, Ryder RW, Batter V, George JR, Hannon WH, Quinn TC. Stability of Human Immunodeficiency Virus Type 1 antibodies in whole blood dried on filter paper and stored under various tropical conditions in Kinshasa, Zaire. *Journal of Clinical Microbiology* 1992;30:1179–1182.

Berry HK. Procedures for testing urine specimens dried on filter paper. *Clinical Chemistry* 1959;5:603–608.

Berry HK, Sutherland B, Guest GM, Warkany J. Simple method for detection of phenylketonuria. *Journal of the American Medical Association* 1958;167:2189–2190.

Berry HK, Scheel C, Marks J. Microbiological test for leucine, valine, isoleucine using urine samples dried on filter paper. *Clinical Chemistry* 1962;8:242–245.

Carter GF. The paper punched disc technique for lead in blood samples with abnormal hemoglobin values. *British Journal of Industrial Medicine* 1978;35:235–240.

Cassol S, Gill MJ, Pilon R, Cormier M, Voigt RF, Willoughby B, Forbes J. Quantification of Human Immunodeficiency Virus Type 1 RNA from dried plasma spots collected on filter paper. *Journal of Clinical Microbiology* 1997;35:2795–2801.

Centers for Disease Control and Prevention (CDC). (n.d.) *Newborn Screening Quality Assurance Program*. http:www.cdc .govlabstandardsnsqap.html (accessed January12, 2014).

Cernik AA, Sayers MHP. Determination of lead in capillary blood using a paper punched disc atomic absorption technique. *British Journal of Industrial Medicine* 1971;28:392–398.

Chace DH, Kalas TA, Naylor EW. Use of tandem mass spectrometry for multianalyte screening of dried blood specimens from newborns. *Clinical Chemistry* 2003;49:1797–1817.

Chaisomchit S, Wichajarn R, Janejai N, Chareonsiriwatana W. Stability of genomic DNA in dried blood spots stored on filter paper. *The Southeast Asian Journal of Tropical Medicine and Public Health* 2005;36:270–273.

Chan K, Puck JM. Development of population based newborn screening for severe combined immunodeficiency. *Journal of Allergy and Clinical Immunology* 2005;115:391–398.

Chapman OD. The complement-fixation test for syphilis: use of patient's whole blood dried on filter paper. *Archives of Dermatology and Syphilology* 1924;9(5):607–611.

Chediak A. The diagnosis of syphilis practiced in a desiccated defibrinated blood drop. *Revista Medico Cubana (Havana)* 1932; 43:953–956.

Chin J, Schmidt NJ, Lennette EH, Hanahoe M. Filter paper disc method of collecting whole blood for serologic studies in children. *American Journal of Epidemiology* 1966;84:74–79.

Civallero G, Michelin K, de Mari J, Viapiana M, Buri M, Coelho JC, Giugliani R. Twelve different enzyme assays on dried-blood filter paper samples for detection of patients with selected inherited lysosomal storage diseases. *Clinica Chimica Acta* 2006;372:98–102.

Cohen AB, Hatgi JN, Wisseman CL Jr. Storage stability of different antibody species against arbovirus and rickettsial antigens in blood dried on filter paper discs. *American Journal of Epidemiology* 1969;89:345–352.

Dussault JH, Laberge C. Thyroxine (T4) Determination in dried blood by radioimmunoassay: a screening method for neonatal hypothyroidism. *Union Medicale du Canada* 1973;102: 2062–2064.

Farrell LN, Reid DBW. Disc plate assay of poliomyelitis antibodies. *Canadian Journal of Public Health* 1959;50:20–26.

Freeble CR, Orsburn B. Use of filter paper microscopic test in a control program. *Public Health Reports (US)* 1952;67:585–588.

Green RH, Opton EMA. Micromethod for determination of poliovirus antibody suitable for mass surveys. *American Journal of Hygiene* 1960;72: 195–203.

Greenland P, Bowley NL, Meiklejohn B, Doane KL, Sparks CE. Blood cholesterol concentration: fingerstick plasma vs venous serum sampling. *Clinical Chemistry* 1990;36:628–630.

Guimaraes MCS, Castilho EA, Celeste BJ, Nakahara OS, Netto VA. Long-term storage of IgG and IgM on filter paper for use in parasitic disease seroepidemiology surveys. *The Bulletin of the Pan American Health Organization* 1985;19:16–28.

Guthrie R. Blood screening for phenylketonuria. *Journal of the American Medical Association* 1961;178:863.

Guthrie R, Susi A. A simple phenylalanine method for detecting phenylketonuria in large populations of newborn infants. *Pediatrics* 1963;32:338–343.

Gwinn M, Pappaioanou M, George JR, Hannon WH, Wasser S, Redus M, Hoff R, Grady GF, Willoughby A, Novello AC, Petersen LR, Dondero TJ Jr, Curran JW. Prevalence of HIV Infection in Childbearing Women in the United States. *Journal of the American Medical Association* 1991;265:1704–1708.

Hamers RL, Smit PW, Stevens W, Schuurman R, Rinke de Wit TF. Dried fluid spots for HIV Type-1 viral load and resistance genotyping: a systematic review. *Antiviral Therapy* 2009;14:619–629.

Hannon WH, De Jesus VR, Chavez MS, Davin BF, Getchell J, Green M, Hopkins PV, Kelm KB, Noorgaard-Pedersen B, Padilla C, Plokhovy E, Mei JV, Therrell BL. *Blood Collection on Filter Paper for Newborn Screening Programs: Approved Standard, 6th edn.* Wayne, PA: Clinical and Laboratory Standards Institute (CLSI); CLSI Document NBS-01-A6; 2013:1–37.

Harvey MA (n.d.). *The Use of Filter Paper to Collect Blood Samples for Diagnostic Applications*. http://www.acefesa.es/bio/fta/903/documentos/903.doc (accessed January 12, 2014).

Hogan, R. Filter paper microscopic test for syphilis. *Journal of Venereal Disease Information* 1950;31:37–45.

Hollegaard MV, Grauholm, J, Børglum A, Grove J, Kreiner-Møller E, Bønnelykke K, Nørgaard M, Benfield TL, Nørgaard-Pedersen B, Mortensen PB, Mors O, Sørensen HT, Harboe1 ZB, Børglum AD, Demontis D, Ørntoft TF, Bisgaard H, Hougaard DM.

Genome-wide scans using archived neonatal dried blood spots samples. *BioMed Central Genomics* 2009;10:297–304.

Holodniy M, Kim S, Katzenstein D, Konrad K, Groves E, Merigan TC. Inhibition of Human Immunodeficiency Virus gene amplification by heparin. *Journal Clinical Microbiology* 1991;29:676–679.

Holtkamp U, Klein J, Sander J, Peter M, Janzen N, Steuerwald U, Blankenstein O. EDTA in dried blood spots leads to false results in neonatal endocrinologic screening. *Clinical Chemistry* 2008;54:602–605.

Holub M, Tuschl K, Ratschmann R, Strnadová KA, Mühl A, Heinze G, Sperl W, Bodamer OA. Influence of hematocrit and localisation of punch in dried blood spots on levels of amino acids and acylcarnitines measured by tandem mass spectrometry. *Clinica Chimica Acta* 2006; 373:27–31.

International Civil Aviation Organization (ICAO). *International Civil Aviation Organization Technical Instructions for the Safe Transport of Dangerous Goods by Air*, Doc 9284, 2005–2006 Edition. http://www.icao.int/publications/Documents/guidance_doc_infectious_substances.pdf (accessed January 12, 2014).

Kalter SS. Disc method for identification and titration of cytopathic viruses and detection of antibodies resulting from their infection. *Journal of Laboratory and Clinical Medicine* 1963;62:525–534.

Karstad L, Spalatin J, Hanson RP. Application of the paper disc technique to the collection of whole blood and serum samples in studies on eastern equine encephalomyelitis. *Journal of Infectious Diseases* 1957;101:295–299.

Kayiran SM, Ozbek N, Turan M, Gürakan B. Significant differences between capillary and venous complete blood counts in the neonatal period. *Clinical and Laboratory Haematology* 2003;25:9–16.

Klas-Bertil A, Heimburger G. The determination of cholinesterase activity in blood samples absorbed on filter paper. *Acta Physiologica Scandinavica* 1953;30:45–54.

Kline MC, Duewer DL, Redman JW, Butler JM, Boyer DA. Polymerase chain reaction amplification of DNA from aged blood stains: quantitative evaluation of the "suitability for purpose" of four papers as archival media. *Analytical Chemistry* 2002;74:1863–1869.

Langer EK, Johnson KJ, Shafer MM, Gorski P, Overdier J, Musselman J, Ross JA. Characterization of the elemental composition of newborn blood spots using sector-field inductively coupled plasma-mass spectrometry. *Journal Exposure Science and Environmental Epidemiology* 2011;21:355–364.

Lindegren ML, Kobrynski L, Rasmussen SA, Moore CA, Grosse SD, Vanderford ML, Spira TJ, McDouga JS, Vogt RF Jr, Hannon WH, Kalman LV, Chen B, Mattson M, Baker TG, Khoury M. Applying public health strategies to primary immunodeficiency disorders: a potential approach to genetic disorders. *The Morbidity and Mortality Weekly Report* 2004;53:1–29.

MacCready RA. Experience with testing for phenylketonuria in Massachusetts. *Pediatrics* 1963;32:308–309.

Mackie IA, Seal DV. (1976). Quantitative tear lysozyme assay in units of activity per microlitre. *British Journal of Ophthalmology* 1976;60:70–74.

McCabe ERB, Huang S-Z, Seltzer WK, Law ML. DNA microextraction from dried blood spots on filter paper blotters: potential applications to newborn screening. *Human Genetics* 1987;75:213–216.

Mei JV, Alexander JR, Adam BW, Hannon WH. Use of filter paper for the collection and analysis of human whole blood specimens. *Journal of Nutrition* 2001;131(5):1631S–1636S.

Mei JV, Zobel SD, Hall EM, De Jesus VR, Adam BW, Hannon WH. Performance properties of filter paper devices for whole blood collection. *Bioanalysis* 2010;2:1397–1403.

Miller M. Maya Masterpiece Revealed at Bonampak. *National Geographic Magazine*, February 1995; pp. 50–69.

Noguchi H. *Serum Diagnosis of Syphilis*, 2nd edn. Philadelphia, PA: JH Lippincott Company; 1911.

Oppelaar L. The use of filter paper as a transport medium for blood and serum. *Tropical and Geographical Medicine* 1966;18:60–66.

Partington MW, Sinnott B. Case finding in phenylketonuria. II. The Guthrie test. *Canadian Medical Association Journal* 1964;91:105–114.

Pellegrino J, Brener, Z. Reacao de fixacao do complement com sangue dessecado no diagnostic do calazar canino. *Revista Brasileira de Malariologia e Doencas Tropicais* 1958;10;39–44.

Rashed MS, Rahbeeni Z, Ozand PT. Application of electrospray tandem mass spectrometry to neonatal screening. *Seminars Perinatology* 1999;23:183–193.

Scheel C, Berry HK. Comparison of serum phenylalanine levels with growth in Guthrie's inhibition assay in newborn infants. *Journal Pediatrics* 1962;61:610–616.

Schmidt V. Ivar Christian Bang (1869–1918), founder of modern clinical microchemistry. *Clinical Chemistry* 1986;32:213–215.

Shaltout AA, Guthrie R, Moussa M, Kandil H, Hassan MF, Dosari L, Hunt MCJ, Fernando NP. Erythrocyte protoporphyrin screening for lead poisoning in bedouin children. A study from Kuwait. *Journal of Tropical Pediatrics* 1989;35:87–91.

Slazyk WE, Phillips DL, Therrell BL Jr, Hannon WH. The effect of lot-to-lot variability in filter paper on the quantitation of thyroxine, thyroid-stimulating hormone, and phenylalanine in dried-blood specimens. *Clinical Chemistry* 1988;34: 53–58.

Songster CL, Barrows GH, Jarrett DD. The fecal smear punch-disc test: a new non-invasive screening test for colorectal cancers. *Cancer* 1980;45:1099–1102.

Strnadova KA, Holub M, Muhl A, Heinze G, Ratschmann R, Mascher H, Stöckler-Ipsiroglu S, Waldhauser F, Votava F, Lebl J, Bodamer OA. Long-term stability of amino acids and acylcarnitines in dried blood spots. *Clinical Chemistry* 2007;53:717–722.

Therrell BL, Hannon WH, Pass KA, Lorey F, Brokopp C, Eckman J, Glass M, Heidenreich R, Kinney S, Kling S, Landenburger G, Meaney FJ, McCabe ERB, Panny S, Schwartz M, Shapira E. Guidelines for the retention, storage, and use of residual dried blood spot samples after newborn screening analysis. *Biochemical and Molecular Medicine* 1996;57:116–124.

Therrell BL Jr, Hannon WH, Bailey DB Jr, Goldman EB, Monaco J, Pedersen BN, Terry SF, Johnson A, Howell RR. Committee report: considerations and recommendations for national guidance regarding the retention and use of residual dried blood spot specimens after newborn screening. *Genetics in Medicine* 2011;13:621–624.

Vaisman A, Hamelin A, Guthe T. Fluorescent antibody technique carried out on dried, eluted blood. I Comparison with FTA, TPI and the tests with lipid antigen carried out on serum. *Bulletin Organization Mondiale de La Sante (Bulletin of the World Health Organization)* 1963;29:1–6.

Van Slyke DD. Ivar Christian Bang. *Scandinavian Journal of Clinical and Laboratory Investigation* 1957;10(Suppl. 31): 18–26.

Verebey K, Eng Y, Davidow B, Ramson A. Rapid, sensitive micro blood lead analysis: a mass screening technique for lead poisoning. *Journal of Analytical Toxicology* 1991;15:237–240.

Walfish PG. Evaluation of three thyroid-function screening tests for detecting neonatal hypothyroidism. *Lancet* 1976;1(7971):1208–1210.

Watson MS, Mann MY, Lloyd-Puryear MA, Rinaldo P, Howell RR. Newborn screening: toward a uniform screening panel and system. *Pediatrics* 2006;117:296–307.

Wolff HL. The filter paper method for shipping blood samples for serological examination. *Tropical and Geographical Medicine* 1958;10:306–308.

World Health Organization (WHO). *Guidance on Regulations for the Transport of Infectious Substances*. Geneva, Switzerland: WHO; 2011–2012 Edition. http://www.who.int/ihr/publications/who_hse_ihr_20100801_en.pdf (accessed January 12, 2014).

Yee HY, Srivuthana K, Elton R, Bhambhani K, Kauffman RE. Capillary blood collection by paper for lead analysis by graphite furnace atomic absorption spectrometry. *Microchemical Journal* 1995; 52(3):370–375.

Zimmermann E. The dried blood test for syphilis. *Munchener Medizinische Wochenschrift* 1939;86:1732–1733.

2

DRIED BLOOD SPOT CARDS

Brad Davin and W. Harry Hannon

2.1 INTRODUCTION

This chapter describes the evolution of the dried blood spot card (DBSC) from refinement of the filter paper to future applications. The chapter describes the importance of the DBSC, the progression of the filter paper, resolution of problems and issues in manufacturing a reproducible homogenous filter paper, and the role of manufacturing the card. The attributes of the DBSC that can enhance the effectiveness for testing are discussed as well as its future applications.

2.2 FILTER PAPER

Filter paper has been used for dried blood spot (DBS) testing for almost a hundred years, as discussed in Chapter 1. The early use of filter paper featured cut sections, paper hole punching, or a precut single sheet (Hogan, 1950; Karstad et al., 1957; Pellegrino and Brener, 1958). It was not until filter paper was used in mass screening that the paper itself was evaluated. Lot-to-lot homogeneity was not monitored and large disparities in the products could be found. The initial filter paper Schleicher & Schuell (S&S) 903 that was chosen by Guthrie in 1960s for blood spot collection featured a rapid adsorption and, therefore, became commonly used in most screening laboratories (Guthrie and Susi, 1963). The filter paper was made solely of cotton linters (Figure 2.1). The manufacturer was able to produce the filter paper by orienting and breaking down the linters into slurry that was of the right density. The filter paper continued to grow in popularity as various industries and laboratories started to experience the utility and benefits of specimen matrix for testing. These benefits included reduced transportation costs; improved sample stability; reduced pathogens; and ease in obtaining samples (Wolff, 1958; Oppelaar, 1966). With this significant increase in use in various testing and inconsistency in lot-to-lot homogeneity of the filter paper, it became quite evident that there was a need to monitor or evaluate the quality of the product to assure the end user that the quality of the spot was consistent. In the late 1970s and early 1980s, the Centers for Disease Control (CDC) began to monitor the production of filter paper utilized in the DBSC (Centers for Disease Control and Prevention, n.d.). Paper manufacturers established the following standard which can be found in the CLSI NBS-01-A6 (Hannon et al., 2013):

1. Filter paper should be made of 100% pure cotton fiber, with no wet-strength additives.
2. Basis weight should be 110 lb ± 5% per ream (179 g/m^2 ± 5%). A ream is defined as 500 sheets 24″ × 36″ (ASTM D646-96).
3. The pH should be 5.7–7.5 (Test method ISO 6588:1981).
4. Ash %: 0.1% maximum (Test method A of ASTM D586-97a).

2.3 FILTER PAPER SOURCES

The CDC then developed and refined the evaluation process as stated in the CLSI standards (Hannon et al., 2013). By developing this testing process and working with the supplier of filter paper, CDC was able to demonstrate that the paper itself was comparable lot to lot. In the early 1990s, a new source of filter paper entered the market. The new filter paper, BFC180, was manufactured by Whatman in the United Kingdom. Whatman proceeded through proper Food and Drug Administration (FDA) channels and obtained a

Dried Blood Spots: Applications and Techniques, First Edition. Edited by Wenkui Li and Mike S. Lee.

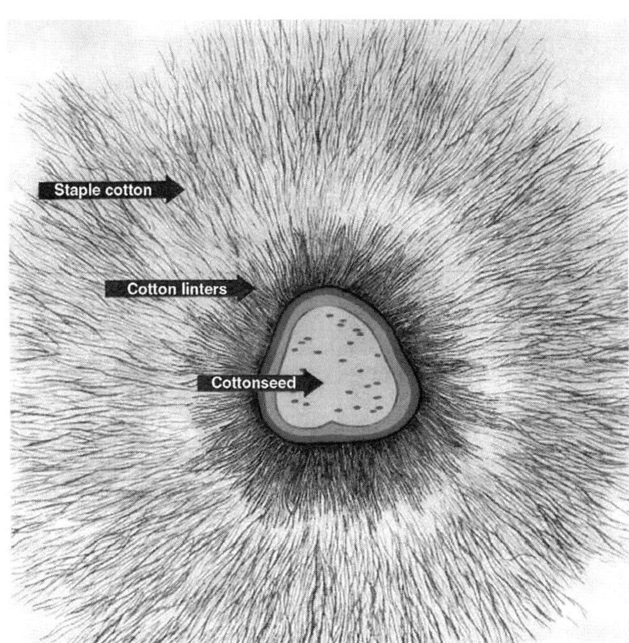

FIGURE 2.1 An illustration of cotton linters.

510K approval from the FDA. Two choices for filter paper product were now available in the market. The FDA clearance on the product was important for a number of reasons. Without this approval the end user had no assurance that the supplier had gone through the required steps necessary to meet the requirements of the product. In addition, the CDC would not include the filter paper in the annual evaluation or its pre-lot certification process unless it was 510K approved. Whatman purchased Schleicher and Schuell in 2005 and removed the BFC180 from the general market. In 2007, a third paper was introduced and approved by the FDA with a 510K approval. This product was labeled 226 and was manufactured by Ahlstrom. The paper is now labeled PerkinElmer 226. Again, two choices of suppliers for filter paper were available for the newborn screening (NBS) community.

In the early 2000 time frame, the pharmaceutical companies began to evaluate filter paper as means of collecting and transporting blood samples for drug discovery and development. This forced the filter paper manufacturers to more closely evaluate the manufacturing processes in order to supply a higher quality product for quantitative assays without concern.

Filter paper has been shown to be a solid product over the years but does require careful scrutiny in order to ensure that it will provide the best transport for DBSs and that the paper itself will not intrude upon the laboratory testing results. Filter paper manufacturers must scrutinize the source of cotton linter for the filter paper, environmental influences in the air and soil such as lead, pesticides, herbicides, radioactive release, or heavy metals during the growth process. Water used to process the filter paper needs to be evaluated closely to make sure that there is no additional minerals or chemicals introduced. The temperature that the raw stocks and final product are exposed to need to be monitored as well, since cotton has been shown to have matrix memory and this memory can affect the results for various compounds.

2.4 GUIDELINES AND PRINTING

The variables in the manufacturing process led to the development of the LA4 (now NBS01) standards by the CLSI (formerly known as NCLS) in the early 1980s. The standards defined the level of integrity necessary for accuracy in DBS testing for NBS (Hannon et al., 2013). The LA4 document is now undergoing its sixth routine revision (revisions happen every 5 years by CLSI policy) and it will become renamed NBS01 at next revision. Standards for evaluation of the raw filter paper were outlined and specific emphasis was given to the manufacturing process for the DBSC, so that the process would not alter the characteristics of the filter paper. These well-founded standards required manufacturers of the DBSC to utilize a procedure that monitors the process and incorporates dedicated ink and glue delivery systems that prevent the type of contact with the paper that alters the flow of the blood or affects the ultimate results of an analytical test. The reason for the dedicated ink and glue systems is well founded. The filter paper is highly absorbent and can easily take up any chemicals used in the day-to-day process of printing. For this reason the applicators (plates) that come in contact with the filter paper need to be free of any other chemicals or additives. The only printing method that is acceptable for use with filter paper is a flexographic method. This method incorporates a raised plate in the area in which the ink is going to be applied to the filter paper. By utilizing this method, no other contact to the filter paper takes place except in the raised area. If the filter paper is being handled as a single sheet during the manufacturing process, then the method that is used to pick up or transport the paper must in no way affect the thickness of the paper by causing compression of the paper. If the manufacturing is a roll-to-roll procedure, then great care must be exercised to maintain minimal tension (stretch) for the filter paper. In this way, the fiber orientation or thickness will not be modified. In addition, the filter paper and card manufacturers must be careful how tight the rolls are re-rolled upon completion of the production run. Rolling of the finished paper can compress the filter paper and, in turn, alter its absorbency characteristics. The production standards that have been developed and incorporated over the years have transformed DBSC to a stable means of collecting spots of blood and transporting and storing them for future applications. While the integrity of the DBS is of paramount importance, the need for additional features became necessary with expansion of their applications. The need to capture demographic information and the ability to accurately track the sample became apparent as mass screening (MacCready, 1963) encompassed linking test results to specific individuals and their captured demographic data. To accomplish this linkage, the DBSC evolved from a single sheet of filter paper to a more complex sample collection device (card) utilizing additional features. The types of features were driven by the higher volumes of cards being processed, the need to validate the quality, the need to collect more patient demographics/data, and the need to accurately track the sample.

2.5 PRINTING ADAPTATIONS

One of the earliest features in the DBSC was the modification of unprinted filter paper to a printed card. The purpose was often to allow the end user to have a specific target such as a printed circle. The size of the printed circle could be varied to allow for the target application of blood and was often controlled by the type of test or number of tests or state or country where sample was collected (local rules and regulations). In the United States, the circle size is 12.7 mm for NBS. In Europe, the size is predominately 10 mm. This small difference can have a significant impact on the sample as well as the provider. A blood spot of 20–30 μL can be applied to the smaller 10 mm circle versus a need for a 55–65 μL requirement for the 12.7 mm circle. As demonstrated, the volume of blood application could affect the distribution of analytes (Mei et al., 2010). The circles started out as solid line circles sometimes printed on both the face and back of the filter paper. This feature created two problems which became evident as the quantifiable attributes became more stringent for the DBSC. The pure physics of printing ink on a paper creates a barrier. If the paper is pictured in its true form—two surfaces (front and back) as well as the center area—then it is evident that by applying blood to the center of the circle, the blood would run into resistance on the surface where the ink has been applied. If observed in slow motion, when the blood's movement makes contact with the printed circle and then flows under the circle and resurfaces on the other side. To compound this issue, if the DBSCs are printed on both sides, then there is a barrier on each side. These barriers create a damming effect that forces blood to pool within the circle and generate a super saturation that alters the volume within a given punch size. Depending on the quality of the manufacturing process, printing the circles can add a margin of compression (calendering) that causes an increased restriction to the blood's movement through the paper. It is for this reason that the preferred marking for the circles should always be a one-side slotted, printed circle. The slotted circles offer less surface resistance allowing the blood movement to proceed freely on the surface as well as in the middle layers with only minimal resistance. The slotted circles also provide a quality assurance indicator in that impermeable ink will create a floret pattern with blood flow.

2.6 FURTHER INNOVATIONS IN MANUFACTURING

In addition to the preprinted circle, the DBSC would often have an area designated for collection of demographic information such as a name and date of sample collection and other patient specific information. Historically, on the original sample collection forms, demographic information was simple and incorporated onto the filter paper sheet. As the number of tests and screening complexity increased, the need to track the samples in a more organized fashion led to the inclusion of Arabic numbering on the filter paper cards.

2.7 FILTER PAPER CARDS CONSTRUCTION

The single-sheet DBSC provided a lot of the basic features required during the early development of the filter paper card and the NBS efforts. However, the number of conditions continued to grow and the need for more information to accompany the samples forced a change in protocols. In an attempt to attach or link the sample to the demographics, it was common to observe the use of a sheet of paper stapled or clipped to the DBS filter paper sample. This procedure did not work well and created numerous compromised samples. The solution to this problem was the development of the tip-on production procedure for DBSCs. This was simply the process of taking a smaller section of filter paper and attaching it by gluing it to the demographic section, not constructed of filter paper. The demographic section could be composed of only one part of heavy-weight paper or with multiple layer removable forms and the attached sample collection filter paper. This type of DBSC would allow for proper chain of custody as well as permanent records at more than one site. Fully automated attachments quickly became the standard as early as the mid-1960s. During this time various glues were tested to allow for the strongest adhesion as well as characteristics that would not adversely affect the analytical test. A number of benefits were allowed by this feature. The end product of the process that now had the filter paper attached allowed the manufacturer of the DBSC to add various complex manufacturing features to the final product while not exposing the filter paper section to adverse environments. The process also permitted the end user (screening laboratories) to reduce their costs by utilizing a smaller quantity of filter paper for DBSC production. By incorporating the filter paper as a "tip-on" attachment to the DBSC, we now have almost unlimited flexibility to capture the demographic information. The "tip-on" also allowed the end user to remove the filter paper section if it was properly identified (barcode, radio frequency identification technology (RFID) chip, label) and store only the DBS sample. This feature was in itself a great cost savings. By attaching the filter paper to the demographic section the manufacturer was able to incorporate a barcode onto the DBSC. Initial attempts to apply the barcode directly

FIGURE 2.2 An illustration of a split barcode used for blood spot collection cards.

to filter paper were problematic with few solutions. If the barcode was applied directly to the filter paper, then the print quality was poor. In addition, the application of the barcode created a new exposure environment for the filter paper and an additional manufacturing step both of which could compromise the quality of the DBSC. An attempt to apply the barcodes directly to the filter paper section was made with adhesive labels but the adhesives would saturate the paper and could cause long-term contamination of the filter paper. By incorporating the "tip-on" method, the filter paper was not compromised and the card itself could have additional features. A split barcode that allows complete separation of the sample from the demographic information without loss of linkage to each piece is commonly used at this time, since it ensures that there is no possibility of numerical mismatch or loss identity (Figure 2.2).

2.8 WRAPAROUND COVER AND CASSETTE CONSTRUCTION

The volume of samples that need to be transported by the postal system or other transportation system continued to grow. Along with this came the need to adapt the DBSC card to allow safe and quality transportation which initiated the need to create the next features, the wraparound cover for the DBS. This feature simply took one of the parts of the multipart card and allowed it to be wrapped around the filter paper sample collection area (Hannon et al., 2013). By adding this feature, the end user was able to add protection to the sample area prior to collection by limiting the exposure of the paper while adding an extra cover for transporting the collected spots. A common issue found when collecting the blood for spotting was that the person obtaining the sample would contaminate the sample area prior to collection by making contact with hands or some other surface. This contact often introduced hand lotions, alcohol, body oil, or DNA contamination. By keeping the sample area covered up to the point of collection and then recovering the collected sample after drying, the potential for contamination is minimized. The wraparound cover helped satisfy future regulations to achieve multiple layered coverage protection in the transport of DBS [World Health Organization (WHO); Guidance on Regulations for the Transport of Infectious Substances (WHO, 2011–2012), the International Civil Aviation Organization's Technical Instructions for Safe Transport of Dangerous Goods by Air (International Civil Aviation, 2012) and the Newborn Screening Quality Assurance Program, Centers for Disease Control and Prevention (CDC)] (Centers for Disease Control and Prevention, n.d.). The next major change in the DBSC took place in the early 1990s when a group of scientists developed a universal DBSC. By locking in on a universal product, the collection, transport, and laboratory process could all be automated. Demand for more efficient methods of distribution of samples from collection to the laboratory as well as within the laboratory for sorting them for a variety of tests led to this change. The solution was the development of a product called the cassette. The cassette in its original format was a three-layered device incorporating a supportive material on the top and bottom with the filter paper sandwiched in the middle section glued together (see Appendix E of CLSI NBS-01-A6; Hannon et al., 2013). The material used to sandwich the filter paper originally was made of a "tag stock" (heavy-weight paper). As equipment was developed to handle the cassettes, it became quite evident that the cassette must be rigid and not subject to distortion in shape. The preferred material at this time is a "tag stock" which can be processed in a rotary-type press. However, the rigidity of the actual cassette comes from the glued multi-layered structure. Distortion in the shape of the cassettes became an issue because of the wet blood application and its drying created a shrinking of the paper that wraps the cassette. For this reason, plastic encasements with flexible glue were evaluated as an option. An additional benefit of the cassette construction was by sandwiching the filter paper it creates an elevated edge to separate cassettes from each other when stacked, which protects the filter paper and samples from cross-contamination. It has been shown that when two DBSs were rubbed abrasively together there could be DNA cross-contamination (Cordovado et al., 2009). Cassettes allow for easy use of automated sampling devices for application of the blood, bulk transport, and automated punching or extraction processes that can be coupled with downstream liquid handling for test reagents. Industries utilized various sizes of cassettes depending on the number of samples required for their specific application.

2.9 FILTER PAPER PRE-TREATMENT

The DBSC has shown over the years that if properly manufactured utilizing an approved filter paper, it can consistently provide high-quality results. In the earlier years of its usage, the DBSC did not employ tight tolerance limits and the product was used primarily as a semiquantitative matrix for mass screening applications. The DBSC has undergone considerable changes and has shown that it can transport blood as well as any other biological sample container. Over time, it has demonstrated high accuracy and capacity to consistently yield high-quality results. DBSC can be used for any type of assay and can be stored for long periods of time under appropriate conditions with reproducible results (Therrell et al.,

1996; 2011). They can also be further modified to allow for spot-specific chemistry. This spot-specific application could allow the end user to apply standards or other reagents to one collection circle while maintaining the non-biased area on the same card. By placing in-line monitoring systems within the manufacturing process, we can ensure the end user that they will receive a high-quality homogenous filter paper for sample collection.

2.10 THE FUTURE OF THE FILTER PAPER CARD

Does advancement of the DBSC and its future applications stop here? The opinion of the authors is that it will not! There is a strong movement toward incorporating the RFID chips into the DBSCs. This feature alone will allow the individuals who are storing the blank card and spotting the DBSC with the sample as well as all subsequent downstream users to imbed information, such as who handled the card, what tests were performed, what the results were to date, what future test should be performed, when results were reported, what were the sample storage environments, and also to document what all the events were associated with storage. The amount and type of information is virtually unlimited. This advancement in DBSC technology will also allow for the ability to gather large amounts of data by screening a collective group of cards at one time. The materials used to hold or contain the filter paper that permit faster automation as well as enhance the security of designed sample carriers are evolving. There has been a strong growth from using the DBSC for blood to using it to collect any biological sample, such as, urine, fecal, synovial fluid, saliva, buccal, epidermal scrapes, and plants tissue (McDade et al., 2007). Where once the DBSC was simply a sample collection device for screening, it is now used in IVD markets such as prenatal screening, vitamin D and steroid testing, medical microbiology, and DNA extraction as well as for home collection of samples for clinical chemistry tests, for infectious diseases, and for therapeutic drug monitoring. In the research use only (RUO) markets we are observing its use in bioanalysis (DMPK) TK and TDM, drug metabolism and pharmacokinetics, and toxicokinetics.

REFERENCES

Centers for Disease Control and Prevention (CDC). *Newborn Screening Quality Assurance Program.* (n.d.) http://www.cdc.gov/labstandards/nsqap.html (accessed January 12, 2014).

Cordovado SK, Earley MC, Hendrix M, Driscoll-Dunn R, Glass M, Mueller PW, Hannon WH. Contamination assessment of DNA from dried blood spots after specimen-to-specimen contact and determination of DNA yield and function using archival neonatal dried blood spots. *Clinica Chimica Acta* 2009:402; 107–113.

Guthrie R, Susi A. A simple phenylalanine method for detecting phenylketonuria in large populations of newborn infants. *Pediatrics* 1963;32:338–343.

Hannon WH, DeJesus VR, Chavez MS, Davin BF, Getchell J, Green M, Hopkins PV, Kelm KB, Noorgaard-Pedersen B, Padilla C, Plokhovy E, Mei JV, Therrell BL. *Blood Collection on Filter Paper for Newborn Screening Programs: Approved Standard, 6th edn.* Wayne, PA: Clinical and Laboratory Standards Institute (CLSI); CLSI Document NBS-01-A6, 2013:1–37.

Hogan, R. Filter paper microscopic test for syphilis. *Journal of Venereal Disease Information* 1950;31:37–45.

International Civil Aviation Organization. *Infectious Substances: Technical Instructions for the Safe Transport of Dangerous Goods by Air, Guidance Document, 2005-2006.* Montreal, Canada: ICAO; 2005. http://www.icao.int/icaonet/dcs/9284.html (accessed November 19, 2012).

Karstad L, Spalatin J, Hanson RP. Application of the paper disc technique to the collection of whole blood and serum samples in studies on eastern equine encephalomyelitis. *Journal of Infectious Diseases* 1957:101;295–299.

MacCready RA. Experience with testing for phenylketonuria in Massachusetts. *Pediatrics* 1963;32:308–309.

McDade TW, Williams S, Snodgrass JJ. What a drop can do: dried blood spots as a minimally invasive method for integrating biomarkers into population-based research. *Demography* 2007;44:899–925.

Mei JV, Zobel SD, Hall EM, De Jesus VR, Adam BW, Hannon WH. Performance properties of filter paper devices for whole blood collection. *Bioanalysis* 2010;2:1397–1403.

Oppelaar L. The use of filter paper as a transport medium for blood and serum. *Tropical and Geographical Medicine* 1966;18:60–66.

Pellegrino J, Brener, Z. Reacao de fixacao do complement com sangue dessecado no diagnostic do calazar canino. *Revista Brasileira de Malariologia e Doencas Tropicais* 1958;10;39–44.

Therrell BL Jr, Hannon WH, Bailey DB Jr, Goldman EB, Monaco J, Pedersen BN, Terry SF, Johnson A, Howell RR. Committee report: considerations and recommendations for national guidance regarding the retention and use of residual dried blood spot specimens after newborn screening. *Genetics in Medicine* 2011;13:621–624.

Therrell BL, Hannon WH, Pass KA, Lorey F, Brokopp C, Eckman J, Glass M, Heidenreich R, Kinney S, Kling S, Landenburger G, Meaney FJ, McCabe ERB, Panny S, Schwartz M, Shapira E. Guidelines for the retention, storage, and use of residual dried blood spot samples after newborn screening analysis. *Biochemical and Molecular Medicine* 1996;57:116–124.

Wolff HL. The filter paper method for shipping blood samples for serological examination. *Tropical and Geographical Medicine* 1958;10:306–308.

World Health Organization (WHO). *Guidance on Regulations for the Transport of Infectious Substances.* Geneva, Switzerland: WHO; 2011–2012. http://biosafety.moh.gov.sg/home/uploadFiles/Common/WHO Guidance on Regulations for Transport of Infectious Substances –PDF (accessed November 20, 2012).

3

DRIED BLOOD SPOT SAMPLE COLLECTION, STORAGE, AND TRANSPORTATION

Joanne Mei

3.1 INTRODUCTION

Dried blood spots (DBS) prepared in the filter paper matrix have straightforward guidelines for collection, storage, and transportation. In the newborn screening community, it has been well recognized that DBS provide the best specimen for testing newborn babies (De Jesús et al., 2010; Therrell, 2010). The collection of DBS is minimally invasive and relies on a heelstick for newborn screening or a fingerstick for older patients. Filter paper used for DBS should be cleared (approved) by the Food and Drug Administration (FDA) for whole blood collection. The paper disk that is punched to aliquot DBS specimens is a volumetric measurement, and just like any collection device, it requires a level of uniformity among and within production lots (Mei et al., 2001). Filter paper has associated with it some level of imprecision that can be characterized so that the device can be standardized to minimize the variation in measurements due to the paper matrix. The ease of collection, transport, and storage make DBS a cost-effective choice for many clinical applications. The Centers for Disease Control and Prevention's (CDC) Newborn Screening Quality Assurance Program (NSQAP), with the cooperation of filter paper manufacturers, routinely evaluates all lots of paper and compares them to the performance of previous lots (NSQAP Current Reports, 2012). The acceptable performance criteria for filter paper are detailed in a Clinical and Laboratory Standards Institute (CLSI) approved standard (CLSI, NBS-01-A6, 2013). The standard describes the minimum criteria for serum absorbance volume for a 1/8-inch (3.2 mm) disk, the diameter of the circle for a 100 μL dried blood aliquot, and the absorption time for the blood aliquot. These endeavors document national and international efforts to standardize the filter paper device for whole blood collection. To that end, proper collection, storage, and transportation efforts ensure that DBS provides an appropriate testing matrix for immediate patient care and can also preserve specimens for future testing (Therrell et al., 1996; Chrysler et al., 2011; Therrell et al., 2011; CLSI NBS-01-A6, 2013).

3.2 GENERAL PROCEDURES AND PRECAUTIONS FOR DBS SAMPLE COLLECTION

3.2.1 Reagents and Materials

The filter paper collection device is made of a pure cotton fiber (100%) paper designed to meet specific criteria for the diameter of the circle and for the absorption time of a 100 μL blood aliquot (Mei et al., 2001; 2010; CLSI NBS-01-A6, 2013). The FDA has registered two commercial sources of filter paper for blood collection as Class II Medical Devices (21 CFR §862.1675). NSQAP analyzed a large array of newborn screening analytes in blood prepared on each FDA-registered filter paper source and found that the difference in analytical results between manufacturers was within 4–5% comparability or, at a minimum, equal to the lot-to-lot variance of a single manufacturer's filter paper products (Mei et al., 2010; NSQAP Current Reports, 2012).

In addition to the CLSI standard for blood collection on filter paper (CLSI NBS-01-A6, 2013), laboratory standards exist for the proper techniques for skin puncture in adults and older children (CLSI H04-A6, 2008). Automated skin puncture devices (retractable lancets) should be used to puncture

Dried Blood Spots: Applications and Techniques, First Edition. Edited by Wenkui Li and Mike S. Lee.

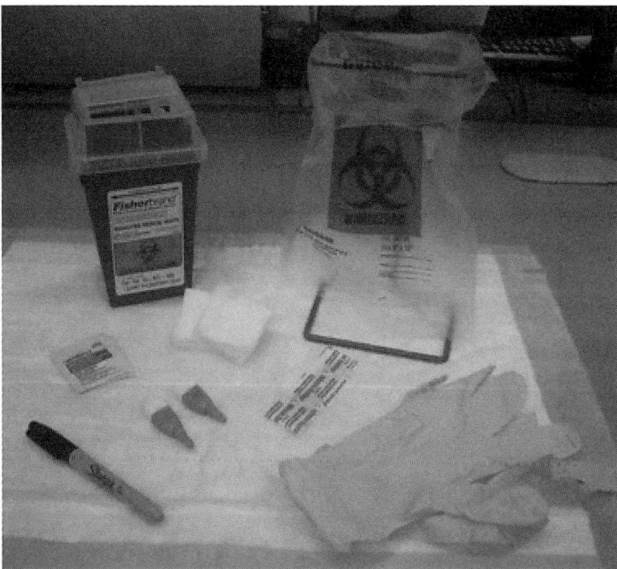

FIGURE 3.1 Supplies needed for the collection of dried blood spots: alcohol wipes (70% isopropanol), gauze pads, sharps disposal container, biohazard bag, gloves, bandages, lancets, and a pen to label specimens.

skin. For newborn heelstick collection, lancets must not puncture the skin deeper than 2.0 mm while lancets that puncture no deeper than 0.85 mm can be used for low birth weight or premature infants (CLSI H04-A6, 2008). Specially made skin puncture devices for collection of DBS are available commercially (Reiner et al., 1990; Program for Appropriate Technology in Health, 2005).

Other materials needed for collecting DBS are as follows: (1) personal protection equipment (PPE); (2) 70% isopropyl alcohol (isopropanol/water: 70/30 by volume) to cleanse the puncture site; (3) sterile gauze to wipe away the first drop of blood; (4) a drying rack or other mechanism that allows the blood spots to air dry in a horizontal position (Figure 3.1); and (5) supplies for storage and shipment of DBS (see Section 3.3).

3.2.2 Biosafety and Infection Control

DBS are prepared from capillary blood which is a combination of venous blood, arterial blood, and tissue fluid (CLSI H04-A6, 2008). All human blood is considered potentially infectious for human immunodeficiency virus, hepatitis B virus, hepatitis C virus, and other blood-borne pathogens. Biosafety and infection control are critical to minimizing exposure to biological fluids and standard precautions should be followed when handling blood specimens (CLSI M29-A3, 2005b; CDC, 2007; www.cdc.gov;). Disposable gloves, a laboratory coat or gown, and protective eyewear should be worn when collecting blood. Gloves should be changed after each patient or when soiled and hands should be washed or sanitized often and between patients. Lancets, needles, or other sharp objects should be disposed in durable, puncture-resistant biohazard waste containers (Figure 3.1). Spills should be cleaned up immediately using a solution of freshly prepared 10% bleach (CLSI M29-A3, 2005b). Follow institutional procedures for safely disposing of solid and liquid biological waste.

3.2.3 Specimen Collection

3.2.3.1 ***General Procedures*** The CLSI standard H04-A6 is an excellent resource for the basic procedures for collecting capillary blood specimens (CLSI H04-A6, 2008). The procedures discussed here will only focus on the preparation of DBS. Follow all manufacturers' instructions for the use and handling of automated skin puncture devices. Before collecting the DBS assemble the needed supplies and put on PPE such as disposable gloves, laboratory coat, and protective eyewear. Confirm the identity of the patient using at least two unique identifiers and ensure that all data elements on the collection form are complete (CLSI H04-A4, 2008; CLSI NBS-01-A6, 2013).

3.2.3.2 ***Newborn Heelsticks*** Confirm the identity of the infant using two unique identifiers. This is done by confirming the infant's name, birthdate, identification number, and/or address with the infant's hospital wrist band, parent, or caregiver. After positively identifying the patient, wash or sanitize hands and put on proper PPE.

For a more successful blood draw, it is best to do preliminary preparation on the puncture site. Pre-warm the site that will be punctured with a commercial warming device or warm towel to increase blood flow to the area. The temperature of the warming device should not be greater than 105°F. This will greatly increase the blood flow for a heelstick in order to collect all the blood that is needed. Massage the patient's leg and heel prior to puncturing to increase blood flow. This can be done by massaging the leg from knee to heel and by massaging the heel from toes to heel. Repeat three to five times.

The puncture site will be on the lateral or medial outer edges of the fleshy part of the heel (see Figure 3.2). Clean the site with 70% isopropyl alcohol and allow to air dry completely. Place the lancet on the heel where the puncture will be made. For best results for creating full drops of blood, make sure the incision will be perpendicular to the heel print so that the blood does not run down the folds of the heel print. After the puncture has been made, wipe away the first drop of blood. This is important because it can contain excess tissue fluids and can cause erroneous newborn screening test results (McCall and Tankersley, 2012). Apply gentle and intermittent pressure to the site to stimulate blood flow. Allow a drop of blood to form then touch the underside of the drop of blood

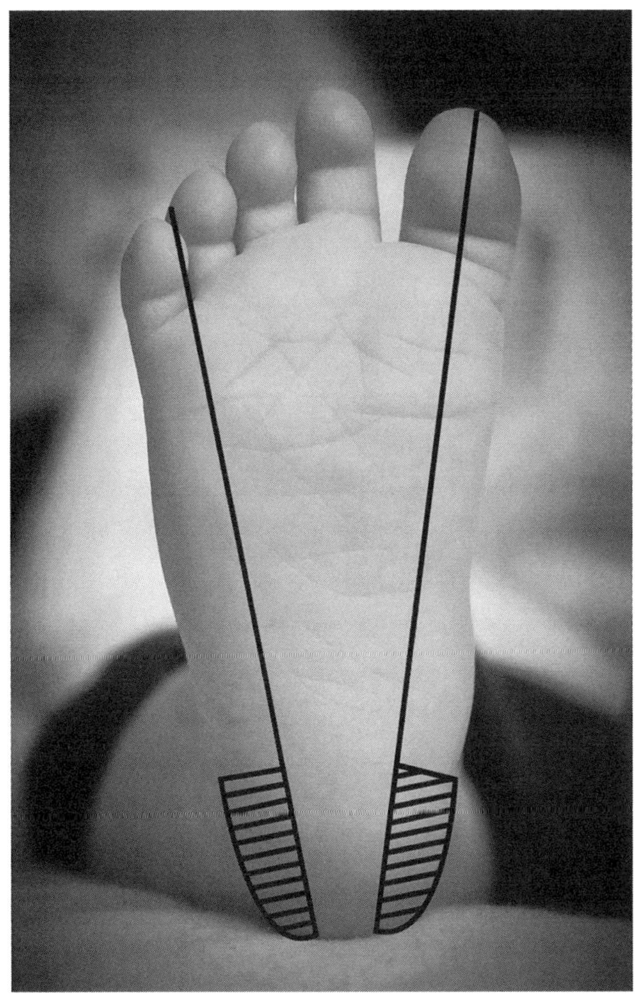

FIGURE 3.2 Skin puncture in newborns should be made within the fleshy portion of the heel (shaded areas). Image provided with permission from the Clinical and Laboratory Standard Institute.

(a)

FIGURE 3.3 (a) A valid DBS specimen has a sufficient quantity of blood that has been allowed to soak through the filter paper to completely fill the preprinted circles on the collection card. Modified and reprinted with permission from the Clinical and Laboratory Standard Institute. (b) Invalid or unsatisfactory specimens. Modified and reprinted with permission from the Clinical and Laboratory Standard Institute. (c-1) DBS specimens showing the back side of a collection card with three fully filled, valid specimens and two specimens with insufficient quantities of blood for testing. Several 1/8 inch punches were removed from the valid spots. It should be noted that the DBS looked acceptable from the front side of the card. Modified and reprinted with permission from Dr Joseph Orsini, New York Department of Health. (c-2) Two DBS specimens appear to be diluted or contaminated compared to the valid specimens on the same collection card. Modified and reprinted with permission from Dr Joseph Orsini, New York Department of Health. (c-3) Serum rings can be seen on two DBS specimens in this example. Modified and reprinted with permission from Dr. Joseph Orsini, New York Department of Health. (c-4) DBS collection card showing four valid specimens and one layered specimen, which was unsatisfactory for testing. Modified and reprinted with permission from Dr Joseph Orsini, New York Department of Health.

to the filter paper and let it soak through the paper and completely fill the preprinted circle. Fill all preprinted circles on the filter paper card. Failure to collect an adequate number of blood spots may invalidate the specimen for some tests and could delay the treatment of affected newborns. Make sure the blood specimen has completely soaked through the card and take care not to layer drops of blood. See Figures 3.3a, 3.3b, and 3.3c for depictions of valid and invalid DBS specimens.

After the specimen has been collected, place gauze over the site and maintain pressure until the bleeding stops. Ensure all specimens are labeled correctly. Remove all collection supplies from the patient area and dispose of the lancet in a biohazard sharps container. Remove and dispose of gloves and wash hands.

Allow the blood spot to dry for a minimum of 3 hours at ambient temperature. See Section 3.2.5 for instructions on correctly drying the blood spots.

3.2.3.3 Older Children and Adults The phlebotomist should first identify the patient by asking the patient (or his/her parent or responsible caregiver) their full name, address, and/or birth date (CLSI H04-A6, 2008). For infants, confirm their identity using two unique identifiers. This is done by asking the patient's name, birthdate, identification number, and/or address or by verifying the information on the patient's hospital wristband if available. After positively identifying the patient, wash or sanitize hands and put on proper PPE.

For fingersticks, which should only be performed on patients 1 year of age or older, the puncture site should be on the palmar surface of the fingertip (fleshy pad) of the middle or ring finger on the non-dominant hand (Figure 3.4) (CLSI H04-A6, 2008). Wipe the site with 70% isopropyl alcohol and allow the skin to air dry. Puncture the site using an automated skin puncture device and discard the device in

Simple Spot Check

Valid specimen:

Allow a sufficient quantity of blood to soak through to completely fill the preprinted circle on the filter paper. Fill all required circles with blood. Do not layer successive drops of blood or apply blood more than once in the same collection circle. Avoid touching or smearing spots.

Invalid specimen: / Possible causes:

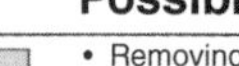

1. Specimen quantity insufficient for testing.

- Removing filter paper before blood has completely filled circle or before blood has soaked through to second side.
- Applying blood to filter paper with a capillary tube.
- Allowing filter paper to come into contact with gloved or ungloved hands or substances such as hand lotion or powder, either before or after blood specimen collection.

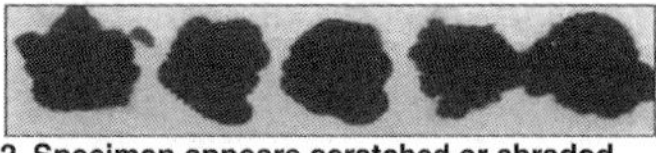

2. Specimen appears scratched or abraded.

- Applying blood with a capillary tube or other device.

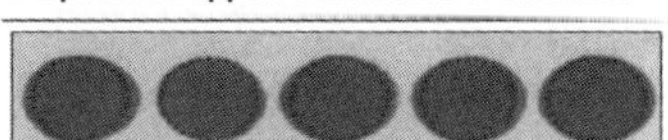

3. Specimen not dry before mailing.

- Mailing specimen before drying for a minimum of 3 hours.

4. Specimen appears supersaturated.

- Applying excess blood to filter paper, usually with a device.
- Applying blood to both sides of filter paper.

5. Specimen appears diluted, discolored or contaminated.

- Squeezing or "milking" of area surrounding the puncture site.
- Allowing filter paper to come into contact with gloved or ungloved hands or substances such as alcohol, formula antiseptic solutions water, and hand lotion or powder either before or after blood specimen collection.
- Exposing blood spots to direct heat.

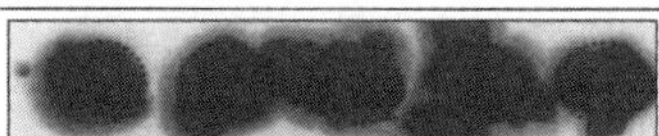

6. Specimen exhibits serum rings.

- Not allowing alcohol at the puncture site to air dry before making skin puncture.
- Allowing filter paper to come into contact with alcohol, and hand lotion.
- Squeezing area surrounding puncture site excessively.
- Drying specimen improperly.
- Applying blood to filter paper with a capillary tube.

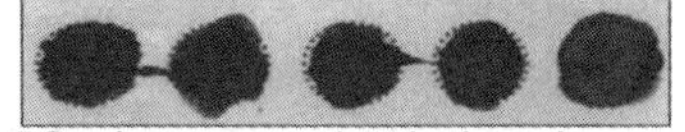

7. Specimen appears clotted or layered.

- Touching the same circle on filter paper to blood drop several times.
- Filling circle on both sides of filter paper.

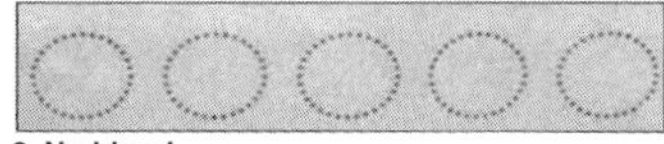

8. No blood.

- Failure to obtain blood specimen.

(b)

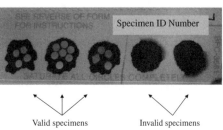

Specimens with insufficient quantity for testing — Valid specimens

(c-1)

Specimen ID Number

Invalid specimens — Valid specimens

(c-3)

Specimen ID Number

Valid specimens — Invalid specimens

(c-2)

Specimen ID Number

Valid specimens — Invalid specimen

(c-4)

FIGURE 3.3 (*Continued*)

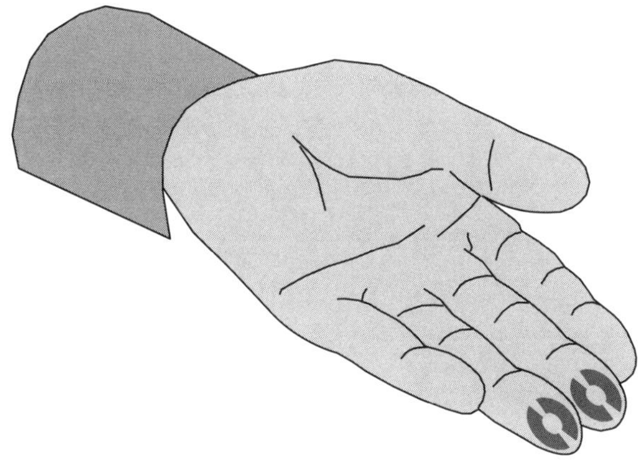

FIGURE 3.4 Skin puncture in older children (greater than 1 year) and adults should be on the planar surface of the middle or ring finger.

a sharps' container. Wipe away the first drop of blood with sterile gauze as the initial drop may contain tissue fluid that could compromise the sample. Gentle pressure should be applied to encourage drops of blood to form but avoid "milking" the site that may squeeze tissue fluid into the drop of blood. Touch the filter paper card to the blood drop and completely fill each preprinted circle with a single application of blood (Figure 3.5) (CLSI H04-A6, 2008). Apply blood only to the printed side of the paper. Do not layer blood drops on top of one another. Observe the back of the card to ensure that blood has completely soaked through the paper (Figure 3.3). For fingerstick specimens, apply an adhesive bandage to the puncture site. Ensure all specimens are labeled correctly. Remove all collection supplies from patient area and dispose of the lancet in a biohazard sharps container. Remove and dispose of gloves and wash hands.

For a more successful blood draw, it is best to do preliminary preparation on the puncture site. Pre-warm the site that will be punctured with a commercial warming device or warm towel to increase blood flow to the area. The temperature of the warming device should not be greater than 105°F. Massage the patient's finger from hand to ring tip to further stimulate blood flow. Repeat three to five times.

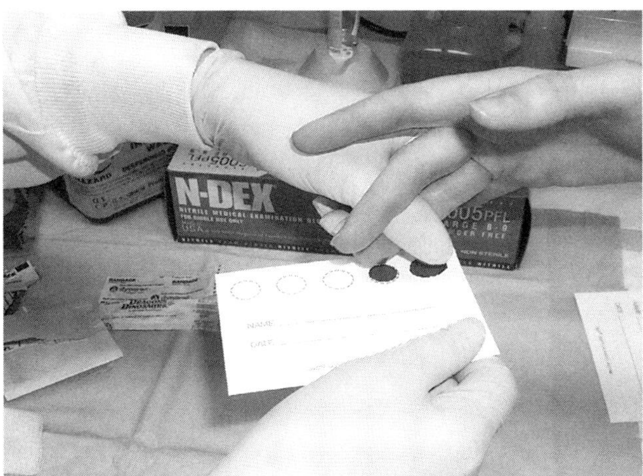

FIGURE 3.5 Application of blood from a fingerstick directly to filter paper. Allow a drop of blood to form and touch the drop to the filter paper.

Allow the blood spot to dry for a minimum of 3 hours at ambient temperature. See Section 3.2.5 for instructions on correctly drying the blood spots.

3.2.4 DBS Prepared by Other Collection Methods

Although they are not the method of choice for preparing DBS, a variety of heparinized microcollection containers are available for capillary blood collection (CLSI NBS-01-A6, 2013). The use of devices containing ethylene diamine tetraacetic acid (EDTA) should be avoided as the EDTA may interfere with some laboratory tests and can give false-negative results (Holtkamp et al., 2008). It should also be noted that heparin can interfere with molecular methods (CLSI MM13-A, 2005a). Glass capillary tubes pose a safety hazard due to the potential for breakage and should be used with caution.

3.2.4.1 Capillary Tube or Microcollection Containers Skin puncture blood specimens may be collected into a capillary tube or into the top of a microcollection container (CLSI H04-A6, 2008). In order to collect blood in a capillary tube, follow fingerstick or heelstick puncture procedures in the previous sections. Hold the tube above or beside the puncture site and touch the end of the tube to the blood drop to allow blood to flow into the tube via capillary action (Figure 3.6a) (McCall and Tankersley, 2012). A fresh capillary tube containing an anticoagulant should be used to fill each circle on the blood collection card (CLSI NBS-01-A6, 2013). Do not remove the tube from the blood drop as this could create an air space which could compromise the volume of blood needed to fill the tube. Fill the tube to the calibration mark and then gently invert the tube two or three times without letting the blood come out of the end to ensure the blood has adequately mixed with the anticoagulant. Lay the full capillary tube aside and collect the next drop in a capillary tube. Repeat this procedure to collect the needed number of tubes. After all tubes have been collected, follow previous procedure on stopping patient's blood flow.

At this time, immediately apply the blood from one capillary tube to the center of a preprinted circle (CLSI NBS-01-A6, 2013). Do not touch the tube to the paper and avoid repeated actions that might scratch or compress the paper. Apply blood to only one side of the paper and use a fresh capillary tube for each circle to be filled (CLSI NBS-01-A6, 2013).

In order to collect blood into a microcollection container, follow fingerstick or heelstick puncture procedures in the

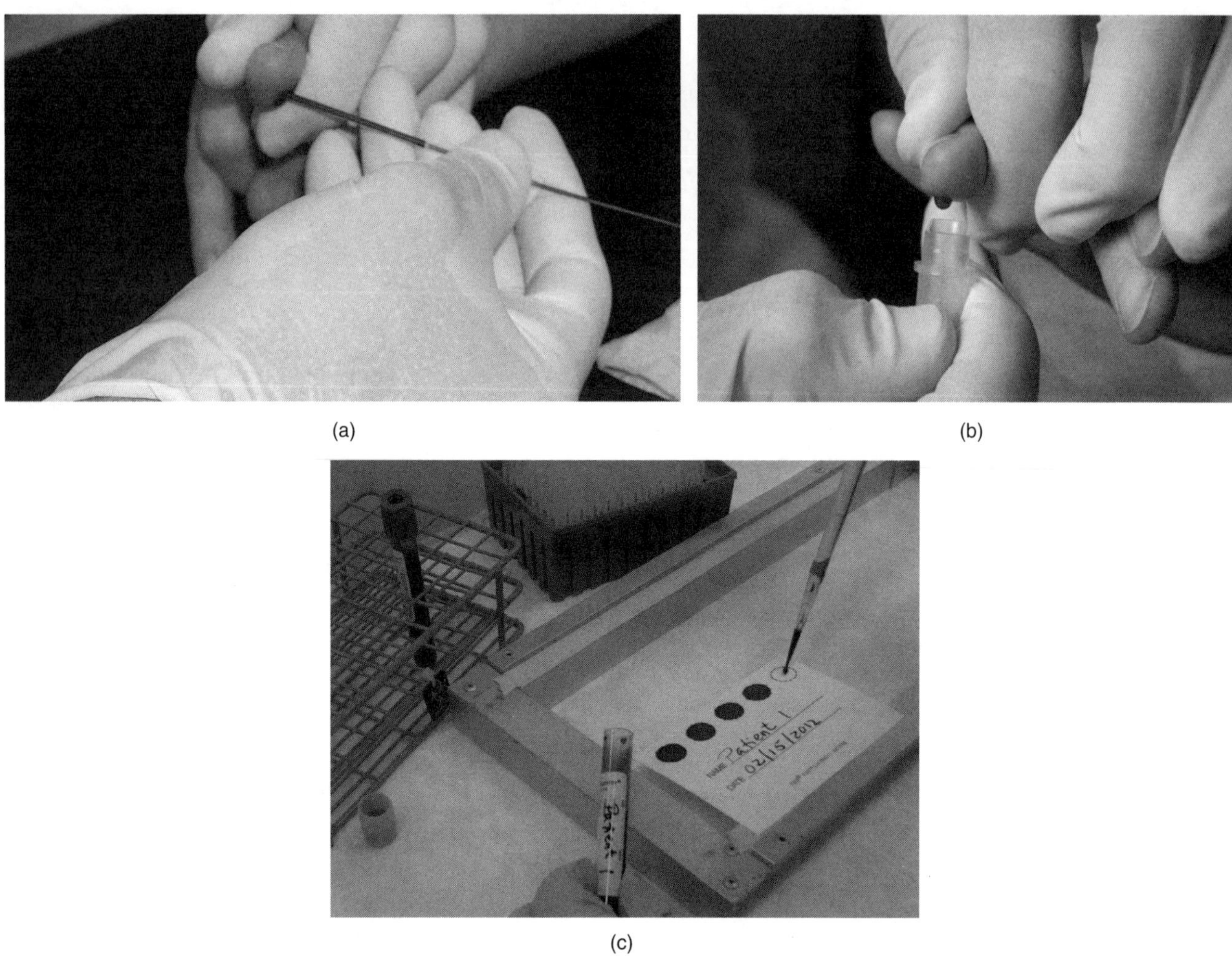

FIGURE 3.6 (a) Collection of blood from a fingerstick into a capillary tube. (b) To collect blood into a microcollection tube, hold the tube below the drop of blood and touch it to the side of the tube. The drop will run down the inside wall. (c) Prepare DBS collected in alternative devices by applying blood to fill preprinted circles on the collection card.

previous sections. Hold the container below the blood drop and touch the collector end of the tube to the blood drop (Figure 3.6b) (McCall and Tankersley, 2012; CDC, http://www.cdc.gov/labstandards/pdf/vitaleqa/Poster_Preparing DBS.pdf). Allow the blood drop to run down the inside wall of the tube. Tap the tube gently to move the blood to the bottom and fill to the appropriate volume to achieve the proper blood to additive ratio. Over- or under-filling of the tube may result in clotting and/or erroneous test results. After capping the tube, gently invert it 8–10 times to mix with the anticoagulant. Lay the full microcollection container aside and continue to collect into other microcollection containers if necessary. Repeat this procedure to collect the needed number of containers. After all containers have been collected, follow previous procedures for stopping the patient's blood flow. Using a pipette, apply drops of blood to the filter paper to completely fill preprinted circles on the collection card (Figure 3.6c).

3.2.4.2 Other Sources of Blood Collection for DBS preparation Venous blood and umbilical cord blood can all be used to collect blood into appropriate vessels followed by direct application to preprinted circles of filter paper cards. None of these sources are the method of choice for DBS preparation and all present challenges to collecting valid specimens for analytical testing due to the following: (a) test results may be different when collected from different vessel sources; (b) veins may be needed by patients for intravenous fluids; (c) venous sampling is more invasive than heelstick or fingerstick; (d) specimens may be contaminated with IV fluids; and (e) lack of anticoagulant in collection devices such as syringes could allow clot formation causing heterogeneous specimens (CLSI NBS-01-A6, 2013). Detailed instructions for collecting blood from venipuncture umbilical catheter or umbilical cord blood are available elsewhere (CLSI H04-A6, 2008; McCall and Tankersley, 2012; CLSI NBS-01-A6, 2013).

3.2.5 Labeling of DBS

3.2.5.1 Newborn Heelsticks DBS collected for newborn screening have specific guidelines for the minimum data elements needed to achieve the goals of screening (CLSI NBS-01-A6, 2013). This information should be captured by individual preprinted data fields on the specimen collection device. Additional information may be collected by newborn screening programs to meet specific needs. Local regulations and institutional policies should be consulted when deviations from the minimum specimen information are encountered. Labeling and collection of data fields should occur before the heel puncture is done.

Minimum data elements for newborn screening include infant's last name; mother's last name; infant's sex and birth date; date of specimen collection; infant's age; infant's birth weight; patient identification number; submitter's identification or birth facility; physician's name and telephone number; name and address of the newborn screening program; unique serial number; expiration date of specimen collection device; appropriate number of preprinted circles for blood collection; manufacturer and lot number of filter paper printed on the filter paper section (CLSI NBS-01-A6, 2013).

3.2.5.2 Older Children and Adults Labeling of capillary DBS specimens for older children and adults occurs after the specimen has been collected (CLSI H04-A6, 2008). The specimen should be labeled immediately after collection but before leaving the patient's side. At a minimum, the label should state the patient's first and last names, identification number, and date of specimen collection.

3.2.6 Drying

Blood spots should be thoroughly air-dried at ambient temperature (18–25°C) for a minimum of 3 hours in a horizontal position over a nonabsorbent, open surface such as drying rack (Figure 3.7a) (CLSI NBS-01-A6, 2013). Do not allow anything to touch the wet spots. Keep the blood spots away from direct sunlight and do not stack, heat, or allow the spots to touch other surfaces during the drying period. A cardboard rack can be used to dry newborn screening collection forms as long as the portion of the card with demographic information and the protective flap are bent back to expose the filter paper strip (Figure 3.7b). After the blood spots are dry, protective flaps may be put back in place over the filter paper strip (Figure 3.7c). Researchers collecting blood spots in the field have found creative ways to dry specimens, such as inserting collection cards between the slats of a bamboo structure (Figure 3.7d).

3.3 SAMPLE STORAGE AND TRANSPORTATION

3.3.1 Packaging and Storage

Dried DBS should be packaged in low gas permeable containers that allow an exchange of air. Suggested supplies for packaging DBS for storage include low gas permeable bags (Bitran®), glassine (weighing paper) or other physical barrier to separate DBS specimens if a fold-over cover (protective flap) is not present on the device, desiccant packs, and an indicator card to monitor humidity (Figures 3.8a and 3.8b). Specimen-to-specimen contact should be avoided to minimize the risk of cross-contamination and maximize the utility of stored DBS specimens for future studies (CDC; http://www.cdc.gov/labstandards/pdf/vitaleqa/Poster_PreparingDBS.pdf).

For storage, DBS should not be packaged in air-tight or leak-proof containers because the lack of air exchange may cause the temperature to rise and moisture to build up inside the sealed, inner environment (CLSI NBS-01-A6, 2013). Long-term (1 year and greater) and accelerated (30 ± 5 days) stability studies conducted at various temperatures (−20°C, 4°C, ambient temperature, and 37°C) with low and high humidity (DBS stored with and without desiccant) indicated that analytes in DBS for newborn screening are stable at −20°C and 4°C for more than 1 year (Behets et al., 1992; Li et al., 2006; Strnadova et al., 2007; Lando et al., 2008; Cordovado et al., 2009; Fingerhut et al., 2009; Adam et al., 2011; Mei et al., 2011). Higher temperatures and humidity (greater than 50%) contributed to accelerated analyte degradation (Adam et al., 2011). Using desiccant packs with DBS during storage and shipping provides some protection from humidity; however, shipping conditions are uncontrolled and desiccant has limited usefulness (CLSI NBS-01-A6, 2013) if the humidity level inside the container rises above the moisture-absorbing capacity of the product.

Newborn screening DBS specimens should be tested upon receipt by the testing facility. DBS can be stored at ambient temperature with low humidity (less than 30%) during the testing process (CLSI NBS-01-A6, 2013). Low humidity and lower temperatures (4–8°C) are suggested for short-term storage (less than 2 years) (CLSI NBS-01-A6, 2013). For storage greater than 2 years, DBS specimens should be kept under low humidity and frozen conditions (−20°C to −70°C) (Therrell et al., 1996; Adam et al., 2011; Brisson et al., 2012; CLSI NBS-01-A6, 2013).

3.3.2 Transportation

Blood spots should be dried for a minimum of 3 hours at ambient temperature before packaging for storage and

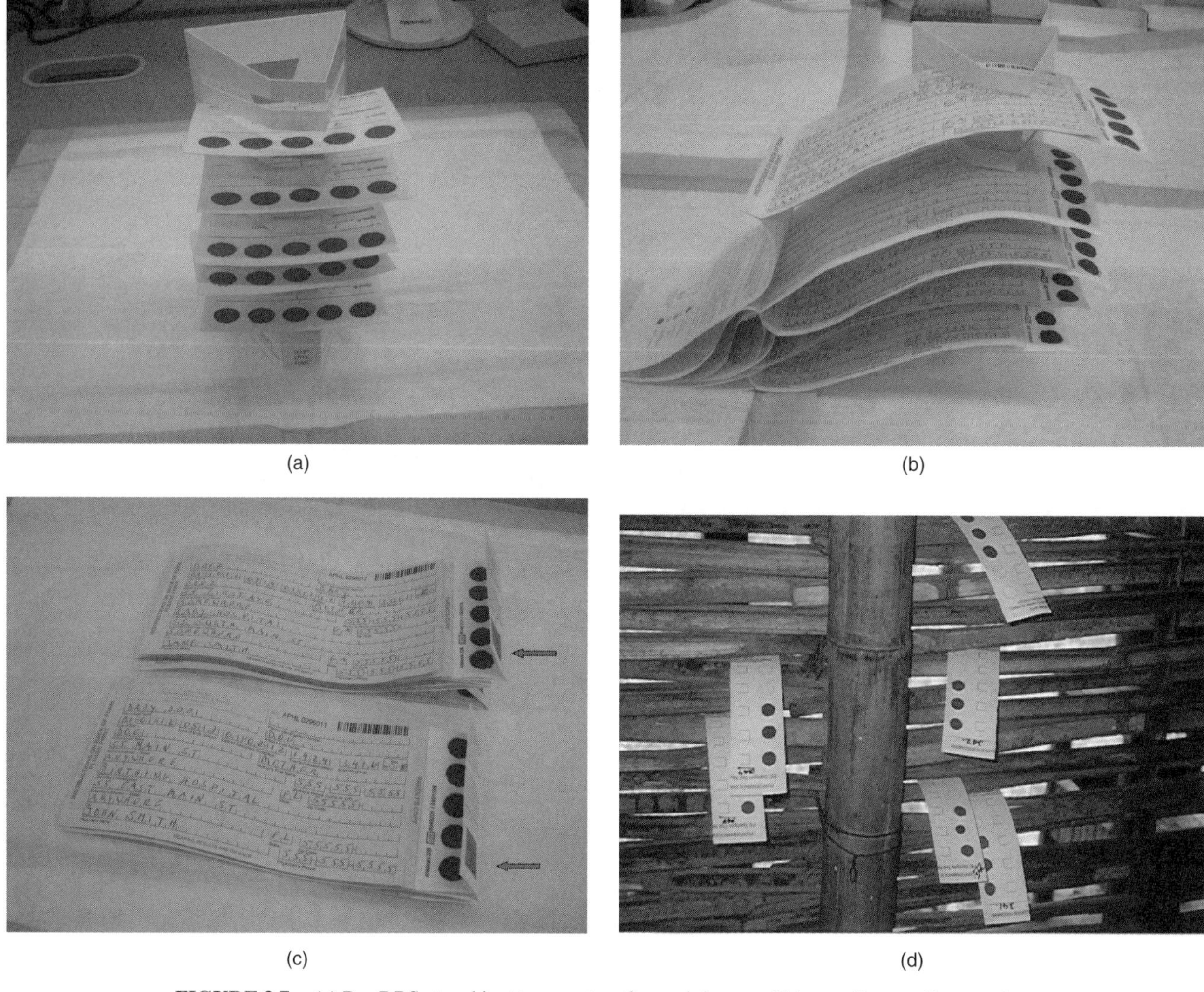

FIGURE 3.7 (a) Dry DBS at ambient temperature for a minimum of 3 hours. Do not allow cards to touch each other or other surfaces. Keep specimens away from direct sunlight. (b) Newborn screening collection devices may be slipped into slots of a cardboard holder to keep them in a horizontal position during the drying process. (c) Newborn screening collection forms can be ordered with a protective flap to serve as a barrier between DBS specimens. The forms can be printed with a biohazard label (arrows) to comply with US Postal Service regulations for shipping non-regulated infectious materials. (d) A creative method of drying DBS in the field: inserted collection cards between bamboo slats of a hut.

transportation (CLSI NBS-01-A6, 2013) and can be transported at ambient temperature. DBS are considered exempt, non-regulated material in the United States (Federal Register, 2006; 2010; United States Postal Services, Publication 52). Non-regulated materials are not subject to regulation as hazardous materials but must be properly packaged for shipping to protect employees as well as transportation personnel so that the Domestic Mail Manual guidelines are met (United States Postal Service 601.10.17.8). The United States Department of Transportation (Federal Register, 2006; 2010) and the United States Postal Service (United States Postal Services, Publication 52) harmonized their regulations with the regulations of the International Air Transporter Association (International Air Transporter Association 2010), the World Health Organization's Guidance on Regulations for the Transport of Infectious Substances (WHO/HSE/IHR/2010.8), and the International Civil Aviation Organization's Technical Instructions for Safe Transport of Dangerous Goods by Air (International Civil Aviation Organization, 2005).

This packaging system meets the basic triple packaging system, that is, the primary container is the filter paper matrix

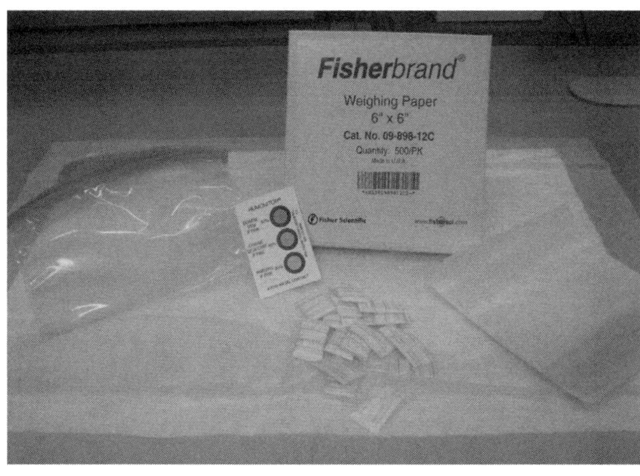

(a)

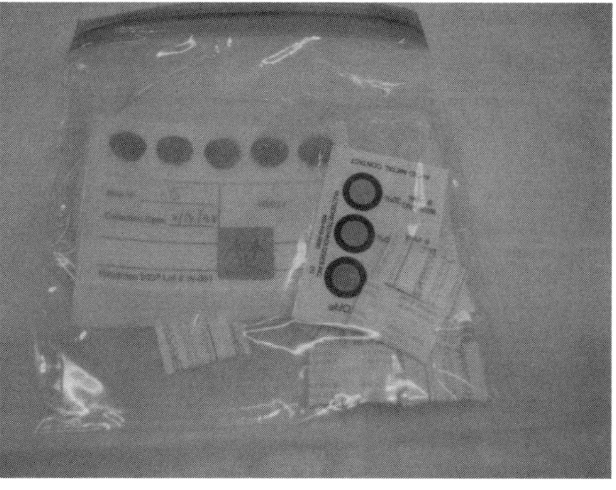

(b)

FIGURE 3.8 (a) Storage supplies for DBS: low gas permeable bag, humidity indicator card, glassine paper, and desiccant packets. (b) DBS cards can be packaged for storage by layering glassine paper between specimens. Desiccant packages are added to keep the internal environment under low humidity conditions (less than 30%). A biohazard label has been placed on the collection card to comply with US Postal Service regulations for shipping non-regulated infectious materials.

that contains the absorbed and dried blood; a fold-over flap (Figure 3.7c) or inner envelope provides the secondary containment; and an outer envelope of sturdy, high-quality paper is the third container used to ship specimens (United States Postal Service 601.10.17.8). A complete return address and delivery address should be on the outer shipping container. No content markings are required on the outer shipping container. The international biohazard symbol must be affixed or printed on either the primary or the secondary container (Figures 3.7c and 3.8b). These levels of containment provide reasonable safety from occupational exposure and maintain optimal specimen integrity and meet the requirements of the United States Occupational Safety and Health Administration (Occupational Safety and Health Administration 29 CFR 1910.1030(g)).

If DBS are collected from a baby's heelstick for newborn screening, then the specimen should be transported to the testing facility as soon as the DBS are dry (minimum of 3 hours) and no later than 24 hours of collection, regardless of weekends. A daily courier service that can provide tracking information should be used. Efforts should be made to avoid transportation delays from the collection site (CLSI NBS-01-A6, 2013).

3.4 COMMON ERRORS IN SAMPLE COLLECTION, STORAGE, AND SHIPMENT

The collection of proper DBS specimens for newborn screening is important to ensure valid testing and to reduce the risk of delayed diagnosis. For other types of testing, such as epidemiological surveys using fingerstick specimens, there may be only one opportunity to collect a valid specimen. A valid specimen consists of a sufficient quantity of blood that has been allowed to soak through the filter paper to completely fill preprinted circles on the collection card and has been thoroughly dried (Figure 3.3a).

All laboratories that test DBS should have procedures in place to address whether DBS specimens that are considered unsatisfactory or invalid (e.g., DBS that do not adequately fill the circle) meet the established acceptance criteria for the intended purposes of the testing (CDC, 2012; CLSI NBS-01-A6, 2013). For all unsatisfactory specimens, a second specimen should be requested. Figures 3.3b and 3.3c provide excellent examples of poorly collected DBS specimens. Figure 3.3c illustrates valid and invalid DBS specimens on the same collection card. The following is a discussion of the common errors made when collecting DBS.

If not enough blood is applied to the preprinted circles, then the resulting specimen will have an insufficient quantity for testing. Figures 3.3b and 3.3c-1 illustrate DBS with so little blood applied that it has not completely soaked through the filter paper. Common causes for this are as follows: (1) removing filter paper before blood has completely filled the circle or before blood has soaked through to the second side; (2) applying blood to filter paper with a capillary tube; and (3) allowing the filter paper to come into contact with gloved or ungloved hands or substances such as hand lotion or powder, either before or after blood specimen collection. Applying blood to the filter paper with a capillary tube or other device can result in a scratched or abraded specimen (Figure 3.3b). Do not touch the capillary tube to the filter paper. This may damage the filter paper fibers (CLSI NBS-01-A6, 2013). Mailing the DBS specimen before the spots are completely dry (minimum of 3 hours at ambient temperature) will result in bright red spots (Figure 3.3b) which could adversely affect newborn screening test results (CLSI NBS-01-A6, 2013). Using a capillary tube to apply too much blood to preprinted circles or applying blood to both sides of the filter paper will result in specimens that appear supersaturated (Figure 3.3b). If the heel or finger is squeezed or "milked," then interstitial fluid may dilute and contaminate the specimen (Figures 3.3b and 3.3c). DBS specimens can also become contaminated if the filter paper collection card comes in contact with substances like alcohol or hand lotion. Serum rings (Figures 3.3b and 3.3c-3) can form if alcohol used at the puncture site is not allowed to dry before applying blood to the filter paper. Improper drying, such as stacking collection cards too close to each other so that air cannot circulate around the blood spots, will cause serum rings (Figures 3.3b and 3.3c-3). Layering (Figures 3.3b and 3.3c-3.4) can occur by repeatedly touching the same filter paper circle to blood drops or when circles are not fully filled initially and additional blood is applied to the filter paper. Occasionally, laboratories receive collection cards with empty circles due to a failure to collect the blood specimen (Figure 3.3b).

3.5 CONCLUSIONS

The use of DBS for biochemical and molecular testing has expanded as filter paper technology has been adapted for the latest analytical methods. Newborn screening alone has pushed the limits of DBS technology by using microliter volumes of whole blood for an ever increasing array of biochemical markers. The collection, storage, and transportation of DBS make them an ideal, noninvasive specimen for population screening and field studies in resource-poor settings. Collecting valid DBS specimens should be the goal of all who use them. For newborn screening, properly collected and transported DBS specimens ensure timely and accurate testing, minimize trauma to the infant and its family, and decrease the burden on the medical system for the recall of newborns for repeat collection and testing. Properly collected and transported DBS specimens for field studies reduce costs and ensure that data from the analyses are of high quality. DBS technology will continue to grow with new applications. Considerations for their use should start with the best techniques for collecting, processing, and storing valid specimens known to be adequate for the intended purpose of the testing.

DISCLAIMER

Use of trade names and commercial sources is for identification only and does not imply endorsement by the Public Health Service or the US Department of Health and Human Services. The findings and conclusions in this report are those of the authors and do not necessarily represent the views of institutions from which the data were obtained or the institutions with which the authors are affiliated.

REFERENCES

Adam BW, Hall EM, Sternberg M, Lim TH, Flores SR, O'Brien S, Simms D, Li LX, De Jesus VR, Hannon WH. The stability of markers in dried-blood spots for recommended newborn screening disorders in the United States. *Clinical Biochemistry* 2011;44:1445–1450.

Behets F, Kashamuka M, Pappaioanou M, Green TA, Ryder RW, Batter V, George JR, Hannon WH, Quinn TC. Stability of human immunodeficiency virus type 1 antibodies in whole blood dried on filter paper and stored under various tropical conditions in Kinshasa, Zaire. *Journal of Clinical Microbiology* 1992;30:1179–1182.

Brisson AR, Matsui D, Rieder MJ, Fraser DD. Translational research in pediatrics: tissue sampling and biobanking. *Pediatrics* 2012;129:153–162.

Centers for Disease Control and Prevention. Guideline for Isolation Precautions: Preventing Transmission of Infectious Agents in Healthcare Settings; 2007. http://www.cdc.gov/hicpac/pdf/isolation/Isolation2007.pdf (accessed January 14, 2014).

Centers for Disease Control and Prevention. Steps for Collecting Fingerstick Blood in a Microtainer® Tube for Preparing Dried Blood Spots. http://www.cdc.gov/labstandards/pdf/vitaleqa/Poster_PreparingDBS.pdf (accessed January 14, 2014).

Centers for Disease Control and Prevention: Good laboratory practices for biochemical genetic testing and newborn screening for inherited metabolic disorders. *Morbidity and Mortality Weekly Report. Recommendations and Reports* 2012; 61.

Chrysler D, McGee H, Bach J, Goldman E, Jacobson PD. The Michigan BioTrust for health: using dried bloodspots for research to benefit the community while respecting the individual. *Journal of Law, Medicine & Ethics* 2011;39:98–101.

Clinical and Laboratory Standards Institute. *Collection, Transport, Preparation, and Storage of Specimens for Molecular Methods; Approved Guideline*. CLSI document MM13-A. Wayne, PA: Clinical Laboratory Standards Institute; 2005a.

Clinical and Laboratory Standards Institute. *Protection of Laboratory Workers from Occupationally Acquired Infections; Approved Guideline-Third Edition*. CLSI document M29-A3. Wayne, PA: Clinical Laboratory Standards Institute; 2005b.

Clinical and Laboratory Standards Institute. *Procedures and Devices for the Collection of Diagnostic Capillary Blood Specimens; Approved Standard-Sixth Edition*. CLSI document H04-A6. Wayne, PA: Clinical and Laboratory Standards Institute; 2008.

Clinical and Laboratory Standards Institute. *Blood Collection on Filter Paper for Newborn Screening Programs; Approved Standard–Fifth Edition*. CLSI document NBS-01-A6. Wayne, PA: Clinical and Laboratory Standards Institute; 2013.

Cordovado SK, Earley MC, Hendrix M, Driscoll-Dunn R, Glass M, Mueller PW, Hannon WH. Assessment of DNA contamination from dried blood spots and determination of DNA yield and function using archival newborn dried blood spots. *Clinica Chimica Acta* 2009;402:107–113.

De Jesús VR, Mei JV, Bell CJ, Hannon WH. Improving and assuring newborn screening laboratory quality worldwide: 30-year experience at the centers for disease control and prevention. *Seminars in Perinatology* 2010;34:125–133.

Federal Register. 39 CFR part 111 New Mailing Standards for Division 6.2 Infectious Substances. 2006; Section 10.17.9(b).

Federal Register. 49 CFR part 173 Shippers General Requirements for Shipments and Packaging. 2010; Section 173.134(b).

Fingerhut R, Ensenauer R, Roschinger W, Arnecke R, Olgemuller B, Roscher AA. Stabilities of acylcarnitines and free carnitine in dried blood samples: implications for retrospective diagnosis of inborn errors of metabolism and neonatal screening for carnitine transporter deficiency. *Analytical Chemistry* 2009;81:3571–3575.

Holtkamp U, Klein J, Sander J, Peter M, Janzen N, Steuerwald U, Blankenstein O. EDTA in dried blood spots leads to false results in neonatal endocrinologic screening. *Clinical Chemistry* 2008;54:602–605.

International Air Transporter Association (IATA). *Transporting and Shipping Infectious Substances A and B*; 2010. http://www.iata.org/whatwedo/cargo/dgr/Documents/DGR52_InfectiousSubstances(DGR362).pdf (accessed January 14, 2014).

International Civil Aviation Organization (ICAO). *International Civil Aviation Organization Technical Instructions for the Safe Transport of Dangerous Goods by Air*; 2005–2006. http://www.icao.int/publications/Documents/guidance_doc_infectious_substances.pdf (accessed January 14, 2014).

Lando VS, Batista MC, Nakamura IT, Mazi CR, Mendonca BB, Brito VN. Effects of long-term storage of filter paper blood samples on neonatal thyroid stimulating hormone, thyroxin, and 17-alpha-hydroxyprogesterone measurements. *Journal of Medical Screening* 2008;15:109–111.

Li L, Zhou Y, Bell CJ, Earley MC, Hannon WH, Mei JV. Development and characterization of dried blood spot materials for the measurement of immunoreactive trypsinogen. *Journal of Medical Screening* 2006;13:79–84.

McCall RE, Tankersley CM. *Phlebotomy Essentials*, 5th edn. Baltimore, MD: Lippincott Williams & Wilkins; 2012.

Mei JV, Alexander JR, Adam, B, Hannon WH. Use of filter paper for the collection and analysis of human whole blood specimens. *Journal of Nutrition* 2001;131:1631S–16316S.

Mei JV, Zobel SD, Hall EM, De Jesús VR, Adam BW, Hannon WH. Performance properties of filter paper devices for whole blood collection. *Bioanalysis* 2010;2:1397–1403.

Mei JV, Li LX, Rasmussen SA, Collier S, Frias JL, Honein MA, Shaw GM, Lorey F, Meyer R, Chaing S, Canfield MA, Jones J, Hannon WH. Effect of specimen storage conditions on newborn dried blood spots used to assess *Toxoplasma gondii* immunoglobulin M (IgM). *Clinica Chimica Acta* 2011;412:455–459.

Newborn Screening Quality Assurance Program. Current Reports; 2012. http://www.cdc.gov/labstandards/nsqap.html (accessed January 14, 2014).

Occupational Health and Safety Administration. 29 CFR 1910.1030(g) (1)(i) Occupational Safety and Health Standards, Bloodborne pathogens. 2009. http://www.osha.gov/pls/oshaweb/owadisp.show_document?p_table=standards&p_id=10051 (accessed January 14, 2014).

Program for Appropriate Technology in Health (PATH). RBP-EIA: Collecting, Processing, and Handling Venous, Capillary, and Blood Spot Samples; 2005. http://www.path.org/publications/files/TS_rbp-eia_blood_collct.pdf (accessed January 14, 2014).

Reiner CB, Meites S, Hayes JR. Optimal sites and depths for skin puncture of infants and children as assessed from anatomical measurements. *Clinical Chemistry* 1990;36:547–549.

Strnadova KA, Holub M, Muhl A, Heinze G, Ratschmann R, Mascher H, Stöckler-Ipsiroglu S, Waldhauser F, Votava F, Lebl J, Bodamer OA. Longterm stability of amino acids and acylcarnitines in dried blood spots. *Clinical Chemistry* 2007;53:717–722.

Therrell BL. An update on public health newborn screening activities. Introduction. *Seminars in Perinatology* 2010;34:103–104.

Therrell BL, Hannon WH, Pass KA, Lorey F, Brokopp C, Eckman J, Glass M, Heidenreich R, Kinney S, Kling S, Landenburger G, Meaney FJ, McCabe ER, Panny S, Schwartz M, Shapira E. Guidelines for the retention, storage, and use of residual blood spot samples after newborn screening analysis: statement of the Council of Regional Networks for Genetic Services. *Biochemical and Molecular Medicine* 1996;57:116–124.

Therrell BL, Hannon WH, Bailey DB, Goldman EB, Monaco J, Norgaard-Pedersen B, Terry SF, Johnson A, Howell RR. Committee report: considerations and recommendations for national guidance regarding the retention and use of residual dried blood spot specimens after newborn screening. *Genetics in Medicine* 2011;13:621–624.

United States Postal Service. Publication 52 Hazardous, Restricted and Perishable Mail; 346 Toxic Substances (Hazard Class 6). 346.234 Nonregulated Materials. 2013. http://pe.usps.com/text/pub52/pub52c3_021.htm (accessed January 14, 2014).

United States Postal Service. 601 Mailability 10.17.8 Packaging Nonregulated Materials. http://pe.usps.gov/text/dmm300/601.htm#1065177 (accessed January 14, 2014).

World Health Organization. WHO/HSE/IHR/2010.8 Guidance on Regulations for the Transport of Infectious Substances; 2011–2012. http://apps.who.int/iris/bitstream/10665/78075/1/WHO_HSE_GCR_2012.12_eng.pdf (accessed January 14, 2014).

4

DRIED BLOOD SPOT SPECIMENS FOR POLYMERASE CHAIN REACTION IN MOLECULAR DIAGNOSTICS AND PUBLIC HEALTH SURVEILLANCE

Chunfu Yang

4.1 INTRODUCTION

The polymerase chain reaction (PCR) technique devised by Saiki et al. (1985) in the mid-1980s and the later automation of a programmable heating block PCR machine by Oste (1989) revolutionized modern molecular biology fields. More recently, the innovation of collecting whole blood on filter paper as dried blood spot (DBS) has simplified the collection, storage, and transportation of blood specimens from collection sites to centralized laboratories for testing (Guthrie and Susi, 1963; McCabe, 1991). The development of DNA microextraction methods for DBS specimens (McCabe, 1991; McCabe et al., 1987) has led to the explosion of DBS-PCR-based technologies as indicated by a number of recent publications that include "a glowing future for dried blood spot sampling" (Spooner, 2010) and several review articles (Hamers et al., 2009; Bertagnolio et al., 2010; Johannessen, 2010; Tanna and Lawson, 2011). Both qualitative and quantitative DBS-PCR-based tests have been applied to many areas that include (1) screening for newborn metabolic disorders (Williams et al., 1988; Jinks et al., 1989; Matsubara et al., 1991; McCabe, 1991; McCabe, 1994; Schwab et al., 2003; Chan and Puck, 2005; Nagy, 2010) and other genetic abnormalities (Evison et al., 1997; Dantonio et al., 2006); (2) detection and monitoring of infectious diseases including cytomegalovirus (CMV) (Barbi et al., 1996), human immunodeficiency virus (HIV) (Cassol et al., 1991; Cassol et al., 1992a; Cassol et al., 1992b; Cassol et al., 1994; Cassol et al., 1996a; Luo et al., 2005; Stevens et al., 2009; Crignis et al., 2010; Huang et al., 2011; Vemu et al., 2011), hepatitis B, C, and E viruses (HBV, HCV, and HEV) (Gupta et al., 1992; Plamondon et al., 2007; Mérens et al., 2009; Crignis et al., 2010), human T-lymphotropic virus (HTLV) and simian T-lymphotropic virus (STLV) (Noda et al., 1993; Sintasath et al., 2009), human herpesvirus 6 (HHV-6) (D'Agaro et al., 2008; Strenger et al., 2011), herpes simplex virus (HSV) (Lewensohn-Fuchs et al., 2003), and measles and rubella (Katz et al., 2002; Centers for Disease Control and Prevention, 2008); and (3) detection of parasitic infections including malaria (Long et al., 1995; Niu et al., 2000; Boonma et al., 2007; Muhamad et al., 2011) and other parasites (Katakura et al., 1997; Tani et al., 2008; Rahmah et al., 2010; Fink et al., 2011; Mumba et al., 2011).

The primary advantages of using DBS for molecular diagnostics and monitoring based on PCR technologies include

- simplified sample collection, storage, and transport;
- ability to ship at ambient temperature to centralized laboratories;
- minimal sample volume for testing;
- suitability for multiplexed platforms to reduce the cost for public health services; and
- reduction of blood-borne pathogen transmission through accidental needle sticks (DBS can be collected using finger or heel pricks).

In this chapter, we provide an overview of the DBS-PCR technologies that have been developed and applied in the past 30 years and give a more detailed description of

Dried Blood Spots: Applications and Techniques, First Edition. Edited by Wenkui Li and Mike S. Lee.

some specific nucleic acid extraction methods that represent significant milestones in the improvement of DBS-PCR technologies.

4.2 PIONEERING DBS NUCLEIC ACID EXTRACTION METHODS FOR PCR

The foundation for using DBS specimens for PCR technique lay with the development of nucleic acid extraction methods from DBS specimens. In 1987, McCabe and colleagues reported the development of a DNA microextraction method from a half-inch diameter blood spot that was approximately equivalent to 50 μL of whole blood. The amount of DNA extracted was just enough for a single Southern blot analysis (McCabe et al., 1987). Concurrent with this development, Saiki and colleagues (1985) devised PCR amplification technique, and the potential for using DBS-extracted DNA for PCR amplification in the diagnosis of sickle cell disease in neonatal screening was realized (Jinks et al., 1989). Almost at the same time, a group of British scientists at St. Mary's Hospital Medical School of London also reported that DNA extracted from one-fifth of a single Guthrie spot that had been stored for more than 15 years could be used for confirmatory diagnosis of cystic fibrosis (Williams et al., 1988). These pioneering efforts using nucleic acid extracted from DBS for PCR amplifications in newborn genetic disorder screenings laid the foundation for the later applications of DBS-PCR technologies in many fields. The improved DNA microextraction technique used by Jinks et al. (1989) and reviewed by McCabe (1991) is summarized in Table 4.1.

TABLE 4.1 DNA Microextraction Method from Dried Blood Spot Specimens on Filter Paper (Jinks et al., 1989; McCabe, 1991) Gold Spring Harbor Laboratory Press

1. A half-inch diameter circle of dried blood spot of interest is cut, washed with dH_2O, and dried.
2. The circle is further cut into small pieces and placed into a tube for incubation with methanol for about 15 min, which is followed by drying it via speed vacuum.
3. 380 μL of 0.15 M NaCl/0.5% SDS solution is added into the tube and incubated at 37°C for 2 h. Then, 50 μg of proteinase K solution is added and the resulting mixture is incubated for about 2 h or overnight at 37°C.
4. The liquid layer is transferred to a new tube and DNA is extracted sequentially with buffered phenol, buffered phenol in combination with chloroform/isoamyl alcohol, and finally chloroform/isoamyl alcohol.
5. A 2.5 × volume of 100% cold pure alcohol and 10% (v/v) of 3 M NaAc (pH 5.2) are added to the tube. The tube is placed in a –20°C freezer overnight and then spun at 14,000 rpm for 15 min. This step is repeated with 75% ethanol.
6. The DNA pellet is air-dried and used for downstream application.

TABLE 4.2 Direct PCR Method of Dried Blood Spot Specimens for Cystic Fibrosis Genotyping (Raskin et al., 1992)

- Cut a 1 mm^2 segment from the testing Guthrie blood spot.
- Place the segment into a 100 μL PCR reaction containing
  a. 1/16 mM of spermidine
  b. 1 mM each of oligonucleotide primers
  c. 10 mM Tris (pH 8.0)
  d. 50 mM KCl
  e. 1.5 mM $MgCl_2$
  f. 0.001% gelatin
  g. 200 mM each of deoxynucleotide phosphate
- Three cycles of heating at 96°C for 3 min and cooling at 55°C for 3 min (to free DNA and minimize inhibition by released proteins and heavy metals).
- Add 1 unit of *Taq* polymerase to the PCR reaction.
- Heat at 72°C for 2 min for an extension.
- Perform 33 cycles of PCR amplification using denaturing, annealing, and extension temperatures and times optimized for each primer set.
- Upon completion, a 10 min extension at 72°C was done.
- The PCR products were cooled on ice and used for downstream analyses.

4.3 DIRECT PCR AMPLIFICATION OF TARGETED GENOMIC DNA FROM DBS

The next DBS advancement for PCR came from the development of direct PCR amplification of targeted genomic DNA from DBS collected on Guthrie cards without any need for DNA extraction (Raskin et al., 1992). To prevent contamination, DBS specimens were placed into individual plastic bags for storage after collection. Prior to testing, a 1 mm^2 segment was cut out from the DBS filter paper and used for the PCR reaction.

Table 4.2 describes the detailed procedure for direct PCR analysis of Guthrie blood spots for the diagnosis of cystic fibrosis in newborns (Raskin et al., 1992). As discussed by authors, the advantages of the direct PCR method included elimination of the time-consuming DNA extraction step and an increased number of potential mutations (diagnoses) that could be identified using a single card.

4.4 EXPANSION OF DBS-PCR FROM NEWBORN SCREENING TO DIAGNOSIS OF INFECTIOUS AND OTHER DISEASES

The simplicity of DBS collection, storage, and transport has attracted many scientists who work in different disciplines including infectious diseases. The first report that used DBS-PCR to diagnose infectious diseases came from a group of Canadian scientists who reported the detection of HIV-1 using DBS-PCR in 1991 (Cassol et al., 1991). Additional

TABLE 4.3 Outline of DNA Extraction Using Roche Amplicor HIV-1 DNA-PCR Kit (Cassol et al., 1996a)

- Excise 0.6 cm circle from a DBS spot.
- Transfer the circle to a 1.5 m screw-cap tube and incubate with 1.0 mL of specimen wash buffer for 30 min at 20°C on an Eppendorf 5436 Thermomixer at 1,000 revolutions/min.
- Aspirate the hemoglobin-containing supernatant.
- Elute DNA by heating the filter at 95°C in 100 μL of 5% Chelex for 1 h.
- Centrifuge at 12,000 g for 1 min.
- 50 μL of the Chelex supernatant is used for downstream analysis.

studies by the same group of scientists (Cassol et al., 1992a; Cassol et al., 1992b) described the use of DBS-PCR for the diagnosis of HIV-1 infections from children born to HIV-1-infected mothers and evaluated the stability of DBS for PCR. Adaptation to a standardized, commercially available microwell plate and amplification and detection kit, Roche HIV-1 Amplicor (Roche Molecular Systems, Somerville, NJ, USA), rendered the assay suitable for use in routine clinical and public health settings (Cassol et al., 1994; Cassol et al., 1996a) and laid the foundation for the recent rapid expansion of this technology for HIV early infant diagnosis (EID) of children born to HIV-1-infected mothers. Use of the Roche Amplicor HIV-1 DNA-PCR kit (Roche Diagnostics, Indianapolis, IN, USA) in the US President's Emergency Plan for AIDS Relief (PEPFAR)-supported programs has since become routine clinical practice for HIV diagnosis in children less than 18 months of age in many resource-limited countries (Stevens et al., 2008) and has saved thousands of lives through early diagnosis and treatment with life-saving antiretroviral therapy (ART) (WHO, 2011). The simplified Roche Amplicor HIV-1 DNA-PCR procedure is shown in Table 4.3.

DBS-PCR has also been used to detect many other infectious diseases. For example, Noda et al. (Noda et al., 1993) described the use of DBS-PCR for the detection of human T-cell lymphotropic virus type 1 infection. This technique has also been used for the detection of genetic risk marker for type 1 diabetes (Dantonio et al., 2006) and in forensic science field (Evison et al., 1997). Table 4.4 summarizes the detection of infectious diseases using DBS-PCR.

4.5 APPLICATION OF DBS-PCR IN SEQUENCING-BASED MOLECULAR EPIDEMIOLOGY ANALYSES

In addition to diagnosis, DBS-PCR technologies have also been applied to sequencing-based analyses in the molecular epidemiology field for surveillance purposes. Systematic collection and analysis of HIV-1 genetic variants from different regions of the HIV pandemic has proved to be a power tool in furthering our understanding of HIV molecular epidemiology and monitoring the HIV pandemic (Robertson et al., 2000; Hemelaar et al., 2011). In 1996, Cassol and colleagues (Cassol et a., 1996b) reported a successful molecular epidemiology surveillance study of HIV viral strains circulating in five Asian countries to confirm the utility of DBS-PCR for large-scale multi-country molecular epidemiology surveys. Following this report, DBS-PCR was used to sequence the extrachromosomal *Plasmodium* DNA within the plastid-like large subunit ribosomal-RNA (Tan et al., 1997). Furthermore, the complete genome of STLV type 3 from a wild *Cercopithecus mona* monkey was analyzed by a genome walking technique using DBS-PCR technique (Sintasath et al., 2009).

4.6 APPLICATION OF DBS-PCR FOR HIV DRUG RESISTANCE SURVEILLANCE AND MONITORING IN RESOURCE-LIMITED SETTINGS

Massive scale-up of ART in resource-limited settings has significantly reduced HIV/AIDS-related morbidity and mortality and decreased HIV transmission. As of December 2011, 6.6 million HIV-infected patients in low- and middle-income countries received ART and represent an increase of 1.35 million over the previous year (WHO, 2011). To maintain the efficacy of the current antiretroviral drug regimens, surveillance and monitoring of the development and transmission of HIV drug resistance (HIVDR) in resource-limited settings is one of the key components in the global HIVDR prevention and assessment strategy recommended by the WHO (Jordan et al., 2008; Bennett et al., 2008). Total nucleic acid extracted from DBS for HIVDR genotyping has become the centerpiece of recent public health research for the implementation of the WHO strategy (Johannessen, 2010; Bertagnolio et al., 2007; McNulty et al., 2007; McNulty et al., 2008; Buckton et al., 2008; Youngpairoj et al., 2008; Garcia-Lerma et al., 2009; Hamers et al., 2009; Monleau et al., 2009; Bertagnolio et al., 2010; Monleau et al., 2010; Yang et al., 2010; Zhou et al., 2011; Rottinghaus et al., 2012). Although DBS-PCR for HIVDR genotyping analysis was first reported in 1996 (Cassol et al., 1996a) the intensive public health research on DBS for HIVDR genotyping in recent years has yielded substantial evidence on how to best preserve viral RNAs in DBS specimens under field conditions as well as during the transport of DBS specimens from collection sites to central genotyping laboratories for testing (Garcia-Lerma et al., 2009; Monleau et al., 2010). For instance, a recent study demonstrated for the first time that DBS specimens stored at ambient temperature for 2 weeks and then transported to

TABLE 4.4 Use of DBS-PCR Technologies in the Detection of Infectious Diseases

Infectious Agent	References	Infectious Agent	References
Cytomegalovirus	(Barbi et al., 1996)	*Plasmodium*	(Long et al., 1995; Niu et al., 2000; Boonma et al., 2007; Muhamad et al., 2011)
Human immunodeficiency virus	(Cassol et al., 1991; Cassol et al., 1992a; Cassol et al., 1992b; Cassol et al., 1994; Cassol et al., 1996a; Luo et al., 2005; Stevens et al., 2008; Stevens et al., 2009; Crignis et al., 2010; Huang et al., 2011; Vemu et al., 2011)	*Babesia*	(Tani et al., 2008)
Human herpesvirus-6 (D'Agaro et al., 2008; Strenger et al., 2011)		*Brugia*	(Rahmah et al., 2010)
Hepatitis viruses B, C, and E	(Gupta et al., 1992; Plamondon et al., 2007; Mérens et al., 2009; Crignis et al., 2010)	*Loa*	(Fink et al., 2011)
Herpes simplex virus	(Lewensohn-Fuchs et al., 2003)	*Trypanosoma*	(Katakura et al., 1997; Mumba et al., 2011)
Human T-cell lymphotropic virus	(Noda et al., 1993)		
Measles and rubella	(Katz et al., 2002; Centers for Disease Control and Prevention, 2008)		

genotyping laboratory at ambient temperature have no statistically significant effect on the genotyping efficiency (Parkin et al., 2012). This and other studies have established that DBS specimens stored under specific ambient conditions and used in combination with a low-cost and broadly sensitive genotyping assay are suitable for HIVDR genotyping in ART-experienced patients (Parkin et al., 2012; Rottinghaus et al., 2012). Table 4.5 describes the DBS sample collection, storage, and shipping conditions that have been demonstrated to be suitable for extracting total nucleic acid for HIVDR genotyping using this assay.

4.7 QUANTITATION OF INFECTIOUS AGENT PARTICLES WITH DBS-PCR

DBS-PCR has traditionally been used for the qualitative detection of DNA materials from target agents. The exploration of DBS-PCR for HIV-1 viral RNA quantitation has extended to quantitative assays that measure viral RNA level from HIV-infected patients (Fiscus et al., 1998). Encouraging results from initial investigations have led to an expansion of this field using different commercially available HIV-1 viral RNA platforms (Alvarez-Munoz et al., 2005;

TABLE 4.5 Impact of Storage and Shipping Conditions on HIV Drug Resistance Genotyping Efficiency (Parkin et al., 2012)

Blood Collection Method	EDTA-Blood	EDTA-Blood	EDTA-Blood	EDTA-Blood	EDTA-Blood	Finger Prick
No. of specimen	103	103	103	103	103	103
Storage	−80°C	−80°C	−80°C	Ambient temperature and then −80°C	Ambient temperature	Ambient temperature and then −80°C
Duration	Study period	Study period	Study period	2 weeks each	4 weeks	2 weeks each
Shipment	Dry ice	Dry ice	Ambient temperature	Ambient temperature	Ambient temperature	Ambient temperature
Genotyping rate (%)	97.1	98.1	97.1	93.2	89.3[a]	77.7[b]

[a]compared to frozen plasma or frozen DBS group; [b]compared to frozen plasma or DBS stored at ambient temperature for two weeks and then at −80°C and shipped at ambient temperature, Fisher's exact test, $P < 0.05$.

Johannessen et al., 2009; Ikomey et al., 2009; Mbida et al., 2009; Andreotti et al., 2010; Pirillo et al., 2011; Rottinghaus et al., 2012), and when combined with second-generation HIV-1 viral RNA quantitation technologies that use real-time detection, DBS-based HIV-1 viral RNA quantitation may soon become a key technology for monitoring patients on ART in resource-limited countries.

4.8 CONCLUSIONS

From its origin in newborn genetic disorder screening, to the detection of infectious agents, and more recent advances in genotyping and quantification of HIV viral RNA, decades of research and innovation have highlighted the versatility and utility of DBS-PCR technologies. Research on further reducing the requirement for hands-on manipulation on DBS-PCR technologies through point-of-care and other novel technologies may further expand the application of DBS-PCR technologies, allowing DBS-PCR technologies to become essential in supporting our efforts to improve public health in resource-limited settings worldwide.

REFERENCES

Alvarez-Munoz MT, Zaragoza-Rodriguez S, Rojas-Montes O, Palacios-Saucedo G, Vazquez-Rosales G, Gomez-Delgado A, Torres J, Munoz O. High correlation of human immunodeficiency virus type-1 viral load measured in dried-blood spot samples and in plasma under different storage conditions. *Archives of Medical Research* 2005;36:382–386.

Andreotti M, Pirillo M, Guidotti G, Ceffa S, Paturzo G, Germano P, Luhanga R, Chimwaza D, Mancini MG, Marazzi MC, Vella S, Palombi L, Giuliano M. Correlation between HIV-1 viral load quantification in plasma, dried blood spots, and dried plasma spots using the Roche COBAS Taqman assay. *Journal of Clinical Virology* 2010;47:4–7.

Barbi M, Binda S, Primache V, Luraschi C, Corbetta C. Diagnosis of congenital cytomegalovirus infection by detection of viral DNA in dried blood spots. *Clinical and Diagnostic Virology* 1996;6:27–32.

Bennett DE, Bertagnolio S, Sutherland D, Gilks CF. The World Health Organization's global strategy for prevention and assessment of HIV drug resistance. *Antiviral Therapy* 2008;13(Suppl 2):1–13.

Bertagnolio S, Soto-Ramirez L, Pilon R, Rodriguez R, Viveros M, Fuentes L, Harrigan PR, Mo T, Sutherland D, Sandstrom P. HIV-1 drug resistance surveillance using dried whole blood spots. *Antiviral Therapy* 2007;12(1):107–113.

Bertagnolio S, Parkin NT, Jordan M, Brooks J, Garcia-Lerma JG. Dried blood spots for HIV-1 drug resistance and viral load testing: a review of current knowledge and WHO efforts for global HIV drug resistance surveillance. *AIDS Review* 2010;12:195–208.

Boonma P, Christensen PR, Suwanarusk R, Price RN, Russell B, Lek-Uthai U. Comparison of three molecular methods for the detection and speciation of Plasmodium vivax and Plasmodium falciparum. *Malaria Journal* 2007;6:124.

Buckton AJ, Bissett SL, Myers RE, Beddows S, Edwards S, Cane PA, Pillay D. Development and optimization of an internally controlled dried blood spot assay for surveillance of human immunodeficiency virus type-1 drug resistance. *Journal of Antimicrobial Chemotherapy* 2008;62(6):1191–1198.

Cassol S, Salas T, Arella M, Neumann P, Schechter MT, O'Shaughnessy M. Use of dried blood spot specimens in the detection of human immunodeficiency virus type 1 by the polymerase chain reaction. *Journal of Clinical Microbiology* 1991;29:667–671.

Cassol SA, Lapointe N, Salas T, Hankins C, Arella M, Fauvel M, Delage G, Boucher M, Samson J, Charest J, Montpetit ML, Oshaughnessy MV. Diagnosis of vertical HIV-1 transmission using the polymerase chain-reaction and dried blood spot specimens. *Journal of Acquired Immune Deficiency Syndromes and Human Retrovirology* 1992a;5:113–119.

Cassol S, Salas T, Gill MJ, Montpetit M, Rudnik J, Sy CT, O'Shaughnessy MV. Stability of dried blood spot specimens for detection of human immunodeficiency virus DNA by polymerase chain reaction. *Journal of Clinical Microbiology* 1992b;30:3039–3042.

Cassol S, Butcher A, Kinard S, Spadoro J, Sy T, Lapointe N, Read S, Gomez P, Fauvel M, Major C. Rapid screening for early detection of mother-to-child transmission of human immunodeficiency virus type 1. *Journal of Clinical Microbiology* 1994;32:2641–2645.

Cassol SA, Read S, Weniger BG, Gomez P, Lapointe N, Ou CY, Babu PG. Dried blood spots collected on filter paper: an international resource for the diagnosis and genetic characterization of human immunodeficiency virus type-1. *Memorias do Instituto Oswaldo Cruz* 1996a;91:351–358.

Cassol S, Weniger BG, Babu PG, Salminen MO, Zheng XW, Htoon MT, Delaney A, Oshaughnessy M, Ou CY. Detection of HIV type 1 env subtypes A, B, C, and E in Asia using dried blood spots: a new surveillance tool for molecular epidemiology. *AIDS Research and Human Retroviruses* 1996b;12:1435–1441.

Centers for Disease Control and Prevention. Recommendations from an ad hoc meeting of the WHO Measles and Rubella Laboratory Network (LabNet) on use of alternative diagnostic samples for measles and rubella surveillance. *Morbidity and Mortality Weekly Report* 2008;57:657–660.

Chan K, Puck JM. Development of population-based newborn screening for severe combined immunodeficiency. *Journal of Allergy and Clinical Immunology* 2005;115:391–398.

Crignis Ed, Re MC, Cimatti L, Zecchi L, Gibellini D. HIV-1 and HCV detection in dried blood spots by SYBR Green multiplex real-time RT-PCR. *Journal of Virological Methods* 2010;165:51–56.

D'Agaro P, Burgnich P, Comar M, Dal Molin G, Bernardon M, Busetti M, Alberico S, Poli A, Campello C, Grp SI. HHV-6 is frequently detected in dried cord blood spots from babies born to HIV-positive mothers. *Current HIV Research* 2008;6:441–446.

Dantonio P, Meredith N, Earley M, Cordovado S, Callan WJ, Rollin D, Morris D, Vogt RF, Hannon WH. A screening system for detecting genetic risk markers of type 1 diabetes in dried blood spots. *Diabetes Technology & Therapeutics* 2006;8:433–443.

Evison MP, Smillie DM, Chamberlain AT. Extraction of single-copy nuclear DNA from forensic specimens with a variety of postmortem histories. *Journal of Forensic Sciences* 1997;42:1032–1038.

Fink DL, Kamgno J, Nutman TB. Rapid molecular assays for specific detection and quantitation of Loa loa microfilaremia. *PLoS Neglected Tropical Diseases* 2011;5:e1299.

Fiscus SA, Brambilla D, Grosso L, Schock J, Cronin M. Quantitation of human immunodeficiency virus type 1 RNA in plasma by using blood dried on filter paper. *Journal of Clinical Microbiology* 1998;36:258–260.

Garcia-Lerma JG, McNulty A, Jennings C, Huang D, Heneine W, Bremer JW. Rapid decline in the efficiency of HIV drug resistance genotyping from dried blood spots (DBS) and dried plasma spots (DPS) stored at 37 degrees C and high humidity. *Journal of Antimicrobial Chemotherapy* 2009;64:33–36.

Gupta BP, Jayasuryan N, Jameel S. Direct detection of hepatitis B virus from dried blood spots by polymerase chain reaction amplification. *Journal of Clinical Microbiology* 1992;30:1913–1916.

Guthrie R, Susi A. A simple phenylalanine method for detecting phenylketonuria in large populations of newborn infants. *Pediatrics* 1963;32:338–343.

Hamers RL, Smit PW, Stevens W, Schuurman R, Rinke de Wit TF. Dried fluid spots for HIV type-1 viral load and resistance genotyping: a systematic review. *Antiviral Therapy* 2009;14:619–629.

Hemelaar J, Gouws E, Ghys PD, Osmanov S. Isolation W-UNfH, characterisation. Global trends in molecular epidemiology of HIV-1 during 2000-2007. *AIDS* 2011;25:679–689.

Huang S, Erickson B, Mak WB, Salituro J, Abravaya K. A novel real time HIV-1 qualitative assay for the detection of HIV-1 nucleic acids in dried blood spots and plasma. *Journal of Virological Methods* 2011;178:216–224.

Ikomey GM, Atashili J, Okomo-Assoumou MC, Mesembe M, Ndumbe PM. Dried blood spots versus plasma for the quantification of HIV-1 RNA using the manual (PCR-ELISA) Amplicor Monitor HIV-1 version 1.5 assay in Yaounde, Cameroon. *Journal of the International Association of Physicians in AIDS Care (JIAPAC)* 2009;8:181–184.

Jinks DC, Minter M, Tarver DA, Vanderford M, Hejtmancik JF, McCabe ER. Molecular genetic diagnosis of sickle cell disease using dried blood specimens on blotters used for newborn screening. *Human Genetics* 1989;81:363–366.

Johannessen A. Dried blood spots in HIV monitoring: applications in resource-limited settings. *Bioanalysis* 2010;2:1893–1908.

Johannessen A, Garrido C, Zahonero N, Sandvik L, Naman E, Kivuyo SL, Kasubi MJ, Gundersen SG, Bruun JN, de MC. Dried blood spots perform well in viral load monitoring of patients who receive antiretroviral treatment in rural Tanzania. *Clinical Infectious Diseases* 2009;49(6):976–981.

Jordan MR, Bennett DE, Bertagnolio S, Gilks CF, Sutherland D. World Health Organization surveys to monitor HIV drug resistance prevention and associated factors in sentinel antiretroviral treatment sites. *Antiviral Therapy* 2008;13(Suppl 2):15–23.

Katakura K, Lubinga C, Chitambo H, Tada Y. Detection of Trypanosoma congolense and T. brucei subspecies in cattle in Zambia by polymerase chain reaction from blood collected on a filter paper. *Parasitology Research* 1997;83:241–245.

Katz RS, Premenko-Lanier M, McChesney MB, Rota PA, Bellini WJ. Detection of measles virus RNA in whole blood stored on filter paper. *Journal of Medical Virology* 2002;67:596–602.

Lewensohn-Fuchs I, Osterwall P, Forsgren M, Malm G. Detection of herpes simplex virus DNA in dried blood spots making a retrospective diagnosis possible. *Journal of Clinical Virology* 2003;26:39–48.

Long GW, Fries L, Watt GH, Hoffman SL. Polymerase chain reaction amplification from Plasmodium falciparum on dried blood spots. *American Journal of Tropical Medicine and Hygiene* 1995;52:344–346.

Luo W, Yang H, Rathbun K, Pau CP, Ou CY. Detection of human immunodeficiency virus type 1 DNA in dried blood spots by a duplex real-time PCR assay. *Journal of Clinical Microbiology* 2005;43:1851–1857.

Matsubara Y, Narisawa K, Tada K, Ikeda H, Yao YQ, Danks DM, Green A, McCabe ER. Prevalence of K329E mutation in medium-chain acyl-CoA dehydrogenase gene determined from Guthrie cards. *Lancet* 1991;338:552–553.

Mbida AD, Sosso S, Flori P, Saoudin H, Lawrence P, Monny-Lobe M, Oyono Y, Ndzi E, Cappelli G, Lucht F, Pozzetto B, Oukem-Boyer OOM, Bourlet T. Measure of viral load by using the Abbott real-time HIV-1 assay on dried blood and plasma spot specimens collected in 2 rural dispensaries in Cameroon. *Journal of Acquired Immune Deficiency Syndromes* 2009;52:9–16.

McCabe ER. Utility of PCR for DNA analysis from dried blood spots on filter paper blotters. *PCR Methods and Applications* 1991;1:99–106.

McCabe ER. DNA techniques for screening of inborn errors of metabolism. *European Journal of Pediatrics* 1994;153:S84–S85.

McCabe ER, Huang SZ, Seltzer WK, Law ML. DNA microextraction from dried blood spots on filter paper blotters: potential applications to newborn screening. *Human Genetics* 1987;75:213–216.

McNulty A, Jennings C, Bennett D, Fitzgibbon J, Bremer JW, Ussery M, Kalish ML, Heneine W, Garcia-Lerma JG. Evaluation of dried blood spots for human immunodeficiency virus type 1 drug resistance testing. *Journal of Clinical Microbiology* 2007;45:517–521.

McNulty A, Diallo K, Zhang J, Titanji B, Kassim S, Bennett D, Abert-Grasse J, Kibuka T, Ndumbe PM, Nkengasong JN, Yang C. Development and application of a broadly sensitive genotyping assay for surveillance of HIV-1 drug resistance in PEPFAR countries. *Antiviral Therapy* 2008;13:A117.

Mérens A, Guérin PJ, Guthmann JP, Nicand E. Outbreak of hepatitis E virus infection in Darfur, Sudan: effectiveness of real-time reverse transcription-PCR analysis of dried blood spots. *Journal of Clinical Microbiology* 2009;47:1931–1933.

Monleau M, Montavon C, Laurent C, Segondy M, Montes B, Delaporte E, Boillot F, Peeters M. Evaluation of different RNA extraction methods and storage conditions of dried plasma or blood spots for human immunodeficiency virus type 1 RNA quantification and PCR amplification for drug resistance testing. *Journal of Clinical Microbiologica* 2009;47(4):1107–1118.

Monleau M, Butel C, Delaporte E, Boillot F, Peeters M. Effect of storage conditions of dried plasma and blood spots on HIV-1 RNA quantification and PCR amplification for drug resistance genotyping. *Journal of Antimicrobial Chemotherapy* 2010;65:1562–1566.

Muhamad P, Chaijaroenkul W, Congpuong K, Na-Bangchang K. SYBR Green I and TaqMan quantitative real-time polymerase chain reaction methods for the determination of amplification of Plasmodium falciparum multidrug resistance-1 gene (pfmdr1). *Journal for Parasitology* 2011;97:939–942.

Mumba D, Bohorquez E, Messina J, Kande V, Taylor SM, Tshefu AK, Muwonga J, Kashamuka MM, Emch M, Tidwell R, Büscher P, Meshnick SR. Prevalence of human African trypanosomiasis in the Democratic Republic of the Congo. *PLoS Neglected Tropical Diseases* 2011;5:e1246.

Nagy AL, Csaki R, Klem J, Rovo L, Toth F, Talosi G, Jori J, Kovacs K, Kiss JG. Minimally invasive genetic screen for GJB2 related deafness using dried blood spots. *International Journal of Pediatric Otorhinolaryngology* 2010;74:75–81.

Niu C, Zhu X, Zhou L, Liu Q. Detection of Plasmodium vivax by nested PCR amplification of dried blood spots on filter papers. *Acta Parasitologica et Medica Entomologica Sinica* 2000;7:7–10.

Noda S, Eizuru Y, Minamishima Y, Ikenoue T, Mori N. Detection of human T-cell lymphotropic virus type 1 infection by the polymerase chain reaction using dried blood specimens on filter papers. *Journal of Virological Methods* 1993;43:111–122.

Oste C. PCR automation. In: Erlich HA (ed.), *PCR Technology-Principles and Applications for DNA Amplification*. New York: Stockton Press; 1989. pp. 23–30.

Parkin N, Diallo K, Parry CR, Mwebaza S, Batamwita R, Devos J, Bbosa B, Lyagoba F, Magambo B, Jordan M, Downing R, Kaleebu P, Yang C, Bertagnolio S. Field study of the utility of dried blood spots (DBS) for HIV-1 drug resistance (HIVDR) genotyping in Kampala, Uganda: storage for 2 weeks and shipping at ambient temperature has no effect on genotyping efficiency. In: International workshop on HIV & hepatitis virus drug resistance and curative strategies, Melia Sitges. Sitges, Spain: International Medical Press; 2012. p. A129.

Pirillo MF, Recordon-Pinson P, Andreotti M, Mancini MG, Amici R, Giuliano M. Quantification of HIV-RNA from dried blood spots using the Siemens VERSANT (R) HIV-1 RNA (kPCR) assay. *Journal of Antimicrobial Chemotherapy* 2011;66:2823–2826.

Plamondon M, Labbe AC, Frost E, Deslandes S, Alves AC, Bastien N, Pepin J. Hepatitis C virus infection in Guinea-Bissau: a sexually transmitted genotype 2 with parenteral amplification? *PLoS One* 2007;2:e372.

Rahmah N, Nurulhasanah O, Norhayati S, Zulkarnain I, Norizan M. Comparison of conventional versus real-time PCR detection of Brugia malayi DNA from dried blood spots from school children in a low endemic area. *Tropical Biomedicine* 2010;27:54–59.

Raskin S, Phillips JA 3rd, Kaplan G, McClure M, Vnencak-Jones C. Cystic fibrosis genotyping by direct PCR analysis of Guthrie blood spots. *PCR Methods and Applications* 1992;2:154–156.

Robertson DL, Anderson JP, Bradac JA, Carr JK, Foley B, Funkhouser RK, Gao F, Hahn BH, Kalish ML, Kuiken C, Learn GH, Leitner T, McCutchan F, Osmanov S, Peeters M, Pieniazek D, Salminen M, Sharp PM, Wolinsky S, Korber B. HIV-1 nomenclature proposal. *Science* 2000;288:55–56.

Rottinghaus EK, Ugbena R, Diallo K, Bassey O, Azeez A, Devos J, Zhang G, Aberle-Grasse J, Nkengasong J, Yang C. Dried blood spot specimens are a suitable alternative sample type for HIV-1 viral load measurement and drug resistance genotyping in patients receiving first-line antiretroviral therapy. *Clinical Infectious Diseases* 2012;54:1187–1195.

Saiki RK, Scharf S, Faloona F, Mullis KB, Horn GT, Erlich HA, Arnheim N. Enzymatic amplification of beta-globin genomic sequences and restriction site analysis for diagnosis of sickle cell anemia. *Science* 1985;230:1350–1354.

Schwab KO, Ensenauer R, Matern D, Uyanik G, Schnieders B, Wanders RA, Lehnert W. Complete deficiency of mitochondrial trifunctional protein due to a novel mutation within the beta-subunit of the mitochondrial trifunctional protein gene leads to failure of long-chain fatty acid beta-oxidation with fatal outcome. *European Journal of Pediatrics* 2003;162:90–95.

Sintasath DM, Wolfe ND, Zheng HQ, LeBreton M, Peeters M, Tamoufe U, Djoko CF, Diffo JLD, Mpoudi-Ngole E, Heneine W, Switzer WM. Genetic characterization of the complete genome of a highly divergent simian T-lymphotropic virus (STLV) type 3 from a wild Cercopithecus mona monkey. *Retrovirology* 2009;6:97.

Spooner N. A glowing future for dried blood spot sampling. *Bioanalysis* 2010;2:1343–1344.

Stevens W, Sherman G, Downing R, Parsons LM, Ou CY, Crowley S, Gershy-Damet GM, Fransen K, Bulterys M, Lu L, Homsy J, Finkbeiner T, Nkengasong JN. Role of the laboratory in ensuring global access to ARV treatment for HIV-infected children: consensus statement on the performance of laboratory assays for early infant diagnosis. *The Open AIDS Journal* 2008;2:17–25.

Stevens WS, Noble L, Berrie L, Sarang S, Scott LE. Ultra-high-throughput, automated nucleic acid detection of human immunodeficiency virus (HIV) for infant infection diagnosis using the Gen-Probe Aptima HIV-1 screening assay. *Journal of Clinical Microbiology* 2009;47:2465–2469.

Strenger V, Pfurtscheller K, Wendelin G, Aberle SW, Nacheva EP, Zohrer B, Zenz W, Nagel B, Zobel G, Popow-Kraupp T. Differentiating inherited human herpesvirus type 6 genome from primary human herpesvirus type 6 infection by means of dried blood spot from the newborn screening card. *Journal of Pediatrics* 2011;159:859–861.

Tan TM, Nelson JS, Ng HC, Ting RC, Kara UA. Direct PCR amplification and sequence analysis of extrachromosomal Plasmodium DNA from dried blood spots. *Acta Tropica* 1997;68:105–114.

Tani H, Tada Y, Sasai K, Baba E. Improvement of DNA extraction method for dried blood spots and comparison of four PCR methods for detection of Babesia gibsoni (Asian genotype) infection in canine blood samples. *Journal of Veterinary Medical Science* 2008;70:461–467.

Tanna S, Lawson G. Analytical methods used in conjunction with dried blood spots. *Analytical Methods* 2011;3:1709–1718.

Vemu L, Talasila S, Dandona R, Anilkumar G, Dandona L. Application of polymerase chain reaction to detect HIV-1 DNA in pools of dried blood spots. *Indian Journal of Microbiology* 2011;51:147–152.

WHO. Global HIV/AIDS response. Epidemic update and health sector progress towards universal access. UNAIDS report on the global AIDS epidemic. Geneva, Switzerland: World Health Organization; 2011.

Williams C, Weber L, Williamson R, Hjelm M. Guthrie spots for DNA-based carrier testing in cystic fibrosis. *Lancet* 1988;2:693.

Yang C, McNulty A, Diallo K, Zhang J, Titanji B, Kassim S, Wadonda-Kabondo N, Aberle-Grasse J, Kibuka T, Ndumbe PM, Vedapuri S, Zhou Z, Chilima B, Nkengasong JN. Development and application of a broadly sensitive dried-blood-spot-based genotyping assay for global surveillance of HIV-1 drug resistance. *Journal of Clinical Microbiology* 2010;48:3158–3164.

Youngpairoj AS, Masciotra S, Garrido C, Zahonero N, de Mendoza C, Garcia-Lerma JG. HIV-1 drug resistance genotyping from dried blood spots stored for 1 year at 4 degrees C. *Journal of Antimicrobial Chemotherapy* 2008;61(6):1217–1220.

Zhou Z, Wagar N, Devos JR, Rottinghaus E, Diallo K, Nguyen DB, Bassey O, Ugbena R, Wadonda-Kabondo N, McConnell MS, Zulu I, Chilima B, Nkengasong J, Yang C. Optimization of a low cost and broadly sensitive genotyping assay for HIV-1 drug resistance surveillance and monitoring in resource-limited settings. *PLoS One* 2011;6:e28184.

5

APPLICATION OF ENZYME IMMUNOASSAY METHODS USING DRIED BLOOD SPOT SPECIMENS

Mireille B. Kalou

5.1 INTRODUCTION

In the late 1960s, two independent groups of researchers in Sweden and the Netherlands reported the successful development of enzyme immunoassays (EIAs) and enzyme-linked immune-sorbent assays (ELISAs) (Engvall and Perlmann, 1972). These assays are laboratory techniques that were originally developed for antibody detection but have since been adapted to capture antigens. They are now widely used for infectious disease diagnosis and clinical chemistry applications. The assay methodology has been adapted and commercialized to create diagnostic tests for a wide range of viral, bacterial, parasitic diseases and metabolic and inflammatory disorders as well as for the detection of hormones (Schuurs and van Weemen, 1980; Van Biervilet et al., 1991; McDade et al., 2004; Baingana et al., 2008; Kapur et al., 2008; Brindle et al., 2010; Lin et al., 2011). Moreover, the ELISA technique has enabled the study of specific biomarkers such as antibodies to HIV, hepatitis virus, herpes simplex virus type 2 (HSV-2) or syphilis (Baqi et al., 1999; Mishra et al., 2006; Swai et al., 2006) or Lyme disease (Kline et al., 1989; Stroffolini et al., 1997; Dehnert et al., 2012), or detection of antigens to parasitic infections such as taenia or malaria (Barton Behravesh et al., 2008; Drakeley and Cook, 2009) in population-based surveys (Hunfeld et al., 2002; Lequin, 2005).

The ELISA technique is a multistep process that is both time-consuming and labor-intensive; however, the use of automated instruments has streamlined the process and enabled the widespread use of the method in laboratories throughout the world. Conventional ELISA protocols have been validated primarily for antibody detection in minimally processed specimens such as serum or plasma (Zaaijer et al., 1992; Filice et al., 1993) and adapted for antigen detection in unprocessed specimens (i.e., stool samples) (Garcia et al., 2000; Grahn et al., 2011).

Challenges with obtaining venous blood specimens in targeted populations, biohazard risks for handlers, and transport, processing, and storage of these specimen types constitute major barriers to use of the ELISA method (Mei et al., 2001; Boemer et al., 2006; Chaillet et al., 2009). To overcome these issues, researchers have evaluated an alternative specimen type, namely dried blood spot (DBS) samples, for use in ELISA assays. Collection of DBS samples is minimally invasive compared to venous blood collection and eliminates the need for venipuncture, especially in infants, young children, and hard-to-reach individuals (Guthrie and Suzi, 1963; Beebe and Briggs et al., 1990; Parker and Cubitt, 1999; Boemer et al., 2006; Sarge-Njie et al., 2006; Mahfoud et al., 2010; Snijdewind et al., 2012). The technique is relatively inexpensive and DBS specimens are easy to transport or ship as a cold chain is not required. The stability of DBS makes it an ideal specimen type for laboratory assays that rely on samples collected from settings with extreme environmental conditions, including elevated ambient temperatures and high humidity (Nyambi et al., 1994; Cassol et al., 1996; Solomon et al., 2004; Alvarez-Muñoz et al., 2005).

The performance of different serological assays has been assessed using DBS samples as an alternative specimen type and comparable results to those of serum and plasma samples have been obtained for a wide range of assays (Table 5.1).

These studies have demonstrated the feasibility of adapting an ELISA by validating DBS as a suitable alternative to conventional samples for use in a variety of settings (Parker

Dried Blood Spots: Applications and Techniques, First Edition. Edited by Wenkui Li and Mike S. Lee.

TABLE 5.1 References for DBS-Based ELISA for the Detection of Major Diseases of Public Health Interest

Type of Disease	Target	References
Viral	HIV	Barin et al. (2006) Beebe and Briggs (1990) Behets et al. (1992) Chaillet et al. (2009) de Castro et al. (2008) Hoxie et al. (1992) Sarge-Njie et al. (2006) Solomon et al. (2002) Thapa et al. (2011) Villar et al. (2011) WHO (2005)
	Hepatitis virus	de Paula (2012) Desbois et al. (2009) Hadlich et al. (2007) Hickman et al. (2008) Judd et al. (2003) Mahfoud et al. (2010) Melgaco et al. (2011) Tuaillon et al. (2010) Villar et al. (2011)
	Herpes virus	Grahn et al. (2011) Hogrefe et al. (2002)
	Dengue virus	Balmaseda et al. (2008) Ruangturakit et al. (1994) Tran et al. (2006)
	Measles/rubella virus	Condorelli et al. (1994) Hardelid et al. (2008) Helfand et al. (2001) Helfand et al. (2007) Ibrahim et al. (2006) Karapanagiotidis et al. (2005) Mercader et al. (2006) Riddell et al. (2002) Riddell et al. (2003) Uzicanin et al. (2011)
Parasitic	*Plasmodium* (malaria)	Corran et al. (2008) Bousema et al. (2010) Drakeley et al. (2005)
	Human African trypanosomiasis	Mumba et al. (2011) Elrayah et al. (2007)
Bacterial	*Treponema* (syphilis)	Coates et al. (1998) Stevens et al. (1992)
Inflammatory and metabolic	Lipoprotein Vitamin A	Van Biervilet et al. (1991) Baingana et al. (2008)
	Insulin Exenatide	Kapur et al. (2008) Lin et al. (2011)
	C-reactive protein	Brindle et al. (2010) McDade et al. (2004)

and Cubitt, 1999; Solomon et al., 2002; Barin et al., 2006; McDade et al., 2007; Brindle et al., 2010). Other researchers have also described the usefulness of DBS samples derived from capillary whole blood collected from a finger prick to identify elevated circulating antibodies such as IgE levels associated with allergic disease and repeated macro-parasitic infections (Tanner and McDade, 2007). Despite the many advantages of using this specimen type, the accuracy of test results may be influenced by factors such as the amount of antibody to be detected, the hematocrit of the blood in the sample spot, the size of the spot, the elution buffer, and the type of filter paper used for blood collection (Villar et al., 2011).

All microbial species have at least one antigen that is unique. These antigens can be purified and used to generate specific monoclonal antibodies. Both purified antigens and monoclonal antibodies have emerged as important components of diagnostic assays based on ELISA principles. ELISA is a biomedical technique in which the analyte to be detected binds either to an antibody or to an antigen which is coated to the solid surface of a microtiter well or similar surface. In this chapter, we review the different ELISA formats for both antibody and antigen detection, describe the most frequent public health applications for DBS-based ELISA methods, and outline considerations for optimizing and validating an ELISA method for use with DBS samples.

5.2 OVERVIEW OF BASIC ELISA SYSTEMS

5.2.1 Antibody Detection

Standard ELISA methods are formatted as 96-well microplate-based assays that use a range of specimen types (i.e., whole blood, plasma, serum, and DBS). This technique is a multistep procedure which includes (1) coating microtiter plate wells with the antigen; (2) adding the specimen which could contain the specific antibody; (3) adding a species-specific antibody conjugated to an enzyme; (4) adding substance containing the enzyme's substrate; and (5) producing a detectable signal translated in a color change of the substrate and indicating a positive reaction. A washing cycle between steps ensures that only specific (high-affinity) binding events are maintained resulting in a signal at the final step (Figure 5.1).

5.2.2 Variation of ELISA Method for Antibody Detection

5.2.2.1 Sandwich ELISA This type of assay format derives its name from the "sandwiching" of the target antigen between two primary antibodies: the capture antibody and the detection antibody. (1) Known quantity of capture antibody is bound to microtiter plate wells. (2) The antigen-containing sample is applied to the microtiter plate. (3) A specific

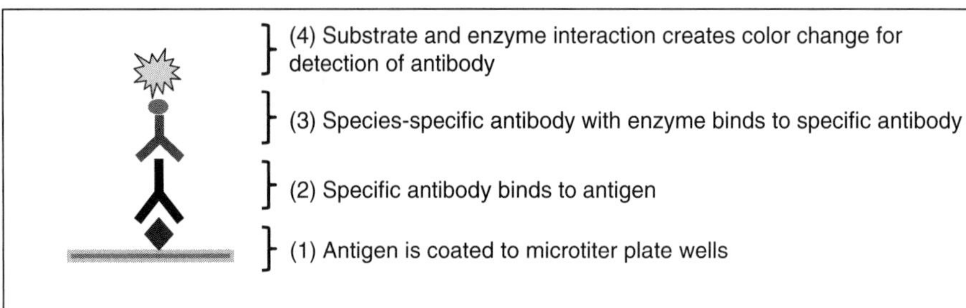

FIGURE 5.1 Basic ELISA principle for antibody detection (indirect ELISA).

antibody is added to bind to the antigen. (4) Enzyme-linked secondary antibodies are applied as detection antibodies to bind specifically to the antibody. (5) After series of washing cycles, the substrate is added and the antibody concentration can be determined by the signal strength generated by the enzyme–substrate reaction (Figure 5.2).

Antigen Detection In antigen detection ELISAs, microtiter plate wells are coated with a capture antibody that recognizes the antigen in the sample under test; any antigen present in the sample binds to the antibodies that coat the well. The bound antigen is then detected by using a second antibody specific for the analyte which is conjugated to an enzyme. A substrate is added to generate a signal; a color change is then noted indicating a positive result. A series of washing cycles between steps removes weakly adherent antigens/antibodies.

5.2.3 Variations of ELISA for Antigen Detection

5.2.3.1 Direct ELISA Direct ELISA uses the method of directly labeling the antibody itself. (1) Microwell plates are coated with a sample containing the target antigen (i.e., patient's serum) and (2) the detecting antibody is applied to the plate for binding to the antigen coated on the plate. The binding of labeled antibody is quantitated by a colorimetric, chemiluminescent, or fluorescent end point (Figure 5.3).

Direct ELISA is relatively quick and avoids potential problems of cross-reactivity of the secondary antibody with components in the antigen sample.

5.2.3.2 Competitive ELISA This ELISA method is based upon any of the above formats (i.e., direct, indirect, or sandwich). This format includes the following steps: (1) incubating unlabeled antibody in the presence of free antigen (sample) for forming antibody/antigen complexes; (2) adding bound antibody/antigen complexes to an antigen-coated well; (3) unbound antibody binds to immobilized antigen, and free antigen/antibody complexes are removed by washing; (4) adding secondary antibody coupled to the enzyme, specific to the primary antibody; and 5) adding substrate, and

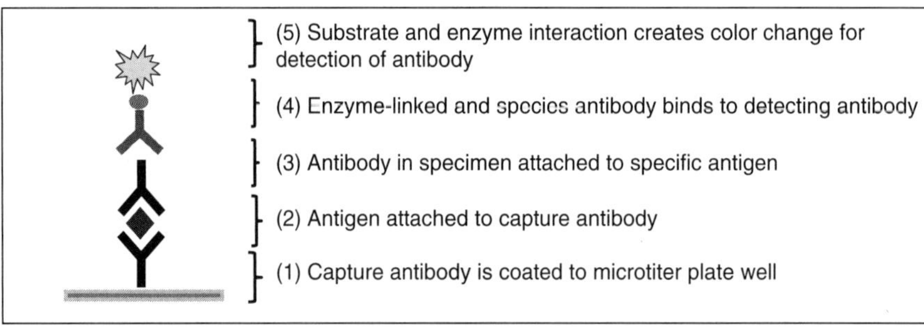

FIGURE 5.2 Sandwich ELISA for antibody detection.

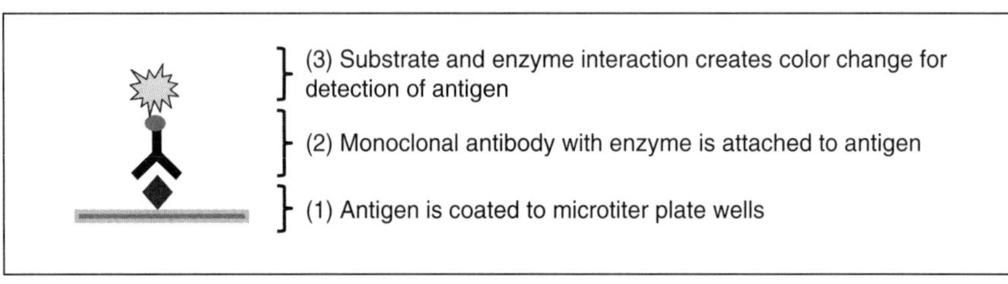

FIGURE 5.3 Direct ELISA for antigen detection. Copyright GenWay Biotech, Inc.

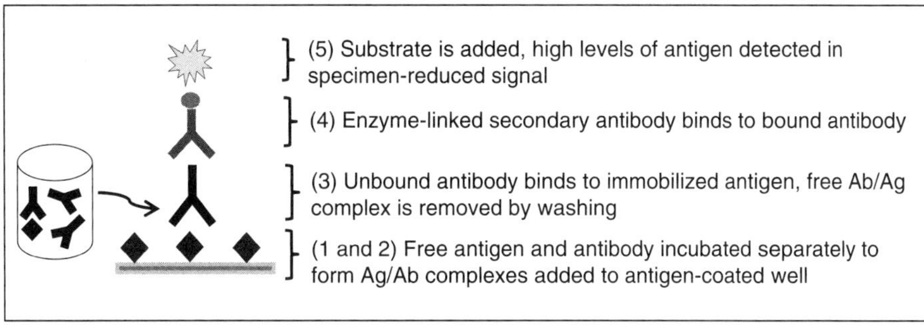
(5) Substrate is added, high levels of antigen detected in specimen-reduced signal
(4) Enzyme-linked secondary antibody binds to bound antibody
(3) Unbound antibody binds to immobilized antigen, free Ab/Ag complex is removed by washing
(1 and 2) Free antigen and antibody incubated separately to form Ag/Ab complexes added to antigen-coated well

FIGURE 5.4 Competitive ELISA for antigen detection.

remaining enzymes elicit a chromogenic or fluorescent signal (Figure 5.4).

For competitive ELISA, the signal intensity is inversely proportional to the target antigen concentration (higher original antigen concentration: weaker signal). It is commonly used for smaller antigens (i.e., hapten) that have only one epitope or antibody-binding site.

5.3 APPLICATIONS OF ELISA METHODS USING DBS

DBS-based ELISA testing has been developed to aid in the diagnosis of infectious diseases metabolic disorders, screen for blood-borne diseases (i.e., blood bank setting), and assess host immune responses following vaccination (Melgaco et al., 2011). The use of DBS-based ELISA methods in clinical laboratories has helped improve screening for blood-borne viral infections such as HIV, hepatitis, measles, and rubella and allowed for targeted prevention interventions and treatment strategies (Table 5.1). Therefore, this section will focus on examples of where DBS-based ELISA methods have been used for the detection of diseases of human public health significance.

Simple approaches have been evaluated for HIV screening in resource-limited setting using DBS as an alternative to serum or plasma to increase testing uptake for diagnostic and HIV sentinel surveillance or surveys (Gwinn et al., 1991). A wide range of commercially available ELISA methods have been adapted for use with DBS specimens for HIV testing. These have been evaluated in many studies and settings with DBS sensitivities and specificities comparable to those recorded for plasma/serum specimens (Sarge-Njie et al., 2006). Despite successful evaluations of modified ELISA assays for HIV diagnosis using DBS, attempts to evaluate single-format ELISA for serotyping HIV in DBS for settings where HIV-1 and HIV-2 types are prevalent have met with limited success (Barin et al., 2006).

Although diagnostic technologies have evolved and enabled the rapid scale-up of HIV testing in laboratory and non-laboratory settings in resource-limited setting, the quality of testing has not kept pace. To address this issue and ensure the accuracy of test results and monitor the performance of testing, the World Health Organization (WHO) has introduced a set of quality assurance activities including national external quality assessment schemes (EQAS) using DBS samples (WHO, 2005). In many developing countries, approximately 10–20% of blood specimens of patients attending HIV clinics or facilities offering routine HIV counseling and testing services are collected onto filter paper and sent to a national or central reference laboratory for quality assurance purposes (Hannon et al., 1989; Chaillet et al., 2009; Parekh et al., 2010; Thapa et al., 2011). This approach has been similarly implemented for sentinel surveillance where DBS samples (specimens of choice) are collected to estimate the HIV prevalence, especially in settings where sample referral systems constitute a great challenge (Hoxie et al., 1992; Van der Loeffa et al., 2003; Solomon et al., 2004). Peripheral testing sites send DBS specimens for retesting at the regional- and central-level laboratories; these laboratories are often enrolled in proficiency testing (PT) programs that allow monitoring the performance of HIV testing performed in the laboratories around the world. Individual-matrix DBS specimens with various levels of sero-reactivity are used as PT or quality control materials and are sent to laboratories where DBS-based ELISA platforms are commonly used. This cost-effective approach for blood collection helps monitor and improve the quality of HIV testing in settings where it may be impractical to refer specimens for additional testing or where there is limited or no access to PT specimens (Hannon et al., 1989; Chaillet et al., 2009; Thapa et al., 2011).

Previous studies have highlighted the use of DBS samples for the detection of serological markers for viral hepatitis by ELISA. Villar et al. (2011) evaluated a modified commercial ELISA's ability to detect three hepatitis B virus (HBV) markers (HBsAg, anti-HBc, and anti-HBs) in DBS samples. The performance of the assays was assessed by comparison with matched serum samples obtained from individuals with or without a past history of HBV infection. Assay specificities ranged between 92.6% and 97.3% for anti-HBc, HBsAg, and anti-HBs assay. Stability studies demonstrated that HBV markers were still detectable in DBS samples for more than 2 months (~63 days) and at low storage temperatures (–20°C), illustrating the feasibility of screening for HBV using DBS specimens stored under extreme conditions.

Commercial ELISAs have also been adapted to successfully diagnose hepatitis A infection by detecting HAV-specific IgM from sera spotted onto filter paper in addition to DBS samples (Desbois et al., 2009; de Paula, 2012). The adjustment method consisted of increasing the sample volume to compensate for the plasma dilution after the DBS elution process, resulting in strong correlations of optical density (OD) ratios between plasma and DBS samples (Karapanagiotidis et al., 2005). Therefore, this technique is reliable for the detection of anti-HAV antibodies after natural infection or vaccination (Melgaco et al., 2011).

Unlike HBV, the accessibility of screening methods for hepatitis C virus (HCV) is generally limited, especially in high-risk populations (i.e., injection drug users) where the prevalence rates are the highest and venous blood collection problematic (Tuaillon et al., 2010). The use of DBS specimens has been demonstrated to increase hepatitis C testing numbers in settings where testing would have been otherwise difficult. Several authors reported a significant increase in hepatitis C testing at sites where DBS testing was offered compared to sites where venipuncture alone was still the common practice (Weinbaum et al., 2005; Sturrock et al., 2007; Hickman et al., 2008; Mahfoud et al., 2010; Rice and Abou-Saleh, 2012). For example, Hickman et al. (2008) described that approximately 70% of individuals in the intervention sites accepted to test for HCV using DBS samples against 21% in the control sites, resulting in a doubling of the total number of HCV antibody tests compared to the preceding 6 months.

Anti-HCV antibody detection is routinely based on highly sensitive second- or third-generation assays that have since been modified and evaluated for use with DBS specimens. Data from previous studies have shown sensitivities and specificities of anti-HCV ELISAs close to 100% for DBS specimens and suggested that DBS-based ELISAs are suitable for HCV epidemiological studies and diagnosis in hard-to-reach populations (McCarron et al., 1999; Tuaillon et al., 2010). However, the determination of the correct dilution factors and storage conditions for DBS specimens is critical to ensure accurate and reproducible ELISA results. For instance, anti-HCV antibodies in DBS samples have been detected in serum dilutions as high as 1:4096 and positive test results have been obtained with HCV-positive DBS specimens even after 12 days of storage at room temperature (Tuaillon et al., 2010).

Modified and optimized ELISA methods are useful to detect hepatitis virus markers and diagnose hepatitis infections using DBS samples that have been stored under optimal conditions. The use of a DBS specimen has been shown to increase testing uptake in settings where testing would have been otherwise difficult. Although studies have demonstrated the usefulness of DBS-based ELISAs in the diagnosis of hepatitis, the performance of ELISA methods varies across hepatitis viruses. This variation in assay performance may be due to the assay validation protocols that are virus specific. Moreover, DBS specimens obtained from immunocompromised individuals have been reported to yield false-negative results, indicating the need to reconfirm DBS specimens with lower anti-HCV titers by methods other than DBS-based ELISA (Fabrizi et al., 2002; Hadlich et al., 2007). Findings from one study showed an increase in the ODs of ELISA in negative HCV DBS specimens stored at room temperature for more than a week, suggesting that inadequate storage conditions could affect the assay performance, resulting in false-positive results and a decrease in the specificity of the assay (Tuaillon et al., 2010).

Dengue virus (DENV) infection is the most prevalent mosquito-borne viral disease worldwide, with over 3 billion people at risk for infection, resulting in 50 million cases of dengue each year (WHO, 2009; Guzman et al., 2010). In many endemic countries, dengue diagnosis is often problematic owing to the lack of reagents, cost, or the delay in obtaining results. To address these challenges, studies have been conducted to determine the usefulness of DBS as a collection method for anti-dengue antibody detection using an ELISA (Vázquez et al., 1991; Balmaseda et al., 2008). In a comparative study conducted in Vietnam, the detection of the anti-dengue IgG antibodies by ELISA in DBS specimens obtained from febrile patients did not seem to have been affected by storage conditions such as high temperature and duration of storage (Ruangturakit et al., 1994). However, the assay performance was impacted by a significant inter-laboratory variability that led to different diagnostic classifications and highlighted the importance of laboratory-specific method validation and optimization for dengue screening (Tran et al., 2006).

HSV-2 infection is believed to be a cofactor for the acquisition of HIV infection. The prevalence of type-specific antibody to HSV-2 is therefore often monitored to identify potential risk for HIV acquisition. To adequately monitor HSV-2 infections worldwide, the standard ELISA technique has been adapted for use with DBS specimens to detect the presence of HSV type-specific antibodies. A dilution protocol for DBS specimens to determine the efficiency of IgG elution has been developed using a recombinant HSV IgG-based ELISA (Hogrefe et al., 2002). A comparison of serum HSV-2 type-specific antibody levels and antibodies in IgG eluted from corresponding DBS by the modified ELISA protocol showed a strong correlation. Study data suggest that using a single dilution (1:4) of DBS eluate in the HSV type-specific ELISA yields results comparable to the serum-based ELISA method.

Nasopharyngeal carcinoma (NPC) is caused by Epstein–Barr virus (EBV) and, although widely distributed, NPC occurs mainly in Southeast Asia (Nor Hashim et al., 2012). NPC is one of the most predominant tumor types in the population in southern China that can be effectively treated when diagnosed early (Sham et al., 1990; Liu et al., 2013). A simple sample collection system such as DBS has therefore been evaluated for screening and diagnosis of NPC in remote populations. DBS-based ELISA protocols for EBV screening have been successfully optimized by changing the volume of the sample buffer, the elution solvent, and the incubation temperature and time. Previous study data showed that combining DBS-based ELISA methods for screening with IgA EBV and IgG EBV immunoblotting methods helped identify more than 98% of healthy donors and confirm all of the NPC samples tested (Fachiroh et al., 2008).

DBS-based ELISA techniques have also been adapted and evaluated to diagnose other viral infections and to monitor the level of antibody production in vaccinated individuals. As an example, commercial ELISAs have been validated for seroprevalence surveys to detect rubella IgG in DBS samples (Hardelid et al., 2008; Snijdewind et al., 2012). Since maternal IgG antibodies are transferred across the placenta, the measurement of rubella IgG antibody in DBS samples collected during newborn screening reflects the maternal antibody prevalence in the population (Gwinn et al., 1991; Helfand et al., 2001, 2007; Karapanagiotidis et al., 2005). Likewise, high sensitivities (100%) and specificities (100%) of DBS-based ELISA have been reported for detection of measles-specific IgM antibody, especially after the onset of rash and in high incidence settings (Uzicanin et al., 2011). Results of DBS-based ELISAs to detect measles IgG and IgM were comparable to those of serum-based ELISAs and suggest that DBS-based ELISAs are useful methods for surveillance and to monitor measles outbreaks (Riddell et al., 2002, 2003; Ibrahim et al., 2006; Mercader et al., 2006). A comparative study conducted in Italy demonstrated that storage conditions (i.e., 15 days at room temperature or several months at 4°C) did not impact the performance of the DBS-based ELISA for IgG detection for measles, rubella, and mumps (Condorelli et al., 1994). Although the performance characteristic of DBS-based ELISAs was close to those of serum-based ELISA, a slight decrease was observed when detecting IgG to measles, rubella, or mumps viruses. Moreover, data indicated the importance of optimizing the sample buffer volume based on the type of filter paper used (Fachiroh et al., 2008).

Bacterial antibody detection in DBS specimens using ELISA has not been extensively described. The most common DBS-based ELISA methods reported are for diagnostic or surveillance purposes; this review will focus on two bacterial infections where DBS-based ELISA has been commonly used. Syphilis surveillance activities target antenatal clinic (ANC) attendees and high-risk populations where syphilis screening is routinely performed using serum specimens collected during prenatal screening or obtained for surveillance or survey purposes. In some cases (i.e., infants) the DBS-based ELISA method has been shown to be appropriate for the detection of treponema IgG which reflects maternal antibody levels. Methods to reduce the concentration of blood substances and the background, ultimately minimizing false-positive results, have been described and shown to be highly specific (>94%) (Stevens et al., 1992). The assay validation protocol requires treating the DBS specimens using buffered saline or treponemal-antibody test diluent to separate high background inherent in DBS specimens from low reactivity specific to low antibody concentrations. However, because this approach is ELISA based, it may constitute a limiting factor in settings where laboratories are not adequately equipped for ELISA testing (Coates et al., 1998). Since the diagnosis of syphilis requires the detection of both treponemal and non-treponemal antibodies, conventional VDRL/RPR testing to detect low levels of non-treponemal antibody will continue to be required even if DBS-based ELISAs expand opportunities for syphilis screening in remote regions.

To implement appropriate public health interventions during outbreaks or routine surveillance activities for meningococcal disease, it is critical to adequately determine the serogroup. DBS-based ELISA methods have also been adapted and validated to detect antibodies to parasitic pathogens such as plasmodia and trypanosomes. As malaria kills over 655,000 people of all ages every year, there is a critical need for identifying accurate early detection mechanisms to ensure timely treatment and care (WHO, 2011). Weekly disease surveillance in sentinel sites as part of the Integrated Disease Surveillance (IDSR) has been established by the WHO in over 15 African countries. Serological approaches using DBS-based ELISA have been validated for detection of antibodies against malaria parasites and have been used as a proxy to measure the intensity of malaria transmission (Drakeley et al., 2005; Corran et al., 2008; Bousema et al., 2010). Cross-reactivity between immune responses to malaria and other parasites has, however, been reported, especially when whole parasite extract is used instead of recombinant proteins which represent single antigens. Although the cross-reactivity was not directly attributed to the use of the DBS specimens, it was noted that using more dilute DBS eluates increased specificity (Bousema et al., 2010).

Untreated human African trypanosomiasis (HAT) cases most often result in death (Mumba et al., 2011); however, study findings suggest that DBS-based serological testing may be appropriate to effectively diagnose individual suspect cases and identify suspect villages for subsequent confirmation (Elrayah et al., 2007). However, the sensitivity and specificity of the assays vary according to the cutoff (CO) point selected based on the OD of the positive control. Increasing the CO point resulted in an increase in the Youden-index and assay specificity with a concurrent decrease in the sensitivity (Hasker et al., 2010). Hence, ELISA method validation is necessary to determine the optimal CO points based on the receiver operating characteristic (ROC) curves when applied to different prevalence settings.

DBS-based ELISA methods have been successfully evaluated to quantify large molecules such as exenatide, a peptide drug used for the treatment of diabetes mellitus type II. The protocol included different extraction solutions, with and without protease inhibitors, and at various incubation times. The findings of this study suggested that exenatide in DBS specimens was still stable after storage at room temperature, 4°C, and 70°C (Lin et al., 2011).

Because quantifying the levels of drug proteins can be challenging in frozen whole blood, ELISA protocols have been evaluated in DBS. A DBS-based ELISA technique appears to be a suitable method for the quantification of other large molecule drugs or biomarkers (Prince et al., 2010). Other studies have also described the usefulness of DBS-ELISA for newborn infant screening of markers such as apo-lipoprotein B and to monitor vitamin A deficiency identified as a major public health problem in resource-poor settings (Van Biervilet et al., 1991; Baingana et al., 2008; Ramakrishnan and Imhoff-Kunsch, 2008).

5.4 OPTIMIZATION AND ELISA METHODS VALIDATION USING DBS SPECIMENS

Since standard ELISA protocols were originally developed for serum or plasma specimens, adaptation of an assay for use with DBS specimens needs to be validated to ensure accuracy, precision, reliability, and to establish detection limits. Evaluation studies have been conducted to optimize specific assays and determine the appropriate storage requirements, the type of filter paper, filter paper disk size and volume of input sample, elution buffer, and CO values to allow detection of the biomarker of interest in DBS specimens (Mei et al., 2001; Dowd et al., 2011).

5.4.1 DBS Elution Protocol

Obtaining a sufficient analyte concentration is critical to target detection and is dependent on the size and number of DBS, as well as the sample volume required for the ELISA. Antibody concentrations in DBS eluates may be lower than in serum specimens, therefore determining the correct volume of DBS eluate necessary for the ELISA method is critical (Hogrefe et al., 2002; Waterboer et al., 2012) to ensure the required assay sensitivity is met. As an example, Croom et al. (2006) reported that despite a 10-fold increase in the elution buffer volume, the modified ELISA method used for the detection of antibodies to HCV in DBS yielded results comparable to those of plasma samples after adjusting for the volume of DBS eluates. Standardization of the dilution scheme is therefore an important consideration for adapting the volume of the DBS eluate for an ELISA method.

5.4.1.1 ***DBS Size*** Prior to the use of DBS specimens in ELISAs, appropriate-sized pieces of filter paper specimens are sampled by punching from the large DBSs. The analyte to be measured is then eluted from the paper and used, albeit in a diluted form, in a normal ELISA (Figure 5.5). Depending on the specific assay protocol, DBS requirements vary from one to five punches with sizes ranging between 2.5 and 8 mm (Hogrefe et al., 2002; Judd et al., 2003; Fachiroh et al., 2008; Brindle et al., 2010; Tuaillon et al., 2010; Melgaco et al., 2011; Villar et al., 2011). Researchers have also described that the location of the punch within the DBS can have an impact on the target analyte concentration. For instance, punches obtained from the periphery of the DBS were shown to reduce the chromatographic effects representing less than 2% of the variation in serum volume obtained after elution of the DBS (Mei et al., 2001; McDade et al., 2004).

5.4.1.2 ***Temperature*** Elution methods frequently require that DBS be soaked in tubes or in microtiter plate wells containing the elution solution at different temperatures and different times (Mercader et al., 2006; Villar et al., 2011). The incubation temperatures for most procedures are method specific and fluctuate between the ambient temperature (23°C) to 4°C for an hour to an 18-hour overnight incubation (Van Biervilet et al., 1991; Mahfoud et al., 2010). Additionally, the optimum incubation time and temperature appear to be dependent on the type of filter paper highlighting the need to evaluate the type of paper when optimizing a DBS-based ELISA method (Fachiroh et al., 2008).

5.4.1.3 ***Elution Buffers*** The effects of different elution buffers on the DBS sample absorbance/cutoff (S/CO) ratios have been extensively evaluated (McCarron et al., 1999; Judd et al., 2003) and need to be taken into consideration when developing a DBS-ELISA method. As blood samples may contain proteins expressed in cells and tissues, failure to adequately select the optimum elution buffer may result in an increase in the assay background interference due to the presence of the lysed cell debris and other proteins ultimately leading to a decreased ability to detect the antibody of interest by the ELISA (Mercader et al., 2006; Villar et al., 2011). Most procedures have used phosphate-buffered saline (PBS) solution (with or without Tween 20) and sodium azide (Hogrefe et al., 2002; Hardelid et al., 2008; Waterboer et al., 2012); however, a variation of this PBS-based solution that contains skim milk powder in either a 10 mM tris and sodium chloride solution or PBS–Tween 20 solution at pH values varying between 7.2 and 7.4 with bovine plasma or serum albumin can be used as elution buffer (Croom et al., 2006; Fachiroh et al., 2008; Melgaco et al., 2011; Villar et al., 2011). Alternatively, researchers have adapted standard ELISA protocols by using the kit specimen diluents provided by the manufacturers, along with other test kit reagents, to elute DBS

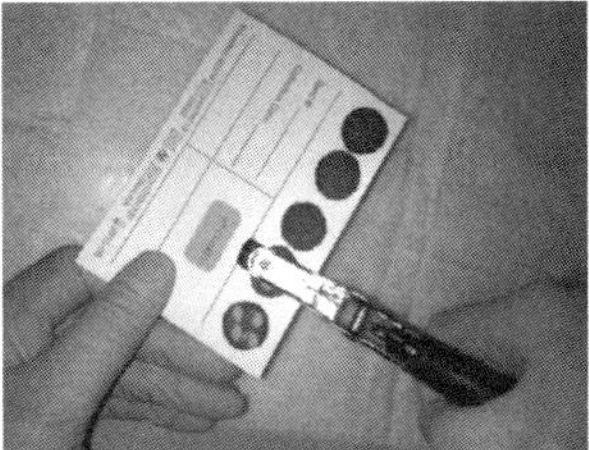

1. DBS punching

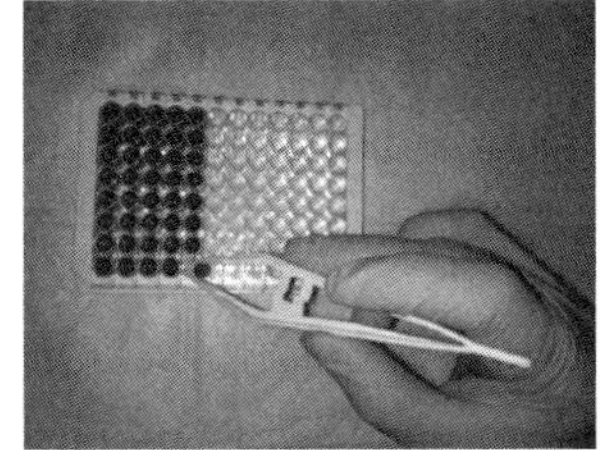

2. Punched DBS samples eluted in microplate wells

3. Eluted material added to ELISA microplate with appropriate dilution and incubated

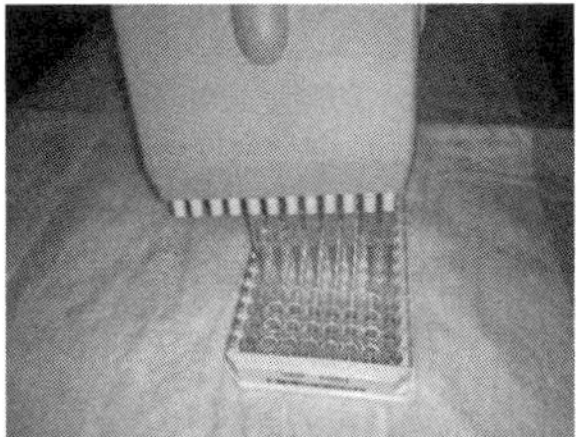

4. After washing step conjugate is added and microplate incubated

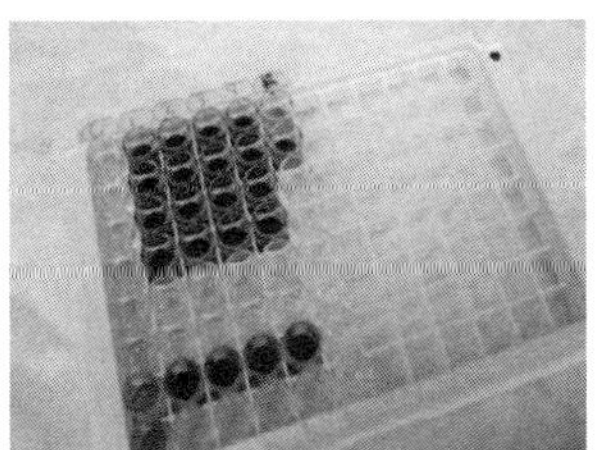

5. After washing step substrate is added and incubated

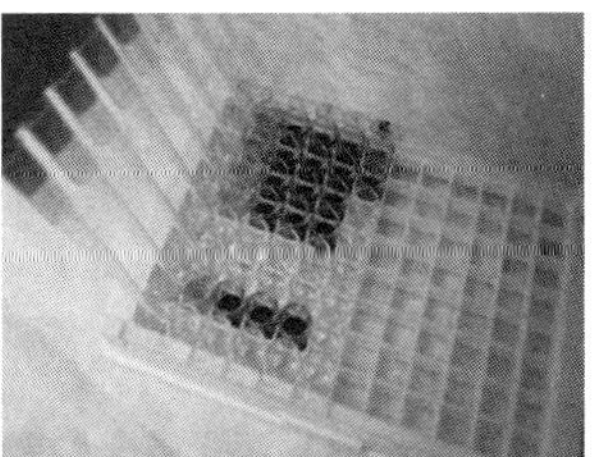

6. Stop solution added

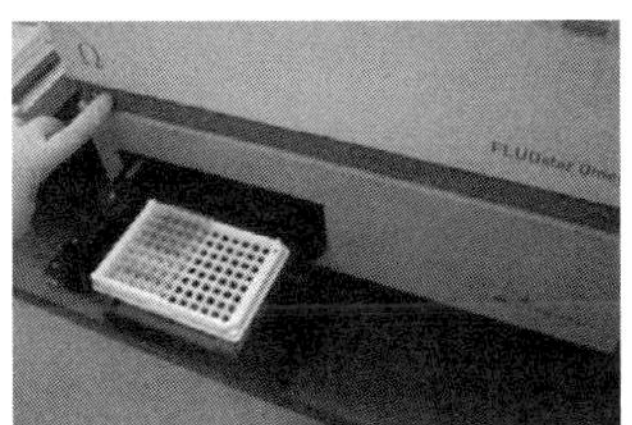

7. Optical densities read

FIGURE 5.5 ELISA technique using DBS specimen.

samples prior to testing (Behets et al., 1992; Tuaillon et al., 2010).

5.4.1.4 Type of Filter Paper Data from several studies have suggested that the optimum volume of sample buffer required for the detection of antibodies by DBS-based ELISA techniques may vary dramatically based on the type of filter paper used to prepare the DBS. For example, one study described that for the detection of EBV IgA by ELISA, 500 μL of sample buffer volume was needed to elute the DBS specimens collected onto Whatman No. 3 whereas 400 μL was used when the blood specimens were spotted onto Schleicher & Schuell (S&S) 903 filter paper (Fachiroh et al., 2008). Similarly, although the OD values of IgA and IgG EBV DBS-based ELISA were comparable to those of plasma-based ELISA, the sensitivity and specificity of DBS-based ELISA varied by the type of filter paper used for DBS sampling, suggesting that the type of paper could impact the performance of the assay if not adequately evaluated.

5.4.2 Defining Cutoff Values

Since ELISA methods were not originally designed for use with DBS specimens, it is critical to determine if the CO values established for plasma samples are suitable for DBS eluates. Several authors have described methods for determining the mean CO values for DBS-based ELISA and calculating assay precision using ODs of positive and negative DBS controls in multiple runs (Judd et al., 2003; Dowd et al., 2011). While studies have reported no significant differences in CO values when using standard deviation (SD) approach to set the CO values, ELISA results need to be interpreted according to the OD and CO ratios, which vary for positive, negative, and indeterminate results and also by analyte tested (Villar et al., 2011). Evaluation of coefficients of variation has therefore been considered as an appropriate method to assess the inter- and intra-assay variability for DBS-based ELISA method (Mercader et al., 2006) and suggest that threshold CO values established by the manufacturer for serum or plasma specimens need to be compared to CO values for different DBS eluate volumes and elution protocols (*DBS elution protocol*).

5.5 FACTORS INFLUENCING THE PERFORMANCE OF DBS-BASED ELISA METHODS

The previous sections provide examples demonstrating the feasibility of using DBS specimens for accurately detecting antibodies to a variety of analytes by ELISA; however, additional factors need to be taken into consideration including the quality of the specimen, the elution protocol, and the optimization process, as these parameters may negatively affect the performance of the assay if not adequately evaluated.

5.5.1 DBS Preparation

Similar to the conventional specimens, the quality of DBS samples may play a critical role in the overall assay precision and performance (Mei et al., 2001; McDade et al., 2004). Since the hematocrit level may influence the blood viscosity and subsequently the distribution of the blood sample on the filter card, it is important that a standard procedure be observed during the application of whole blood to the filter paper. Uneven sample coverage due to inappropriate application methods or variable hematocrit levels can cause variable ELISA results. For example, blood samples blotted or smeared onto the paper or dropped on the top of a previously collected drop of blood may lead to significant concentration differences between and within spots and ultimately impact the accuracy of the assay based upon the location of the punched spot.

5.5.2 Anticoagulant

In some instances, DBS cards are prepared from leftover whole blood collected in tubes containing anticoagulant. As anticoagulants affect blood viscosity and therefore the thickness and homogeneity of the spot, proper evaluations of DBS cards prepared from blood collected in anticoagulant tubes should be conducted prior to using these samples in an ELISA.

5.5.3 Punching

To maximize the assay performance, blood spotting and punching procedures should be well defined and consistent. In a high-throughput setting, manual punching can be labor-intensive and can impact the consistency of the size of the spots, leading to issues with assay reproducibility. Thus, the use of semi-automated punching in high-throughput laboratories can ensure that adequate sample volume is obtained.

5.5.4 Cross-Contamination

As with most laboratory techniques, cross-contamination typically results in false results and ultimately decreases the performance of the assay. Contamination may be due to carryover during the punching process if paper punchers are not adequately cleaned between punches. Current procedures require that the blood spotted onto filter paper cards be air dried for several hours prior to shipping or storage; however, this method may lead to contamination from circulating air and from exposed surfaces. As an example, de Castro et al. (2008) reported a false-positive sample which failed to be confirmed even after testing several times, suggesting a cross-contamination of the actual DBS sample leading to a decrease in the specificity and positive predictive value of the HIV diagnostic ELISA under evaluation (de Castro et al., 2008).

5.6 CONCLUSIONS

As described in the sections above, ELISA methods have been widely evaluated using DBS specimens and shown to be suitable for a wide range of biomarkers. The assay performance depends on the optimization protocol and may be negatively impacted if the spotting technique, sample volume, punching method, type of filter paper, and elution protocol are not adequately evaluated. Moreover, determining the optimum CO values specific to each analyte is critical for both the inter- and intra-assay variability and reproducibility, and the presence or absence of anticoagulant in the blood sample used to prepare DBS specimens should be considered as it may affect the detection of certain analytes. The potential limitations of DBS can, however, be overcome through careful optimization and validation procedures leading to increased testing uptake in settings where this would not have been possible, ultimately impacting public health programs worldwide.

DISCLAIMER

The findings and conclusions in this article are those of the author and do not necessarily represent the views of the US Centers for Disease Control and Prevention. The use of trade names is for identification purposes only and does not constitute endorsement by the US Centers for Disease Control and Prevention or the Department of Health and Human Services.

REFERENCES

Alvarez-Muñoz MT, Zaragoza-Rodríguez S, Rojas-Montes O, Palacios-Saucedo G, Vázquez-Rosales G, Gómez-Delgado A, Torres J, Muñoz O. High correlation of human immunodeficiency virus type-1 viral load measured in dried-blood spot samples and in plasma under different storage conditions. *Archives of Medical Research* 2005;36:382–386.

Baingana RK, Matovu DK, Garrett D. Application of retinol-binding protein enzyme immunoassay to dried blood spots to assess vitamin A deficiency in a population-based survey: the Uganda Demographic and Health Survey 2006. *Food and Nutrition Bulletin* 2008;29:297–305.

Balmaseda A, Saborio S, Tellez Y, Mercado JC, Pérez L, Hammond SN, Rocha C, Kuan G, Harris E. Evaluation of immunological markers in serum, filter-paper blood spots, and saliva for dengue diagnosis and epidemiological studies. *Journal of Clinical Virology* 2008;43:287–291.

Baqi S, Shah SA, Baig MA, Mujeeb SA, Memon A. Seroprevalence of HIV, HBV, and syphilis and associated risk behaviours in male transvestites (Hijras) in Karachi, Pakistan. *International Journal of STD and AIDS* 1999;10:300–304.

Barin F, Plantier JC, Brand D, Brunet S, Moreau A, Liandier B, Thierry D, Cazein F, Lot F, Semaille C, Desenclos JC. Human immunodeficiency virus serotyping on dried serum spots as a screening tool for the surveillance of the AIDS epidemic. *Journal of Medical Virology* 2006;78:S13–S18.

Barton Behravesh C, Mayberry LF, Bristol JR, Cardenas VM, Mena KD, Martínez-Ocaña J, Flisser A, Snowden KF. Population-based survey of taeniasis along the United States-Mexico border. *Annals of Tropical Medicine and Parasitology* 2008;102:325–333.

Beebe JL, Briggs LC. Evaluation of enzyme-linked immunoassay systems for detection of human immunodeficiency virus type 1 antibody from filter paper disks impregnated with whole blood. *Journal of Clinical Microbiology* 1990;28:808–810.

Behets F, Kashamuka M, Pappaioanou M, Green TA, Ryder RW, Batter V, George JR, Hannon WH, Quinn TC. Stability of human immunodeficiency virus type 1 antibodies in whole blood dried on filter paper and stored under various tropical conditions in Kinshasa, Zaire. *Journal of Clinical Microbiology* 1992;30:179–182.

Boemer F, Vanbellinghen JF, Bours V, Schoos R. Screening for sickle cell disease on dried blood: a new approach evaluated on 27,000 Belgian newborns. *Journal of Medical Screening* 2006;13:132–136.

Bousema T, Youssef RM, Cook J, Cox J, Alegana VA, Amran J, Noor AM, Snow RW, Drakeley C. Serologic markers for detecting malaria in areas of low endemicity, Somalia, 2008. *Emerging Infectious Diseases* 2010;16:392–399.

Brindle E, Fujita M, Shofer J, O'Connor KA. Serum, plasma, and dried blood spot high-sensitivity C-reactive protein enzyme immunoassay for population research. *Journal of Immunology Methods* 2010;362:112–120.

Cassol SA, Read S, Weniger BG, Gomez P, Lapointe N, Ou CY, Babu PG. Dried blood spots collected on filter paper: an international resource for the diagnosis and genetic characterization of human immunodeficiency virus type-1. *Memórias do Instituto Oswaldo Cruz* 1996;91:351–358.

Chaillet P, Zachariah R, Harries K, Rusanganwa E, Harries AD. Dried blood spots are a useful tool for quality assurance of assurance of rapid HIV testing in Kigali, Rwanda. *Transactions of the Royal Society of Tropical Medicine and Hygiene* 2009;103:634–637.

Coates GL, Guarenti L, Parker SP, Willumsen JF, Tomkins AM. Evaluation of the sensitivity and specificity of a Treponema pallidurn dried blood spot technique for use in the detection of syphilis. *Transactions of the Royal Society of Tropical Medicine and Hygiene* 1998;92:44.

Condorelli F, Scalia G, Stivala A, Gallo R, Marino A, Battaglini CM, Castro A. Detection of immunoglobulin G to measles virus, rubella virus, and mumps virus in serum samples and in microquantities of whole blood dried on filter paper. *Journal of Virological Methods* 1994;49:25–36.

Corran PH, Cook J, Lynch C, Leenderyse H, Manjurano A, Griffin J, Cox J, Abeku T, Bousema T, Ghani AC, Drakeley C, Riley E. Dried blood spots as a source of anti-malarial antibodies for epidemiological studies. *Malaria Journal* 2008;7:195–207.

Croom HA, Richards KM, Best SJ, Francis BH, Johnson EI, Dax EM, Wilson KM. Commercial enzyme immunoassay adapted for the detection of antibodies to hepatitis C virus in dried blood spots. *Journal of Clinical Virology* 2006;36:68–71.

de Castro AC, Borges LG, Souza Rda S, Grudzinski M, D'Azevedo PA. Evaluation of the human immunodeficiency virus type 1 and 2 antibodies detection in dried whole blood spots (DBS) samples. *Revista do Instituto de Medicina Tropical de São Paulo* 2008;50:151–156.

de Paula VS. Laboratory diagnosis of hepatitis A: alternative samples for diagnosis. *Future Virology* 2012;7:461–472.

Dehnert M, Fingerle V, Klier C, Talaska T, Schlaud M, Krause G, Wilking H, Poggensee G. Seropositivity of Lyme borreliosis and associated risk factors: a population-based study in children and adolescents in Germany (KiGGS). *PLoS One* 2012;7:e41321.

Desbois D, Roque-Afonso AM, Lebraud P, Dussaix E. Use of dried serum spots for serological and molecular detection of hepatitis a virus. *Journal of Clinical Microbiology* 2009;47:1536–1542.

Dowd JB, Aiello A, Chyu L, Huang YY, McDade TW. Cytomegalovirus antibodies in dried blood spots: a minimally invasive method for assessing stress, immune function, and aging. *Immunity and Ageing* 2011;8:3.

Drakeley C, Cook J. Potential contribution of sero-epidemiological analysis for monitoring malaria control and elimination: historical and current perspectives. *Advances in Parasitology* 2009;69:299–352.

Drakeley CJ, Corran PH, Coleman PG, Tongren JE, McDonald SL, Carneiro I, Malima R, Lusingu J, Manjurano A, Nkya WM, Lemnge MM, Cox J, Reyburn H, Riley EM. Estimating medium- and long-term trends in malaria transmission by using serological markers of malaria exposure. *Proceedings of the National Academy of Sciences of the United States of America* 2005;102:5108–5113.

Elrayah IE, Rhaman MA, Karamalla LT, Khalil KM, Buscher P. Evaluation of serodiagnostic tests for T.b. gambiense human African trypanosomiasis in southern Sudan. *Eastern Mediterranean Health Journal* 2007;13:1098–1107.

Engvall E, Perlmann P. Enzyme-linked immunosorbent assay, ELISA III. Quantitation of specific antibodies by enzyme-labeled anti-immunoglobulin in antigen-coated tubes. *Journal of Immunology* 1972;109:129–135.

Fabrizi F, Poordad FF, Martin P. Hepatitis C infection and the patient with end-stage renal disease. *Hepatology* 2002;36:3–10.

Fachiroh J, Prasetyanti PR, Paramita DK, Prasetyawati AT, Anggrahini DW, Haryana SM, Middeldorp JM. Dried-blood sampling for Epstein-Barr virus immunoglobulin G (IgG) and IgA serology in nasopharyngeal carcinoma screening. *Journal of Clinical Microbiology* 2008;46:1374–1380.

Filice G, Patruno S, Campisi D, Chiesa A, Orsolini P, Debiaggi M, Bruno R, Tinelli M. Specificity and sensitivity of 3rd generation EIA for detection of HCV antibodies among intravenous drug-users. *The New Microbiologica* 1993;16: 35–42.

Garcia LS, Shimizu RY, Bernard CN. Detection of Giardia lamblia, Entamoeba histolytica/Entamoeba dispar, and Cryptosporidium parvum antigens in human fecal specimens using the triage parasite panel enzyme immunoassay. *Journal of Clinical Microbiology* 2000;38:3337–3340.

Grahn A, Studahl M, Nilsson S, Thomsson E, Bäckström M, Bergström T. Varicella-zoster virus (VZV) glycoprotein E is a serological antigen for detection of intrathecal antibodies to VZV in central nervous system infections, without cross-reaction to herpes simplex virus 1. *Clinical and Vaccine Immunology* 2011;18:1336–1342.

Guthrie R, Susi A. A simple phenylalanine method for detecting phenylketonuria in large populations of newborn infants. *Pediatrics* 1963;32:338–343.

Guzman MG, Halstead SB, Artsob H, Buchy P, Farrar J, Gubler DJ, Hunsperger E, Kroeger A, Margolis HS, Martínez E, Nathan MB, Pelegrino JL, Simmons C, Yoksan S, Peeling RW. Dengue: a continuing global threat. *National Reviews: Microbiology* 2010;8:S7–S16.

Gwinn M, Pappaioanou M, George JR, Hannon WH, Wasser SC, Redus MA, Hoff R, Grady GF, Willoughby A, Novello AC, Petersen LR, Dondero TJ, Curran JW. Prevalence of HIV infection in childbearing women in the United States. Surveillance using newborn blood samples. *Journal of the American Medical Association* 1991;265:1704–1708.

Hadlich E, Alvares-Da-Silva MR, Dal Molin RK, Zenker R, Goldani LZ. Hepatitis C virus (HCV) viremia in HIV-infected patients without HCV antibodies detectable by third-generation enzyme immunoassay. *Journal of Gastroenterology and Hepatology* 2007;22:1506–1509.

Hannon WH, Lewis DS, Jones WK, Powell MK. A quality assurance program for human immunodeficiency virus seropositivity screening of dried-blood spot specimens. *Infection Control and Hospital Epidemiology* 1989;10:8–13.

Hardelid P, Williams D, Dezateux C, Cubitt WD, Peckham CS, Tookey PA, Cortina-Borja M. Agreement of rubella IgG antibody measured in serum and dried blood spots using two commercial enzyme-linked immunosorbent assays. *Journal of Medical Virology* 2008;80:360–364.

Hasker E, Lutumba P, Mumba D, Lejon V, Büscher P, Kande V, Muyembe JJ, Menten J, Robays J, Boelaert M. Diagnostic accuracy and feasibility of serological tests on filter paper samples for outbreak detection of T.b. gambiense human African trypanosomiasis. *American Journal of Tropical Medicine and Hygiene* 2010;83:374–379.

Helfand RF, Keyserling HL, William I, Murray A, Mei J, Moscatiello C, Icenogle J, Bellini WJ. Comparative detection of measles and rubella IgM and IgG derived from filter paper blood and serum samples. *Journal of Medical Virology* 2001;65:751–757.

Helfand RF, Cabezas C, Abernathy E, Castillo-Solorzano C, Ortiz AC, Sun H, Osores F, Oliveira L, Whittembury A, Charles M, Andrus J, Icenogle J. Dried blood spots versus sera for detection of rubella virus-specific immunoglobulin M (IgM) and IgG in samples collected during a rubella outbreak in Peru. *Clinical and Vaccine Immunology* 2007;14:1522–1555.

Hickman M, McDonald T, Judd A, Nichols T, Hope V, Skidmore S, Parry JV. Increasing the uptake of hepatitis C virus testing among injecting drug users in specialist drug treatment and prison settings by using dried blood spots for diagnostic testing: a cluster randomized controlled trial. *Journal of Viral Hepatology* 2008;15:250–254.

Hogrefe WR, Ernst C, Su X. Efficiency of reconstitution of immunoglobulin g from blood specimens dried on filter paper and utility in herpes simplex virus type-specific serology screening. *Clinical and Diagnostic Laboratory Immunology* 2002;9:1338–1342.

Hoxie NJ, Vergeront JM, Pfister JR, Hoffman GF, Markwardt-Elmer PA, Davis JP. Improving estimates of HIV-1 seroprevalence among childbearing women: use of smaller blood spots. *American Journal of Public Health* 1992;82:1370–1373.

Hunfeld KP, Ernst M, Zachary P, Jaulhac B, Sonneborn HH, Brade V. Development and laboratory evaluation of a new recombinant ELISA for the serodiagnosis of Lyme disease. *Die Wiener klinische Wochenschrift* 2002;114:580–585.

Ibrahim SA, Abdallah A, Saleh EA, Osterhaus AD, De Swart RL. Measles virus-specific antibody levels in Sudanese infants: a prospective study using filter-paper blood samples. *Epidemiology and Infection* 2006;134:79–85.

Judd A, Parry J, Hickman M, McDonald T, Jordan L, Lewis K, Contreras M, Dusheiko G, Foster G, Gill N, Kemp K, Main J,

Murray-Lyon I, Nelson M. Evaluation of a modified commercial assay in the detecting antibody to hepatitis C virus in oral fluids and dried blood spots. *Journal of Medical Virology* 2003;71:49–55.

Kapur S, Kapur S, Zava D. Cardiometabolic risk factors assessed by a finger stick dried blood spot method. *Journal of Diabetes Science and Technology* 2008;2:236–241.

Karapanagiotidis T, Riddell M, Kelly H. Detection of rubella immunoglobulin M from dried venous blood spots using a commercial enzyme immunoassay. *Diagnostic Microbiology and Infectious Disease* 2005;53:107–111.

Kline RL, Brothers TA, Brookmeyer R, Zeger S, Quinn TC. Evaluation of human immunodeficiency virus seroprevalence in population surveys using pooled sera. *Journal of Clinical Microbiology* 1989;27:1449–1452.

Lequin RM. Enzyme immunoassay (EIA)/enzyme-linked immunosorbent assay (ELISA). *Clinical Chemistry* 2005;51: 2415–2418.

Lin Y-Q, Khetarpal R, Zhang Y, Song H, Li SS. Combination of ELISA and dried blood spot technique for the quantification of large molecules using exenatide as a model. *Journal of Pharmacological and Toxicological Methods* 2011;64:124–128.

Liu Z, Ji MF, Huang QH, Fang F, Liu Q, Jia WH, Guo X, Xie SH, Chen F, Liu Y, Mo HY, Liu WL, Yu YL, Cheng WM, Yang YY, Wu BH, Wei KR, Ling W, Lin X, Lin EH, Ye W, Hong MH, Zeng YX, Cao SM. Two Epstein-Barr virus-related serologic antibody tests in nasopharyngeal carcinoma screening: results from the initial phase of a cluster randomized controlled trial in Southern China. *American Journal of Epidemiology* 2013;177:242–250.

Mahfoud Z, Kassak K, Kreidieh K, Shamra S, Ramia S. Prevalence of antibodies to human immunodeficiency virus (HIV), hepatitis B and hepatitis C and risk factors in prisoners in Lebanon. *Journal of Infection in Developing Countries* 2010;3:144–149.

McCarron B, Fox R, Wilson K, Cameron S, McMenamin J, McGregor G, Pithie A, Goldberg D. Hepatitis C antibody detection in dried blood spots. *Journal of Viral Hepatitis* 1999;6: 453–456.

McDade TW, Burhop J, Dohnal J. High-sensitivity enzyme immunoassay for C-reactive protein in dried blood spots. *Clinical Chemistry* 2004;50:652–654.

McDade TW, Williams S, Snodgrass JJ. What a drop can do: dried blood spots as a minimally invasive method for integrating biomarkers into population-based research. *Demography* 2007;44:899–925.

Mei JV, Alexander JR, Adam BW, Hannon WH. Use of filter paper for the collection and analysis of human whole blood specimens. *Journal of Nutrition* 2001;131:1631S–1636S.

Melgaco JG, Pinto MA, Rocha AM, Freire M, Gaspar LP, Lima SMB, Cruz OG, Vitral CL. The use of dried blood spots for assessing antibody response to hepatitis A virus after natural infection and vaccination. *Journal of Medical Virology* 2011;83:208–217.

Mercader S, Featherstone D, Bellini WJ. Comparison of available methods to elute serum from dried blood spot samples for measles serology. *Journal of Virological Methods* 2006;137:140–149.

Mishra V, Vaessen M, Boerma JT, Arnold F, Way A, Barrere B, Cross A, Hong R, Sangha J. HIV testing in national population-based surveys: experience from the demographic and health surveys. *Bulletin of the World Health Organization* 2006;84:537–545.

Mumba D, Bohorquez E, Messina J, Kande V, Taylor SM, Tshefu AK, Muwonga J, Kashamuka MM, Emch M, Tidwell R, Büscher P, Meshnick SR. Prevalence of human African trypanosomiasis in the Democratic Republic of the Congo. *PLoS Neglected Tropical Diseases* 2011;5:e1246.

Nor Hashim NA, Ramzi NH, Velapasamy S, Alex L, Chahil JK, Lye SH, Munretnam K, Haron MR, Ler LW. Identification of genetic and non-genetic risk factors for nasopharyngeal carcinoma in a Southeast Asian population. *Asian Pacific Journal of Cancer Prevention* 2012;13:6005–6010.

Nyambi PN, Fransen K, De Beenhouwer H, Chomba EN, Temmerman M, Ndinya-Achola JO, Piot P, Van der Groen G. Detection of human immunodeficiency virus type 1 (HIV-1) in heel prick blood on filter paper from children born to HIV-1-seropositive mothers. *Journal of Clinical Microbiology* 1994;32:2858–2860.

Parekh BS, Kalou M, Alemnji G, Ou CY, Gershy-Damet GM, Nkengasong JN. Scaling up HIV rapid testing in developing countries. Comprehensive approach for implementing quality assurance. *American Journal of Clinical Pathology* 2010;134: 573–584.

Parker SP, Cubitt WD. The use of the dried blood spot sample in epidemiological studies. *Journal of Clinical Pathology* 1999;52:633–639.

Prince PJ, Matsuda KC, Retter M, Scott G. Assessment of DBS technology for the detection of therapeutic antibodies. *Bioanalysis* 2010;2:1449–1460.

Ramakrishnan U, Imhoff-Kunsch B. *Anemia and Iron Deficiency in Developing Countries. Nutrition and Health: Handbook of Nutrition and Pregnancy*. Totowa, NJ: Humana Press/Springer; 2008.

Rice P, Abou-Saleh MT. Detecting antibodies to hepatitis C in injecting drug users: a comparative study between saliva, serum, and dried blood spot tests. *Addictive Disorders and Their Treatment* 2012;11:76–83.

Riddell MA, Leydon JA, Catton MG, Kelly HA. Detection of measles virus-specific immunoglobulin M in dried venous blood samples by using a commercial enzyme immunoassay. *Journal of Clinical Microbiology* 2002;40:5–9.

Riddell MA, Byrnes GB, Leydon JA, Kelly HA. Dried venous blood samples for the detection and quantification of measles IgG using commercial enzyme immunoassay. *Bulletin of World Health Organization* 2003;81:701–707.

Ruangturakit S, Rojanasuphot S, Srijuggravanvong A, Duangchanda S, Nuangplee S, Igarashi A. Storage stability of dengue IgM and IgG antibodies in whole blood and serum dried on filter paper strips detected by ELISA. *Southeast Asian Journal of Tropical Medicine and Public Health* 1994;25:560–564.

Sarge-Njie R, Schim Van Der Loeff M, Ceesay S, Cubitt D, Sabally S, Corrah T, Whittle H. Evaluation of the dried blood spot filter paper technology and five testing strategies of HIV-1 and HIV-2 infections in West Africa. *Scandinavian Journal of Infectious Diseases* 2006;38:1050–1056.

Schuurs AH, van Weemen BK. Enzyme-immunoassay: a powerful analytical tool. *Journal of Immunoassay* 1980;1:229–249.

Sham JS, Wei WI, Zong YS, Choy D, Guo YQ, Luo Y, Lin ZX, Ng MH. Detection of subclinical nasopharyngeal carcinoma by fibreoptic endoscopy and multiple biopsy. *Lancet* 1990;335:371–374.

Snijdewind IJ, van Kampen JJ, Fraaij PL, van der Ende ME, Osterhaus AD, Gruters RA. Current and future applications of dried blood spots in viral disease management. *Antiviral Research* 2012;93:309–321.

Solomon SS, Solomon S, Rodriguez II, et al. Dried blood spots (DBS): a valuable tool for HIV surveillance in developing/tropical countries. *International Journal of STD and AIDS* 2002; 13:25–28.

Solomon SS, Pulimi S, Rodriguez II, Chaguturu SK, Satish Kumar SK, Mayer KH, Solomon S. Dried blood spots are an acceptable and useful HIV surveillance tool in a remote developing world setting. *International Journal of STD and AIDS* 2004;15:658–661.

Stevens R, Pass K, Fuller S, Wiznia A, Noble L, Duva S, Neal M. Blood spot screening and confirmatory tests for syphilis antibody. *Journal of Clinical Microbiology* 1992;30:2353–2358.

Stroffolini T, Pretolani S, Miglio F, Rapicetta M, Villano U, Bonvicini F, Baldini L, Sampogna F, Giulianelli G, Stefanelli ML, Carloni A, Sorcinelli A, Ghironzi G, Gasbarrini G. Population-based survey of hepatitis A virus infection in the Republic of San Marino. *European Journal of Epidemiology* 1997;13:687–689.

Sturrock CJ, Currie MJ, Vally H, O'keefe EJ, Primrose R, Habel P, Schamburg K, Bowden FJ. Community-based sexual health care works: a review of the ACT outreach program. *Sexual Health* 2007;4:201–204.

Swai RO, Somi G, Matee MI, Killewo J, Lyamuya EF, Kwesigabo G, Tulli T, Kabalimu TK, Ng'ang'a L, Isingo R, Ndayongeje J. Surveillance of HIV and syphilis infections among antenatal clinic attendees in Tanzania-2003/2004. *BioMed Central Public Health* 2006;6:91–100.

Tanner S, McDade TW. Enzyme immunoassay for total immunoglobulin E in dried blood spots. *American Journal of Human Biology* 2007;19:440–442.

Thapa B, Koirala S, Upadhaya BP, Mahat K, Malla S, Shakya G. National external quality assurance scheme for HIV testing using dried blood spot: a feasibility study. *SAARC Journal of Tuberculosis, Lung Diseases, and HIV/AIDS* 2011;8:23–27.

Tran TN, de Vries PJ, Hoang LP, Phan GT, Le HQ, Tran BQ, Vo CM, Nguyen NV, Kager PA, Nagelkerke N, Groen J. Enzyme-linked immunoassay for dengue virus IgM and IgG antibodies in serum and filter paper blood. *BioMed Central Infectious Disease* 2006;6:13–20.

Tuaillon E, Mondain AM, Meroueh F, Ottomani L, Picot MC, Nagot N, Van de Perre P, Ducos J. Dried blood spot for hepatitis C virus serology and molecular testing. *Hepatology* 2010;51:752–758.

Uzicanin A, Lubega I, Nanuynja M, Mercader S, Rota P, Bellini W, Helfand R. Dried blood spots on filter paper as an alternative specimen for measles diagnostics: detection of measles immunoglobulin M antibody by a commercial enzyme immunoassay. *Journal of Infectious Diseases* 2011;204:S564–S569.

Van Biervilet JP, Michiels G, Rosseneu M. Quantifications of lipoprotein(a) in dried blood spots and screening for above-normal lipoprotein(a) concentrations in newborns. *Clinical Chemistry* 1991;37:706–708.

Van der Loeffa MFS, Sarge-Njie R, Ceesayc S, Awasanaa AA, Jayce P, Samc O, Jairehc KO, Cubittd D, Milligana P, Whittl IIC. Rcgional diffcrcnccs in HIV trends in the Gambia: results from sentinel surveillance among pregnant women. *AIDS* 2003;17:1841–1846.

Vázquez S, Fernández R, Llorente C. Usefulness of blood specimens on paper strips for serologic studies with inhibition ELISA. *Revista do Instituto de Medicina Tropical de São Paulo* 1991;33:309–311.

Villar LM, de Oliveira JC, Cruz HM, Yoshida CF, Lampe E, Lewis-Ximenez LL. Assessment of dried blood spot samples as a simple method for detection of hepatitis B virus markers. Infectious diseases: HIV, STI, hepatitis. *Journal of Medical Virology* 2011;83:1522–1529.

Waterboer T, Dondog B, Michael KM, Michel A, Schmitt M, Vaccarella S, Franceschi S, Clifford G, Pawlita M. Dried blood spot samples for seroepidemiology of infections with human papillomaviruses, Helicobacter pylori, hepatitis C virus, and JC virus. *Cancer Epidemiology, Biomarkers and Prevention* 2012;21:287–293.

Weinbaum CM, Sabin KM, Santibanez SS. Hepatitis B, hepatitis C, and HIV in correctional populations: a review of epidemiology and prevention. *AIDS* 2005;19:S41–S46.

World Health Organization. Guidelines for Assuring the Accuracy and Reliability of HIV Rapid Testing: Applying a Quality System Approach; 2005. http://hinfo.humaninfo.ro/gsdl/healthtechdocs/documents/s15126e/s15126e.pdf (accessed June 18, 2012).

World Health Organization. *Dengue: Guidelines for Diagnosis, Treatment, Prevention and Control.* Geneva: WHO Press; 2009. http://whqlibdoc.who.int/publications/2009/9789241547871_eng.pdf (accessed November 20, 2012).

World Health Organization. World Malaria Report 2011; 2011. www.who.int/malaria/world_malaria_report_2011/en/index.html (Retrieved November 20, 2012).

Zaaijer HL, von Exel-Oehlers P, Kraaijeveld T, Altena E, Lelie PN. Early detection of antibodies to HIV-1 by third-generation assays. *Lancet* 1992;340:770–772.

6

APPLICATIONS OF DRIED BLOOD SPOTS IN NEWBORN AND METABOLIC SCREENING

DONALD H. CHACE, ALAN R. SPITZER, AND VÍCTOR R. DE JESÚS

6.1 INTRODUCTION

6.1.1 History and Chronology in the Use of DBS in NBS

Dried blood spots (DBS) on filter paper are the biological specimens used by laboratories for purposes of screening newborns for inherited metabolic or congenital diseases. Robert Guthrie in the early 1960s established the routine use of DBS for phenylketonuria (PKU) testing (Guthrie, 1969). The PKU test marked the beginning of worldwide newborn screening (Guthrie, 1992). As PKU testing expanded to all US states, Europe, and other nations throughout the 1960s and early 1970s, new disease screens were added to newborn screening programs. In the 1990s, tandem mass spectrometry (MS/MS) was introduced and shown to screen more than 30 disorders of amino acid, organic acid, and fatty acid metabolism (Chace et al., 2003). The concept of screening one disease with one suitable test was replaced by modern multiplexed approaches such as MS/MS that could screen for multiple metabolites from a single punch from a blood spot (Chace et al., 2002). Today, this platform continues to expand to include other metabolic disorders including lysosomal storage diseases (Li et al., 2004; Gelb et al., 2006; Duffner et al., 2009; Mechtler et al., 2012). Multiplex platforms based on molecular analysis for cystic fibrosis (CF), sickle cell disease (hemoglobinopathies), and severe combined immunodeficiency syndrome (SCID) are being implemented in many laboratories (Gregg et al., 1997; Lorey et al., 2001; Comeau et al., 2004, 2010; Chan and Puck, 2005; Baker et al., 2010; Hoppe, 2011).

6.1.1.1 The First DBS Assay The analytical test (Guthrie and Susi, 1963) developed by Robert Guthrie for detection of PKU was quite interesting in that it was based on the response of a living organism to the presence of phenylalanine (Phe), the primary marker for detection of PKU. It was known as the bacterial inhibition assay (BIA). A special colony of bacteria that required Phe for growth was utilized in this assay. Punched disks from blood spots were arrayed on a sheet of agar containing these special bacteria. The Phe present in the blood punch would be used by the bacteria to grow until it was consumed. Therefore, the amount of growth of the bacteria was correlated with the amount of Phe present in a DBS. After staining the agar gel, the bacterial growth could be measured by the diameter of the growth zone surrounding the DBS. The method was simple, could accommodate many spots on an agar gel, and was easily adapted by public health labs that had ample expertise in this type of assay. This platform could also be adapted to other metabolites by growing special bacteria that were sensitive to other metabolites such as leucine (Leu) for detection of maple syrup urine disease (MSUD) (Naylor and Guthrie, 1978).

6.1.1.2 Expansion of Newborn Screening Tests The expansion of newborn screening for the first 20 years was relatively modest for a variety of reasons that included requirements to developing infrastructure for screening from sample collection to follow-up of positive results, availability of new technologies, and complex politics and health issues around which new candidate disorders should be

Dried Blood Spots: Applications and Techniques, First Edition. Edited by Wenkui Li and Mike S. Lee.

screened (see Chapter 1). The policies for addition of new disorders have always included the rules that the new screening test must be for a treatable disorder or health benefits could be derived from early detection. In addition, the analytical test had to be highly accurate and sensitive (low false result rates, very predictive of disease), relatively easy to use, have a high capacity for population screening, and cost effective. The tools that met the analytical needs were based on fluorometry, immunoassays, enzyme assays, and some rapid chromatographic methods and electrophoresis in the first few decades following the introduction of BIA to screening.

The second most prominent inherited metabolic disorder that was implemented nearly universally was an endocrine disorder, congenital hypothyroidism (CH). CH is characterized by a low thyroxine (T4) concentration and a high thyroid-stimulating hormone (TSH) concentration. Initially, the test was primarily based on the measurement of total thyroxine (total T4) using a variety of techniques, but improvements of the analysis came with the use of TSH analysis that either replaced or supplemented T4 analysis (Mitchell et al., 1978). Other disorders that were introduced into many screening programs soon after PKU and T4 included congenital adrenal hyperplasia (CAH), galactosemia (GAL), sickle cell disease, biotinidase deficiency, and other amino acid disorders such as MSUD, homocystinuria (HCys), and tyrosinemia types I and II (Tyr I, Tyr II) (De Jesus et al., 2010a). In terms of universality of screening, additional disorders created a disparity between states with some screening seven to nine disorders while others only screening two or three disorders. This disparity became even more apparent with the introduction of new and existing disorders screened by MS/MS (Levy, 1998; Levy and Albers, 2000).

A more modern approach to DBS analysis of newborns was developed in the early 1990s that involved multiplex platforms. These new analytical systems include MS/MS and molecular screening (Witchel et al., 1997; Muhl et al., 2001; Moslinger et al., 2003). Although MS/MS started as a supplemental screening program (Ziadeh et al., 1995; Chace and Naylor, 1999), it has become nearly universal 20 years after its development. Molecular techniques, historically important in disease confirmation and characterization of disorders detected by other analytical techniques (Dobrowolski et al., 1999; McCabe and McCabe, 1999), is presently undergoing a change, with rapid growth in newborn screening laboratories as a routine second-tier analysis and new growth as a primary screening platform (see Chapters 4 and 22) for many diseases. A multiplexed platform like MS/MS (Chace, 2001) transformed the efficiency of newborn screening because it (1) improved the methods of measurement for many metabolites such as Phe and methionine (Met) with a higher degree of analytical selectivity; (2) measured more than one biomarker, which improved clinical selectivity for metabolic diseases such as PKU (measurement of both Phe and tyrosine (Tyr) in the same analysis enables calculation of highly PKU-selective Phe/Tyr ratio); and (3) enabled screening for a variety of new conditions (e.g., MCAD deficiency) by detecting a new family of metabolites (i.e., acylcarnitines) in a single analysis. Furthermore, the platform could be used to expand or improve screening of disorders by either the addition of more specific markers (succinylacetone (SUAC) in Tyr I) or through the use of the same platform (Turgeon et al., 2008; De Jesus et al., 2010a) for other markers in other assays such as lysosomal storage disorders (LSD) (De Jesus et al., 2009; Duffner et al., 2009). The key point is that all assays are based on using the dried filter paper blood spot without special treatment or modification of the sample during the post-collection and pre-analysis.

6.1.2 Chapter Organization and Perspectives

Describing the use of DBS in newborn screening is best served by describing primarily the methods used in screening (an assay perspective), while briefly summarizing which DBS assays are the best choices for detection of specific disorders (a disease detection perspective). Historically, many different methods have been used to detect and quantify metabolites or biomarkers that were most closely correlated with the presence of an inherited disease. Which method was chosen to analyze DBS in NBS laboratories was and still is based on cost, laboratory expertise, assay selectivity and sensitivity, clinical selectivity and sensitivity, and logistics (i.e., laboratory size, facility support, technical support). This chapter will primarily focus on methods that were either historically significant, most commonly used in current screening programs, the newest multiplex methods such as MS, and a very new approach to multiple platform multiple analyte analysis that combines two different methods such as primary tier and secondary tier screening with MS/MS and LC/MS, or DNA confirmation using blood spots from the same DBS card. This approach is best illustrated in the chart shown in Figure 6.1, which gives an overview of DBS methods used in the past and present as well as the relationship to metabolites and disorders screened. Various patterned lines connect methods to key metabolites measured and solid lines connect metabolites to disorders detected. Different patterned lines differentiate multiplex (multiple metabolites) methods from single metabolite analysis. Some lines connect assays directly to the disorder for technologies that detect mutations rather than metabolites, that is, molecular methods, while some connections are for methods used to confirm disorders using a secondary screen (separate and different tests) on a smaller subset of samples that are presumptive positive for a disorder. Other aspects of this chapter will focus on new applications, approaches, and issues related to DBS and metabolic disease screening that includes newborns and other

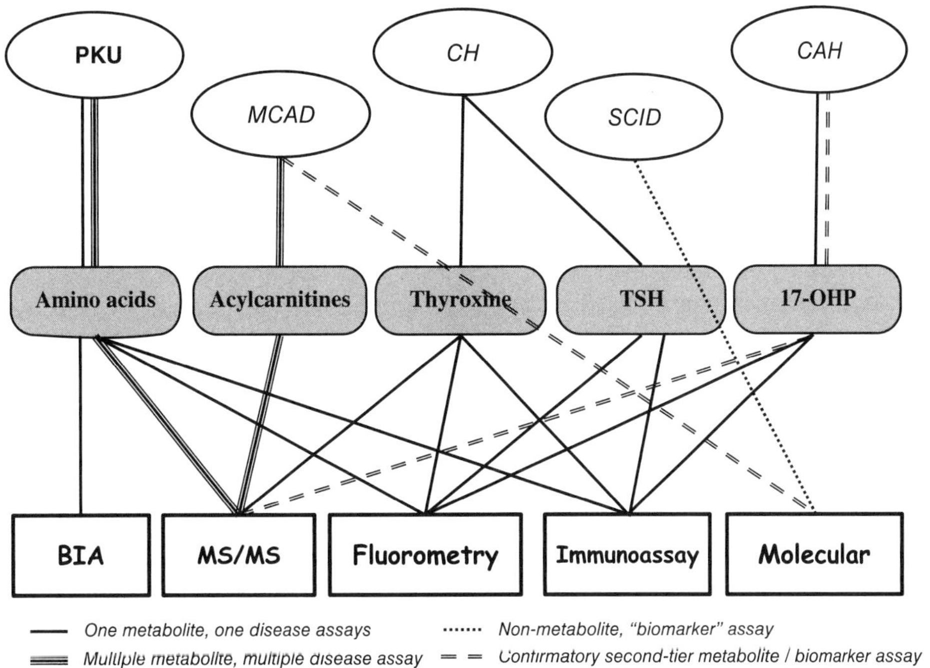

FIGURE 6.1 Illustration of the relationship between analytical method, biomarker quantified, and disease detected in newborn screening using DBS. Rectangular boxes show the analytical method utilized. Rounded boxes show the single or group of metabolites. Ovals represent specific diseases. Various formatted lines connect each of the analytical methods with the metabolite they detect and, in a single case connects the method (molecular) to the disease it detects (SCID, MCAD) because no metabolite is measured rather than a specific mutation for each disease (mutation not shown). Many methods connect to the same metabolite, illustrating multiple approaches to measuring a single biomarker. Complex formatted lines demonstrate multiplex methods such as MS/MS and certain multiplexed immunoassays. In the case of PKU or MCAD deficiency, multiple markers are used to detect a single disease. Some methods may use two different groups of metabolites to detect that disease (CH) or for confirmation (MCAD) or CAH using multiple steroid metabolites.

potential screening populations (i.e., very low birth weight infants).

6.2 FUNDAMENTALS OF DRIED BLOOD SPOTS

6.2.1 A Class II Medical Device

The most challenging concept regarding DBS is the issue of volume and its relationship to a liquid sample. A punch from a dried blood specimen has a specific volume. Although it appears flat, a DBS is actually a disk defined by an area (πr^2) and a width or height (πhr^2) which is the volume of a cylinder. What is challenging is the association between this dried disk volume and a liquid volume counterpart. If the absorptivity of the paper is controlled, then volume equivalence between an actual liquid and a dried specimen can be made. As a result, the volume of blood present in a punched disk in terms of cubic centimeters (cm^3) can be expressed in microliters (μL). Note that 1 mL is also equal to 1 cm^3. The Clinical and Laboratory Standards Institute (CLSI) has developed an approved standard (Hannon et al., 2007) that describes the analytical process needed to determine how much serum is present in DBS punches prepared from whole blood at a hematocrit of 55%. The Centers for Disease Control and Prevention's Newborn Screening Quality Assurance Program (NSQAP) has evaluated filter paper lots of Grade 903® paper for more than 30 years (Bell, 2010; DeJesus, 2010). The published acceptable serum absorption volume per 1/8 inch disk punched from a 100 μL spot of intact-cell blood of 55% hematocrit is 1.54 ± 0.17 μL (Hannon et al., 2007). Manufacturers who produce this highly specialized paper must meet these guidelines, as they are regulated by the Food and Drug Administration (FDA) as a Class II Medical Device (21 CFR §862.1675). Unfortunately, both the hematocrit of newborns and the method of collecting blood and applying it to filter paper are variable. These factors affect the analysis to a greater potential error than measuring liquid blood. There are also differences between how a liquid specimen can be used versus a dried specimen, such as the ability to mix a standard or additive uniformly in a liquid. The coefficients of variation associated with DBS analyses are often within the range of many established assays

(<15% CV), especially if the comparative volumes between a liquid specimen and a DBS are the same, that is, 3–8 μL of liquid versus 1/8th–3/16th punches.

6.2.2 Liquid versus Dried Blood Specimens

The most common specimens derived from blood and used in most assays in clinical chemistry laboratories are plasma or serum. Plasma is obtained after centrifugation from blood collected in Vacutainer® tubes containing an anticoagulant (e.g., heparin, ethylenediaminetetraacetic acid (EDTA), citrate, oxalate) while serum is obtained after centrifugation without anticoagulant. Whole blood is rarely used in the clinical lab for routine metabolic analysis. In fact, liquid whole blood is rarely used in any clinical laboratory for metabolic analysis because its characteristics are quite different than a DBS. It is important to distinguish the differences between liquid and dried biological specimens regardless of whether they are used in clinical or newborn screening assays. Working with liquids is quite different than dried specimens, especially as they relate to the volume sampled and quantitative approaches that are needed to accurately determine the concentration of the metabolite or biomarker measured.

A typical volume of blood collected by a phlebotomist ranges from 5 to 10 mL. Blood or plasma samples are sent either internally to a hospital-based laboratory or externally to a commercial/specialty/reference lab under refrigerated conditions to preserve sample integrity. A typical volume of plasma sampled is anywhere from 100 to 500 μL. Top left panel of Figure 6.2 illustrates this concept. The quantity of

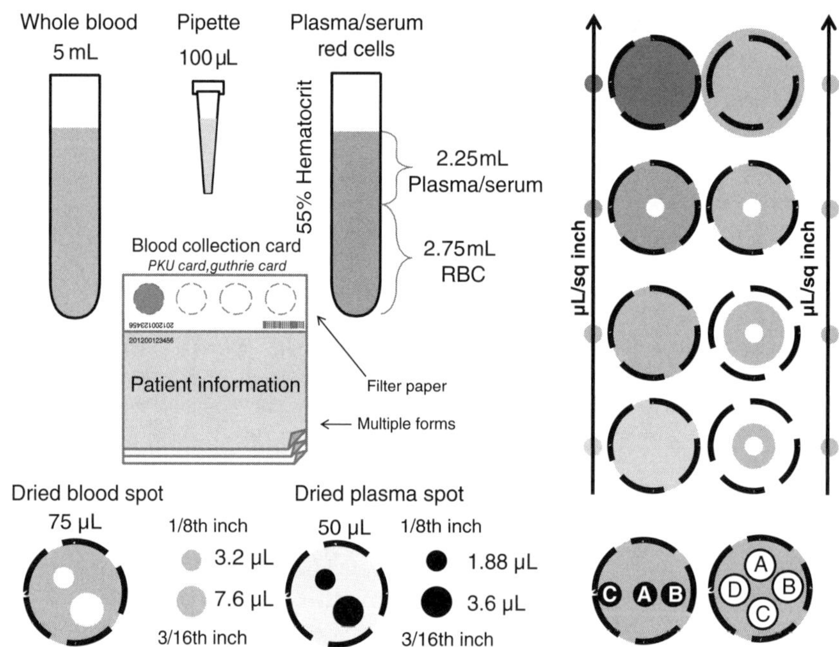

FIGURE 6.2 Illustration of DBS and volume relationships. The left panel shows volumes of different blood collection schemes in tubes, pipettes, and dried blood spots (DBS). Blood collected in tubes usually requires 5–10 mL per tube. From this blood, approximately half is available as plasma or serum depending upon the patient's hematocrit. A typical clinical chemistry analysis utilizes 100 μL or more of plasma as represented by the pipette. A dried blood spot obtained from the heel of a newborn or a finger prick can contain various volumes of blood but volume can usually be gauged by how full the dashed circle is. A full dashed circle contains approximately 80–100 μL of blood. Plasma, which has a hematocrit of zero, is less viscous than blood and the volume contained within the same diameter is less. A filled dashed circle contains approximately 40–50 μL of plasma. The volume of sample based on punch size ranges from 3.2 to 7.6 μL for blood depending on size while a plasma punched sample contains slightly more than half as much as shown. The panel on the right shows a series of spots with various intensities of color. The left column shows the volume of blood increasing with increasing hematocrit (arrow pointing to top of page, μL/sq in). The right column shows the volume of blood applied to filter paper for a given punch size. The volume present in a punch is greatest for the highest volume of blood applied to the paper and is also shown with the arrow pointing up (μL/sq in). The blood spot in the lower right represents punch location. Studies have been described suggesting there is no difference between locations A, B, or C for DBS. Other matrices such as plasma and serum have not been investigated.

blood applied to a DBS card for newborn screening comes from a free flowing droplet that forms on the heel of an infant after lancing. The paper is applied to the outer surface of the drop that formed on the heel, which leads to the absorption of blood onto the paper. For convenience, the filter paper is marked with dashed circles to provide a target and an approximate volume of application (with the goal to fill the circle) per spot (as shown in Figure 6.2, middle left). The filter paper is attached to a form that contains patient information. Both filter paper and form are numbered and barcoded for patient identification (see Chapter 2). Some collection sites for NBS occasionally use capillary pipettes to withdraw the blood that formed on the heel. The blood is then applied to the paper from the capillary. This method has the disadvantage of the pipette scratching the surface of the paper during application, which changes the absorptive characteristics of the paper or affects blood clotting, lysing, or mixing with anticoagulant or other contaminants within the pipette prior to spotting on paper. The volume of blood that can be obtained from a heel stick is approximately 0.3–0.5 mL for four to six spots, approximately 1/10th less than is collected in a typical venipuncture. Unlike liquid blood which requires refrigeration, blood spots are dried at room temperature for at least 3 hours (Hannon et al., 2007) and then typically shipped in a regular envelope via regular mail or express courier to a newborn screening laboratory for analysis.

The volume characteristics of DBS are as follows: approximately 75 μL of blood will completely fill the dashed circle on a filter paper card designed for newborn screening applications. Contact with the paper will lyse the red blood cells. This hemolysate has been shown to distribute evenly across the blood spot from the center to the edges. This even distribution has been demonstrated using radiolabeled thyroxine (Mei et al., 2001). Blood hemolysate contains both red cell contents and plasma contents so there should be no compartmentalization of metabolites. The typical zones punched by screening labs are illustrated in the blood spot shown in Figure 6.2 (bottom right, in the center, middle, or edge punches). Studies suggest punching DBS in the middle zone and at N, E, S, and W distribution are not different. Generally, laboratories do not punch the center of a blood spot because more sample can be obtained using the B location at N, E, S, and W (four punches) versus a center punch and extreme edge where there is an increasing likelihood of there being less blood due to incomplete absorption.

The size of a DBS punch (diameter) is directly related to the volume of dried blood (cm^3) or area (cm^2), since the width (height, thickness) is constant. The volume of dried blood is related to the viscosity of the original liquid and is directly related to the hematocrit as well as the volume applied. Based on the CLSI standard, the volume of blood in a 1/8th inch punch is calculated to be 3.42 μL at 55% hematocrit. Presuming a fixed width of the paper, calculating the ratio of the area of an 1/8th inch punch, and relating it to the serum volume presented in the CLSI standard, one derives a factor than can be used to calculate the volume of blood in any size blood punch such as a 3/16th inch punch, which would have a 7.69-μL volume.

The impact of hematocrit on blood volume in a DBS is illustrated in Figure 6.2. The higher the hematocrit, as represented by darker spots, the higher the volume of blood in a specified punch area for a fixed volume of blood applied to filter paper. It is also worth noting that for a fixed volume of blood, very low hematocrits will produce larger blood spots, more similar to that of plasma or serum. A zero hematocrit is plasma or serum and these produce the largest spots for a specific volume (Figure 6.2 bottom panel), where a comparison of the estimated volumes for different punch sizes for blood (55% hematocrit) and plasma (no hematocrit) is estimated.

For fixed hematocrits, the size of blood spots for samples is related to the volume of blood in a DBS as well. The larger the volume of blood applied using a pipette, the greater the blood volume in a DBS as shown in Figure 6.2, right panel. This finding is presumably due to a greater lateral flow relative to complete saturation/penetration of the layers or spaces within the sandwich/sheets/fibers of cellulose that make up the filter paper (see below). Plasma spreads to a greater extent as described above. Most interestingly, even different plasma samples have different viscosities. Deproteinized plasma that comes from patients on dialysis or other biological fluids such as urine has even smaller volumes per unit area. In fact, special papers are often used for urine that are thicker, hence having a greater volume per same diameter punch. More rigorous studies on the uniformity of plasma, urine, and other biological matrices have yet to be completed. It is presumed, however, that the impact of variable hematocrit will be much less on volume in these biological matrices and may be a better choice for quantitative applications.

6.2.3 Interaction of Paper with Biological Matrix and Substrates

The filter paper used in newborn screening is prepared from cotton fibers. Cotton is comprised of cellulose, a polymer of glucose (Figure 6.3, top panel). The glucose subunits of the long polymeric stands of cellulose interact with other cellulose polymeric strands by extensive hydrogen bonding. Hydrogen bonding also occurs within the same polymeric strand of cellulose between glucose subunits. In the dry form, the polymer is tightly bound to other strands but addition of water disrupts the hydrogen bonding and interlaces between polymeric strands (Figure 6.3, center panels). This causes an expansion of the distance between polymers as compared to dry material. However, upon drying, the cellulose polymers contract back to their original tightly packed form as shown in Figure 6.3, middle panels. If small molecules are present

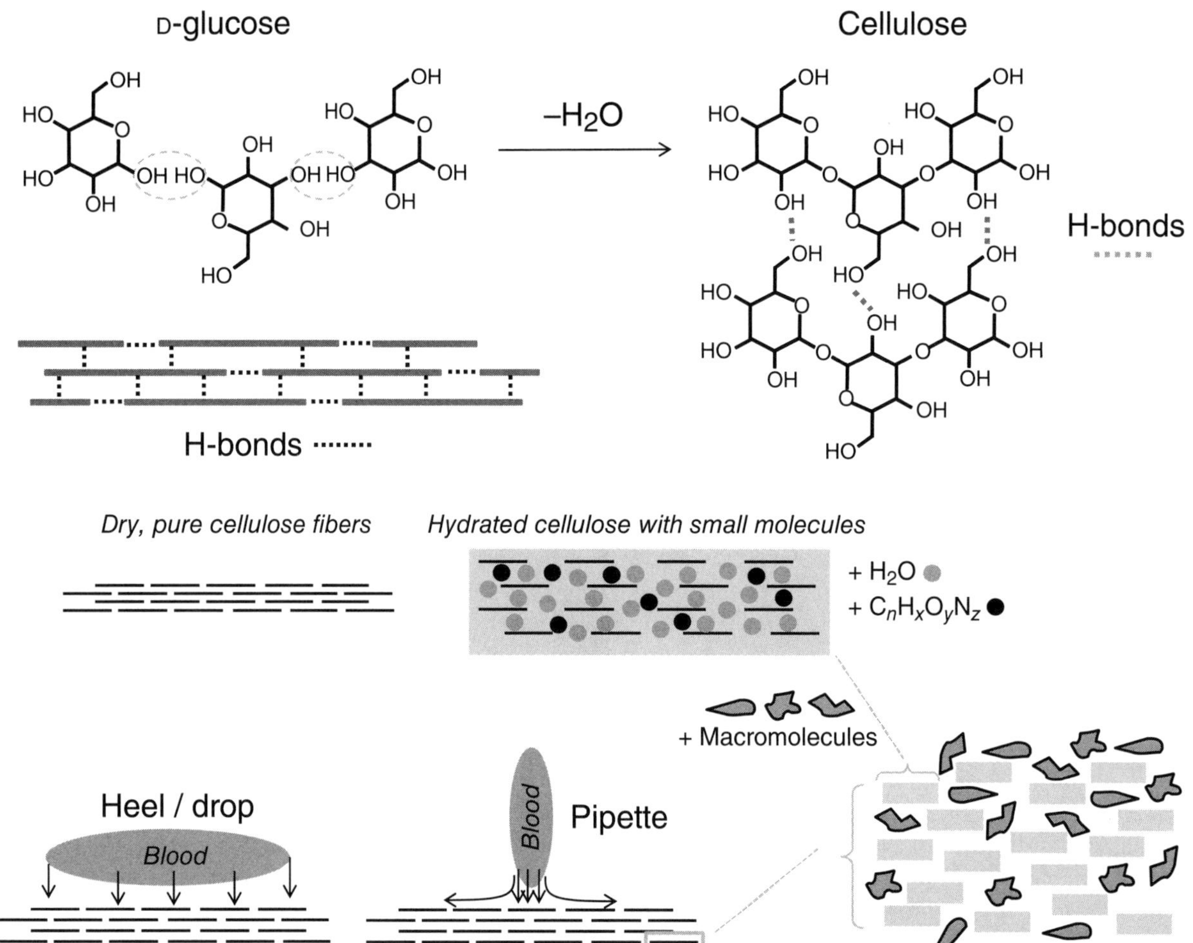

FIGURE 6.3 Interaction of matrix and metabolites with cellulose-based filter paper. The structure of D-glucose and its polymeric form, cellulose, is shown in the top panel. Extensive hydrogen bonding of cellulose polymers to each other, which leads to formation of cotton fibers, is illustrated. Paper is manufactured from the fibers with varying degrees of compression that results in pores of varying sizes forming within the dry matrix of pure cellulose fibers. Addition of water or other hydrophilic solvents to the paper causes it to swell due to a disruption of hydrogen bonds between cellulose polymers, and the dispersion of solvent within the gaps and pores increases its net volume and apparent swelling. Organic molecules within the solvent may also interact with the hydrogen bonds of cellulose or simply fill the gaps. On a larger macro level, macromolecules such as proteins and DNA intersperse within the gaps between fibrous layers. Drying of the paper causes it to return to its previous state but with molecules trapped within the cellulose matrix. The bottom panel shows the potential impact of blood application, using a passive absorption method such as a drop of blood from a heel or a more active stream of liquid applied from a pipette.

in the aqueous matrix, then they may interact directly with cellulose through hydrogen bonding or remain dissolved in the aqueous solvent within the pores of the paper. On drying, the small nonvolatile molecules become trapped in the cellulose matrix. It could be argued that this is one reason why molecules dried on filter paper are more stable, because they are "shielded" from the environment and stabilized by the cellulose backbone. Macromolecules such as DNA, RNA, proteins, and complex carbohydrates presumably also interlace between fibers but perhaps on a larger macro level, with groups of glucose polymer strands enveloping these molecules and stabilizing them (Figure 6.3, lower right panel). It is known that enzymes retain their activity when reconstituted so the cellulose does not chemically interact or degrade proteins (Garrick et al., 1973; Fujimoto et al., 2000).

The fibers in the filter paper can be compressed to make the paper more or less absorptive. Manufacturing standards are primarily designed to ensure that the compression and ultimate absorptivity are reproducible. Highly compressed paper—consider paper used for writing—is quite thin and much less permeable to water while the filter paper used in newborn screening is much less compressed and quite permeable to water, yet not so loosely packed as to fall apart. From a macro viewpoint, there are likely large gaps between cotton fibers enabling the blood or other matrix to pass through the various channels (Figure 6.3, bottom panel). What is interesting is how the type of liquid may interact with the surface when applied to the paper. If the paper surface is considered to be similar to large drains or strainers, only so much water can penetrate these gaps while some water will move laterally. More viscous material may pool at the surface and penetrate slowly but with little lateral flow, compared to less viscous samples (i.e., blood versus plasma or urine). The method of application may also have an impact on the absorption. In newborn screening, the paper is held to the surface of the heel of an infant where a drop is present. The drop is allowed to absorb slowly into the paper. This technique can be compared to adding blood to paper using a pipette where there is a large quantity of blood at the surface very quickly—causing more lateral flow and perhaps less penetration into the paper.

Our discussion into the possible dynamics of fluid and filter paper composition may help in the understanding of many of the characteristics of DBS used in newborn screening and why it has been so successful in this application. But the discussion also assists in the design of new assays with different biomarkers, different biological matrices, and new types or uses for filter paper.

6.2.4 Sample Collection and DBS

The method or means of collecting blood samples for newborn screening has a direct impact on the quality of any newborn screening assay in terms of accuracy and precision. Current guidelines for collection, storage, and shipment have been described (see Chapter 3). What is important to note is that standardized collection procedures reduce variation in blood volume and contamination, which improves the analytical precision and accuracy of newborn screening assays.

6.3 ANALYTICAL METHODS IN NEWBORN SCREENING

6.3.1 Fundamental Premise

The one common aspect of all newborn screening methods is the DBS. Clinical methods were either adapted to the DBS or created for the DBS platform. All assays share the following requirements. First, the size of the DBS punch is either one or two punches of a diameter of 1/8th or 3/16th inch. Because most laboratories are interested in detecting multiple biomarkers and diseases, as many as 10 punches may be obtained from the filter paper card, which typically has four to six circles, depending on the requirement of the private, regional, or national government policy. Because spots are not always completely filled, and assay repeats are common for abnormal results, care is taken to use the least amount of DBS "real estate" in performing initial screening. From a practical perspective, this approach requires that the assays that analyze the greatest number of diseases in a single analysis be punched first (typically MS/MS), followed by other assays like CH and CAH. Second, the assay must be able to perform the analysis, with sufficient detection limits and wide enough analytical range, no more than two 3/16th inch DBS punches (approximately 15.4 μL). Often, however, it is necessary to analyze samples sizes equivalent to a single 1/8th inch DBS punch (approximately 3.42 μL). Third, the analyte must be extractable (recovered) from the DBS matrix with high precision, accuracy, and reproducibility. A high recovery is one of the most important aspects in newborn screening analysis of DBS, since internal standards are always added to the extracted matrix and not to the DBS itself. Fourth, the combined analytical approaches must be able to analyze hundreds of samples per day in a cost-effective manner, typically a few dollars (US) per test using 2012 standards. This last requirement translates to assays that typically use the 96 microtiter well plate format (for which automated punchers are adapted) and for which the sample rate (i.e., time between sample injections) is approximately 3 minutes or less. The analytical preparation is commonly less than 24 hours (overnight incubations meet this requirement since results are usually obtained the following morning after noon receipt).

6.3.2 Classical Methods Overview

6.3.2.1 BIA Although chromatographic methods (GC, LC, thin layer chromatography (TLC)) were available in the 1960s when Dr Guthrie developed the BIA assay, they were not sufficiently simple or accurate enough to measure amino acids. Furthermore, staffing newborn screening laboratories at this early stage in NBS history generally came from existing public health laboratory professionals skilled in microorganism analysis and characterizations. As described in Section 6.1.1.1, the BIA relies on special bacterial colonies that have a unique affinity for the target analyte (i.e., Phe). Blood spots are placed on agar gel with Phe-free media in an array. Because the blood spots contain the target analyte, the bacteria will grow until Phe is completely consumed. Therefore, the extent of growth, as measured by a growth zone or diameter from the center of the spot to the edge of the colony, is related to the quantity of Phe in the blood spot. The correlation between the diameter of the growth zone and Phe concentration is surprisingly linear. This assay, in addition to being simple to use, could be expanded to other metabolites such as Leu (Naylor and Guthrie, 1978), Met (Naughten et al., 1998), and galactose (Badawi et al., 1996) by growing these metabolite-specific bacterial colonies.

Limitations of the BIA assays included the fact that the bacterial colonies must be kept in good supply (which was never a problem when only a few labs were using this assay). Also, the assays failed if patient blood samples contained antibiotics. The analysis was imprecise because it required growth of an organism, which is always subject to changes in the environment. Because of this lack of precision, the cutoff concentration of Phe was higher than most of today's modern methods. In general, higher cutoffs reduce false-positive rates. In addition, the time of collecting a DBS when using the BIA test was no less than 2 and often 3 days (48–72 hours) after birth. For many amino acid disorders (i.e., PKU), longer waiting times to obtain a blood sample allowed time for the metabolite to accumulate in an affected newborn's blood. This resulted in better differentiation between normal and abnormal metabolite levels, thus helping to establish cutoff levels, especially for less precise methods. Because of improvements in analytical precision of techniques such as MS/MS, laboratories can now reliably analyze blood for detection of PKU for samples collected at less than 24 hours due to early hospital discharge (Chace et al., 1998).

6.3.2.2 ***Chromatography*** TLC and paper chromatography were considered as potential methods for quantifying Phe to detect PKU in the early days of newborn screening. A more recent example of TLC analysis, as well as high-performance liquid chromatography (HPLC) for amino acids, has been described (Wu, 1991). Although the use of higher resolution chromatography techniques such as GC and LC had the potential to be the first multianalyte approach in newborn screening, they were primarily restricted to clinical labs and urine analysis for confirming amino acid metabolism disorders and, more recently, fatty acid and organic acid metabolic screening. The real hindrance to LC or GC techniques are the long run times that make them less than ideal for very high throughput screening. The cost of having a sufficient number of instruments running in parallel to achieve high throughput also becomes an issue. Even modern LC analytical approaches, which use multiple LC systems for separation and MS for detection of 17-hydroxyprogesterone from other steroids for screening of CAH, have not yet been used routinely. HPLC and GC together with MS (LC-MS, GC-MS) are used extensively in the confirmation of presumptive positive newborns for metabolic disorders in urine and blood (Chalmers and Lawson, 1982; Reilly et al., 1998; Lehotay, 1991). Today, LC coupled to MS has emerged as a key technique in newborn screening from a second-tier testing perspective (see subsequent sections).

6.3.3 Fluorometry

Methods based in fluorometry are perhaps the most numerous and widely used methods in newborn screening throughout the past three decades. The principle of this assay is that biomarkers/metabolites fluoresce upon excitation by an ultraviolet beam of light and fluorometers can then quantify the intensity of light emitted after excitation by this beam of UV, correlating with concentration. Fluorometric assays that detect Phe were perhaps the most common assays that initially replaced the Guthrie BIA test. Improvements were made to fluorometric methods that cause increased excitation of target metabolites, enabling enhanced sensitivity and a higher number of metabolites to be analyzed. One common technique used in newborn screening laboratories is based upon immunoassays with time-resolved fluorescence (Diamandis, 1988).

One of the major drawbacks of fluorometry is that different molecules may fluoresce similarly in a particular assay. One such example is the measurement of Phe extracted from DBS. Tyrosine has a very similar structure and will also contribute a fluorescence signal at the same wavelength used to quantify Phe. The net result is a positive bias in an assay based on the concentration of Phe. This bias has been estimated to be as much as 1 mg/dL (60 μM) from the presence of Tyr. However, the assay still easily detects PKU, because Tyr is usually not elevated or is slightly depressed. Nevertheless, in one study comparing fluorometric analysis to MS/MS analysis, the bias and false results were quite apparent for fluorometry (Chace et al., 1993). Methods for detecting amino acids using fluorometry have largely been replaced by MS/MS (see below).

6.3.4 Immunoassays

The assays described here essentially include all of the assays used in newborn screening that require antigen antibody complex formation for detection. The most common assay is based on Luminex xMAP® Technology. Luminex color-codes tiny beads, called microspheres, and they have produced 500 distinct color-coded sets. Each bead set can be coated with a reagent specific to a particular bioassay, allowing the capture and detection of specific analytes from a sample. Inside the Luminex analyzer, a light source excites the internal dyes that identify each microsphere particle, as well as any reporter dye captured during the assay. Many readings are made on each bead set, which further validate the results. Using this process, xMAP® Technology allows multiplexing of up to 500 unique bioassays within a single sample. This technology can be used for the analysis of the CF transmembrane conductance regulator gene (Dunbar and Jacobson, 2000; Lacbawan et al., 2012), T4 and TSH for CH (Lukacs et al., 2003), and SCID (Janik et al., 2010). For further discussion on enzyme immunoassays see Chapter 5.

6.3.5 Electrophoresis

There are numerous gel electrophoretic methods available for detection of proteins. Electrophoresis is considered the best method for screening and diagnosis of hemoglobinopathies. Isoelectric focusing (IEF) is the most common technique for the analysis of hemoglobin variants in DBS (Kleman et al., 1989). Routine testing of adults and newborns consists of alkaline electrophoresis followed by citrate agar electrophoresis, in order to confirm the presence of sickle cell (HbS) or HbS-like variants. However, the resolving power of these systems is rather poor for many abnormal hemoglobins, and unequivocal identification must be performed by other methods (e.g., globin chain electrophoresis; Schneider, 1974).

6.3.6 Tandem Mass Spectrometry (MS/MS)

6.3.6.1 Overview The number of metabolites currently detected using MS/MS from a single DBS and a single sample injection is more than 50 (Chace et al., 2003). However, the actual number of biomarkers, which include metabolites and their molar ratios, markers for assay quality assurance (signal intensity of standards, markers formed if incomplete derivatization, or excessive acylcarnitines hydrolysis; Chace et al., 2003), special markers that detect sample contamination from anticoagulants such as EDTA (Fingerhut et al., 2009), total parenteral nutrition (TPN) solutions (Chace et al., 2010, 2011), or drugs and their metabolites that form acylcarnitines (Millington et al., 1985; Abdenur et al., 1998; Silva et al., 2001), is far more extensive. Although it might be argued that some newer immunoassays have the potential to detect 100 or more antigen–antibody complexes, the current MS/MS assay for amino acids, acylcarnitines, and SUAC does not require expensive, specialized reagents other than stable isotope-labeled internal standards and easily obtained derivatizing reagents such as hydrazine or 3N hydrochloric acid (HCl) in butanol.

MS/MS is a highly selective technique for measuring two primary compound classes (Chace et al., 2003), α-amino acids and free carnitine/acylcarnitines, as well as SUAC (Turgeon et al., 2008; Chace et al., 2009b). Most newborn screening laboratories that have access to MS/MS instrumentation have replaced their other classical methods for measuring amino acids with this technology. Furthermore, the only practical rapid method for screening for carnitine and acylcarnitines is through MS/MS and flow injection sample introduction techniques. This approach makes for one of the most cost-effective methods in newborn screening with respect to metabolites measured and diseases detected from a single DBS. Furthermore, the same platform is now used to detect up to nine LSD (Li et al., 2004; De Jesus et al., 2009; Duffey et al., 2010; Khaliq et al., 2011; Spacil et al., 2011; Wolfe et al., 2011). These assays rely on functional enzymatic reactions in which multiple probes are used to measure lysosomal enzyme activities, and their products are easily detected in a single flow injection analysis by MS/MS. Even more interesting is the fact that the MS/MS platform can be coupled to HPLC to perform specialized second-tier analysis to confirm several diseases in which another more selective assay is typically required to rule out presumptive positives from either disorders detected by MS/MS (i.e., amino acids and organic acids in MSUD, methylmalonic acid (MMA) by chromatographic separation; Matern et al., 2007; Turgeon et al., 2010) or disorders detected by time-resolved fluorometric analysis (i.e., multiple steroids for CAH; Lacey et al., 2004; Minutti et al., 2004; De Jesus et al., 2010c). In addition, MS/MS is sufficiently sensitive so that these modern instruments can be used to measure compounds that are present in blood at low concentrations and those in which concentration decreases in the disease state (e.g., thyroxine; Chace et al., 2004). All of these assays or potential assays are based on the analysis of a DBS from the same original filter paper card obtained from a newborn.

6.3.6.2 Amino Acids The principles behind MS/MS and its ability to detect multiple metabolites extracted from a DBS are based on the structural commonalities of certain classes of compounds. All essential amino acids have an amine and carboxylic acid attached to the α-carbon to which the variable R group is attached. Because MS/MS is a method for detecting the original intact molecular ions (precursor ions) and subsequent fragment ions (products ions) after breaking them up into reproducible patterns in the collision chamber, it is well suited to selectively detect compound classes like amino acids. In the case of α-amino acids, the carboxylic acid attached to the α-carbon fragments forms a nonionized (neutral) formic acid molecule or butyl formate if the amino acids are derivatized as butylesters. The masses of these molecules are 46 and 102 respectively, and only α-amino acids that have an α-carbon share these two unique products. An extensive discussion of amino acids analysis using MS/MS including the fragmentation process has been described (Chace, 2001; Chace et al., 2003).

The specific type of analysis of amino acids using MS/MS is called a neutral loss experiment, since the fragments (whether formic acid or its butylester) are neutral uncharged molecules and cannot be detected in the second mass analyzer. The product ion that differs by the mass of these neutral molecules is, in fact, detected. The MS/MS can link the products and precursors in a neutral loss scan that is quite selective for many α-amino acids. The amino acids that are in key in the detection of amino acid disorders in newborn screening are Phe, Leu, Ile, Val, Met, Tyr, Glu, Ala, Cit, Arg, and Orn. The analysis is imperfect, however, in that some amino acids like Leu and Ile cannot be separated by mass and, therefore, the ion signal of their mass represents the sum of their individual ion intensities. Other amino acids such as cysteine or cystine are not easily detected because they either form multiple charges or are not extracted efficiently because of protein binding to thiol (-SH) functional groups. These compounds must be analyzed using a different sample preparation method (Accinni et al., 2003; Turgeon et al., 2010) and are most useful as a confirmatory, second-tier test.

The classical approach to sample preparation of amino acid prior to MS/MS is derivatization of carboxylic acids to butylesters using 3N HCl in butanol. Dicarboxylic acids such as glutamate and aspartate have two carboxylic acid sites. Because MS/MS is based on the detection of positive ions that arises from molecules like protonated amines, the presence of carboxylic acids that form negative ions may cause a net neutral or net negative charge (i.e., one positive ion from the amine group and two negative ions from the carboxylic acid group). By making esters of the carboxylic acids, the negative ion is neutralized and the molecule is insured to have a net positive charge. This process represents a major difference between derivatized and nonderivatized techniques, as well as the mass differences measured for amino acids. Although derivatization techniques have improved sensitivity of analysis as compared to nonderivatized techniques, many laboratories have opted for a nonderivatized approach (Zoppa et al., 2006) for a variety of reasons. A more extensive discussion of the advantages and disadvantages of derivatization has been described (De Jesus et al., 2010a).

6.3.6.3 Acylcarnitines The other family of compounds detected in the MS/MS analysis of DBS is the carnitine class of molecules that include both esterified (acylcarnitines) and free un-esterified carnitine (Chace, 2001, 2005; Chace et al., 2003). The molecules are quite similar to amino acids in that they have a terminal carboxylic acid, a terminal amino group (quaternary ammonium), and an R group (fatty acids of varying chain length from 2 to 20 carbons of various levels of unsaturation and functional groups) attached to a secondary alcohol (Chace, 2001) at position 3 in the carnitine backbone. Carnitine, a free hydroxyl group (no fatty acid ester attached), is L-carnitine (free carnitine) and is written as C0 or FC. The notation for the other carnitines generally includes the number of carbons in the fatty acid chain and the number of double bonds, whether it contains hydroxyl units or carboxyl units (e.g., C3, C18:2, C16OH, C5DC for propionic, linoleic, hydroxypalmitoic, glutaric acids). Unlike amino acids, the MS/MS analysis of carnitines and acylcarnitines shares a common product ion of m/z 85 Da rather than a neutral molecule. Therefore, the MS/MS can focus on this single common ion in the second mass analyzer and detect all acylcarnitines and free carnitine precursor ions. The net result of the precursor ion scan as it is called and the neutral loss scan for amino acids is selectivity for detection without the need for chromatography.

Sample preparation is the same for amino acids and acylcarnitines. Both families are extracted in 90–100% methanol from a DBS, can be derivatized as butylesters, or are left underivatized (free acids). It is interesting to note that both free acids and butylester derivatives of acylcarnitines and carnitine share the same product ion of 85 Da. Like amino acids, dicarboxylic acids are less sensitive when they are not derivatized. In addition, the mass shifts from dicarboxylic acids overlap with other important acylcarnitines, which can be an issue in screening as previously described (De Jesus et al., 2010a).

6.3.6.4 Succinylacetone (SUAC) The tyrosinemias are different diseases that all share a common elevation of tyrosine. One of them, tyrosinemia type I, may not result in an elevation of Tyr during the collection period (1–2 days), increasing the potential for false-negative results (Chace et al., 2002). Biochemical methods were developed previously using a spectrophotometric microassay for δ-aminolevulinate dehydratase in DBS (Schulze et al., 2001) to detect SUAC, a primary marker for this specific tyrosinemia. More recently, MS/MS has been used to detect SUAC in DBS using hydrazine as a derivatization agent (Magera et al., 2006; Turgeon et al., 2008; Chace et al., 2009b). Early methods utilized DBS that had previously been extracted with methanol for analysis of acylcarnitines and amino acids. Most recent methods actually derivatize SUAC during the extraction process for amino acids and acylcarnitines, increasing the efficiency of a combined amino acid and acylcarnitines method, while reducing false-positive results (Dhillon et al., 2011).

6.3.6.5 Thyroxine Thyroxine (T4) is a hormone whose biochemical precursors include tyrosine and four iodine atoms. Thyroxine is normally tightly bound to thyroid-binding globulin (TBG), less tightly bound to nonspecific binding proteins in blood, and in its free, unbound form (free T4) that is considered the critical clinical marker. The molecule, interestingly, has an α-amino acid function group (because of its precursor heritage in tyrosine) and therefore can be detected using MS/MS with a neutral loss scan function in the same manner as other α-amino acids. The concentration of both free and bound (total) T4 is still quite low in blood, making it difficult to detect without sensitive mass spectrometers, especially when the disease is characterized by a decrease in T4 levels in affected newborns with CH. The recovery of T4 using MS techniques is less than 50%, using the standard methods for sample preparation for amino acids and acylcarnitines (Chace et al., 2009c). It is uncertain whether this low recovery is due to the measurement of free T4 and loosely bound T4 (T4 bound to albumin and aromatic binding proteins) or from incomplete extraction from TBG in the DBS. The impact of protein binding and metabolites in DBS on extraction efficiency is one area of research that has yet to be investigated.

6.3.6.6 Pseudo Isotope Dilution MS The use of stable isotope-labeled internal standards is unique to MS because the differentiating feature of natural and stable isotopes is mass. Stable isotope-labeled internal standards are also the ideal quantification standards for mass spectrometry-based assays due to the fact that these standards are nearly chemically identical to native metabolites. The design of internal standards in terms of choice of isotope, the number of isotopes, their position in the molecule, and purity and enrichment is critical to MS/MS analysis. During a period of two decades, standards have been introduced to meet the quantitative requirements of MS/MS analysis in DBS for amino acids and acylcarnitines, often available as sets of amino acid and acylcarnitines standards enriched with deuterium (D or ^{2}H), carbon-13 (^{13}C), and nitrogen-15 (^{15}N). There are over 20 acylcarnitines and amino acid stable isotope-enriched standards used in metabolic screening of DBS (Chace, 2001; Chace et al., 2003). Figure 6.4 illustrates the MS/MS profile of the methanol extraction solvent containing only the stable isotope internal standards.

In clinical or pharmaceutical applications of mass spectrometry, stable isotopes are typically added to a liquid matrix (typically plasma or urine) at concentrations that are within the range of the concentration of the metabolite expected (De Brabandere et al., 1998). The ion intensity ratio at the mass of the unlabeled target metabolite to the mass of the added stable isotope is calculated. The concentration of the unlabeled metabolite is then calculated from the ion ratio and the concentration of stable isotope in the liquid matrix. This is known as isotope dilution MS (IDMS). DBS is not a liquid and the internal standard cannot be mixed with it. Alternatively, internal standards are prepared as part of the extraction solvent of the DBS which for NBS is pure methanol (classical method). They can then be mixed together and diluted with pure methanol to the concentration needed (i.e., working solution). In each assay, a specific volume of the stable isotope extraction solvent is mixed with the DBS punch. Only the extracted metabolites can be quantified with the internal standard in the methanol mixture. Because accurate quantification relies on the extraction efficiency, the technique is somewhat less ideal that IDMS. The terms pseudo IDMS or

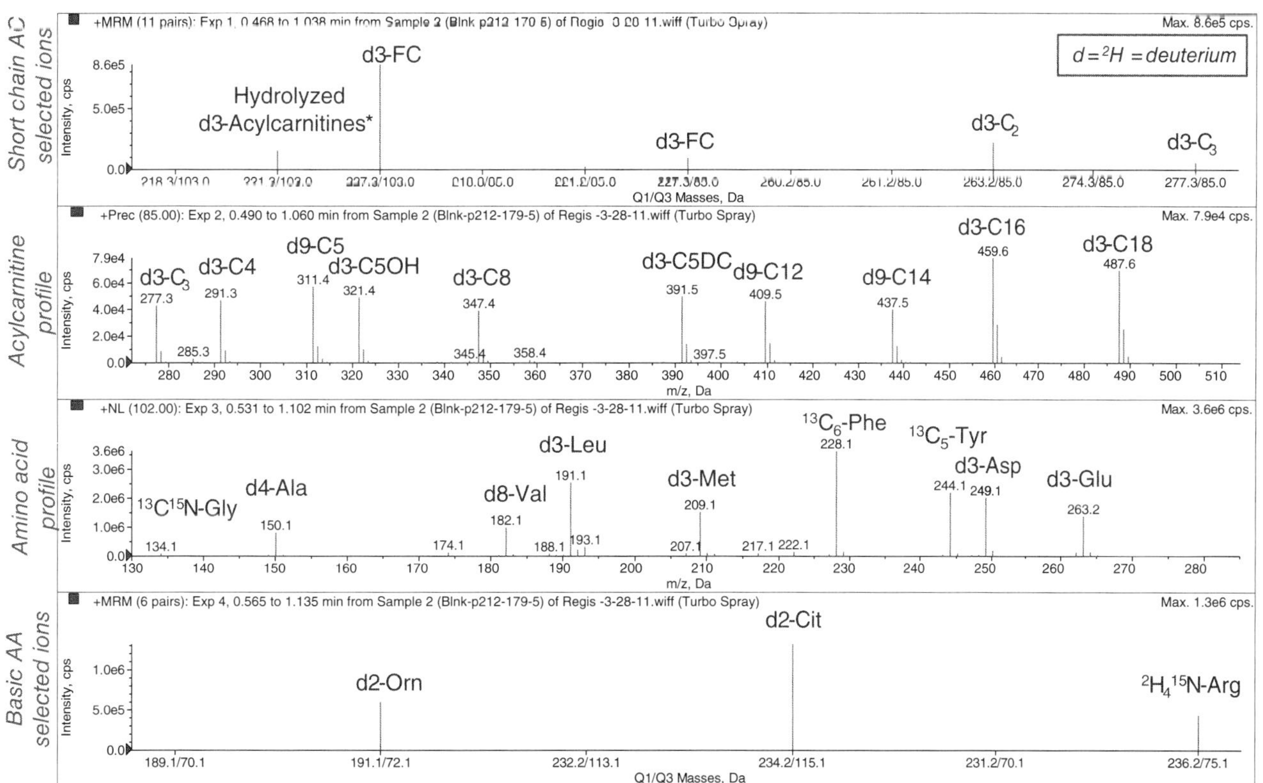

FIGURE 6.4 MS/MS profile of stable isotope internal standards mixture used in the extraction of DBS for acylcarnitines and amino acids. The top panel shows the selection ion profiles with ion transitions for the internal standards of free carnitine and short-chain acylcarnitines. The second panel shows a full scan profile using precursor 85 Da scan function showing the set of acylcarnitine stable isotope internal standards, ranging from short chain to long chain and including hydroxyl and dicarboxylic acid internal standards. The bottom two panels show an amino acid full scan profile using a NL 102 scan function and a selection ion profile of the transitions for basic amino acids. The internal standards are labeled in gray and include isotopic compositions. All compounds are derivatized as butylesters.

*a marker to estimate free carnitine produce from hydrolysis of acylcarnitines during sample preparation.

pIDMS is used instead (Chace et al., 2009a). Fortunately, for most acylcarnitines and amino acids, the extraction efficiency is greater than 90% and the differences in reduction in analytical accuracy are not significant.

6.3.6.7 Metabolic Profiles versus Quantification The concept of multiplexing has been used previously in molecular analysis and some immunoassays to represent the ability of an assay to measure multiple metabolites in a single microtiter well or tube. Generally, multiplex analysis would have specific reagents that could be mixed in the same well that would target a specific metabolite and generate a product that would be unique to that metabolite. Rather than use separate wells for each targeted compound the assay is designed such that the reagents, products, and detection do not overlap or interact, thus saving time and improving efficiency. MS/MS is unique in that special reagents are not required for every compound measured, and the assay is multiplexed so that more than one compound can be distinctly detected without interference from others. Also, some of the reagents added (e.g., stable isotope-labeled internal standards) are unique to many of the compounds quantified, but unlike other assays do not interact with the target analyte. In addition to measuring multiple metabolites from a single microtiter well, MS/MS can produce visual profiles that show the relative concentrations of all the metabolites in the analysis similar to chromatographic techniques such as GC and LC. The one major difference in the *x*-axis is the mass to charge ratio in MS/MS versus time for chromatographic techniques. A visual profile is essential in some chromatographic assays, where a clinical interpretation of metabolic profiles (Husek, 1995; Duez et al., 1996; Clayton, 2001; Pitt and Eggington, 2002) (e.g., organic acids in urine) is required, and is also important in the interpretation of newborn screening MS/MS results as well.

A visual profile is the qualitative assay of MS/MS of a dried blood spot. However, with the addition of internal standards the MS/MS profile becomes quantitative. As shown in Figure 6.5 for DBS from pooled adult blood, the relative concentration of endogenous metabolites can be compared to the known amount of internal standards that serve as a visual cue to concentration (Figure 6.4), and quantitative ion

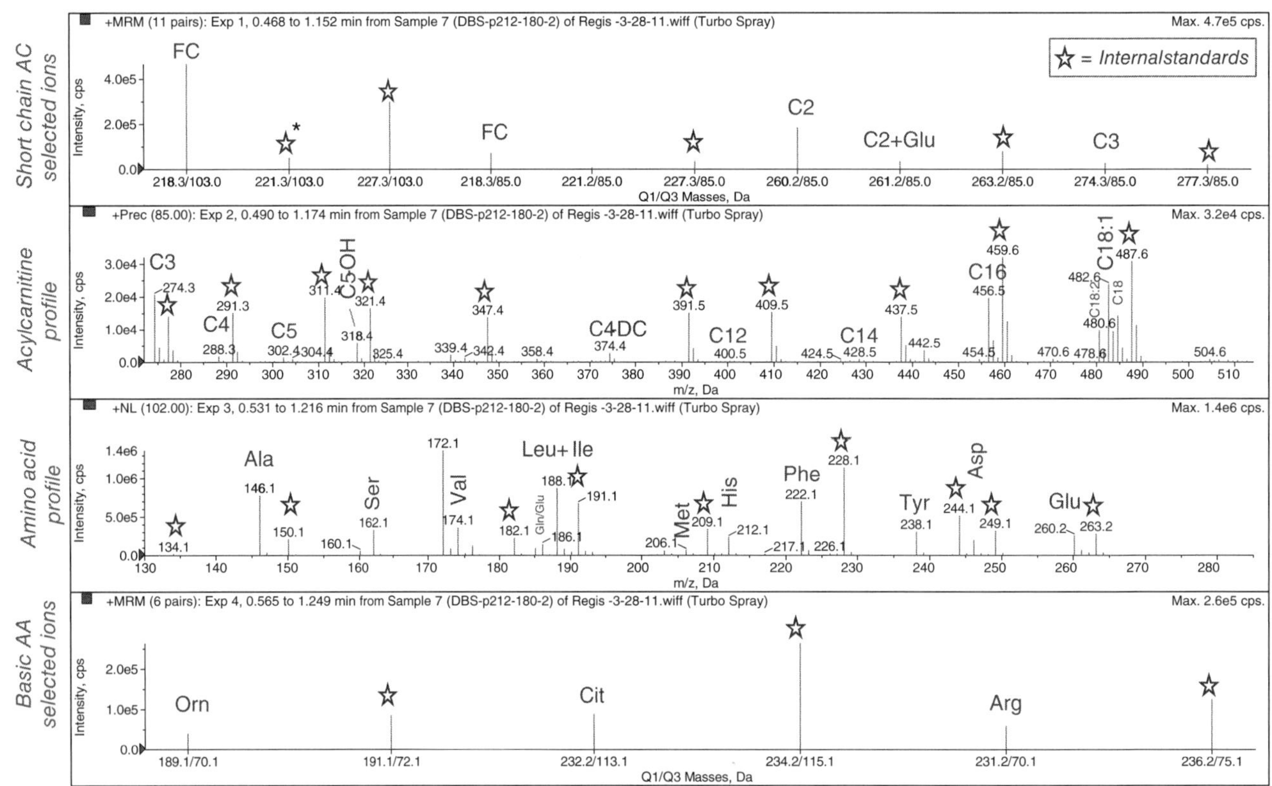

FIGURE 6.5 MS/MS profile of a DBS from pooled adult blood. The first through four panels are selected ion profiles of free and short-chain acylcarnitines, full scan acylcarnitine profile, full scan amino acid profile, and selected ion profiles of basic amino acids for pooled blood from adults. The endogenous metabolites are labeled with internal standards marked with a star. See Figure 4 for a more detailed profile of the internal standards only mixture. All compounds are prepared as butylesters.

*a marker to estimate free carnitine produce from hydrolysis of acylcarnitines during sample preparation.

ratios can be calculated and exported to quantitative programs designed to calculate concentrations from ion intensity data of endogenous and internal standards as described below.

Calculation of concentrations using MS/MS is somewhat less traditional in newborn screening applications than many other MS- and non-MS-based assays that utilize traditional standard curves for each batch of analysis. The reason for this difference is that it is impractical to have a typical standard curve in each analysis for at least 20 key metabolites measured. In the MS/MS analysis, we rely on a set of standards provided by a vendor that has been validated for concentration. A typical NBS laboratory verifies these standards using unlabeled material they purchased from independent sources. The internal standard stock solutions contain 10% of stated concentrations depending on the source (i.e., commercially prepared individually, in groups or as part of a kit, or synthesized independently from in-house synthesis). Therefore, it is possible to know with good accuracy how much standard is added to each well containing the dried blood specimen.

Figure 6.6 illustrates the process of obtaining a concentration value for Phe in a DBS from pIDMS principles that compare traditional IDMS methods that use a liquid blood specimen. The volume of blood collected and added to a tube can be measured accurately, while the volume of blood in a DBS is somewhat less as described previously. For illustrative purposes however, as shown in Figure 6.6, the volume of blood sampled is presumed to be accurate. We also know that the concentration of Phe in the punched disk or the tube is the same at 75 μM. A known quantity of standard is added (known volume of and concentration of $^{13}C_6$-Phe or *Phe) to either the tube or to the well containing the dried

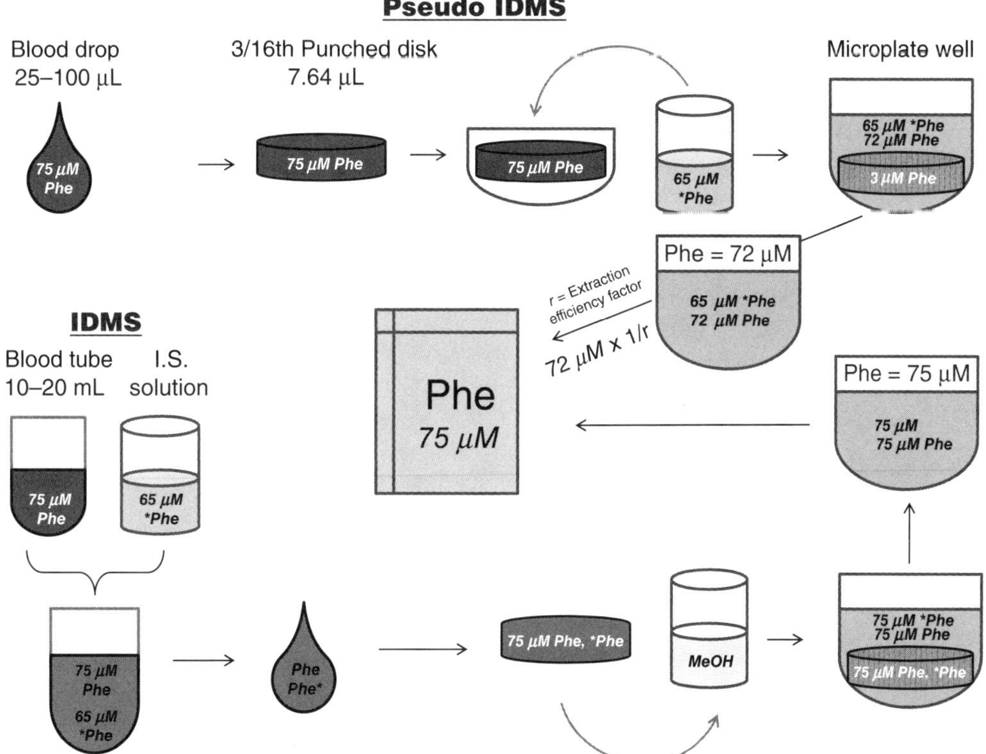

FIGURE 6.6 Determination of concentration of metabolites in a blood using IDMS and a DBS using pIDMS principles. Illustration of measuring metabolite concentration in whole blood or DBS using IDMS and pIDMS techniques, respectively. The concentration in blood for Phe is 75 μM as shown in the blood drop representing the sample. This blood can be added to a tube (IDMS) or filter paper (pseudo IDMS). The concentration of Phe is measured by quantifying the known amount of added internal standard *Phe. The * represent stable isotopically labeled Phe which is most frequently in the form of $^{13}C_6$.The internal standard is diluted in whole blood as shown in the IDMS illustration, while the blood disk is added to a solution of extraction solvent containing the internal standard pIDMS. The ion ratios measured by MS/MS reflect the concentration ratios of the metabolites as shown for both IDMS and pIDMS. Blood containing internal standard can be spotted prior to analysis as shown in the bottom line. pIDMS measurements require correction to account for the extraction efficiency of Phe. Both procedures should produce the final accurate concentration measurement of 75 μM for Phe.

blood disk. The *Phe standard is diluted and mixed with the Phe in liquid blood while only the "extracted Phe" from the DBS is mixed or diluted with the internal standard. Perhaps the term pIDMS should have been renamed "reverse isotope dilution," since it is the metabolite measured that is diluted in the internal standard mixture. The amount of Phe extracted may be slightly less, as shown in the illustration, such that the relative concentration in the DBS is different than the relative concentration measured in a liquid specimen. However, this "extraction efficiency" or recovery can be corrected, since that value can be determined experimentally for most metabolites. Note in the illustrations that liquid blood that already contains internal standard can be spotted to take advantage of many other positive aspects of DBS such as sample clean-up.

In terms of concentration calculation, the first step to obtain this value is to identify the metabolite and its respective reference standard. Consider an example using Phe and $^{13}C_6$-Phe, as shown in Figure 6.6. An ion ratio of the ion counts (or relative ion intensity) for Phe and $^{13}C_6$-Phe is calculated from the data obtained in the MS/MS analysis for the specific *m/z* values of each compound. Since the concentration of the $^{13}C_6$-Phe added to each well is known, the concentration of Phe can be determined. If there was a 100% extraction of Phe from the DBS, then the formula would simply be what is shown in Equation 1 (note that cps is counts per second):

$$\text{Phe}\,(\mu\text{mol/L}) = \text{Phe(cps)}/{}^{13}\text{C}_6\text{-Phe}\,(\text{cps}) \times {}^{13}\text{C}_6\text{-Phe}\,(\mu\text{mol/L}). \quad (6.1)$$

However, the extraction of metabolites is not always 100% and it is necessary to introduce a constant that is based on the extraction efficiency of metabolite recovery as shown in Equation 2, where *r* is the extraction efficiency or recovery:

$$\text{Phe}\,(\mu\text{mol/L}) = \text{Phe(cps)}/{}^{13}\text{C}_6\text{-Phe}\,(\text{cps}) \times {}^{13}\text{C}_6\text{-Phe}\,(\mu\text{mol/L}) \times 1/r. \quad (6.2)$$

6.3.6.8 Improving Precision and Quantification Using Molar Ratios One of the advantages to analytical platforms that measure many metabolites in a single assay is that the relative concentrations of each metabolite are useful in two major ways. First, metabolites that are linked biochemically, as in the case of Phe and Tyr and detection of PKU, enable calculation of mathematical relationships between the two connected metabolites. In PKU, the conversion of Phe to Tyr by phenylalanine hydroxylase is either reduced in activity or absent. The result is an increase in Phe proximal to the enzyme block with an associated decrease in Tyr distal to the enzyme. A ratio of Phe to Tyr should be an even more sensitive indicator for the disease because the numerator increases while the denominator decreases. This finding has been reported in one study (Chace et al., 1993), which showed that MS/MS and molar ratios enabled both more accurate and earlier detection of PKU in NBS as compared to a classical screening method. Second, a ratio of biomarker or metabolite to a closely related structural biomarker or metabolite with a similar fluid compartmentalization may help reduce errors due to volume changes that occur during spotting or extraction. Phe and Tyr molar ratios are the ideal metabolites for molar ratios since they both have a metabolic linkage and are compartmentalized similarly. It has been shown that molar ratios improve the utility of measurement in DBS and should be utilized together with the concentrations of key screening metabolites.

6.3.7 Molecular Methods

Analysis of DBS using molecular techniques is expanding rapidly. There are a variety of approaches that are discussed in more detail in other chapters (see Chapters 4 and 22). It is noteworthy, however, to point out that molecular analysis detects the defective genes, rather than abnormal metabolites. Therefore, many of the issues of blood volume are mute and not an issue. Sample contamination with DNA material from other specimens can be problematic, however. Further, detection of a single gene for a specific disorder may not be conclusive and failure to detect a specific gene in association with a disorder does not completely rule out the possibility of disease. Nevertheless, the use of both metabolic and molecular testing schemes for particular disorders is a very powerful disease screening approach that can substantially reduce false results and improve screening efficacy.

6.4 DETECTION OF METABOLIC DISEASE IN A DBS

6.4.1 Overview

The previous section described the methods used in newborn screening to detect a variety of disorders. The disorders currently screened in most newborn screening programs are disorders of amino acid, organic acid, and fatty acid metabolism, CH, CAH, GAL, hemoglobinopathies, biotinidase deficiency, CF, and, most recently, SCID. Some diseases are almost all detected by a specific technology such as amino acid, organic acid, and fatty acid disorders using MS/MS, while SCID requires molecular methodology. Most other disorders can be detected using one or more methods based on the preferred approach of the laboratories. The diseases selected for screening must meet certain accepted criteria that include the following: it is a treatable condition; an assay is available to test for the disease in the newborn and has acceptable false-positive and near-zero false-negative rates (Wilson and Jungner, 1968). One understated requirement is that the disorder biomarker(s) must be obtained from a DBS for neonatal screening purposes. Therefore, the analytical methods chosen are important as described in the previous section. In 2006, the American College of Medical Genetics recommended a uniform panel of conditions for inclusion by all NBS programs in the United States (Watson et al., 2006). The expert working group recommended 54 conditions (29 core conditions—including hearing screening—and 25 secondary targets) for implementation in all state NBS programs. This "uniform panel" was subsequently endorsed by the Secretary of Health's Advisory Committee on Heritable Disorders and Genetic Diseases in Newborns and Children and has subsequently been amended to include two new conditions, SCID and congenital heart disease (CHD). To date, all states have expanded to include at least the core biochemical tests. Note that two tests, newborn hearing screening and screening for CHD, are physical tests performed at the bedside in newborn nurseries and do not utilize DBS.

6.4.2 Results and Interpretation of Newborn Screening of DBS Perspective

When Dr Guthrie's PKU test became the first test in the evolution of public health newborn screening programs, he relied on reporting the concentration of a single analyte for the inherited disease PKU. This single metabolite assay is not only Guthrie's approach to the detection of PKU, but perhaps still the most common clinical chemistry diagnostic approach to disease detection to this day. An abnormal result whether from a newborn screening test or from a diagnostic assay relies on whether the concentration of a diagnostic or screening marker is above a certain concentration threshold. Many questions surround how that threshold is determined. For many labs, it is based on pure statistical values obtained from analysis of a large number of "normal" samples; for others, it is based on thresholds that are adjusted to accommodate fewer false-positive or false-negative results. One concept is certain in single metabolite analytical approaches with regard to abnormal results: the number of false-positive results is inversely proportional to false-negative results for methods that utilize a single metabolite analysis and a single threshold cutoff concentration.

For Dr Guthrie's BIA test, the precision of the DBS analysis was poor relative to other more modern techniques. Since the goal of any newborn screening test is to reduce false-negative results, Dr Guthrie had two choices; to lower the threshold concentration for Phe in the PKU test with a concomitant increase in false positives, or change the conditions of the test by shifting the collection time by one extra day. This latter approach enabled abnormal patients to accumulate Phe further, distinguishing them from the normal population. Some more modern tests like fluorometry have better precision but their accuracy is poorer, having higher concentrations in normal newborns from a higher bias, due to cross-reactivity of other metabolites that can fluoresce like Tyr. This approach results in a higher false-positive rate if the cutoffs are not adjusted as described above, and increasing cutoffs can result in an increased frequency of false-negative results.

Raising or lowering thresholds to accommodate a method is not always a valid solution. There is little room for error when the blood spot itself has a significant impact on quantitative variation. Historically, false-positive rates were 0.1–1% for many assays and when all tests were combined the total false-positive rate for a newborn screening program could easily exceed 5%. Methods that could improve accuracy of the measurement with little bias and little interference, while also adding multiple metabolite cutoffs, could, in fact, reduce both the false-positive and false-negative rates. This is clearly the case with MS/MS and the detection of PKU. With MS/MS not only Phe is measured, but also Tyr and other amino acids. Molar ratios such as Phe/Tyr and Phe/Leu, and detection of biomarkers that suggest contamination such as intravenous nutrition containing amino acids (Chace et al., 1998, 2011), can also be determined. Molar ratios improve precision for the measurement of certain metabolites like Phe while also improving the diagnostic efficacy because of the metabolic relationship of Phe and Tyr. By combining all of these relationships into an interpretation scheme of variable cutoffs and evaluation of the results, the idea of false positives is actually questionable. An elevation of Phe, a low Tyr, and a high Phe/Tyr ratio are no longer a false positive, but rather a true result for detection of a sample contaminated with an amino acid solution given to premature infants (TPN), for which further proof is obtained by a normal Phe/Leu ratio. That result does not suggest PKU but rather simply reflects an infant whose sample may have been collected improperly. The result is not indicative of a laboratory test failure and becomes useful information for the laboratory. This modern approach to interpretation using MS/MS as an example is illustrated in Figure 6.7.

6.4.3 Chemical Interferences in DBS Analysis in NBS

Any clinical assay for metabolites must identify or attempt to find any potential interference that could result in a false-positive or false-negative finding. Interferences with DBS analyses include common contaminants such as metabolites of antibiotics, steroids, and other drugs used in the treatment of sick newborns or passed from mom to infant at birth or through breast milk. These contaminants can also be introduced during the collection process. Detection of these interferences and avoidance of them will reduce analytical error. There are certain common sources of error in newborn screening using DBS, and these are described below.

6.4.3.1 Anticoagulants One of the first known interferences in newborn screening that is detected by MS/MS is

Analytical results with upper cutoff (classical) — Enhanced with molar ratios

Patient	Leu	Met	Phe	Tyr	Flag	Leu/Phe	Phe/Tyr	Flag
$Cutoff_{(high)}$	300	80	180	240		4	2.5r	
1001	113	25	62	91	OK	1.8	0.7	OK
1002	125	18	74	86	OK	6.9	0.9	OK
1003	<u>424</u>	31	97	103	**Leu**	<u>4.4</u>	0.9	**Leu/Phe**
1004	<u>502</u>	<u>110</u>	<u>240</u>	140	**Leu,Met, Phe**	2.1	1.7	OK
QA high	<u>434</u>	<u>162</u>	<u>299</u>	<u>354</u>	**Leu,Met, Phe, Tyr**	1.4	0.8	OK

Laboratory interpretation guidelines

Classical Chemistry Report:

1003: *Elevated Leu, Leu/Phe possible MSUD, alert physician obtain another specimen*

1004: *Elevated Leu, Met, Phe, Leu/Phenormal alert physician rule out MSUD, PKU, homocystinuria obtain another specimen*

Modern Chemistry Report:

1003: *Moderately elevated Leu(assign code, score, level) abnormal Leu/Phe ratio (assign code, score, level) probable MSUD. Review profile adjust interpretation based on term date, collection date, infant age, QA/QC, lab bias ... order second-tier Leuanalysis Positive second-tier increase odds of MSUD Urgent report, follow-up, repeat*

1004: *Moderately elevated Leu, Met, Phebut no ratios flagged. VLBW infant, TPN administration confirmed Profile consistent with TPN contaminationincluding markers suggestive of TPN. Review profile Specimen questionable – second specimen required after TPN discontinued*

FIGURE 6.7 Illustration of result interpretation scheme for multiple metabolite analysis. The upper table shows a classical interpretation using specific upper abnormal cutoff values for four amino acids (row 1), with an example of possible results from four patients and one quality control DBS. Concentrations above the cutoffs are underlined. A flag is a notation in many interpretation programs that states what metabolites were found to be abnormal (above or below cutoff or within normal range marked as OK). The table on the right shows molar ratio values which are usually part of modern interpretation programs in MS/MS systems. Abnormal results are underlined and any comments for abnormal results are shown in bold. The lower left panel illustrates a typical report based on the concentration of metabolites and molar ratios from cutoff results. The lower right panel shows more extensive interpretation that may indicate results associated with a code, score, or level that provides information of the likelihood of a true-positive result compared to the normal population. New interpretation programs in newborn screening also include instructions for additional recommended tests, when another sample is requested, and the level of urgency for patient follow-up, evaluation, or care.

EDTA. This molecule is known to interfere with some time-resolved fluorometry assays (Holtkamp et al., 2008) and its detection is important in order to reduce false results (Fingerhut et al., 2009). EDTA exhibits four distinct peaks in the acylcarnitine profile, yet it does not interfere with quantification of any key metabolites in screening by MS/MS. The presence of EDTA, however, alerts the screening laboratory to a potential problem with other assays, some of which have known issues. Knowledge of EDTA in a sample also suggests an important piece of clinical information, namely that the sample was collected in a Vacutainer® tube containing EDTA or a capillary pipette containing the same, rather than a heel stick. This information could indicate that an infant was sick, or premature, or another circumstance that makes the case unusual. Although collection using a capillary pipette containing EDTA is obtained from a heel stick, its use could result in scratching the surface of the filter paper collection device and thus change its absorption properties, possibly altering the quantity of blood analyzed. Awareness of this issue could assist in a borderline result interpretation.

6.4.3.2 Total Parenteral Nutrition (TPN) and Premature or Sick Infants Premature and very sick infants in pediatric or neonatal intensive care units have the highest false-positive result rates within the newborn screening sample population. This observation is due, in part, to immature metabolism, treatments with steroids, antibiotics, transfusions, and other therapies needed by these very low birth weight infants or very sick term infants. The major problem with premature infants are abnormal metabolic profiles, especially elevated amino acids. This finding, which is usually due to administration of high concentration of amino acids provided for nutrition, may also be due to contamination of blood with the TPN (intravenous) feeding solutions if improper collection technique occurs (Chace et al., 2010, 2011). TPN solutions contain high concentrations of dextrose and amino acids. An increase in amino acid concentrations in blood drawn through a line may lead to false results for several diseases. Because TPN also contains dextrose that is detectable in the acylcarnitine profile, these markers can serve as flags to indicate that the blood spot may have been contaminated with TPN solution, rendering the amino acid profile invalid.

6.4.3.3 Other Contaminants Other contaminants occur not from the detection of high concentrations of the unaltered chemicals themselves, but as a result of metabolite that either interferes with the metabolic process (elevating the concentration of important key screening metabolites) or form metabolites themselves that are detected in the analysis. Two examples are valproic acid, used in the treatment of patients at risk for seizures, and pivalic acid, a counter ion in some drug formulations in certain countries. Pivalic acid is metabolized by β-oxidation and forms an acylcarnitine whose mass cannot be differentiated from C5 (isovalerylcarnitine), causing a potential false-positive result for diseases characterized by elevated C5 such as isovaleric acidemia (Abdenur et al., 1998). Valproic acid is also metabolized by β-oxidation to form medium-chain acylcarnitines species including C8 (octanoylcarnitine) and C10 (decanoylcarnitine) (Van Hove et al., 1993; Silva et al., 2001). Detection of these metabolites could result in a false-positive result for MCAD deficiency, although the ratio of C8/C10 can be used to assist in differentiating MCAD deficiency from valproate treatment. However, the use of valproate in newborns is rare. The other potential contamination that may occur include therapies that made provide high concentrations of fatty acids such as medium-chain triglyceride (Donnell et al., 2002) or carnitine supplementation in IV solutions (Chace et al., 2003).

6.4.4 From Lab Test to Diagnosis (Bench to Bedside)

The purpose of screening, unlike diagnostics, is to identify infants at birth that are at risk for a metabolic disease that can be treated early and reduce any damage that is produced by the disease and to improve the health of an affected infant. Although the goal of screening is to achieve 100% detection with no false results, in practice, it can never be achieved. The considerable variation in the metabolic profiles of affected and normal newborns that often overlap between disease and normal metabolite ranges and variations in how newborn screening labs set cutoffs and interpret data make perfection in this area an impossibility. Finally, no diagnosis of a patient can be made unless that result is communicated to and acted upon by a clinician who runs a series of tests to confirm the presence or absence of a disorder.

Timely collection, analysis, and reporting of a DBS are key to a positive outcome for many affected newborns. Specimens are now collected on the second day of life or between 24 and 48 hours. They are often shipped via courier or regular mail and arrive in the newborn screening laboratory within an additional 24–72 hours. Although DBS can be shipped via regular mail, this "convenience of a DBS" does not necessarily result in a quicker collection to report time period. Many labs therefore use daily couriers or overnight delivery services. Some disorders, especially those that are fatty acid oxidation defects like MCAD deficiency can result in death at any time after birth without symptoms, and therefore, rapid turnaround is critical in some instances. Other disorders such as PKU produce greater damage with higher concentrations of Phe but generally are treatable if detected and treatment is initiated by the first week of life. Although laboratories cannot control the time it takes to receive a sample, or the time it takes to contact an infant affected by a disorder, the time it takes to process a DBS in the laboratory and produce an accurate result is something that can be affected by choice of test, quality of test, and number of repeats for accurate diagnosis.

Even though the analysis of DBS is a screening test, it may be the only opportunity to detect a disorder. As discussed in the previous section, interpretation of results with as much information as possible can assist the clinician in the time it takes to make a diagnosis and to get the child under his/her care. Communication of what the test results mean and what steps are necessary is often important in cases of rare metabolic disorders. This step may be especially problematic when the physicians of record may never have encountered such a disease in their career previously. Often, metabolic referral centers are set up in regions to provide both patients and physicians support and treatment of an affected newborn. Today, newborn screening laboratories are improving the certainty of results by adding confirmatory tests of initial positive results from the same DBS that the original analysis was performed. This process can reduce false positives and improve the certainty of a positive newborn screening finding.

6.4.5 Improving DBS Screening Accuracy Using Second-Tier Tests

While the appropriate interpretation of the metabolic profiles can reduce false-positive results in NBS, there are a few instances where interpretive skills are not sufficient to avoid unnecessary clinical follow-up of an abnormal NBS result (Matern et al., 2007). Newborns with false-positive screening results are more often hospitalized than healthy children with normal screening results (Gurian et al., 2006); thus, there is a great need to improve screening specificity without sacrificing diagnostic sensitivity. Screening assay specificity can be enhanced by the use of second-tier testing using DBS. These laboratory assays have a lower sample throughput but can measure additional metabolites that either strongly support the presumption of a true-positive case or refute the notion that the patient has the disorder (Chace and Hannon, 2010). Furthermore, second-tier tests use the original DBS sample after the initial primary screening test has made a presumptive positive identification. The use of second-tier tests is an effective approach to reducing false-positive rates (Matern et al., 2007) in NBS activities.

There are several second-tier assays reported in the peer-reviewed literature for disorders in which a "positive" metabolic profile may require expensive and time-consuming follow-up activities. Perhaps the most common application is second-tier testing for CAH. The primary screening assay for CAH relies on the detection of 17-hydroxyprogesterone (17-OHP) using a fluoroimmunoassay (or fluorometric-based detection). However, elevated levels of 17-OHP can result from cross-reacting metabolites found in low weight birth as well as in premature babies. A second-tier MS-based test was developed to assay other steroid metabolites in the biochemical pathway of sterol synthesis, resulting in an 89% reduction of false-positive results in one study (Lacey et al., 2004). A DBS proficiency testing program has been established to help laboratories achieve technical proficiency when performing second-tier CAH testing (De Jesus et al., 2010c). The second-tier CAH test requires a chromatographic separation of five different analytes, thus making the assay unsuitable as a primary screen test due to the time it takes to complete the assay. This is a key feature of second-tier tests; however, this approach can be successfully introduced at a regional level in order to facilitate greater cost savings for laboratories.

Another example of a second-tier approach occurs with branched chain amino acids (BCAAs), to enhance NBS for MSUD (Oglesbee et al., 2008). Newborn screening for MSUD relies on finding increased concentrations of the BCAAs leucine, isoleucine, allo-isoleucine, and valine by MS/MS. Current NBS MS/MS assays cannot differentiate between the three leucine species since they are isobaric, thus necessitating chromatographic separation for unequivocal identification and quantitation. This method has been successfully employed to monitor MSUD patients in a hospital-based laboratory, where it has been shown to generate DBS analyte results comparable to plasma amino acid analysis (Yu, 2010).

A third example is the analysis of total homocysteine (tHCY), MMA, and 2-methylcitric acid (MCA) in DBS by MS/MS to enhance the identification of propionate, methionine, and cobalamin metabolism disorders (la Marca et al., 2008; Turgeon et al., 2010). While current NBS MS/MS can detect methionine and propionylcarnitine in the primary screen, these analytes are not specific for these conditions and can lead to a higher number of false-positive results. Three specific markers are tHCY, MMA, and MCA, which are detectable in DBS. The ultimate combination of DBS specimens and robust analytical approaches to analyte detection by MS-based assays presents limitless possibilities to enhancing NBS assay specificity.

6.5 NEWBORN SCREENING AS A SUBSET OF METABOLIC SCREENING

"Beyond newborn screening" is an alternative title for this section because one of the most exciting areas for the use of DBS may soon be developed, using the same tests in newborn screening for alterations of metabolism that are not a result of genetic abnormalities. Essentially, newborn screening is a smaller piece of the "metabolic screening pie", and there is an entire rapidly expanding area of metabolic testing and research. But much of what we have learned understanding many biomarkers and metabolic interactions came from newborn screening of inherited diseases.

A natural extension of newborn screening includes metabolic screening of a similar population of infants whose metabolic profiles have greater variation as a result of gestational age, weight, nutrition, drug treatments, and so on such as that which occurs in premature and very low birth weight infants. Metabolic screening can also be extended to include older infants and children, adolescents, adults, seniors, forensic (see Chapter 12), disease status such as diabetes and infectious disease, obesity, pharmacology/toxicology, and diet in children and adults. In metabolic screening, the number of metabolites that are detected in a DBS will likely expand significantly. But many of the most recent "non-newborn screening" DBS test relies on the detection of the same metabolites using the same analytical methods but with a different interpretation. The applications include the metabolic studies in critically premature infants to assess nutritional status (Clark et al., 2007; Kelleher et al., 2008; Steinbach et al., 2008),

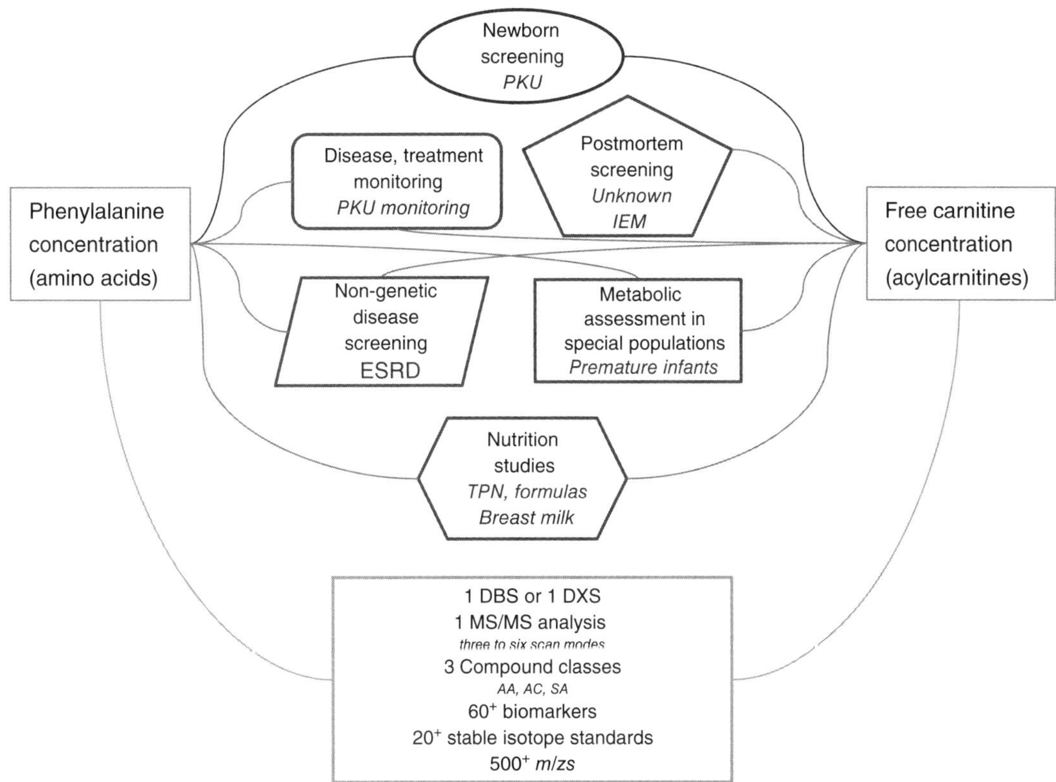

FIGURE 6.8 Applications of metabolic screening of two compound classes in multiple areas of medicine and medical research. An illustration connecting the primary MS/MS panels (amino acids and acylcarnitines) with other metabolic applications beyond the primary screen of a DBS.

postmortem screening (Chace et al., 2001; Dott et al., 2006) (see Chapter 12), analysis of carnitine in dialysis patients (Chazot et al., 2003; Reuter et al., 2005), obesity (Mihalik et al., 2010) and nutrition research, and disease/treatment monitoring (Spitzer and Chace, 2008). An example of how measurement of amino acids and acylcarnitines can apply to many metabolic studies is illustrated in Figure 6.8.

6.6 PROGNOSTICATIONS AND PREDICTIONS

That the past predicts the future in the case of newborn and metabolic screening using DBS represents another way to misquote Shakespeare: "Past is Prologue and Epilogue." The analysis of DBS in newborn screening has established a foundation, or perhaps a footprint (appropriate analogy for an assay based on a heel stick), rather than a blue print, for how to develop and apply screening tests for new diseases and new populations of patients. An analysis of a DBS is an analytical challenge compared to other sample matrices, but newborn screening scientists have harnessed this matrix platform successfully, such that hundreds of millions of infants have been screened and thousands of lives improved as a result of screening. It is only natural to believe that extending DBS analysis to other endogenous metabolic screening studies and assays as well as exogenous drug metabolite studies will validate the technique even further. It is not just the DBS matrix that has had an impact on how metabolic and newborn screening has been done. DBS analysis in NBS has opened new doors to improving lives through clinical chemistry and laboratory science, with the science of DBS experiencing a renaissance (Levy, 1998; Chace, 2003) in both clinical chemistry and pharmaceutical analyses.

DISCLAIMER

The findings and conclusions in this report are those of the authors and do not necessarily represent the views of the Centers for Disease Control and Prevention.

REFERENCES

Abdenur JE, Chamoles NA, Guinle AE, Schenone AB, Fuertes AN. Diagnosis of isovaleric acidaemia by tandem mass spectrometry: false positive result due to pivaloylcarnitine in a newborn screening programme. *Journal of Inherited Metabolic Disease* 1998;21:624–630.

Accinni R, Campolo J, Parolini M, De Maria R, Caruso R, Maiorana A, Galluzzo C, Bartesaghi S, Melotti D, Parodi O. Newborn screening of homocystinuria: quantitative analysis of total homocyst(e)ine on dried blood spot by liquid chromatography with fluorimetric detection. *Journal of Chromatography B - Analytical Technologies in the Biomedical and Life Sciences* 2003;785:219–226.

Badawi N, Cahalane SF, McDonald M, Mulhair P, Begi B, O'Donohue A, Naughten E. Galactosaemia–a controversial disorder. Screening & outcome. Ireland 1972–1992. *Irish Medical Journal* 1996;89:16–17.

Baker MW, Laessig RH, Katcher ML, Routes JM, Grossman WJ, Verbsky J, Kurtycz DF, Brokopp CD. Implementing routine testing for severe combined immunodeficiency within Wisconsin's newborn screening program. *Public Health Reports* 2010;125(Suppl 2):88–95.

Bell C. In: Bell C (ed.), *Newborn Screening Quality Assurance Program Annual Report*. Atlanta, GA: Centers for Disease Control and Prevention; 2010.

Chace DH. Mass spectrometry in the clinical laboratory. *Chemical Reviews* 2001;101:445–477.

Chace DH. Mass spectrometry-based diagnostics: the upcoming revolution in disease detection has already arrived. *Clinical Chemistry* 2003;49:1227–1228; author reply 8–9.

Chace DH. Clinical analysis: inborn errors of metabolism. In: Worsfold P, Townshend A, Poole, C (eds), *Encyclopedia of Analytical Science,* 2nd edn. Elsevier Academic Press; 2005.

Chace DH, Hannon WH. Impact of second-tier testing on the effectiveness of newborn screening. *Clinical Chemistry* 2010;56: 1653–1655.

Chace DH, Naylor EW. Expansion of newborn screening programs using automated tandem mass spectrometry. *Mental Retardation and Developmental Disabilities Research Reviews* 1999;5: 150–154.

Chace DH, Millington DS, Terada N, Kahler SG, Roe CR, Hofman LF. Rapid diagnosis of phenylketonuria by quantitative analysis for phenylalanine and tyrosine in neonatal blood spots by tandem mass spectrometry. *Clinical Chemistry* 1993;39: 66–71.

Chace DH, Sherwin JE, Hillman SL, Lorey F, Cunningham GC. Use of phenylalanine-to-tyrosine ratio determined by tandem mass spectrometry to improve newborn screening for phenylketonuria of early discharge specimens collected in the first 24 hours. *Clinical Chemistry* 1998;44:2405–2409.

Chace DH, DiPerna JC, Mitchell BL, Sgroi B, Hofman LF, Naylor EW. Electrospray tandem mass spectrometry for analysis of acylcarnitines in dried postmortem blood specimens collected at autopsy from infants with unexplained cause of death. *Clinical Chemistry* 2001;47:1166–1182.

Chace DH, Kalas TA, Naylor EW. The application of tandem mass spectrometry to neonatal screening for inherited disorders of intermediary metabolism. *Annual Review of Genomics and Human Genetics* 2002;3:17–45.

Chace DH, Kalas TA, Naylor EW. Use of tandem mass spectrometry for multianalyte screening of dried blood specimens from newborns. *Clinical Chemistry* 2003;49:1797–1817.

Chace DH, Lim T, Hansen CR, Adam BW, Hannon WH. Quantification of malonylcarnitine in dried blood spots by use of MS/MS varies by stable isotope internal standard composition. *Clinica Chimica Acta* 2009a;402:14–18.

Chace DH, Lim T, Hansen CR, De Jesus VR, Hannon WH. Improved MS/MS analysis of succinylacetone extracted from dried blood spots when combined with amino acids and acylcarnitine butyl esters. *Clinica Chimica Acta* 2009b;407:6–9.

Chace DH, Singleton S, Diperna J, Aiello M, Foley T. Rapid metabolic and newborn screening of thyroxine (T4) from dried blood spots by MS/MS. *Clinica Chimica Acta* 2009c;403:178–183.

Chace DH, De Jesus VR, Lim TH, Hannon WH, Spitzer AR. Tandem mass spectrometric identification of dextrose markers in dried-blood spots from infants receiving total parenteral nutrition. *Clinica Chimica Acta* 2010;411:1806–1816.

Chace DH, De Jesus VR, Lim TH, Hannon WH, Clark RH, Spitzer AR. Detection of TPN contamination of dried blood spots used in newborn and metabolic screening and its impact on quantitative measurement of amino acids. *Clinica Chimica Acta* 2011;412:1385–1390.

Chalmers RA, Lawson AM. *Organic Acids in Man: Analytical Chemistry, Biochemistry, and Diagnosis of the Organic Acidurias*. London; New York: Chapman and Hall; 1982.

Chan K, Puck JM. Development of population-based newborn screening for severe combined immunodeficiency. *Journal of Allergy and Clinical Immunology* 2005;115:391–398.

Chazot C, Blanc C, Hurot JM, Charra B, Jean G, Laurent G. Nutritional effects of carnitine supplementation in hemodialysis patients. *Clinical Nephrology* 2003;59:24–30.

Clark RH, Chace DH, Spitzer AR. Effects of two different doses of amino acid supplementation on growth and blood amino acid levels in premature neonates admitted to the neonatal intensive care unit: a randomized, controlled trial. *Pediatrics* 2007;120:1286–1296.

Clayton PT. Applications of mass spectrometry in the study of inborn errors of metabolism. *Journal of Inherited Metabolic Disease* 2001;24:139–150.

Comeau AM, Parad RB, Dorkin HL, Dovey M, Gerstle R, Haver K, Lapey A, O'Sullivan BP, Waltz DA, Zwerdling RG, Eaton RB. Population-based newborn screening for genetic disorders when multiple mutation DNA testing is incorporated: a cystic fibrosis newborn screening model demonstrating increased sensitivity but more carrier detections. *Pediatrics* 2004;113:1573–1581.

Comeau AM, Hale JE, Pai SY, Bonilla FA, Notarangelo LD, Pasternack MS, Meissner HC, Cooper ER, DeMaria A, Sahai I, Eaton RB. Guidelines for implementation of population-based newborn screening for severe combined immunodeficiency. *Journal of Inherited Metabolic Disease* 2010; 33:S273–S281.

De Brabandere VI, Hou P, Stockl D, Thienpont LM, De Leenheer AP. Isotope dilution-liquid chromatography/electrospray ionization-tandem mass spectrometry for the determination of serum thyroxine as a potential reference method. *Rapid Communications in Mass Spectrometry* 1998;12:1099–1103.

De Jesus VR, Zhang XK, Keutzer J, Bodamer OA, Muhl A, Orsini JJ, Caggana M, Vogt RF, Hannon WH. Development and evaluation of quality control dried blood spot materials in newborn screening for lysosomal storage disorders. *Clinical Chemistry* 2009;55:158–164.

De Jesus VR, Chace DH, Lim TH, Mei JV, Hannon WH. Comparison of amino acids and acylcarnitines assay methods used in newborn screening assays by tandem mass spectrometry. *Clinica Chimica Acta* 2010a;411:684–689.

De Jesus VR, Mei JV, Bell CJ, Hannon WH. Improving and assuring newborn screening laboratory quality worldwide: thirty-year experience at the centers for disease control and prevention. *Seminars in Perinatology* 2010b;34:125–133.

De Jesus VR, Simms DA, Schiffer J, Kennedy M, Mei JV, Hannon WH. Pilot proficiency testing study for second tier congenital adrenal hyperplasia newborn screening. *Clinica Chimica Acta* 2010c;411:1684–1687.

Dhillon KS, Bhandal AS, Aznar CP, Lorey FW, Neogi P. Improved tandem mass spectrometry (MS/MS) derivatized method for the detection of tyrosinemia type I, amino acids and acylcarnitine disorders using a single extraction process. *Clinica Chimica Acta* 2011;412:873–879.

Diamandis EP. Immunoassays with time-resolved fluorescence spectroscopy: principles and applications. *Clinical Biochemistry* 1988;21:139–150.

Dobrowolski SF, Banas RA, Naylor EW, Powdrill T, Thakkar D. DNA microarray technology for neonatal screening. *Acta Paediatrica. Supplement* 1999;88:61–64.

Donnell SC, Lloyd DA, Eaton S, Pierro A. The metabolic response to intravenous medium-chain triglycerides in infants after surgery. *Journal of Pediatrics* 2002;141:689–694.

Dott M, Chace D, Fierro M, Kalas TA, Hannon WH, Williams J, Rasmussen SA. Metabolic disorders detectable by tandem mass spectrometry and unexpected early childhood mortality: a population-based study. *American Journal of Medical Genetics Part A* 2006;140:837–842.

Duez P, Kumps A, Mardens Y. GC-MS profiling of urinary organic acids evaluated as a quantitative method. *Clinical Chemistry* 1996;42:1609–1615.

Duffey TA, Sadilek M, Scott CR, Turecek F, Gelb MH. Tandem mass spectrometry for the direct assay of lysosomal enzymes in dried blood spots: application to screening newborns for mucopolysaccharidosis VI (Maroteaux-Lamy syndrome). *Analytical Chemistry* 2010;82:9587–9591.

Duffner PK, Caggana M, Orsini JJ, Wenger DA, Patterson MC, Crosley CJ, Kurtzberg J, Arnold GL, Escolar ML, Adams DJ, Andriola MR, Aron AM, Ciafaloni E, Djukic A, Erbe RW, Galvin-Parton P, Helton LE, Kolodny EH, Kosofsky BE, Kronn DF, Kwon JM, Levy PA, Miller-Horn J, Naidich TP, Pellegrino JE, Provenzale JM, Rothman SJ, Wasserstein MP. Newborn screening for Krabbe disease: the New York State model. *Pediatric Neurology* 2009;40:245–252; discussion 53–55.

Dunbar SA, Jacobson JW. Application of the luminex LabMAP in rapid screening for mutations in the cystic fibrosis transmembrane conductance regulator gene: a pilot study. *Clinical Chemistry* 2000;46:1498–1500.

Fingerhut R, Dame T, Olgemoller B. Determination of EDTA in dried blood samples by tandem mass spectrometry avoids serious errors in newborn screening. *European Journal of Pediatrics* 2009;168:553–558.

Fujimoto A, Okano Y, Miyagi T, Isshiki G, Oura T. Quantitative Beutler test for newborn mass screening of galactosemia using a fluorometric microplate reader. *Clinical Chemistry* 2000;46:806–810.

Garrick MD, Dembure P, Guthrie R. Sickle-cell anemia and other hemoglobinopathies. Procedures and strategy for screening employing spots of blood on filter paper as specimens. *New England Journal of Medicine* 1973;288:1265–1268.

Gelb MH, Turecek F, Scott CR, Chamoles NA. Direct multiplex assay of enzymes in dried blood spots by tandem mass spectrometry for the newborn screening of lysosomal storage disorders. *Journal of Inherited Metabolic Disease* 2006;29:397–404.

Gregg RG, Simantel A, Farrell PM, Koscik R, Kosorok MR, Laxova A, Laessig R, Hoffman G, Hassemer D, Mischler EH, Splaingard M. Newborn screening for cystic fibrosis in Wisconsin: comparison of biochemical and molecular methods. *Pediatrics* 1997;99:819–824.

Gurian EA, Kinnamon DD, Henry JJ, Waisbren SE. Expanded newborn screening for biochemical disorders: the effect of a false-positive result. *Pediatrics* 2006;117:1915–1921.

Guthrie R. Screening for phenylketonuria. *Triangle* 1969;9:104–109.

Guthrie R. The origin of newborn screening. *Screening* 1992;1:5–15.

Guthrie R, Susi A. A simple phenylalanine method for detecting phenylketonuria in large populations of newborn infants. *Pediatrics* 1963;32:338–343.

Hannon WH , Davin B, Fernhoff P, Halonen T, Lavochkin M, Miller J, Ojodu J, Therrell BL. *Blood Collection on Filter Paper for Newborn Screening Programs; Approved Standard-Fifth Edition*. Wayne, PA: Clinical Laboratory Standards Institute; 2007.

Holtkamp U, Klein J, Sander J, Peter M, Janzen N, Steuerwald U, Blankenstein O. EDTA in dried blood spots leads to false results in neonatal endocrinologic screening. *Clinical Chemistry* 2008;54:602–605.

Hoppe CC. Newborn screening for hemoglobin disorders. *Hemoglobin* 2011.

Husek P. Simultaneous profile analysis of plasma amino and organic acids by capillary gas chromatography. *Journal of Chromatography B - Analytical Technologies in the Biomedical and Life Sciences* 1995;669:352–357.

Janik DK, Lindau-Shepard B, Comeau AM, Pass KA. A multiplex immunoassay using the Guthrie specimen to detect T-cell deficiencies including severe combined immunodeficiency disease. *Clinical Chemistry* 2010;56:1460–1465.

Kelleher AS, Clark RH, Steinbach M, Chace DH, Spitzer AR. The influence of amino-acid supplementation, gestational age and time on thyroxine levels in premature neonates. *Journal of Perinatology* 2008;28:270–274.

Khaliq T, Sadilek M, Scott CR, Turecek F, Gelb MH. Tandem mass spectrometry for the direct assay of lysosomal enzymes in dried blood spots: application to screening newborns for mucopolysaccharidosis IVA. *Clinical Chemistry* 2011;57:128–131.

Kleman KM, Vichinsky E, Lubin BH. Experience with newborn screening using isoelectric focusing. *Pediatrics* 1989;83:852–854.

Lacbawan FL, Weck KE, Kant JA, Feldman GL, Schrijver I; Biological and Molecular Genetic Resource Committee of the College of American Pathologists. Verification of performance specifications of a molecular test: cystic fibrosis carrier testing using the Luminex liquid bead array. *Archives of Pathology & Laboratory Medicine* 2012;136:14–19.

Lacey JM, Minutti CZ, Magera MJ, Tauscher AL, Casetta B, McCann M, Lymp J, Hahn SH, Rinaldo P, Matern D. Improved specificity of newborn screening for congenital adrenal hyperplasia by second-tier steroid profiling using tandem mass spectrometry. *Clinical Chemistry* 2004;50:621–625.

la Marca G, Malvagia S, Casetta B, Pasquini E, Donati MA, Zammarchi E. Progress in expanded newborn screening for metabolic conditions by LC-MS/MS in Tuscany: update on methods to reduce false tests. *Journal of Inherited Metabolic Disease* 2008;31(Suppl 2):S395–S404.

Lehotay DC. Chromatographic techniques in inborn errors of metabolism. *Biomedical Chromatography* 1991;5:113–121.

Levy HL. Newborn screening by tandem mass spectrometry: a new era. *Clinical Chemistry* 1998;44:2401–2402.

Levy HL, Albers S. Genetic screening of newborns. *Annual Review of Genomics and Human Genetics* 2000;1:139–177.

Li Y, Scott CR, Chamoles NA, Ghavami A, Pinto BM, Turecek F, Gelb MH. Direct multiplex assay of lysosomal enzymes in dried blood spots for newborn screening. *Clinical Chemistry* 2004;50:1785–1796.

Lorey F, Cunningham G, Vichinsky EP, Lubin BH, Witkowska HE, Matsunaga A, Azimi M, Sherwin J, Eastman J, Farina F, Waye JS, Chui DH. Universal newborn screening for Hb H disease in California. *Genetic Testing* 2001;5:93–100.

Lukacs Z, Mordac C, Kohlschutter A, Kruithof R. Use of microsphere immunoassay for simplified multianalyte screening of thyrotropin and thyroxine in dried blood spots from newborns. *Clinical Chemistry* 2003;49:335–336; author reply 6.

Magera MJ, Gunawardena ND, Hahn SH, Tortorelli S, Mitchell GA, Goodman SI, Rinaldo P, Matern D. Quantitative determination of succinylacetone in dried blood spots for newborn screening of tyrosinemia type I. *Molecular Genetics and Metabolism* 2006;88:16–21.

Matern D, Tortorelli S, Oglesbee D, Gavrilov D, Rinaldo P. Reduction of the false-positive rate in newborn screening by implementation of MS/MS-based second-tier tests: the Mayo Clinic experience (2004–2007). *Journal of Inherited Metabolic Disease* 2007;30:585–592.

McCabe ER, McCabe LL. State-of-the-art for DNA technology in newborn screening. *Acta Paediatrica. Supplement* 1999;88:58–60.

Mechtler TP, Stary S, Metz TF, De Jesus VR, Greber-Platzer S, Pollak A, Herkner KR, Streubel B, Kasper DC. Neonatal screening for lysosomal storage disorders: feasibility and incidence from a nationwide study in Austria. *Lancet* 2012;379:335–341.

Mei JV, Alexander JR, Adam BW, Hannon WH. Use of filter paper for the collection and analysis of human whole blood specimens. *Journal of Nutrition* 2001;131:1631S–1636S.

Mihalik SJ, Goodpaster BH, Kelley DE, Chace DH, Vockley J, Toledo FG, DeLany JP. Increased levels of plasma acylcarnitines in obesity and type 2 diabetes and identification of a marker of glucolipotoxicity. *Obesity (Silver Spring)* 2010;18:1695–1700.

Millington DS, Bohan TP, Roe CR, Yergey AL, Liberato DJ. Valproylcarnitine: a novel drug metabolite identified by fast atom bombardment and thermospray liquid chromatography-mass spectrometry. *Clinica Chimica Acta* 1985;145:69–76.

Minutti CZ, Lacey JM, Magera MJ, Hahn SH, McCann M, Schulze A, Cheillan D, Dorche C, Chace DH, Lymp JF, Zimmerman D, Rinaldo P, Matern D. Steroid profiling by tandem mass spectrometry improves the positive predictive value of newborn screening for congenital adrenal hyperplasia. *Journal of Clinical Endocrinology and Metabolism* 2004;89:3687–3693.

Mitchell ML, Larsen PR, Levy HL, Bennett AJ, Madoff MA. Screening for congenital hypothyroidism. Results in the newborn population of New England. *Journal of American Medical Association* 1978;239:2348–2351.

Moslinger D, Muhl A, Suormala T, Baumgartner R, Stockler-Ipsiroglu S. Molecular characterisation and neuropsychological outcome of 21 patients with profound biotinidase deficiency detected by newborn screening and family studies. *European Journal of Pediatrics* 2003;162(Suppl 1):S46–S49.

Muhl A, Moslinger D, Item CB, Stockler-Ipsiroglu S. Molecular characterisation of 34 patients with biotinidase deficiency ascertained by newborn screening and family investigation. *European Journal of Human Genetics* 2001;9:237–243.

Naughten ER, Yap S, Mayne PD. Newborn screening for homocystinuria: Irish and world experience. *European Journal of Pediatrics* 1998;157(Suppl 2):S84–S87.

Naylor EW, Guthrie R. Newborn screening for maple syrup urine disease (branched-chain ketoaciduria). *Pediatrics* 1978;61:262–266.

Oglesbee D, Sanders KA, Lacey JM, Magera MJ, Casetta B, Strauss KA, Tortorelli S, Rinaldo P, Matern D. Second-tier test for quantification of alloisoleucine and branched-chain amino acids in dried blood spots to improve newborn screening for maple syrup urine disease (MSUD). *Clinical Chemistry* 2008;54:542–549.

Pitt JJ, Eggington M, Kahler SG. Comprehensive screening of urine samples for inborn errors of metabolism by electrospray tandem mass spectrometry. *Clinical Chemistry* 2002;48:1970–1980.

Reilly AA, Bellisario R, Pass KA. Multivariate discrimination for phenylketonuria (PKU) and non-PKU hyperphenylalaninemia after analysis of newborns' dried blood-spot specimens for six amino acids by ion-exchange chromatography. *Clinical Chemistry* 1998;44:317–326.

Reuter SE, Evans AM, Faull RJ, Chace DH, Fornasini G. Impact of haemodialysis on individual endogenous plasma acylcarnitine concentrations in end-stage renal disease. *Annals of Clinical Biochemistry* 2005;42:387–393.

Schneider RG. Differentiation of electrophoretically similar hemoglobins–such as S, D, G, and P; or A2, C, E, and O–by electrophoresis of the globin chains. *Clinical Chemistry* 1974; 20:1111–1115.

Schulze A, Frommhold D, Hoffmann GF, Mayatepek E. Spectrophotometric microassay for delta-aminolevulinate dehydratase in dried-blood spots as confirmation for hereditary tyrosinemia type I. *Clinical Chemistry* 2001;47:1424–1429.

Silva MF, Selhorst J, Overmars H, van Gennip AH, Maya M, Wanders RJ, de Almeida IT, Duran M. Characterization of plasma acylcarnitines in patients under valproate monotherapy using ESI-MS/MS. *Clinical Biochemistry* 2001;34:635–638.

Spacil Z, Elliott S, Reeber SL, Gelb MH, Scott CR, Turecek F. Comparative triplex tandem mass spectrometry assays of lysosomal enzyme activities in dried blood spots using fast liquid chromatography: application to newborn screening of Pompe, Fabry, and Hurler diseases. *Analytical Chemistry* 2011;83:4822–4828.

Spitzer AR, Chace D. Proteomics- and metabolomics-based neonatal diagnostics in assessing and managing the critically ill neonate. *Clinics in Perinatology* 2008;35:695–716, vi.

Steinbach M, Clark RH, Kelleher AS, Flores C, White R, Chace DH, Spitzer AR; Pediatrix Amino-Acid Study Group. Demographic and nutritional factors associated with prolonged cholestatic jaundice in the premature infant. *Journal of Perinatology* 2008;28:129–135.

Turgeon C, Magera MJ, Allard P, Tortorelli S, Gavrilov D, Oglesbee D, Raymond K, Rinaldo P, Matern D. Combined newborn screening for succinylacetone, amino acids, and acylcarnitines in dried blood spots. *Clinical Chemistry* 2008;54:657–664.

Turgeon CT, Magera MJ, Cuthbert CD, Loken PR, Gavrilov DK, Tortorelli S, Raymond KM, Oglesbee D, Rinaldo P, Matern D. Determination of total homocysteine, methylmalonic acid, and 2-methylcitric acid in dried blood spots by tandem mass spectrometry. *Clinical Chemistry* 2010b;56:1686–1695.

Van Hove JL, Zhang W, Kahler SG, Roe CR, Chen YT, Terada N, Chace DH, Iafolla AK, Ding JH, Millington DS. Medium-chain acyl-CoA dehydrogenase (MCAD) deficiency: diagnosis by acylcarnitine analysis in blood. *American Journal of Human Genetics* 1993;52:958–966.

Watson MS, Mann MY, Lloyd-Puryear MA, Rinaldo P, Howell RR. Newborn screening: toward a uniform screening panel and system. *Genetics in Medicine* 2006;8:12S–252S.

Wilson JM, Jungner YG. [Principles and practice of mass screening for disease]. *Boletín de la Oficina Sanitaria Panamericana* 1968;65:281–393.

Witchel SF, Nayak S, Suda-Hartman M, Lee PA. Newborn screening for 21-hydroxylase deficiency: results of CYP21 molecular genetic analysis. *Journal of Pediatrics* 1997;131:328–331.

Wolfe BJ, Blanchard S, Sadilek M, Scott CR, Turecek F, Gelb MH. Tandem mass spectrometry for the direct assay of lysosomal enzymes in dried blood spots: application to screening newborns for mucopolysaccharidosis II (Hunter Syndrome). *Analytical Chemistry* 2011;83:1152–1156.

Wu JT. Screening for inborn errors of amino acid metabolism. *Annals of Clinical and Laboratory Science* 1991;21:123–142.

Yu C. Monitoring PKU and MSUD in one method from dried blood spots using LC/MS/MS. *2010 Newborn Screening and Genetics Testing Symposium*. Orlando, FL: Association of Public Health Laboratories; 2010.

Ziadeh R, Hoffman EP, Finegold DN, Hoop RC, Brackett JC, Strauss AW, Naylor EW. Medium chain acyl-CoA dehydrogenase deficiency in Pennsylvania: neonatal screening shows high incidence and unexpected mutation frequencies. *Pediatric Research* 1995;37:675–678.

Zoppa M, Gallo L, Zacchello F, Giordano G. Method for the quantification of underivatized amino acids on dry blood spots from newborn screening by HPLC-ESI-MS/MS. *Journal of Chromatography B - Analytical Technologies in the Biomedical and Life Sciences* 2006;831:267–273.

7

DRIED BLOOD SPOTS FOR USE IN HIV-RELATED EPIDEMIOLOGICAL STUDIES IN RESOURCE-LIMITED SETTINGS

SRIDHAR V. BASAVARAJU AND JOHN P. PITMAN

7.1 BACKGROUND

Of the approximately 34 million persons infected with the human immunodeficiency virus (HIV) globally, an estimated 22.9 million reside in sub-Saharan Africa (WHO, 2011). Another 4.8 million HIV-infected persons live in South or Southeast Asia (WHO, 2011). In sub-Saharan Africa, nearly all countries meet the World Health Organization (WHO) criteria for a generalized HIV epidemic (Figure 7.1). Monitoring the HIV epidemic through epidemiological studies is a priority for nearly all levels of the public health infrastructure in affected areas, including policy and program planners, clinicians, and laboratory scientists. Population-based and other (e.g., antenatal clinics (ANC)) HIV surveillance, viral load monitoring, drug resistance genotyping, and prevention of mother-to-child transmission studies in resource-limited settings rely on laboratory testing of blood or serum samples for HIV infection. However, in resource-limited settings where much of this work occurs, efforts to monitor the epidemic face several major challenges in the collection, storage, and transport of blood and serum samples, including

- *Occupational exposure*. During the blood collection process, occupational HIV exposure among healthcare workers, particularly during phlebotomy, poses a substantial risk of HIV transmission from infected patients to phlebotomists via accidental needle sticks (Alvarado-Ramy et al., 2003; Castella et al., 2003; Shariati et al., 2007). These risks are magnified in resource-limited settings where healthcare workers are often inadequately trained in safe phlebotomy practices, experience stock outs of important protective equipment including gloves, and do not use retractable or other devices designed to prevent needle-stick injuries (Nsubuga and Jaakkola, 2005; Sadoh et al., 2006; Taegtmeyer et al., 2008).
- *Cost and logistical considerations*. The collection of plasma specimens traditionally used for HIV testing requires syringes and glass or plastic tubes, as well as centrifuges and other capacity for initial specimen processing (Johannessen, 2010). Wet specimens also require storage and transport under cold-chain. Infrastructure limitations including a lack of reliable power sources can preclude such studies before they even begin; delays in transportation or cold-chain failures during the course of a study can also result in compromised specimens (Biggar et al., 1997). Transportation of potentially HIV-infected biohazardous specimens by commercial courier services is expensive and may pose another occupational HIV exposure risk to couriers (Biggar et al., 1997).

All told, the financial costs and logistical barriers associated with the collection, storage, and transport of wet blood or plasma specimens can prohibit facilities, Ministries of Health, and international donors from pursuing such studies on a regular or routine basis.

For these reasons, dried blood spot (DBS) specimens have played an increasingly important role in the conduct of large-scale and *routine* HIV-related epidemiological

Dried Blood Spots: Applications and Techniques, First Edition. Edited by Wenkui Li and Mike S. Lee.

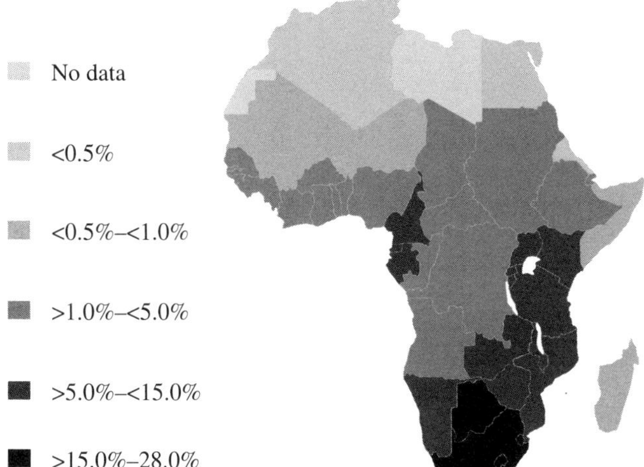

FIGURE 7.1 Estimated HIV/AIDS prevalence among persons aged 15–49 years in Africa by country, 2009. (*Note*: Data are 2009 estimates obtained from Joint United Nations Program on HIV/AIDS 2010 global view of HIV infection. Available at: http://www.unaids.org/globalreport/global_report.htm.)

studies in resource-limited settings. There are two advantages associated with DBS specimens in these settings: safety and stability. DBS specimens can be prepared via heel or finger stick with little training requirements and minimal risk for occupational HIV exposure to healthcare workers (Biggar et al., 1997). Once the samples are dried, the HIV viral particles are no longer infectious (Yang et al., 2010). As a result, these samples can be packaged in lightweight envelopes and transported with low biohazard risk at ambient temperatures (Biggar et al., 1997; Yang et al., 2010). Due to reduced capital investment costs for centrifuges and cold-chain equipment, as well as the lower cost of consumables, DBS specimens present a considerable cost savings compared to plasma specimens (Toledo et al., 2005; Sarge-Njie et al., 2006).

7.2 INITIAL STUDIES DESCRIBING HIV DETECTION IN DRIED BLOOD SPOTS

The concept of using dried whole blood on filter paper for laboratory monitoring of the HIV epidemic was first suggested in studies published in the late 1980s and early 1990s. The detection of antibodies to HIV using DBS specimens was first described in 1987 (Farzadegan et al., 1987). The following year, the sensitivity and specificity of HIV antibody detection using an enzyme-linked immunoadsorbent assay (ELISA) and Western blot were found to be comparable between DBS and serum specimens (Varnier et al., 1988). In 1991, Cassol et al. described the first highly reproducible polymerase chain reaction (PCR) assay for HIV detection in DBS specimens (Cassol et al., 1991). This study described an adapted PCR assay employing duplicate deoxyribonucleic acid (DNA) testing with three primer sets and demonstrated high sensitivity and specificity (Cassol et al., 1991). The following year, Yourno and Conroy described a two-step, nested PCR procedure designed to detect a single copy of HIV proviral DNA in DBS (Yourno and Conroy, 1992). In these studies, the authors described the potential application of DBS for conducting large HIV-related serosurveys:

> Use of the nested PCR with blood spots from clinical specimens showed the assay to be both sensitive and specific. The nested assay is capable of unequivocally detecting infection in individuals when suboptimal amounts of mononuclear cell DNA are present (Yourno and Conroy, 1992).

Stability studies—including several that have documented HIV antibody and viral nucleic acid detection under various storage conditions (Farzadegan et al., 1987; Behets et al., 1992; Cassol et al., 1992, 1997; Alvarez-Munoz et al., 2005; Uttayamakul et al., 2005; Castro et al., 2008; Kane et al., 2008; Leelawiwat et al., 2009)—have provided important data to support the utilization of DBS for HIV-related epidemiological studies (Table 7.1). This evidence base, which includes descriptions of the conditions for preparing, storing, and transporting DBS, has been essential to promote the use of DBS in HIV-related epidemiological studies. Many of these studies have included storage conditions to account for temperature and humidity conditions frequently encountered in resource-limited regions of sub-Saharan Africa and South and Southeast Asia for extended time periods. From these studies, parameters for storage conditions have been established for conducting HIV-related epidemiological studies. These parameters include the following:

- Storage and transport of samples must include desiccants and a humidity indicator card (Bertagnolio et al., 2010).
- For short-term storage (several days to 2 weeks), HIV antibody and nucleic acid are stable if stored at room temperature, or even at 37°C, while for longer periods (beyond 2 weeks up to 1 year), optimal storage is under frozen conditions (see Table 7.1).

7.2.1 Surveillance

With the recognition of HIV in sub-Saharan Africa in the early 1980s, public health surveillance systems, which historically had focused on vaccine-preventable diseases such as polio, smallpox, and measles, were, in some countries, expanded to detect, describe, and monitor the emerging HIV epidemic. In other countries, parallel HIV-only surveillance systems were put in place. Worldwide, the main goals of HIV surveillance are to understand sources of incident infections, monitor disease transmission, observe trends among

TABLE 7.1 Summary of Studies Evaluating HIV Antibody or Viral Nucleic Acid Stability in Dried Blood Spots under Different Storage Conditions

Study Authors	Specimen Source Country and Year	Number and Source Patients	Antibody/Nucleic Acid Detection	Storage Time and Conditions	Assay	Study Findings and Conclusions
Alvarez-Munos et al.	Mexico, 2005	127 (3 children, 124 adults) HIV-1-positive patients	HIV-1 RNA	Time: 1, 3, and 7 days Conditions: 4°C, 22°C, and 37°C	NucliSENS HIV-1 RNA QT	Strong agreement of HIV-1 RNA concentration between DBS and plasma at 7 days, even at high temperature (37°C)
Behets et al.	Zaire, 1992	Five HIV positive with AIDS, 1 HIV-1-infected patient with seronegative blood, 1 HIV-negative blood donor, and CDC Quality Assurance Program specimens	HIV-1 antibody	Time: 20 weeks Conditions: metal cargo direct sunlight; non-air-conditioned house; 4°C, and −20°C	Western blot (Miniblotter system)	Hot and humid storage conditions result in progressive decline in optical density ratios in samples with low antibody titer. Recommend storage in gas-impermeable bags with dessicant
Brambilla et al.	Viral Quality Assurance panels (USA), 2003	Three panels: Panel A: HIV-1 stock diluted in plasma and whole blood Panel B: 6 HIV-1-positive patients and 1 HIV negative Panel C: 4 HIV-1-positive patients	HIV-1 RNA	RT, −30°C, −70°C stored for up to 1 year; −30°C then thawed stored at RT for 3 months	1) NucliSENS HIV-1 RNA QT 2) modified Amplicor HIV-1 monitor	No significant loss of HIV-1 RNA concentration despite storage for 1 year at RT
Cassol et al.	Canada, 1992	Two specimen types: 1) Specimens prepared by diluting 8E5 cells containing 1 HIV proviral DNA copy in CEM cells 2) 24 field specimens: 5 patients with AIDS-related complex, 9 with AIDS, 10 asymptomatic HIV infection	HIV DNA	Time: 12 weeks Conditions: −20°C, 22°C, 37°C; 60% humidity and freeze thawing	In-house PCR	HIV DNA detectable under all conditions. Some variability in detection noted at low HIV DNA levels, which is overcome with ethanol fixation
Castro et al.	Brazil, 2008	190 HIV-positive and 267 HIV-negative patients	HIV-1 and -2 antibody	Time: 6 weeks Conditions: −70°C, −20°C, 4°C, RT	UMELISA HIV-1 and -2 recombinant test	Strong antibody stability at low temperatures for 6 weeks. Drop in stability at RT at 6 weeks

Farzadegan et al.	Africa (unspecified country), USA, 1987	35 HIV-infected patients or hospital staff (Africa), 9 HIV-positive patients (USA), and 1 HIV-negative patient (USA)	HIV antibody	Time: 8 weeks Conditions: RT, refrigeration and 37°C; dry conditions and 90% humidity	2 ELISA (Litton Bionetics and ElectroNucleonics); and Western blot	ELISA results remain consistent at all storage conditions for 8 weeks. Slight decline in Western blot band clarity at 8 weeks after storage at 37°C and 90% humidity
Kane et al.	Senegal, 2008	20 HIV-1-infected infectious disease clinic patients and 41 HIV-1-infected (5 children, 36 adults) suburban dispensary patients	HIV-1 RNA	Time: 15 days Conditions: 37°C	NucliSENS EasyQ HIV-1	No HIV-1 RNA degradation noted after 15 days of storage at 37°C
Leelawiwat et al.	Thailand, 2009	56 HIV-1-infected infants and 106 HIV-1-negative infants	HIV-1 DNA and RNA	Time: 12 months Conditions: −70°C, −20°C, 25°C, and 37°C	Amplicor HIV-1 DNA version 1.0; modified manual NucliSENS RNA NASBA; Amplicor Monitor version 1.5	DNA more stable than RNA. Decline in detection noted at low viral loads when stored at 37°C for 12 months. DBS should be stored at −70°C
Mitchell et al.	USA, 2008	Two HIV-1-infected antiretroviral naive patients	HIV-1 DNA	Time: 12 months Conditions: −20°C, 37°C, and 85% humidity	Amplicor HIV-1 DNA version 1.5	Significant decline in HIV-1 DNA detection when stored at high temperature and humidity for 1 year. DBS should be stored frozen
Uttayamakul et al.	Thailand, 2005	109 HIV-1-infected (80 asymptomatic and 29 symptomatic) patients and 100 HIV-1-negative (60 healthy adults and 40 HIV-1-uninfected infants born to HIV-1-infected mothers) patients	HIV-1 DNA and RNA	Time: unspecified Condition: −20°C	Amplicor HIV-1 DNA version 1.0; NucliSENS RNA NASBA	Strong correlation between HIV-1 RNA detection between plasma and DBS

RT, room temperature; NASBA, nucleic acid sequence-based amplification.

specific subgroup populations, and measure associated morbidity (Diaz et al., 2009). At the national and subnational level, HIV epidemic surveillance frequently involves estimating prevalence and incidence of the disease through annual or biennial surveillance studies (WHO, 2001). These studies can be conducted via household surveys, testing at sentinel sites (e.g., ANC), or through special subpopulation-based studies (WHO, 2001; WHO, 2003; Diaz et al., 2009). From 2001 to 2008, approximately 30 low- and middle-income countries conducted national surveillance studies that included HIV testing (Diaz et al., 2009).[1]

As noted above, surveillance studies in sub-Saharan Africa and other resource-limited settings have faced substantial logistical challenges related to the preparation, storage, and transport of wet plasma and blood specimens. Shortages of serum tubes, inadequate quality assurance in cold-chain preservation, and delays in sample delivery from collection sites to a laboratory have resulted in challenges to surveillance programs (WHO, 2001; Solomon et al., 2002). These challenges have resulted in the adoption of DBS as the preferred sample medium by national population-based HIV surveillance programs in many African settings (Bezuidenhout et al., 2002; Garcia-Calleja et al., 2006; INS and ICF Macro, 2010; KNBS and ICF Macro, 2010). Garcia-Calleja et al. described 16 of 20 population-based surveillance studies in sub-Saharan Africa between 2001 and 2005 that used DBS as the specimen medium of choice (Table 7.2) (Garcia-Calleja et al., 2006). These studies have collectively involved the transport and testing of over 100,000 DBS specimens and resulted in important insight into HIV prevalence and trends across the region (Garcia-Calleja et al., 2006).

Among sentinel surveillance methods, one of the most common relies on HIV testing of pregnant females attending ANC (Diaz et al., 2009). When using ANC sentinel sites, blood samples are frequently tested on site without transport to a centralized laboratory. Likely due to the on-site testing feature of ANC surveillance, few reports describe DBS-based sample collection in ANC sentinel sites. However, some limited ANC-based surveillance studies involving a centralized laboratory have used DBS specimens (Taylor et al., 1996; Dandona et al., 2006).

In specific subgroup based surveillance studies (e.g., among most-at-risk populations), DBS specimens have played a more prominent role. For HIV surveillance among female sex workers, DBS specimens have been advocated due to the ease of collection (Ghys et al., 2001) and have allowed for important characterization of HIV infection among clients of female sex workers in resource-limited settings (Couture et al., 2008). Additional subgroup-specific surveillance studies involving injecting drug users, conflict zones, remote rural areas, and migrant workers have successfully used DBS specimens (Solomon et al., 2004; Hesketh et al., 2006; Platt et al., 2006; Gutierrez et al., 2007; Emmanuel et al., 2009; Judd et al., 2009; Charpentier et al., 2011).

7.2.2 Viral Load Monitoring

In 2010, an estimated 6.6 million HIV-infected persons in low- and middle-income countries (approximately 5 million of whom were in sub-Saharan Africa) were receiving antiretroviral therapy (ART) (UNAIDS, 2010). For proper disease monitoring and clinical management, regular quantification of HIV-1 viral load using PCR-based assays is essential. Early and sustained suppression of viral load has been shown to reduce morbidity and mortality associated with HIV infection (Rouet et al., 2005; Lundgren et al., 2008; Anema et al., 2009; Isaakidis et al., 2010) and to significantly cut the risk of vertical (Chasela et al., 2010) and sexual transmission among serodiscordant sex partners (Cohen et al., 2011). Furthermore, early recognition of treatment failure and emergence of drug resistance is imperative for the prevention of the accumulation of drug resistance mutations and modification of ART regimens, including switching to second-line regimens (Cozzi-Lepri et al., 2007; Hammer et al., 2008; Johannessen, 2010). Viral load monitoring is also known to be superior to cluster of differentiation 4 (CD4) lymphocyte cell count as a method of recognizing ART treatment failure (Badri et al., 2008; Keiser et al., 2009). With the continued scale-up of public health activities related to ART services, notably under the emerging "treatment as prevention" strategy (Cohen, 2011), and the continuing reduction in prices for these medications, viral load monitoring in resource-limited settings is a major global health program priority. However, while ART enrollment programs have been substantially expanded across sub-Saharan Africa, laboratory capacity has lagged (Johannessen et al., 2009). For example, in South Africa where approximately 1 million persons are receiving ART (van Zyl et al., 2011), the National Health Laboratory Service operates 18 laboratories with viral load monitoring capacity

[1]Often, surveillance studies rely on anonymous testing where a collected specimen is de-linked from the patient (WHO, 2001; Diaz et al., 2009). However, even when test results may be de-linked from an individual patient, study design may allow for linkage to individual demographic or behavioral characteristics (Garcia-Calleja et al., 2006). Surveillance studies are not intended to identify individual HIV cases, but rather to generate estimates about prevalence and/or incidence among a target population. Estimates from even routine surveillance sites can be difficult to generalize to larger groups, such as a national population. Another role for surveillance studies involves using molecular epidemiology to identify and track changes in circulating HIV subtypes and detect emerging viral strains or mutations (Cassol et al., 1996). These studies enhance knowledge related to identifying more easily transmitted viral strains or those having altered virulence (Cassol et al., 1996). Molecular epidemiologic surveillance further informs vaccine development by identifying circulating subtypes within populations targeted for clinical vaccine trials (Kalish et al., 1995; Cassol et al., 1996; Arroyo et al., 2006; Carr, 2006; Arroyo et al., 2009).

TABLE 7.2 National Population-Based HIV Surveillance Studies Conducted in Sub-Saharan Africa between 2001 and 2005 Using Dried Blood Spots

Country	Year of Survey	Type of Survey	Age Group (Years)		Sample Size	Type of Specimen	Adult HIV Prevalence (%)[a]
			Females	Males			
Burkina Faso	2003	DHS[b]	15–49	15–59	7515	DBS	1.8
Burundi	2002	Household survey	>12	>12	5569	Venous	3.6
Cameroon	2004	Household survey, restricted to urban areas	15–49	15–59	9900	DBS	5.5
Congo	2003	Household survey	15–49	15–49	3453	Venous	4.2
Equatorial Guinea	2004	DHS	15–49	15–49	1449	DBS	3.2
Ghana	2003	DHS	15–49	15–59	9144	DBS	2.2
Guinea	2005	DHS	15–49	15–59	6377	DBS	1.5
Kenya	2003	DHS	15–49	15–59	6002	DBS	6.7
Lesotho	2004	DHS	15–49	15–59	5043	DBS	23.5
Mali	2001–2002	DHS	15–49	15–59	6846	DBS	1.7
Niger	2002	Household survey	15–49	15–49	6056	DBS	0.9
South Africa	2003	Household survey	15–24	15–24	11,904	Oral fluids	3.0
South Africa	2005	Household survey	>2	>2	15,851	DBS	10.2
Rwanda	2005	DHS	15–49	15–59	10,020	DBS	16.2
Senegal	2005	DHS	15–49	15–59	7524	DBS	0.7
Sierra Leone	2005	Household survey	15–49	15–49	8308	DBS	1.5
Tanzania	2004	AIS[c]	15–49	15–49	10,747	DBS	7.0
Uganda	2004–2005	AIS	15–49	15–49	16,714	Venous	7.1
Zambia	2002	DHS	15–49	15–59	3807	DBS from venous draw	15.6
Zimbabwe	2001–2002	Household survey	15–29	15–29	10,744	DBS	16.5

Source: Reproduced with permission from Garcia-Calleja et al. (2006). Copyright 2006 BMJ Publishing Group Ltd.

[a]HIV prevalence rates are calculated using the respective survey data and are typically among persons aged 15–49 years; Mali includes males 15–59; Burundi for 15 +, South Africa 15–24; Zimbabwe 15–29; Congo for urban areas only.

[b]Demographic Health Survey (DHS)—DHSs are funded by the US Agency for International Development and follow a standard methodology (including standard survey questionnaire, biomarker testing, and anonymous HIV testing) in conducting a nationally representative survey collecting data on fertility, family planning, maternal and child health, gender, HIV/AIDS, malaria, and nutrition.

[c]AIDS Indicator Survey (AIS) follows a standardized survey protocol to collect nationally representative data related to HIV/AIDS in a country. The standardized methodology allows for comparison of findings across countries and over time.

(Stevens, 2010). Limited laboratory infrastructure in remote areas and resulting transport challenges have prompted the exploration of DBS specimens for viral load monitoring in resource-limited settings.

Several previous studies based in industrialized countries have noted good correlation in viral load estimates between DBS and plasma specimens with high viral loads (e.g., greater than 3.5 log copies of HIV RNA/mL). These studies have also found that when compared to plasma specimens, sensitivity is lower in DBS samples with low viral loads below 3 log copies of HIV RNA/mL (O'Shea et al., 1999; Brambilla et al., 2003; Garrido et al., 2009). Additional discrepancies in viral load quantification between DBS and plasma specimens have been described, and additional research has been proposed to describe optimal RNA extraction methods and DBS specimen size. Some published reports have observed higher viral load measurements in DBS specimens than in corresponding plasma samples (Monleau et al., 2009; Reigadas et al., 2009; Steinmetzer et al., 2010), likely due to the contribution of cell-associated HIV-1 proviral DNA (Monleau et al., 2009). Due to the small sample volume input (75 μL with DBS versus 500–1000 μL for plasma), other studies have observed lower detection thresholds when using DBS in comparison to plasma specimens (Garrido et al., 2009; Andreotti et al., 2010; Johannessen, 2010). As noted above, these studies estimate that the HIV viral load detection limit with DBS specimens is around 3 $\log_{10}$ copies/mL (Garrido et al., 2009; Andreotti et al., 2010; Johannessen, 2010). In the United States, treatment guidelines call for switching to second-line ART regimens with any sustained detectable viremia in the setting of treatment adherence (Hammer et al., 2008; Johannessen, 2010). However, for resource-limited settings, WHO recommends continuing first-line regimens as long as viral load remains below 5000 copies/mL, as clinical disease progression is unlikely below this threshold (WHO, 2010b). HIV-related epidemiological studies in resource-limited settings, including sub-Saharan Africa, have focused on the performance of DBS specimens in viral load monitoring in conformity with WHO treatment guidelines.

Epidemiological studies set in resource-limited settings have described viral load measurements in DBS samples from patients receiving ART, which were collected and stored under field conditions (Table 7.3). These studies paid specific attention to existing WHO treatment guidelines (Alvarez-Munoz et al., 2005; Waters et al., 2007; Kane et al., 2008; Johannessen et al., 2009; Mbida et al., 2009; Viljoen et al., 2010; van Zyl et al., 2011). The primary goal of viral load epidemiological studies has been to ensure that patients experiencing treatment failure (viral load >5000 copies/mL) are identified when using DBS specimens. Several epidemiological studies set in resource-limited settings have aimed to establish DBS as an acceptable alternative to plasma on samples collected, stored, and transported under field conditions (Alvarez-Munoz et al., 2005; Waters et al., 2007; Kane et al., 2008; Mbida et al., 2009; Viljoen et al., 2010). While one study by Waters et al. using samples collected and stored under field conditions in Uganda suggested an overestimate of viral load when using DBS (Waters et al., 2007), remaining field-based studies have largely shown DBS to be suitable for the viral load monitoring recommendations set by WHO for resource-limited settings, though with generally lower sensitivity than plasma (Alvarez-Munoz et al., 2005; Kane et al., 2008; Mbida et al., 2009; Viljoen et al., 2010). Acceptable correlation between viral load estimates in plasma and DBS has been shown on field samples, whether the testing was conducted abroad (in a laboratory located in an industrialized country) (Johannessen et al., 2009; Mbida et al., 2009) or in a resource-limited setting (Alvarez-Munoz et al., 2005; Kane et al., 2008; Viljoen et al., 2010). Most field studies examining viral load quantification in DBS specimens have used real-time PCR technology (Kane et al., 2008; Johannessen et al., 2009; Mbida et al., 2009; Viljoen et al., 2010). While both laboratory and field studies have noted lower sensitivity in viral load detection using DBS in comparison to plasma samples, an important observation is that the performance of DBS when using real-time PCR technology is sufficient to meet current WHO treatment guidelines for resource-limited settings (Johannessen et al., 2009; Viljoen et al., 2010). However, if the viral load threshold recommended by WHO for switching to second-line ART regimens is reduced to below 5000 copies/mL, DBS may not be suitable for detecting treatment failure in resource-limited settings.

7.2.3 Drug Resistance Testing

Monitoring HIV-infected patients for drug resistance mutations is a major priority for public health programs in resource-limited settings. Testing for drug resistance can be performed phenotypically with cell culture assays or genotypically by characterizing HIV-coding regions corresponding to drug targets (Bertagnolio et al., 2010). Due to technical ease in comparison to cell culture assays, most drug resistance testing in resource-limited settings occurs by genotyping assays. For patients receiving ART treatment, monitoring for drug resistance may provide insight into treatment failure and whether regimens should be changed to second-line therapy (Bertagnolio et al., 2010). Monitoring for drug resistance is also useful in untreated, recently infected patients as this information may inform initial selection of treatment regimens (Bertagnolio et al., 2010). From a broader epidemiological perspective, population-based surveillance for drug-resistant HIV infections contributes to ART policy decisions and HIV postexposure prophylaxis guidelines (Bertagnolio et al., 2010).

While HIV drug resistance genotyping has historically been performed with plasma samples, commercially available drug resistance genotyping assays have been evaluated with DBS (Masciotra et al., 2007; Hallack et al., 2008). Several studies have been conducted to investigate the effect of time, temperature, and humidity on drug resistance genotyping using DBS specimens (Bertagnolio et al., 2007; McNulty et al., 2007; Buckton et al., 2008; Garcia-Lerma et al., 2009; Bertagnolio et al., 2010) and have identified similar parameters for using DBS in the stability studies described in Table 7.1. Genomic sequence concordance between plasma and DBS specimens has been evaluated in several studies and shows an overall 94.5–99.9% concordance in drug resistance genotyping (Ziemniak et al., 2006; Bertagnolio et al., 2007; Masciotra et al., 2007; McNulty et al., 2007). Efforts to detect HIV drug resistance using DBS specimens in sub-Saharan Africa have been aided by the development of two DBS-specific, broadly sensitive, drug resistance genotyping assays (Buckton et al., 2008; Yang et al., 2010). These DBS assays detect a wider range of HIV-1 subtypes and circulating recombinant factors and have improved HIV drug resistance genotyping. The commercial available assays are designed and evaluated primarily for only HIV-1 subtype B, which predominates in Western Europe and North America (Yang et al., 2010). The two DBS-specific, broadly sensitive, drug resistance genotyping assays are particularly valuable and directly applicable in sub-Saharan Africa due to the region's very high diversity of circulating HIV strains (Yang et al., 2010).

Given the current emphasis on scale-up of ART in resource-limited settings and the global importance of monitoring the emergence and changes in HIV drug resistance patterns, WHO has convened a subgroup panel of experts to develop standardized methods for using DBS specimens for use in drug resistance testing in resource-limited settings and particularly in HIV drug resistance epidemiological surveys (Bertagnolio et al., 2010). HIV drug resistance epidemiological surveys following standard WHO methodologies are classified as those studying drug-naïve patients and those studying patients receiving ART (Bennett et al., 2008; Jordan et al., 2008; Bertagnolio et al., 2010). Due to the lower sensitivity of detection using DBS specimens (described in Section 7.2.2), WHO advises against using DBS in drug

TABLE 7.3 Epidemiological Studies Describing HIV Viral Load Quantification from Dried Blood Spots Collected from HIV-1-Infected Patients in Resource-Limited Settings and Stored under Field Conditions

Study Authors	Country of Specimen Collection	Country of Specimen Testing	Study Design	Assay	Study Findings and Conclusions
Alvarez-Munoz et al.	Mexico	Mexico	Random sample of patients attending ART clinic for routine monitoring; sample intended to be representative of Mexico City and three states with humid, tropical climates	NucliSENS HIV-1 RNA QT and COBAS Amplicor HIV-1 Monitor 1.5	Statistically significant correlation in $\log_{10}$ HIV-1 RNA copies/mL between DBS and plasma; however, mean $\log_{10}$ RNA copies/mL in DBS were generally lower than paired plasma samples
Johannessen et al.	Tanzania	Plasma: Tanzania DBS: Spain	Cross-sectional survey of HIV-infected patients receiving antiretroviral therapy (November 2007 to June 2008)	COBAS TaqMan real-time (plasma) and NucliSENS EasyQ HIV-1 assay (DBS)	Reduced sensitivity of DBS with low-level viremia (<1000 copies/mL); Optimal DBS threshold is 5000 copies/mL (within WHO guidelines for switching to second-line ART regimen)
Kane et al.	Senegal	Senegal	Consecutive samples obtained from patients receiving routine monitoring at a district hospital at day 1 or month 1 following antiretroviral therapy initiation	NucliSENS EasyQ HIV-1 assay	No significant variation of viral load between DBS and plasma specimens

(continued)

TABLE 7.3 ***(Continued)***

Study Authors	Country of Specimen Collection	Country of Specimen Testing	Study Design	Assay	Study Findings and Conclusions
Mbida et al.	Cameroon	France	Random sample of patients attending two Cameroonian dispensaries representative of a rural area (November 2007)	Abbott real-time HIV-1 assay with combined automated extraction and amplification	Good correlation between viral load quantified from plasma compared with DBS; estimated detection threshold with DBS 3 log copies/mL[3] Whole blood can be stored on Whatman 903 filter paper cards for up to 3 months at room temperature, before testing HIV-1 RNA viral load
Viljoen et al.	Burkina Faso and South Africa	Burkina Faso and South Africa	Two-part study involving HIV-1-infected mothers and infants (some mothers were receiving ART): 1. Part 1—South Africa and Burkina Faso: large-scale retrospective survey 2. Part 2—prospective cohort study in South Africa only	Combination NucliSENS extraction method/generic HIV viral load assay on two distinct open, real-time PCR instruments using either oneor two DBS circles	DBS sensitivity reduced compared with plasma; sensitivities of the assays with one or two DBS circles are sufficient to meet WHO guidelines (5000 copies/mL) for switching to second-line ART regimen
Waters et al.	Uganda	Netherlands	Cohort of HIV-infected patients receiving regular viral load monitoring	Manual RNA extraction; quantification with COBAS TaqMan real time	Significant correlation in viral load s in DBS and plasma; high rates of false-positive detectable viral load results with DBS

[2]Study protocol was not specifically designed for determination of detection thresholds. However, authors estimated 3 log copies per mL.

resistance surveys for 12-month endpoint monitoring among patients receiving ART (Bertagnolio et al., 2010). However, for HIV drug resistance surveys among infected patients not yet receiving treatment, DBS are suitable for use in settings where plasma specimens may not be feasible (Bertagnolio et al., 2010). These epidemiological surveys are intended to study drug resistance among patients at a sentinel site prior to ART enrollment and provide insight into circulating drug-resistant subtypes and allow for projection of ART requirements by program and policy planners (Bertagnolio et al., 2010). To date, multiple resource-limited countries have conducted HIV drug resistance epidemiological surveys among specific populations or sentinel sites using DBS specimens (Bertagnolio et al., 2007; Garrido et al., 2008; Kamoto and Aberle-Grasse, 2008; Somi et al., 2008; Mousavi et al., 2010; Zhang et al., 2010; Holguin et al., 2011).

7.3 PREVENTION OF MOTHER-TO-CHILD HIV TRANSMISSION (PMTCT) AND EARLY DIAGNOSIS OF HIV INFECTION AMONG INFANTS

Since initial reports in sub-Saharan Africa recognized that HIV can be transmitted from mother to infant during pregnancy, at childbirth, and/or through breastfeeding (Hira et al., 1990; Nicoll et al., 1990; Lombardi et al., 1991; Ryder and Temmerman, 1991), the prevention of vertical HIV transmission has been a public health priority in the region. Additionally, studies have described that untreated HIV infection may lead to rapid death among infants (Scott et al., 1989) and early initiation of treatment can dramatically reduce morbidity and mortality related to HIV infection among infants (Violari et al., 2008). Given the success of PMTCT programs to reduce rates of vertical transmission worldwide, the Joint United Nations Programme on HIV/AIDS (UNAIDS) has called for the virtual elimination of mother-to-child transmission by 2015 (UNAIDS, 2011).

Following the initial publication describing the effectiveness of zidovudine (AZT) in the prevention of mother-to-child HIV transmission (PMTCT) (CDC, 1994), reducing vertical transmission became a focus of public health programs in resource-limited settings (Cohen, 1995; Mansergh et al., 1996; Morris, 1998). WHO has released PMTCT recommendations which include ART for pregnant females (WHO, 2010a). Implementation of WHO guidelines in resource-limited settings has been shown to dramatically reduce vertical transmission (Dabis et al., 2005; WHO, 2010a). In 2009, an estimated 50% of pregnant women in eastern and southern Africa received an HIV test during pregnancy, and 68% of HIV-positive females received ART to prevent vertical transmission (UNICEF, 2009). With considerable support from the US President's Emergency Plan for AIDS Relief (PEPFAR) and the Global Fund to Fight AIDS, Tuberculosis and Malaria (Global Fund), efforts are underway to scale up PMTCT programs in resource-limited settings with generalized HIV epidemics to ensure 80% coverage of testing for pregnant women at the national level and 85% coverage of antiretroviral drug prophylaxis and treatment, as indicated, for pregnant women found to be HIV infected (PEPFAR, 2009b).

Similarly, WHO has released recommendations related to early infant HIV diagnosis programs in order to rapidly identify new infant infections and ensure universal access to care and treatment among children and infants (WHO, 2006). These guidelines recommend virological testing (DNA, RNA, or p24 antigen[3]) of HIV-exposed infants at 6 weeks of age (WHO, 2006). With PEPFAR and Global Fund support, early infant diagnosis programs in resource-limited settings with generalized HIV epidemics are attempting to ensure 65% coverage of testing for HIV-exposed infants (PEPFAR, 2009a). Given the difficulty in collecting blood from young infants, WHO has endorsed the use of DBS in early infant diagnosis programs in an effort to improve access to HIV testing in resource-limited settings (PEPFAR, 2009a; WHO, 2006).

Early epidemiological studies related to vertical transmission, including some conducted in resource-limited settings, involved placebo-controlled clinical trials (Connor et al., 1994; Sperling et al., 1996; Shaffer et al., 1999; Leroy et al., 2002). However, recent studies in resource-limited settings have focused on monitoring and evaluation of coverage and impact of PMTCT programs (Moodley et al., 2011), identifying appropriate drug regimens (Thomas et al., 2011), tracking drug resistance transmission among mothers and infants (Zeh et al., 2011), and evaluating morbidity and mortality (Heidari et al., 2011). DBS specimens are especially valuable in resource-limited settings for use in PMTCT and early infant diagnosis studies due to the small sample volume requirements and ease of obtaining specimens. Even in the United States, seroepidemiological programs to monitor HIV infection among infants and mothers using DBS were first instituted in the late 1980s and early 1990s (Hannon et al., 1989; Gwinn et al., 1991). Similar seroepidemiological studies among neonates involving DBS specimens were implemented in the United Kingdom during the same time (Peckham et al., 1990; Ades et al., 1991; Dadswell et al., 1992).

In sub-Saharan Africa, the first published studies describing the use of DBS specimens in perinatal transmission studies were conducted in Malawi and South Africa (Biggar et al., 1996, 1997; Taylor et al., 1996). Following these studies, Rollins et al. in South Africa advanced the use of DBS in HIV-related PMTCT epidemiological studies by describing a combined HIV antibody and RNA test for DBS specimens.

[3]HIV antibody testing of exposed infants is considered unreliable due to the possible presence of maternal antibodies which may persist beyond the first 12 months of age.

This assay could be used to simultaneously measure prevalence, incidence, and mother-to-child transmission (Rollins et al., 2002). As a result of this work, monitoring the impact of PMTCT and early infant diagnosis programs through epidemiological studies has been facilitated by implementation of DBS specimens. These studies have included HIV surveillance among infants presenting to immunization clinics (Rollins et al., 2007), evaluation of multi-drug regimens in reducing transmission to infants (Namukwaya et al., 2011; Thomas et al., 2011), and description of impact and program evaluation of early HIV diagnosis among infants programs implemented on the national level (Creek et al., 2008; Nuwagaba-Biribonwoha et al., 2010).

7.4 EMERGING STRATEGIES AND APPLICATIONS OF DBS IN HIV-RELATED EPIDEMIOLOGICAL STUDIES IN RESOURCE-LIMITED SETTINGS

Other than the key areas presented in this chapter, surveillance, viral load monitoring, drug resistance testing, prevention of mother-to-child transmission, and some additional strategies and applications of DBS in HIV-related epidemiological activities are emerging. These include specimen pooling strategies for viral load monitoring and HIV screening of donor blood intended for transfusion in resource-limited settings.

Interest in pooling strategies has been driven by concerns about the high cost of viral load monitoring, regardless of the specimen type used. In South Africa, for example, a viral load test, inclusive of labor, reagents, and instrumentation, has been estimated to cost approximately $40 (van Zyl et al., 2011). Recent studies using plasma samples have shown that pooled testing, either in matrices or minipools, with a clinically appropriate viral load threshold may accurately detect ART failure among HIV-infected patients (Smith et al., 2009). Significant cost savings may be realized if pooling strategies are implemented based on a predetermined pretest probability of ART failure (May et al., 2010). Unfortunately, using similar methodology, pooled testing using DBS specimens has been evaluated in a resource-limited setting (van Zyl et al., 2011) and found to produce increased variability in efficiency and decreased negative predictive value when used with larger eluant volumes (van Zyl et al., 2011). However, excellent efficiency has been observed when pooling dried plasma spots (DPS) (van Zyl et al., 2011). These findings suggest that pooled viral load testing with DBS requires further optimization and that DPS may also be a suitable sample medium for pooled viral load monitoring in resource-limited settings. By reducing the number of tests needed to evaluate viral loads in multiple patients, both sample mediums hold promise for reducing costs associated with viral load monitoring in resource-limited settings.

Another emerging area where DBS specimens may be applicable for use in resource-limited settings is in HIV screening of donor blood by national blood transfusion services. Previous studies have described a high risk of transfusion-transmitted HIV infection in sub-Saharan Africa (Consten et al., 1997; Moore et al., 2001; Baggaley et al., 2006). For resource-limited settings, the WHO advocates universal screening of all donor blood for HIV, hepatitis B and C, and syphilis (WHO, 2010c). The minimum evaluated sensitivity and specificity levels of blood screening assays should be at least 99.5% (WHO, 2010c). In sub-Saharan Africa, many blood services use only third-generation assays to screen donor blood for HIV infection, while several use fourth-generation assays. Previous studies have shown that adding nucleic acid testing (NAT) may reduce the window period of HIV detection in donor blood units to 6–15 days following infection (Corfec et al., 1999; Busch et al., 2005), compared to 15–22 days with fourth-generation p24 and HIV antibody assays (Ly et al., 2004; Busch et al., 2005). Furthermore, the WHO recommends that donor blood screening in resource-limited settings be centralized in one or few national laboratories in order to optimize the cost-effective use of resources and standardize laboratory testing by eliminating variations between laboratory practices and testing results (WHO, 2010c). A substantial focus of recent epidemiological studies related to blood transfusion safety in sub-Saharan Africa has been on quantifying the burden of transfusion-transmitted HIV infection following implementation of blood safety initiatives in line with WHO recommendations (Heyns et al., 2006; Ouattara et al., 2006; Basavaraju et al., 2010; Lefrère et al., 2010; Tagny et al., 2011). These studies have typically involved testing plasma samples in a centralized laboratory (Heyns et al., 2006; Ouattara et al., 2006).

Given logistical challenges associated with transport of plasma samples, one study evaluating the blood safety in Kenya and the potential additional benefit of adding NAT implemented a pooled screening methodology using DBS specimens (Basavaraju et al., 2010). Typically, NAT screening for transfusion-transmitted infections by blood services is conducted on pooled specimens to reduce personnel time and testing costs. While the findings from the Kenya study did not suggest an additional safety benefit of NAT in Kenya (Basavaraju et al., 2010), the study served as a proof of concept for centralized blood bank screening with NAT using DBS specimens rather than traditional liquid plasma or gel tubes.

Several previous studies have already demonstrated acceptable sensitivity and specificity of fourth-generation serological assays with DBS specimens (Lakshmi et al., 2007; Dandona et al., 2008). Building on this work, Okonji et al. investigated the suitability of pooled DBS specimens for centralized NAT screening by blood transfusion services in resource-limited settings, by testing a standard dilution panel in 24 replicates and performing a probit analysis to

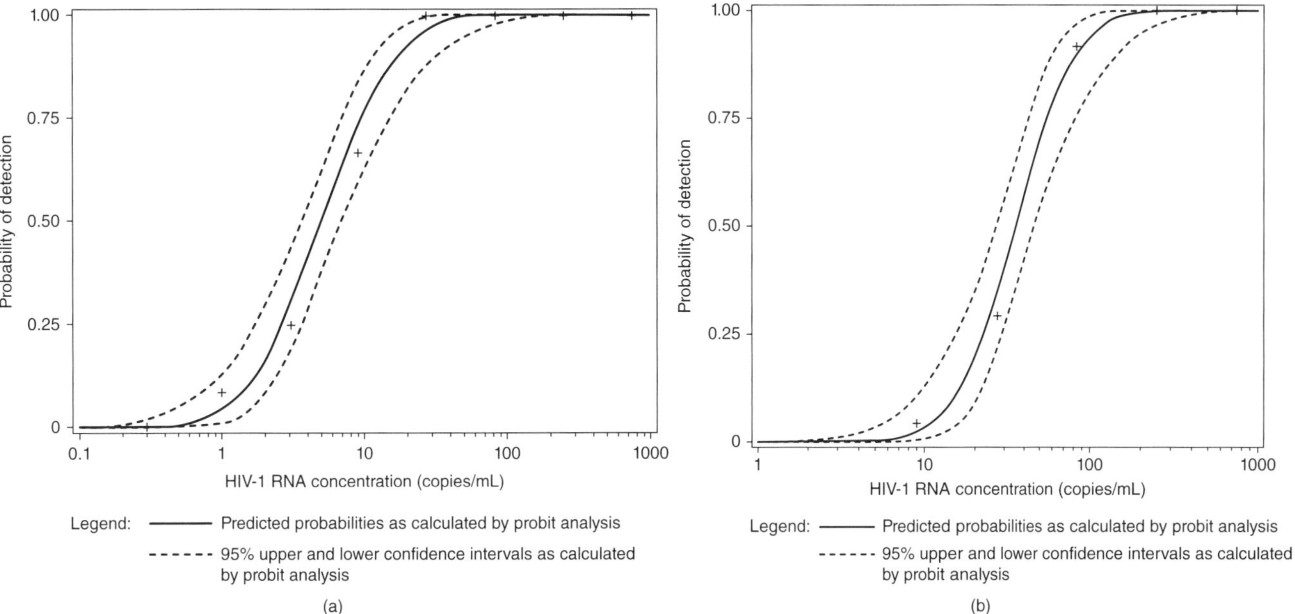

FIGURE 7.2 (a) Probability of detection of HIV-1 RNA in plasma samples using the Roche COBAS Ampliscreen test version 1.5 as calculated by probit analysis on the proportion of positive results on a standard dilution series prepared from a Viral Quality Assurance laboratory stock standard. (b) Probability of detection of HIV-1 RNA in dried blood spots using the Roche COBAS Ampliscreen test version 1.5 as calculated by probit analysis on the proportion of positive results on a standard dilution series prepared from a Viral Quality Assurance laboratory stock standard. (*Note.* Viral Quality Assurance laboratory HIV-1 RNA stock standard is described by Yen-Lieberman et al. (1996). A 70,000 copies/mL specimen from this laboratory was obtained to prepare these plasma and DBS dilution panels.)

establish 50% and 95% detection thresholds (Okonji et al., 2011). DBS and paired plasma dilution series were tested with the Roche COBAS Ampliscreen, a NAT assay approved by the US Food and Drug Administration (FDA) for qualitative screening of individual and pooled samples of whole blood and components intended for transfusion (Okonji et al., 2011). The COBAS Ampliscreen is intended to test pools of up to 24 specimens (Roche, 2003). According to FDA guidelines, HIV-1 NAT blood screening assays should have a 95% detection limit at or below 100 copies/mL for pooled specimens and 5000 copies/mL for individual donations (Lelie et al., 2002). Results of the probit analysis are presented in Figure 7.2. The 95% detection threshold with DBS specimens when pooling specimens and testing with the COBAS Ampliscreen is 106.7 copies/mL (Okonji et al., 2011). However, for individual specimens that detection threshold is 2561 copies/mL. The COBAS Ampliscreen detection threshold with DBS suggests acceptability for individual donations, but additional optimization is required for pooled specimens. With further improvement in detection sensitivity, DBS specimens may eventually be utilized for reduced cost HIV screening by national blood services in resource-limited settings.

7.5 CONCLUSION

The use of DBS specimens has substantially improved the capacity to monitor the HIV epidemic in resource-limited settings. Stability studies have adequately outlined collection, storage, and transport conditions for the specimens. In particular, HIV epidemic surveillance, viral load monitoring, drug resistance testing, and prevention of mother-to-child programs have been assisted by the use of DBS specimens. While technological advances related to DBS specimens have expanded their use in some areas of HIV epidemiology (e.g., surveillance and early infant diagnosis), other improvements are necessary to further the role of DBS for other screening activities (e.g., viral load monitoring and blood donor testing.) To increase the role of DBS in future epidemiological applications in resource-limited settings, additional work is needed to improve HIV detection sensitivity using DBS specimens.

DISCLAIMER

The findings and conclusions in this article are those of the authors and do not necessarily represent the views of the

US Centers for Disease Control and Prevention. The use of trade names is for identification purposes only and does not constitute endorsement by the US Centers for Disease Control and Prevention or the Department of Health and Human Services.

REFERENCES

Ades AE, Parker S, Berry T, Holland FJ, Davison CF, Cubitt D, Hjelm M, Wilcox AH, Hudson CN, Briggs M, et al. Prevalence of maternal HIV-1 infection in Thames regions: results from anonymous unlinked neonatal testing. *Lancet* 1991;337:1562–1565.

Alvarado-Ramy, F, Beltrami EM, Short LJ, Srivastava PU, Henry K, Mendelson M, Gerberding JL, Delclos GL, Campbell S, Solomon R, Fahrner R, Culver DH, Bell D, Cardo DM, Chamberland ME. A comprehensive approach to percutaneous injury prevention during phlebotomy: results of a multicenter study, 1993–1995. *Infection Control and Hospital Epidemiology* 2003;24:97–104.

Alvarez-Munoz MT, Zaragoza-Rodrıguez S, Rojas-Montes O, Palacios-Saucedo G, Vazquez-Rosales G, Gomez-Delgado A, Torresa J, Munoza O. High correlation of human immunodeficiency virus type-1 viral load measured in dried-blood spot samples and in plasma under different storage conditions. *Archives of Medical Research* 2005;36:382–386.

Andreotti, M, Pirillo M, Guidotti G, Ceffa S, Paturzo G, Germano P, Luhanga R, Chimwaza D, Mancini MG, Marazzi MC, Vella S, Palombi L, Giuliano M. Correlation between HIV-1 viral load quantification in plasma, dried blood spots, and dried plasma spots using the Roche COBAS Taqman assay. *Journal of Clinical Virology* 2010;47:4–7.

Anema A, Lima VD, Johnston K, Levy A, Montaner JS. Expanded highly active antiretroviral therapy coverage – a powerful strategy to curb progression to aids, death and new infections. *European Infectious Diseases* 2009;3:41–43.

Arroyo MA, Sateren WB, Serwadda D, Gray RH, Wawer MJ, Sewankambo NK, Kiwanuka N, Kigozi G, Wabwire-Mangen F, Eller M, Eller LA, Birx DL, Robb ML, McCutchan FE. Higher HIV-1 incidence and genetic complexity along main roads in Rakai district, Uganda. *Journal of Acquired Immune Deficiency Syndromes* 2006;43:440–445.

Arroyo MA, Sateren WB, Foglia G, Kibaya R, Langat L, Wasunna M, Bautista CT, Scott PT, Shaffer DN, Robb ML, Michael NL, Birx DL, McCutchan FE. Short communication: HIV type 1 genetic diversity among tea plantation workers in Kericho, Kenya. *AIDS Research and Human Retroviruses* 2009;25:1061–1064.

Badri, M, Lawn SD, Wood R. Utility of CD4 cell counts for early prediction of virological failure during antiretroviral therapy in a resource-limited setting. *BMC Infectious Diseases* 2008;8: 89.

Baggaley, RF, Boily MC, White RG, Alary M. Risk of HIV-1 transmission for parenteral exposure and blood transfusion: a systematic review and meta-analysis. *AIDS* 2006;20:805–812.

Basavaraju SV, Mwangi J, Nyamongo J, Zeh C, Kimani D, Shiraishi RW, Madoda R, Okonji JA, Sugut W, Ongwae S, Pitman JP, Marum LH. Reduced risk of transfusion-transmitted HIV in Kenya through centrally co-ordinated blood centres, stringent donor selection and effective p24 antigen-HIV antibody screening. *Vox Sanguinis* 2010;99:212–219.

Behets F, Kashamuka M, Pappaioanou M, Green TA, Ryder RW, Batter V, George JR, Hannon WH, Quinn TC. Stability of human immunodeficiency virus type 1 antibodies in whole blood dried on filter paper and stored under various tropical conditions in Kinshasa, Zaire. *Journal of Clinical Microbiology* 1992;30:1179–1182.

Bennett DE, Myatt M, Bertagnolio S, Sutherland D, Gilks CF. Recommendations for surveillance of transmitted HIV drug resistance in countries scaling up antiretroviral treatment. *Antiviral Therapy* 2008;13(Suppl 2):25–36.

Bertagnolio S, Soto-Ramirez L, Pilon R, Rodriguez R, Viveros M, Fuentes L, Harrigan PR, Mo T, Sutherland D, Sandstrom P. HIV-1 drug resistance surveillance using dried whole blood spots. *Antiviral Therapy* 2007;12:107–113.

Bertagnolio S, Parkin NT, Jordan M, Brooks J, Garcia-Lerma JG. Dried blood spots for HIV-1 drug resistance and viral load testing: a review of current knowledge and WHO efforts for global HIV drug resistance surveillance. *AIDS Reviews* 2010;12:195–208.

Bezuidenhout F, Brookes HJ, Chauveau J, Covlin M, Connolly C, Ditlopo P, Kelly K, Moatti JP, Loundou DA, Parker W, Richter L, Schwabe C, Shisana O, Simbayi LC, Stoker D, Toefy Y, Zyl Jv. *Nelson Mandela/HSRC Study of HIV/AIDS*. Capetown, South Africa: Human Sciences Research Council Publishers; 2002.

Biggar RJ, Miotti PG, Taha TE, Mtimavalye L, Broadhead R, Justesen A, Yellin F, Liomba G, Miley W, Waters D, Chiphangwi JD, Goedert JJ. Perinatal intervention trial in Africa: effect of a birth canal cleansing intervention to prevent HIV transmission. *Lancet* 1996;347:1647–1650.

Biggar RJ, Miley W, Miotti P, Taha TE, Butcher A, Spadoro J, Waters D. Blood collection on filter paper: a practical approach to sample collection for studies of perinatal HIV transmission. *Journal of Acquired Immune Deficiency Syndromes and Human Retrovirology* 1997;14:368–373.

Brambilla D, Jennings C, Aldrovandi G, Bremer J, Comeau AM, Cassol SA, Dickover R, Jackson JB, Pitt J, Sullivan JL, Butcher A, Grosso L, Reichelderfer P, Fiscus SA. Multicenter evaluation of use of dried blood and plasma spot specimens in quantitative assays for human immunodeficiency virus RNA: measurement, precision, and RNA stability. *Journal of Clinical Microbiology* 2003;41:1888–1893.

Buckton, AJ, Bissett SL, Myers RE, Beddows S, Edwards S, Cane PA, Pillay D. Development and optimization of an internally controlled dried blood spot assay for surveillance of human immunodeficiency virus type-1 drug resistance. *Journal of Antimicrobial Chemotherapy* 2008;62:1191–1198.

Busch MP, Glynn SA, Stramer SL, Strong DM, Caglioti S, Wright DJ, Pappalardo B, Kleinman SH. A new strategy for estimating risks of transfusion-transmitted viral infections based on rates of detection of recently infected donors. *Transfusion* 2005;45:254–264.

Carr JK. Viral diversity as a challenge to HIV-1 vaccine development. *Current Opinion in HIV and AIDS* 2006;1:294–300.

Cassol S, Salas T, Arella M, Neumann P, Schechter MT, O'Shaughnessy M. Use of dried blood spot specimens in the detection of human immunodeficiency virus type 1 by the polymerase chain reaction. *Journal of Clinical Microbiology* 1991;29:667–671.

Cassol S, Salas T, Gill MJ, Montpetit M, Rudnik J, Sy CT, O'Shaughnessy MV. Stability of dried blood spot specimens for detection of human immunodeficiency virus DNA by polymerase chain reaction. *Journal of Clinical Microbiology* 1992;30:3039–3042.

Cassol S, Weniger BG, Babu PG, Salminen MO, Zheng X, Htoon MT, Delaney A, O'Shaughnessy M, Ou CY. Detection of HIV type 1 env subtypes A, B, C, and E in Asia using dried blood spots: a new surveillance tool for molecular epidemiology. *AIDS Research and Human Retroviruses* 1996;12:1435–1441.

Cassol S, Gill MJ, Pilon R, Cormier M, Voigt RF, Willoughby B, Forbes J. Quantification of human immunodeficiency virus type 1 RNA from dried plasma spots collected on filter paper. *Journal of Clinical Microbiology* 1997;35:2795–2801.

Castella A, Vallino A, Argentero PA, Zotti CM. Preventability of percutaneous injuries in healthcare workers: a year-long survey in Italy. *Journal of Hospital Infection* 2003;55:290–294.

Castro AC, Borges LG, Souza Rda S, Grudzinski M, D'Azevedo PA. Evaluation of the human immunodeficiency virus type 1 and 2 antibodies detection in dried whole blood spots (DBS) samples. *Revista do Instituto de Medicina Tropical de Sao Paulo (Sao Paulo)* 2008;50:151–156.

Centers for Disease Control and Prevention (CDC). Zidovudine for the prevention of HIVtransmission from mother to infant. *Morbidity and Mortality Weekly Report* 1994;43:285–287.

Charpentier C, Koyalta D, Ndinaromtan M, Tchobkreo B, Jenabian MA, Day N, Si-Mohamed A, Weiss H, Belec L. Distribution of HIV-1 and HSV-2 epidemics in Chad revealing HSV-2 hot-spot in regions of high-risk HIV spread. *Journal of Infection in Developing Countries* 2011;5:64–67.

Chasela CS, Hudgens MG, Jamieson DJ, Kayira D, Hosseinipour MC, Kourtis AP, Martinson F, Tegha G, Knight RJ, Ahmed YI, Kamwendo DD, Hoffman IF, Ellington SR, Kacheche Z, Soko A, Wiener JB, Fiscus SA, Kazembe P, Mofolo IA, Chigwenembe M, Sichali DS, van der Horst CM. Maternal or infant antiretroviral drugs to reduce HIV-1 transmission. *New England Journal of Medicine (Boston, MA)* 2010;362:2271–2281.

Cohen J. Bringing AZT to poor countries. *Science* 1995;269:624–626.

Cohen J. Breakthrough of the year. HIV treatment as prevention. *Science* 2011;334:1628.

Cohen MS, Chen YQ, McCauley M, Gamble T, Hosseinipour MC, Kumarasamy N, Hakim JG, Kumwenda J, Grinsztejn B, Pilotto JH, Godbole SV, Mehendale S, Chariyalertsak S, Santos BR, Mayer KH, Hoffman IF, Eshleman SH, Piwowar-Manning E, Wang L, Makhema J, Mills LA, de Bruyn G, Sanne I, Eron J, Gallant J, Havlir D, Swindells S, Ribaudo H, Elharrar V, Burns D, Taha TE, Nielsen-Saines K, Celentano D, Essex M, Fleming TR. Prevention of HIV-1 infection with early antiretroviral therapy. *New England Journal of Medicine (Boston, MA)* 2011;365:493–505.

Connor EM, Sperling RS, Gelber R, Kiselev P, Scott G, O'Sullivan MJ, VanDyke R, Bey M, Shearer W, Jacobson RL, et al. Reduction of maternal-infant transmission of human immunodeficiency virus type 1 with zidovudine treatment. Pediatric AIDS Clinical Trials Group Protocol 076 Study Group. *New England Journal of Medicine (Boston, MA)* 1994;331:1173–1180.

Consten EC, van der Meer JT, de Wolf F, Heij HA, Henny PC, van Lanschot JJ. Risk of iatrogenic human immunodeficiency virus infection through transfusion of blood tested by inappropriately stored or expired rapid antibody assays in a Zambian hospital. *Transfusion* 1997;37:930–934.

Corfec EL, Pont FL, Tuckwell HC, Rouzioux C, Costagliola D. Direct HIV testing in blood donations: variation of the yield with detection threshold and pool size. *Transfusion* 1999;39:1141–1144.

Couture MC, Soto JC, Akom E, Labbe AC, Joseph G, Zunzunegui MV. Clients of female sex workers in Gonaives and St-Marc, Haiti characteristics, sexually transmitted infection prevalence and risk factors. *Sexually Transmitted Diseases* 2008;35:849–855.

Cozzi-Lepri A, Phillips AN, Ruiz L, Clotet B, Loveday C, Kjaer J, Mens H, Clumeck N, Viksna L, Antunes F, Machala L, Lundgren JD. Evolution of drug resistance in HIV-infected patients remaining on a virologically failing combination antiretroviral therapy regimen. *AIDS* 2007;21:721–732.

Creek T, Tanuri A, Smith M, Seipone K, Smit M, Legwaila K, Motswere C, Maruping M, Nkoane T, Ntumy R, Bile E, Mine M, Lu L, Tebele G, Mazhani L, Davis MK, Roels TH, Kilmarx PH, Shaffer N. Early diagnosis of human immunodeficiency virus in infants using polymerase chain reaction on dried blood spots in Botswana's national program for prevention of mother-to-child transmission. *Pediatric Infectious Disease Journal* 2008;27:22–26.

Dabis F, Bequet L, Ekouevi DK, Viho I, Rouet F, Horo A, Sakarovitch C, Becquet R, Fassinou P, Dequae-Merchadou L, Welffens-Ekra C, Rouzioux C, Leroy V, ADPS Group. Field efficacy of zidovudine, lamivudine and single-dose nevirapine to prevent peripartum HIV transmission. *AIDS* 2005;19:309–318.

Dadswell JV, Dowding B, Fletcher M, Pinney GJ, Sellwood J, Williams DL. A pilot study of dried blood spot testing for HIV antibody in neonates. *Communicable Disease Report CDR Review* 1992;2:R126–127.

Dandona L, Lakshmi V, Sudha T, Kumar GA, Dandona R. A population-based study of human immunodeficiency virus in South India reveals major differences from sentinel surveillance-based estimates. *BMC Medicine* 2006;4:31.

Dandona L, Dandona R, Kumar GA, Reddy GB, Ameer MA, Ahmed GM, Ramgopal S, Akbar M, Sudha T, Lakshmi V. Risk factors associated with HIV in a population-based study in Andhra Pradesh state of India. *International Journal of Epidemiology* 2008;37:1274–1286.

Diaz T, Garcia-Calleja JM, Ghys PD, Sabin K. Advances and future directions in HIV surveillance in low- and middle-income countries. *Current Opinion in HIV and AIDS* 2009;4:253–259.

Emmanuel F, Archibald C, Razaque A, Sandstrom P. Factors associated with an explosive HIV epidemic among injecting drug users in Sargodha, Pakistan. *Journal of Acquired Immune Deficiency Syndromes* 2009;51:85–90.

Farzadegan H, Quinn T, Polk BF. Detecting antibodies to human immunodeficiency virus in dried blood on filter papers. *Journal of Infectious Diseases* 1987;155:1073–1074.

Garcia-Calleja JM, Gouws E, Ghys PD. National population based HIV prevalence surveys in sub-Saharan Africa: results and implications for HIV and AIDS estimates. *Sexually Transmitted Diseases* 2006;82(Suppl 3):iii64–70.

Garcia-Lerma JG, McNulty A, Jennings C, Huang D, Heneine W, Bremer JW. Rapid decline in the efficiency of HIV drug resistance genotyping from dried blood spots (DBS) and dried plasma spots (DPS) stored at 37 degrees C and high humidity. *Journal of Antimicrobial Chemotherapy* 2009;64:33–36.

Garrido C, Zahonero N, Fernandes D, Serrano D, Silva AR, Ferraria N, Antunes F, Gonzalez-Lahoz J, Soriano V, Mendoza Cd. Subtype variability, virological response and drug resistance assessed on dried blood spots collected from HIV patients on antiretroviral therapy in Angola. *Journal of Antimicrobial Chemotherapy* 2008;61:694–698.

Garrido C, Zahonero N, Corral A, Arredondo M, Soriano V, de Mendoza C. Correlation between human immunodeficiency virus type 1 (HIV-1) RNA measurements obtained with dried blood spots and those obtained with plasma by use of Nuclisens EasyQ HIV-1 and Abbott RealTime HIV load tests. *Journal of Clinical Microbiology* 2009;47:1031–1036.

Ghys PD, Jenkins C, Pisani E. HIV surveillance among female sex workers. *AIDS* 2001;15(Suppl 3):S33–40.

Gutierrez JP, Conde-Gonzalez CJ, Walker DM, Bertozzi SM. Herpes simplex virus type 2 among Mexican high school adolescents: prevalence and association with community characteristics. *Archives of Medical Research* 2007;38:774–782.

Gwinn M, Pappaioanou M, George JR, Hannon WH, Wasser SC, Redus MA, Hoff R, Grady GF, Willoughby A, Novello AC, et al. Prevalence of HIV infection in childbearing women in the United States. Surveillance using newborn blood samples. *Journal of the American Medical Association* 1991;265:1704–1708.

Hallack R, Doherty LE, Wethers JA, Parker MM. Evaluation of dried blood spot specimens for HIV-1 drug-resistance testing using the Trugene HIV-1 genotyping assay. *Journal of Clinical Virology* 2008;41:283–287.

Hammer SM, Eron JJ Jr, Reiss P, Schooley RT, Thompson MA, Walmsley S, Cahn P, Fischl MA, Gatell JM, Hirsch MS, Jacobsen DM, Montaner JS, Richman DD, Yeni PG, Volberding PA. Antiretroviral treatment of adult HIV infection: 2008 recommendations of the International AIDS Society-USA Panel. *Journal of the American Medical Association* 2008;300:555–570.

Hannon WH, Lewis DS, Jones WK, Powell MK. A quality assurance program for human immunodeficiency virus seropositivity screening of dried-blood spot specimens. *Infection Control and Hospital Epidemiology* 1989;10:8–13.

Heidari S, Mofenson L, Cotton MF, Marlink R, Cahn P, Katabira E. Antiretroviral drugs for preventing mother-to-child transmission of HIV: a review of potential effects on HIV-exposed but uninfected children. *Journal of Acquired Immune Deficiency Syndromes* 2011.

Hesketh T, Li L, Ye X, Wang H, Jiang M, Tomkins A. HIV and syphilis in migrant workers in eastern China. *Sexually Transmitted Diseases* 2006;82:11–14.

Heyns AdP, Benjamin RJ, Swanevelder JPR, Laycock ME, Pappalardo BL, Crookes RL, Wright DJ, Busch MP. Prevalence of HIV-1 in blood donations following implementation of a structured blood safety policy in South Africa. *Journal of the American Medical Association* 2006;295:519–526.

Hira SK, Mangrola UG, Mwale C, Chintu C, Tembo G, Brady WE, Perine PL. Apparent vertical transmission of human immunodeficiency virus type 1 by breast-feeding in Zambia. *Journal of Pediatrics* 1990;117:421–424.

Holguin A, Erazo K, Escobar G, de Mulder M, Yebra G, Martin L, Jovel LE, Castaneda L, Perez E. Drug resistance prevalence in human immunodeficiency virus type 1 infected pediatric populations in Honduras and El Salvador during 1989–2009. *Pediatric Infectious Disease Journal* 2011;30:e82–87.

INS and ICF Macro. *2009 National Survey on Prevalence, Behavioral Risks, and Information About HIV and AIDS in Mozambique (INSIDA).* Calverton, MD: INS and ICF Macro; 2010.

Isaakidis P, Raguenaud ME, Te V, Tray CS, Akao K, Kumar V, Ngin S, Nerrienet E, Zachariah R. High survival and treatment success sustained after two and three years of first-line art for children in Cambodia. *Journal of the International AIDS Society* 2010;13:11.

Johannessen A. Dried blood spots in HIV monitoring: applications in resource-limited settings. *Bioanalysis* 2010;2:1893–1908.

Johannessen A, Garrido C, Zahonero N, Sandvik L, Naman E, Kivuyo SL, Kasubi MJ, Gundersen SG, Bruun JN, de Mendoza C. Dried blood spots perform well in viral load monitoring of patients who receive antiretroviral treatment in rural Tanzania. *Clinical Infectious Diseases* 2009;49:976–981.

Jordan MR, Bennett DE, Bertagnolio S, Gilks CF, Sutherland D. World Health Organization surveys to monitor HIV drug resistance prevention and associated factors in sentinel antiretroviral treatment sites. *Antiviral Therapy* 2008;13(Suppl 2):15–23.

Judd A, Rhodes T, Johnston LG, Platt L, Andjelkovic V, Simic D, Mugosa B, Simic M, Zerjav S, Parry RP, Parry JV. Improving survey methods in sero-epidemiological studies of injecting drug users: a case example of two cross sectional surveys in Serbia and Montenegro. *BMC Infectious Diseases* 2009;9:14.

Kalish ML, Baldwin A, Raktham S, Wasi C, Luo CC, Schochetman G, Mastro TD, Young N, Vanichseni S, Rubsamen-Waigmann H, et al. The evolving molecular epidemiology of HIV-1 envelope subtypes in injecting drug users in Bangkok, Thailand: implications for HIV vaccine trials. *AIDS* 1995;9:851–857.

Kamoto K, Aberle-Grasse J. Surveillance of transmitted HIV drug resistance with the World Health Organization threshold survey method in Lilongwe, Malawi. *Antiviral Therapy* 2008;13(Suppl 2):83–87.

Kane CT, Ndiaye HD, Diallo S, Ndiaye I, Wade AS, Diaw PA, Gaye-Diallo A, Mboup S. Quantitation of HIV-1 RNA in dried blood spots by the real-time Nuclisens EasyQ HIV-1 assay in Senegal. *Journal of Virological Methods* 2008;148:291–295.

Keiser O, MacPhail P, Boulle A, Wood R, Schechter M, Dabis F, Sprinz E, Egger M. Accuracy of who CD4 cell count criteria for virological failure of antiretroviral therapy. *Tropical Medicine & International Health* 2009;14:1220–1225.

KNBS and ICF Macro. *Kenya Demographic and Health Survey 2008–09*. Calverton, MD: KNBS and ICF Macro; 2010.

Lakshmi V, Sudha T, Bhanurekha M, Dandona L. Evaluation of the Murex HIV Ag/Ab combination assay when used with dried blood spots. *Clinical Microbiology and Infection* 2007;13:1134–1136.

Leelawiwat W, Young NL, Chaowanachan T, Ou CY, Culnane M, Vanprapa N, Waranawat N, Wasinrapee P, Mock PA, Tappero J, McNicholl JM. Dried blood spots for the diagnosis and quantitation of HIV-1: stability studies and evaluation of sensitivity and specificity for the diagnosis of infant HIV-1 infection in Thailand. *Journal of Virological Methods* 2009;155:109–117.

Lefrère J-J, Dahourouh H, Dokekias AE, Kouao MD, Diarra A, Diop S, Tapko J-B, Murphy EL, Laperche S, Pillonel J. Estimate of the residual risk of transfusion-transmitted human immunodeficiency virus infection in sub-Saharan Africa: a multinational collaborative study. *Transfusion* 2010; [Epub ahead of print].

Lelie PN, Drimmelen HAJv, Cuypers HTM, Best SJ, Stramer SL, Hyland C, Allain J-P, Moncharmont P, Defer C, Nubling M, Glauser A, Cardoso MdS, Viret J-F, Lankinen MH, Grillner L, Wirthmuller U, Coste J, Schottstedt V, Masecar B, Dax EM. Sensitivity of HCV RNA and HIV RNA blood screening assays. *Transfusion* 2002;42:527–536.

Leroy V, Karon JM, Alioum A, Ekpini ER, Meda N, Greenberg AE, Msellati P, Hudgens M, Dabis F, Wiktor SZ. Twenty-four month efficacy of a maternal short-course zidovudine regimen to prevent mother-to-child transmission of HIV-1 in West Africa. *AIDS* 2002;16:631–641.

Lombardi V, Moschese V, Scarlatti G, Wahren B, Jansson M, Wigzell H, Rossi P. Maternal markers affecting the risk of vertical transmission of HIV. *Archives of Aids Research* 1991;5:49–52.

Lundgren JD, Babiker A, El-Sadr W, Emery S, Grund B, Neaton JD, Neuhaus J, Phillips AN. Inferior clinical outcome of the CD4+ cell count-guided antiretroviral treatment interruption strategy in the SMART study: role of CD4+ cell counts and HIV RNA levels during follow-up. *Journal of Infectious Diseases* 2008;197:1145–1155.

Ly TD, Laperche S, Brennan C, Vallari A, Ebel A, Hunt J, Martin L, Daghfal D, Schochetman G, Devare S. Evaluation of the sensitivity and specificity of six HIV combined p24 antigen and antibody assays. *Journal of Virological Methods* 2004;122:185–194.

Mansergh G, Haddix AC, Steketee RW, Nieburg PI, Hu DJ, Simonds RJ, Rogers M. Cost-effectiveness of short-course zidovudine to prevent perinatal HIV type 1 infection in a sub-Saharan African developing country setting. *Journal of the American Medical Association* 1996;276:139–145.

Masciotra S, Garrido C, Youngpairoj AS, McNulty A, Zahonero N, Corral A, Heneine W, de Mendoza C, Garcia-Lerma JG. High concordance between HIV-1 drug resistance genotypes generated from plasma and dried blood spots in antiretroviral-experienced patients. *AIDS* 2007;21:2503–2511.

May S, Gamst A, Haubrich R, Benson C, Smith DM. Pooled nucleic acid testing to identify antiretroviral treatment failure during HIV infection. *Journal of Acquired Immune Deficiency Syndromes* 2010;53:194–201.

Mbida AD, Sosso S, Flori P, Saoudin H, Lawrence P, Monny-Lobe M, Oyono Y, Ndzi E, Cappelli G, Lucht F, Pozzetto B, Oukem-Boyer OO, Bourlet T. Measure of viral load by using the Abbott Real-Time HIV-1 assay on dried blood and plasma spot specimens collected in 2 rural dispensaries in Cameroon. *Journal of Acquired Immune Deficiency Syndromes* 2009;52:9–16.

McNulty A, Jennings C, Bennett D, Fitzgibbon J, Bremer JW, Ussery M, Kalish ML, Heneine W, Garcia-Lerma JG. Evaluation of dried blood spots for human immunodeficiency virus type 1 drug resistance testing. *Journal of Clinical Microbiology* 2007;45:517–521.

Monleau M, Montavon C, Laurent C, Segondy M, Montes B, Delaporte E, Boillot F, Peeters M. Evaluation of different RNA extraction methods and storage conditions of dried plasma or blood spots for human immunodeficiency virus type 1 RNA quantification and PCR amplification for drug resistance testing. *Journal of Clinical Microbiology* 2009;47:1107–1118.

Moodley D, Srikewal J, Msweli L, Maharaj NR. A bird's eye view of PMTCT coverage at two regional hospitals and their referral clinics in a resource-limited setting. *South African Medical Journal* 2011;101:122–125.

Moore A, Herrera G, Nyamongo J, Lackritz E, Granade T, Nahlen B, Oloo A, Opondo G, Muga R, Janssen R. Estimated risk of HIV transmission by blood transfusion in Kenya. *Lancet* 2001;358:657–660.

Morris K. Short course of AZT halves HIV-1 perinatal transmission. *Lancet* 1998;351:651.

Mousavi SM, Hamkar R, Gouya MM, Safaie A, Zahraei SM, Yazdani Z, Asbaghi Namini S, Bertagnolio S, Sutherland D, Sandstrom P, Brooks J. Surveillance of HIV drug resistance transmission in Iran: experience gained from a pilot study. *Archives of Virology* 2010;155:329–334.

Namukwaya Z, Mudiope P, Kekitiinwa A, Musoke P, Matovu J, Kayma S, Salmond W, Bitarakwate E, Mubiru M, Maganda A, Galla M, Byamugisha J, Fowler MG. The impact of maternal highly active antiretroviral therapy and short-course combination antiretrovirals for prevention of mother-to-child transmission on early infant infection rates at the Mulago National Referral Hospital in Kampala, Uganda, January 2007 to May 2009. *Journal of Acquired Immune Deficiency Syndromes* 2011;56:69–75.

Nicoll A, Killewo JZ, Mgone C. HIV and infant feeding practices: epidemiological implications for sub-Saharan African countries. *AIDS* 1990;4:661–665.

Nsubuga FM and Jaakkola MS. Needle stick injuries among nurses in sub-Saharan Africa. *Tropical Medicine & International Health* 2005;10:773–781.

Nuwagaba-Biribonwoha H, Werq-Semo B, Abdallah A, Cunningham A, Gamaliel JG, Mtunga S, Nankabirwa V, Malisa I, Gonzalez LF, Massambu C, Nash D, Justman J, Abrams EJ. Introducing a multi-site program for early diagnosis of HIV infection among HIV-exposed infants in Tanzania. *BMC Pediatrics* 2010;10:44.

Okonji JA, Basavaraju SV, Mwangi J, Shiraishi RW, Odera M, Ouma K, Pitman JP, Marum LH, Ou CY, Zeh C. Comparison of HIV-1 detection in plasma specimens and dried blood spots using the Roche COBAS Ampliscreen HIV-1 test in Kisumu, Kenya. *Journal of Virological Methods* 2011;179(1):21–25.

O'Shea S, Mullen J, Corbett K, Chrystie I, Newell ML, Banatvala JE. Use of dried whole blood spots for quantification of HIV-1 RNA. *AIDS* 1999;13:630–631.

Ouattara H, Siransy-Bogui L, Fretz C, Diane KM, Konate S, Koidio A, Minga KA, Hyda J, Koffi-Abe N, Offoumou AM, Abissey S. Residual risk of HIV, HVB and HCV transmission by blood transfusion between 2002 and 2004 at the Abidjan National Blood Transfusion Center. *Transfusion Clinique et Biologique* 2006;13:242–245.

Peckham CS, Tedder RS, Briggs M, Ades AE, Hjelm M, Wilcox AH, Parra-Mejia N, O'Connor C. Prevalence of maternal HIV infection based on unlinked anonymous testing of newborn babies. *Lancet* 1990;335:516–519.

Platt L, Bobrova N, Rhodes T, Uuskula A, Parry JV, Ruutel K, Talu A, Abel K, Rajaleid K, Judd A. High HIV prevalence among injecting drug users in Estonia: implications for understanding the risk environment. *AIDS* 2006;20:2120–2123.

Reigadas S, Schrive MH, Aurillac-Lavignolleb V, Fleury HJ. Quantitation of HIV-1 RNA in dried blood and plasma spots. *Journal of Virological Methods* 2009;161:177–180.

Roche COBAS Ampliscreen HIV-1 Test Version 1.5 Manufacturer's Package Insert. 2003. Branchburg, NJ: RM Systems; 1.

Rollins NC, Dedicoat M, Danaviah S, Page T, Bishop K, Kleinschmidt I, Coovadia HM, Cassol SA. Prevalence, incidence, and mother-to-child transmission of HIV-1 in rural South Africa. *Lancet* 2002;360:389.

Rollins N, Little K, Mzolo S, Horwood C, Newell ML. Surveillance of mother-to-child transmission prevention programmes at immunization clinics: the case for universal screening. *AIDS* 2007;21:1341–1347.

Rouet F, Ekouevi DK, Chaix ML, Burgard M, Inwoley A, Tony TD, Danel C, Anglaret X, Leroy V, Msellati P, Dabis F, Rouzioux C. Transfer and evaluation of an automated, low-cost real-time reverse transcription-PCR test for diagnosis and monitoring of human immunodeficiency virus type 1 infection in a West African resource-limited setting. *Journal of Clinical Microbiology* 2005;43:2709–2717.

Ryder RW and Temmerman M. The effect of HIV-1 infection during pregnancy and the perinatal period on maternal and child health in Africa. *AIDS* 1991;5(Suppl 1): S75–85.

Sadoh WE, Fawole AO, Sadoh AE, Oladimeji AO, Sotiloye OS. Practice of universal precautions among healthcare workers. *Journal of the National Medical Association* 2006;98:722–726.

Sarge-Njie R, Schim M, Van Der Loeff, Ceesay S, Cubitt D, Sabally S, Corrah T, Whittle H. Evaluation of the dried blood spot filter paper technology and five testing strategies of HIV-1 and HIV-2 infections in West Africa. *Scandinavian Journal of Infectious Diseases* 2006;38:1050–1056.

Scott GB, Hutto C, Makuch RW, Mastrucci MT, O'Connor T, Mitchell CD, Trapido EJ, Parks WP. Survival in children with perinatally acquired human immunodeficiency virus type 1 infection. *New England Journal of Medicine (Boston, MA)* 1989;321:1791–1796.

Shaffer N, Chuachoowong R, Mock PA, Bhadrakom C, Siriwasin W, Young NL, Chotpitayasunondh T, Chearskul S, Roongpisuthipong A, Chinayon P, Karon J, Mastro TD, Simonds RJ. Short-course zidovudine for perinatal HIV-1 transmission in Bangkok, Thailand: a randomised controlled trial. Bangkok Collaborative Perinatal HIV Transmission Study Group. *Lancet* 1999;353:773–780.

Shariati B, Shahidzadeh-Mahani A, Oveysi T, Akhlaghi H. Accidental exposure to blood in medical interns of Tehran University of Medical Sciences. *Journal of Occupational Health* 2007;49:317–321.

Smith DM, May SJ, Perez-Santiago J, Strain MC, Ignacio CC, Haubrich RH, Richman DD, Benson CA, Little SJ. The use of pooled viral load testing to identify antiretroviral treatment failure. *AIDS* 2009;23:2151–2158.

Solomon SS, Solomon S, Rodriguez II, McGarvey ST, Ganesh AK, Thyagarajan SP, Mahajan AP, Mayer KH. Dried blood spots (DBS): a valuable tool for HIV surveillance in developing/tropical countries. *International Journal of STD and AIDS* 2002;13:25–28.

Solomon SS, Pulimi S, Rodriguez II, Chaguturu SK, Satish Kumar SK, Mayer KH, Solomon S. Dried blood spots are an acceptable and useful HIV surveillance tool in a remote developing world setting. *International Journal of STD and AIDS* 2004;15:658–661.

Somi GR, Kibuka T, Diallo K, Tuhuma T, Bennett DE, Yang C, Kagoma C, Lyamuya EF, Swai RO, Kassim S. Surveillance of transmitted HIV drug resistance among women attending antenatal clinics in Dar es Salaam, Tanzania. *Antiviral Therapy* 2008;13(Suppl 2):77–82.

Sperling RS, Shapiro DE, Coombs RW, Todd JA, Herman SA, McSherry GD, O'Sullivan MJ, Van Dyke RB, Jimenez E, Rouzioux C, Flynn PM, Sullivan JL. Maternal viral load, zidovudine treatment, and the risk of transmission of human immunodeficiency virus type 1 from mother to infant. Pediatric Aids Clinical Trials Group Protocol 076 Study Group. *New England Journal of Medicine* 1996;335:1621–1629.

Steinmetzer K, Seidel T, Stallmach A, Ermantraut E. HIV load testing with small samples of whole blood. *Journal of Clinical Microbiology* 2010;48:2786–2792.

Stevens W. Point of Care Technology: Perspective from South Africa. INTEREST Workshop, Maputo, Mozambique; 2010.

Taegtmeyer M, Suckling RM, Nguku PM, Meredith C, Kibaru J, Chakaya JM, Muchela H, Gilks CF. Working with risk:

occupational safety issues among healthcare workers in Kenya. *AIDS Care* 2008;20:304–310.

Tagny CT, Mbanya D, Leballais L, Murphy E, Lefrere JJ, Laperche S. Reduction of the risk of transfusion-transmitted human immunodeficiency virus (HIV) infection by using an HIV antigen/antibody combination assay in blood donation screening in Cameroon. *Transfusion* 2011;51:184–190.

Taylor MB, Parker SP, Crewe-Brown HH, McIntyre J, Cubitt WD. Seroepidemiology of HTLV-I in relation to that of HIV-1 in the Gauteng region, South Africa, using dried blood spots on filter papers. *Epidemiology and Infection* 1996;117:343–348.

Thomas TK, Masaba R, Borkowf CB, Ndivo R, Zeh C, Misore A, Otieno J, Jamieson D, Thigpen MC, Bulterys M, Slutsker L, De Cock KM, Amornkul PN, Greenberg AE, Fowler MG. Triple-antiretroviral prophylaxis to prevent mother-to-child HIV transmission through breastfeeding-the Kisumu Breastfeeding Study, Kenya: a clinical trial. *PLoS Medicine* 2011;8:e1001015.

Toledo Jr AC, Januario JN, Rezende RM, Siqueira AL, Mello BF, Fialho EL, Ribeiro RA, Silva HL, Pires EC, Simoes TC, Greco DB. Dried blood spots as a practical and inexpensive source for human immunodeficiency virus and hepatitis C virus surveillance. *Memorias do Instituto Oswaldo Cruz* 2005;100:365–370.

UNAIDS. UNAIDS Report on the Global AIDS Epidemic 2010; 2010. http://www.unaids.org/documents/20101123_GlobalReport_em.pdf (accessed May 23, 2011).

UNAIDS. Global Plan Towards the Elimination of New HIV Infections among Children by 2015 and Keeping Their Mothers Alive 2011–2015; 2011. http://www.unaids.org/en/media/unaids/contentassets/documents/unaidspublication/2011/20110609_JC2137_Global-Plan-Elimination-HIV-Children_en.pdf (accessed January 9, 2012).

UNICEF. Prevention of Mother-to-Child Transmission; 2009. http://www.unicef.org/esaro/factsonchildren_5797.html (accessed June 12, 2011).

U.S. President's Emergency Plan for AIDS Relief. Annex: PEPFAR and Prevention, Care, and Treatment; 2009a. http://www.pepfar.gov/about/strategy/prevention_care_treatment/index.htm (accessed January 15, 2012).

U.S. President's Emergency Plan for AIDS Relief. Prevention: Priority Interventions; 2009b. http://www.pepfar.gov/press/strategy_briefs/138399.htm (accessed July 12, 2011).

Uttayamakul S, Likanonsakul S, Sunthornkachit R, Kuntiranont K, Louisirirotchanakul S, Chaovavanich A, Thiamchai V, Tanprasertsuk S, Sutthent R. Usage of dried blood spots for molecular diagnosis and monitoring HIV-1 infection. *Journal of Virological Methods* 2005;128:128–134.

van Zyl GU, Preiser W, Potschka S, Lundershausen AT, Haubrich R, Smith D. Pooling strategies to reduce the cost of HIV-1 RNA load monitoring in a resource-limited setting. *Clinical Infectious Diseases* 2011;52:264–270.

Varnier OE, Lillo FB, Reina S, De Maria A, Terragna A, Schito G. Whole blood collection on filter paper is an effective means of obtaining samples for human immunodeficiency virus antibody assay. *AIDS Research and Human Retroviruses* 1988;4:131–136.

Viljoen J, Gampini S, Danaviah S, Valea D, Pillay S, Kania D, Meda N, Newell ML, Van de Perre P, Rouet F. Dried blood spot HIV-1 RNA quantification using open real-time systems in South Africa and Burkina Faso. *Journal of Acquired Immune Deficiency Syndromes* 2010;55:290–298.

Violari A, Cotton MF, Gibb DM, Babiker AG, Steyn J, Madhi SA, Jean-Philippe P, McIntyre JA. Early antiretroviral therapy and mortality among HIV-infected infants. *New England Journal of Medicine* 2008;359:2233–2244.

Waters L, Kambugu A, Tibenderana H, Meya D, John L, Mandalia S, Nabankema M, Namugga I, Quinn TC, Gazzard B, Reynolds SJ, Nelson M. Evaluation of filter paper transfer of whole-blood and plasma samples for quantifying HIV RNA in subjects on antiretroviral therapy in Uganda. *Journal of Acquired Immune Deficiency Syndromes* 2007;46:590–593.

WHO. *HIV Surveillance Report for Africa 2000*. Harare, Zimbabwe: WHO; 2001.

WHO. *Reconciling Antenatal Clinic-Based Surveillance and Population-Based Survey Estimates of HIV Prevalence in Sub-Saharan Africa*. Lusaka, Zambia: WHO; 2003.

WHO. *Antiretroviral Therapy of HIV Infection in Infants and Children* (2006 revision). Geneva, Switzerland: WHO; 2006. http://whqlibdoc.who.int/publications/2010/9789241599764_eng.pdf (accessed January 15, 2012).

WHO. *Antiretroviral Drugs for Treating Pregnant Women and Preventing HIV Infection in Infants: Recommendations for a Public Health Approach*. Geneva, Switzerland: WHO; 2010a.

WHO. *Antiretroviral Therapy for HIV Infection in Adults and Adolescents: Recommendations for a Public Health Approach*. 2010 revision. Geneva, Switzerland: WHO; 2010b.

WHO. *Screening Donated Blood for Transfusion-Transmissible Infections: Recommendations*. Geneva, Switzerland: WHO; 2010c. http://www.who.int/bloodsafety/ScreeningTTI.pdf (accessed March 19, 2012).

WHO. Global HIV/AIDS Response: Epidemic Update and Health Sector Progress Towards Universal Access Progress Report. http://whqlibdoc.who.int/publications/2011/9789241502986_eng.pdf (accessed January 15, 2012).

Yang C, McNulty A, Diallo K, Zhang J, Titanji B, Kassim S, Wadonda-Kabondo N, Aberle-Grasse J, Kibuka T, Ndumbe PM, Vedapuri S, Zhou Z, Chilima B, Nkengasong JN. Development and application of a broadly sensitive dried-blood-spot-based genotyping assay for global surveillance of HIV-1 drug resistance. *Journal of Clinical Microbiology* 2010;48:3158–3164.

Yen-Lieberman B, Brambilla D, Jackson B, Bremer J, Coombs R, Cronin M, Herman S, Katzenstein D, Leung S, Lin HJ, Palumbo P, Rasheed S, Todd J, Vahey M, Reichelderfer P. Evaluation of a quality assurance program for quantitation of human immunodeficiency virus type 1 RNA in plasma by the AIDS Clinical Trials Group virology laboratories. *Journal of Clinical Microbiology* 1996;34:2695–2701.

Yourno J, Conroy J. A novel polymerase chain reaction method for detection of human immunodeficiency virus in dried blood spots on filter paper. *Journal of Clinical Microbiology* 1992;30:2887–2892.

Zeh C, Weidle PJ, Nafisa L, Lwamba HM, Okonji J, Anyango E, Bondo P, Masaba R, Fowler MG, Nkengasong JN, Thigpen MC, Thomas T. HIV-1 drug resistance emergence among breastfeeding infants born to HIV-infected mothers during a single-arm trial of triple-antiretroviral prophylaxis for prevention of mother-to-child transmission: a secondary analysis. *PLoS Medicine* 2011;8:e1000430.

Zhang J, Kang D, Fu J, Sun X, Lin B, Bi Z, Nkengasong JN, Yang C. Surveillance of transmitted HIV type 1 drug resistance in newly diagnosed HIV type 1-infected patients in Shandong province, China. *AIDS Research and Human Retroviruses* 2010;26:99–103.

Ziemniak C, George-Agwu A, Moss WJ, Ray SC, Persaud D. A sensitive genotyping assay for detection of drug resistance mutations in reverse transcriptase of HIV-1 subtypes B and C in samples stored as dried blood spots or frozen RNA extracts. *Journal of Virological Methods* 2006;136:238–247.

8

USE OF DRIED BLOOD SPOT SAMPLES IN HCV-, HBV-, AND INFLUENZA-RELATED EPIDEMIOLOGICAL STUDIES

Harleen Gakhar and Mark Holodniy

8.1 HEPATITIS C VIRUS

8.1.1 Introduction

Viral hepatitis is one of the leading causes of chronic liver diseases. Hepatitis C virus (HCV) was first identified as a separate viral entity in the late 1980s. The HCV genome was successfully cloned in 1989 by Choo et al. (Choo et al., 1989). Since then, major strides in the understanding of its life cycle, disease pathogenesis, and treatment options have been made. HCV is a member of the family Flaviviridae, within a separate genus, *Hepacivirus.* HCV is a positive strand RNA virus of approximately 9.6 kb in length. Phylogenetic evaluation has identified at least six major genotypes (referred to as genotypes 1–6) around the world with as much as 35% genetic variation in between different genotypes. Less diverse types have been classified as subtypes. Although HCV genotypes 1, 2, and 3 appear to have worldwide distribution, their relative prevalence varies from one geographic area to another. Lack of proofreading by the HCV polymerase enzyme and high replication rate lead to frequent mutations in the genome resulting in a complex of genetic variants within individual isolates called quasispecies. Although, no differences in pathogenesis and transmissibility between different genotypes have been found, response rates to treatment vary between them. Genotype 1 is usually less responsive to treatment compared to genotypes 2 and 3.

8.1.2 Epidemiology

An estimated 130–170 million persons are infected with HCV worldwide (EASL International Consensus Conference on Hepatitis C, 1999; Shepard et al., 2005). Incidence of new infections has dramatically decreased since the introduction of blood screening measures in developed countries, but new cases still occur mainly as a result of injection drug use. Prevalence of HCV varies from 0.5% to 2% in developed countries like North America, Northern and Western Europe, and Australia. In Egypt, 6–28% of the population is infected with HCV (Abdel-Wahab et al., 1994). Injection therapy for schistosomiasis infection is thought to be responsible for this high rate of prevalence (Frank et al., 2000). Other countries with high prevalence rates include China, Japan, Pakistan, and other countries in Asia and Africa.

The parenteral route is the most common and efficient mode of transmission of HCV. Infections through blood transfusion rapidly decreased after blood screening for HCV started in early 1990s. However, transfusion is still a significant cause of new infections in the developing world where screening measures are not as robust. Maternal–fetal transmission does occur but is less efficient. Similarly, sexual transmission can be an important transmission route, particularly in men who have sex with men, where it has also been associated with HIV transmission. Lastly, occupational exposure by needle stick injury is another risk factor, although prevalence of HCV infection is similar in healthcare workers as in the general population.

8.1.3 Diagnosis

Diagnostic tests for HCV are divided into serological and nucleic acid tests (Pawlotsky, 2002, 2003). Serological tests are used for screening and are based on antibody detection.

Dried Blood Spots: Applications and Techniques, First Edition. Edited by Wenkui Li and Mike S. Lee.

Over the last few years, significant improvement has been made in the performance of these tests with a third generation of enzyme immunoassays (EIAs) using different recombinant proteins and with increasing sensitivity and specificity. The US FDA has approved two EIA assays for HCV diagnosis: Abbott HCV EIA 2.0 (Abbott Laboratories, Abbott Park, IL, USA) and ORTHO HCV Version 3.0 ELISA (Ortho-Clinical Diagnostics, Raritan, NJ, USA), as well as a chemiluminescence immunoassay (VITROS Anti-HCV assay, Ortho-Clinical Diagnostics, Raritan, NJ, USA). A positive EIA test is confirmed with an immunoblot assay and the only FDA-licensed supplemental test is the RIBA 3.0 (Chiron RIBA® HCV 3.0 SIA, Chiron Corp., Emeryville, CA, USA). This test is considered positive if at least two out of three different antibodies are detected and indeterminate if only one is detected. It is important to confirm a positive EIA test, because of occasional false-positive tests, especially if patients are coinfected with HIV or HTLV or have renal failure.

HCV nucleic acid tests can be qualitative or quantitative. Qualitative polymerase chain reaction (PCR) assays confirm viremia and are the most specific test for HCV diagnosis. They are also used for screening blood donors. The introduction of nucleic acid testing (NAT) for blood donor screening significantly reduced the incidence of post-transfusion HCV infection. Quantitation of serum HCV RNA (viral load) is important for treatment purposes because the rate of virologic response is an important treatment monitoring parameter and predictor of final outcome. HCV treatment guidelines suggest checking HCV viral loads at 4 and 12 weeks during a treatment regimen, at end of treatment, and 6 months after treatment has been completed to determine whether a sustained virologic response (SVR) has been achieved. HCV can be quantified using either target amplification techniques, which include PCR and transcription-mediated amplification (TMA), or signal amplification with the branched DNA (bDNA) assay. The bDNA VERSANT assay (Siemens) has a range of 615–7,690,000 IU/mL and real-time PCR assays (Abbott, Roche) have a range of detection of 10–100 million IU/mL.

HCV genotype determination is also important in treatment decisions and influences the specific regimens and duration of treatment. HCV genotype can be determined by either testing for type-specific antibodies with a competitive EIA (Murex HCV serotyping assay; Murex Diagnostics, Dartford, UK) or by molecular genotyping. Molecular genotyping can be performed by direct sequence analysis (TruGene HCV 5′NC genotyping kit, Siemens), by reverse hybridization using genotype-specific oligonucleotide probes (line probe assay, INNO-LiPA HCV, Innogenetics, Belgium) and Versant™ HCV Genotype 2.0 System (LiPA), or by restriction fragment length polymorphism (RFLP). No HCV genotyping assay has been approved by the US FDA.

Another pretreatment predictor of SVR is interleukin-28b (IL28b) genotype. Molecular-based assays are used to detect a single nucleotide polymorphism near the IL28b gene on chromosome 19. Patients with a CC genotype achieve SVR twice as frequently as patients with genotype CT or TT.

Since the approval of direct acting antivirals (DAAs) for HCV treatment such as HCV protease inhibitors (PIs), antiviral resistance to PIs has also been described. In PI registration studies, some patients had PI resistance even at baseline. Although there are no US FDA-approved assays for HCV PI resistance, several laboratories have developed RT-PCR and sequencing assays for the detection of mutations in the HCV NS3 gene that confer resistance to the currently approved PIs. PI resistance testing is selectively recommended in patients who have a virologic breakthrough on treatment or in patients who do not achieve an SVR. Another application of this test is to look for cross-resistance between PIs, as decisions to use another PI after treatment failure is controversial.

8.1.4 Use of Dried Blood for HCV Diagnosis and Treatment

Several studies have evaluated the use of dried blood spots (DBS) made from whole blood, in the diagnosis of HCV infection and some have also attempted to use them for HCV quantification and genotyping. Although the ideal sample for testing HCV antibodies and RNA is blood plasma or serum, in resource-limited settings, it can be inconvenient to collect, separate, store, and transport these specimens. Dried spots can also be made applying plasma (DPS) or serum (DSS) on filter paper. While no study has used DPS for HCV testing, Abe et al. utilized DSS to investigate its stability at room temperature. Although HCV RNA was detectable after 4 weeks at room temperature, a 10-fold reduction in viral load was found (Abe and Konomi, 1998). There are some studies in the HIV literature, which have compared DBS and DPS (Andreotti et al., 2010; Monleau et al., 2010). It is much more convenient to make DBS than DPS or DSS which would require separation of whole blood into plasma and serum. Table 8.1 summarizes the studies conducted to validate DBS for HCV testing.

The majority of studies using DBS have been performed in Europe. One of the first papers published in this field was by Parker et al. who developed an in-house EIA assay for the detection of HCV antibodies in DBS obtained from neonates in London and South Africa (Parker et al., 1997). HCV RIBA was used for confirmation. The in-house EIA test detected all positive and negative serum samples with 100% accuracy and had 100% sensitivity and specificity when DBS from the London cohort were used. Among DBS from 569 South African neonates, 4.9% were false positive by initial EIA screening. Retesting these samples using antigen-coated and -uncoated wells showed only 2 of 28 to be positive (confirmed by RIBA). The third-generation HCV EIA tests have >99%

TABLE 8.1 Summary of Studies Conducted to Validate DBS for HCV Testing

Study	Sample Size	Specimen Site	Filter Paper and Preparation	Assay Used	Temperature Stability or Duration	Sensitivity and Specificity/LOD
Parker et al. (1997)	448 HCV– newborns 569 South African neonates (25 with HIV and HTLV-1 Ab)	Newborn DBS	Guthrie card dried @ RT, stored at 4°C	In-house ELISA RIBA 3.0 immunoblot	ND	4.9% false + ELISA in South African RIBA sensitivity 99.8%
(Abe and Konomi, 1998)	8 HCV-positive and 4 HCV-negative patients	Serum samples	Whatman GF/C glass microfiber filter air dried, stored at RT	One-step RT-PCR	PCR + in all DSS stored at RT for 4 weeks (serial dilution of one frozen serum sample). DSS stored for 4 weeks @ RT—10-fold reduction in RNA yield in 6/8 samples	100% sensitivity and specificity
McCarron et al. (1999)	108 paired samples, HCV-infected group; 112 paired samples unexposed persons	Whole blood	Guthrie filter paper DBS stored at 4°C until tested	Serum EIA (Abbott); RIBA (Chiron) HCV RNA (Roche Amplicor) Monolisa HCV Ab (Sanofi Pasteur)	ND	Sensitivity 100% and specificity 87.5% if T/CO 0.99 Sensitivity 97.2% and specificity 100% if T/CO 1.99
Solmone et al. (2002)	124 pairs from 34 HCV + pts 24 pairs from 19 HCV– pts	EDTA whole blood by venipuncture	S&S card	In-house RT-PCR, TMA Genotype-INNO-LiPA (Siemens)	• 16 HCV RNA-replicate DBS stored at RT, tested at 2–4 weeks for 11 months—100% detection • 8 HCV genotype-replicate DBS stored at RT for 11 months—100% agreement	100% sensitivity and specificity (both tests) Limiting dilution analysis Two viremic patients with VL 604,000 and 807,000 IU/mL were used; 75% samples positive with 53 IU and 100% samples with 1328 IU.
Judd et al. (2003)	381 HCV negative 253 HCV positive	Capillary blood	Guthrie cards stored @ RT for <15 days and then at –25°C	HCV Ab- Ortho HCV 3.0 SAVe ELISA (serum), modified Ortho HCV—oral and DBS RIBA HCV 3.0 (Ortho)	ND	>99% sensitivity and 100% specificity when using multiple CO values
Croom et al. (2006)	75 anti-HCV positive, 108 anti-HCV negative	Whole blood	S&S card air dried at RT, stored at –20°C	Monolisa anti-HCV plus version EIA Murex anti-HCV EIA	ND	100% sensitivity and specificity (Δ for + DBS: 5.89, –DBS: –6.64) Genetic systems elution buffer lowest cross-reactivity

(*continued*)

TABLE 8.1 ***(Continued)***

Study	Sample Size	Specimen Site	Filter Paper and Preparation	Assay Used	Temperature Stability or Duration	Sensitivity and Specificity/LOD
De Crignis et al. (2010)	For validation HIV and HCV calibrated virus panels 25 paired samples for analysis • 13 coinfected • 4 HIV-1 • 3 HCV • 5 healthy donors	Whole blood	Whatman no 3 filter paper Dried at 37°C for 1 h	1. Extraction-QIAamp DNA Mini kit 2. Multiplex Sybr Green RT-PCR	ND	Sensitivity 94% Specificity 100% LOD—(using standard panels) HCV 2500 copies/mL HIV 400 copies/mL
Tuaillon et al. (2010)	Paired samples of 100 HCV Ab+ (62 HCV RNA+) and 100 HCV Ab– serum samples	Whole blood	Whatman 903 Dried for 18 h, stored at –20°C	1. HCV Ab titers—Ortho HCV 3.0 ELISA 2. INNO-LIA HCV Score for confirmation 3. HCV RNA extraction—Cobas Ampliprep (Roche) 4. HCV RNA quantification—Cobas Taqman HCV test (Roche) 5. Genotyping—RT-PCR and sequenced	1. 8 DBS at RT for 2–12 days HCV neg DBS—OD reached threshold 4/5 DBS after 6 days HCV pos DBS—No difference 2. Decrease in HCV RNA in DBS at 6 days at RT	EIA—99% sensitivity, 98% specificity PCR—97% concordance Correlation coefficient 0.94 LOD—15 IU/mL for serum samples DBS—477–25,000,000 IU/mL.
Santos et al. (2011)	• 100 pts 4 weeks after HCV treatment • 68 pts after 24 weeks 2039 DBS samples	Capillary blood	SS903 collection cards Dried for 2 h	In-house quantitative PCR	ND	Sensitivity 98% Specificity 94.3% 10/2039 DBS tested needed recollection (invalid results) LOD—DBS 1000 IU/mL (assay limit of 50 RNA IU/mL)
Bennett and Gunson (2012)	Mock DBS using HCV RNA international standard 80 HCV Ab-positive paired samples	Whole blood	Whatman Protein Saver TM903 Cards Dried at RT for 1 h, stored at 4°C	Extraction—Abbott M2000 In-house RT-PCR Real-Time TaqMan RT-PCR	One patient—2 DBS samples tested in duplicate. DBS kept at RT, 4°C, 20°C, –80°C at BL and multiple times for 1 year No significant difference in CT	100% sensitivity, 95.8% specificity; LOD—150—250 IU/mL Inter-assay—little variation (SD 0.83); intra-assay—little variation (SD 0.73)

ND, not done; SD, standard deviation; RT, room temperature; RT-PCR, reverse transcriptase PCR; DBS, dried blood spots; T/CO, optical density of eluate test/cutoff ratio; LOD, limit of detection; S&S, Schleicher and Schuell; CT, cycle threshold; TMA, transcription-mediated amplification.

specificity but they can have as much as a 35% false-positive rate if the prevalence of HCV infection is <10% in the population being studied. However, in the study conducted by Parker et al., the DBS negative samples from London had no false positives (Parker et al., 1997). The phenomenon of a relatively higher false-positive rate by EIA has been described before in tropical populations. The higher false-positive rate is thought to be due to some nonspecific binding of antigens. To examine the performance of HCV RIBA, 40 positive samples were tested and 6 were found to be indeterminate; and after doubling the volume of the DBS eluate, only one sample with low reactivity remained unconfirmed. The authors did not describe how the eluate volume was determined, thereby highlighting a common problem. All the assays using DBS need to be validated, due to lack of standardization.

Halfon et al. evaluated DBS for the detection of IL28B SNP DNA using real-time PCR (Halfon et al., 2012). This study also investigated multiple other sample sources like plasma, serum, and buccal epithelial cells. They found 100% concordance using all the above-mentioned sample collection methods. This is the only study which has evaluated DBS for this assay.

8.1.4.1 Sample Collection and Drying The majority of studies have used anticoagulated whole blood obtained from venipuncture or capillary blood from a finger stick and applied that blood on filter paper. Few studies have also made DBS from anti-HCV-positive frozen serum samples by mixing it with HCV-negative packed red blood cells (PRBC) or whole blood. Bennett et al. made mock DBS using NIBSC HCV RNA (international standard) in negative blood for the validation of the assay (Bennett and Gunson, 2012).

The volume of blood used per spot ranged from 50 to 80 μL. DBS should be completely dry before storage or transportation. Most studies have air-dried DBS at room temperature anywhere from 1 hour to overnight. Drying times usually depend on the humidity of the surrounding environment and should be evaluated before using DBS for clinical purposes.

Most DBS studies for HCV testing used Whatman 903 filter paper (previously called Schleicher & Schuell). Whatman filter paper is a highly absorbent cotton linter, and the most commonly used type in HCV studies is the Whatman 903 protein saver card. The sample collection area of the 903 protein saver card contains five half-inch circles. Each circle holds 75–80 μL of sample. The card fits into Whatman foil barrier Ziploc bags for storage. However, Abe et al. used GF/C glass microfiber filter (Whatman Ltd, UK) to show stability of HCV RNA on DSS (Abe and Konomi, 1998). This kind of filter is typically used for cell harvesting or filtering fine particulate matter. It is not clear why this filter paper was used in the study.

Although the technique of collecting DBS samples is relatively simple, there are many steps where potential error can compromise assay results. The field personnel should be acquainted with the correct way to make the spots. During the process of making DBS, if the drop of blood is not evenly distributed on the pre-drawn circle, the results could vary depending on where the filter paper is punched.

In a study done by Solmone, the question of risk of cross-contamination while preparing DBS was evaluated (Solmone et al., 2002). DBS were prepared in close proximity using 15 HCV-positive and 15 HCV-negative blood samples. RT-PCR was performed and no evidence of cross-contamination was seen. Although this is the only HCV study that has addressed this question, other studies have evaluated cross-contamination risk with HIV testing. Driver et al. in South Africa evaluated the risk of cross-contamination using automated and manual excision of DBS for HIV PCR testing using the Amplicor HIV-1 DNA version 1.5 assay and assessed three different cleaning protocols with manual and automated punch (Driver et al., 2007). They found one false positive out of 574 HIV-negative DBS using an automated punch, which may have been secondary to punching process. Overall, the risk of cross-contamination in the process of making DBS appears to be very low. A recent study showed a 3% risk of false positives, which was attributed to the tube caps used in PCR, and this risk was decreased by spinning the tubes prior to opening (Mitchell et al., 2010). This result was only found in the group who had very high HCV viral loads (5000 copies/punch) and there were no false positives in the group with lower viral load (500 copies/punch). None of the studies have found a significant risk of false positive secondary to sample preparation. Establishing a standardized protocol for DBS preparation will be helpful for people in resource-limited settings using DBS.

ViveST (ViveBio, Buford, GA, formally Sample Tanker) is an alternative sample collection system that has been studied extensively for HIV sample collection, and minimally for HCV. It consists of a sample collection matrix and desiccant within a cryovial tube. This sealed system reduces the risk of contamination. One study evaluated ViveST for collection and transportation of dried plasma in HIV and HCV coinfected samples (Lloyd et al., 2009). The advantage of ViveST over the standard filter paper DBS is the amount of plasma/whole blood that can be utilized to prepare the sample. DBS are limited to 50–100 μL of serum or whole blood, whereas ViveST matrix can retain 1 mL of plasma or serum, which can be eluted off in its entirety and thus one sample can be used for a variety of tests.

8.1.4.2 Storage Processing and storage are very critical elements to ensure optimum testing. There is no consensus regarding the appropriate temperature to store DBS and the length of time for which it can be stored at ambient temperature. Tuaillon et al. evaluated eight of their samples for 2–12 days and tested them periodically using HCV antibody EIA and RNA quantitative assays (Tuaillon et al., 2010).

The absorbance (optical density) of the negative samples increased after 6 days and almost reached the cutoff value. Also, decrease in HCV RNA levels was seen at day 6 of storage at room temperature as compared to –20°C. Abe et al. stored DSS at room temperature to check the stability of HCV RNA after 1, 2, 3, and 4 weeks and found that HCV RNA was detectable in all eight positive samples even after 4 weeks (Abe and Konomi, 1998). However, a 10-fold reduction in viral load using DSS was seen by serial dilution when one frozen serum sample (stored at –80°C) was compared to DSS stored at room temperature for 4 weeks. In contrast, Solmone et al. tested a replicate DBS set from 16 HCV RNA-positive patients stored at room temperature at intervals of 2–4 weeks over an 11-month period and demonstrated a 100% detection rate (Solmone et al., 2002). Only qualitative analysis was done in this study. In addition, eight DBS samples with known genotypes were tested after storage at room temperature for 11 months with 100% concordance for genotype detection.

Bennett et al. made a set of DBS from the blood sample of one known HCV RNA-positive patient and evaluated them at 21°C, 4°C, –20°C, and –80°C for viral load (Bennett and Gunson, 2012). The samples were tested in duplicate using two spots each at baseline and then at multiple time points for the next 12 months. Minimal variation was seen in the threshold cycle (C_T) values at all time points and all temperature conditions (standard deviation (SD) 0.49–0.64). To our knowledge, this is the only study demonstrating no HCV RNA degradation in 1-year duration at varying temperatures.

Although few studies have demonstrated the long-term stability of HCV antibody or RNA in DBS, several studies have assessed the stability of HIV RNA in DBS and have reported favorably on the durability of these specimens over a range of temperatures and storage conditions.

One of the major advantages of DBS over plasma or serum is its ability to be stored at room temperature. A common limitation of these studies is their small sample sizes. The available data do not allow us to come to a definitive conclusion about storage conditions. More research needs to be done to answer this question before DBS can be recommended as the preferred specimen collection and storage method. Larger HCV stability studies should be performed looking at different analytes, particularly HCV RNA, and at different temperatures to simulate various transport and storage conditions.

8.1.4.3 Elution Elution is the process of extracting the analyte from DBS. As the amount of whole blood applied to the dried spots is limited, this process becomes very crucial for accurate results. Various studies have employed different elution buffers. Croom et al. compared four elution buffers using a modified protocol of Monolisa EIA (80 μL of eluate used with 20 μL of diluent vs. 20 μL of serum with 80 μL of diluent) (Croom et al., 2006). Buffers compared were PBS/T, 0.3% skim milk powder in 10 mM tris pH 7.4 containing 150 mM sodium chloride, Monolisa EIA specimen diluent, and Genetic systems rLAV EIA elution buffer (BioRad, France). Although the elution capacity (measured by S/CO ratios) of all the buffers was equivalent, there was less nonspecific reactivity with Genetic Systems elution buffer (evident in negative samples). When this buffer was used to compare paired DBS and plasma samples of 75 HCV antibody-positive and 108 HCV antibody-negative samples, there was a significant difference between the S/CO ratio of the positive and negative samples, thereby reducing the chances of false negatives or positives. Some groups have employed in-house elution buffers without offering details. No studies have compared elution buffers for use in NAT for HCV.

8.1.4.4 Antibody Detection Assays A number of different EIA assays have been evaluated using DBS samples. Some studies have described test validation results by comparing plasma samples with DBS (paired samples), while others have utilized DBS as their specimen collection method to conduct epidemiological studies.

Parker et al. developed an in-house HCV antibody test and conducted a study using DBS samples from multiple ethnic groups and also included samples positive for HIV and HTLV-1 antibodies to assess specificity (Parker et al., 1997). HTLV-1-positive samples showed cross-reactivity (6/28) with HCV antibody, 5 of which were confirmed negative with HCV RIBA. Overall, it was a well-conducted study, but as with any in-house assay, the reproducibility is questionable and the comparison with different laboratories is not feasible.

Two studies used the standard Ortho HCV 3.0 SAVe ELISA (Ortho-Clinical Diagnostics, Raritan, NJ, USA) (Judd et al., 2003; Tuaillon et al., 2010). Tuaillon et al. made a slight modification by using 100 μL of eluate instead of 20 μL of serum plus 200 μL of specimen diluent. Two hundred DBS samples (100 HCV positive and 100 HCV negative) were tested using the manufacturer's (Ortho HCV) OD cutoff value. They reported 99% sensitivity and 98% specificity with one false positive (HIV-infected patient) and one false negative.

Judd et al. looked at oral samples and DBS as alternate sample collection methods for HCV diagnosis (Judd et al., 2003). For this study, Ortho HCV was used according to the standard protocol for the serum samples. For oral samples and DBS, modifications were made to the original assay, which included altering the cutoff value. It is unclear from the publication whether any modification was made in the reagents used. Although DBS showed 100% specificity and >99% sensitivity using the preliminary cutoff value derived from validation of the assay and using alternate cutoff values, performance of the assay with oral samples demonstrated 80% sensitivity with the preliminary cutoff value. By lowering the cutoff value, test sensitivity for the oral samples improved.

Toledo et al. conducted a study using Umelisa HCV test, which has not been validated, to assess the prevalence of HIV and HCV in Minas Gerais, Brazil (Toledo et al., 2005). They tested 4211 DBS for HCV and found 30 positive and 4 borderline results. HCV prevalence ranged from 0% to 2.86% (mean 0.71%, 95% CI of 0.46–0.97%) depending on the geographic region tested. Borderline test results were discarded, as they did not perform confirmatory tests. As stated earlier, it is important to consider the prevalence of the population as the sensitivity and specificity of any test vary by prevalence. McCarron used the HCV Monolisa ELISA test in Glasgow, Scotland, where the HCV seroprevalence among intravenous drug users (IDU) is as high as 77% (McCarron et al., 1999). Using this test on DBS samples with the standard cutoff, their results demonstrated 100% sensitivity and 87.5% specificity. By increasing the cutoff value, the sensitivity decreased to 97.2% and specificity increased to 100%. Judd et al. also demonstrated that altering the cutoff value could change sensitivity and specificity of a test (Judd et al., 2003). This result emphasizes the need for validating the assay prior to using DBS for reliable results.

The US FDA has not approved any commercial HCV EIA test for use with DBS. Overall, it seems that DBS can reliably be used for HCV serological testing although it would be helpful if the groups evaluating the use of DBS clearly explain their rationale for the modifications they apply. This explanation will be helpful for future research so that definitive protocols for DBS use can be established in the field beyond research.

8.1.4.5 HCV Viral Load Assay A limited number of studies have assessed DBS and HCV viral loads assays. Abe et al. was the first group that used RT-PCR for HCV detection on DSS and demonstrated stability of HCV RNA at room temperature (Abe and Konomi, 1998). A one-step PCR method was used to detect HCV RNA in patients with hepatocellular carcinoma (HCC). Solmone et al. compared an in-house RT-PCR assay and TMA for qualitative HCV RNA detection using replicate DBS pairs from 34 HCV-positive patients and detected all positive samples using both assays with no false positives (Solmone et al., 2002). However, contradicting their results, they reported >99% sensitivity and >95% specificity. A recent well-done study by Tuaillon et al. used the COBAS Taqman HCV test (Roche Diagnostics, Mannheim, Germany) with a limit of detection (LOD) of 15 IU/mL, to detect and quantify HCV RNA from 62 HCV-infected paired DBS and serum samples. HCV RNA was detected in 60 of 62 samples. The two false negatives had low viral loads of 178 and 331 IU/mL in the corresponding serum samples (Tuaillon et al., 2010). Five samples were detectable but not quantifiable because they were below the LOD. This study used regression analysis and found good correlation between the HCV RNA levels in serum and DBS ($R^2 = 0.94$, $P < 0.0001$). The mean difference between the viral load obtained from DBS and serum samples was –2.27 log 10 IU/mL ± 0.47. DBS samples with viral loads below 1000 IU/mL had lower sensitivity compared to serum samples. Sybr Green RT-PCR was evaluated in one study to detect HIV and HCV in the same sample (De Crignis et al., 2010). Sybr Green is a fluorescent nucleic acid binding dye used after PCR. It binds nonspecifically to all PCR amplicons. The test was validated using standard virus panels and the lower LOD was calculated for each virus alone and in combination. Twenty-five known patient samples were tested, which included 13 HIV coinfected samples. The PCR test detected 15/16 HCV samples and 12/17 HIV-positive samples. The false negatives were below the LOD.

Santos et al. used an in-house quantitative PCR assay on patient samples to monitor HCV RNA in a clinical trial (Santos et al., 2012). Their test had a lower LOD of 50 IU/mL with a range of detection from 50 to 10 million IU/mL. Paired plasma and DBS samples were collected from 100 patients undergoing HCV treatment with peg-interferon alfa-2b and ribavirin after 4 weeks of treatment. Samples from 68 of these patients were also collected after 24 weeks on treatment. DBS and plasma HCV viral loads were compared in these two sample sets. There were two false negatives (viral loads 456, 2614 IU/mL) and four false positives from patients tested after 4 weeks. No discordant results after 24 weeks of treatment were found, which could be because all the samples were divided into either HCV RNA <200 IU/mL (below the LOD) or >5000 IU/mL. However, in the other group, the HCV RNA levels were distributed in all ranges. The reported sensitivity of 98% is acceptable but specificity of 94.3% is rather low for PCR. The plasma samples of the four false positives were repeatedly negative and the authors explain this by the possibility of HCV being present in B-lymphocytes contained in whole blood and absent in plasma. The presence of intracellular HCV in peripheral blood mononuclear cells (PBMCs) has been described in several publications, and so is a plausible explanation for these discordant results, as whole blood spots could be expected to include PBMCs harboring HCV after successful treatment, whereas cell-free plasma or serum would not.

Bennett et al. used an in-house RT-PCR assay and Real-Time TaqMan RT-PCR assay in their study (Bennett and Gunson, 2012). LOD using serum samples for the in-house assay was 50 IU/mL. The expected LOD using DBS was calculated as 250 IU/mL, but when serial dilutions of mock DBS were tested, LOD was demonstrated to be between 150 and 250 IU/mL. Both assays were used on paired samples of 80 HCV antibody-positive patients; 100% sensitivity and 95.8% specificity were seen (one false positive using DBS). Intra-assay and inter-assay variability was also examined using a mock DBS with concentration of 1000 IU/mL and very little variation was seen (SD 0.83 for inter-assay and 0.73 for intra-assay).

Few papers have calculated the lower LOD by serial dilution of blood samples and applying it on dried spots to determine what dilution level is detectable. It is difficult to come to a reliable conclusion regarding the performance of DBS for HCV viral load assays, as none of these studies are comparable. All the studies have employed different PCR techniques and methods of calculating lower LODs. The sample size used to calculate the lower LOD is small. More studies are needed evaluating various PCR techniques using multiple replicates of DBS to produce reproducible and reliable results. Although overall there seems to be little doubt that DBS can be used for HCV viral detection, reliable quantification is still questionable. The above-mentioned studies lack consistency and are variable in their methods of DBS utilization. Thus, there is a need to standardize these assays for quality control issues and for comparison between different laboratories.

8.1.5 DBS Utility

Utilizing DBS samples in epidemiologic studies shows great promise especially in resource-limited settings. Several pilot studies are being conducted in the United Kingdom utilizing DBS to detect HCV-positive patients and DBS has been incorporated as a sample collection method in the Strategy against HCV in UK and Scotland. Hope et al. assessed the incidence of HCV in IDUs in Bristol, UK, by testing for HCV antibodies and HCV RNA using DBS (Hope and Hickman, 2011). Incidence was defined as patients with presence of HCV RNA but absence of antibody. It is difficult to measure the incidence of HCV as the majority of cases are asymptomatic and come to medical attention late in their clinical course. Although this is not the most effective way to calculate incidence their results were very similar to prior studies done in the same population (Judd et al., 2005; Maher et al., 2007). Craine and colleagues conducted an audit in a substance misuse service to see the impact of DBS introduction on HCV testing (Craine et al., 2009b). The option of testing by a finger prick was given to clients who did not want venipuncture. Samples were tested for HCV antibody using the Ortho HCV test that was used by Judd et al. in their DBS study. A sixfold increase was seen in HCV testing as compared to the previous year when only venipuncture was available. Craine et al. conducted another study in an IDU population in South Wales, UK, to study the incidence of HCV using DBS (Craine et al., 2009a). Patients were tested for HCV antibody at baseline and again at 1 year follow-up demonstrating that 17 of 286 had seroconverted resulting in an incidence rate of 5.9%.

Bravo and colleagues have used DBS to study the effect of supervised injection facilities (SIF) on young heroin drug users in Spain (Bravo et al., 2009). HCV was diagnosed using Monolisa anti-HCV plus version 2 (BioRad) on a group of 249 injectors recruited from the ITINERE cohort. SIF users were more likely to be HCV positive (43.2% vs. 19.5%). The same group has also conducted additional HCV prevalence and behavior studies in drug user populations using DBS (Vallejo et al., 2008; Brugal et al., 2009). Costa et al. conducted a study to look at prevalence of HCV and HIV infection in pregnant women in Central Brazil (Costa et al., 2009). Unlike HIV, HCV screening is not mandatory in prenatal screening. Blood samples were collected by finger prick onto S&S 903 filter paper from 28,561 pregnant females and allowed to dry for 4 hours before sending it to reference laboratory. The DETECT (ADALTIS Inc, Canada) assay was used for initial HCV antibody screening and Hepanostika HCV Ultra (Beijing United Biomedical Co. Ltd, China) for confirmatory testing. Samples were also tested for HCV RNA using AMPLICOR HCV test (Roche, USA). Confirmatory tests were done on whole blood samples. In the initial screen, 65 positive samples were found. However, there were 17 false positives (28.3%) after confirmatory tests were performed. It was not clear why the false-positive rate was so high. The true prevalence rate in this population was 0.15%. Finally, DBS have also been used to test HCV prevalence in prison populations (Hickman et al., 2008; Mahfoud et al., 2010).

8.1.6 Summary

Although, significant advancements have been made in recent years utilizing DBS for HCV diagnosis and management, many areas require more research. Comprehensive studies regarding various storage conditions and their impact on the various assays are of prime importance. HCV antibody and qualitative HCV RNA assays are likely to yield more reliable results with DBS than HCV RNA quantitative assays. Whether DPS or DSS offer better performance for HCV assays is not known. There are a limited number of studies, which have used DBS to measure HCV viral load for HCV management. Thus, it is very important to better understand the effect of DBS storage and transport conditions on HCV viral load. The question of how much loss of HCV viral load in a DBS clinical sample is acceptable for treatment purposes is a very difficult question. It is important to quantify that loss, if any, for various assays by conducting well-done studies using replicate samples and doing multiple test runs to assess multiple potential sources of variability.

8.2 HEPATITIS B VIRUS

8.2.1 Epidemiology

Hepatitis B virus (HBV) is one of the leading causes of chronic liver disease in the world. Approximately 350 million people are chronically infected with HBV worldwide

(Lavanchy 2004). HBV is a 3200 kb DNA virus with a partially double-stranded structure. HBV infection produces large quantities of subviral filamentous surface antigen (HBsAg), used for the detection of HBV infection. HBV is divided into eight genotypes (A–H) based on at least 8% genetic diversity between them. There are also subtypes recognized based on a minimum 4% diversity. Additional data support the observation that genotypic differences can be translated into differences in disease severity and treatment outcomes.

Prevalence of HBV varies geographically. According to the World Health Organization (WHO), the low prevalence rate (0.1–2%) countries include the United States, New Zealand, and parts of Western Europe. Intermediate prevalence (2–6%) areas include Central and South America, Eastern and Southern Europe, and Japan; and high prevalence (>7%) areas include Southeast Asia, China, Middle East except Israel, Haiti, Dominican Republic, and Africa (Te and Jensen, 2010).

Acute HBV infection can be asymptomatic or symptomatic or even rarely present with fulminant hepatitis. The rate of conversion to chronic infection is inversely proportional to the age of disease acquisition. In high prevalence countries where vertical transmission is common, more than 90% of the infants get chronic HBV infection. On the other hand, healthy immunocompetent adults are able to clear the infection and chances of progression to chronic HBV infection are less than 5%. Chronic HBV infection is also a common cause of HCC worldwide.

8.2.2 Transmission

HBV is transmitted parenterally via apparent or unapparent percutaneous and mucosal exposure to infected blood and other body fluids. It is much more infectious than HCV and HIV. Risk factors of infection include blood transfusion, sexual promiscuity, sharing or reusing syringes between injection drug users, and tattooing. In low prevalence areas, HBV is typically a disease of young adults who acquire the infection through risky behavior. In countries with high prevalence, the mode of transmission is mainly perinatal infection or horizontal transmission in early childhood. Other high-risk groups include healthcare workers and hemodialysis patients.

8.2.3 Diagnosis

The presence of HBsAg in serum is the hallmark of HBV infection. HBsAg is detected in blood by radioimmunoassay (RIA) and EIA. It is detectable in the blood about 2–10 weeks after exposure. Persistence of HBsAg for more than 6 months suggests chronic infection while seroconversion (loss) is a sign of recovery. Acute infection is diagnosed by the presence of HBV core IgM antibody (HBcAb). Chronic infection is diagnosed by the presence of HBsAg, HBc IgG Ab. Presence of HBeAg usually suggests a state of high viral replication and infectivity. On the other hand, its loss and development of HBeAb suggests low viral replication and low infectivity except with HBe mutants where the absence of HBeAg is the result of a mutation, which prevents its formation. It has in fact been associated with poorer prognosis and higher chances of progression to cirrhosis and HCC. PCR-based assays for serum HBV DNA (HBV viral load) are used for quantitative detection and therapeutic monitoring of HBV infection. Newer PCR assays are very sensitive and have a range of 30 to 1.1×10^8 IU/mL. It has also been suggested that the level of viremia correlates with the development of cirrhosis and HCC. The gold standard test for HBV genotyping is nucleic acid sequencing, but other alternatives include RFLP, INNO-LIPA, and serotyping (using antibodies against HBsAg).

One of the main treatment approaches for HBV is inhibition of the viral polymerase using nucleoside analogs that terminate viral DNA synthesis. Lamivudine is a widely used nucleoside analog for HBV treatment and lamivudine resistance is a major problem, especially when used as monotherapy. An amino acid substitution of the methionine residue of the conserved tyrosine (Y), methionine (M), aspartate (D), aspartate (D) (YMDD) motif of RNA-dependent DNA polymerase confers resistance to lamivudine. Assays used to detect this codon change include population sequencing, RFLP, line probe assays, and real-time PCR.

Although most patients who have been exposed to HBV and cleared the infection have evidence of HBcAb and HBsAb, other scenarios exist where patients demonstrate HBcAb alone or HBcAb and HBsAb and demonstrate low-level viremia. In almost all cases, presence of HBsAb alone is usually the result of HBV vaccination.

8.2.4 Dried Blood Spots for HBV Diagnosis and Management

The use of DBS samples for HBV infection diagnosis was first described by Farzadegan et al. in 1978 in Iran (Farzadegan et al., 1978). One drop of capillary whole blood from 10 healthy HBsAg-positive carriers was applied on Whatman 4 filter paper. Different assays including immunoelectroosmophoresis (IEOP) (serological test used for HBV in early 1970s), hemagglutination inhibition (HAI), reverse passive hemagglutination (RPHA by Abbott's Auscell), and RIA (Abbott's Ausria II) were used for comparison. Samples were allowed to stand at 4°C and 37°C for 1, 3, 7, 14, and 30 days to assess the stability of the samples. RIA assay was found to be the best method for detecting HBsAg because it tested positive on all the samples at baseline and even after storing at room temperature for 30 days. The authors mention that there was a reduction of RIA counts per minute at 25°C and 37°C and the best storage temperature was 4°C, although there is no mention that the samples were stored at that temperature.

Zoulek et al. conducted a study in 1985 and concluded that DBS should only be used to screen HBV carriers and not otherwise (Zoulek et al., 1985). Paired samples of 86 patients were tested for all HBV serological markers (RIA assay). All DBS samples were positive for HBcAb when the serum titer was >1:100, while 13/25 were positive when serum titer was lower. Similarly for HBsAb, DBS were positive in all 17 samples with serum titer >30 and it was negative in 1 of 4 samples with antibody titers between 10 and 30. The only sample with titer <10 was not positive using DBS. Three of 86 were positive for HBeAb and DBS was positive for all of them. Similarly, DBS was positive in all 24 samples positive for HBsAg. Almost 50% of the samples with HBcAb titer <1/100 were missed in this study. The authors do not comment on false positives from DBS obtained from negative serum samples. Similarly, Forbi et al. tested 300 paired samples for HBsAg using EIA (Shantest^{TM}-HBsAg, India) and the manufacturer's cutoff and found the test to be 78.6% sensitive and 88.6% specific (Forbi et al., 2010).

There has been significant advancement in the field of DBS testing since then and now we understand that rigorous testing and standardization is required to get accurate results. In the study done by Zoulek and Forbi, none of the steps were optimized for the use of DBS, so the results are difficult to interpret. Table 8.2 summarizes the studies done to validate HBV testing in DBS.

8.2.4.1 ***Specimen Collection*** Most HBV studies have also used Whatman filter paper (previously called Schleicher & Schuell). Gupta et al. used Whatman 3MM filter paper or a BA85 nitrocellulose membrane for amplification (Gupta et al., 1992). In order to make sure that PCR was not affected by solid support, they used DBS from one HBV-positive and one HBV-negative patient to perform PCR analysis and compared it with whole blood. Southern blot hybridization was used to confirm that the 140 bp product was HBV specific. No difference was seen in the efficacy with or without filter paper. The reason they used BA85 nitrocellulose membrane is not clearly defined in the study. Blood volumes ranging from 5 to 80 μL have been used in various studies to make DBS. Most of the studies have dried the spots at room temperature for about 2–4 hours before storing or using them further.

Studies have evaluated the effect of different humidity conditions on DBS and have demonstrated that there can be degradation of the analyte in high humidity conditions. This has been tested primarily in the context of HIV and newborn screening tests. However, there are no studies showing the importance of complete drying of the spots before storage or the effect of incomplete drying of the spots on HBV results.

8.2.4.2 ***Storage and Stability*** One of the key advantages of using DBS in resource-limited setting is the ability to transport them relatively easily as compared to a serum/plasma specimen which requires wet or dry ice to preserve analytes such as antibody, antigens, or nucleic acid if they are not analyzed within a few hours. A number of studies have evaluated different DBS storage conditions for HBV testing and whether storing the samples at room temperature has any adverse effect on the results. Lira et al. analyzed 10 DBS samples from their cohort of 47 patients after storing them at 4°C, 25°C, and 37°C for 7 days before storage at –20°C (Lira et al., 2009). Viral load and quality of the DNA extracted were not significantly different as compared to the gold standard in plasma stored at –70°C. Gupta et al. also found similar results and noted that samples were stable by PCR detection for up to 5 months at 37°C (Gupta et al., 1992). However, they concluded that it is preferable to keep the specimen at room temperature as there is less denaturation of proteins and, hence, less interference in gel analysis after amplification.

Jardi et al. also studied the stability of DBS samples stored at room temperature for 3 days and 1, 2, and 3 weeks (Jardi et al., 2004). They took samples with viral loads of 10^4, 10^5, and 10^7 copies/mL and tested them at varying intervals and reported no significant decrease in DNA levels and no degradation (using nested PCR on the amplification product). However, details about these results are not provided in the study.

Overall the existing data indicate that there is potential in the use of DBS for HBV DNA testing after storage at ambient temperature but additional work is needed to make definite conclusions. None of the studies are comparable and they all have limitations of small sample size and incomplete data.

Villa and colleagues included two patient groups in their study, one HBsAg and HBcAb positive and the second group HBsAg negative but HBcAb and HBsAb positive with known antibody titers (Villa et al., 1981). The samples were stored at room temperature and tested using RIA after 1, 7, 15, 30, 60, and 180 days. In the first group, HBsAg gave positive results after each time interval regardless of the volume used for elution (250 or 600 μL of saline). In the same group, HBcAb also gave positive results although they report some loss of titer as compared to serum samples after 30, 60, and 180 days (with both 150 and 600 μL of saline). HBeAg and HBeAb also showed a 20% decrease in activity after 15 days of storage in the same sample group. In the second group, when testing for HBsAb, the results were always positive when the serum Ab titer was ≥1/1000 but when the titer was ≤1/100, the results were borderline or sometimes negative after 15 days of storage at room temperature. Interestingly, when they stored eight samples from the second group at 4°C and –20°C, there was no difference in results. These inconsistent results appear to be a function of antigen or antibody titer

TABLE 8.2 Summary of Studies Conducted to Validate HBV Testing in DBS

Study	Sample Size	Specimen Site	Filter Paper and Preparation	Assays Used	Temperature Stability or Duration	Sensitivity and Specificity/LOD
Farzadegan et al. (1978)	10 HBsAg +	Capillary whole blood/ spot venipuncture	Whatman 4	RIA, IEOP, HAI and RPHA	Stored @ 1, 3, 7, 14, 30 days @ 4°C and 37°C. Marginal reduction in cpm @ 25°C and 37°C	100% sensitivity using RIA
Villa et al. (1981)	1. 12 HBsAg +, HBcAb + 2. 12 HBsAg–, HBsAb +, HBcAb +	Capillary blood	Filter paper (not described) air dried, stored at RT	RIA (Ausria 11, Abbott) Serial dilution to detect titer	Stored @ 1, 7, 15, 30, 60, 180 days. Ab titer >1/1000—stable at RT; titers <1/100—negative after 15 days of storage: temp did not affect results; slight loss of titer after 30, 60, 180 days	100% specific Sensitivity decreased when titer <1/10. Negative if antibody was detected on undiluted serum only
Zoulek et al. (1985)	86 persons	Venous blood	Filter paper stored at RT for 2 weeks, then –20°C	RIA (Abbott)	ND	HBcAb—All DBS + when serum Ab >1:100; 13/25 + in lower titers. HBsAb—serum titer >30–; all 17 +; titer 10–30—1/4 +; titer <10—1/1 sample negative. HBeAb/HBsAg 100% concordance HBeAg—1 false negative out of 86 samples
Gupta et al. (1992)	Validation using reconstituted HBsAg + serum: 60 HBsAg +	Capillary blood	Whatman 3MM; nitrocellulose membrane (BA85: Schleicher & Schuell).	In-house PCR	Stable 5 months @ 37°C	56/60 samples concordant; $10–10^4$ copies/5 μL DBS depending on PCR cycle#
Jardi et al. (2004)	82 HBsAg +	Capillary or whole blood	Whatman 903 protein saver, dried 2 h at RT; stored @ –20°C	QIAamp extraction Light cycler RT-PCR Genotype—RFLP on S gene Precore mutant—nested PCR Lamivudine resistance—INNO-LiPA HBV DR assay	HBV DNA @ 10^4, 10^5, and 10^7 copies/mL; RT and –20°C; tested in duplicate @ day 3, weeks 1, 2, 3 No effect on RT for 7 days Stable @ –20°C after 3 weeks	Correlation coefficient[a] 0.96 LOD—2000 DNA copies/mL Precore—90% concordance Lamivudine resistance 97% concordance

(continued)

TABLE 8.2 (*Continued*)

Study	Sample Size	Specimen Site	Filter Paper and Preparation	Assays Used	Temperature Stability or Duration	Sensitivity and Specificity/LOD
Mendy et al. (2005)	45 HBsAg+	Venous and capillary blood	Whatman grade BFC180	HBsAg—Determine (Abbott) AFP—CIE and RIA	ND	Sensitivity 96% and specificity 100%
Lira et al. (2009)	47 HBsAg+	Whole blood	Whatman 903 protein saver, air dried for 4 h @ RT, stored at –20°C	DNA extraction—QIAamp DNA microkit PCR—Cobas Amplicor v1.5 (Roche)	10 samples stored @ 4°C, 25°C, 37°C for 7 days prior to storing at –20°C. No temp effect on DNA quality or viral load	Sensitivity 100% Correlation coefficient[a] 0.93
Forbi et al. (2010)	300 paired samples	Venous blood	Whatman no. 3 air dried, stored at RT	Third-generation EIA—Shantest, India)	ND	DBS—sensitivity 78.6% and specificity of 88.6%.
Villar et al. (2011)	1. Elution parameters—10 paired samples 2. Cutoff value—155 paired samples 3. Stability—1 positive and 1 negative	Whole blood and paired samples tested	Whatman 903 protein saver, dried for 4 h @ RT	Modified EIA (ETI-AB-COREK PLUS, HBcAb; ETI-AB-AUK-3, HBsAb; ETI-MAK-4, HBsAg, Diasorin)	HBcAb—DBS detectable after 183 days but storage at –20°C resulted in lowest variation. HBsAg—false positive seen after 63 days at 22–25°C	Varied for different analytes using different cutoff values

RT, room temperature; RIA, radioimmunoassay; CIE, counterimmunoelectrophoresis; ND, not done; DBS, dried blood spots; LOD, limit of detection.
[a]Correlation of the viral load in plasma versus DBS samples, expressed as Pearson's correlation coefficient or R^2.

rather than storage temperature. The results provided in this study are purely descriptive and hence the conclusions are difficult to verify. More in line with other work done with HIV or HCV, Villar et al. tested replicates (number not specified) of one positive and one negative DBS sample and found that HBcAb was positive even after 183 days, but storage at –20°C resulted in the lowest variation of OD values (Villar et al., 2011). For HBsAg, there were false negatives after 63 days of storage at 22–25°C, as compared to 4–8°C and –20°C and also false-positive results at all temperatures except –20°C. For HBsAb, one false-negative result was seen at 183 days after storage at 22–25°C. The number of discordant results was not specified. Although some variation in OD was seen in all environmental conditions over time, variation was least at –20°C.

With this limited data and small sample size studies, it is difficult to draw a conclusion about the optimum storage conditions. Studies with bigger sample numbers, multiple replicates, and better statistical analysis are needed to verify these findings before a definitive storage condition for HBV testing using DBS can be specified.

8.2.4.3 Elution As the volume of blood used to make DBS is very small, it is imperative that the yield from the DBS be very good. Villa et al. evaluated different elution volumes for all serologic markers. Even after optimizing the elution volume, when HBV assays were used on a group of samples made from HBsAg-negative but HBcAb- and HBsAb-positive patients, positive results were obtained only when the antibody titer was >1/100. Some studies have used PBS for elution in varying volumes. Mendy et al. used 100 μL of PBS for testing HBsAg and alpha-fetoprotein (AFP) (Mendy et al., 2005). Zoulek et al. and Forbi et al. used 400 and 50 μL of PBS, respectively, as the elution buffer (Zoulek et al., 1985; Forbi et al., 2010). Jardi et al. incubated the DBS in 200 μL of water and kept it at 37°C overnight before centrifugation (Jardi et al., 2004). Results of the above studies are described below. Similarly Gupta et al. used 75 μL of water but they exposed the microcentrifuge tubes to radiation in microwave oven at full power for 10 minutes (Gupta et al., 1992). The hypothesis behind this was that radiation helps to amplify the signal about two- to threefold by denaturing the inhibitory factors in the mixture. Cheyrou et al. first used this method for HBV PCR testing in 1991 (Cheyrou et al., 1991). Mahfoud et al. used PBS/Tween and sodium azide for HBsAg and HBcAb detection with ELISA but did not compare paired samples to determine the sensitivity and specificity of the assay (Mahfoud et al., 2010). Villar et al. attempted to test all these parameters in a very detailed study (Villar et al., 2011). Parameters evaluated included type of elution buffer, volume of the buffer, and also the diameter of the DBS circle used to extract the material. Ten paired samples were tested and the results showed that PBS/BSA 0.5% buffer gave the lowest level of nonspecific reactivity. Other buffers tested were PBS alone, PBS/Tween 20, PBS/Tween 20/sodium azide, PBS/Tween 20/BSA 5%. The investigators used 700 μL of buffer and 6 mm filter paper for HBsAg and HBcAb whereas 300 μL of buffer with 12.5mm filter paper for HBsAb. No details are provided as to how these parameters were optimized. Different cutoff values for HBV serological markers were then evaluated using these numbers (described below).

Lira et al., while validating PCR for HBV viral load testing, diluted the DBS eluate to 1:100 in saline when the plasma viral load was $>10^6$ IU/mL and 1:1000 when viral load was $>10^7$ IU/mL (Lira et al., 2009). Two dried spots were used for elution when plasma viral load was less than $\log 10^4$ IU/mL. Because they used one DBS spot per patient, which is 50 μL of blood as compared to 100 μL of blood, a dilution factor of 2 was used. Using this scheme, the average viral load in DBS was 5.29 $\log_{10}$ with SD of 1.46 as compared to 5.48 with SD of 1.32 in plasma. Good linear correlation was seen (Pearson correlation coefficient of 0.93). Although these modifications helped optimize the assay, it is difficult to utilize them in field where the serum sample may be unavailable.

There are multiple important steps involved in eluting the analyte from DBS, which need to be standardized. Different groups have used different buffers, varying volumes of buffers, and used different protocols for elution without much overlap. More research is needed in this area with studies like the one by Villar et al. comparing various options for elution. The use of microwave by Gupta et al. is also very interesting and relatively easy to do, provided the equipment is available, and should be explored for DNA extraction in resource-limited settings.

8.2.4.4 HBV Serologic Assays Different serological assays have been used in various studies to evaluate DBS in diagnosing HBV infection. Villa et al. validated the RIA (Ausria II, Ausab, Corab, and anti e RIA kit from Abbott Laboratories) for DBS samples (Villa et al., 1981). Their sample size was small (24 patients and no mention of replicates) and they did not recalculate the cutoff value of the test, using the serum cutoff value. They did see loss of activity after 30, 60, and 180 days of storage at room temperature when compared to frozen serum for HBcAb, HBeAg, and HBeAb. Farzadegan et al. used RIA (Abbott's Ausria II methods) and showed some loss of activity at 25°C and 37°C as compared to 4°C. Tappin et al. conducted a study to determine HBcAb positivity in neonatal DBS samples to assess past and present infection in mothers (Tappin et al., 1998). They used a passive hemagglutination assay (CORE-CELL by Green Cross Laboratories) to detect HBcAb. The test was validated using four panels of samples of varying serological combinations. Cutoff value was established using results from one of the panels. None of the negative samples were reactive above a dilution of 1:2, so 1:4 was chosen as the cutoff and using this cutoff, the sensitivity was 93% and specificity was 100% in a group of samples comprising 14 HBsAg-/HBeAb- and HBcAb-positive patients; 15 HBsAg-negative, but HBsAb- and HBcAb-positive patients; and 15 HBV negative patients. Another panel of patient samples consisting of 56 HBV immune patients, who were HBcAb positive, demonstrated a sensitivity of 79% and specificity was maintained at 100%. The hemagglutination assay was then used for 14,044 Guthrie card samples to screen for evidence of HBV exposure. Although 75 samples were reactive, 66 samples had a titer of >1:4. The authors concluded that the prevalence of infection was 0.5% (66/14,044), which is similar to the prevalence of HBV exposure documented among blood donors in Scotland. Mendy et al. studied HBsAg detection using Determine HBsAg assay (Abbott Laboratories, Dartford, UK) on a cohort of 45 known HBsAg-positive patients (Mendy et al., 2005). Sensitivity obtained for DBS as compared to plasma was 96% (two false negatives) and specificity was 100%. This assay is a qualitative and visually read immunoassay, which can use 50 μL of whole blood directly on the test card.

Villar et al. used manual commercial EIAs (ETI-AB-COREK PLUS, Diasorin, Italy for HBcAb; ETI-AB-AUK-3, Diasorin for HBsAb; and ETI-MAK-4, Diasorin for HBsAg) for various HBV serological markers to compare the results between DBS and serum samples (Villar et al., 2011). Three different cutoff values were evaluated in the study: (1) manufacturer's recommendation; (2) 3 SDs above the mean DBS absorbance of HBV seronegative samples; and (3) area under the receiver operating characteristic (AUROC) curve analysis was calculated for DBS absorbance value. The sensitivity and specificity of all the serological markers varied with different cutoff values and demonstrated that a commercial assay is adaptable to DBS but may require modifications.

8.2.4.5 Viral Load Assay There are limited number of studies that have assessed NAT for HBV DNA (viral load) detection and quantification using DBS. Two studies (Jardi et al. and Lira et al.) compared DBS and serum and used the QIAamp kit for DNA extraction (Jardi et al., 2004; Lira et al., 2009). Quantitative PCR analysis was done and the HBV core gene was amplified. Seven duplicate serial dilutions were made using a standard virus panel containing 10^9 HBV DNA copies/mL and mixing it with whole blood negative for HBV. A linear relationship was observed between C_T values and DNA copy number in the range of 10^2–10^8 copies/mL for serum samples and 10^2–10^7 copies/mL for DBS samples. Significant correlation was seen by linear regression analysis when 82 paired patient samples were tested ($P <$ 0.0001). Using this analysis, the LOD of the DBS assay was 2000 HBV DNA copies/mL. Out of eight samples with serum DNA between 10^3 and 10^4 copies/mL, seven were detectable using DBS but none of the samples with DNA levels $<10^3$ copies/mL were detectable. They also calculated intra- and inter-assay CVs in two patients. The intra-assay CVs were 5.9% and 3.2% and inter-assay CVs were 8.3% and 6.3%. Jardi et al. also assessed HBV genotyping, precore mutants, and YMDD HBV polymerase gene motif in their study. HBV genotypes were analyzed by PCR and RFLP on the S gene sequence. Nested PCR was used to detect the presence of precore mutants at position 1896. In 20 chronically HBV-infected patients, matching results were found using DBS and serum samples for 18 patients. One patient was genotype A and one was genotype D by DBS sample but by serum analysis they were both a combination of A and D. Sample types were 100% concordant for the detection of precore mutants. A total of 15 patients were tested for YMDD mutation, and none were found at baseline in both sample types. However, after 1 year of lamivudine treatment, 14 patients had concordant serum and DBS results and 1 patient had a combination of YMDD, YVDD, and YIDD detected in the serum, whereas only YMDD in the DBS sample. Thus, overall, there was 97% concordance between the two sample types although the sample size was very small.

8.2.4.6 ***DBS and Alpha-fetoprotein*** AFP is a glycoprotein that is secreted by the liver and is used as a screening tool for HCC in patients with chronic HBV and HCV. There are some inherent limitations of this test, which include poor sensitivity and specificity but there is some evidence of survival benefit from utilizing this test for HCC surveillance. American Academy for the Study of Liver (AASLD) guidelines recommend HCC screening every 6–12 months using abdominal ultrasound (US) alone or both US and AFP. Tsao et al. used EIA to determine the AFP levels in a group of healthy and HCC patients (Tsao et al., 1986). The validation of the test was done using known standard levels of AFP to make DBS (25 μL of blood on each spot). The spot was incubated overnight in alkaline phosphatase conjugated to rabbit anti-AFP antibody in a tube, which had polystyrene beads coated with mouse monoclonal antibody to AFP. Results were measured colorimetrically and using this method a range of 9–900 μg/L of AFP could be measured. The authors report that the test could pick up values up to 3 mg/mL, but the data regarding number of samples tested and the degree of variation between blood and DBS samples were not described in the study. A group of 242 healthy donors and 60 HCC patients were tested using the above assay and results were compared to serum samples tested by RIA. There were no false positives and the sensitivity of the test was 72% by both methods. A more recent study comparing DBS and serum for AFP testing was performed by Mendy et al. using counterimmunoelectrophoresis (CIE) (qualitative assay) and RIA (quantitative assay) (Mendy et al., 2005). Sensitivity of 95% and specificity of 100% were seen with the qualitative CIE as the gold standard, but performance was significantly decreased with a quantitative RIA assay (sensitivity 71% and specificity of 75%). It is difficult to optimize AFP testing using DBS when the clinical utility is suboptimal using serum specimen. More research will be needed to optimize the assay and cutoff value to at least make it comparable to current standards.

8.2.4.7 ***Use of DBS in HBV Epidemiologic Studies*** Vallejo and group conducted a cross-sectional study to look at the prevalence and risk factors for HBV in an injection and non-injection heroin user population <30 years of age in three cities in Spain (Vallejo et al., 2008). DBS samples were collected from 949 people and tested for HBsAg (Monolisa anti-HBs Plus EIA, BioRad) and HBcAb (Monolisa anti-HBc Plus EIA, BioRad). There are no specific details provided regarding the assay. Prevalence of HBV was significantly higher in injection drug users (22.5%) as compared to non-injection heroin users (7.4%) and they strongly recommend reinforcing the need for HBV vaccination in this population. Mahfoud et al. used DBS to study the prevalence of HIV, HCV, and HBV in a prisoner population in Lebanon (Mahfoud et al., 2010). HBV prevalence in the general population in Lebanon is extremely low (<1%). About 250 prisoners were tested for HBsAg and HBcAb using modified ELISA (Monolisa HBsAg ULTRA and Monolisa anti-HBc PLUS, BioRad). HBV prevalence in this population was 2.4%. Most of these prisoners were from the same cell suggesting that there was probably an outbreak of HBV.

Komas et al. conducted a study to look at the prevalence of HBV in Bangui, Central African Republic, in a cohort of 801 healthy high school adolescents and young adults (Komas et al., 2010). DBS were obtained by finger prick and stored at −20°C. HBsAg, HBsAb, and HBcAb were tested (Abbott-Murex Biotech Ltd, Dartford, Kent, UK). The test was first validated using paired DBS and serum from 15 HBsAg-positive samples. The cutoff value was the same as recommended by the manufacturer. One false-positive sample was found using DBS, which was later confirmed negative with the Murex HBsAg Confirmatory test, Version 3. The prevalence in this population was found to be 42.3% and

15.5% were HBsAg positive. Dried blood spots can be a very valuable tool in large-scale HBV epidemiological studies to help target at-risk and difficult-to-reach populations. DBS are being used for these studies in many places, but the results may not be fully reliable unless assays have been optimized for use with DBS.

8.2.5 Summary

As with HCV, HBV antibody, antigen, and qualitative HBV DNA assays are likely to yield more reliable results with DBS than HBV RNA quantitative assays. Whether DPS or DSS offer better performance for HBV assays is not known. There are a limited number of studies, which have used DBS to measure HBV viral load, genotype, and antiviral resistance for HBV management.

8.3 INFLUENZA

8.3.1 Introduction

Influenza virus affects millions of people yearly and is a major cause of morbidity and mortality. According to the WHO, there are about 3–5 million cases of severe influenza illness worldwide and about 250,000–500,000 deaths annually. Most cases of influenza are undiagnosed, as it is a relatively self-limiting respiratory illness in healthy adults and many times is indistinguishable from other respiratory viruses. In addition, influenza is also a major burden on a nation's economy due to absenteeism and productivity losses during epidemics.

Influenza virus is a single-stranded RNA virus belonging to the Orthomyxoviridae family. It includes influenza virus A, B, and C. Influenza types A and B cause most human diseases. There is only one influenza B subtype, but influenza A can be classified into numerous subtypes depending on the combination of their hemagglutinin (HA) and neuraminidase (NA) proteins. These proteins are on the surface of the virus and help in the entry and exit of viral particles from the respiratory epithelial cell.

Influenza exhibits a seasonal pattern and peaks between the months of December and March. Most of the epidemic and pandemics in history have been caused by influenza A. It is spread by direct person-to-person contact and also via contaminated surfaces. The virus is also transmitted by droplet mode. Historically, the worst influenza pandemic was in 1918 that resulted in millions of deaths. Most recently, the pandemic in 2009 was attributed to an H1N1 influenza strain of swine origin.

8.3.2 Clinical Features

Influenza typically causes an acute febrile illness associated with rhinorrhea, sore throat, and constitutional symptoms (malaise, fatigue, and myalgias). Influenza is usually self-limiting in an otherwise healthy patient with complete recovery within 7–10 days. Certain groups of people are more vulnerable to complications including infants, the elderly, HIV-infected patients, pregnant women, and other immunocompromised patients. Primary viral pneumonia and secondary bacterial pneumonia are the two most common complications of influenza. Other less common but potentially devastating complications include encephalitis, rhabdomyolysis, and pericarditis. A vaccine is available and is administered annually to healthcare workers and the public. Antiviral medications such as oseltamivir can be used to decrease symptoms and possibly spread of infection.

8.3.3 Diagnosis

Diagnosis of influenza infection is dependent on adequate specimen collection, timing of specimen collection, and the assay used. Nasopharyngeal (NP) swab samples are commonly tested, but oral and other respiratory secretion samples can be used. The rapid diagnostic assays have a turnaround time of 1–2 hours and are based on detecting viral antigen in the specimen. There are numerous rapid diagnostic assays that have been approved by the US FDA. However, these assays have limited sensitivity, which can vary anywhere from 40% to 80%. These assays have to be interpreted with caution because of the risk of getting false-positive results in low prevalence season and false-negative results in high prevalence season. Direct and indirect immunofluorescence assays are very operator dependent. They also suffer from low sensitivity and are unable to subtype influenza A viruses. Viral culture used to be the gold standard for influenza diagnosis before PCR. However, it does not have perfect sensitivity and the turnaround time is longer. Rapid shell vial cultures have reduced this time and can yield positive results in about 48 hours. RT-PCR has now replaced viral culture as the gold standard. It has high sensitivity and can yield positive results after many days of symptom initiation. The sample should ideally have a high viral load for the assay to be highly sensitive and be collected soon after becoming symptomatic. NP swab and bronchoalveolar lavage (BAL) samples have much better yield than throat swab. FDA-approved PCR assays are now available and are being increasingly used in clinical laboratories.

8.3.4 Dry Respiratory Sample

According to the Centers for Disease Control and Prevention (CDC) recommendations, respiratory samples can be collected from NP, nasal wash, BAL, or throat swab and need to be put in viral transport medium (VTM) where they can be stored at 4°C for not more than 4 days and later need to be frozen. Transportation of these specimens requires dry ice, which is not always practical in all settings. Alternatives to this collection method, which would require less infrastructure, but at the same time be efficient in delivering the sample, are required.

Similar to utilizing DBS for other viruses, some work has been done to evaluate dry respiratory samples for diagnosing influenza. There have been a few studies that have used FTA paper for avian respiratory viruses (avian influenza, infectious bronchitis virus, hemorrhagic enteritis virus, infectious bursal disease virus) and also mycoplasma.

There is increasing interest in developing relatively easy and cheap alternatives for diagnosing avian influenza since the emergence of highly pathogenic avian influenza virus (HPAIV) (H5N1) in South East Asia in the 1990s. The natural reservoir of AIV is wild aquatic birds and AIV can cause severe disease in domestic poultry. AIV usually does not infect humans but sporadic cases do occur. Researchers are interested in understanding the ecology of these viruses so that outbreaks of AIV can be predicted in advance.

Abdelwhab and group compared FTA paper with cellulose filter paper for infectivity and RT-PCR after storing these samples at room temperature for 5 months (Abdelwhab et al., 2011). They used HPAIV (H5N1) and low pathogenic avian influenza virus (LPAIV) H6N2 in the study. The eluate from the card was inoculated into the allantoic sac of embryonated chicken eggs and the AAF obtained after 5 days of incubation was used for hemagglutination and RT-PCR. Virus spotted on FTA cards was not infective when tested as early as 1 hour after sampling. On the other hand, virus on cellulose filter paper retained its infectivity for at least 3 days. All the tests were run in triplicate. Stability and sensitivity were evaluated by comparing FTA samples with direct examination of swab fluid. RT-PCR from FTA cards was positive even after 5 months of storage at room temperature although sensitivity was slightly lower. The 3 positive samples out of 20 which were negative using FTA cards but positive with direct swab testing had low C_T value and hence most likely had low viral loads.

Nobuto filter paper has been used to detect AIV antibody in blood (Dusek et al., 2011). This is a cellulose filter paper, which is used to collect whole blood or serum and was first used by Dr Kenzo Nobuto for diagnosis of toxoplasmosis. It can usually absorb about 100 μL of blood. The study used blocking ELISA to detect AIV antibodies at different storage conditions. Blood was collected before and 15 days after inoculation of AIV in a mallard. Serum stored at –30°C was compared with Nobuto paper samples stored at room temperature (days 38 and 99) and stored at –30°C (days 9, 38, and 99). All 24 positive samples were identified in serum and from Nobuto paper stored at room temperature for 38 days after collection. One sample in all other groups was false negative. No false positives were seen.

Kraus et al. have developed a protocol of using FTA cards for sample collection and avian influenza RNA extraction from the cloaca of ducks (Kraus et al., 2011). The authors suggested people who are not well trained in this field could easily use this method for AIV surveillance in remote areas.

Our group evaluated ViveST, a novel dried-sample collection system based on a similar principle as FTA filter paper, in order to assess its utility for influenza RNA testing (Winters et al., 2011). This is the only study done to date that evaluates dry specimen collection for influenza testing in humans. We collected samples from patients (Dacron swabs from oropharynx or nasopharynx or sputum samples) and placed them in VTM. These samples were stored at –80°C until use. Samples containing VTM were then applied to the ViveST matrix and allowed to dry for 16 hours at room temperature. Influenza RNA was eluted with a PBS buffer and extracted from the eluate using MagMax Viral RNA isolation kit. Presence of influenza was confirmed using the CDC RT-PCR assay for 2009 pandemic influenza A.

We also assessed quantitative recovery from the dried samples by serially diluting 12 samples and comparing them with frozen diluted samples. These samples were tested at 1 and 3 weeks to test their stability. Quantitative measurement was performed by calculating the mean C_T. We found that the mean threshold cycles were slightly more for the sample tanker (ST) samples as compared to the frozen counterparts but not significantly different (median difference of 2.75 cycles). C_T was not significantly different among the 1- and 3-week samples as well. In these diluted samples, there were 9 of 48 which gave discordant results and the median C_T of these were 35.5 indicating that the viral burden in these samples was very low. Thirty-six samples were tested for influenza A using the CDC assay to see if the slight loss of RNA in the dry specimen collection method influences the overall recovery rate. We found 100% concordance with the frozen samples.

Moore at al. conducted a study on respiratory viruses in Wales, UK, comparing dry cotton-tipped swabs with and without VTM and also compared nucleic acid sequence-based amplification (NASBA) versus immunofluorescence and traditional cell culture technique (Moore et al., 2008). They tested the specimen for four different respiratory viruses including influenza A and B viruses and found that the detection rate was the same with both methods of specimen collection. The detection rate of influenza from dry cotton-tipped swabs stored for 5 days at room temperature was higher than rates from paired samples collected in VTM and assayed by cell culture and immunofluorescence.

8.3.5 Summary

Dried respiratory samples for diagnosing influenza have great potential as a collection method in resource-limited settings. At this time there is minimal data to support its widespread use. Moore et al. have stated that dried respiratory swabs are routinely used in Wales for community influenza surveillance. More studies are required to evaluate and compare various filter papers or swabs and detection tests (antigen tests vs. influenza nucleic acid) that are best supported by dried samples, and more clinical studies are needed to confirm the above findings.

REFERENCES

Abdel-Wahab MF, Zakaria S, Kamel M, Abdel-Khaliq MK, Mabrouk MA, Salama H, Esmat G, Thomas DL, Strickland GT. High seroprevalence of hepatitis C infection among risk groups in Egypt. *American Journal of Tropical Medicine and Hygiene* 1994;51(5):563–567.

Abdelwhab EM, Lüschow D, Harder TC, Hafez HM. The use of FTA(R) filter papers for diagnosis of avian influenza virus. *Journal of Virological Methods* 2011;174(1–2):120–122.

Abe K, Konomi N. Hepatitis C virus RNA in dried serum spotted onto filter paper is stable at room temperature. *Journal of Clinical Microbiology* 1998;36(10):3070–3072.

Andreotti M, Pirillo M, Guidotti G, Ceffa S, Paturzo G, Germano P, Luhanga R, Chimwaza D, Mancini MG, Marazzi MC, Vella S, Palombi L, Giuliano M. Correlation between HIV-1 viral load quantification in plasma, dried blood spots, and dried plasma spots using the Roche COBAS Taqman assay. *Journal of Clinical Virology* 2010;47(1):4–7.

Bennett S, Gunson RN, McAllister GE, Hutchinson SJ, Goldberg DJ, Cameron SO, Carman WF. Detection of hepatitis C virus RNA in dried blood spots. *Journal of Clinical Virology* 2012;54(2):106–9.

Bravo MJ, Royuela L, De la Fuente L, Brugal MT, Barrio G, Domingo-Salvany A; Itínere Project Group. Use of supervised injection facilities and injection risk behaviours among young drug injectors. *Addiction* 2009;104(4):614–619.

Brugal MT, Pulido J, Toro C, de la Fuente L, Bravo MJ, Ballesta R, Soriano V, Barrio G, Vallejo F, Domingo-Salvany A, Castellano Y; Project Itinere Group. Injecting, sexual risk behaviors and HIV infection in young cocaine and heroin users in Spain. *European Addiction Research* 2009;15(3):171–178.

Cheyrou A, Guyomarc'h C, Jasserand P, Blouin P. Improved detection of HBV DNA by PCR after microwave treatment of serum. *Nucleic Acids Research* 1991;19(14):4006.

Choo QL, Kuo G, Weiner AJ, Overby LR, Bradley DW, Houghton M. Isolation of a cDNA clone derived from a blood-borne non-A, non-B viral hepatitis genome. *Science* 1989;244(4902):359–362.

Costa ZB, Machado GC, Avelino MM, Gomes Filho C, Macedo Filho JV, Minuzzi AL, Turchi MD, Stefani MM, de Souza WV, Martelli CM. Prevalence and risk factors for Hepatitis C and HIV-1 infections among pregnant women in Central Brazil. *BMC Infectious Diseases* 2009;9:116.

Craine N, Hickman M, Parry JV, Smith J, Walker AM, Russell D, Nix B, May M, McDonald T, Lyons M. Incidence of hepatitis C in drug injectors: the role of homelessness, opiate substitution treatment, equipment sharing, and community size. *Epidemiology and Infection* 2009a;137(9):1255–1265.

Craine N, Parry J, O'Toole J, D'Arcy S, Lyons M. Improving blood-borne viral diagnosis; clinical audit of the uptake of dried blood spot testing offered by a substance misuse service. *Journal of Viral Hepatitis* 2009b;16(3):219–222.

Croom HA, Richards KM, Best SJ, Francis BH, Johnson EI, Dax EM, Wilson KM. Commercial enzyme immunoassay adapted for the detection of antibodies to hepatitis C virus in dried blood spots. *Journal of Clinical Virology* 2006;36(1):68–71.

De Crignis E, Re MC, Cimatti L, Zecchi L, Gibellini D. HIV-1 and HCV detection in dried blood spots by SYBR Green multiplex real-time RT-PCR. *Journal of Virological Methods* 2010;165(1):51–56.

Driver GA, Patton JC, Moloi J, Stevens WS, Sherman GG. Low risk of contamination with automated and manual excision of dried blood spots for HIV DNA PCR testing in the routine laboratory. *Journal of Virological Methods* 2007;146(1–2):397–400.

Dusek RJ, Hall JS, Nashold SW, TeSlaa JL, Ip HS. Evaluation of Nobuto filter paper strips for the detection of avian influenza virus antibody in waterfowl. *Avian Diseases* 2011;55(4):674–676.

EASL International Consensus Conference on Hepatitis C. Paris, 26–28, February 1999, Consensus Statement. European Association for the Study of the Liver. *Journal of Hepatology* 30(5):956–961.

Farzadegan H, Noori KH, Ala F. Detection of hepatitis-B surface antigen in blood and blood products dried on filter paper. *Lancet* 1978,1(8060):362–363.

Forbi JC, Obagu JO, Gyar SD, Pam CR, Pennap GR, Agwale SM. Application of dried blood spot in the sero-diagnosis of hepatitis B infection (HBV) in an HBV hyper-endemic nation. *Annals of African Medicine* 2010;9(1):44–45.

Frank C, Mohamed MK, Strickland GT, Lavanchy D, Arthur RR, Magder LS, El Khoby T, Abdel-Wahab Y, Aly Ohn ES, Anwar W, Sallam I. The role of parenteral antischistosomal therapy in the spread of hepatitis C virus in Egypt. *Lancet* 2000;355(9207):887–891.

Gupta BP, Jayasuryan N, Jameel S. Direct detection of hepatitis B virus from dried blood spots by polymerase chain reaction amplification. *Journal of Clinical Microbiology* 1992;30(8):1913–1916.

Halfon P, Ouzan D, Khiri H, Pénaranda G, Castellani P, Oulès V, Kahloun A, Amrani N, Fanteria L, Martineau A, Naldi L, Bourlière M. Detection of IL28B SNP DNA from buccal epithelial cells, small amounts of serum, and dried blood spots. *PLoS One* 2012;7(3):e33000.

Hickman M, McDonald T, Judd A, Nichols T, Hope V, Skidmore S, Parry JV. Increasing the uptake of hepatitis C virus testing among injecting drug users in specialist drug treatment and prison settings by using dried blood spots for diagnostic testing: a cluster randomized controlled trial. *Journal of Viral Hepatitis* 2008;15(4):250–254.

Hope VD, Hickman M,Ngui SL, Jones S, Telfer M, Bizzarri M, Ncube F, Parry JV. Measuring the incidence, prevalence and genetic relatedness of hepatitis C infections among a community recruited sample of injecting drug users, using dried blood spots. *Journal of Viral Hepatitis* 2011;18(4):262–270.

Jardi R, Rodriguez-Frias F, Buti M, Schaper M, Valdes A, Martinez M, Esteban R, Guardia J. Usefulness of dried blood samples for quantification and molecular characterization of HBV-DNA. *Hepatology* 2004;40(1):133–139.

Judd A, Parry J, Hickman M, McDonald T, Jordan L, Lewis K, Contreras M, Dusheiko G, Foster G, Gill N, Kemp K, Main J, Murray-Lyon I, Nelson M. Evaluation of a modified commercial assay in detecting antibody to hepatitis C virus in oral fluids and dried blood spots. *Journal of Medical Virology* 2003;71(1):49–55.

Judd A, Hickman M, Jones S, McDonald T, Parry JV, Stimson GV, Hall AJ. Incidence of hepatitis C virus and HIV among new injecting drug users in London: prospective cohort study. *British Medical Journal* 2005;330(7481):24–25.

Komas NP, Baï-Sepou S, Manirakiza A, Léal J, Béré A, Le Faou A. The prevalence of hepatitis B virus markers in a cohort of students in Bangui, Central African Republic. *BMC Infectious Diseases* 2010;10:226.

Kraus RH, van Hooft P, Waldenström J, Latorre-Margalef N, Ydenberg RC, Prins HH. Avian influenza surveillance with FTA cards: field methods, biosafety, and transportation issues solved. *Journal of Visualized Experiments* 2011;August 2(54):2832.

Lavanchy D. Hepatitis B virus epidemiology, disease burden, treatment, and current and emerging prevention and control measures. *Journal of Viral Hepatitis* 2004;11(2):97–107.

Lira R, Maldonado-Rodriguez A, Rojas-Montes O, Ruiz-Tachiquin M, Torres-Ibarra R, Cano-Dominguez C, Valdez-Salazar H, Gomez-Delgado A, Muñoz O, Alvarez-Muñoz MT. Use of dried blood samples for monitoring hepatitis B virus infection. *Virology Journal* 2009;6:153.

Lloyd RM Jr, Burns DA, Huong JT, Mathis RL, Winters MA, Tanner M, De La Rosa A, Yen-Lieberman B, Armstrong W, Taege A, McClernon DR, Wetshtein JL, Friedrich BM, Ferguson MR, O'Brien W, Feorino PM, Holodniy M. Dried-plasma transport using a novel matrix and collection system for human immunodeficiency virus and hepatitis C virus virologic testing. *Journal of Clinical Microbiology* 2009;47(5):1491–1496.

Maher L, Li J, Jalaludin B, Chant KG, Kaldor JM. High hepatitis C incidence in new injecting drug users: a policy failure? *Australian and New Zealand Journal of Public Health* 2007;31(1):30–35.

Mahfoud Z, Kassak K, Kreidieh K, Shamra S, Ramia S. Prevalence of antibodies to human immunodeficiency virus (HIV), hepatitis B and hepatitis C and risk factors in prisoners in Lebanon. *Journal of Infection in Developing Countries* 2010;4(3):144–149.

McCarron B, Fox R, Wilson K, Cameron S, McMenamin J, McGregor G, Pithie A, Goldberg D. Hepatitis C antibody detection in dried blood spots. *Journal of Viral Hepatitis* 1999;6(6):453–456.

Mendy M, Kirk GD, van der Sande M, Jeng-Barry A, Lesi OA, Hainaut P, Sam O, McConkey S, Whittle H. Hepatitis B surface antigenaemia and alpha-foetoprotein detection from dried blood spots: applications to field-based studies and to clinical care in hepatitis B virus endemic areas. *Journal of Viral Hepatitis* 2005;12(6):642–647.

Mitchell C, Kraft K, Peterson D, Frenkel L. Cross-contamination during processing of dried blood spots used for rapid diagnosis of HIV-1 infection of infants is rare and avoidable. *Journal of Virological Methods* 2010;163(2):489–491.

Monleau M, Butel C, Delaporte E, Boillot F, Peeters M. Effect of storage conditions of dried plasma and blood spots on HIV-1 RNA quantification and PCR amplification for drug resistance genotyping. *Journal of Antimicrobial Chemotherapy* 2010;65(8):1562–1566.

Moore C, Corden S, Sinha J, Jones R. Dry cotton or flocked respiratory swabs as a simple collection technique for the molecular detection of respiratory viruses using real-time NASBA. *Journal of Virological Methods* 2008;153(2):84–89.

Parker SP, Cubitt WD, Ades AE. A method for the detection and confirmation of antibodies to hepatitis C virus in dried blood spots. *Journal of Virological Methods* 1997;68(2):199–205.

Pawlotsky JM. Use and interpretation of virological tests for hepatitis C. *Hepatology* 2002;36(5 Suppl 1):S65–S73.

Pawlotsky JM. Use and interpretation of hepatitis C virus diagnostic assays. *Clinics in Liver Disease* 2003;7(1):127–137.

Santos C, Reis A, Dos Santos CV, Damas C, Silva MH, Viana MV, Ferraz ML, Carnauba D, El-Far F, Serra F, Diaz RS. The use of real-time PCR to detect hepatitis C virus RNA in dried blood spots from Brazilian patients infected chronically. *Journal of Virological Methods* 2012;179(1):17–20

Shepard CW, Finelli L, Alter MJ. Global epidemiology of hepatitis C virus infection. *Lancet Infectious Diseases* 2005;5(9):558–567.

Solmone M, Girardi E, Costa F, Pucillo L, Ippolito G, Capobianchi MR. Simple and reliable method for detection and genotyping of hepatitis C virus RNA in dried blood spots stored at room temperature. *Journal of Clinical Microbiology* 2002;40(9):3512–3514.

Tappin DM, Greer K, Cameron S, Kennedy R, Brown AJ, Girdwood RW. Maternal antibody to hepatitis B core antigen detected in dried neonatal blood spot samples. *Epidemiology and Infection* 1998;121(2):387–390.

Te HS, Jensen DM. Epidemiology of hepatitis B and C viruses: a global overview. *Clinical of Liver Diseases* 2010;14(1):1–21, vii.

Toledo AC Jr, Januário JN, Rezende RM, Siqueira AL, Mello BF, Fialho EL, Ribeiro RA, Silva HL, Pires EC, Simões TC, Greco DB. Dried blood spots as a practical and inexpensive source for human immunodeficiency virus and hepatitis C virus surveillance. *Memorias do Instituto Oswaldo Cruz* 2005;100(4):365–370.

Tsao D, Hsiao KJ, Wu JC, Chou CK, Lee SD. Two-site enzyme immunoassay for alpha-fetoprotein in dried-blood samples collected on filter paper. *Clinical Chemistry* 1986;32(11):2079–2082.

Tuaillon E, Mondain AM, Meroueh F, Ottomani L, Picot MC, Nagot N, Van de Perre P, Ducos J. Dried blood spot for hepatitis C virus serology and molecular testing. *Hepatology* 2010;51(3):752–758.

Vallejo F, Toro C, de la Fuente L, Brugal MT, Soriano V, Silva TC, Bravo MJ, Ballesta R, Barrio G. Itinere Project Group. Prevalence of and risk factors for hepatitis B virus infection among street-recruited young injection and non-injection heroin users in Barcelona, Madrid and Seville. *European Addiction Research* 2008;14(3):116–124.

Villa E, Cartolari R, Bellentani S, Rivasi P, Casolo G, Manenti F. Hepatitis B virus markers on dried blood spots. A new tool for epidemiological research. *Journal of Clinical Pathology* 1981;34(7):809–812.

Villar LM, de Oliveira JC, Cruz HM, Yoshida CF, Lampe E, Lewis-Ximenez LL. Assessment of dried blood spot samples as a simple method for detection of hepatitis B virus markers. *Journal of Medical Virology* 2011;83(9):1522–1529.

Winters M, Lloyd R Jr, Shahidi A, Brown S, Holodniy M. Use of dried clinical samples for storing and detecting influenza RNA. *Influenza and Other Respiratory Viruses* 2011;5(6):413–417

Zoulek G, Bürger P, Deinhardt F. Markers of hepatitis viruses A and B: direct comparison between whole serum and blood spotted on filter-paper. *Bulletin of the World Health Organization* 1985;63(5):935–939.

9

APPLICATIONS OF DRIED BLOOD SPOTS IN GENERAL HUMAN HEALTH STUDIES

Eleanor Brindle, Kathleen A. O'Connor, and Dean A. Garrett

9.1 INTRODUCTION

Simplified specimen collection and assay methods offer great promise for expanding our understanding of human health and its correlates in population and public health fields (McDade et al., 2007; Weir, 2007; Steptoe, 2011). Less invasive and more efficient biomarker measurements facilitate the sampling of a broader range of settings outside of the clinic, encompassing more human biological variation and including both high- and low-resource settings. Additionally, when coupled with mixed-methods approaches, simpler biomarker measurement fosters new research directions, examining ways that the social environment and biological health intersect.

In this chapter, we explore public health and population health research recently made possible by dried blood spot collection and assay methods, and discuss some of the logistics and mechanics of DBS collection and DBS biomarker measurement for population samples collected outside of the clinic (Section 9.1). We discuss DBS collection protocols and tips, the importance of understanding DBS assay performance, specimen treatment issues, disease transmission risks, new technological developments (point-of-care testing, multiplexing), continuing logistical and technological challenges, and developing new directions in DBS biomarkers of health (Section 9.2). We use two of our own projects, one from a high-resource job-site setting and one from a multinational collection effort, the Demographic and Health Survey (DHS), undertaken in low-resource settings, as examples to illustrate some of these mechanics (Section 9.3).

9.1.1 Clinical and Non-Clinical Health Research

A majority of studies that form our knowledge of human health are based on clinical research. Clinically based research takes advantage of the best diagnostic tools available and existing healthcare infrastructure that makes them practical. In that setting, significant volumes of blood are collected by practiced phlebotomists in a controlled environment and specimens are processed appropriately at a dedicated facility following established procedures. These routine, well-established procedures are the gold standard for biomarker measurement and yield reliable results for a broad range of markers. However, these standard venous blood collection procedures are not feasible for research undertaken outside clinical settings (Global Health Diagnostics Forum, 2006).

Research outside of clinical settings in public and population health has contributed a great deal to our understanding of human biological variation (McDade, 2003; Valeggia, 2007) and health status (National Research Council, 2008; Crimmins et al., 2010), and has a continuing rich future ahead to inform us of how physical, social, and biological environments enhance and hinder health and well-being. It is critical to continue expanding our knowledge of the determinants of human health status outside of the clinic and in populations other than those enjoying the best health and longest life. Even in wealthy countries, much clinical research has focused on pathology, rather than the full range of human variation, leaving us lacking in our understanding of pathways and linkages to behavior and ecological context that contribute to the morbidity and mortality risk of populations, communities, and individuals.

Dried Blood Spots: Applications and Techniques, First Edition. Edited by Wenkui Li and Mike S. Lee.

Areas of human health research that stand to benefit from simplified methods of biological data collection span a number of academic disciplines. Recent research increasingly combines perspectives across social, biological, and biomedical areas, helping to form a more complete picture of the influences on individual and population health. Social workers, psychologists, epidemiologists, anthropologists, sociologists, and demographers have increasingly turned to study designs that link physiology and behavior. As social scientists and biologists enter into research together using biological measures (Butz and Torrey, 2006), they need methods compatible with existing techniques for subject recruitment and participation, which occur in the community, rather than the clinic. Communities range from a rural village near a community health center to a job site in a resource-rich employer setting. An advantage of this approach is less bias selection toward participants engaged in the healthcare system. This can allow for detection of subclinical and preclinical indicators of micronutrient or metabolic syndrome status, for example, as well as provide population ranges for biomarkers (Zimmermann et al., 2006; Fujita et al., 2011; Wander et al., 2012a).

Blood is a more invasive, but also much richer data source than urine or saliva for many health biomarkers. Dried blood spot methodology minimizes invasiveness while still providing access to rich biological information in broader contexts than was previously possible. DBS can also be more directly compared to clinical serum and plasma levels than urine or saliva specimens. Additionally, recent technology has made possible increasing the number of analytes measured (e.g., multiplex assay platforms) in the very low volume specimens that DBS provide.

Urine and saliva specimens have been widely used for research requiring repeated sampling and research conducted in nonclinical population settings (Lu et al., 1999; Ellison et al., 2002; Goldman et al., 2004; Ferrell et al., 2005; Lindau and McDade, 2008; Trumble et al., 2010; Trumble et al., 2012). In particular, saliva specimens have been useful in characterizing diurnal patterns of hormones within individuals across time (Ruttle et al., 2011; Skinner et al., 2011). Urine specimens have been used in a number of human health studies requiring repeated longitudinal sampling designs (Ferrell et al., 2007; Woods et al., 2007; Santoro et al., 2008; O'Connor et al., 2009). While these fluids are limited in the range of biomarkers available, they have low invasiveness and suitability for repeated sampling strategies, even those requiring multiple samples across the day. DBS, like venous blood collection generally, are not ideal for very frequent repeated sampling designs, but can allow more frequent sampling than venous blood draws.

DBS have reduced, but not negligible, invasiveness and risk of disease transmission as compared to venous blood draws. Thus, while DBS collection can more easily be done in nonclinical settings, sharps disposal, biohazard waste disposal, and participant discomfort still require safe and ethical procedures conducted by trained personnel. As many studies have demonstrated, and as we show here, these can be accomplished in a range of community settings, including at work sites, small urban and rural community health clinics, and even in settings more remote from a health clinic (Shell-Duncan and McDade, 2005).

9.1.2 DBS Use in Public and Population Health Settings

DBS are a good choice for studies with the challenges of large sample size and longitudinal assessment, including in hard-to-reach populations. These are two key advantages that DBS collection brings to human health research. Examples from nutrition, particularly in micronutrient malnutrition, and HIV research illustrate the impact of the availability of DBS techniques thus far. Some of these studies may straddle the line between clinical and nonclinical research, particularly those that include HIV testing, and may take place in field clinics where laboratory facilities are limited.

DBS are seen frequently in nutrition and supplementation studies, as challenging research design is a common feature in that field, either because of the remoteness or vulnerability of the subject population (Bhutta et al., 2009; Kabahenda et al., 2011) or because of a need for large-scale, nationally representative population samples (Baingana et al., 2008; Thomas and Frankenberg, 2008; Garrett et al., 2011), as will be discussed in greater detail later in this chapter. Zimmerman et al (2003) used a DBS measure of thyroid function to evaluate efficacy of introducing fortified salt in a community with iodine deficiency. This prospective study of 377 Moroccan children in rural mountain villages demonstrated a significant decrease in mean thyroglobulin, indicating improved thyroid function in DBS collected 5 and 12 months after iodine supplementation was introduced, as compared to baseline thyroglobulin levels in DBS collected before the intervention was implemented (Zimmermann et al., 2003).

DBS are widely used in HIV research as well. Spielberg et al. demonstrated that DBS could be collected at home and sent to a lab for HIV testing, and that this could be done with excellent adherence to a regimen of self-collecting DBS at 4, 12, and 20 weeks into the study in four extremely high-risk populations in the United States (Spielberg et al., 2000). Kane et al. demonstrated that DBS can be used to regularly monitor viral load in patients receiving ARV HIV treatment in Senegal (Kane et al., 2008). Other studies report on the benefits of DBS for infant screening (Sherman et al., 2005; Patton et al., 2008), determining HIV subtype variability and resistance to antiretroviral therapy (Garrido et al., 2008), and simultaneous screening for HIV and hepatitis (Lukacs et al., 2005). As in the case of national-level screening for micronutrient deficiency, DBS are an invaluable tool for country-wide infectious disease screening (Parker and Cubitt, 1999; Solomon et al., 2002) and have been used extensively for nationwide HIV surveillance in both developed (Barin et al., 2006) and developing (Mishra et al., 2006; Bärnighausen et al., 2008) country settings.

DBS have contributed to much recent research in a wide and still growing range of applications in the more traditionally social sciences to facilitate effective bridging of the social and biological sciences as well as yield new insights into human biology and health and how they are influenced by the social world. Most of the leading causes of morbidity and mortality, both chronic and acute, in both developed and (very rapidly) developing countries are believed to be in significant part a function of individual behavior, social dynamics, and social structures. Integrating biomarkers into biobehavioral health has been a productive way to address this priority in health-related research (Giles, 2011).

Many panel studies of sociobehavioral health incorporating DBS have taken place in the United States and other industrialized countries, addressing areas such as, for example, social and spatial network effects on health, health disparities, biobehavioral health, health economics, and biodemography (i.e., the Health and Retirement Study, National Longitudinal Study of Adolescent Health, Los Angeles Family and Neighborhood Survey, National Social Life, Health, and Aging Project). Outside of higher resource settings, DBS panel studies have been conducted in less wealthy nations (i.e., Mexican Family Life Survey, Work and Iron Status Evaluation), and more are expected, as non-communicable diseases emerge as a growing problem in the developing world (World Health Organization, 2010; Wagner and Brath, 2011; Lakshmy et al., 2012). Anthropological life history and health research using DBS has fostered an even broader perspective on our understanding of human health, biology, environment, and social structures in populations and settings even farther removed from industrial infrastructure and lifeways; an active theme in much recent work in this area has been on immunological function, which has been particularly facilitated by DBS (see, e.g., McDade et al., 2008; Sharrock et al., 2008; Blackwell et al., 2011; Wander et al., 2012b;). DBS methods have thus contributed to a large and growing number of interdisciplinary social and biological studies to understand and improve individual, community, and population health.

9.2 METHODOLOGICAL CONSIDERATIONS IN DBS HEALTH RESEARCH

9.2.1 DBS Collection Strategies

9.2.1.1 Venous Blood Collection Three requirements of venipuncture create the most daunting obstacles from a researcher's perspective: venipuncture skill, maintaining a cold chain, and time constraints for sample processing. Even experienced phlebotomists may not be successful in obtaining blood from every patient. Children, the elderly, and those with compromised veins due to illness or drug abuse present the greatest challenges. Phlebotomists must be relied upon to follow strict protocols to avoid hemolysis of specimens and to ensure that any collection tube additives are properly mixed with the blood. Once blood has been drawn, it must be kept cold continuously until analysis.

Most clinical and research assays have been, and continue to be, based on venous blood, separated into serum or plasma components, and not whole blood. There are tight time constraints for this task, which must generally be completed in less than 2 hours for serum and plasma, depending on the collection tubes used (NCCLS, 2004). Additionally, to separate serum or plasma from whole blood requires a centrifuge and is a task that involves significant risk of disease transmission, as a technician must process large volumes of blood via procedures that aerosolize the samples, increasing the chance of accidental exposure.

9.2.1.2 DBS Blood Collection Strategies In contrast to the above requirements for venous blood collection, there is minimal processing of DBS until time of assay, lower biohazard risk, and specimens can often tolerate less than optimal collection conditions (McDade et al., 2007), although this varies by analyte of interest in DBS.

One important benefit to using DBS specimens as compared to other less invasive sample types is the range of markers that have been measured successfully from DBS. Any marker that is measured in blood can in theory be measured in DBS, with few known exceptions (Mei et al., 2001; McDade et al., 2007), and DBS results can be expected to correspond more directly to circulating blood levels of an analyte than saliva or urine results (i.e., Wong et al., 2004).

Other advantages of DBS are primarily logistical. As compared to venous blood samples, DBS can be collected easily in nonclinical settings and can even be collected by research participants themselves. Puncture of the finger or heel with a sterile lancet requires little practice, and research staff or study participants can be taught proper DBS collection technique in a matter of minutes. Obtaining blood from infants, the elderly, and those with fragile veins can be difficult even for highly skilled phlebotomists, and capillary blood collection offers an excellent alternative for those groups. Costs for DBS collection supplies are comparable to those for venipuncture.

Particularly when specimens are collected in challenging field settings, collecting an adequate volume of specimen may be difficult. In our experience, it is more difficult to collect a sufficient amount of blood from a finger prick when participants are cold or dehydrated or when they have very calloused fingers. We suggest using the largest lancet size available, ideally at least 1.5 mm wide by 2.0 mm deep, in order to create enough blood flow for collecting a minimum of five spots and to avoid the need to perform a second finger puncture. In addition, DBS collection is not without risk. Individuals can faint and many are not comfortable with having blood taken; these individuals should not be encouraged to participate in the protocol of DBS collection if they are not comfortable with it.

A number of creative methods can be used to increase blood flow and increase the number of spots that can be collected from a single finger prick, and to make the subject more comfortable while collection is happening. Wander et al. collected DBS in a study using immunological measures to test predictions of the hygiene hypothesis among young children in Tanzania (Wander et al., 2008, 2012a, 2012b). To both increase blood flow and decrease anxiety, Wander offered soccer balls, tennis balls, and balloons and encouraged children to play while awaiting their turn for DBS collection. In addition to enough blood to fill a hemoglobin test cuvette (HemoCue) and a rapid HIV test kit (SD BioLine HIV-1/2 3.0 rapid HIV-1/2 test), this yielded a minimum of three to four blood spots from each child, sufficient for the five ELISAs to be conducted for the primary study, with enough left-over in most cases for additional future use (K. Wander, personal communication). Other sources have suggested that warming the skin with a warm towel or other warming device for 3–5 minutes can stimulate blood flow (Clinical and Laboratory Standards Institute, 2008). Simply rubbing the hands may also be an effective method of warming the fingers to stimulate blood flow.

Methods for minimizing discomfort and maximizing the volume of blood obtained for a population of adults would, of course, be very different. Strategies used in a population of men recruited in their workplace for the Health Initiatives for Men (HIM) study, discussed in detail below, included having men drink water and wave their arms or do a few knee bends to help them to relax and increase blood flow. We also found that it was helpful to maintain physical contact by having the researcher keep their hand on the participants' arm during the DBS collection process to assess temperature and any potential participant stress and to encourage compliance with requests to keep the arm relaxed.

9.2.2 Specimen Treatment, Biohazard, and Safety Issues

When research is conducted far from the laboratory, maintenance of a cold chain and transport of the specimens are greatly simplified by DBS as compared to liquid samples. While articles commonly claim that DBS require no refrigeration, this is not strictly true. Stability varies by analyte: for example, DBS retinol decreases after 1 week of storage then seems to stabilize (Erhardt et al., 2002), whereas DBS retinol-binding protein (RBP) remains stable even at high ambient temperatures for up to 6 weeks (Fujita et al., 2007). Other analytes have been tested for their short-term durability in DBS with varying results (i.e., Worthman and Stallings, 1994; O'Broin et al., 1997; Cook et al., 1998; McDade and Shell-Duncan, 2002; Miller et al., 2006; Kane et al., 2008; Brindle et al., 2010). However, a number of analytes are significantly more stable in DBS than in liquid form, and often protection from extreme ambient temperatures and humidity is all that is required. Instead of a cumbersome liquid nitrogen dewar, a simple cooler will generally be sufficient for days or weeks of field storage, depending on the analytes of interest. DBS must also be protected from humidity and insects, but these requirements can be easily achieved in field settings using desiccant and plastic storage containers.

Risk of disease transmission, a particularly important consideration in human health research, is reduced in DBS as compared to venous blood. Eliminating liquid handling reduces the risk of contact with samples through either spills or aerosolization during sample processing. The weight and bulk of DBS are significantly less than even small liquid specimens and can further simplify transport from the field to the lab and decrease the specimen storage space demands for large studies.

The advantages of DBS methods have helped simplify two challenges that have introduced limitations in studies of human health and biology. First, simplifying sample collection logistics allows for more representative sampling, particularly for research conducted in hard-to-reach populations, whether in wealthy or impoverished nations. As will be illustrated in the two examples given in Section 9.3, simplified sample collection is essential to obtaining an unbiased sample appropriate to the population under study. With conventional techniques, research subjects may be understudied because social or economic factors make them less likely to engage in the healthcare system, and therefore less likely to be captured in clinical research. Or, potential participants may be physically hard to reach, as with residents of remote rural areas not served by roads or other infrastructure. This exacerbates another potential source of bias in a sample: those least able to travel because of lack of means or physical capacity will be underrepresented in studies that require a visit to a field clinic.

Second, DBS methods facilitate studies requiring repeated sampling to understand patterns of temporal variability in biomarkers. Such studies have included examinations of the changes in markers over long time scales, such as an ovarian cycle (Wander et al., 2008), or over seasons of a year (Chiriboga et al., 2009) and shorter term fluctuations, like diurnal variation in cortisol (Wong et al., 2004) or the effects of short-term intensive physical training (Nindl et al., 2003).

9.2.3 Storing DBS for Future Use

Much discussion has been published about the ethical considerations and potential benefits of long-term storage of specimens left over from large-scale collection efforts (Vaught, 2006; Sgaier et al., 2007; Willett et al., 2007). Ethical considerations include responsibilities to specimen donors, and measures for insuring identity and data protection are necessary for archiving specimens. Institutional review boards (IRBs) have established policies regarding confidentiality and consent for storing the specimens for uses other than the primary purpose for which they were collected. The use of stored specimens when subject identity is not known and human subjects are no longer engaged in the project may require minimal human subjects' approval.

DBS specimens can be much more efficiently stored in a repository than the larger volumes of blood components, urine, or saliva collections. A key advantage for storing DBS specimens is their use for pilot research exploring new research directions. Storage facility requirements for long-term preservation are a concern often overlooked, but the costs of maintaining an archive for serum or other conventional specimens can be considerable; this cost should be considered for both study duration and for years after a study and its funding have concluded. Consideration should be given to the amount of DBS left over after a primary study is complete and whether the DBS are worth archiving. This is a complicated question, however, as the volume of specimen required for assay has been rapidly decreasing as laboratory technology improves and multiplexing techniques advance (Gordon and Michel, 2012).

Long-term stability of analytes in DBS specimens is not often well characterized, and optimal procedures may vary across environmental settings (Lindau and McDade, 2008; Brindle et al., 2010). The quality of biomarker data from DBS hinges upon assumptions on the stability of analytes under various treatment regimes, including long-term storage. The dearth of data to evaluate long-term stability of DBS also translates into a lack of information about appropriate short- and long-term storage temperatures. We found DBS C-reactive protein (CRP) to be stable at –20°C for up to about 1 year, followed by a slow decline in CRP values over roughly 2 years of testing (Brindle et al., 2010). If not previously documented, stability parameters should be evaluated, if possible, for a study's key biomarkers and expected storage conditions.

9.2.4 Comparability of DBS and Venous Blood Assay Results

A set of matched specimens, such as paired DBS and venous blood draws, is valuable for examining the nature of the correlation between DBS and serum/plasma measure of a given analyte. This is critical for understanding the strengths and limitations of the DBS method for an analyte or biomarker. As new biomarkers come on-line, assays can be adapted for DBS, and good documentation of the relationship of a biomarker in the two different matrices is critical. A good

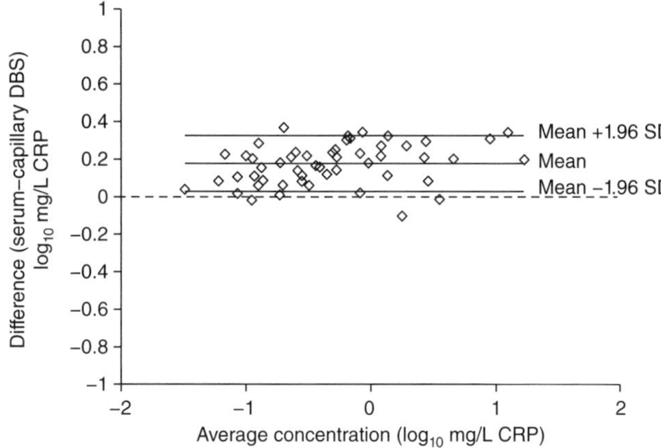

FIGURE 9.1 Bland–Altman plot of differences between log serum CRP and log CRP in capillary DBS plotted against average log CRP concentration; $n = 52$ matched specimen sets. Center line indicates mean difference; upper and lower lines indicate the 95% confidence interval. Broken line indicates the line of equality. Redrawn from Brindle et al. (2010).

gold standard test is to compare clinical or kit assays in both serum and DBS.

Bland–Altman comparisons (Bland and Altman, 1999) are a useful tool for such comparisons, as they reveal the magnitude of differences between measures, the consistency of the degree of difference from sample to sample, and any concentration-dependent bias in the difference. Figure 9.1 illustrates the use of Bland–Altman plots comparing CRP in 52 paired serum and DBS specimens from US adults. While the Pearson correlation for two sample types was very high (0.979, $p < 0.001$), the Bland–Altman plot reveals that the serum values are consistently higher than DBS values. This allows a simple adjustment to serum equivalents to be made when necessary (McDade and Shell-Duncan, 2002; Brindle et al., 2010).

The comparability of DBS to serum or plasma results is particularly important for biological markers that have cutoff values or clinical diagnostic reference ranges established in serum or plasma; it is important to know how these ranges may change when DBS are used. Even excellent correlations between DBS and conventional blood values may conceal subtle differences in absolute values. These differences can, and should, be characterized thoroughly, and DBS results should be adjusted to serum/plasma equivalents when necessary before standard cutoff values and reference ranges are applied (Melby et al., 2005; Brindle et al., 2010; Dowd et al., 2011).

9.2.5 DBS Biomarker Assays

Other limitations of DBS are related to the technical problems that they can introduce to assays designed for serum or plasma. These issues are discussed in greater detail in other chapters, so will be mentioned only briefly here. When DBS are eluted from the filter paper, the specimens are diluted. For analytes that are at very low concentrations, further dilution of already small volumes may lead to results below the assay limits of detection. Attempts to mitigate this problem may require additional steps in preparing specimens for assay, such as extraction or concentration, or may involve increasing the amount of DBS sample consumed for the test.

DBS assay methodology, in concurrence with changing assay technology, increasingly allows for multiple assay results from relatively small volumes of blood. Skogstrand et al. demonstrated this principle by measuring 25 analytes in two 3 mm DBS punches, or the equivalent of about 3.1 μL of serum, using a bead-based multi-analyte assay system (Skogstrand et al., 2005). In theory, these same two small punches of DBS used to measure 25 analytes can now be used with the same bead-based multiplexing technology that allows up to 500 analytes to be measured simultaneously. These techniques are still relatively new, and continuously improving (Chowdhury et al., 2009; Gordon and Michel, 2012; Salvante et al., 2012).

However, methods that measure analytes singly are still the most easily available and widely used, and most assays already validated for use with DBS use conventional single analyte techniques, such as ELISA. The abundance of the marker of interest in the blood is the most important determinant of the amount of DBS required for an assay, and this varies by orders of magnitude among analytes (Anderson, 2002). For example, C-reactive protein is relatively abundant and can easily be measured using a single 3 mm diameter punch for an assay in the high-sensitivity range (Brindle et al., 2010), while four punches of the same size may be required to assay progesterone (Edelman et al., 2007). The number of assays that can be done from a single blood spot therefore varies greatly by exactly which analytes are to be measured and which assays are used. In addition, the size of the blood spot can vary, yielding as few as two, or as many as seven, 3 mm diameter punches per spot, in spite of best efforts to keep spot size consistent.

DBS biomarker assays are still in the early phases of application. Consequently, it is important to characterize assay performance and relationships to serum values for each new assay that comes on-line in a rapidly moving field of technological advances for measuring more things in ever smaller specimens.

9.2.6 Limitations of DBS

For justifiable health and safety reasons, frequent repeated DBS sampling may be prohibited by some IRBs, which have policies limiting the amount of blood a subject may provide within a particular time frame. Consequently, studies requiring repeated blood sampling, such as studies of the time course of variation in biomarkers either in normal physiology or in disease states, or longitudinal studies using high sampling frequency to explore behavioral or environmental effects on biology are not particularly feasible. Small-scale validation studies of serial sampling can be conducted with DBS under some circumstances, for example, one study collected paired DBS and urine across 1.5 menstrual cycles, a challenging protocol requiring fairly highly motivated participants and requiring a nearby lab setting (Wander et al., 2008).

Diagnostic tests designed for use with serum or plasma may or may not yield comparable results in DBS. This is particularly relevant for quantitative tests that rely on cutoff values to identify a disease state or micronutrient deficiency, because DBS results may be highly correlated with serum or plasma results, yet differ in absolute value (Brindle et al., 2010; Wander et al., 2012a).

9.3 CASE STUDIES

We describe two of our own studies whose motivations and target populations illustrate the breadth of population and public health research using dried blood spots. In the first, we discuss use of dried blood spot collection methods in resource-poor settings for micronutrient deficiency screening of large, nationally representative samples as part of the DHSs in Uganda in 2006 and in Tanzania in 2010 (Baingana et al., 2008; Garrett et al., 2011). In a second example, we describe a smaller biomarker collection study, a community pilot study on men's health and health behaviors, conducted with participants at a local job site—the HIM study. This biodemographic mixed-methods pilot study measured a panel of biomarkers relevant to men's health and health behaviors in a sample of 94 US men.

9.3.1 DBS in Demographic and Health Surveys

The DHS program has been at the forefront of the movement to incorporate biomarker testing in large-scale national household surveys. To date, DHS has conducted biomarker tests on millions of respondents, measuring 16 different biomarkers, in more than 80 countries (see Table 9.1).

Since 1986, these surveys have generated a rich body of information related to the health of populations in developing countries and are the leading source of data worldwide on the health of populations in developing countries. These data are critical to researchers and policy makers alike: they help to better understand the occurrence and distribution of a disease or condition in a population, serve as a tool for advocacy, and guide the development of policies and programs to improve the health of these populations.

Biomarkers are collected mainly from three main age groups: children 6–59 months; women age 15–49 years,

TABLE 9.1 Biomarker Tests Done by the Demographic and Health Surveys (DHSs)

Biomarker	Year Test First Done	Number of Surveys Including Test	Population Typically Tested[a]	Sampling Method and Equipment Used
Anthropometry (weight, height, age)	1987	147	Women 15–49 years Children 0(6)–59 months Men 15–59 years	Noninvasive, measuring board and scale
Hemoglobin (for anemia)	1995	62	Women 15–49 years Children 0–59 months Men 15–59 years	Capillary blood, HemoCue™ portable analyzer
HIV	2001	32	Women 15–49 years Men 15–59 years	Capillary blood, DBS
Blood pressure	1998	8	Women 15–49 years Men 15–59 years	Noninvasive, automatic cuff
Syphilis	1996	6	Women 15–49 years Men 15–59 years	Venous blood, RPR
Vitamin A	1996	4	Women 15–49 years Children 0(6)(12)–59 months	Capillary blood, DBS, HPLC, RBP-EIA
Malaria	2006	3	Women 15–49 years Children 0(6)–59 months	Capillary blood, RDT and thick/thin slides

Other, less common tests: hepatitis B, hepatitis C, herpes, measles, tetanus, chlamydia, diabetes, lipids, CRP, transferrin. DBS, dried blood spots; RPR, rapid plasma reagin; HPLC, high-performance liquid chromatography; RBP-EIA, retinol-binding protein enzyme immunoassay; RDT, rapid diagnostic test.
[a]Varies by country, for example, men 15–54 or 15–64; children under 3 years.

and men 15–49 years. Because there is a priority to collect biomarkers at the household level and the surveys are conducted primarily in resource-poor countries, tests for measuring or monitoring biomarkers in population-based surveys should be the least intrusive as possible to respondents and should, preferably, provide results on the spot. The equipment used need to be rugged, portable, resilient to extreme temperatures and degrees of humidity, and relatively inexpensive. Meeting all of these conditions can be challenging. When it' is not feasible to collect the biomarker at the household, DHS often collects blood samples and processes blood samples as dried blood spots, then tests the samples later in a centralized laboratory.

9.3.1.1 Logistical Considerations of Sample Collection

DBS are the sample of choice in DHS because the adverse conditions under which DHS operates can pose considerable challenges in the collection, handling, and storage of samples. In the past, many tests required the collection of venous whole blood which further complicated the survey operations, making it necessary to have specially trained phlebotomists accompany the interviewers in the field. Apart from the logistical issues, venous blood collection makes the survey more expensive as it calls for more intensive and longer periods of training for health workers and requires specialized equipment for the preservation of samples. However, the main challenge in collecting venous whole blood is the preservation of the integrity of the blood components prior to analysis in a central laboratory. The preservation of venous whole blood samples requires having a "cold chain" in the field setting.

DBS samples have several advantages over venous blood (Parker and Cubitt, 1999; McDade et al., 2007). The use of DBS samples has eliminated the need for a cold chain and refrigeration of specimens in the field, reducing considerably the complexity of storage in remote areas and transport to the laboratory. DBS samples are obtained from the prick of a finger (or a heel in the case of very young children), thus reducing the need for syringes and needles, test tubes, and racks, and reducing the invasiveness of the procedure. DBS samples are collected in preprinted circles on a filter paper card, dried overnight, protected with glassine paper, sealed tightly in commercially available plastic bags with desiccants and humidity indicators, and then transported with relatively little effort to the central laboratory for testing.

The increased application of tests based on the collection of DBS has revolutionized the capacity to measure biomarkers in field settings, and the DHS has taken advantage of this collection medium to assess the prevalence of vitamin A (VAD) and iron deficiency (ID) in three surveys, the details of which have been given elsewhere (Baingana et al., 2008; Garrett et al., 2011).

9.3.1.2 Biomarkers of Nutrition in the Uganda and Tanzania Demographic and Health Surveys

In addition to collecting anthropometric data and screening for the prevalence of anemia in women and children, the DHS collected biomarker data on the prevalence of VAD and ID in the

Uganda (UDHS) and Tanzania (TDHS) DHSs in 2006 and 2010, respectively.

9.3.1.3 Subjects and Methods The UDHS was conducted in 2006 and the Tanzania DHS in 2010. In the Uganda DHS and the Tanzania DHS, levels of RBP (a proxy for retinol) and transferrin receptor (sTfR) were measured to assess the prevalence of VAD and ID, respectively. We also measured CRP to "correct" the RBP data for the effect of inflammation/infection. DBS samples were collected from women and children and transferred to a central laboratory for testing. All testing was carried out by local personnel using commercially obtained test kits. Hemoglobin was also measured in the household in women and children using a portable device (HemoCue, Angelhom, Sweden) to assess the prevalence of anemia.

9.3.1.4 Sample Size Calculations In DHS surveys, sample size is determined for various indicators including fertility and mortality of a given population. Typically, all households or half of the sampled households are included for anemia testing. Sample size determinations are usually made for smaller regions and extended to the entire population. Depending on the country, a sample of 300–1500 households comprises a smaller region.

Although serum is the preferred sample for most biochemical assays, DHS typically collects blood samples as DBS because of the reasons mentioned previously. For the measurement of CRP, RBP, and sTfR in the UDHS and TDHS, DBS were collected for the analysis of these markers. The sample collection procedure has been described in detail elsewhere (Baingana et al., 2008). Briefly, blood was collected directly from a finger stick into preprinted circles on special filter paper cards (Whatman 903), dried overnight, packaged, and sent to the laboratory for testing.

9.3.1.5 Laboratory Analysis of DBS for CRP, RBP, and sTfR All DBS samples collected in the UDHS 2006 and TDHS 2010 were checked individually for stability (gauged by humidity indicator card), adequacy of spots, blood clots, smudges, contamination with dirt, and so on, or incompletely filled circles. Once the samples were deemed adequate for testing, they were allowed to come to room temperature and tested using commercially available test kits with the exception of CRP testing in the UDHS, which was tested using an in-house assay (CSDE Biodemography Core, University of Washington; Brindle et al., 2010). In the TDHS, CRP was measured using the Bender MedSystems instant enzyme immunoassay (Bender MedSystems, Campus Vienna Biocenter 2, Vienna, Austria), RBP with the Scimedx RBP Assay (Scimedx Corporation, Denville, NJ, USA; Hix et al., 2004) and sTfR was determined using a commercial enzyme immunoassay (TF-94, Ramco Laboratories, Stafford, TX, USA) adapted by McDade and Shell-Duncan (McDade and Shell-Duncan, 2002; Shell-Duncan and McDade, 2005). Because these assays were not developed for DBS, it is necessary to assess the elution profile of the analytes of interest from the filter paper matrix. We validated the DBS as a sample matrix for CRP, RBP, and sTfR. The procedure for validation of DBS for these markers has been discussed previously (Baingana et al., 2008).

9.3.1.6 Discussion Vitamin A and iron testing were successfully included in the Uganda and Tanzania DHS, an achievement that was due largely to the application of DBS for the measurement of biomarkers of nutritional status. In total, over 23,000 DBS samples were collected in the two surveys and tested in the respective country using local staff. This achievement demonstrates that it is feasible to effectively introduce biomarker collection into large, population-based surveys. Nonetheless, there were a number of logistical challenges to overcome. Specifically, there were challenges related to sample collection, sample transport to a central testing facility, supervision of staff, management of large number of samples and management of large volumes of data, and testing for three biomarkers as part of a nationally representative survey.

The successful outcome of these two surveys—Tanzania and Uganda—demonstrates that it is feasible to measure biomarkers in challenging environments, even where local technical capacity may be limited, if adequate training of staff and sustained supervision of the project are made a priority of stakeholders. Despite the extensive training of staff and capacity building that go into each DHS survey, there are still numerous challenges within the survey country that have a direct bearing on the outcome of the testing. These challenges must be not only recognized prior to implementation of the survey, but addressed head-on if the survey is to be successful. For example, in most DHS countries, infrastructure is a serious hindrance to testing large numbers of samples. There is usually inadequate space for sample analysis, lack of refrigerators and freezers for storage of samples, unstable supply of electricity, and a lack of facilities to produce deionized water. In many laboratories, the equipment needed to conduct testing is not available or if it is, has not been serviced and/or is not working properly.

Added-value can be built into these surveys if left-over samples (DBS) are preserved for future use. In many surveys, a specific request is made to respondents for their samples to be used for related studies to which a large majority of respondents consent. Because the DHS is the leading source of demographic and health data, the assessment of vitamin and mineral disorders (VMD) in the context of a DHS provides a very unique blend of demographic, health, and nutrition data that are invaluable not only to ministries of health, but to academicians, researchers, program managers, and public health workers.

9.3.2 DBS in the Health Initiatives for Men Study

The HIM study collected DBS and other biometrics paired with survey data from a community sample of working men in a pilot project to assess feasibility of DBS and point-of-care measurement assessment in a workplace setting and to preliminarily test hypotheses relating biomarkers to men's health beliefs, behaviors, and social environments. The interdisciplinary team consisted of physicians, social scientists, and employer/employee representatives from a local employer. This mixed-methods (qualitative and quantitative) pilot study was designed to inform future directions for exploring biological, behavioral, and structural factors influencing a fundamental health disparity: the higher mortality in males than females at all ages of the lifespan, including in utero (Hazzard, 1986).

We aimed to (1) identify common incentives and barriers to health that men perceive and experience; (2) examine familial, social, biological, structural, and community factors influencing men's health-related behaviors; and (3) explore biological and sociocultural factors simultaneously to see where and how they intersect. In particular, we were interested in how marital and parental status were associated with various biomarkers, in light of the strong sociological and demographic evidence supporting a consistent survival advantage for married over unmarried men (Weden and Brown, 2010; Rendall et al., 2011). This study was also motivated by clinician's concerns with men delaying health care, and thus we focused on identifying incentives and barriers to engaging men more fully with their health.

9.3.2.1 Participants and Study Objectives The study sampled men across the job gradient from hourly employees to salaried executives of a single company, in a setting where all employees had good health insurance and access to healthcare related information. In this near-optimal health setting, we examined how incentives and barriers for men to engage in healthy behaviors were shaped by norms related to health and by individual perceptions and intentions with regard to health-related behaviors. Our aims were to assess key health behavior domains: health norms; healthcare intentions and behaviors; health-related social networks; incentives and barriers to healthy behaviors; and use of the healthcare system. Biomarker goals of the pilot project are to (1) optimize the biomarker data collection and assay methods for minimally invasive, accurate, point-of-care, and high-throughput data collection and processing; (2) demonstrate that sufficient variability in the measures of interest could be captured in this setting; and (3) test whether key variables in the conceptual model were correlated in predicted ways.

In this community-based participatory research project we worked closely with the community at the workplace to implement the study. The employer partnered with us because of their interest in improving men's health and offered significant support in allowing participants to complete the study protocol at the workplace during work hours. In addition to gathering data on men's health behaviors and perceptions in a survey, we collected biometric data on site. Procedures were approved by the IRBs of the University of Washington and the Pennsylvania State University and by the Human Resources Office of PBC. We did not share individual-level biomarker or survey data with any employees of our business partner; they have access only to aggregate-level data.

9.3.2.2 Logistics of DBS Collection and Assay at a Job-Site Setting The men who participated in the study came to a set of four small conference rooms provided for biomarker and survey collection procedures. The research team obtained consent, and then collected basic demographic data, including age, information on medications the men were taking (they were advised in advance to bring this information along), time of wake-up that morning (to control for diurnal effects in biomarkers such as cortisol and testosterone), and time of last meal. Anthropometrics (height, weight, percent body fat, and waist circumference) were measured as well as blood pressure and pulse using portable devices. Blood was collected from a finger prick with a sterile lancet; blood drops were then collected in a small plastic weigh boat (about 1.5 inches wide) and then (1) used in a number of on-site point-of-care devices, and (2) spotted onto Whatman 903 filter paper cards using a pipette, for later assay in the laboratory.

Blood collection and point-of-care testing and preparation of DBS were expedited by collecting blood drops in the small plastic tray. Additionally, it made it simpler for two researchers to work together during blood collection so that one could maintain contact with the participant and ensure their comfort, while the other could manipulate blood specimens out of the participant's view. Other advantages included allowing the researcher to touch the participant's finger to the collection tray to continue a flow of blood (which you cannot do with filter paper) and facilitating a few extra drops of blood to be reserved for a few minutes to allow repeat testing in the event of a failure of the point-of-care test devices.

The entire biometric data collection phase took an average of 16 minutes. The participants received $20 for participation in the biomarker and survey collection (a survey followed the biomarker collection) and, perhaps more importantly, were given a brochure with their results from the point-of-care tests, along with reference range information and contacts who could answer any questions regarding their test results. Many of the men were interested in getting this information and appreciated the take-away documentation. We had high participation in this phase of the study; in fact we expanded beyond our initial sample size by 14 participants ($N = 80$ increased to $N = 94$).

From our experience with this pilot study, as well as others, we learned that participants whose hands are cool have potential for difficulty collecting blood; warm hands and active circulation assist in the collection of DBS spots. Adequate or recent hydration facilitates better blood flow as well. As discussed above, hydration and activity status appeared to be related to blood flow in the same ways among 2- to 7-year-old Tanzanian children (Wander et al., 2012b).

In the HIM study, for about one-third of the men it was necessary to perform more than one finger prick to get sufficient blood flow to complete all the tests and collect an adequate number of blood spots. Participants were asked for permission before another lancet stick was attempted, and none of the men refused. Total blood volume required for the point-of-care testing was about 65 μL. The average number of DBS collected was 5.3, or approximately 265 μL of whole blood. The six DBS assays we performed required nine punches, 3 mm each, or about 1.5–2 blood spots. The remaining bloods spots were archived at –20°C.

Although there were no problems with participants fainting in this study, we have in other settings seen individuals faint in response to finger pricks. Individuals who are low on sleep, have not eaten recently, or are not well hydrated may be more likely to faint. Smelling salts should be kept at hand during the collection process. It is also important to make sure participants are in a chair that provides good support so that they will not fall in the event that they do faint. The technician collecting DBS needs to be sensitive to and prepared for such events.

Partly out of this concern for fainting, and partly given the schedule for specimen collection across the day from 8 AM to 5 PM, we avoided asking participants to fast before providing specimens. Instead, we asked the men when they last ate. Time since last meal should be noted when collecting specimens, as some clinical reference ranges apply only after fasting (e.g., glucose) and other analytes may increase briefly following food intake. Men were also asked to provide their time of wake-up, and sample collection time was noted; these are included as control variables in data analyses.

9.3.2.3 HIM DBS Assay Methods for Point-of-Care and Laboratory Biomarker Measures The combination of DBS collection methods with point-of-care testing was well suited to this study. The point-of-care tests were FDA approved for diagnostic use and not subject to Clinical Laboratory Improvement Amendments (CLIA) regulations for diagnostic labs, meaning we could feel confident in providing the results to the participants. They had the added benefit of offering rapid testing, with results available from all analyzers in approximately 10 minutes.

Point-of-care tests included (1) a Cholestech LDX for measurement of total cholesterol, HDL cholesterol, triglycerides, calculated LDL cholesterol, and glucose; (2) a Bio-Rad In2it for measuring glycosylated hemoglobin (HbA1c), an indicator of blood glucose control over the previous 3–4 months; and (3) a Chempaq XBC which provided measurements of hemoglobin, white blood cell count, and a three-part white blood cell count differential (lymphocytes, monocytes, and granulocytes).

Laboratory DBS assays were used for analytes that are less relevant for individual's health status (using assays designed for research but not approved for diagnostic purposes), but valuable in addressing the research goals. Three of the assays—testosterone, cortisol, and CRP—were developed and/or validated in-house at the CSDE Biodemography Lab. The other three assays—sex hormone-binding globulin (SHBG), alpha-1-acid glycoprotein (AGP), and Epstein–Barr virus antibody (EBV)—were serum kits adapted for use with DBS. Assay details are provided in Table 9.2.

For all of the laboratory DBS assays, performance in DBS was evaluated by comparing assay results using matched serum and DBS collected simultaneously. As demonstrated in Figure 9.1 for CRP and in Figure 9.2 and Table 9.2, serum and DBS may be highly correlated while not providing identical results. To evaluate this, we looked at both the correlation between serum and DBS paired specimens and the analyte recovered from DBS as a percent of the serum measurement (Table 9.2). The number of specimen pairs tested varied by assay. These data suggest good but varied recovery of most of the analytes in DBS compared to serum. However, the Bland–Altman plot for SHBG shows both consistently lower values in DBS than serum and a DBS dose-dependent relationship with serum values (Figure 9.2); moreover, the recovery of SHBG in DBS is not particularly high (Table 9.2).

We were able to demonstrate the ability to collect DBS in a job setting with good participation and rich point-of-care and laboratory assay biomarker results. Among other findings, testosterone and SHBG were correlated with age, and CRP was positively correlated with body fat, as expected. Despite good access to healthcare and information resources, 68% of participants showed biomarkers (BMI, cholesterol, blood pressure, glycosylated hemoglobin) suggestive of a need for health intervention. This was a convenience sample of men, so results may reflect some selection bias toward men concerned with their health. Controlling for time of wake-up (diurnal variation), testosterone in single men tended to be higher than in partnered men, a finding observed in several previous recent studies using salivary specimens (e.g., Gray et al., 2002; Gettler et al., 2011a).

Coupling these hormonal findings with qualitative analysis results from the HIM study, we find that family is an important effector and correlate of men's health in *both* hormonal and behavioral domains. These data have implications for the evolutionary biology of men's health, as well as for

TABLE 9.2 Comparisons of Assay Results from Matched Serum and Dried Blood Spot (DBS) Specimens Collected Simultaneously

Assay	Assay Source	*N* (Matched Serum and DBS Pairs)	Pearson Correlation DBS and Serum	DBS/Serum (%) ± SD[a]
Testosterone	In-house ELISA (Muir et al., 2001)	69	0.954 ($p < 0.001$)	92% ± 10%
Cortisol	In-house ELISA (Munro and Stabenfeldt, 1985)	51	0.912 ($p < 0.001$)	102% ± 16%
SHBG	ELISA kit, DRG International, Marburg, Germany, Cat # EIA-2996	66	0.986 ($p < 0.001$)	63% ± 12%
CRP	In-house ELISA (Brindle et al., 2010)	52	0.979 ($p < 0.001$)	67% ± 15%
AGP	ELISA kit, GenWay Biotech, Inc., San Diego, CA, Cat # GWB-505E5B (Wander et al., 2012a)	9	0.913 ($p < 0.001$)	99% ± 10%
EBV antibody	ELISA kit, DiaSorin, Stillwater, MN, ETI VCA-G, cat # P001606A (McDade et al., 2000)	30	0.743 ($p < 0.001$)	88% ± 22%

DBS were prepared from venous whole blood. SHBG, sex hormone-binding globulin; CRP, C-reactive protein; AGP, alpha-1-acid glycoprotein; EBV, Epstein–Barr virus antibody.

[a]DBS/serum (%) ± SD refers to recovery of analyte in DBS as percent of that measured in serum.

identifying health behaviors and social structures to target to help improve men's health. The hormonal and behavioral data together support the well-established sociodemographic finding of the benefits of marriage for men's health and merit further exploration. The biodemography and evolution of male biology has been an active area of research in anthropology, in particular, with a rapidly growing body of work supporting biomarker links with men's social, energetic, and immunologic environments (Ellison et al., 2002; Gray et al., 2006; Kuzawa et al., 2009; Vitzthum et al., 2009; Muehlenbein et al., 2010; Trumble et al., 2010; Gettler et al., 2011b; Trumble et al., 2012). A range of specimen types and biomarkers of health and biology have been used in these studies (serum, saliva, urine), and DBS stand to offer much new and additional information to male health and male reproductive ecology and biodemography. In contrast, in the clinical research on men's health there has been great emphasis on declining testosterone with age, testosterone supplementation, and the relationships of these with comorbidities and mortality risk (Khaw et al., 2007; Laughlin et al., 2008).

The health status biomarker results for the HIM project suggest a need for better health management in this sample of (motivated) men despite their being in a generally high health resource job setting. The survey and qualitative data can be used to point ways forward for improving men's health motivations and actions. In addition to shedding light on human evolution and ways for improving healthcare access for men, a better understanding of the biosocial factors associated with the mortality gender disparity may help address a potential public health problem: as developing countries transition to highly industrialized market economies, they may follow the demographic transition trend of a widening gender gap in mortality (Gage, 1994; Drevenstedt et al., 2008; Weden and Brown, 2010). In the United States and elsewhere, better recognition of the role of family in men's health could help justify and inform policy aiming to support healthier families and healthier men.

9.4 SUMMARY AND CONCLUSIONS

Availability of biomarker measurements in dried blood spots has propelled forward several areas of human health research previously hindered by challenges of specimen collection by venipuncture or limitations of alternative specimen types. Underrepresented areas in our body of knowledge, including research targeting those groups least able or least likely to come for an appointment in a traditional research setting and studies of the normal range of human biological variation between individuals and within individuals across time, stand to benefit immensely from availability of DBS methods. This is evident in the recent explosion of work in which socioeconomic and behavioral factors intersect with biomarker measurements and studies of hard-to-reach populations coming from the global health community.

While DBS methods do have a few limitations as compared to conventional venous blood collection, their logistical advantages for research in nonclinical settings often greatly outweigh the disadvantages, and through broader use, the limitations of DBS can be expected to diminish. DBS assays may require additional testing and validation, but this minor drawback will decrease in importance as more and more researchers develop and share methods. As more data collection efforts are undertaken in more populations,

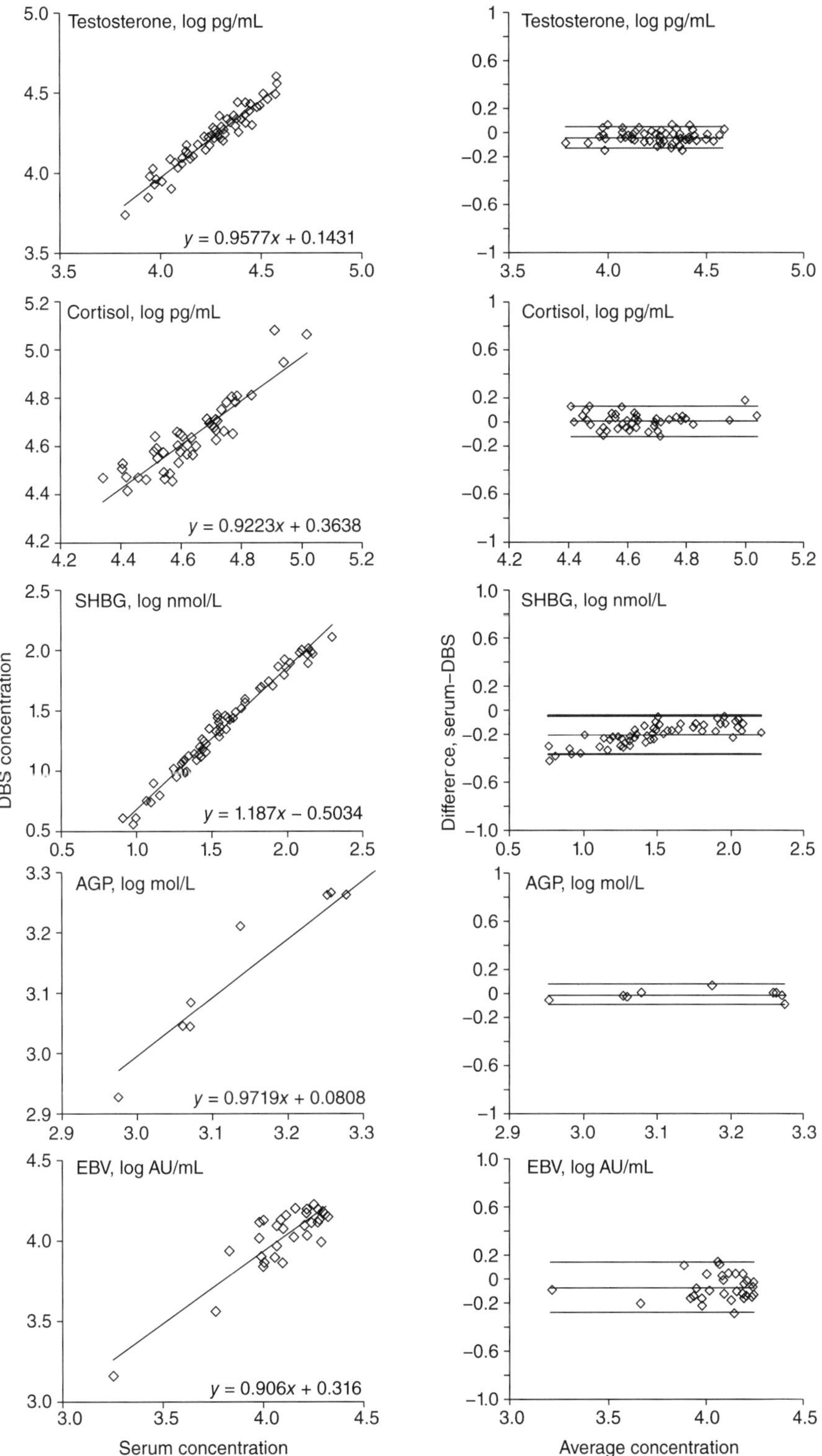

FIGURE 9.2 Scatter plots (left) and Bland–Altman plots (right) for paired serum and DBS specimens measured by immunoassay for testosterone (n = 69 pairs), cortisol (n = 51 pairs), sex hormone-binding globulin (SHBG; n = 66 pairs), alpha-1-acid glycoprotein (AGP; n = 9 pairs), and Epstein–Barr virus antibody (EBV; n = 30 pairs). In scatter plots, solid line = linear regression. In Bland–Altman plots, center line indicates mean difference between serum and DBS measures; upper and lower lines indicate the 95% confidence interval.

comparability of DBS with serum or plasma diagnostic tests will be better estimated and thus less of a concern.

In summary, DBS are particularly well suited to studies of human health. Their convenience expands the potential repertoire of measurements that can be easily incorporated into a number of field settings and makes it possible to attempt larger and more complex research designs. Combining this simpler sample collection tool with continually improving assay methods makes a more comprehensive view of human health accessible. This simplified tool facilitates entry from research disciplines not previously engaged in biological measurement, thus broadening the concept of health and allowing inclusion of contextual data not previously considered in clinical or biological studies of health parameters.

REFERENCES

Anderson, NL. The human plasma proteome: history, character, and diagnostic prospects. *Molecular & Cellular Proteomics* 2002;1(11):845–867. doi:10.1074/mcp.R200007-MCP200

Baingana RK, Matovu DK, Garrett DA. Application of retinol-binding protein enzyme immunoassay to dried blood spots to assess vitamin A deficiency in a population-based survey: the Uganda Demographic and Health Survey 2006. *Food and Nutrition Bulletin* 2008;29(4):297–305.

Barin F, Plantier J-C, Brand D, Brunet S, Moreau A, Liandier B, Thierry D, Cazein F, Lot F, Semaille C, Desenclos JC. Human immunodeficiency virus serotyping on dried serum spots as a screening tool for the surveillance of the AIDS epidemic. *Journal of Medical Virology* 2006;78(Suppl 1): S13–S18. doi:10.1002/jmv.20600

Bärnighausen T, Wallrauch C, Welte A, McWalter TA, Mbizana N, Viljoen J, Graham N, Tanser F, Puren A, Newell ML. HIV incidence in rural South Africa: comparison of estimates from longitudinal surveillance and cross-sectional cBED assay testing. *PloS One* 2008;3(11):e3640. doi:10.1371/journal.pone.0003640

Bhutta Z, Klemm R, Shahid F, Rizvi A, Rah JH, Christian P. Treatment response to iron and folic acid alone is the same as with multivitamins and/or anthelminthics in severely anemic 6- to 24-month-old children. *Journal of Nutrition* 2009;139(8):1568–1574. doi:10.3945/jn.108.103507

Blackwell AD, Gurven MD, Sugiyama LS, Madimenos FC, Liebert MA, Martin MA, Kaplan HS, Snodgrass JJ. Evidence for a peak shift in a humoral response to helminths: age profiles of IgE in the Shuar of Ecuador, the Tsimane of Bolivia, and the U.S. NHANES. *PLoS Neglected Tropical Diseases* 2011;5(6):e1218. doi:10.1371/journal.pntd.0001218.

Bland JM, Altman DG. Measuring agreement in method comparison studies. *Statistical Methods in Medical Research* 1999;8(2):135–160.

Brindle E, Fujita M, Shofer JB, O'Connor KA. Serum, plasma, and dried blood spot high-sensitivity C-reactive protein enzyme immunoassay for population research. *Journal of Immunological Methods* 2010;362(1–2):112–120. doi:10.1016/j.jim.2010.09.014

Butz WP, Torrey BB. Some frontiers in social science. *Science (New York, N.Y.)* 2006;312(5782):1898–1900. doi:10.1126/science.1130121

Chiriboga DE, Ma Y, Li W, Stanek EJ, Hébert JR, Merriam PA, Rawson ES, Ockene IS. Seasonal and sex variation of high-sensitivity C-reactive protein in healthy adults: a longitudinal study. *Clinical Chemistry* 2009;55(2):313–321. doi:10.1373/clinchem.2008.111245

Chowdhury F, Williams A, Johnson P. Validation and comparison of two multiplex technologies, Luminex and Mesoscale Discovery, for human cytokine profiling. *Journal of Immunological Methods* 2009;340(1):55–64. doi:10.1016/j.jim.2008.10.002

Clinical and Laboratory Standards Institute. *Procedures and Devices for the Collection of Diagnostic Capillary Blood Specimens; Approved Standard*, 6th edn, Vol. 28. Clinical and Laboratory Standards Institute; 2008.

Cook JD, Flowers CH, Skikne BS. An assessment of dried blood-spot technology for identifying iron deficiency. *Blood* 1998;92(5):1807–1813.

Crimmins EM, Kim JKI, Vasunilashorn S. Biodemography: new approaches to understanding trends and differences in population health and mortality. *Demography* 2010;47: 41–64.

Dowd JB, Aiello AE, Chyu L, Huang YY, McDade TW. Cytomegalovirus antibodies in dried blood spots: a minimally invasive method for assessing stress, immune function, and aging. *Immunity & Ageing* 2011;8(1):3. doi:10.1186/1742-4933-8-3

Drevenstedt GL, Crimmins EM, Vasunilashorn S, Finch CE. The rise and fall of excess male infant mortality. *Proceedings of the National Academy of Sciences of the United States of America* 2008;105(13):5016–5021. doi:10.1073/pnas.0800221105

Edelman A, Stouffer R, Zava DT, Jensen JT. A comparison of blood spot vs. plasma analysis of gonadotropin and ovarian steroid hormone levels in reproductive-age women. *Fertility and Sterility* 2007;88(5):1404–1407. doi:10.1016/j.fertnstert.2006.12.016

Ellison PT, Bribiescas RG, Bentley GR, Campbell BC, Lipson SF, Panter-Brick C, Hill K. Population variation in age-related decline in male salivary testosterone. *Human Reproduction* 2002a;17(12):3251–3253.

Erhardt JG, Craft NE, Heinrich F, Biesalski HK. Rapid and simple measurement of retinol in human dried whole blood spots. *Journal of Nutrition* 2002;132(2):318–321.

Ferrell RJ, O'Connor KA, Rodríguez G, Gorrindo T, Holman DJ, Brindle E, Miller RC, Schechter DE, Korshalla L, Simon JA, Mansfield PK, Wood JW, Weinstein M. Monitoring reproductive aging in a 5-year prospective study: aggregate and individual changes in steroid hormones and menstrual cycle lengths with age. *Menopause* 2005;12(5):567–577. doi:10.1097/01.gme.0000172265.40196.86

Ferrell RJ, O'Connor KA, Holman DJ, Brindle E, Miller RC, Rodriguez F, Simon JA, Mansfield PK, Wood JW,

Weinstein M. Monitoring reproductive aging in a 5-year prospective study: aggregate and individual changes in luteinizing hormone and follicle-stimulating hormone with age. *Menopause* 2007;14(1):29–37. doi:10.1097/01.gme.0000227859.50473.20

Fujita M, Brindle E, Shofer JB, Ndemwa P, Kombe Y, Shell-Duncan B, O'Connor KA. Retinol-binding protein stability in dried blood spots. *Clinical Chemistry* 2007;53(11):1972–1975. doi:10.1373/clinchem.2007.093104

Fujita M, Shell-Duncan B, Ndemwa P, Brindle E, Lo Y-J, Kombe Y, O'Connor KA. Vitamin A dynamics in breastmilk and liver stores: a life history perspective. *American Journal of Human Biology* 2011;23(5):664–673. doi:10.1002/ajhb.21195

Gage TB. Population variation in cause of death: level, gender, and period effects. *Demography* 1994;31(2):271–296.

Garrett DA, Sangha JK, Kothari MT, Boyle D. Field-friendly techniques for assessment of biomarkers of nutrition. *American Journal of Clinical Nutrition* 2011;94:685–690. doi:10.3945/ajcn.110.005751.Am

Garrido C, Zahonero N, Fernándes D, Serrano D, Silva AR, Ferraria N, Antúnes F, González-Lahoz J, Soriano V, de Mendoza C. Subtype variability, virological response and drug resistance assessed on dried blood spots collected from HIV patients on antiretroviral therapy in Angola. *Journal of Antimicrobial Chemotherapy* 2008;61(3):694–698. doi:10.1093/jac/dkm515

Gettler LT, McDade TW, Feranil AB, Kuzawa CW. Longitudinal evidence that fatherhood decreases testosterone in human males. *Proceedings of the National Academy of Sciences of the United States of America* 2011a;108(39):16194–16199. doi:10.1073/pnas.1105403108

Gettler LT, McDade TW, Kuzawa CW. Cortisol and testosterone in Filipino young adult men: evidence for co-regulation of both hormones by fatherhood and relationship status. *American Journal of Human Biology* 2011b;23(5):609–620. doi:10.1002/ajhb.21187

Giles J. Social science lines up its biggest challenges. *Nature* 2011;470(7332):18–19. doi:10.1038/470018a

Global Health Diagnostics Forum. The right tools can save lives. *Nature* 2006;444(7120):681. doi:10.1038/444681a

Goldman N, Weinstein M, Cornman J, Singer B, Seeman TE, Chang MC. Sex differentials in biological risk factors for chronic disease: estimates from population-based surveys. *Journal of Women's Health* 2004;13(4):393–403. doi:10.1089/154099904323087088

Gordon J, Michel G. Discerning trends in multiplex immunoassay technology with potential for resource-limited settings. *Clinical Chemistry* 2012;58(4):690–698. doi:10.1373/clinchem.2011.176503

Gray PB, Kahlenberg SM, Barrett ES, Lipson SF, Ellison PT. Marriage and fatherhood are associated with lower testosterone in males. *Evolution and Human Behavior* 2002;23(3):193–201. doi:10.1016/S1090-5138(01)00101-5

Gray PB, Yang C-FJ, Pope HG. Fathers have lower salivary testosterone levels than unmarried men and married non-fathers in Beijing, China. *Proceedings of the Royal Society B: Biological Sciences* 2006;273(1584):333–339. doi:10.1098/rspb.2005.3311

Hazzard WR. Biological basis of the sex differential in longevity. *Journal of the American Geriatrics Society* 1986;34(6):455–471.

Hix J, Martinez C, Buchanan I, Morgan J, Tam M, Shankar A. Development of a rapid enzyme immunoassay for the detection of retinol-binding protein. *American Journal of Clinical Nutrition* 2004;79(1):93–98.

Kabahenda M, Mullis RM, Erhardt JG, Northrop-Clewes CA, Nickols SY. Nutrition education to improve dietary intake and micronutrient nutriture among children in less-resourced areas: a randomised controlled intervention in Kabarole district, western Uganda. *South African Journal of Clinical Nutrition* 2011;24(2):83–88.

Kane CT, Ndiaye HD, Diallo S, Ndiaye I, Wade AS, Diaw PA, Gaye-Diallo A, Mboup S. Quantitation of HIV-1 RNA in dried blood spots by the real-time NucliSENS EasyQ HIV-1 assay in Senegal. *Journal of Virological Methods* 2008;148(1–2):291–295. doi:10.1016/j.jviromet.2007.11.011

Khaw K-T, Dowsett M, Folkerd E, Bingham S, Wareham N, Luben R, Welch A, Day N. Endogenous testosterone and mortality due to all causes, cardiovascular disease, and cancer in men: European prospective investigation into cancer in Norfolk (EPIC-Norfolk) Prospective Population Study. *Circulation* 2007;116(23):2694–2701. doi:10.1161/CIRCULATIONAHA.107.719005

Kuzawa W, Gettler LT, Muller MN, McDade TW, Feranil AB. Fatherhood, pairbonding and testosterone in the Philippines. *Hormones and Behavior* 2009;56(4):429–435. doi:10.1016/j.yhbeh.2009.07.010

Lakshmy R, Mathur P, Gupta R, Shah B, Anand K, Mohan V, Desai NG, Mahanta J, Joshi PP, Thankappan KR. Measurement of cholesterol and triglycerides from a dried blood spot in an Indian Council of Medical Research-World Health Organization multicentric survey on risk factors for noncommunicable diseases in India. *Journal of Clinical Lipidology* 2012;6(1):33–41. doi:10.1016/j.jacl.2011.10.021

Laughlin GA, Barrett-Connor E, Bergstrom J. Low serum testosterone and mortality in older men. *The Journal of Clinical Endocrinology and Metabolism* 2008;93(1):68–75. doi:10.1210/jc.2007-1792

Lindau ST, McDade TW. Minimally invasive and innovative methods for biomeasure collection in population-based research. In: Weinstein M, Vaupel JW, Wachter KW (eds) *Biosocial Surveys*. Washington, DC: National Academies Press; 2008, pp. 251–277.

Lu Y, Bentley GR, Gann PH, Hodges KR, Chatterton RT. Salivary estradiol and progesterone levels in conception and nonconception cycles in women: evaluation of a new assay for salivary estradiol. *Fertility and Sterility* 1999;71(5):863–868.

Lukacs Z, Dietrich A, Ganschow R, Kohlschütter A, Kruithof R. Simultaneous determination of HIV antibodies, hepatitis C antibodies, and hepatitis B antigens in dried blood spots–a feasibility study using a multi-analyte immunoassay. *Clinical Chemistry and Laboratory Medicine* 2005;43(2):141–145. doi:10.1515/CCLM.2005.023

McDade TW. Life history theory and the immune system: steps toward a human ecological immunology. *American*

Journal of Physical Anthropology 2003;Suppl 37:100–125. doi:10.1002/ajpa.10398

McDade TW, Shell-Duncan B. Whole blood collected on filter paper provides a minimally invasive method for assessing human transferrin receptor level. *Journal of Nutrition* 2002;132(12):3760–3763.

McDade TW, Stallings JF, Angold A, Costello EJ, Burleson M, Cacioppo JT, Glaser R, Worthman CM. Epstein-Barr virus antibodies in whole blood spots: a minimally invasive method for assessing an aspect of cell-mediated immunity. *Psychosomatic Medicine* 2000;62(4):560–567.

McDade TW, Williams S, Snodgrass JJ. What a drop can do: dried blood spots as a minimally invasive method for integrating biomarkers into population-based research. *Demography* 2007;44(4):899–925.

McDade TW, Reyes-García V, Tanner S, Huanca T, Leonard WR. Maintenance versus growth: investigating the costs of immune activation among children in lowland Bolivia. *American Journal of Physical Anthropology* 2008;136(4):478–484. doi:10.1002/ajpa.20831

Mei JV, Alexander JR, Adam BW, Hannon WH. Use of filter paper for the collection and analysis of human whole blood specimens. *Journal of Nutrition* 2001;131(5):1631S–1636S.

Melby MK, Watanabe S, Whitten PL, Worthman CM. Sensitive high-performance liquid chromatographic method using coulometric electrode array detection for measurement of phytoestrogens in dried blood spots. *Journal of Chromatography. B, Analytical Technologies in the Biomedical and Life Sciences* 2005;826(1–2):81–90. doi:10.1016/j.jchromb.2005.08.009

Miller AA, Sharrock KCB, McDade TW. Measurement of leptin in dried blood spot samples. *American Journal of Human Biology* 2006;18(6):857–860. doi:10.1002/ajhb.20566

Mishra V, Vaessen M, Boerma JT, Arnold F, Way A, Barrere B, Cross A, Hong R, Sangha J. HIV testing in national population-based surveys: experience from the Demographic and Health Surveys. *Bulletin of the World Health Organization* 2006;84(7):537–545.

Muehlenbein MP, Hirschtick JL, Bonner JZ, Swartz AM. Toward quantifying the usage costs of human immunity: altered metabolic rates and hormone levels during acute immune activation in men. *American Journal of Human Biology* 2010;22(4):546–556. doi:10.1002/ajhb.21045

Muir C, Spironello-Vella E, Pisani N, DeCatanzaro D. Enzyme immunoassay of 17 beta-estradiol, estrone conjugates, and testosterone in urinary and fecal samples from male and female mice. *Hormone and Metabolic Research* 2001;33(11):653–658. doi:10.1055/s-2001-18692

Munro CJ, Stabenfeldt G. Development of a cortisol enzyme immunoassay in plasma. *Clinical Chemistry* 1985;31(6):956.

National Research Council. Biosocial surveys. In: Weinstein M, Vaupel JW, Wachter KW (eds), *Social Sciences*. Washington, DC: National Academies Press; 2008.

NCCLS. *Procedures for the Handling and Processing of Blood Specimens; Approved Guideline*, 3rd edn. NCCLS document H18-A3 (ISBN 1-56238-555-0). NCCLS, 940 West Valley Road, Suite 1400, Wayne, Pennsylvania 19087-1898 USA, 2004.

Nindl BC, Kellogg MD, Khosravi MJ, Diamandi A, Alemany JA, Pietila DM, Young AJ, Montain SJ. Measurement of insulin-like growth factor-I during military operational stress via a filter paper blood spot assay. *Diabetes Technology & Therapeutics* 2003;5(3):455–461. doi:10.1089/152091503765691974

O'Broin SD, Kelleher BP, Davoren A, Gunter EW. Field-study screening of blood folate concentrations: specimen stability and finger-stick sampling. *American Journal of Clinical Nutrition* 1997;66(6):1398–1405.

O'Connor KA, Ferrell RJ, Brindle E, Shofer JB, Holman DJ, Miller RC, Schechter DE, Singer B, Weinstein M. Total and unopposed estrogen exposure across stages of the transition to menopause. *Cancer Epidemiology, Biomarkers & Prevention* 2009;18(3):828 836. doi:10.1158/1055-9965.EPI-08-0996

Parker SP, Cubitt WD. The use of the dried blood spot sample in epidemiological studies. *Journal of Clinical Pathology* 1999;52(9):633–639.

Patton JC, Coovadia AH, Meyers TM, Sherman GG. Evaluation of the ultrasensitive human immunodeficiency virus type 1 (HIV-1) p24 antigen assay performed on dried blood spots for diagnosis of HIV-1 infection in infants. *Clinical and Vaccine Immunology* 2008;15(2):388–391. doi:10.1128/CVI.00265-07

Rendall MS, Weden MM, Favreault MM, Waldron H. The protective effect of marriage for survival: a review and update. *Demography* 2011;48(2):481–506. doi:10.1007/s13524-011-0032-5

Ruttle PL, Shirtcliff EA, Serbin LA, Fisher DB-D, Stack DM, Schwartzman AE. Disentangling psychobiological mechanisms underlying internalizing and externalizing behaviors in youth: longitudinal and concurrent associations with cortisol. *Hormones and Behavior* 2011;59(1):123–132. doi:10.1016/j.yhbeh.2010.10.015

Salvante KG, Brindle E, McConnell D, O'Connor KA, Nepomnaschy PA. Validation of a new multiplex assay against individual immunoassays for the quantification of reproductive, stress, and energetic metabolism biomarkers in urine specimens. *American Journal of Human Biology* 2012;86:81–86. doi:10.1002/ajhb.21229

Santoro N, Crawford SL, Lasley BL, Luborsky JL, Matthews Ka, McConnell D, Randolph JF, Gold EB, Greendale GA, Korenman SG, Powell L, Sowers MF, Weiss G. Factors related to declining luteal function in women during the menopausal transition. *Journal of Clinical Endocrinology and Metabolism* 2008;93(5):1711–1721. doi:10.1210/jc.2007-2165

Sgaier SK, Jha P, Mony P, Kurpad A, Lakshmi V, Kumar R, Ganguly NK. Public health. Biobanks in developing countries: needs and feasibility. *Science* 2007;318(5853):1074–1075. doi:10.1126/science.1149157

Sharrock KCB, Kuzawa CB, Leonard WR, Tanner S, Reyes-García VE, Vadez V, Huanca T, McDade TW. Developmental changes in the relationship between leptin and adiposity among Tsimané children and adolescents. *American Journal of Human Biology* 2008;20(4):392–398. doi:10.1002/ajhb.20765

Shell-Duncan B, McDade TW. Cultural and environmental barriers to adequate iron intake among northern Kenyan schoolchildren. *Food and Nutrition Bulletin* 2005;26(1):39–48.

Sherman GG, Stevens G, Jones SA, Horsfield P, Stevens WS. Dried blood spots improve access to HIV diagnosis and care for infants in low-resource settings. *Journal of Acquired Immune Deficiency Syndromes (1999)* 2005;38(5):615–657.

Skinner ML, Shirtcliff EA, Haggerty KP, Coe CL, Catalano RF. Allostasis model facilitates understanding race differences in the diurnal cortisol rhythm. *Development and Psychopathology* 2011;23(4):1167–1186. doi:10.1017/S095457941100054X

Skogstrand K, Thorsen P, Nørgaard-Pedersen B, Schendel DE, Sørensen LC, Hougaard DM. Simultaneous measurement of 25 inflammatory markers and neurotrophins in neonatal dried blood spots by immunoassay with xMAP technology. *Clinical Chemistry* 2005;51(10):1854–1866. doi:10.1373/clinchem.2005.052241

Solomon SS, Solomon S, Rodriguez II, McGarvey ST, Ganesh AK, Thyagarajan SP, Mahajan AP, Mayer KH. Dried blood spots (DBS): a valuable tool for HIV surveillance in developing/tropical countries. *International Journal of STD & AIDS* 2002;13(1).25–28.

Spielberg F, Critchlow C, Vittinghoff E, Coletti AS, Sheppard H, Mayer KH, Metzgerg D, Judson FN, Buchbinder S, Chesney M, Gross M. Home collection for frequent HIV testing: acceptability of oral fluids, dried blood spots and telephone results. HIV Early Detection Study Group. *AIDS* 2000;14(12):1819–1828.

Steptoe A. Psychosocial biomarker research: integrating social, emotional and economic factors into population studies of aging and health. *Social Cognitive and Affective Neuroscience* 2011;6(2):226–233. doi:10.1093/scan/nsq032

Thomas D, Frankenberg E. Comments on collecting and utilizing biological indicators in social science surveys. In: Weinstein M, Vaupel J, Wachter K (eds), *Biosocial Surveys*. Washington, DC: National Academies Press; 2008. pp. 149–155.

Trumble BC, Brindle E, Kupsik M, O'Connor KA. Responsiveness of the reproductive axis to a single missed evening meal in young adult males. *American Journal of Human Biology* 2010;22(6):775–781. doi:10.1002/ajhb.21079

Trumble BC, Cummings D, von Rueden C, O'Connor KA, Smith EA, Gurven M, Kaplan H. Physical competition increases testosterone among Amazonian forager-horticulturalists: a test of the 'challenge hypothesis'. *Proceedings of the Royal Society B: Biological Sciences* 2012;279(1739):2907–2912. doi:10.1098/rspb.2012.0455

Valeggia CR. Taking the lab to the field: monitoring reproductive hormones in population research. *Population and Development Review* 2007;33(3):525–542. doi:10.1111/j.1728-4457.2007.00183.x

Vaught JB. Blood collection, shipment, processing, and storage. *Cancer Epidemiology, Biomarkers & Prevention* 2006;15(9):1582–1584. doi:10.1158/1055-9965.EPI-06-0630

Vitzthum VJ, Worthman CM, Beall CM, Thornburg J, Vargas E, Villena M, Soria R, Caceres E, Spielvogel H. Seasonal and circadian variation in salivary testosterone in rural Bolivian men. *American Journal of Human Biology* 2009;21(6):762–768. doi:10.1002/ajhb.20927

Wagner K-H, Brath H. A global view on the development of non communicable diseases. *Preventive Medicine* 2011;54(Suppl):S38–S41. doi:10.1016/j.ypmed.2011.11.012

Wander K, Brindle E, O'Connor KA. C-reactive protein across the menstrual cycle. *American Journal of Physical Anthropology* 2008;136(2):138–146. doi:10.1002/ajpa.20785

Wander K, Brindle E, O'Connor KA. Sensitivity and specificity of C-reactive protein and α1-acid glycoprotein for episodes of acute infection among children in Kilimanjaro, Tanzania. *American Journal of Human Biology* 2012a;24(4):565–568. doi:10.1002/ajhb.22261

Wander K, O'Connor K, Shell-Duncan B. Expanding the hygiene hypothesis: early exposure to infectious agents predicts delayed-type hypersensitivity to candida among children in Kilimanjaro. *PLoS ONE* 2012b;7(5):e37406. doi:10.1371/journal.pone.0037406

Weden MM, Brown RA. Historical and life course timing of the male mortality disadvantage in Europe: epidemiologic transitions, evolution, and behavior. *Social Biology* 2010;53(1-2):61–80.

Weir D. Elastic powers: the integration of biomarkers into the health and retirement study. In: Weinstein M, Vaupel JW, Wachter KW (eds), *Biosocial Surveys*. Washington, DC: National Academies Press; 2007. pp. 78–95.

Willett WC,Blot WJ, Colditz GA, Folsom AR, Henderson BE, Stampfer MJ. Merging and emerging cohorts: not worth the wait. *Nature* 2007;445(7125):257–258. doi:10.1038/445257a

Wong V, Yan T, Donald A, McLean M. Saliva and bloodspot cortisol: novel sampling methods to assess hydrocortisone replacement therapy in hypoadrenal patients. *Clinical Endocrinology* 2004;61(1):131–137. doi:10.1111/j.1365-2265.2004.02062.x

Woods NF, Mitchell ES, Tao Y, Viernes H-M, Stapleton PL, Farin FM. Polymorphisms in the estrogen synthesis and metabolism pathways and symptoms during the menopausal transition: observations from the Seattle Midlife Women's Health Study. *Menopause (New York, N.Y.)* 2007;13(6):902–910. doi:10.1097/01.gme.0000227058.70903.9f

World Health Organization. *Global Status Report on Noncommunicable Diseases*. World Health Organization; 2010.

Worthman CM, Stallings JF. Measurement of gonadotropins in dried blood spots. *Clinical Chemistry* 1994;40(3):448–453.

Zimmermann MB, Moretti D, Chaouki N, Torresani T. Development of a dried whole-blood spot thyroglobulin assay and its evaluation as an indicator of thyroid status in goitrous children receiving iodized salt. *American Journal of Clinical Nutrition* 2003;77(6):1453–1458.

Zimmermann MB, de Benoist B, Corigliano S, Jooste PL, Molinari L, Moosa K, Pretell EA, Al-Dallal ZS, Wei Y, Zu-Pei C, Torresani T. Assessment of iodine status using dried blood spot thyroglobulin: development of reference material and establishment of an international reference range in iodine-sufficient children. *Journal of Clinical Endocrinology and Metabolism* 2006;91(12):4881–4887. doi:10.1210/jc.2006-1370

10

APPLICATIONS OF DRIED BLOOD SPOTS IN ENVIRONMENTAL POPULATION STUDIES

ANTONIA M. CALAFAT AND KAYOKO KATO

10.1 APPLICATION IN ENVIRONMENTAL POPULATION STUDIES

10.1.1 Introduction

Technological advances in analytical chemistry have made it possible to measure trace levels of multiple environmental chemicals in biological fluids and tissues. These advances have contributed to the increased use of biomonitoring methods to determine the exposure of target chemicals in humans. Specimens such as blood or urine are analyzed to determine the actual concentration of the environmental chemicals, their corresponding metabolites, or adducts of the chemicals with endogenous biomolecules (Angerer et al., 2007).

Biomonitoring data permit accurate assessments of human exposure because concentrations of these biomarkers represent an integrative measure of exposure to the target chemicals from multiple sources and routes. Therefore, biomonitoring data in combination with indirect measures of exposure (e.g., environmental monitoring, questionnaire information) are the most appropriate tools for exposure assessment and can provide useful information about differences in exposures across geography, race/ethnicity, and socioeconomic status.

For the most part, little information exists about the extent of human exposure to many environmental chemicals. Furthermore, the potential toxic health effects of these chemicals in humans are largely unknown. Therefore, studies to investigate the prevalence of these exposures are warranted. Pregnant women, infants, and young children are recognized as subpopulations susceptible to the effects of potentially harmful environmental chemicals (Arbuckle, 2010). However, these segments of the population can be poorly represented in general population biomonitoring surveys such as the National Health and Nutrition Examination Survey (NHANES). NHANES is an ongoing survey conducted by the Centers for Disease Control and Prevention (CDC) and is designed to evaluate the health and nutrition status of the civilian noninstitutionalized US population (CDC, 2003). Since 1999, NHANES has included an ongoing exposure assessment of the US population to select environmental chemicals. However, the amount of biological specimens collected from preadolescents is limited; therefore, environmental chemicals exposure data in NHANES for young children are also rather limited. Specifically, blood is collected on NHANES participants 1 year of age and older, and urine specimens are requested of all participants 6 years of age and older. Therefore, biomonitoring urinary measurements are restricted to participants 6 years of age and older. Similarly, environmental chemical measurements in blood or blood serum are available only from participants 12 years and older, with the exception of lead, cadmium, and mercury—measured in all NHANES participants—and cotinine—measured in every person 3 years of age or older (CDC, 2003).

Biomonitoring studies among pregnant women, infants, and young children are of interest because exposures during these critical time periods, including exposures to some environmental chemicals, may impact health later in life (Arbuckle, 2010). Appropriate biological matrices for biomonitoring during such critical time windows will depend upon matrix availability and the different classes of environmental chemicals to be monitored (Barr et al., 2005). For

Dried Blood Spots: Applications and Techniques, First Edition. Edited by Wenkui Li and Mike S. Lee.

example, biomonitoring matrices unique to the fetus include amniotic fluid and meconium. Other matrices potentially useful for biomonitoring are cord blood, placenta, umbilical cord, and dried blood spots (DBS).

Since Dr Robert Guthrie first began collecting heel prick blood spot samples from newborns to detect phenylketonuria in the 1960s, DBS have been collected routinely from newborns within 48 hours of birth in the United States. In fact, the collection of DBS from approximately 4 million infants each year for routine newborn screening purposes is one of the largest US public health programs (Therrell and Adams, 2007).

In the United States and in many other countries worldwide (Borrajo, 2007; Jaques et al., 2008; Kerruish et al., 2008; Fujii et al., 2010; Rothwell et al., 2011), DBS are screened for metabolic and other disorders, and state public health departments store residual newborn DBS for potential reanalysis. Residual specimens can be used for quality control (QC) evaluation of new screening tests, forensic testing, therapeutic drug monitoring, and research purposes. For example, DBS may be used for population-level survey research, specifically for measuring clinical biomarkers of neuroendocrine, cardiovascular, metabolic, and immune/inflammatory function (Mcdade et al., 2007; van Amsterdam and Waldrop, 2010) and genotyping purposes after extracting DNA from the DBS (Olshan, 2007).

Residual DBS may also provide a valuable resource to assess exposures to environmental chemicals at birth, with the possibility of extrapolating to gestational exposures. The use of these residual DBS, however, even without any personal identifiers, to evaluate exposure to environmental chemicals has been rather limited (Olshan, 2007). Because the simple finger or heel prick to collect a few drops of whole blood on filter paper is less invasive than conventional venipuncture, DBS can be a potentially useful biomonitoring matrix not only for infants and young children but also for pregnant women and other adults. In this chapter, we will provide a general overview of the use of DBS for biomonitoring.

10.1.2 Analytical Considerations for Biomonitoring Using DBS

The technique for collecting DBS has been validated for a number of pharmaceutical compounds, and numerous studies have shown good correlation between analyses of DBS and conventional blood plasma (Boy et al., 2008; Edelbroek et al., 2009; Li and Tse, 2010). However, biomonitoring generally requires measuring the target analytes at relatively low concentrations (e.g., at or below parts per billion). Therefore, biomonitoring with DBS must employ suitable, state-of-the-art analytical methods that can provide adequate sensitivity and selectivity at trace concentrations. Furthermore, the volume of whole blood in the entire DBS is generally small (~75 μL), and other biomolecules often are present at concentrations higher than the trace levels common for the target environmental chemicals. Thus, extraction, separation, and quantification of these target chemicals from DBS will require highly sensitive, specific, and selective methods.

Below are some analytical considerations relevant for the successful use of DBS for biomonitoring purposes.

- *Sample volume.* DBS have a limited amount of available serum/whole blood for testing. A typical drop of capillary blood collected from a finger/heel stick includes ~75 μL of whole blood. Most newborn screening assays use only a punch from the DBS (~1.5 μL serum) (Mei et al., 2001; CLSI, 2007; CDC, 2011), but other assays may require an entire fully filled DBS. Of interest, although punch sizes are quite reproducible, the amount of blood contained within each punch can differ significantly and act as a source of variance (Mei et al., 2001). Furthermore, the entire residual DBS may not be available for biomonitoring due to punches taken for newborn screening tests. Therefore, pooling several DBS may be a viable strategy to obtain the needed volume of blood to achieve adequate sensitivity (Spliethoff et al., 2008). However, limitations related to the use of pools for population studies, applicable to other matrices, would also apply to DBS pools (Caudill, 2011). Regardless of the approach used, analytical methods that use DBS for biomonitoring purposes must provide sufficient sensitivity to accurately detect trace levels of the target compounds in relatively small (<100 μL) volumes of blood.
- *Contributions from the filter paper.* Filter paper variables, such as potential differences among lots and manufacturers, uneven distribution of the blood, and potential presence of the target chemicals, need to be evaluated and defined. Further, for ubiquitous environmental chemicals, such as certain compounds used in personal care and consumer products (e.g., bisphenol A [BPA], triclosan, parabens), care is needed to avoid contaminating DBS during collection or storage. Strict collection, handling, and storage protocols are particularly important if the chemicals monitored as exposure biomarkers are ubiquitous environmental contaminants or environmental degradates. Some of these chemicals (e.g., phthalates, polybrominated diphenyl ethers, polyfluoroalkyl chemicals [PFCs], BPA) have been detected in indoor air and dust (Wilson et al., 2003; Hwang et al., 2008; Volkel et al., 2008; Fromme et al., 2009; Rudel and Perovich, 2009; Weschler, 2009). DBS for newborn screening purposes are collected and handled according to the Clinical and Laboratory Standards Institute guidelines (CLSI, 2007) to avoid specific contamination (e.g., cross-contamination with maternal blood if cord blood is evaluated, extraneous DNA if

used for DNA testing), but these DBS will not necessarily be stored under controlled conditions (Therrell et al., 2011) and will not be stored to avoid environmental contaminants. Therefore, care must be taken during collection and processing of DBS specimens that may be used for biomonitoring to ensure that sampling materials do not contain detectable levels of the target chemicals and also that these materials are dust free. Field or procedural filter paper blanks can be used to assess potential contamination during sample collection, storage, processing, and/or transport (NIOSH, 1994). Recommended practices for sampling and processing applicable to ongoing and future biomonitoring initiatives that use conventional matrices (e.g., blood and urine) are also applicable with DBS (Calafat and Needham, 2009). Finally, production lots of DBS paper and whole blood should be screened for background levels of the target chemicals and of any potential analytical interferences before preparation of standards and QC materials.

- *Extraction of biological material from the filter paper.* The blood is allowed to dry on a filter paper for DBS. Therefore, the target analytes must be extracted from the paper. For this purpose, the required amount of DBS can be placed into a buffer solution. The extracted DBS can be reconstituted as hemolyzed whole blood and used for biomonitoring. The choice of extraction technique will depend on the physicochemical properties of the target analyte and availability of analytical instrumentation. Also, because of the uneven distribution of analytes in the whole blood onto the DBS, extraction efficiencies of the environmental contaminants may vary. A reference molecule proportional to blood volume may be needed to normalize the concentrations of some analytes (Olshan, 2007). For example, the normalization of DBS concentrations of minor and trace elements using the concentrations of major elements such as magnesium or calcium may be necessary for the measurement of metals and will need to be further evaluated.
- *Stability.* Degradation of some analytes on DBS may occur and likely will vary by analyte. Moreover, the time between DBS collection and analysis could be days, months, or even years if residual DBS newborn screening specimens are used for biomonitoring studies. Last, the storage conditions of DBS specimens might vary depending upon the newborn screening programs (see Section 1.4 below). Therefore, stability of the target analytes and the potential impact of storage conditions must be evaluated because the validity of the biomonitoring data will strongly depend on the integrity of the DBS specimen. For example, storage at low temperature and low humidity (Therrell et al., 1996) may be advisable, particularly if DBS will be analyzed for thermally labile compounds. Special storage and also shipping conditions may be required for light-sensitive compounds, although some chemicals may gain photostability on DBS compared to liquid matrices like plasma or blood (Bowen et al., 2010).
- *Comparison data between DBS and serum, plasma, or whole blood.* DBS results must be compared with results obtained from matched, simultaneously collected serum or plasma samples via venipuncture using a previously established protocol to confirm the validity of DBS for biomonitoring (Edelbroek et al., 2009). Furthermore, although hematocrit can be variable, the plasma content of whole blood is approximately 55%. Therefore, a correction factor must be applied to compare concentrations of target chemicals in DBS with concentrations of the same compounds in plasma or serum (i.e., multiply the whole blood/DBS level by 2 to make the results comparable to serum/plasma levels).

10.1.3 Selection of Biomarkers of Environmental Chemicals

A good understanding of the toxicokinetics of the environmental chemicals (e.g., distribution among body compartments, metabolism) and their bioactivity at environmental exposure levels is required to properly interpret any biomonitoring measurements. For persistent chemicals, the biomarker of exposure is generally the parent compound. On the other hand, nonpersistent chemicals can be partially or fully metabolized (e.g., phase I biotransformations) to increase their hydrophilic character, and both parent compound and Phase I metabolites can be excreted unchanged or can undergo phase II biotransformations. Generally, blood (or its components) is the preferred matrix for assessing exposure to persistent compounds. Therefore, in principle, DBS, like blood or its components, would be an adequate matrix for assessing exposure to metals and persistent organic chemicals (e.g., PFCs and dichlorodiphenyltrichloroethane [DDT]). For nonpersistent chemicals (e.g., perchlorate; see Section 2.5.), blood levels may vary considerably in response to episodic exposures and rapid clearance. Therefore, the analytical procedure has to be sensitive enough to allow for the detection of the target biomarker at the trace levels expected in DBS.

Furthermore, for ubiquitous environmental chemicals, such as phthalates, BPA, triclosan, or sunscreen agents (e.g., 2-hydroxy-4-methoxybenzophenone, 2-ethylhexylsalicylate, 2-ethylhexyl-*p*-methoxycinnamate), attention to potential contamination during sample collection and storage is important. For example, contamination of biological specimens with diesters of phthalic acid, commonly known as phthalates—industrial chemicals used in the manufacture of many consumer and personal care products—is difficult to avoid. In humans, phthalates quickly metabolize into their hydrolytic monoacid (commonly referred to as "monoester") (Calafat and Needham, 2008, 2009). In some cases, the monoesters can be oxidized into oxidative metabolites. Hydrolytic and oxidative metabolites are then primarily excreted in the urine as phase II conjugates (e.g., glucuronides) and/or as the unchanged or free species. Measuring the urinary concentrations of the total (conjugated plus free) species of these metabolites is the most common biomonitoring approach for assessing human exposure to phthalates. Esterases present in blood can hydrolyze phthalate diesters to their hydrolytic monoesters, beginning immediately after sample collection (Kato et al., 2003). Therefore, the use of hydrolytic monoester concentrations in blood (Kato et al., 2003) or DBS should be avoided as biomarkers of exposure even though they can be determined accurately. By contrast, the concentrations of oxidative metabolites can potentially be used as biomarkers of exposure to phthalates because these metabolites are not formed from the simple hydrolysis of phthalate diesters by esterases. Unfortunately, because the concentrations of the oxidative metabolites in blood are much lower than the corresponding urinary concentrations, methods with increased analytical sensitivity are needed to maximize the utility of the oxidative metabolites as biomarkers in matrices other than urine.

10.1.4 Storage of Residual DBS and Their Potential Impact for DBS Use in Biomonitoring

Residual DBS represent a potentially valuable resource for research. However, several factors may hinder their use for biomonitoring purposes. First, the cost of storage, retrieval, and preparation of DBS for biomonitoring projects could be substantial (e.g., under monitored conditions or subfreezing temperatures), particularly in times of limited budgets. Second, two important considerations related to storage of residual DBS are their duration and the physical conditions (i.e., temperature, humidity). Storage guidelines do not exist. In fact, both factors vary considerably from state to state within the United States and, presumably, also differ by countries and localities worldwide. For example, duration can range from more than 6 months for about half of 47 states to over a year for more than 40% of states (Olney et al., 2006; Therrell et al., 2011). Temperature and humidity conditions also can be quite varied, from room temperature to freezing and from storage in sealed plastic bags containing desiccants to humidity-monitored environments.

Lack of written storage guidelines regulating physical conditions, privacy, and consent issues of the potential use of DBS for biomonitoring purposes might affect future use of these specimens.

10.1.5 Advantages of Using DBS for Biomonitoring

The current clinical standard for blood collection involves venipuncture to obtain several milliliters of whole blood, centrifugation under specified conditions, and transfer of the upper plasma or serum layer into another tube, followed by predetermined storage conditions. The current clinical standard for DBS, by contrast, is much less complicated: after a simple finger or heel prick, a few drops of blood are collected, the blood is placed on a filter paper or card and allowed to dry, and the DBS cards are stacked and stored (CLSI, 2007). Reports exist of some analytes relevant to drug monitoring remaining stable at room temperature for at least 1 week (Edelbroek et al., 2009), providing considerable flexibility in procedures for sample storage and transport.

The ease of collection and the ease of transport are two major advantages of DBS blood sampling over venipuncture. For DBS, blood collection is relatively painless and noninvasive, can happen at home, and can be performed by persons without medical training. This factor is a particular advantage for research with infants, children, and the elderly, for whom venipuncture may be problematic, as well as for research in remote or underserved communities, where the logistics of venipuncture may limit access to willing participants. At the same time, controlling for potential external contamination under these collection conditions could be a limitation of using DBS for biomonitoring of certain chemicals. This is particularly critical for chemicals with widespread commercial and industrial use when the target biomarker to be measured in the DBS is the chemical used in commercial applications (Calafat and Needham, 2009).

For conventional venipuncture, transportation of the biological samples requires use of dry ice and is therefore very costly. Moreover, shipping of hazardous materials such as blood is regulated under the International Air Transport Association (http://www.iata.org/whatwedo/cargo/dangerous_goods/Pages/infectious_substances.aspx). By contrast, DBS cards can be shipped in a sealed bag with desiccant and although they need to be labeled with a biohazard symbol, the symbol does not need to be displayed on the outside of the shipping package (CLSI, 2007).

10.1.6 Disadvantages of Using DBS

Despite the appealing qualities of DBS collection for biomonitoring purposes, it has several drawbacks. First, the blood volume collected on a single DBS is rather small (~75 μL). Therefore, achieving adequate sensitivity is challenging and often requires costly hyphenated analytical techniques (e.g., liquid or gas chromatography coupled to mass spectrometry). Furthermore, the target analytes will need to be extracted from the filter paper, thus increasing the complexity and cost of the analytical method. One such technique, micro solid phase extraction, may provide additional cleanup of the matrix-depleted DBS extract but cannot be used for volatile compounds and certain unstable metabolites. Another consequence of the limited sample volume available in DBS is that repeat analyses may not be possible.

Another important limitation is the fact that DBS samples are not homogeneous (Edelbroek et al., 2009; Liang et al., 2009; Denniff and Spooner, 2010). Unlike liquid whole blood, plasma, or serum, the distribution of some analytes between the peripheral and central parts of a DBS differs. Additionally, although this is not an issue for newborn screening applications, the quality of DBS specimens for absolute quantification for biomonitoring may be questionable because of possible DBS paper quality issues and lack of uniformity in sample collection procedures (e.g., blood volume on the spot, degree of hemolysis) (Li and Tse, 2010).

DBS offer the promise of making biomonitoring available to a wide range of populations but at the same time represent substantial analytical challenges for both method development and quantitation of analytes. Specifically, before DBS can be successfully used as part of routine biomonitoring programs, in addition to addressing the factors described above, comparison of results obtained with conventional venipuncture techniques and DBS sampling must be conducted. Studies that use DBS as the biomonitoring matrix should provide sufficient background information on all steps involved during the collection, processing, storage, and transport of DBS. This includes recording information on the sampling time and location (e.g., home, hospital), whether sample collection will be part of prescheduled or *ad hoc* health visits (e.g., child well-being, prenatal care appointments, delivery), and detailed description of all procedures, utensils, and materials used during collection (e.g., wipes, containers) and storage (e.g., plastic bags, temperature) of DBS before arrival to the laboratory for analysis.

10.2 APPLICATION OF DBS FOR BIOMONITORING OF ENVIRONMENTAL CHEMICALS

Several environmental chemicals have been measured in DBS (Table 10.1). In this section, we provide an overview of current options for measuring select biomarkers in DBS specimens and discuss issues related to methods evaluation that can help investigators decide on the appropriateness of a method for their research.

10.2.1 Metals

Of interest are major elements such as calcium and magnesium; minor elements such as mercury, iron, and zinc; and trace elements such as lead, arsenic, cadmium, chromium, and copper. Concentrations of lead, mercury, and cadmium in 18 DBS from Utah and three other Rocky Mountain states in the United States were reported by Chaudhuri et al. (Chaudhuri et al., 2009). The metals were extracted with acid, followed by detection with inductively coupled plasma-mass spectrometry (ICP-MS). The authors suggested that the filter paper may contain cadmium, and to a lesser extent lead but not mercury, and that the levels of these chemicals vary considerably among different lots of filter paper and manufacturers. Unfortunately, the high background levels for cadmium precluded the measurement of this element in DBS. The authors reported adequate stability of lead and mercury in DBS for 8.5 months of storage at room temperature. Therefore, the method was successfully applied to the measurement of lead and mercury: concentrations ranged from 0.64 to 1.84 μg/dL for lead (limit of detection [LOD] = 0.36 μg/dL) and from <0.65 to 2.83 μg/L for mercury (LOD = 0.65 μg/L).

10.2.2 Dichlorodiphenyldichloroethylene

Burse et al. conducted a study in Texas to assess prenatal exposure to organochlorine compounds. They examined DBS from 2-day-old infants to find the concentration of dichlorodiphenyldichloroethylene (DDE), a metabolite of the pesticide DDT (Burse et al., 1997). One whole DBS was cut and placed in a tube that contained 2.0 mL of a 1:16 methanol:phosphate buffer solution (pH 7.4) containing 0.1% v/v of Tween 20. DDE was extracted from the filter paper with hexane, and the DBS extracts were evaporated under a stream of nitrogen. DDE was quantified by capillary gas chromatography with electron capture detection (GC-ECD). Concentrations in 10 DBS ranged from 0.13 to 1.87 pg/μL with a median value of 0.69 pg/μL. Recently, Patterson et al. reported a method to quantify DDE and polybrominated biphenyl flame-retarding chemicals (e.g., BB-153) in DBS with two-dimensional GC high-resolution magnetic sector MS (Patterson et al., 2011).

10.2.3 Hexachlorohexane and DDT

In a study conducted in India which involved 37 persons from the general population and 47 workers occupationally exposed to hexachlorohexane (HCH) and DDT, these compounds were measured in DBS. The target compounds were extracted with a hexane:acetone (1:1) solution, and the extract was concentrated by evaporation and analyzed with GC-ECD (Dua et al., 1996). The recoveries ranged from 84 to 92% with standard deviations of 2.7–8.4 at 2 μg/L for the HCH isomers and for DDT and its metabolites; the LODs were 0.003 μg/L (HCH isomers) and 0.01 μg/L (DDT and metabolites). DBS and venipuncture whole blood concentrations showed statistically significant correlations for HCH ($r = 0.93$, $p < 0.001$) and DDT ($r = 0.95$, $p < 0.001$). The HCH and DDT spiked onto the DBS were stable up to 1 month, and no concentration difference was observed between venous blood and DBS. The mean concentrations for the general population group ($n = 37$) were 21.6 μg/L for the sum of HCH isomers (α, β, γ, and δ) and 20.79 μg/L for total DDT (o,p' and p,p'-DDE). For the exposed group

TABLE 10.1 Examples of Environmental Chemical Studies Using DBS

Use of DBS for Human Studies							
Target Chemicals	Detection Technique	DBS Sample Size/Volume	Extraction Solvent	Sample Cleanup Approach	DBS median Concentrations	Reference	Median Serum Concentrations from the US General Population
Select heavy metals	ICP/MS	1/4 in. (11.5 μL)	2% HCl[a]		Utah (USA), (n = 18, newborn) Lead: 0.64–1.84 μg/dL Mercury: <0.65–2.83 μg/L	(Chaudhuri et al., 2009)	Lead: 1.27 pg/dL[b] Mercury: 0.769 μg/L[b]
DDE	GC/ECD	75–100 μL	Hexane		Texas (USA) (n = 10, newborn) 0.69 pg/μL	(Burse et al., 1997)	238 ng/g lipid[c]
HCH, DDT	GC/ECD		Hexane:acetone (1:1)		n = 37 (general population) HCH: 21.6 μg/L DDT: 20.79 μg/L	(Dua et al., 1996)	HCH: <7.8 (ng/g lipid)[c] DDT: <7.8 (ng/g lipid)[c]
PFCs	HPLC/MS/MS	~75 μL	75% methanol	On-line SPE	Texas (USA) 2007 (n = 98, newborn) PFHxS: 0.3 μg/L PFNA: 0.3 μg/L PFOA: 0.9 μg/L PFOS: 2.2 μg/L	(Kato et al., 2009)	PFHxS: 2.0 μg/L serum[b] PFNA: 1.23 μg/L serum[b] PFOA: 4.3 μg/L serum[b] PFOS: 13.6 μg/L serum[b]
	HPLC/MS/MS	24 × 1/4 in. (~32.6 μL)	Water	Ion-pair extraction	New York (USA) 2003–2004 (n = 20 pooled, newborn) PFOSA: 0.28 μg/L PFHxS: 1.64 μg/L PFNA: 0.35 μg/L PFOA: 0.73 μg/L PFOS: 1.59 μg/L	(Spliethoff et al., 2008)	PFOSA: <.2 μg/L serum[c] PFHxS: 1.9 μg/L serum[c] PFNA: 1.0 μg/L serum[c] PFOA: 4.1 μg/L serum[c] PFOS: 21.2 μg/L serum[c]
Perchlorate	IC/MS/MS	10 × 3.2 mm	Methanol		New York (n = 10, pooled) 3.03 μg/L	(Otero-Santos et al., 2009)	N/A in serum 3.0 μg/L urine[c]

[a] 2% HCl containing 0.05% 2-mercaptoethanol, 0.001% L-cystein, 10 ppb iridium and rhodium.
[b] The National Health and Nutrition Examination Survey 2007–2008 (CDC, 2013).
[c] The National Health and Nutrition Examination Survey 2003–2004 (CDC, 2013).

($n = 47$), the concentrations were about three times higher: 68.01 μg/L for total HCH and 58.43 μg/L for total DDT. The authors suggested that DBS provided a great advantage because venipunctures were often difficult to standardize for the handling of blood samples in tropical field work.

10.2.4 PFCs

In DBS, PFCs have been measured by isotope dilution high-performance liquid chromatography-tandem mass spectrometry (HPLC-MS/MS) in two different studies from the United States (Spliethoff et al., 2008; Kato et al., 2009). Of interest, the concentrations of PFCs in DBS from these two studies were quite similar, suggesting that exposure to some PFCs may occur in utero. Spliethoff et al. measured perfluorooctane sulfonate (PFOS), perfluorooctane sulfonamide (PFOSA), perfluorohexane sulfonate (PFHxS), perfluorooctanoate (PFOA), and perfluorononanoate (PFNA) in residual blood spots collected in New York State's newborn screening program between 1997 and 2007. The analyses were conducted with 110 composite DBS (representing 2640 infants). The PFCs in the DBS were extracted by an ion-pair extraction method using methyl *tert*-butyl ether, 0.5 M tetrabutyl ammonium hydrogen sulfate (adjusted to pH 10) and 0.25 M sodium carbonate/sodium bicarbonate buffer. The DBS extracts were centrifuged, evaporated, and reconstituted in methanol. The recoveries (60–112%) and precision (RSD <10–26%) were adequate. The LODs, expressed in units of whole blood equivalents, were 0.04–0.4 μg/L. All analytes were detected in >90% of the specimens. Concentrations of PFOS, PFOSA, PFHxS, and PFOA exhibited significant exponential declines after the year 2000, coinciding with the phase-out production of PFOS in the United States (Spliethoff et al., 2008).

More recently, Kato et al. measured PFHxS, PFOS, PFOA, and PFNA in 98 single residual DBS collected in 2007 from infants in Texas (Kato et al., 2009). To extract the target analytes from the DBS, the authors added methanol, 0.1 M formic acid, and a stable isotope-labeled internal standard solution. After vortex mixing and ultrasonication, the four PFCs in the DBS extracts were detected and quantified by on-line solid phase extraction coupled to HPLC-MS/MS. The analytical LODs using one whole DBS (~75 μL of blood) were <0.4 μg/L for all four compounds. The accuracy and precision of the method were acceptable (recovery: 52.9–145.9%, interday precision (RSD, 3 weeks, 50 points): 13.6–22.1%; intraday precision: 2.3–12.6%). PFOS and PFOA were detected in all DBS at concentrations ranging from 0.4 to 11.3 μg/L (PFOS) and 0.4 to 3.5 μg/L (PFOA); PFNA was detected in almost all specimens (98%) and PFHxS was detected less frequently (70%). The median concentrations were 2.2 μg/L (PFOS), 0.9 μg/L (PFOA), and 0.3 μg/L (PFHxS, PFNA). These concentrations are lower than the serum concentrations reported among NHANES participants (Kato et al., 2009). Note, however, that because most PFCs bind to proteins in serum/plasma and the serum/plasma content of whole blood is approximately 55%, concentrations of PFCs in DBS (expressed as whole blood) are expected to be lower than in serum (see Section 1.2).

10.2.5 Perchlorate

Environmental exposure to perchlorate has been associated with changes in thyroid hormone levels in US women, leading to concerns about perchlorate exposure and potential thyroid dysfunction in newborns (CDC, 2011). Newborn DBS provide a convenient matrix for assessing both neonatal thyroid function and perchlorate exposure. In one study, perchlorate was measured in DBS collected from 100 infants by ion chromatography-MS/MS (IC-MS/MS) with stable isotope dilution (Otero-Santos et al., 2009) with a LOD of 0.1 μg/L. To each sample, consisting of 10 DBS punches of 3.2 mm, methanol was added and the sample vortexed. The methanol extract solutions were evaporated and transferred to microcentrifuge filtration tubes, and the stable isotope-labeled internal standard solution was added, followed by spin-filtrate centrifugation. The filtered samples were analyzed by IC-MS/MS. The recoveries (97–112%), precision (7.7–16.2% RSD), and accuracy (4.8–12.1% difference) at three different concentrations (0.33, 3.3, 33 μg/L) were adequate. The QC materials were made by adding 2.27 and 27.3 μg/L of perchlorate to whole blood, then spotted on filter paper cards (100 μL/spot), and allowed to dry overnight. The stability of perchlorate in the QC materials was examined up to 24 months; results showed it would be stable in DBS stored at −20°C for a prolonged period. Perchlorate was detected in 100% of the DBS examined. Traces of perchlorate were also detected in the filter paper and required the subtraction of filter paper blanks from DBS perchlorate values.

10.2.6 Measurement of Protein Adducts

Adult and infant DBS were used to measure hemoglobin or serum albumin adducts of carcinogens, primarily 1,4-benzoquinone (1,4-BQ). First, globin was selectively isolated from DBS, and the cysteinyl adducts of 1,4-BQ were detected by GC-MS. Other than these preliminary results (Olshan, 2007), no additional data exist in the published peer-reviewed literature for this specific use of DBS.

10.2.7 Novel Analytical Technologies with Potential Applications in DBS Biomonitoring

In the last few years, several analytical approaches have offered the possibility of analyzing DBS spiked with pharmaceutical compounds with minimal sample preparation and extraction steps. Some of these techniques are increased automation, adding derivatizing reagents on the DBS card (Ingels et al., 2010), direct analysis without sample preparation (Abu-Rabie and Spooner, 2009; Wiseman et al., 2010; Heinig et al., 2011; Manicke et al., 2011), on-line desorption of DBS (Deglon et al., 2009; Thomas et al., 2010; Deglon et al., 2011), chip-based infusion nanoelectrospray ionization (Kertesz and Van Berkel, 2010), desorption electrospray ionization (Costa and Cooks, 2007; Ranc et al., 2008), paper spray ionization (Manicke et al., 2011), and thin-layer chromatography mass spectrometry (Abu-Rabie and Spooner, 2009). These novel quantitative methods, already applicable in the pharmaceutical field for the analysis of whole blood, offer incredible promise for the analysis of DBS for biomonitoring purposes.

10.2.8 Biomonitoring of Nonhuman Species

The ease of collection and transport of DBS could also facilitate the use of DBS for biomonitoring in nonhuman species, such as wild birds (Shlosberg et al., 2011; Shlosberg et al., 2012). For example, DBS could be used to establish reference values of potentially harmful environmental agents in avian species. Such reference values could be useful for the diagnosis of environmental contaminant-mediated ill-health (such as manifestations of sickness, increased mortality, reduction in population, poor breeding success, abnormal behavior) in individual birds or in bird populations by comparing DBS concentrations of the target compounds in investigated birds to the reference values.

In the case of rodent experiments, tail pricks are the preferred mode of bleeding. Graduated EDTA-coated capillaries can be used to collect the blood from tail pricks, followed by direct spotting onto DBS filter paper (Uyeda et al., 2011). As with human biomonitoring, analytical method parameters (e.g., accuracy, precision, linearity range, robustness/ruggedness) should be well established. Moreover, these methods must have high sensitivity; some of these animals are small, and the volume of blood that can be obtained without harming the animal could be rather low (25–500 μL) (Clark et al., 2011; Uyeda et al., 2011).

10.3 CONCLUSIONS

In population-based studies, biomonitoring data can be useful to establish background and/or reference levels of select environmental chemicals and can also be used to identify segments of the population that may be the most highly exposed. Urine, blood, and breast milk are the most commonly used biomonitoring matrices in population-based studies. Adding DBS to data-collection protocols for population-based studies could provide biomonitoring data from infants, young children, pregnant women or the elderly, for whom venipuncture may not be possible. These populations represent some of the most vulnerable and susceptible in terms of the potential adverse health effects from environmental chemical exposures.

DBS have the potential to be a potential matrix for assessing exposure to select environmental chemicals among both human populations and nonhuman species. High-quality analytical methods already permit the determination of target environmental chemicals in DBS. However, limited data exist on the suitability of DBS for biomonitoring purposes, particularly for chemicals with widespread commercial and industrial use. Validated protocols for the collection, handling, shipping, and storage procedures to preserve the integrity of both DBS and the target analytes must be developed first to guarantee valid generation and interpretation of biomonitoring data. Further research is needed to determine the suitability of DBS for epidemiologic studies to assess exposures to environmental chemicals.

DISCLAIMER

The findings and conclusions in this chapter are those of the authors and do not necessarily represent the views of the Centers for Disease Control and Prevention.

REFERENCES

Abu-Rabie P, Spooner N. Direct quantitative bioanalysis of drugs in dried blood spot samples using a thin-layer chromatography mass spectrometer interface. *Analytical Chemistry* 2009;81:10275–10284.

Angerer J, Ewers U, Wilhelm M. Human biomonitoring: state of the art. *International Journal of Hygiene and Environmental Health* 2007;210:201–228.

Arbuckle TE. Maternal-infant biomonitoring of environmental chemicals: the epidemiologic challenges. *Birth Defects Research. Part A, Clinical and Molecular Teratology* 2010;88: 931–937.

Barr DB, Wang RY, Needham LL. Biologic monitoring of exposure to environmental chemicals throughout the life stages: requirements and issues for consideration for the National Children's Study. *Environmental Health Perspectives* 2005;113:1083–1091.

Borrajo GJC. Newborn screening in Latin America at the beginning of the 21st century. *Journal of Inherited Metabolic Disease* 2007;30:466–481.

Bowen CL, Hemberger MD, Kehler JR, Evans CA. Utility of dried blood spot sampling and storage for increased stability of photosensitive compounds. *Bioanalysis* 2010;2:1823–1828.

Boy RG, Henseler J, Mattern R, Skopp G. Determination of morphine and 6-acetylmorphine in blood with use of dried blood spots. *Therapeutic Drug Monitoring* 2008;30:733–739.

Burse VW, DeGuzman MR, Korver MP, Najam AR, Williams CC, Hannon WH, Therrell BL. Preliminary investigation of the use of dried-blood spots for the assessment of in utero exposure to environmental pollutants. *Biochemistry and Molecular Medicine* 1997;61:236–239.

Calafat AM, Needham LL. Factors affecting the evaluation of biomonitoring data for human exposure assessment. *International Journal of Andrology* 2008;31:139–143.

Calafat AM, Needham LL. What additional factors beyond state-of-the-art analytical methods are needed for optimal generation and interpretation of biomonitoring data? *Environmental Health Perspectives* 2009;117:1481–1485.

Caudill SP. Important issues related to using pooled samples for environmental chemical biomonitoring. *Statistics in Medicine* 2011;30:515–521.

CDC. *National Health and Nutrition Examination Survey*. Atlanta, GA: Centers for Disease Control and Prevention; National Center for Health Statistics; 2003. http://www.cdc.gov/nchs/nhanes.htm (accessed January 16, 2014).

CDC. *National Report on Human Exposure to Environmental Chemicals*. Updated Tables. Atlanta, GA: Centers for Disease Control and Prevention; National Center for Environmental Health; Division of Laboratory Sciences; 2013. http://www.cdc.gov/exposurereport/ (accessed January 17, 2014).

CDC. *Newborn Screening Quality Assurance Program. Program Reports*. Atlanta, GA: Centers for Disease Control and Prevention; National Center for Environmental Health; Division of Laboratory Sciences; 2011. http://www.cdc.gov/labstandards/nsqap_reports.html (accessed January 18, 2014).

Chaudhuri SN, Butala SJM, Ball RW, Braniff CT. Pilot study for utilization of dried blood spots for screening of lead, mercury and cadmium in newborns. *Journal of Exposure Science and Environmental Epidemiology* 2009;19:298–316.

Clark GT, Giddens G, Burrows L, Strand C. Utilization of dried blood spots within drug discovery: modification of a standard DiLab (R) AccuSampler (R) to facilitate automatic dried blood spot sampling. *Laboratory Animals* 2011;45:124–126.

CLSI. In: Hannon WH, Whitley RJ, Davin B, Fernhoff P, Halonen T, Lavochkin M, Miller J, Ojodu J, Therrell BL (eds), *Blood Collection on Filter Paper for Newborn Screening Programs*; Approved Standard-Fifth Edition. CLSI document LA4-A5. Wayne, PA: Clinical and Laboratory Standards Institute; 2007.

Costa AB, Cooks RG. Simulation of atmospheric transport and droplet-thin film collisions in desorption electrospray ionization. *Chemical Communications* 2007;(38):3915–3917.

Deglon J, Thomas A, Cataldo A, Mangin P, Staub C. On-line desorption of dried blood spot: a novel approach for the direct LC/MS analysis of mu-whole blood samples. *Journal of Pharmaceutical and Biomedical Analysis* 2009;49:1034–1039.

Deglon J, Thomas A, Daali Y, Lauer E, Samer C, Desmeules J, Dayer P, Mangin P, Staub C. Automated system for on-line desorption of dried blood spots applied to LC/MS/MS pharmacokinetic study of flurbiprofen and its metabolite. *Journal of Pharmaceutical and Biomedical Analysis* 2011;54:359–367.

Denniff P, Spooner N. The effect of hematocrit on assay bias when using DBS samples for the quantitative bioanalysis of drugs. *Bioanalysis* 2010;2:1385–1395.

Dua VK, Pant CS, Sharman VP, Pathak GK. Determination of HCH and DDT in finger-prick whole blood dried on filter paper and its field application for monitoring concentrations in blood. *Bulletin of Environmental Contamination and Toxicology* 1996;56:50–57.

Edelbroek PM, van der Heijden J, Stolk LML. Dried blood spot methods in therapeutic drug monitoring: methods, assays, and pitfalls. *Therapeutic Drug Monitoring* 2009;31:327–336.

Fromme H, Tittlemier SA, Volkel W, Wilhelm M, Twardella D. Perfluorinated compounds – exposure assessment for the general population in western countries. *International Journal of Hygiene and Environmental Health* 2009;212:239–270.

Fujii C, Sato Y, Harada S, Kakee N, Gu YH, Kato T, Shintaku H, Owada M, Hirahara F, Umehashi H, Yoshino M. Attitude to extended use and long-term storage of newborn screening blood spots in Japan. *Pediatrics International* 2010;52:393–397.

Heinig K, Wirz T, Bucheli F, Gajate-Perez A. Determination of oseltamivir (Tamiflu (R)) and oseltamivir carboxylate in dried blood spots using offline or online extraction. *Bioanalysis* 2011;3:421–437.

Hwang HM, Park EK, Young TM, Hammock BD. Occurrence of endocrine-disrupting chemicals in indoor dust. *Science of the Total Environment* 2008;404:26–35.

Ingels ASME, Lambert WE, Stove CP. Determination of gamma-hydroxybutyric acid in dried blood spots using a simple GC-MS method with direct "on spot" derivatization. *Analytical and Bioanalytical Chemistry* 2010;398:2173–2182.

Jaques AM, Collins VR, Pitt J, Halliday JL. Coverage of the Victorian newborn screening programme in 2003: a retrospective population study. *Journal of Paediatrics and Child Health* 2008;44:498–503.

Kato K, Silva MJ, Brock JW, Reidy JA, Malek NA, Hodge CC, Nakazawa H, Needham LL, Barr DB. Quantitative detection of nine phthalate metabolites in human serum using reversed-phase high-performance liquid chromatography-electrospray ionization-tandem mass Spectrometry. *Journal of Analytical Toxicology* 2003;27:284–289.

Kato K, Wanigatunga AA, Needham LL, Calafat AM. Analysis of blood spots for polyfluoroalkyl chemicals. *Analytica Chimica Acta* 2009;656:51–55.

Kato K, Wong L-Y, Jia LT, Kuklenyik Z, Calafat AM. Trends in exposure to polyfluoroalkyl chemicals in the U.S. population: data from the national health and nutrition examination survey (NHANES) 1999–2008. *Environmental Science & Technology* 2011; 45:8037–8045.

Kerruish NJ, Webster D, Dickson N. Information and consent for newborn screening: practices and attitudes of service providers. *Journal of Medical Ethics* 2008;34:648–652.

Kertesz V, Van Berkel GJ. Fully automated liquid extraction-based surface sampling and ionization using a chip-based robotic nanoelectrospray platform. *Journal of Mass Spectrometry* 2010;45:252–260.

Li W, Tse FLS. Dried blood spot sampling in combination with LC-MS/MS for quantitative analysis of small molecules. *Biomedical Chromatography* 2010;24:49–65.

Liang XR, Li YH, Barfield M, Ji QC. Study of dried blood spots technique for the determination of dextromethorphan and its metabolite dextrorphan in human whole blood by LC-MS/MS. *Journal of Chromatography B* 2009;877:799–806.

Manicke NM, Yang QA, Wang H, Oradu S, Ouyang Z, Cooks RG. Assessment of paper spray ionization for quantitation of pharmaceuticals in blood spots. *International Journal of Mass Spectrometry* 2011;300:123–129.

Mcdade TW, Williams S, Snodgrass JJ. What a drop can do: dried blood spots as a minimally invasive method for integrating biomarkers into population-based research. *Demography* 2007;44:899–925.

Mei JV, Alexander JR, Adam BW, Hannon WH. Use of filter paper for the collection and analysis of human whole blood specimens. *Journal of Nutrition* 2001;131:1631S–1636S.

NIOSH. NIOSH Manual of Analytical Methods, 4th ed.; 1994. http://www.cdc.gov/niosh/nmam/chaps.html (accessed January 16, 2014).

Olney RS, Moore CA, Ojodu JA, Lindegren ML, Hannon WH. Storage and use of residual dried blood spots from state newborn screening programs. *Journal of Pediatrics* 2006;148:618–622.

Olshan AF. Meeting report: the use of newborn blood spots in environmental research: opportunities and challenges. *Environmental Health Perspectives* 2007;115:1767–1769.

Otero-Santos SM, Delinsky AD, Valentin-Blasini L, Schiffer J, Blount BC. Analysis of perchlorate in dried blood spots using ion chromatography and tandem mass spectrometry. *Analytical Chemistry* 2009;81:1931–1936.

Patterson DG, Welch SM, Turner WE, Sjodin A, Focant JF. Cryogenic zone compression for the measurement of dioxins in human serum by isotope dilution at the attogram level using modulated gas chromatography coupled to high resolution magnetic sector mass spectrometry. *Journal of Chromatography A* 2011;1218:3274–3281.

Ranc V, Havlicek V, Bednar P, Lemr K. Nano-desorption electrospray and kinetic method in chiral analysis of drugs in whole human blood samples. *European Journal of Mass Spectrometry* 2008;14:411–417.

Rothwell EW, Anderson RA, Burbank MJ, Goldenberg AJ, Lewis MH, Stark LA, Wong B, Botkin JR. Concerns of Newborn Blood Screening Advisory Committee members regarding storage and use of residual newborn screening blood spots. *American Journal of Public Health* 2011;101:2111–2116.

Rudel RA, Perovich LJ. Endocrine disrupting chemicals in indoor and outdoor air. *Atmospheric Environment* 2009;43:170–181.

Shlosberg A, Rumbeiha WK, Lublin A, Kannan K. A database of avian blood spot examinations for exposure of wild birds to environmental toxicants: the DABSE biomonitoring project. *Journal of Environmental Monitoring* 2011;13:1547–1558.

Shlosberg A, Wu Q, Rumbeiha WK, Lehner A, Cuneah O, King R, Hatzofe O, Kannan K, Johnson M. Examination of Eurasian griffon vultures (Gyps fulvus fulvus) in Israel for exposure to environmental toxicants using dried blood spots. *Archives of Environmental Contamination and Toxicology* 2012;62: 502–511.

Spliethoff HM, Tao L, Shaver SM, Aldous KM, Pass KA, Kannan K, Eadon GA. Use of newborn screening program blood spots for exposure assessment: declining levels of perfluorinated compounds in New York state infants. *Environmental Science and Technology* 2008;42:5361–5367.

Therrell BL, Adams J. Newborn screening in North America. *Journal of Inherited Metabolic Disease* 2007;30:447–465.

Therrell BL, Hannon WH, Pass KA, Lorey F, Brokopp C, Eckman J, Glass M, Heidenreich R, Kinney S, Kling S, Landenburger G, Meaney FJ, Mccabe ERB, Panny S, Schwartz M, Shapira E. Guidelines for the retention, storage, and use of residual dried blood spot samples after newborn screening analysis: statement of the Council of Regional Networks for Genetic Services. *Biochemistry and Molecular Medicine* 1996;57:116–124.

Therrell BL, Hannon WH, Bailey DB, Goldman EB, Monaco J, Norgaard-Pedersen B, Terry SF, Johnson A, Howell RR. Committee report: considerations and recommendations for national guidance regarding the retention and use of residual dried blood spot specimens after newborn screening. *Genetics in Medicine* 2011;13:621–624.

Thomas A, Deglon J, Steimer T, Mangin P, Daali Y, Staub C. On line desorption of dried blood spots coupled to hydrophilic interaction/reversed-phase LC/MS/MS system for the simultaneous analysis of drugs and their polar metabolites. *Journal of Separation Science* 2010;33:873–879.

Uyeda C, Pham R, Fide S, Henne K, Xu GF, Soto M, James C, Wong P. Application of automated dried blood spot sampling and LC-MS/MS for pharmacokinetic studies of AMG 517 in rats. *Bioanalysis* 2011;3:2349–2356.

van Amsterdam P, Waldrop C. The application of dried blood spot sampling in global clinical trials. *Bioanalysis* 2010;2:1783–1786.

Volkel W, Genzel-Boroviczeny O, Demmelmair H, Gebauer C, Koletzko B, Twardella D, Raab U, Fromme H. Perfluorooctane sulphonate (PFOS) and perfluorooctanoic acid (PFOA) in human breast milk: results of a pilot study. *International Journal of Hygiene and Environmental Health* 2008;211:440–446.

Weschler CJ. Changes in indoor pollutants since the 1950s. *Atmospheric Environment* 2009;43:153–169.

Wilson NK, Chuang JC, Lyu C, Menton R, Morgan MK. Aggregate exposures of nine preschool children to persistent organic pollutants at day care and at home. *Journal of Exposure Analysis and Environmental Epidemiology* 2003;13:187–202.

Wiseman JM, Evans CA, Bowen CL, Kennedy JH. Direct analysis of dried blood spots utilizing desorption electrospray ionization (DESI) mass spectrometry. *Analyst* 2010;135:720–725.

11

THE USE OF DRIED BLOOD SPOTS AND STAINS IN FORENSIC SCIENCE

Donald H. Chace and Nicholas T. Lappas

11.1 INTRODUCTION

Dried blood spot (DBS) most commonly refers to blood that has been applied to a special filter paper and dried. Its use was pioneered in the early 1960s by Dr Robert Guthrie for quantitation of phenylalanine and the subsequent screening of phenylketonuria, an inherited metabolic disorder of amino acid metabolism (Guthrie and Susi, 1963). The scope of DBS screening for inherited metabolic disorders increased in the past five decades to include galactosemia, congenital hypothyroidism, congenital adrenal hyperplasia, cystic fibrosis, sickle cell disease, and disorders of amino acid, organic acid, and fatty acid metabolism (Chace et al., 2003b) (see Chapter 6). Several of these disorders, especially aberrations of organic acid metabolism and fatty acid oxidation, have been correlated with sudden and unexplained death (Bennett et al., 1991; Boles et al., 1994, 1998; Rashed et al., 1995; Keppen and Randall, 1999; Matern et al., 1999). In many of these cases, findings such as fatty liver or cardiomyopathy remained inconclusive results until a biochemical test was developed to detect markers clearly associated with abnormal metabolism. The biochemical test that was key to this recognition was MS/MS with detection of acylcarnitine metabolites, the same metabolites used to detect inherited disorders in newborns. This test, based on the analysis of blood or other fluids collected at autopsy and applied to filter paper, was aptly named the metabolic autopsy (Bennett and Rinaldo, 2001; Rinaldo et al., 2004).

Collection of DBS and examination of metabolic profiles is also useful for other medicolegal purposes such as ruling out shaken baby syndrome by the confirmation of a metabolic disease such as glutaric acidemia (Gago et al., 2003). This application in the forensic sciences might be termed metabolic pathology to distinguish it from forensic toxicology, which includes the detection of drugs and other toxins in blood and other bodily fluids and tissues (Figure 11.1). DNA identification, another common forensic use of DBS, will not be presented here (see Chapter 4).

It would be a somewhat limiting discussion of DBS to presume that only dried blood on the special cellulose filter paper used in newborn screening (NBS) or drug detection would be of value in forensic analysis. In addition to blood, forensic scientists utilize a variety of both antemortem and postmortem samples. Antemortem samples such as urine and saliva are collected from persons for a number of purposes (e.g., pre-employment screening, the investigation of industrial accidents). Postmortem samples are collected as a matter of routine at autopsy and depending on the condition of the body and the circumstance of death (i.e., exposure to fire, drowning, postmortem decomposition) and may include any of the available samples (e.g., urine, bile, and vitreous humor and tissues such as liver and brain). In addition to the biological samples collected for the purpose described, forensic investigations often rely on stains found at the scenes of crimes or other legal interest. These DBSs, which may be found on a variety of substrates, may also be used for the detection of endogenous metabolites as well as drugs and metabolites. Although these DBS substrates may be less than ideal as compared to the use of filter paper with defined absorptive characteristics (see other chapters), they are important from a qualitative perspective in a forensic investigation.

The use of DBS for NBS tests is perhaps the most common metabolic screening method in use today. However,

Dried Blood Spots: Applications and Techniques, First Edition. Edited by Wenkui Li and Mike S. Lee.

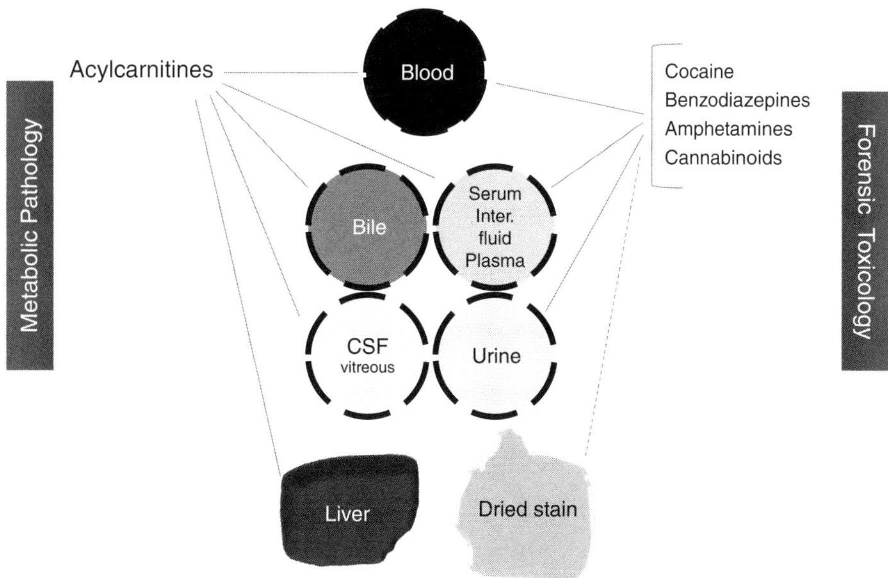

FIGURE 11.1 Examples of analytes detectable in DBS in metabolic pathology and forensic toxicology.

the forensic use of DBS will likely expand as new methods for the detection of endogenous metabolites and drugs and their metabolites in both ante- and postmortem medicolegal and forensic cases are developed. This expansion of DBS as a new collection matrix presents several advantages such as reduced storage requirements, diminished shipped costs, reduced potential for contraction of infectious diseases, enhanced stability of drugs and their metabolites, and smaller sample requirements.

This chapter on forensic application of DBSs in both antemortem and postmortem cases is divided into two sections: the metabolic analysis of DBS in forensic metabolic pathology and the use of DBS in forensic toxicology for the detection of drugs and their metabolites.

11.2 FORENSIC METABOLIC PATHOLOGY

Early research suggested that metabolic disorders were thought to account for 1–4% of all infant deaths in the first years of life (Howat et al., 1985; Leonard et al., 1988; Bennett et al., 1990, 2000; Holton et al., 1991; Boles et al., 1998; Rinaldo et al., 1999; Bennett and Rinaldo, 2001). Retrospective analysis revealed that many of these metabolic disorders were enzymatic defects in fatty acid oxidation or organic acid metabolism. With the introduction of MS/MS and acylcarnitine analysis, it was observed that acylcarnitines were biomarkers for fatty acid and organic acid metabolic disorders in sick or postmortem infants. Early studies showed that many infants died of these disorders without the benefit of NBS and it was thought that early detection could prevent these deaths. Much of the early metabolic screening of postmortem infants was performed using bile specimens and MS/MS in cases that were recognized as most likely metabolic diseases based on various signs such as, fatty liver, enlarged heart, and family history of unexpected deaths of siblings. Subsequently, the development of methods for the detection of acylcarnitines in DBS in NBS revealed several advantages including ease of collection, requirement for small samples, and ease of shipment; these advantages would allow a greater number of routine analyses in cases of infant deaths. Postmortem screening of DBS collected at autopsy paralleled the expansion of NBS analysis of DBS using MS/MS and, in fact, the findings of this screening led to the more rapid expansion and acceptance of MS/MS in NBS (Chace et al., 2001; Dott et al., 2006). Both NBS and postmortem screening were clearly intertwined. However, postmortem metabolic screening required a greater expertise in interpretation as described below.

11.2.1 The Metabolic Autopsy

Disorders of fatty acid oxidation and organic acid metabolism are associated with sudden death of an affected newborn but also in less common instances, an older child, adolescent, or adult (Wilcox et al., 2002; Rinaldo et al., 2004). A pathological exam does not always produce findings associated with a death due to metabolic disease (e.g., fatty liver or enlarged heart). Generally, toxicological analyses do not result in abnormal findings except an occasional incidental finding of an elevated concentration of specific fatty acids and organic acid metabolites (Ross et al., 1996) in urine. However, a biochemical analysis of blood for acylcarnitines can reveal a probable cause of death for a variety of metabolic disorders including medium-chain acyl CoA dehydrogenase (MCAD) deficiency. The methodology for detection of acylcarnitine in postmortem DBS is described below.

11.2.1.1 Postmortem DBS and Other Dried Biological Samples (DXS) Blood is collected at autopsy most commonly using a syringe, and often directly from the heart and stored in a tube. Due to possible alterations after death, blood collected from autopsy usually differs from blood collected from newborns. The condition of postmortem blood varies widely depending on the state of the body of the deceased (e.g., it may be hemolyzed, clotted, or contaminated with interstitial fluid). Typically, 25–50 μL of blood is pipetted to the same type of filter paper used in NBS and allowed to dry before shipping to a laboratory specialized in postmortem metabolic screening. DBSs are shipped to the lab using similar procedures as that used in NBS. One issue unique to forensic science is the need for an intact "chain of custody" of the sample that requires the maintenance of a written record of receipts of every transfer of the sample from person to person.

If blood is not available at autopsy, then bile is satisfactory. Bile is applied to filter paper in the same manner as blood, as has been demonstrated for postmortem metabolic screening (Maurer, 1998; Rinaldo et al., 1999), although the volume of the punched sample will be less. Other specimens may also be used and applied to filter paper. These DXS specimens may include urine, CSF, vitreous fluids, and other fluids. In addition, tissue samples have been analyzed by placing them on the filter paper and allowing the fluid surrounding the tissue (liver most commonly) to be absorbed in the paper. Essentially, in an attempt to investigate a possible cause of death, even imperfect specimens can be analyzed and the results interpreted appropriately. If there is any indication of a metabolic disorder present in the DXS or DBS sample, then an original newborn blood spot will be requested from the medical examiner or submitting physician. The original blood spot serves to confirm the presence of a metabolic disease if also positive. In addition, the original DBS specimens are easier to analyze and interpretation of results is more reliable. Unfortunately, many DBSs are being destroyed by public health laboratories before they can be retrieved in cases of presumptive positives in postmortem screening.

11.2.1.2 Tandem Mass Spectrometry of Acylcarnitines The analysis of DBS for "metabolic autopsy" utilizes tandem mass spectrometry to detect primarily acylcarnitines although it also includes amino acids. Acylcarnitine profiles together with free carnitine concentrations are useful in the detection of many fatty acid oxidation disorders including MCAD deficiency. MCAD deficiency is a disorder of a medium-chain length fatty acid (octanoic acid) oxidation, specifically the enzyme medium-chain acyl CoA dehydrogenase deficiency. Historically, it has been the most common metabolic disorder found in infants and children who have died without explanation. A brief description of the analysis, the metabolites, and disorders detected by MS/MS is described below and compared to MS/MS as used in NBS.

11.2.1.3 Acylcarnitines, Fatty Acid Metabolism, and MS/MS As described in Chapter 6, MS/MS of DBS of acylcarnitines and amino acids is most commonly used in NBS. The postmortem screening assay is nearly identical to that for a DBS, with only two major differences. First, the specimen source is blood from a deceased newborn although this is not exclusive since a sample may be collected from an infant in an ER who dies shortly after sample collection or is an original newborn blood spot requested by the medical examiner. Second, interpretation of the MS/MS spectra, the so-called postmortem metabolic profile, is quite different than that for NBS (Chace et al., 2001; Dott et al., 2006). Metabolic profiles of acylcarnitines (an example can be found in Chapter 6) are different in newborns as compared to the blood obtained from a deceased individual. A postmortem DBS profile is characterized by a significant increase in the concentration of two short-chain acylcarnitines (C_4 and C_4OH). These elevations would be considered abnormal in NBS and would be suggestive of short-chain acyl CoA dehydrogenase (SCHAD) deficiency. However, their presence at elevated concentrations is normal in postmortem screening. Further, the amino acid concentrations are greater in postmortem profiles than in newborns and with few exceptions are not useful for determining the cause of death from a metabolic disease due to the generalized high concentration. Presumably cell lysis and protein hydrolysis are responsible for this "normal" postmortem MS/MS profile. The elevated concentrations of amino acids are helpful in confirming that the specimen is indeed postmortem or has not been diluted. With regard to DXS specimens, their profiles vary by the manner by which acylcarnitines are eliminated. Bile contains much higher concentration of acylcarnitines as compared to blood, whereas urine contains very high dicarboxylic acid acylcarnitines making that matrix difficult to interpret. In fact, the challenges of interpretation of "normal" and abnormal postmortem profiles can be difficult without the experience of having analyzed many hundreds of specimens, although these results of numerous postmortem screens have been characterized (Chace et al., 2001).

11.2.1.4 Impact of Routine Postmortem DBS Screening Determination of the cause of death due to metabolic disease can directly impact mortality of current or future siblings of the deceased. Many metabolic diseases such as MCAD deficiency have no symptoms and can only be detected by analysis of blood. Detection of a death due to a metabolic disease can result in screening other children and parents for the disorder, preventing their possible unexpected death due to the treatability of these disorders. The number of metabolic disorders thought to have caused the deaths of affected newborns is 1% of all infants screened. The number of infants screened has varied by state, but it increased substantially when DBS was utilized in part due to ease of collection and in part because DBS could be analyzed using the same technology and in some of the same labs performing NBS. The cost of the postmortem screening could be reduced substantially when it is part of a program that has a much larger screening population, that is, routine metabolic newborn screening and postmortem screening. For these reasons, in some states nearly all infant deaths have been screened for a metabolic disease (Chace et al., 2003a; Dott et al., 2006) as described in one study in Virginia. The number of cases of undetected metabolic disease and subsequent metabolic deaths has been reduced substantially due to the routine NBS for these disorders. However, death is still possible in affected infants between the time of birth and sample reporting which can take 3–5 days. It is thought about 5–10% of all metabolic deaths will occur during this time period. Therefore, postmortem screening remains an active program within pathology laboratories and medical examiners' offices.

11.2.1.5 Forensic Analysis of Metabolic Profiles Other Than Metabolic Disease The analysis of DBS of acylcarnitines is not restricted to infants who have died of a metabolic disorder. DBS is analyzed routinely from infants suspected of child abuse in order to rule out a cause attributable to a metabolic disease. The most common reason for a metabolic profile of a newborn is shaken baby syndrome since, historically, some cases of suspected abuse were determined to be due to glutaric acidemia type I (GA-I). Both GA-I and shaken baby syndrome may show cerebral hemorrhages upon examination (Gago et al., 2003). An analysis that shows the presence of GA-I rules out abuse. It is also possible that metabolic profiles may be abnormal in cases of malabsorption due to an existing pathology or malnutrition resulting in an increased turnover of fat and subsequent increased beta oxidation. The profiles may resemble a metabolic disease or profiles associated with administration of medium-chain fatty acids. These data may be supportive of other forensic or pathology findings, but are likely conclusive by themselves. Finally, many metabolic disorders produce high concentrations of small molecules, many of which are similar to toxins such as ethylene glycol. Methylmalonic acid (MMA) disorders produce small volatile organic acids that are quite similar to toxins such as ethylene glycol. These two compounds can be differentiated by the use of GC/MS, but a false-positive identification of ethylene glycol may occur as a result of improper methods of identification. An example of this occurred in the case of a suspected ethylene glycol poisoning in an infant based on an unconfirmed GC analysis (Shoemaker et al., 1992). The erroneous identification of ethylene glycol by this method in blood obtained from the infant resulted in the arrest of the infant's mother on a first-degree murder charge and her conviction and imprisonment. As a result of subsequent findings, it was determined that the infant had not been poisoned, but suffered from methylmalonic acidemia (Fraser, 2002). This case exemplifies the need for confirmatory analysis of those results obtained from methods such as GC and RIA in cases of forensic metabolic profiling and drug detection. The Forensic Toxicology Laboratory Guidelines prepared by the Society of Forensic Toxicology and the American Academy of Forensic Sciences state that "... the detection or initial identification of drugs and other toxins should be confirmed whenever possible by a second technique based on a different principle" (SOFT/AAFS, 2006). The guidelines further recommend that the confirmatory test should have greater specificity for the analyte than the initial test method and that MS confirmation is appropriate for results obtained by means of detection methods such as colorimetry, immunoassay, and GC. The significance of false-positive and false-negative results is presented schematically in Figure 11.2. As is depicted, false-positive results are generally considered to be of greater significance than false-negative results in forensic toxicology, whereas greater significance is attached to false-negative results in metabolic screening.

11.3 FORENSIC TOXICOLOGY AND DBS

Although there are a number of reports in the forensic literature of drug detection in DBS, they are not as numerous as the reports of the use of DBS for the clinical purposes of neonatal testing (Henderson et al., 1993, 1997; Sosnoff et al., 1996) and therapeutic drug monitoring (TDM) (Edelbroek et al., 2009). Several of the clinical findings are adaptable to forensic use, but the nature and condition of the samples at collection and the purpose to which the results are put differ from forensic applications in which samples often are degraded or present in small and limited quantities. In forensic toxicology, several circumstances can be envisaged in which drug detection in DBS would be of value. These include the detection of drugs in DBS found on objects collected from persons or incident scenes of interest, prepared from blood samples collected for the purposes of detecting and/or monitoring drug use, and prepared from postmortem blood samples for the purposes of increasing drug stability and ease of transport and storage.

Early suggestions were made that nongenetic markers, "environmental in origin," such as metals and drugs (Kind, 1961) and drugs of abuse (Curry, 1965), might be of forensic value in the individualization of bloodstains. However, it was recognized that the limits of the methods of detection then available presented a problem to the implementation of this suggestion and that the development of newer methods of detection with lower detection limits would be required. This

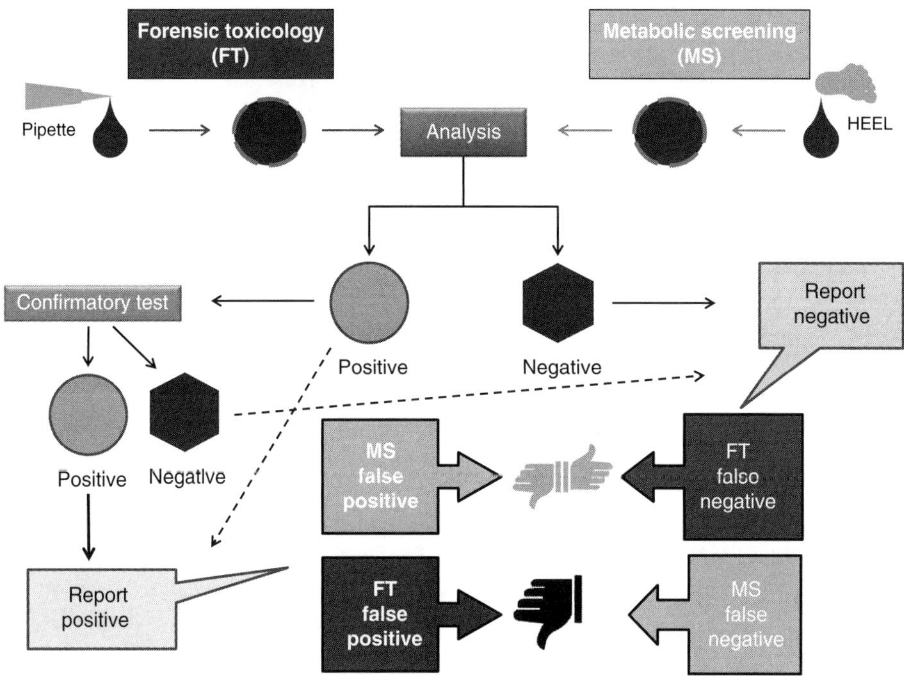

FIGURE 11.2 The relative significance of false-positive and false-negative results in metabolic screening and forensic toxicology.

review discusses several common drugs of abuse reported to have been detected in dry stains and the various parameters involved in their detection.

Subsequent to the development and widespread use of immunoassays and mass spectrometry, forensic interest in the detection of drugs in DBS was stimulated largely in the 1970s and 1980s and interest once again has emerged more recently as evidenced by the increasing number of articles of the detection of drugs of abuse in DBS reported in the refereed literature. A summary of selected methodological parameters of studies from the refereed literature is presented in Table 11.1.[1]

As is apparent from the data in Table 11.1, several of the methods have elements in common: DBS is generally prepared on filter paper, commonly Whatman 903 paper, using blood volumes of 100 μL or less; extraction of drugs from DBS was often achieved by the use of detergents or simple solvent, although LLE and SPE were also employed; the earlier methods of detection employed RIA and GC/MS, whereas more recent methods have relied on LC/MS/MS, which permits the detection of nonvolatile and polar analytes at nanogram and sub-nanogram quantities (Li and Tse, 2010) without extensive sample preparation such as extensive extractions and derivatization (Sosnoff et al., 1996; Boy et al., 2008; Li and Tse, 2010).

[1]The detection of several additional drugs in DBS has been reported at a recent meeting of the International Association of Forensic Toxicologists (Stove et al., 2012).

Several reports of drug detection in DBS were published in the forensic literature in the late1970s by Shaler and Smith and their coworkers. They published a series of articles describing the detection, by means of RIA, of several drugs—diphenylhydantoin (Shaler et al., 1978), morphine (Smith et al., 1980), digoxin (Smith, 1981), phenobarbital (Smith and Pomposini, 1981), and cocaine metabolites (Smith and Liu, 1986)—in DBS. They suggested that the detection of these drugs in DBS, when used in conjunction with or in place of the identification of genetic markers, could aid in the differentiation of bloodstain evidence since the stability of drugs in DBS might be greater than that of the genetic markers employed at that time. The DBSs that were analyzed in all but one of these reports were prepared with blood from persons who were being treated with the drugs and, therefore, contained therapeutic drug concentrations. By means of a fluorometric method, it was estimated that salicylate could be detected in a 50 μL DBS obtained from a person for as long as 16 hours after the ingestion of 600 mg of aspirin (King, 1979).

Although methodological parameters were not a major focus of these early reports, there were several important findings that generally have been corroborated subsequently, namely that satisfactory extraction of drugs from DBS did not require liquid–liquid extraction (LLE) or solid-phase extraction (SPE), but was sufficiently efficient with the use of surfactants or organic solvents and that drugs in DBS were detectable for extended periods of time (e.g., digoxin for 14 days and diphenylhydantoin for 6 months).

TABLE 11.1 Methods for the Detection of Drugs of Abuse in DBS

Size of Analytical Sample; Stain Substrate	Extraction Solvent	Detection Method	Detection Limit	Reference
Cocaine				
100 μL; Guthrie card 903	pH 6 phosphate buffer in ultrasonic bath	LC/MS/MS	0.74 ng/spot	Alfazil and Anderson (2008)
10 μL; Whatman 903 Protein Saver card	Methanol	HPLC/SF	LOD, 7 ng/mL and LOQ 20 in a 10 μL spot	Mercolini et al., (2010)
20 μL; Sartorius TFN card	Methanol/acetone	LC/MS	0.001 ng/20 μL spot	Thomas et al. (2012)
Benzoylecgonine				
$\frac{1}{4}$ inch section of bloodstained tampon	0.05% SDS	RIA	0.57 ng of BE equivalents	Smith and Liu (1986)
$\frac{1}{4}$ inch diameter spots; Schleicher & Schuell Grade 903 filter paper	Tween 20/phosphate buffer, pH 7.4	RIA	10 ng/mL	Henderson et al. (1993)
$\frac{1}{4}$ inch diameter spots (~12 μL); filter paper	2 mM CH_3COONH_4, followed by methanol	LC/MS/MS	~2 ng/mL	Sosnoff et al. (1996)
50 μL; NR	Methanol/SPE	GC/MS	<2 ng in a 50 μL DBS	Schütz et al. (2002)
10 μL; Whatman 903 Protein Saver card	Methanol	HPLC/SF	LOD, 7 ng/mL and LOQ 20 in a 10 μL spot	Mercolini et al. (2010)
30 μL; Whatman 903	Water with ultrasonication by online extraction	Online LC/MS/MS	~3 μL <10 ng/mL	Sausserau et al. (2012)
Flunitrazepam and its metabolites				
~50 μL; unbleached cotton cloth	Acetic acid/SPE	GC/MS	10 ng	Elian (1999)
Several benzodiazepines				
50–100 μL; cotton	pH 10.3 Trizma Ethyl acetate	GC	NR	Hammond et al. (1979)
100 μL; Guthrie card 903	pH 6 phosphate buffer in ultrasonic bath	LC/MS/MS	0.3–0.62 ng	Alfazil and Anderson (2008)
Alprazolam				
100 μL; Whatman 903 filter paper	Org LLE	LC/MS/MS	LOD, 0.02ng LOQ, 0.07 ng	Jantos and Skopp (2011)

(continued)

TABLE 11.1 ***(Continued)***

Size of Analytical Sample; Stain Substrate	Extraction Solvent	Detection Method	Detection Limit	Reference
Amphetamine/methamphetamine				
NR	Methanol/SPE	GC/MS	<2 ng in a 50 μL DBS	Schütz et al. (2002)
MDMA				
100 μL; Whatman 903 paper	Ethyl acetate	LC/MS/MS	0.16 ng/mL MDMA in an 18 mm punch	Jantos et al. (2011)
20 μL; Sartorius TFN card	Methanol/acetone	LC/MS	0.005 ng/20 μL spot	Thomas et al. (2012)
MDA				
100 μL; Whatman 903 paper	Ethyl acetate	LC/MS/MS	0.12 ng/mL MDA in an 18 mm punch	Jantos et al. (2011)
20 μL; Sartorius TFN card	Methanol/acetone	LC/MS	0.01 ng/20 μL spot	Thomas et al. (2012)
Morphine and heroin				
50 μl; cloth sheet	Triton-X or SDS	RIA	14 ng/mL morphine detected in a DBS prepared from a volunteer given a 5 mg dose of morphine	Smith et al. (1980)
Curtain and tissue from the scene of a homicide	Methanol SPE	GC/MS	NR for stains	Schütz et al. (2002)
Morphine				
100 μL; Whatman BFC 180 paper	Borate buffer, pH 8.5/LLE	LC/MS/MS	0.4 ng/100 μL spot	Boy et al. (2008)
6-Monoacetylmorphine				
100 μL; Whatman BFC 180 paper	Borate buffer, pH 8.5/LLE	LC/MS/MS	0.8 ng/100 μL spot	Boy et al. (2008)
Cannabinoids				
Paper or textile	Methanol	GC/MS	0.82 ng THC/50 μL DBS 0.27 ng THC-COOH/50 μL DBS 0.13 ng 11-OH THC/50 μL DBS	Schütz et al. (2002)
20 μL; Sartorius TFN card	Methanol/acetone	LC/MS	0.005 ng/20 μL spot THC and THC-COOH	Thomas et al. (2012)

11.4 COMMON DRUGS OF ABUSE DETECTED IN DRIED BLOODSTAINS

11.4.1 Cocaine and Its Metabolites

There have been a number of reports of the detection of cocaine and its metabolites in dried bloodstains—generally, cocaine metabolites have been detected since the half-life ($t\frac{1}{2}$) of these metabolites in blood is greater than that of cocaine.

Various limits of detection (LODs) have been reported for cocaine: less than 2 ng cocaine in 50 μL DBS by GC/MS (Schütz et al., 2002); less than10 ng/mL by means of LC/MS/MS (Alfazil and Anderson, 2008); 0.07 ng in 10 μL DBS by HPLC with fluorescence detection (Mercolini et al., 2010); and 0.001 ng in 20 μL DBS by LC/MS (Thomas et al., 2012). There have been several reports of cocaine stability in DBS: benzoylecgonine (BE), a product of cocaine hydrolysis, was not detected in ~12 μL spiked DBS, containing cocaine at a concentration of 1000 ng/mL and stored at 45°C for 108 hours (Henderson et al., 1993); 75% of the initial concentration of cocaine was detected after 17 days of DBS storage (Skopp and Potsch, 2001); the loss of cocaine in DBS stored at freezer, refrigerator, and room temperatures for 1 month was less than 10%, 11%, and 19.9%, respectively (Alfazil and Anderson, 2008); less than 50% and more than 80% of the initial concentration of cocaine (10 or 50 ng/mL) in DBS was detected after 6 months of storage at 4°C and –20°C, respectively (Sausserau et al., 2012); and cocaine was detectable after 7 days of storage at 2–8°C (Thomas et al., 2012). Generally, the stability of cocaine in DBS was greater than in liquid blood stored under the same conditions.

Cocaine and/or its metabolites were detected in menstrual blood on a tampon[2] obtained from an alleged sexual assault victim (Smith and Liu, 1986). Subsequently, an RIA method for the detection of BE in DBS, which had <1% cross-reactivity for cocaine or any of its other metabolites, was determined to have a limit of quantitation (LOQ) of 10 ng/mL (Henderson et al., 1993). BE was detectable at a concentration of <40 ng/mL (50 μL) by means of GC/MS. The LOD of BE in DBS by means of LC-MS/MS was determined to be approximately 2 ng/mL (~12 μL spot) (Sosnoff et al., 1996), 7 ng/mL (10 μL spot) (Mercolini et al., 2010), and <10 ng/mL (~3 μL) (Sausserau et al., 2012). The stability of BE was estimated to be several years in desiccated DBS stored at either refrigerated or room temperatures (Henderson et al., 1993) and greater than 90% in DBS after 6 months of storage at 4°C or –20°C (Sausserau et al., 2012). An excellent correlation was determined between the BE concentrations in DBS and whole blood (Sausserau et al., 2012).

[2]It is not clear whether the blood was dried prior to analysis. The detection relied on the use of unconfirmed RIA; therefore, the specific compounds detected were not identified.

Cocaethylene, the unique metabolite produced by the use of cocaine and ethanol, has been detected in DBS at concentrations of 4 ng/mL (Mercolini et al., 2010). The storage stability of ecgonine methyl ester was temperature dependent: less than 15% of the initial concentration of ecgonine methyl ester in DBS was detected after 6 months of storage at 4°C, whereas more than 85% was detected after the same storage period at –20°C (Sausserau et al., 2012).

The concentrations of cocaine and its metabolites in DBS, after correcting for hematocrit, and in plasma samples were in close agreement (Mercolini et al., 2010).

11.4.2 Benzodiazepines

Several benzodiazepines (BDZ), present at therapeutic concentrations, were detected in DBS by GC (Hammond et al., 1979). An extraction employing subtilisin Carlsberg was reported to improve analyte recoveries from DBS. A GC/MS method developed for the detection of flunitrazepam and its metabolites, 7-aminoflunitrazepam and desmethylflunitrazepam, in DBS from victims of reported sexual assault was reported to be linear from 0.1 to 50 μg/dL (Elian, 1999). By means of LC/MS/MS the LOD and the LOQ for six BDZ in DBS were decreased from less than 1 ng to between approximately 1 and 2 ng, respectively (Alfazil and Anderson, 2008), and 0.02 and 0.07 ng for alprazolam, respectively (Jantos and Skopp, 2011).

BDZ have been shown to be stable under various storage conditions. Approximately 40–80% of BDZ and their metabolites were detectable after 4–6 months of storage in the dark at room temperature. The amount of BDZ in 100 μL DBS (100 ng) was found to decrease by approximately 10–15%, 5–11%, and less than 10% after being stored for 1 month at room temperature, 4°C, or –20°C, respectively (Alfazil and Anderson, 2008).

There was excellent correlation between the alprazolam concentrations, over the range of 3.7–20.7 ng/mL, detected in DBS and in the whole blood from which the DBSs were prepared (Jantos and Skopp, 2011).

11.4.3 Amphetamines

Fewer than 2 ng of amphetamine and methamphetamine were detected by means of GC/MS in 50 μL DBS prepared from spiked blood samples (Schütz et al., 2002). 3,4-Methylenedioxymethamphetamine (MDMA) and its metabolite 3,4-methylenedioxyamphetamine (MDA) were detected at concentrations of 1.6 and 0.12 ng/mL, respectively, in 100 μL spiked DBS (Jantos et al., 2011) and at concentrations of 0.25 and 0.5 ng/mL, respectively, in 20 μL spiked DBS (Thomas et al., 2012). Concentrations of MDMA and MDA in whole blood and DBS prepared from the whole blood have been shown to be sufficiently similar that the differences would not cause a difference in interpretation (Jantos et al., 2011).

11.4.4 Opioids

Morphine was detected by RIA in a 50 μL DBS collected from a volunteer to whom a single dose of 5 mg of morphine had been administered (Smith et al., 1980); LOD for morphine detection in DBS was less than 5 ng morphine in 50 μL DBS by means of GC/MS (Schütz et al., 2002) and 0.4 ng in 100 μL DBS by means of LC/MS/MS (Boy et al., 2008).

Codeine and the morphine metabolite 6-monoacetylmorphine (6-MAM) were detected at LOD of 4.91 ng in 50 μL DBS by GC/MS (Schütz et al., 2002) and 0.8 ng in 100 mL DBS by LC-MS/MS (Boy et al., 2008).

Only 57% of the morphine in a DBS was detected after a storage period of 9 months (the storage temperature was not reported) (Smith et al., 1980). Significant degradation was not reported in DBS stored at –20°C and 4°C for up to 7 days and at 40°C for up to 5 days (Boy et al., 2008). Although only 50% of 6-MAM was detected in DBS after 5 days of storage at 40°C, this was greater stability than in whole blood; it was estimated that the $t\frac{1}{2}$ of 6-MAM at 4°C was 138 hours and at 40°C 96 hours, respectively, whereas the corresponding values in whole blood were 42 and 13 hours (Boy et al., 2008).

A significant correlation in morphine concentrations was reported between authentic whole sample and DBS prepared from those samples (Boy et al., 2008).

11.4.5 Cannabinoids

The first reported detection of cannabinoids in DBS was by means of GC-MS: THC and two of its metabolites were detected at LOD of less than 1 ng in 50 μL DBS (Schütz et al., 2002). Recently, the detection of Δ^9-tetrahydrocannabinol (THC) and tetrahydrocannabinol-9-carboxylic acid (THC-COOH) with an LOD of 0.005 ng in 20 μL DBS was reported (Thomas et al., 2012). These drugs were detectable in DBS stored at 2–8°C for 7 days, the longest period evaluated.

11.5 CONCLUSIONS

This review of the literature allows several conclusions to be drawn as to the advantages and disadvantages of drug detection in DBS for forensic purposes, as well as the shortcomings of the work thus far reported.

The value in forensic toxicology of drug detection in DBS may be summarized as follows:

1. DBSs may be collected by finger-prick and consequently are easier to collect than whole blood samples by venipuncture. Since this collection may not require medical personnel (Boy et al., 2008), samples can be collected closer to the time of incident by nonmedical personnel (Jantos et al., 2011), and therefore may be of value in cases of individuals who are suspected of driving under the influence of drugs (Skopp and Potsch, 2001; Mercolini et al., 2010).
2. The use of DBS offers the benefits of ease of storage and shipping (Jantos et al., 2011). In addition, DBS affords increased safety (Jantos et al., 2011) since certain disease vectors (e.g., the HIV-1 and -2 viruses, the human T-cell leukemia/lymphoma virus-I and -II, and the hepatitis C virus) are inactivated as a result of drying, although hepatitis B virus in high titer may remain active in dried stains for several days (Parker and Cubitt, 1999).
3. Although certain methods relied on the use of LLE (Boy et al., 2008) or SPE (Henderson et al., 1993) of drugs from DBS, in many cases drugs were extracted easily from DBS with water (Sausserau et al., 2012), methanol (Mercolini et al., 2010), or other organic solvents (Jantos et al., 2011). The ease and efficiency with which drugs are extracted from DBS may be due in part to the retention of certain matrix components of blood on the stain substrate (Mercolini et al., 2010).
4. The most commonly employed method of detection was LC-MS/MS by which drugs were detected and quantitated at sub-nanogram amounts.
5. Generally, drugs and their metabolites have been reported to have equal or greater stability and be detected for longer periods in DBS than in whole blood samples (Skopp and Potsch, 2001; Alfazil and Anderson, 2008; Sausserau et al., 2012), in many cases due to the inhibition of enzymatic and/or nonenzymatic hydrolysis, for example, cocaine (Alfazil and Anderson, 2008) and 6-MAM (Boy et al., 2008).
6. Drug concentrations detected in DBS have been reported to be well correlated with concentrations in whole blood. Some of these drugs include several amphetamines (Jantos et al., 2011; Sausserau et al., 2012), cocaine and its metabolites (Mercolini et al., 2010; Sausserau et al., 2012), and morphine (Boy et al., 2008).

Although the use of DBS in forensic toxicology appears to offer several advantages, many of the investigations of their use were method development or proof-of-principle exercises, and not sufficiently evaluated for use in either ante- or postmortem forensic toxicology. Also, the number of drugs and drug groups that have been evaluated is small and should be extended to several other drugs of interest to forensic toxicologists. Therefore, additional work, including the following, is needed before the full advantages of DBS are realized.

1. There are limited reports of the detection of drugs in DBS that were prepared using authentic blood samples. Those reported include blood obtained from neonates (Henderson et al., 1993, 1997), from volunteers to whom therapeutic drug doses had been administered (Jantos et al., 2011), from drivers suspected of driving under the influence of drugs (Sausserau et al., 2012), and from known cocaine (Mercolini et al., 2010) and opioid users (Boy et al., 2008). There is a need for extensive testing of postmortem samples including hemolyzed and putrefied samples since the detection of drugs in DBS prepared with postmortem blood samples is rare (Henderson et al., 1993).
2. Rarely (Schütz et al., 2002) were the DBSs analyzed found on surfaces other than filter paper (e.g., from the scene of a crime).
3. Due to the small volume of blood in DBS, detection of multiple drug groups may not be possible, although several DBS could be used (Boy et al., 2008).
4. Although studies have reported significant correlation between the drug concentrations in DBS with those in whole blood, the detection limits of cocaine and metabolites was determined to be greater in DBS than in plasma (Mercolini et al., 2010), possibly due to the method of detection.
5. Additional stability studies are necessary. There have been only a few studies of extended storage, and generally in those studies blood samples were obtained from volunteers and contained preservatives or anticoagulants (Boy et al., 2008) that may have influenced the stability of drugs.

REFERENCES

Alfazil AA, Anderson RA. Stability of benzodiazepines and cocaine in blood spots stored on filter paper. *Journal of Analytical Toxicology* 2008;32:511–515.

Bennett MJ, Rinaldo P. The metabolic autopsy comes of age. *Clinical Chemistry* 2001;47:1145–1146.

Bennett MJ, Allison F, Pollitt RJ, Variend S. Fatty acid oxidation defects as causes of unexpected death in infancy. *Progress in Clinical and Biological Research* 1990;321:349–364.

Bennett MJ, Rinaldo P, Millington DS, Tanaka K, Yokota I, Coates PM. Medium-chain acyl-CoA dehydrogenase deficiency: postmortem diagnosis in a case of sudden infant death and neonatal diagnosis of an affected sibling. *Pediatric Pathology* 1991;11:889–895.

Bennett MJ, Rinaldo P, Strauss AW. Inborn errors of mitochondrial fatty acid oxidation. *Critical Reviews in Clinical and Laboratory Science* 2000;37:1–44.

Boles RG, Martin SK, Blitzer MG, Rinaldo P. Biochemical diagnosis of fatty acid oxidation disorders by metabolite analysis of postmortem liver. *Human Pathology* 1994;25:735–741.

Boles RG, Buck EA, Blitzer MG, Platt MS, Cowan TM, Martin SK, Yoon H, Madsen JA, Reyes-Mugica M, Rinaldo P. Retrospective biochemical screening of fatty acid oxidation disorders in postmortem livers of 418 cases of sudden death in the first year of life [see comments]. *Journal of Pediatrics* 1998;132:924–933.

Boy RG, Henseler J, Mattern R, Skopp G. Determination of morphine and 6-acetylmorphine in blood with use of dried blood spots. *Therapeutic Drug Monitoring* 2008;30:733–739.

Chace DH, DiPerna JC, Mitchell BL, Sgroi B, Hofman LF, Naylor EW. Electrospray tandem mass spectrometry for analysis of acylcarnitines in dried postmortem blood specimens collected at autopsy from infants with unexplained cause of death. *Clinical Chemistry* 2001;47:1166–1182.

Chace, DH, Kalas TA, Fierro M, Hannon H, Rasmussen SA, Wolf K, Williams J, Dott M, Rinaldo P. Contribution of selected metabolic diseases to early childhood deaths–Virginia, 1996-2001. *Morbidity and Mortality Weekly Report* 2003a;52:677–679.

Chace DH, Kalas TA, Naylor EW. Use of tandem mass spectrometry for multianalyte screening of dried blood specimens from newborns. *Clinical Chemistry* 2003b;49.1797–1817.

Curry AS. Science against crime. *International Science and Technology* 1965;47:39–48.

Dott M, Chace D, Fierro M, Kalas TA, Hannon WH, Williams J, Rasmussen SA. Metabolic disorders detectable by tandem mass spectrometry and unexpected early childhood mortality: a population-based study. *American Journal of Medical Genetics A* 2006;140:837–842.

Edelbroek PM, van der Heijden J, Stolk ML. Dried blood spot methods in therapeutic drug monitoring: methods, assays and pitfalls. *Therapeutic Drug Monitoring* 2009;31:327–336.

Elian AA. Detection of low levels of flunitrazepam and its metabolites in blood and blood stains. *Forensic Science International* 1999;101:107–111.

Fraser AD. Clinical toxicologic implications of ethylene glycol and glycolic acid poisoning. *Therapeutic Drug Monitoring* 2002;24:232–238.

Gago LC, Wegner RK, Capone A Jr, Williams GA. Intraretinal hemorrhages and chronic subdural effusions: glutaric aciduria type 1 can be mistaken for shaken baby syndrome. *Retina* 2003;23:724–726.

Guthrie R, Susi A. A simple phenylalanine method for detecting phenylketonuria in large populations of newborn infants. *Pediatrics* 1963;32:338–343.

Hammond MD, Osselton MD, Moffat AC. The extraction and analysis of benzodiazepine drugs in bloodstains. *Journal of the Forensic Science Society* 1979;19:193–198.

Henderson LO, Powell MK, Hannon WH, Miller BB, Martin ML, Hanzlick RL, Vroon D, Sexson WR. Radioimmunoassay screening of dried blood spot materials for benzoylecgonine. *Journal of Analytical Toxicology* 1993;17:42–47.

Henderson LO, Powell MK, Hannon WH, Bernert JT Jr, Pass KA, Fernhoff P, Ferre CD, Martin L, Franko E, Rochat RW, Brantley MD, Sampson E. An evaluation of the use of dried blood spots from newborn screening for monitoring the prevalence of cocaine use among childbearing women. *Biochemical and Molecular Medicine* 1997;61:143–151.

Holton JB, Allen JT, Green CA, Partington S, Gilbert RE, Berry PJ. Inherited metabolic diseases in the sudden infant death syndrome [see comments]. *Archives of Diseases in Children* 1991;66:1315–1317.

Howat AJ, Bennett MJ, Variend S, Shaw L, Engel PC. Defects of metabolism of fatty acids in the sudden infant death syndrome. *British Medical Journal* (Clinical Research Edition) 1985;290:1771 1773.

Jantos R, Skopp G. Comparison of drug analysis in whole blood and dried blood spots. *Toxichem + Krimtech* 2011;78:268–275.

Jantos R, Veldstra JL, Mattern R, Brookhuis KA, Skopp G. Analysis of 3,4-methylenemethamphetamine: whole blood versus blood spots. *Journal of Analytical Toxicology* 2011;35:269–273.

Keppen LD, Randall B. Inborn defects of fatty acid oxidation: a preventable cause of SIDS. *South Dakota Journal of Medicine* 1999;52:187–188; discussion 188–189.

Kind SS. The individuality of human bloodstaining. *Journal of the Forensic Science Society* 1961;2:75.

King LA. The fluorometric detection of salicylate in a bloodstain. *Journal of Forensic Sciences* 1979;24:317–318.

Leonard JV, Green A, Holton JB, Bartlett K. Inborn errors of metabolism and unexpected infant deaths [letter]. *Lancet* 1988;2:854.

Li W, Tse FLS. Dried spot sampling in combination with LC-MS/MS for quantitative analysis of small molecules. *Biomedical Chromatography* 2010;24:49–65.

Matern D, Strauss AW, Hillman SL, Mayatepek E, Millington DS, Trefz FK. Diagnosis of mitochondrial trifunctional protein deficiency in a blood spot from the newborn screening card by tandem mass spectrometry and DNA analysis. *Pediatric Research* 1999;46:45–49.

Maurer HH. Liquid chromatography-mass spectrometry in forensic and clinical toxicology. *Journal of Chromatography B* 1998;713:3–25.

Mercolini L, Mandrioli R, Gerra G, Raggi MA. Analysis of cocaine and two metabolites in dried blood spots by liquid chromatography with fluorescence detection: a novel test for cocaine and alcohol intake. *Journal of Chromatography A* 2010;1217:7242–7248.

Parker SP, Cubitt WD. The use of the dried blood spot sample in epidemiological studies. *Journal of Clinical Pathology* 1999;52:633–639.

Rashed MS, Ozand PT, Bennett MJ, Barnard JJ, Govindaraju DR, Rinaldo P. Inborn errors of metabolism diagnosed in sudden death cases by acylcarnitine analysis of postmortem bile. *Clinical Chemistry* 1995;41:1109–1114.

Rinaldo P, Yoon HR, Yu C, Raymond K, Tiozzo C, Giordano G. Sudden and unexpected neonatal death: a protocol for the postmortem diagnosis of fatty acid oxidation disorders. *Seminars in Perinatology* 1999;23:204–210.

Rinaldo P, Hahn S, Matern D. Clinical biochemical genetics in the twenty-first century. *Acta Paediatrics Supplement* 2004;93:22–26; discussion 27.

Ross KF, Guileyardo JM, Bennett MJ, Barnard JJ. Comment on whole blood levels of dodecanoic acid, a routinely detectable forensic marker for a genetic disease often misdiagnosed as sudden infant death syndrome (SIDS): MCAD deficiency [letter; comment]. *American Journal of Forensic Medicine and Pathology* 1996;17:349–350.

Sausserau E, Lacroix C, Gaulier JM, Goulle JP. On-line liquid chromatography/tandem mass spectrometry simultaneous determination of opiates, cocainics and amphetamines in dried blood spots. *Journal of Chromatography B* 2012;885-886:1–7.

Schütz H, Gotta JC, Erdmann F, Risse M, Weiler G. Simultaneous screening and detection of drugs in small blood samples and bloodstains. *Forensic Science International* 2002;126:191–196.

Shaler RC, Smith FP, Mortimer CE. Detection of drugs in a bloodstain. I: diphenylhydantoin. *Journal of Forensic Sciences* 1978;23:701–706.

Shoemaker JD, Lynch RE, Hoffmann JW, Sly WS. Misidentification of proprionic acid as ethylene glycol in a patient with methylmalonic acidemia. *Journal of Pediatrics* 1992;120:417–421.

Skopp G, Potsch L. Detection of cocaine in blood stains (abstract). *Archiv fur kriminologie* 2001;207:81–88.

Smith FP. Detection of digoxin in bloodstains. *Journal of Forensic Sciences* 1981;26:193–197.

Smith FP, Liu RH. Detection of cocaine metabolite in perspiration stain, menstrual bloodstain and hair. *Journal of Forensic Sciences* 1986;31:1269–1273.

Smith FP, Pomposini DA. Detection of phenobarbital in bloodstains, semen, seminal stains, saliva stains, perspiration stains, and hair. *Journal of Forensic Sciences* 1981;26:582–586.

Smith FP, Shaler RC, Mortimer CE, Errichetto LT. Detection of drugs in bloodstains. II: morphine. *Journal of Forensic Sciences* 1980;25:369–373.

SOFT/AAFS. Forensic Toxicology Laboratory Guidelines; 2006. http://www.soft-tox.org/files/Guidelines_2006_Final.pdf (accessed April 6, 2012).

Sosnoff CS, Ann Q, Bernert JT Jr, Powell MK, Miller BB, Henderson LO, Hannon WH, Fernhoff P, Sampson EJ. Analysis of benzoylecgonine in dried blood spots by liquid chromatography-atmospheric pressure chemical ionization tandem mass spectrometry. *Journal of Analytical Toxicology* 1996;20:179–184.

Stove CP, Ingels AME, DeKesel PMM, Lambert WE. Dried blood spots in toxicology: from the cradle to the grave. *Critical Reviews in Toxicology* 2012;42:230–243.

Thomas A, Geyer H, Schänzer W, Crone C, Kellmann M, Moehring T, Thevis M. Sensitive determination of prohibited drugs in dried blood spots (DBS) for doping controls by means of a benchtop quadrupole/Orbitrap mass spectrometer. *Analytical and Bioanalytical Chemistry* 2012;403(5):1279–1289.

Wilcox RL, Nelson CC, Stenzel P, Steiner RD. Postmortem screening for fatty acid oxidation disorders by analysis of Guthrie cards with tandem mass spectrometry in sudden unexpected death in infancy. *Journal of Pediatrics* 2002;141:833–836.

PART II

PHARMACEUTICAL APPLICATIONS

12

PHARMACEUTICAL PERSPECTIVES OF USE OF DRIED BLOOD SPOTS

CHRISTOPHER EVANS AND NEIL SPOONER

12.1 EARLY DEVELOPMENTS

The concept of collecting blood by blotting/spotting onto paper was first introduced by Robert Guthrie in 1963 (Guthrie and Suzi, 1963). The benefits of using this simple "dried blood spot" (DBS) technology for screening of newborn infants were quickly realized and this technology now plays a critical role in routinely screening for a variety of metabolic diseases in neonates (Pitt, 2010). However, there was less immediate interest in the adoption of this technique for collecting blood samples for the quantitative assessment of circulating drug concentrations in clinical and nonclinical samples derived from pharmaceutical discovery and development studies, despite some early key investigations which highlighted a number of the potential advantages in discovery stage animal pharmacokinetic studies (Beaudette and Bateman, 2004). It was not until the preliminary investigations by Barfield, Spooner, and co-workers that highlighted the practicalities and potential benefits of the technology that the interest of investigators in this area was peeked (Barfield et al., 2008; Spooner et al., 2009). A number of factors had combined to make the time right for further investigation of DBS sampling for quantitative drug bioanalysis:

- The widespread availability of increasingly sensitive mass spectrometers, enabling assays to be developed with the required analytical sensitivity for the small blood volumes used for DBS.
- The requirement of regulatory agencies for pharmaceutical companies to develop medicines for children as a first intent (European Union, 2006), which needed a mechanism for handling small blood volumes.
- The increasing expectations from the public, governments, and regulators around higher standards of animal welfare for drug development, which led to the 3Rs in animal welfare: replacement, reduction, and refinement. This highlighted the need for techniques to be developed to enable fewer rodents to be used in drug development and toxicology studies and for simplified, less stressful, procedures to be used.
- Performing clinical drug trials in remote locations with no, or limited access to centrifuges for the production of plasma, freezers for plasma storage, and facilities for shipping samples to the bioanalytical laboratory in the frozen state.
- The desire to generate drug exposure data where typically they were not previously readily available, for example, investigator sites and home monitoring (therapeutic drug monitoring applications).

Pharmaceutical researchers quickly realizing the potential benefits and impact to drug discovery and development embraced this high-potential concept of microsampling for supporting both nonclinical and clinical studies. The advantages of DBS sampling over conventional wet whole blood, plasma, or serum sample collection are now well understood and widely accepted, namely, a simple and practical way of handling reduced sample volumes. From a nonclinical perspective, this leads to improved data quality through serial sampling or the ability to sample from main study animals, rather than TK satellite animals. Simplicity in sample collection is derived by not having to centrifuge blood to derive plasma, followed by transfer to a separate tube and freezing

Dried Blood Spots: Applications and Techniques, First Edition. Edited by Wenkui Li and Mike S. Lee.

before shipment and storage. In addition, it was theorized that an accurate volume of blood need not be applied to the DBS collection card substrate, if a fixed diameter punch was used to subsample the sample at the time of analysis (although there is more on this story, later in this introduction). Further, the technique enables simplified sample collection (elimination or reduction of rodent warming) and processing procedures (where appropriate, the use of human finger prick sampling (Spooner et al., 2010)), reduced animal use, ambient temperature sample shipment and storage, cost savings for multi-site clinical trials, and so on, and thus, has the potential to have a profound impact on drug development. Furthermore, the technique has enabled quantitative bioanalytical data to be generated in situations and studies where it was previously difficult or impossible to do so, for example, the use of DBS sampling at home or local doctors surgery for therapeutic drug monitoring of drugs with a narrow therapeutic index (Edelbroek et al., 2009); the determination of drug exposures in small and vulnerable children (Pandya et al., 2011; Patel et al., 2013); the ability to perform clinical studies in locations without access to a centrifuge to produce plasma, or freezing conditions for the storage and shipment of the derived samples.

12.2 IN-DEPTH INVESTIGATIONS

While these potential benefits are undeniable, potential practitioners of DBS sampling have wanted to further understand the limitations and potential pitfalls of the technique, particularly when the drug exposure data are to be used for regulatory purposes. Hence, over the last 5 years there have been a large number of publications from researchers in both industry and academia investigating the fundamentals associated with this technique. These include but are certainly not limited to

- understanding that control blood can be used when up to 14 days old for the preparation of DBS samples for use as quantitative calibration standards and quality control samples (Abu-Rabie and Spooner, 2010);
- demonstrating that different spotting devices do not result in variability in the quantitative data generated when the sub-punching approach is used (Spooner et al., 2009);
- definition that DBS samples are dry within 2 hours when dried under standard laboratory conditions (Denniff and Spooner, 2010a);
- understanding the effect of ambient conditions on sample integrity (Denniff and Spooner, 2010a);
- demonstrating greater on-card stability for a variety of analytes as DBS samples, rather than conventional liquid storage (Bowen et al., 2010; D'Arienzo et al., 2010; Heinig et al., 2010; Liu et al., 2011; Bowen et al., 2012);
- understanding the alternative approaches to the incorporation of internal standard into the bioanalytical process (van Baar et al., 2013);
- illustrating that when using the sub-punch method prior to sample extraction and analysis, precise and accurate quantitative data could be obtained if the volume of blood spotted was actually not that precise (Spooner et al., 2009);
- defining the temperature and humidity conditions likely to be encountered by a DBS sample during ambient shipping across international borders and storage (Bowen et al., 2011);
- illustrating that DBS samples are not necessarily homogenous and that the nature of the spot distribution can vary between different substrate types (Clark et al., 2010; Ren et al., 2010; O'Mara et al., 2011; de Vries et al., 2013; Cobb et al., 2013);
- demonstrating that the degree of blood hematocrit can have an effect on the size of the DBS samples derived and hence, potentially give a bias in the measured concentration when the sub-punch method is used (Denniff and Spooner, 2010b; O'Mara et al., 2011; de Vries et al., 2013);
- demonstrating that validated DBS methods result in equivalent precision and accuracy as liquid-based methods (Evans, 2010);
- demonstrating that good quality incurred sample reproducibility (ISR) data can be derived for DBS samples for a wide variety of small molecule drugs (Barfield et al., 2011; Meesters et al., 2012a); and
- application of DBS to large molecules (peptides and proteins) using either ELISA, or LC-MS/MS detection, with and without enzymatic digestion (Kehler et al., 2010; Prince et al., 2010; Kehler et al., 2011; Burns et al., 2012; Sleczka et al., 2012; Kaendler et al., 2013).

12.3 VALIDATION REQUIREMENTS

There are currently no regulatory guidances around the validation of bioanalytical methods for DBS sampling. However, based upon the increasing understanding of DBS samples and their similarities and differences to conventional wet matrix samples, researchers have published opinions and recommendations on how the validation of a regulated quantitative bioanalytical method using DBS sampling might be performed (Spooner et al., 2009; Timmerman et al., 2011a; Xu et al., 2013; Timmerman et al., 2013). Outputs are pending from various industry-sponsored consortia (EBF and IQ) that will provide specific recommendations for method validation and study design.

12.4 PUBLICATION EXPLOSION!

DBS-related literature and content at scientific gatherings has exploded in recent years, where "more than 2300 papers have been published since 1936, while the number in the last three years alone (since 2010) is more than 1200" (Demirev, 2013). DBS sampling is now a commonplace topic in the pharmaceutical industry!

Additionally, over the last 5 years there have been numerous dedicated DBS conferences and workshops, not to mention a plethora of technical sessions contained within more general national and international meetings. This interest has spurred the formation of industry focus groups, adopting a collaborative approach toward increasing the understanding and acceptance of the DBS approach. These groups are generating and sharing data and knowledge, not keeping proprietary knowledge to any one company, and are sharing the output with the wider scientific community through conference presentations and peer-reviewed publications (IQ Consortium Microsampling Working Group, in preparation) (Abbott et al., 2010; Timmerman et al., 2011b; Timmerman et al., 2012a; Timmerman et al., 2012b, 2013; EBF, 2014).

12.5 OTHER USES OF DBS SAMPLING FOR PHARMACEUTICAL DEVELOPMENT

In addition to their use in the regulated quantitative bioanalysis of drugs, interest in DBS technology has also spread to other aspects of pharmaceutical development. A number of manuscripts have demonstrated the successful implementation of the technology for drug discovery study designs, resulting in reduced animal usage, simpler procedures, and higher quality serial exposure data in rodents (Beaudette and Bateman, 2004; Discenza et al., 2012; Rahavendran et al., 2012). The technology has also proved to have utility for the collection of samples for the determination of biomarkers (McDade et al., 2007) and for simplified sampling for therapeutic drug monitoring (Edelbroek et al., 2009). The latter area is particularly important for marketed pharmaceuticals with a narrow therapeutic index, for which dose adjustment is required for individual patients to ensure efficacy, while minimizing potential adverse events, or to indicate whether prescribed medication is actually being administered. In this context, DBS technology offers the possibility of immediate sampling in a doctor's surgery, or possibly in the patient's home, provided suitable safeguards are taken to ensure the integrity of the collected sample.

The use of dried sampling has also been applied to matrices other than blood, including plasma (Barfield et al., 2011). The technique has also shown utility for sampling any volume-limited fluids, that is, synovial fluid, tears, and even cerebrospinal fluid (Christianson et al., 2010a; Christianson et al., 2010b). One of the single largest secondary benefits of the exploration into DBS has been the increased interest in microsampling techniques; looking to leverage the benefits of blood microsampling, researchers are exploring a variety of approaches (Musteata et al., 2009; Jonson et al., 2012; Bowen et al., 2013).

12.6 HURDLES

While there seems to have been some cautious optimism and acceptance of the DBS technique from some regulators, particularly where implemented in nonclinical study support (Beharry et al., 2010; Viswanathan et al., 2012a), the data from some of these investigations have revealed some hurdles and obstacles. The impact of hematocrit on spot size and recovery, spot homogeneity, and best practices for inclusion of bioanalytical internal standard currently seem to be the largest sticking points. In addition, there are further complexities for investigators wishing to change from determining systemic exposure in plasma to that in blood (Emmons et al., 2010), particularly if their intent is to do so within the same regulatory submission package. These complexities have led to a more conservative approach from regulators when it comes to implementing DBS sampling for use in clinical study support, and it seems that they are not yet ready to accept DBS-only exposure data (Xu et al., 2013). Recently, there have been requests from regulators to provide exposure data for both wet and dry samples from clinical patients in order to demonstrate concordance between the matrices, before proceeding with the use of DBS alone for further clinical studies. At the time of writing, sponsors are doing this in a sparse-sampling approach. An upcoming publication from the IQ Consortium Microsampling Working Group will suggest these concordance determinations should occur in a number of samples capable of providing an acceptable representation of the study in the intended population and at relevant doses. Consideration should be given to the expected correlation; plasma to blood may not result in a 1:1 relationship (similarly for a venous to peripheral comparison), while wet blood to DBS should (IQ Consortium Microsampling Working Group, in preparation). The hope remains that as more DBS concordance data are generated and published by increasing numbers of practitioners, the level of confidence placed on DBS methods will increase and the requirement for the comparison of wet and dry samples will diminish, particularly for drug development programs where the entirety of the clinical and nonclinical program is performed in the same matrix.

Another important hurdle to the more widespread adoption of DBS sampling is the added complexity in developing, validating, and applying the bioanalytical methods in the quantitative bioanalytical laboratory. The small sample size (typically equivalent to <3 μL blood) means that it may be difficult to develop robust assays, particularly for drugs with low systemic exposure. Further, there are a number of different DBS substrate cards commercially available. These have different properties and there do not appear to be any consistent rules as to which analytes will work best with which substrate. In addition, the solvent used to extract the analyte from the DBS samples often needs to be optimized. A further complexity in the bioanalytical laboratory is that the DBS sample needs to be punched prior to extraction, which is considered a more laborious process than simply sub-aliquoting a liquid sample. Thankfully, these issues are beginning to be addressed. Mass spectrometers are becoming increasingly sensitive and automation, either for punching the DBS sample or direct liquid elution, is increasingly becoming available (Abu-Rabie et al., 2011).

12.7 THE WAY AHEAD!

Despite these hurdles, users and vendors still recognize the numerous benefits of DBS sampling listed above and are therefore developing strategies to minimize or eliminate the issues. These focus on careful consideration as to how DBS sampling is deployed in drug development and the investigation and implementation of novel consumables.

With regard to how DBS sampling is deployed, two strategies for implementation have evolved and are being adopted within various organizations. One approach is to exclusively use DBS sampling from an early stage of development, where DBS data may be the only data type collected for pharmacokinetic/toxicokinetic assessment. This provides consistency in sample type and does not require any cross-comparison between, for example, plasma and blood exposures. Clear guidance has been provided by Emmons and Rowland on the considerations around the determination of drug exposures in blood or plasma (Emmons et al., 2010). Alternatively, an asset could progress through nonclinical and clinical development in a liquid matrix, and DBS be implemented later in development to address specific populations or study designs, that is, pediatric, remote areas, and home monitoring. In using this approach, DBS sampling may offer the ability to generate exposure data that were difficult, or impossible to obtain using traditional PK sampling techniques.

Another important consideration is the potential for blood hematocrit to impact the quantitative drug exposure data. It is likely that this will only be a major concern if there are subjects with a broad range of hematocrit values in a given study and the blood spot sub-punch approach is used. If, however, the hematocrit values encountered for subjects are different to what is considered normal, but do not cover a wide range, then a good approach might be to adjust the hematocrit of the control blood used to prepare calibration standards and quality control samples. This should minimize any assay bias effects encountered. In any case, the effect of hematocrit should be carefully explored as part of the validation of the bioanalytical method. A further consideration is that for some analytes the recovery can also vary with the blood hematocrit and this in turn has the potential to mask the bias effects. The authors therefore also recommend an investigation of analyte recovery with hematocrit as part of the quantitative assay validation.

Another approach to overcome the hematocrit effect and to eliminate any homogeneity issues is to punch-out and extract the whole DBS sample. However, this relies on an accurate volume of blood being spotted at the point of sample origin, something that is not required when the sub-punch method is utilized. It is the authors' belief that this is not a preferred approach with currently available spotting equipment. For example, if pipettes were to be used, they are an expensive item, require regular calibration, and would necessitate considerable training of the practitioners. Not something that is practical for multi-site clinical studies! An alternative would be to use affordable and disposable micropipettes and capillaries. However, these would again require considerable training of medical practitioners and a concern that there is no way of confirming the quality and accuracy of the blood volume spotted.

In addition to the above, there have been a number of recent publications that point the way to alternative technologies that may help to minimize or overcome the hematocrit issues. These include facilitating the spotting of an accurate volume of blood, separating the plasma component on the sample substrate, and determining the hematocrit value of the test samples (Li et al., 2011; Fan and Lee, 2012; Li et al., 2012; Meesters et al., 2012b; Capiau et al., 2013). It is our firm belief that the broad understanding of the issues and willingness to realize the potential of dried sampling will lead to further in-depth investigations and publications and that this will lead to a new generation of dried sampling technologies which will be more acceptable to the regulated quantitative bioanalytical community and to the regulators themselves (Viswanathan et al., 2012b).

This desire will also likely result in the discovery of additional ingenious vendor solutions to circumvent the negative influence of hematocrit. Solutions to the hematocrit effects and more widespread acceptance of the technique will only serve to propel this technology forward once again. The other analytical "hurdles" to the technique are considered by the authors to have less impact on the quality of data obtained from DBS samples and therefore do not serve as that great of an impediment to the use of this technique for regulated bioanalysis.

12.8 CONCLUSIONS

It is our belief that the early optimism around the DBS technique was completely founded (Spooner, 2010a; Spooner,

2010b) and that it shows great promise to revolutionize our industry. Several secondary benefits have already been demonstrated. However, considering our understanding around the technological hurdles and cautious optimism from regulators, more justification is needed for use in its current form—widespread implementation is still a ways off. However, once vendors solve some of the analytical challenges and these new approaches have been thoroughly investigated, we anticipate another flurry of related publications followed by increasing implementation and adoption of the technology.

The enthusiasm and scientific investigations into DBS are readily demonstrated in the various chapters within this book. The editors have looked to capture a diverse series of topics representing the current state of the art associated with technology and some forward looking concepts in Parts II and III. The compilation of this information by so many thought leaders in the area will no doubt lead to further understanding of the issues and many proven benefits of the DBS technique, particularly for its use in regulated quantitative drug bioanalysis.

REFERENCES

Abbott R, Smeraglia J, White S, Luedtke S, Brunet L, Thomas E, Globig S, Timmerman P. Conference report: connecting strategies on dried blood spots. *Bioanalysis*, 2010;2:1809–1816.

Abu-Rabie P. Direct analysis of DBS: emerging and desirable technologies. *Bioanalysis* 2011;3:1675–1678.

Abu-Rabie P, Spooner N. Study to assess the effect of age of control human and animal blood on its suitability for use in quantitative bioanalytical DBS methods. *Bioanalysis* 2010;2:1373–1384.

Barfield M, Wheller R. Use of dried plasma spots in the determination of pharmacokinetics in clinical studies: validation of a quantitative bioanalytical method. *Analytical Chemistry* 2011;83:118–124.

Barfield M, Spooner N, Lad R, Parry S, Fowles S. Application of dried blood spots combined with HPLC-MS/MS for the quantification of acetaminophen in toxicokinetic studies. *Journal of Chromatography B* 2008;870:32–37.

Barfield M, Ahmad S, Busz M. GlaxoSmithKline's experience of incurred sample reanalysis for dried blood spot samples. *Bioanalysis* 2011;3:1025–1030.

Beaudette P, Bateman KP. Discovery stage pharmacokinetics using dried blood spots. *Journal of Chromatography B* 2004;809:153–158.

Beharry M. DBS: a UK (MHRA) regulatory perspective. *Bioanalysis* 2010;2:1363–1364.

Bowen C, Hemberger M, Kehler J, Evans C. Utility of dried blood spot sampling and storage for increased stability of photosensitive compounds. *Bioanalysis* 2010;2:1823–1828.

Bowen C, Dopson W, Kemp D, Lewis M, Lad R, Overvold C. Investigations into the environmental conditions experiences during ambient sample transport: impact to dried blood spot sample shipments. *Bioanalysis* 2011;3:1625–1633.

Bowen CL, Volpatti J, Cades J, Licea-Perez H, Evans CA. Evaluation of glucuronide metabolite stability in dried blood spots. *Bioanalysis* 2012;4:2823–2832.

Bowen CL, Licea-Perez H, Karlinsey MZ, Jurusik K, Pierre E, Siple J, Kenney, J, Stokes A, Spooner N, Evans C. A novel approach to capillary plasma microsampling for quantitative bioanalysis. *Bioanalysis* 2013;5(9):1131–1135.

Burns D, Brunner L, Rajendran S, Johnson B, Ma M, Wang J. Validation of a ligand binding assay using dried blood spot sampling. *AAPS Journal* 2012;15:123–131.

Capiau S, Stove VV, Lambert WE, Stove CP. Prediction of the hematocrit of dried blood spots via potassium measurement on a routine clinical chemistry analyzer. *Analytical Chemistry* 2013;85:404–410.

Christianson CD, Laine DF, Zimmer JSD, Johnson CJL, Sheaff CN, Carpenter A, Needham SR. Development and validation of an HPLC–MS/MS method for the analysis of dexamethasone from pig synovial fluid using dried matrix spotting. *Bioanalysis*, 2010a;2:1829–1837.

Christianson CD, Johnson CJ, Sheaff CN, Laine DF, Zimmer JSD, Needham SR. Dried Matrix Spot Analysis – Analysis of Drugs from Tears and Other Translucent Fluids Using a Novel Color Indicating Technology. Presented at CPSA, Langhorne, PA; 2010b.

Clark GT, Haynes JJ, Baylis MA, Burrows L. Utilization of DBS within drug discovery: development of a serial microsampling pharmacokinetic study in mice. *Bioanalysis* 2010;2:1477–1488.

Cobb Z, de Vries R, Spooner N, Williams S, Staelens L, Doig M, Broadhurst R, Barfield M, van de Merbel N, Schmid B, Siethoff C, Ortiz J, Verheij E, van Baar B, White S, Timmerman P. In-depth study of homogeneity in DBS using two different techniques: results from the EBF DBS-microsampling consortium. *Bioanalysis* 2013;5(17): 2161–2169.

D'Arienzo CJ, Li QC, Discenza L, Cornelius G, Hynes J, Cornelius L, Santella JB, Olah T. DBS sampling can be used to stabilize prodrugs in drug discovery rodent studies without the addition of esterase inhibitors. *Bioanalysis* 2010;2:1415–1422.

Demirev PA. Dried blood spots: analysis and applications. *Analytical Chemistry* 2013;85:779–789.

Denniff P, Spooner N. Effect of storage conditions on the weight and appearance of dried blood spot samples on various cellulose-based substrates. *Bioanalysis* 2010a;2:1817–1822.

Denniff P, Spooner N. The effect of hematocrit on assay bias when using DBS samples for the quantitative analysis of drugs. *Bioanalysis* 2010b;2:1385–1395.

de Vries R, Barfield M, van de Merbel N, Schmid B, Siethoff C, Ortiz EV, van Baar B, Cobb Z, White S, Timmerman P. The effect of hematocrit on bioanalysis of DBS: results from the EBF DBS-microsampling consortium. *Bioanalysis* 2013;5(17):2147–2160.

Discenza L, Obermeier MT, Westhouse R, Olah TV, D'Arienzo CJ. A bioanalytical strategy utilizing dried blood spots and LC-MS/MS in discovery toxicology studies. *Bioanalysis* 2012;4:1057–1064.

EBF. 5th Open Meeting, Barcelona, 14–16 November 2012, Day 2, Breakout Session I http://bcn2012.europeanbioanalysis forum.eu/Slides/ (accessed 20 January 2014).

Edelbroek PM, van der Heijden J, Stolk LM. Dried blood spot methods in therapeutic drug monitoring: methods assays and pitfalls. *Therapeutic Drug Monitoring* 2009;31:327–336.

Emmons G, Rowland M. Pharmacokinetic considerations as to when to use dried blood spot sampling. *Bioanalysis* 2010;2:1791–1796.

European Union. Regulation (EC) No 1901/2006 of the European Parliament and of the Council of 12 December 2006 on medicinal products for paediatric use and amending Regulation (EEC) No 1768/92, Directive 2001/20/EC, Directive 2001/83/EC and Regulation (EC) No 726/2004. Official Journal of the European Union 27.12.2006, L378/1–378/19; 2006.

Evans CA. Method Reproducibly of Dried Blood Spot Methods – Four years of Experience. Presented at the Applied Pharmaceutical Analysis Annual Symposium, Baltimore, MD, September 2010.

Fan L, Lee JA. Managing the effect of hematocrit on DBS analysis in a regulated environment. *Bioanalysis* 2012;4:345–347.

Guthrie R, Suzi A. A simple phenylalanine method for detecting phenylketonuria in large populations of newborn infants. *Pediatrics* 1963;32:338–343.

Heinig K, Bucheli F, Hartenbach R, Gajate-Perez A. Determination of mycophenolic acid and its phenyl glucuronide in human plasma, ultrafiltrate, blood, DBS and dried plasma spots. *Bioanalysis* 2010;2:1423–1436.

Jonson O, Palma Villar R, Nilsson LB, Erikson M, Konigsson K. Validation of a bioanalytical method using capillary microsampling of 8 μl plasma samples: application to a toxicokinetic study in mice. *Bioanalysis* 2012;4:1989–1998.

Kaendler K, Warren A, Lloyd P, Sims J, Sickert D. Evaluation of dried blood spots for the quantification of therapeutic monoclonal antibodies and detection of anti-drug antibodies. *Bioanalysis* 2013;5:613–622.

Kehler JR, Bowen CL, Boram SL, Evans CA. Application of DBS for quantitative assessment of the peptide Exendin-4; comparison of plasma and DBS method by UHPLC-MS/MS. *Bioanalysis* 2010;2:1461–1468.

Kehler J, Akella N, Citerone D, Szapacs M. Application of DBS for the quantitative assessment of a protein biologic using on-card digestion LC-MS/MS or immunoassay. *Bioanalysis* 2011;3:2283–2290.

Li F, Zulkoski J, Fast D, Michael S. Perforated dried blood spots: a novel format for accurate microsampling. *Bioanalysis* 2011;3:2321–2333.

Li Y, Henion J, Abbott R, Wang P. The use of a membrane filtration device to form dried plasma spots for the quantitative determination of guanfacine in whole blood. *Rapid Communications in Mass Spectrometry* 2012;26:1208–1212.

Liu G, Ji QC, Jemal M, Tymiak AA, Arnold ME. Approach to evaluating dried blood spot sample stability during drying process and discovery of a treated card to maintain analyte stability by rapid on-card pH modification. *Analytical Chemistry* 2011;83:9033–9038.

McDade TW, Williams S, Snodgrass JJ. What a drop can do: dried blood spots as a minimally invasive method for integrating biomarkers into population-based research. *Demography* 2007;44:899–925.

Meesters RJW, Hoof GP, Gruters R, van Kampen JJA, Luider TM. Incurred sample reanalysis comparison of dried blood spots and plasma samples on the measurement of lopinavir in clinical samples. *Bioanalysis* 2012a;4:237–240.

Meesters RJW, Zhang J, vanHuizen NA, Hoof GP, Gruters RA, Luider TM. Dried matrix on paper discs: the next generation DBS microsampling technique for managing hematocrit effect in DBS analysis. *Bioanalysis* 2012b;4:2027–2035.

Musteata F. Pharmacokinetic applications of microdevices and microsampling techniques. *Bioanalysis* 2009;1:171–185.

O'Mara M, Hudson-Curtis B, Olson K, Yueh Y, Dunn J, Spooner N. The effect of hematocrit and punch location on assay bias during quantitative bioanalysis of dried blood spot samples. *Bioanalysis* 2011;3:2335–2347.

Pandya HC, Spooner N, Mulla H. Dried blood spots, pharmacokinetic studies and better medicines for children. *Bioanalysis* 2011;3:779–786.

Patel P, Mulla H, Kairamkonda V, Spooner N, Gade S, Della-Pasqua O, Field D, Pandya HC. Dried blood spots and sparse sampling: a practical approach to estimating pharmacokinetic parameters of caffeine in preterm infants. *British Journal of Clinical Pharmacology* 2013;75:805–813.

Pitt J. Newborn screening. *Clinical Biochemistry Review* 2010;31:57–68.

Prince PJ, Matsuda KC, Retter MW, Scott G. Assessment of DBS technology for the detection of therapeutic antibodies. *Bioanalysis*, 2010;2:1449–1460.

Rahavendran SV, Vekich S, Skor H, Batugo M, Nguyen L, Shetty B, Shen Z. Discovery pharmacokinetic studies in mice using serial microsampling, dried blood spots and microbore LC-MS/MS. *Bioanalysis* 2012;4:1077–1095.

Ren X, Paehler T, Zimmer M, Guo Z, Zane P, Emmons G. Impact of various factors on radioactivity distribution in different DBS papers. *Bioanalysis* 2010;2:1469–1475.

Sleczka B, D'Arienzo CJ, Tymiak AA, Olah TV. Quantitation of therapeutic proteins following direct trypsin digestion of dried blood spot samples and detection by LC-MS-based bioanalytical methods in drug discovery. *Bioanalysis* 2012;4:29–40.

Spooner N. A glowing future for dried blood spot sampling. *Bioanalysis* 2010a;2:1343–1344.

Spooner N. Dried blood spot sampling for quantitative bioanalysis: time for a revolution? *Bioanalysis* 2010b;2:1781.

Spooner N, Lad R, Barfield M. Dried blood spots as a sample collection technique for the determination of pharmacokinetics in clinical studies: considerations for the validation of a quantitative bioanalytical method. *Analytical Chemistry* 2009;81:1557–1563.

Spooner N, Ramakrishnan Y, Barfield M, Dewit O, Miller S. Use of DBS sample collection to determine circulating drug concentrations in clinical trials: practicalities and considerations. *Bioanalysis* 2010;2:1515–1522.

Timmerman P, White S, Globig S, Ludtke S, Brunet L, Smeraglia J. EBF recommendation on the validation of bioanalytical methods for dried blood spots. *Bioanalysis* 2011a;3:1567–1575.

Timmerman P, White S, Globig S, Lüdtke S, Brunet L, Smith C, Smeraglia J. EBF and dried blood spots: from recommendations to potential resolution. *Bioanalysis*, 2011b;3:1787–1789.

Timmerman P, White S, Cobb Z, de Vries R, Thomas E, van Baar B, Smith C, Zimmer D, Delrat P, Smeraglia J, Lüdtke S. Updates from the EBF DBS-microsampling consortium. *Bioanalysis* 2012a;4:1969–1970.

Timmerman P, White S, Cobb Z, de Vries R, Thomas E, van Baar B. Update of the EBF recommendation for the use of dried blood spots in regulated bioanalysis towards the conclusions from the EBF DBS microsampling consortium. *Bioanalysis* 2012b;5:2129–2136.

Timmerman P, White S, Cobb Z, de Vries R, Thomas E, van Baar B. Update of the EBF recommendation for the use of DBS in regulated bioanalysis integrating the conclusions from the DBS-microsampling consortium. *Bioanalysis* 2013;5(17):2129–2136.

van Baar BLM, Verhaeghe T, Heudi O, Rohde M, Wood S, de Vries R, White S, Timmerman P. IS addition in bioanalysis of DBS: results from the EBF DBS-microsampling consortium. *Bioanalysis* 2013;5(17):2137–2145.

Viswanathan CT. Perspectives on microsampling: DBS. *Bioanalysis* 2012a;4:1417–1419.

Viswanathan CT. Bringing new technologies into regulatory space. *Bioanalysis* 2012b;4:2763–2764.

Xu Y, Woolf E, Agrawal NGB, Kothare P, Pucci V, Bateman KP. Merck's perspective on the implementation of dried blood spot technology in clinical drug development – what, when and how. *Bioanalysis* 2013;5:341–350.

13

PUNCHING AND EXTRACTION TECHNIQUES FOR DRIED BLOOD SPOT SAMPLE ANALYSIS

PHILIP WONG AND CHRISTOPHER A. JAMES

13.1 INTRODUCTION

Dried blood spot (DBS) sampling begins with the collection of small volumes of whole blood that are dried onto paper cards. Due to the unique characteristics of this type of sample matrix relative to the more familiar technique of handling liquid biofluids during analysis, sampling and extraction of DBS samples present specific and unique challenges. The general approach to the analysis of DBS samples is to punch a small disk from the blood spot followed by extraction of the disk with a solvent or buffer solution to yield a liquid sample. For small molecule compounds, the liquid sample can either be analyzed directly or undergo further sample processing steps (e.g., liquid–liquid extraction (LLE) or solid-phase extraction (SPE)) followed by LC-MS/MS analysis or other analytical techniques. Recently, analytical methods for DBS analysis have been reviewed (Li and Tse, 2010; Tanna and Lawson, 2011); although a number of analytical techniques, such as direct desorption (Wiseman et al., 2010; Crawford et al., 2011) and direct elution (Loppacher et al., 2011; Ganz et al., 2012), are reported which do not require punching or extraction procedures, these techniques are not commonly used and require special equipment in addition to conventional bioanalytical LC-MS/MS systems. These approaches will be discussed in separate chapters of this book. The main focus of this chapter is on punching and extraction of DBS samples of small molecule compounds prior to analysis; extraction of samples of large molecules (LM) will also be briefly discussed, but currently there are only a limited number of publications on LM bioanalysis with DBS.

The punch size provides a volumetric measure of the sample for quantitative analysis for most DBS methods. The punch size presents two analytical challenges. First, punching a small disk (e.g., 3 mm diameter) from the DBS samples only utilizes a small volume of the sample for analysis (e.g., ~3 μL for a 3 mm diameter disk). The small volume can potentially limit assay sensitivity. When sensitivity is an issue, punching a larger diameter of blood spot using multiple spots or concentrating the final extract may increase the sensitivity of an assay to some degree. A second potential issue is that the effective volume of sample is dependent on both the size of the spot (punch diameter) and how the sample is distributed across the spot when applied to the card. In particular, the diameter of the spot can be affected by the viscosity of the sample (related to the hematocrit level); poor spotting technique can also render analytical results unreliable.

13.2 PUNCHING

13.2.1 Punching Devices

There are a number of manual punching tools which can be used to sample the DBS cards. Punching tools are available in sizes from as small as 0.35 mm to as large as 12.0 mm diameter (Figure 13.1a). These punching tools are of very low cost and are commonly used in pharmaceutical industry for DBS analysis. During the punching process, a DBS card is placed on top of a punching mat (Figure 13.1b). The punching tool is then placed on the center of the blood spot on the card and is pressed down. By twisting the tool, the

Dried Blood Spots: Applications and Techniques, First Edition. Edited by Wenkui Li and Mike S. Lee.

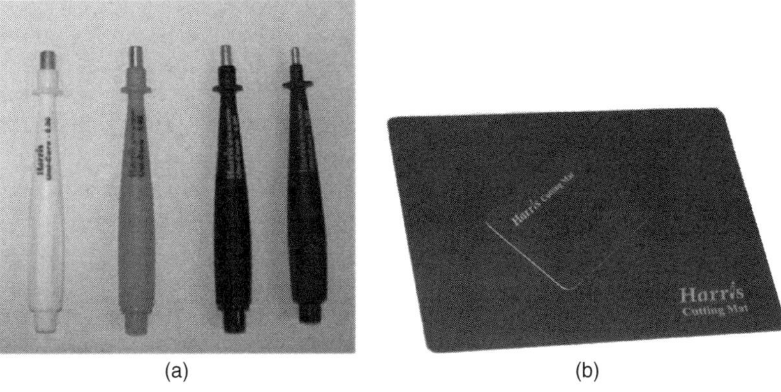

FIGURE 13.1 (a) Manual punching tools and (b) punching mats.

blood spot is cut and is then dispensed into a cluster tube or 96-well plate for extraction. This punching process is then repeated with another DBS sample. Although carryover does not appear to be a major issue, the samples can be punched in order from low to high concentration, or a blank card (without blood spot) can be punched by the tool before the next sample is punched. After a number of punching cycles, these tools become blunt and can be disposed or the cutting heads replaced. Although simple and inexpensive, manual punching processes are time consuming and labor intensive and can also lead to the risk of repetitive strain injuries. A hand held electric punching device (e.g., Harris e-Core™) can be used to reduce labor during the manual punching process. This device, similar in size to an electronic pipette, cuts the disk by rotating the cutting head.

The punching process can also be performed by robotic systems in a fully or semi-automated fashion. Both fully and semi-automatic punching systems can be equipped with punching heads of variable sizes (e.g., 1.2, 1.5, 2.0, 3.2, 4.7, and 6.0 mm diameter). In a semi-automated system, sample cards are fed in manually. A barcode reader can be added for sample identification to streamline the automation process. In an advanced model of a semi-automated system, such as BSD 600 Duet as shown in Figure 13.2, a light targeting system guides the user by highlighting the specific area to be punched helping to ensure accurate sampling. Additional features include an anti-static system (humidifier) and a punch cleaning system (vacuum) to remove particles during the punching process. The anti-static feature avoids potential problems with static electricity buildup that can prevent the disk from falling directly into the sample well; in addition when punching a large number of samples, dust particles from DBS cards rapidly accumulate in the system and might cause contamination. When the DBS sample is ready to be punched, a hand/foot switch is triggered by the user or by a time delay auto-triggered system. A disk detection system confirms that the punched disk is passed through the chute and placed in the receiving well/tube. This semi-automated system can handle cards of various formats and is ideal for medium-throughput analysis. A fully automated punching system can be used for high-throughput DBS sample analysis. The automated punching system can communicate with Laboratory Information Management System (LIMS) software and determine what and how many samples to be punched. In a fully automated punching system, the DBS cards are fed by grippers and thus require no operator attention. A robotic arm is used to pick up the sample plates containing the punched disks and place them onto a liquid-handling workstation for the downstream processing. There are a number of fully and semi-automated card

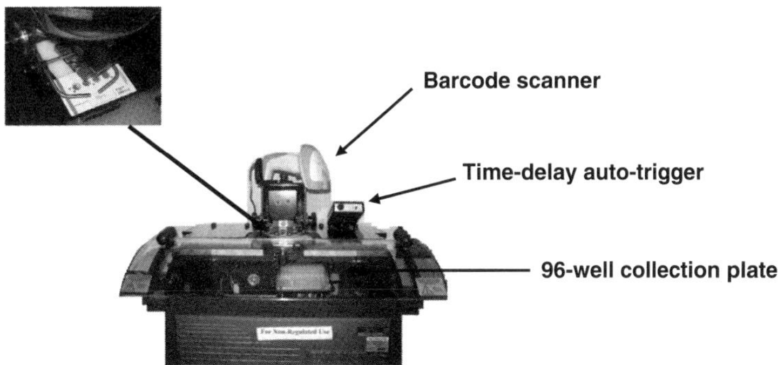

FIGURE 13.2 A semi-automated punching system, BSD 600 duet. The insert shows the DBS card.

punching systems commercially available. These include instruments from Luminex (BSD Robotics), PerkinElmer, Hudson Robotics, Tomtec, Hamilton Robotics, and other manufacturers. Some of the automation systems can accommodate different card formats. For example, the BSD-1000 can handle card sizes of 50 × 50, 80 × 50, and 108 × 50 mm. The options for automation systems are likely to increase as DBS collection becomes more widely adopted for bioanalysis. Alternative approaches to automation involving direct elution from the cards will be discussed in separate chapters.

13.2.2 Volumetric Issue

As indicated in the introduction, the effective volume of the sample analyzed is dependent on both the punch diameter and how a given volume of the blood sample spreads on the card and is distributed within the DBS formed. Hematocrit is a particular issue for DBS analysis which has been discussed in multiple publications (Denniff and Spooner, 2010; Arora et al., 2011; Li et al., 2011; O'Mara et al., 2011; Youhnovski et al., 2011). Hematocrit is the percentage of blood cells in whole blood by volume and the ranges in adult humans are between 41% and 50% for males and 36% and 44% for females. The hematocrit range is about 28–55% for infants of less than 2 years old (Denniff and Spooner, 2010). The hematocrit level of a diseased population may be significantly different from that of a healthy population and can also be altered by specific drug treatments. A major concern for DBS sampling is variation in hematocrit which affects the overall viscosity of whole blood and, thus, the spot size. Blood samples with higher hematocrit levels are more viscous, and therefore, spread less and have a smaller spot size than the same volume of blood with a low hematocrit level. This variability in hematocrit levels could lead to inaccuracy in the determination of the concentration of analyte in the DBS sample. The effects of hematocrit on the area of DBS have been studied (Denniff and Spooner, 2010). The experiments involved spotting 15 μL of human blood with hematocrit levels ranging from 20% to 80% on three types of paper substrate and determining the area from eight replicate blood spots. Figure 13.3 illustrates the relationship between human blood hematocrit and the area of the DBSs obtained on different paper substrates. The decrease in spot area with increasing hematocrit is linear on the three types of paper substrate tested. The results also showed that the blood spot area at a given hematocrit value depends on the type of cards used. In order to overcome the effect of variable hematocrit, the entire blood spot can be sampled for analysis (Li et al., 2011; Youhnovski et al., 2011). Punching the entire blood spot can also be useful in method development and validation as a way to assess extraction recovery. However, the volume analyzed is then dependent on accuracy of the volume spotted onto the DBS cards when the blood sample was collected. This introduces an additional requirement during sample collection and an additional potential source of error.

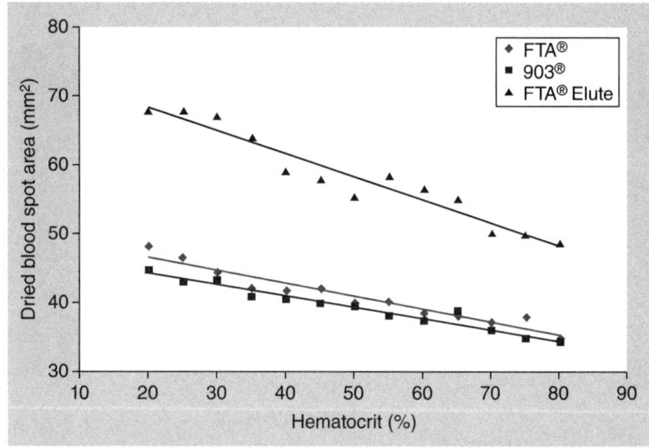

FIGURE 13.3 Relationship between human blood hematocrit and the area of the dried blood spots obtained on different paper substrate. Reproduced with permission from Denniff and Spooner, 2010. Copyright 2010 Future Science Ltd.

13.2.3 Effect of Punch Positions

One assumption of the DBS sampling technique is that the analyte is evenly distributed throughout the DBS. However, recent studies have shown that there may be significant differences in analyte distribution between the perimeter and the center of the blood spot (O'Mara et al., 2011). One study involved spotting 20 μL of blood at control hematocrit level (45%) containing a radiolabeled compound on four types of DBS cards. After drying, the sample was punched using a 3 mm diameter punch at two locations (center and perimeter). The punched samples were combusted followed by scintillation counting analysis. The analyte distribution was measured as P/C where P is the analyte concentration in the perimeter while C is the analyte concentration in the center of the DBS. A P/C value of 1.0 indicates an even distribution of analyte on the DBS. The comparison of perimeter to center punch ratios (P/C) for five compounds and four blood spot card types at control hematocrit level (45%) is shown in Table 13.1. A value of 1.40 was measured for compound GSK4 in DBS on both Ahlstrom 226 and Whatman DMPK C cards, indicating the analyte concentration measured in the perimeter punches was greater than that measured in the center punch. However, for Whatman FTA Elute card type, all compounds tested were close to being homogeneously distributed across the spot. These results indicate that the homogeneous distribution of analyte can depend on both the card type and the nature of the analyte. It is unclear if these results represent how most compounds behave on the cards or if the results are specific for the compounds studied. Another study also showed inhomogeneous distribution of three analytes on five types of DBS cards by measuring the radioactivity

TABLE 13.1 Comparison of perimeter to center punch ratios (P/C) for various compounds and different blood spot card types at control hematocrit level (45%)

	Mean perimeter to center ratios ($n = 4$) for four card types			
Compound	Ahlstrom 226	Whatman FTA	Whatman FTA Elute	Whatman DMPK C
Acetaminophen	1.17	1.11	0.93	1.09
GSK1	1.22	1.07	0.94	1.23
GSK2	1.25	1.02	0.98	1.18
GSK3	1.46	1.07	0.89	1.37
GSK4	1.40	1.17	0.99	1.40

Source: Reproduced with permission from O'Mara et al., 2011.

distribution on the DBS (Ren et al., 2010). It is recommended to test the influence of punch position for every new method. The effect can, however, be minimized by consistently sampling blood from the center of the blood spot.

13.3 EXTRACTION TECHNIQUES

13.3.1 Solvent Extraction

Water-miscible organic solvents are most commonly used to extract analytes for DBS analysis of small molecule compounds. Solvents commonly used are methanol (MeOH), acetonitrile (ACN), or a mixture of MeOH, ACN, and water (e.g., 75% MeOH/water). The proportion of the aqueous content can be adjusted to optimize the extraction recovery. Based on visual observation, it appears that the higher the aqueous content, the more the dissolution of blood cells and other endogenous components from the DBS (West et al., 2010), resulting in more potential for matrix effect. Typically, an internal standard (IS) is added to the extraction solvent for DBS analysis to compensate for matrix effect, loss due to sample handling, and the drift in detector response. After the extraction solvent is added to the DBS samples, the mixture is then vortex mixed from 0.5 to 60 minutes. The vortexing time may be analyte dependent. Longer extraction times may not increase the extraction recovery but may improve reproducibility (Wong et al., 2011). Sonication also assists the dissolution of analytes into the extraction solvent. After centrifugation, the supernatants are transferred for LC-MS/MS analysis. For example, in the analysis of acetaminophen in dog DBS (Barfield et al., 2008), MeOH (100 μL) containing an IS ($^2H_4$-acetaminophen) was added to DBS samples (3 mm diameter disk) which were then vortex mixed for approximately 30 seconds followed by centrifugation. The supernatant was transferred for LC-MS/MS analysis. Although organic solvent/water extraction is simple and easy to implement, it has been reported to suffer from severe matrix effects (Liu et al., 2010).

Although the majority of reports currently in the literature are focused on small molecules, there is interest in the use of DBS for LM (Brindle et al., 2010; Prince et al., 2010). For the analysis of LM in DBS, organic solvent/water extraction will potentially denature the protein. Aqueous buffers (e.g., mixture of phosphate-buffered saline and Tween 20) are, therefore, commonly used in the extraction of LM. In a typical extraction procedure, aqueous buffer is added to DBS and vortex mixed for an hour. The samples are incubated overnight at 4°C. After incubation, the samples are vortex mixed for another 30–60 minutes and centrifuged. The supernatants are transferred for immunoassay. There are also reports of LC-MS/MS methods being used to quantitate protein therapeutics, where an enzyme solution (e.g., Lys-C or trypsin) is added to the DBS samples to perform digestion (Kehler et al., 2011; Olah, 2012) prior to analysis of a signature peptide to quantitate the molecule.

13.3.1.1 Effect of Hematocrit on Extraction Recovery

The effect of hematocrit on extraction recovery has been studied (Li et al., 2011). The experiments involved the analysis of the entire blood spot such that a known amount of analyte is present in the DBS regardless of the hematocrit levels. The analyte and IS used were lansoprazole and lansoprazole-d_4, respectively. A blood spotting volume of 5 μL and 200 μL of an extraction solvent (90% ACN/water) containing 2 ng/mL IS were used. After vortex mixing (10 minutes) and centrifugation, the extracts were analyzed by LC-MS/MS. Figure 13.4a illustrates that the normalized responses (analyte peak area and the analyte/IS ratio) decreased with increasing hematocrit levels, indicating a lower analyte recovery at high hematocrit levels. Pretreatment with water, sonication, and prolonged vortexing time appeared to help release the analyte from the DBS samples with high hematocrit, as the normalized responses became relatively constant (Figure 13.4b). These results indicated that extraction recovery and quantitative results may be affected at high hematocrit levels unless the variability in extraction recovery is compensated by the IS; however, in this example, the IS added to the extraction solvent did not fully compensate for changes in recovery of the analyte. A number of IS addition methods have been investigated in order to compensate for variability in extraction recovery (Meesters et al., 2011). In the study, nevirapine (NVP) and its stable isotope-labeled compound (NVP-d_3) were used as a model compound and an IS, respectively. A blood spotting volume of 25 μL (three blood spots), a 7 mm diameter punch, and a 200 μL of an extraction solvent (80% ACN/water) were used. After sonication for 30 minutes, the extracts (20 μL) were mixed with a MALDI matrix solution (20 μL) and then subjected to MALDI-QqQ-MS/MS analysis. The IS addition methods investigated were (i) adding IS to liquid blood samples before spotting, (ii) adding IS to DBS cards before spotting (i.e., IS-treated cards), and (iii) adding IS to DBS (i.e., adding IS after blood spotting). For method

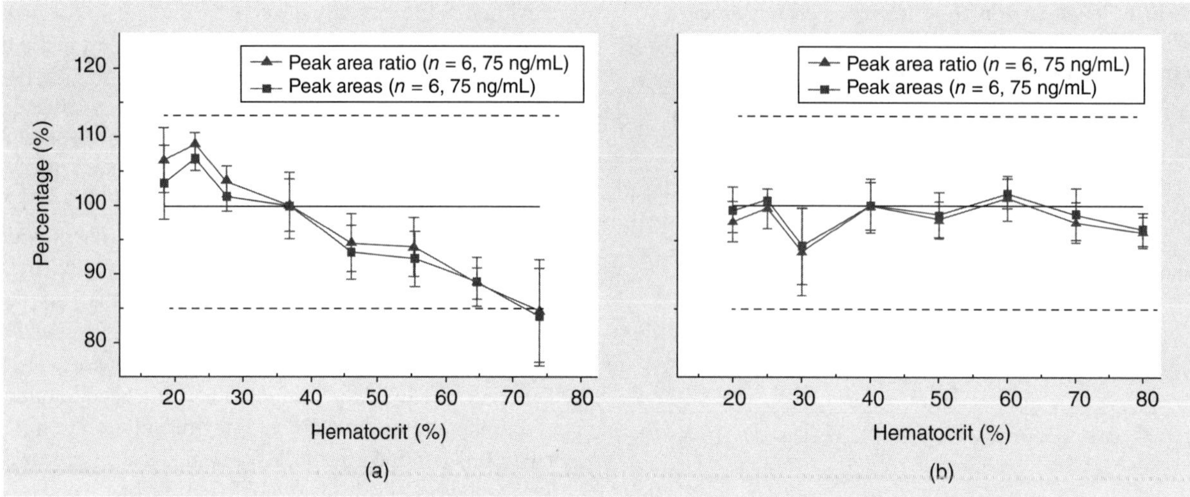

FIGURE 13.4 The average lansoprazole responses obtained from the entire blood spot samples prepared at 75 ng/mL using absolute peaks areas and peak area ratios of lansoprazole to lansoprazole-d4 as a function of hematocrit from 18.5% to 74.0%. (a) Optimized extraction conditions for method validation were used. (b) A combination of pretreatment with water, sonication, and prolonged vortex was employed for sample extraction. Reproduced with permission from Li et al., 2011. Copyright 2011 Future Science Ltd.

(i), adding IS to blood samples before spotting yielded good accuracy (7.9% bias) and precision (4.2% CV). However, this method of IS addition is not practical and requires accurate spiking of IS to blood samples during blood collection. For method (ii), pretreating the cards with IS yielded acceptable accuracy (–12.4% bias) and precision (8.5% CV). It has been suggested that spotting of whole blood onto IS-treated cards might cause displacement of the IS from the center of the spot due to possible chromatographic effects (Heinig et al., 2010a); however, no clear evidence of this effect was found by Meesters et al. (2011). This IS displacement effect may be dependent on the chemical nature of the IS, the DBS card material, and solvent used. For method (iii), the method of adding IS in 50% MeOH/water (2.5 μL) to DBS yielded poor accuracy (–45.0% bias) and precision (14.2%). This is likely a suboptimal method for adding IS, and the poor accuracy and precision may be due to the small volume of IS added or the partial re-dissolving of the dried blood, resulting in uneven distribution of the analyte and the IS. Other techniques of adding IS to DBS have also been investigated such as spraying IS onto DBS. This technique is being developed for application to on-line DBS analysis in a fully automated fashion (Zimmer et al., 2011).

13.3.1.2 Effect of Paper Substrate on Extraction Recovery A thorough study of the effect of paper substrate on the extraction recovery has been reported (Simeone et al., 2011). In the study, a mixture of salmeterol, nefazodone, and verapamil in rat DBS using four types of DBS cards (DMPK A, DMPK B, DMPK C, and Whatman 903) were extracted with various organic solvent/water mixtures (MeOH, ACN, 50% MeOH/water, 0.05% formic acid in 70% ACN/water, and methyl-*tert*-butyl ether (MTBE)). DMPK A and B cards are chemically treated cards. The results indicate that the extraction recovery of an analyte depends on paper substrate and the extraction solvent. DMPK B cards consistently gave the highest recoveries, regardless of the analyte studied or extraction solvent used. Another study also showed that paper impregnated with chemicals (ammonium acetate, formic acid, and pasteurized plasma protein) showed higher recovery than the paper without impregnation (van der Heijden et al., 2009). Higher extraction recovery on chemically treated cards was also reported by Heinig et al. (2010b). This higher recovery may be due to reduced polar–polar interactions between the paper substrate and the analyte in the DBS on the chemically treated card. Hence, different paper substrates (chemically treated vs. untreated cards) could lead to different analyte recoveries. Although chemically treated cards may yield higher recovery, ion suppression and ion-pair chromatographic effects have been reported in chemically treated cards (Chen et al., 2012). Difference in paper/card lot could also cause variability in analyte recovery. Therefore, DBS paper material must be manufactured with high uniformity to yield reproducible concentration results, and the potential for lot-to-lot variability of DBS cards should be considered when validating methods, especially when supporting long-term studies.

13.3.2 Protein Precipitation

DBS samples can also be prepared by protein precipitation (PPT) in which samples are first extracted with water or an aqueous solution. The dried blood from the blood spot will appear to almost completely dissolve in water upon vortex mixing and sonication. The resulting solution (reddish brown color) can then be subjected to PPT by the addition of ACN, MeOH, or other PPT reagents. For example, in the analysis of the compound BMS-X in rat DBS, a 50 μL aliquot of IS (d_9-BMS-X) in aqueous solution was added to the DBS samples (3 mm diameter) in a 96-well plate (Liu et al., 2010). The plate was then centrifuged to make sure the DBS disk completely submerged in the aqueous solution. The 96-well plate was sonicated for 5 minutes, and then 250 μL of ACN or MeOH was added to the samples. The plate was vortexed for 1 minute and sonicated again for 15 minutes. It was then centrifuged and 200 μL of the supernatant was transferred to a new plate, dried down, and reconstituted in 100 μL of 20% ACN/water. The extraction recoveries using ACN or MeOH were greater than 90%. In the analysis of morphine and its active metabolites (morphine 3β-glucuronide and morphine 6β-glucuronide) in DBS (Clavijo et al., 2011), water was used to extract the analytes followed by PPT with methanol:0.2 M $ZnSO_4$ (7:3, v/v). The recoveries of morphine and its active metabolites were all greater than 93%. The benefit of using water or aqueous solution as the extraction solvent is that high extraction recovery is usually obtained as dried blood appears to be highly soluble in water. However, PPT extracts can suffer from significant matrix effects in LC-MS/MS analysis (Liu et al., 2010). Alternatively, after extraction with water, the aqueous solution can undergo LLE or SPE for sample clean-up.

13.3.3 Liquid–Liquid Extraction

Compared with solvent extraction and PPT, LLE has the potential to provide a cleaner extract for LC-MS/MS analysis. In a typical LLE, DBS samples were first extracted with an aqueous solution followed by the addition of a water-immiscible organic solvent. Common organic solvents used are chloroform, hexane, MTBE, and ethyl acetate (Liang et al., 2009, 2010; Newman et al., 2009; Niggebrugge et al., 2011). For example, in the LLE of dextromethorphan and its metabolite dextrorphan in DBS samples (Liang et al., 2009), 50 μL of a mixture of two stable isotope-labeled ISs in 50% ACN/water and 100 μL of 2% ammonium hydroxide were added to the DBS samples followed by the addition of 700 μL MTBE. The purpose of the addition of ammonium hydroxide is to make the charge state of the analytes neutral and thus more efficient for extraction of the analytes from aqueous phase to organic phase. The mixture was then vortex mixed for 3 minutes. After centrifugation, the organic layer was transferred to a 96-well collection plate. The samples were then dried down with a stream of nitrogen and reconstituted with 150 μL of 20% ACN/water followed by vortex mixing for approximately 1 minute. The extracts (10–20 μL) were then injected into the LC-MS/MS system for analysis. The matrix effects (B/A – 1) found for the analytes and IS were negligible (<3.2%), where A is the sample with no matrix and no extraction and B is the sample with matrix but without extraction.

Although LLE provides cleaner extracts than solvent extraction and PPT of DBS samples, it has a number of disadvantages which include low extraction recovery for hydrophilic and polar compounds, large solvent consumption, the labor-intensive nature of the process, and the difficulty of automating the process. Environmental and safety concerns now also limit the choice of solvents that may be used.

13.3.4 Solid-phase Extraction

SPE is probably the most versatile of all of the sample extraction techniques. It potentially can provide an additional clean-up procedure after solvent extraction of DBS samples, and the wide range of sorbents available for SPE makes this technique applicable to almost all classes of analytes. Recent advances in sample preparation techniques including LLE and SPE have been reviewed (Kole et al., 2011). There are a number of reports of using SPE in conjunction with solvent extraction of DBS samples (Blessborn et al., 2007; Lejeune et al., 2007; Suyagh et al., 2010; Römsing et al., 2011). For example, canrenone in pediatric DBS samples (6 mm diameter disk) was first extracted with methanol (2 mL) containing 17α-methyltestosterone (IS) for 60 minutes (Suyagh et al., 2010). The DBS disks were removed and the extracts were then dried down and reconstituted in 1 mL of 10% ACN/water followed by SPE using Oasis HLB cartridges. The SPE cartridges were first conditioned with 1 mL MeOH and then 1 mL water. The reconstituted sample (1 mL) was loaded onto the cartridges followed by washing with 1 mL of 5% methanol/water and then 1 mL of 60% methanol/water. Finally, 1 mL methanol was used for the elution step. The eluate was evaporated to dryness and then reconstituted in 100 μL of mobile phase (60% MeOH/water). The matrix effect found for both the analyte and IS was less than 10%. The advantages of SPE are cleaner extracts, ease of automation, high recoveries for polar compounds, and lower solvent consumption. SPE can also be performed on-line and was reported for the analysis of everolimus in human DBS using Oasis HLB cartridges (van der Heijden et al., 2009) and the analysis of mycophenolic acid and its phenyl glucuronide in human DBS using Hypersil gold C_{18} trapping column (Heinig et al., 2010b).

Although solvent extraction of DBS samples (without additional sample clean-up steps) has a greater risk of being affected by matrix effects, it is still widely used for sample preparation particularly during the early stages of drug development due to its simplicity and low cost of analysis. The use of procedures such as LLE and SPE that provide superior sample clean-up will likely to be a more appropriate option for bioanalysis during the clinical drug development stage or for high sensitivity and other challenging methods.

13.4 CONCLUDING REMARKS

Punching and extraction of disks from DBS samples are key and critical steps in the application of most DBS analytical methods. Punching of a fixed size DBS disk provides a simple and relatively robust quantitative sampling technique and avoids critical volumetric steps during sample collection. However, the benefits of the DBS brings with it potential issues unique to this format. Notably, the nature of the sample can affect spot size (e.g., hematocrit) formed for a given volume of blood. Additionally, there may be the possibility of nonhomogeneous distribution of the analyte through the spot. A DBS sample is also effectively a solid sample, and therefore, there are greater concerns about achieving good extraction recovery of the analyte from the DBS, particularly as the most practical and commonly used technique of adding IS does not always compensate effectively for extraction recovery losses.

In spite of these potential issues, it should be noted that many methods using DBS have been successfully developed and validated. However, these potential issues with DBS require that the analyst must investigate the analytical conditions thoroughly to have a full understanding of how the analyte behaves in various steps in the method, and also to understand how variability of study sample might affect the analytical results.

REFERENCES

Arora R, Hudson WC, Yong B, Boguszewski P. Dried Blood Spot Analysis—Consistent Spot Homogeneity with Variable Spot Punch Locations. Presented at the 59th ASMS Conference on Mass Spectrometry and Allied Topics, Denver, CO, June 5–9, 2011.

Barfield M, Spooner N, Lad R, Parry S, Fowles S. Application of dried blood spots combined with HPLC-MS/MS for the quantification of acetaminophen in toxicokinetic studies. *Journal of Chromatography B* 2008;870:32–37.

Blessborn D, Römsing S, Annerberg A, Sundquist D, Björkman A, Lindegardg N, Bergqvist Y. Development and validation of an automated solid-phase extraction and liquid chromatographic method for determination of lumefantrine in capillary blood on sampling paper. *Journal of Pharmaceutical and Biomedical Analysis* 2007;45:282–287.

Brindle E, Fujita M, Shofer J, O'Connor KA. Serum, plasma, and dried blood spot high-sensitivity C-reactive protein enzyme immunoassay for population research. *Journal of Immunological Methods* 2010;362:112–120.

Chen X, Zhao H, Hatsis P, Amin J. Investigation of dried blood spot card-induced interferences in liquid chromatography/mass spectrometry. *Journal of Pharmaceutical and Biomedical Analysis* 2012;61:30–37.

Clavijo CF, Hoffman KL, Thomas JJ, Carvalho B, Chu LF, Drover DR, Hammer GB, Christians U, Galinkin JL. A sensitive assay for the quantification of morphine and its active metabolites in human plasma and dried blood spots using high-performance liquid chromatography-tandem mass spectrometry. *Analytical and Bioanalytical Chemistry* 2011;400:715–728.

Crawford E, Gordon J, Wu JT, Musselman B, Liu R, Yu S. Direct analysis in real time coupled with dried blood spot sampling for bioanalysis in a drug-discovery setting. *Bioanalysis* 2011;3:1217–1226.

Denniff P, Spooner N, The effect of hematocrit on assay bias when using DBS samples for quantitative bioanalysis of drugs, *Bioanalysis* 2010;2:1385–1395.

Ganz N, Singrasa M, Nicolas L, Gutierrez M, Dingemanse J, Döbelin W, Glinski M. Development and validation of a fully automated online human dried blood spot analysis of bosentan and its metabolites using the sample card and prep DBS system. *Journal of Chromatography B* 2012;885–886:50–60.

Heinig K, Bucheli F, Hartenbach R, Gajate-Perez A. Determination of mycophenolic acid and its phenyl glucuronide in human plasma, ultrafiltrate, blood, DBS and dried plasma spots. *Bioanalysis* 2010;2:1423–1435.

Heinig K, Wirz T, Gajate-Perez A. Sensitive determination of a drug candidate in dried blood spots using a TLC-MS interface integrated into a column-switching LC-MS/MS system. *Bioanalysis* 2010;2:1873–1882.

Kehler J, Akella N, Citerone D, Szapacs M. Application of DBS for the quantitative assessment of a protein biologic using on-card digestion LC-MS/MS or immunoassay. *Bioanalysis* 2011;3:2283–2290.

Kole PL, Venkatesh G, Kotecha J, Sheshala R. Recent advances in sample preparation techniques for effective bioanalytical methods. *Biomedical Chromatography* 2011;25:199–217.

Lejeune D, Souletie I, Houzé S, Le bricon T, Le bras J, Gourmel B, Houzé P. Simultaneous determination of monodesethylchloroquine, chloroquine, cycloguanil and proguanil on dried blood spots by reverse-phase liquid chromatography. *Journal of Pharmaceutical and Biomedical Analysis* 2007;43:1106–1115.

Li W, Tse FLS. Dried blood spot sampling in combination with LC-MS/MS for quantitative analysis of small molecules. *Biomedical Chromatography* 2010;24:49–65.

Li F, Zulkoski J, Fast D, Michael S. Perforated dried blood spots: a novel format for accurate microsampling. *Bioanalysis* 2011;3:2321–2333.

Liang X, Li Y, Barfield M, Ji QC. Study of dried blood spots technique for the determination of dextromethorphan and its metabolite dextrorphan in human whole blood by LC-MS/MS. *Journal of Chromatography B* 2009;877:799–806.

Liang X, Jiang H, Chen X. Human DBS sampling with LC-MS/MS for enantioselective determination of metoprolol and its metabolite O-desmethyl metoprolol. *Bioanalysis* 2010;2:1437–1448.

Liu G, Patrone L, Snapp HM, Batog A, Valentine J, Cosma G, Tymiak A, Ji QC, Arnold ME. Evaluating and defining sample preparation procedures for DBS LC-MS/MS assays. *Bioanalysis* 2010;2:1405–1414.

Loppacher M, Fankhauser C, Schetter K, Schranz U, Altmeyer M, Koller A, Mueller B, Walpen S. Direct Extraction/Analysis of Dried Blood Spots (DBS): A Fully Automated System Including Spot Localization, Internal Standard (IS) Application and Multiple Batch Analysis. Presented at the 59th ASMS Conference on Mass Spectrometry and Allied Topics, Denver, CO, June 5–9, 2011.

Meesters R, Hooff G, van Huizen N, Gruters R, Luider T. Impact of internal standard addition on dried blood spot analysis in bioanalytical method development. *Bioanalysis* 2011;3:2357–2364.

Newman MS, Brandon TR, Groves MN, Gregory WL, Kapur S, Zava DT. A liquid chromatography/tandem mass spectrometry method for determination of 25-hydroxy vitamin D2 and 25-hydroxy vitamin D3 in dried blood spots: a potential adjunct to diabetes and cardiometabolic risk screening. *Journal of Diabetes Science and Technology* 2009;3:156–162.

Niggebrugge AE, Podagatlapalli RK, Sanduja R, Khadang A, Pamidi C, Hayes R. LC-MS/MS Quantitation of Isradipine by Supported Liquid Extraction (SLE) and Dried Blood Spot (DBS) Sampling. Presented at the 59th ASMS Conference on Mass Spectrometry and Allied Topics, Denver, CO, June 5–9, 2011.

Olah TV. Quantitation of therapeutic proteins following direct trypsin digestion of dried blood spot samples and detection by LC-MS-based bioanalytical methods in drug discovery. *Bioanalysis* 2012;4:29–40.

O'Mara M, Hudson-Curtis B, Olson K, Yueh Y, Dunn J, Spooner N. The effect of hematocrit and punch location on assay bias during quantitative bioanalysis of dried blood spot samples. *Bioanalysis* 2011;3:2335–2347.

Prince PJ, Matsuda KC, Retter MW, Scott G. Assessment of DBS technology for the detection of therapeutic antibodies. *Bioanalysis* 2010;2:1449–1460.

Ren X, Paehler T, Zimmer M, Guo Z, Zane P, Emmons GT. Impact of various factors on radioactivity distribution in different DBS papers. *Bioanalysis* 2010;2:1469–1475.

Römsing S, Lindegardh N, Bergqvist Y. Determination of tafenoquine in dried blood spots and plasma using LC and fluorescence detection. *Bioanalysis* 2011;3:1847—1853.

Simeone J, Mather J, Plumb RS, Rainville P, Bowen C. Comprehensive Study to Determine Interfering Compounds From Various Dried Blood Spot Cards Using a Range of Extraction Solvents. Presented at the 59th ASMS Conference on Mass Spectrometry and Allied Topics, Denver, CO, June 5–9, 2011.

Suyagh MF, Laxman KP, Millership J, Collier P, Halliday H, McElnay JC. Development and validation of a dried blood spot-LC-APCI-MS assay for estimation of carenone in pediatric samples. *Journal of Chromatography B* 2010;878:769–776.

Tanna S, Lawson G. Analytical methods used in conjunction with dried blood spots. *Analytical Methods* 2011;3:1709–1718.

van der Heijden J, de Beer Y, Hoogtanders K, Christiaans M, de Jong GJ, Neef C, Stolk L. Therapeutic drug monitoring of everolimus using the dried blood spot method in combination with liquid chromatography-mass spectrometry. *Journal of Pharmaceutical and Biomedical Analysis* 2009;50:664–670.

West R, Burrows M, Ellis G. Solvent Effects on Dried Blood Spot (DBS) Extractions. Presented at the 3rd EBF Open Symposium, Barcelona, Spain, December 1–3, 2010.

Wiseman JM, Evans CA, Bowen CL, Kennedy JH. Direct analysis of dried blood spots utilizing desorption electrospray ionization (DESI) mass spectrometry. *Analyst* 2010;135:720–725.

Wong P, Pham R, Whitely C, Soto M, Salyers K, James C, Bruenner B. Application of automated serial blood sampling and dried blood spot technique with liquid chromatography-tandem mass spectrometry for pharmacokinetic studies in mice. *Journal of Pharmaceutical and Biomedical Analysis* 2011;56: 604–608.

Youhnovski N, Bergeron A, Furtado M, Garofolo F. Pre-cut dried blood spot (PCDBS): an alternative to dried blood spot (DBS) technique to overcome hematocrit impact. *Rapid Communications in Mass Spectrometry* 2011;25:2951–2958.

Zimmer D, Hassler S, Betschart B, Sack S, Fankhauser C, Loppacher M. Internal Standard (IS) Application to Dried Blood Spots (DBS) – Evaluation of a Novel Automated Application System. Presented at the 59th ASMS Conference on Mass Spectrometry and Allied Topics, Denver, CO, June 5–9, 2011.

14

CONSIDERATIONS IN DEVELOPMENT AND VALIDATION OF LC-MS/MS METHOD FOR QUANTITATIVE ANALYSIS OF SMALL MOLECULES IN DRIED BLOOD SPOT SAMPLES

Wenkui Li

14.1 INTRODUCTION

The use of dried blood spots (DBS) obtained from heel, toe or finger pricks and spotted onto filter paper for the collection and analysis of human blood dates back to the early 1960s, when Dr Robert Guthrie used the specimens to measure phenylalanine in newborns for the detection of phenylketonuria (Guthrie and Susi, 1963). This novel approach of blood collection led to the population screening of newborns and other clinical diagnosis, epidemiological studies (bacterial and viral, etc.), general human health studies (DNA and RNA testing, etc.), therapeutic drug monitoring (since 1980s), and pharma R&D (since 2000s) (Li and Tse, 2010).

DBS sampling offers a number of advantages over conventional whole blood, plasma, or serum sample collection (Mci et al., 2001; Li and Tse, 2010). Firstly, it is a less-invasive sampling method (finger, toe or heel prick, rather than conventional venous cannula) and requires a smaller blood volume (less than 100 μL, compared to more than 0.5 mL blood volume in conventional blood sample collection). This can help recruitment of study subjects. Secondly, as there is no requirement for freezer and dry ice in most applications, DBS sampling offers a simpler storage and easier transfer of study samples compared to aqueous study sample handling. Thirdly, DBS samples can be conveniently collected by patients themselves or guardians with minimum training. The samples can be sent by regular mail to the assigned laboratory for analysis. This can not only void unnecessary costs but also enable effective PK/PD or PK/safety sample collection at home setting in clinical trials and/or therapeutic drug monitoring wherever conventional blood/plasma/serum sample collection is deemed impractical. Fourth, DBS sampling reduces the infection risk of HIV/AIDS and other infectious pathogens to a minimum level (Parker and Cubit, 1999; Li and Tse, 2010). The combined advantage of above allows for a significant simplification of blood sample collection and handling for newborns, infants, pediatrics, and other special patient populations for various studies. DBS sampling is relatively new to the pharmaceutical community. However, the impressive growth in the evaluation and implementation of the technology in drug discovery and development and post-approval therapeutic drug monitoring has been well demonstrated in recent years (Li and Tse, 2010). There is no doubt that DBS sampling will play an increasing role in support of “fit-for purpose” preclinical or clinical studies.

The analysis of DBS samples using tandem mass spectrometry (MS/MS) could be traced back to the early 1990s when Drs Millington and Chace et al. employed the precursor ion scan of m/z 99 for the detection of acylcarnitines and the constant neutral loss scan of 102 Da and product ion scan for the analysis of phenylalanine and tyrosine in newborn screening (Millington et al., 1990; Chace et al., 1993). In LC-MS/MS analysis, assay selectivity can be readily achieved by three stages of separation of the analyte(s) of interest from unwanted components in DBS samples. The three stages of separation include (1) sample extraction (protein precipitation (PPT), liquid–liquid extraction (LLE), solid-phase extraction (SPE), etc.), (2) liquid chromatography via

Dried Blood Spots: Applications and Techniques, First Edition. Edited by Wenkui Li and Mike S. Lee.

various columns, and (3) tandem mass spectrometric detection in selected reaction monitoring (SRM) or multiple reaction monitoring (MRM) mode as demonstrated by numerous applications (Li and Tse, 2010). Rapid advance in mass spectrometer technology in the past decades has produced MS/MS systems with significantly enhanced assay sensitivity in LC-MS/MS quantitation of various small molecules in DBS samples. This allows for a broader implementation of the technology in various studies.

Despite the many potential benefits of DBS sampling and numerous applications using the technique as highlighted above, there are several DBS-associated challenges in quantitative LC-MS/MS bioanalysis:

- A single 3/8″ (or 3 mm) disc punched from DBS card is only equivalent to approximately 2.5 μL of blood sample. Apparently, LC-MS/MS assay sensitivity is a challenge.
- Unlike aqueous samples for which sample homogeneity for quantitative analysis could be readily ensured by completely thawing and thoroughly vortex-mixing the content in study sample vials/tubes, DBS is a special matrix in which the analyte of interest distributes after blood spotting. Although each individual DBS disc for analysis should theoretically represent a specific homogenous blood volume, for example, 2.5 μL, DBS sample quality might vary from spot to spot and from the spots of standards/QC samples to those of study samples after spots are made and dried (Li and Tse, 2010). Unfortunately, re-mixing the dried blood prior to analysis is not possible. The degree of inhomogeneity across a set of DBS study samples can be affected by a number of factors, including, but not limited to, card type (chemically treated vs. untreated), physicochemical properties of the analyte, blood hematocrit (HT or HCT), and spotting technique (Cobb et al., 2013). It is necessary to investigate the homogeneity for the analyte of interest on the intended DBS samples, perhaps using a range of HT levels, during method development.
- In LC-MS/MS analysis of aqueous samples, internal standard (IS) can be thoroughly mixed with the analyte of interest in study sample aliquots to ensure any variation due to sample extraction (PPT, LLE, or SPE) and analysis (LC and MS/MS) can be compensated for to the highest level by the IS (typically stable isotope-labeled one. Unfortunately, in DBS-LC-MS/MS, the use of IS might not provide such compensation in sample extraction, although it may well track the variation of the analyte of interest during LC-MS/MS. This is because the analyte of interest is in the dried form of blood, while the IS is commonly in solution unless otherwise specified and validated.
- Depending on the physicochemical properties of the analyte(s) of interest and its interaction with blood matrix components and card materials, some analytes might stay on the surface of DBS card at lower concentration and permeate the spot at higher concentration, resulting in a higher extraction recovery at lower concentration level but a lower recovery at higher concentration level, while other compounds might behave in the opposite. In either case, assay performance will be problematic if extraction method is not optimal (van der Heijden et al., 2009). Pretreatment, for example, impregnation, of the paper/card with pasteurized plasma proteins has been tested useful in deactivating the sites of active adsorption on the card paper and making desorption of analyte(s) easier (van der Heijden et al., 2009). Pretreatment of DBS card can also be made by adding chemicals to the card during manufacturing. For example, FTA (DMPK-A) and FTA Elute (DMPK-B) cards, manufactured by Whatman (part of GE Healthcare), are chemically treated with proprietary reagents that, upon contact, lyse cells, denature proteins, and prevent the growth of bacteria and other microorganisms. However, the presence of the additional chemicals might be a source of matrix effect; some extra effort (compared to aqueous samples) is needed not only to optimize extraction method to ensure the analyte of interest is off the card during extraction, but also to minimize the possible matrix interference in LC-MS/MS.

14.2 DBS-LC-MS/MS METHOD DEVELOPMENT

The above challenges have to be assessed and/or addressed properly during development stage of a DBS-LC-MS/MS assay method prior to study sample analysis in support of a preclinical or clinical study.

14.2.1 DBS Card Selection and Assay Sensitivity Assessment

DBS cards are commercially available from a range of manufacturers. One important step of implementing DBS sampling in a preclinical or clinical study is to identify a DBS card that is the best for the analyte of interest in the intended blood matrix. This requires some bioanalytical development work.

The experiment of DBS card selection could be conducted by spotting the middle QC samples of the analyte(s) of interest onto at least one chemically treated cards (e.g., DMPK-A or DMPK-B cards) and one untreated cards (e.g., DMPK-C cards or PE 226 cards) or other available cards as appropriate. This is followed by proper drying (~2 hours) and extraction of punched discs with various extraction solvents in triplicates. The resulting extracts along with neat solution(s) with equivalent analyte concentrations are assayed using LC-MS/MS. In general, colored DBS sample extract should be avoided, if possible, to minimize matrix effect. A comparison of analyte peak area (or peak area ratio to IS) of the extracted QC samples versus those of neat solution should lead to a conclusion on which DBS card is superior to the other for the analyte of interest.

Shifting of chromatographic retention time and distortion of chromatographic peak shape along with suppression of MS signal intensity have been reported when using DBS cards pretreated with sodium dodecyl sulfate (SDS), an anionic surfactant. SDS forms ion pairs with compounds containing basic amine groups, resulting in increased retention on a C(18) stationary phase and peak shape distortion. However, the interference can be greatly alleviated and/or completely overcome with the use of nonacidic mobile phases and/or DBS cards without SDS coating (Chen et al., 2012).

Once the card type and extraction solvent are established, the LLOQ can be estimated and/or improved, if needed, by choosing a larger punch, more punches/sample, and/or using more sensitive MS/MS system. Assay accuracy and precision at the LLOQ level should be confirmed in combination with other assay parameters during assay validation.

14.2.2 Impact Assessment of DBS Sample Properties on Assay Performance

***14.2.2.1 Effect of Hematocrit* (HT)** HT, packed cell volume (PCV), or erythrocyte volume fraction (EVF) is the proportion of blood volume that is occupied by red blood cells. Independent of body size, HT is considered an integral part of a subject's complete blood count results, along with hemoglobin concentration, white blood cell count, and platelet count. HT is normally about 0.41–0.51 for men and 0.37–0.47 for women (Wilhelm et al., 2009). The percentage might be outside of the above ranges in certain populations, for example, 0.28–0.67 for neonates (0–1 year old) and 0.35–0.42 for children (2–12 years old). Capillary blood tends to have a higher HT (e.g., 0.61) than venous blood (Kayiran et al., 2003).

Since HT is directly proportional to the viscosity of blood, it affects flux and diffusion properties of the blood spotted on the paper/card. At a high HT value, the distribution of blood sample through the paper/card might be poor, resulting in smaller blood spots when compared to the blood spots with a low HT. This means that more blood sample per punch would be taken for analysis, leading to a higher measured analyte concentration than that from the low HT DBS samples (Adam et al., 2000; Mei et al., 2001). Holub et al. tested the effect of five increasing HT values (0.2, 0.3, 0.4, 0.5, and 0.6) in combination with two different punching positions for 31 amino acids and acylcarnitines. For all test analytes that showed a correlation with HT, concentrations were found to be significantly higher in samples with high HT and significantly lower in those with low HT regardless of the position of the punch (Holub et al., 2006).

It seems that HT effect is case dependent. Little or no change in 25(OH)-vitamin D_3 concentrations was seen by Newman et al. from the DBS samples with HT ranging from 0.4 to 0.6 (Newman et al., 2009). In a separate test conducted by Wilhelm et al. for assessing HT impact on DBS results, cyclosporine A QC samples were prepared at concentrations of 60 and 800 μg/L using human blood with various HT values (0.20, 0.29, 0.35, 0.44, 0.59, and 0.72). The QC samples were analyzed in duplicate. The measured cyclosporine A concentrations were compared to those obtained from the DBS samples prepared using blood with HT of 0.36 (similar to the target patient population of 0.35). No apparent (difference from −12% to 14%) impact of HT was observed on the quantification of cyclosporine A (Wilhelm et al., 2009). In another case, Li et al. (2011) evaluated the impact due to HT on the accuracy of determination of NIM811, a cyclosporine A analogue, in human DBS samples. No difference (i.e., outside of ±15% of nominal concentrations) in the accuracy of determination was observed when comparing the measured results to the nominal values for all QC samples (low, middle, and high) prepared using blank blood with HT values between 0.3 and 0.5, a range for a majority of human subjects. However, HT effect is noticeable for the middle and high QC samples but not for the low QC samples prepared using blood with a HT value of ≥0.5. The higher HT value of the blood was associated with a higher bias (%) for the obtained higher concentration QC sample results. This phenomenon suggested the presence of some kind of "blood–analyte" interaction. The magnitude of this interaction is proportional to both the blood HT value and the concentration of the analyte in the blood. This interaction might have resulted in an altered spreadability or diffusion of the blood with a higher HT and a higher analyte concentration across the DBS card compared to "normal" blood with regular HT values (0.37–0.51) (Li and Tse, 2010).

In any case, during a DBS-LC-MS/MS assay development, the influence of HT on assay performance needs to be assessed. In this assessment, fresh blood samples with adjusted low (e.g., 0.30) and high (e.g., 0.60) HT values are used to prepare DBS QCs at low and high or low, middle, and high concentrations. After drying, each of the above DBS QC samples was analyzed in three replicates along with calibration standards that are prepared using fresh blood with a middle HT value (e.g., 0.40). A difference outside ±15% of the nominal values for the measured analyte concentrations in the QC samples should serve as a sight of concern, which might trigger further evaluation.

14.2.2.2 Influence of Blood Spot Volume Different blood volumes spotted onto DBS card/paper might result in different analyte concentrations to be measured from a fixed punch size and HT value. Different card types might also give different blood spot sizes when the same volume of blood is spotted. Adam et al., by using HPLC, reported that the mean of the measured phenylalanine concentrations from 35 μL spots (85.1 ± 4.7 mg/L blood) was lower than that from 100 μL spots (95.5 ± 9.1 mg/L blood) (Adam et al., 2000). In contrast, no apparent impact due to blood volume has been seen in many recent LC-MS/MS applications for quantitative analysis of small molecules in DBS samples (Li and Tse, 2010). Nonetheless, during development of a DBS-LC-MS/MS method, the relationship between dried blood spot area/weight and the volume of blood spotted on the paper/card should be examined by spotting increasing volumes of blood onto the intended paper/card (at least in duplicate) and measuring the areas of the obtained spots (ter Heine et al., 2008) or weighing the obtained spots (Hoogtanders et al., 2007). A linear relationship represents an even distribution of blood on the card/paper. A more practical approach is to spot QC samples at two (e.g., low and high) or three (low, middle, and high) concentration levels onto the paper/card at least in three replicates with three or more increasing volumes (e.g., 10, 20, 30, and 40 μL). After drying, a single punch (e.g., 3 mm) is taken from the center of each DBS sample and analyzed along with a set of calibration standards, for which the spotted blood volume could be at middle or at a different level (ter Heine et al., 2008; Damen et al., 2009; Liang et al., 2009; Spooner et al., 2009). The obtained QC sample results by LC-MS/MS outside ±15% of the nominal values would serve as a sight of concern for further evaluation.

14.2.2.3 Impact due to Sample Inhomogeneity and Chromatographic Effect In addition to the effect due to HT and blood volume that needs to be examined as above, sample inhomogeneity and chromatographic effect (or distribution effect) (Mei et al., 2001) due to the possible interaction of blood and/or analyte with the materials of DBS card/paper is another factor that might result in a significant difference in the measured analyte concentrations between the central and peripheral areas within a spot or from a spot to the other. Two common inhomogeneity issues are (1) decreased analyte concentrations along the edge of a spot and (2) volcano effect by which the analyte concentrations tend to be lower in the center than at the edge of a spot (Ren et al., 2010; Cobb et al., 2013).

There are two approaches, namely radiolabeled (e.g., 14C-labeled) and non-radiolabeled, to measure homogeneity of DBS samples. For the former, a radiolabeled analyte of interest is added to control blood at low and high or low, middle, and high QC concentrations. Then, replicates of spots with two or more different blood volumes are made onto the intended DBS cards (e.g., GE DMPK-A, DMPK-B and DMPK-C cards, PE 226 card, and Agilent Bond Elut card). Once the spots are dried (2 hours at room temperature), the radioactivity across the spots can be measured using quantitative autoradiography. Ren et al. (2010) reported that a smaller spotting volume (e.g., 30 μL) showed slightly better homogeneity, especially with FTA elute (DMPK-B card) paper, compared to the one with a relatively large blood volume (e.g., 100 μL).

For the non-radiolabeled approach, DBS QC samples can be prepared at low and high or low, middle, and high concentration levels on the intended DBS cards. Upon complete dryness of the QC spots, discs (×3) are punched from the center and edge area of DBS QC samples at each level. This is followed by analysis along with calibration standards punched from the center area of the respective spots prepared using the same blood volume (e.g., 20 μL) as for the QCs. A good reproducibility (<15% CV of the mean measured values) and accuracy (within ±15% of nominal values and between the centers and edges) of the QC samples (n = 3) would suggest that chromatographic effect or possible inhomogeneity is negligible (Li and Tse, 2010; Li et al., 2011, 2012).

Another approach is to use well rotor-mixed incurred (fresh) blood samples. A fixed volume of the blood is spotted onto the card/paper. After drying, a minimum of five replicates of punches are made from different locations of the same spot. The samples are extracted and LC-MS/MS responses are compared to demonstrate the comparability of each punch (Koal et al., 2005).

In summary, HT, blood volume, and sample inhomogeneity (blood distribution) might have a direct impact on the use of DBS sampling for quantitative LC-MS/MS bioanalysis. Unless these factors are carefully evaluated, accurate pipetting using a calibrated pipette followed by cutting the whole blood spot from the card/paper appears to be the only option for accurate and precise quantification of the analyte of interest (Li and Tse, 2010) in DBS samples.

14.2.3 Sample Extraction

A common extraction procedure for DBS samples is to punch one or more DBS discs from DBS card into tubes or 96-well plates. This is followed by addition of a fixed volume of extraction solvent containing appropriate IS(s). As a general recommendation, the intended extraction solution should be able to interrupt the binding of analyte to the matrix proteins and the paper materials. The analyte of interest is then extracted with gentle shaking or vortex-mixing. Sonication may be applied to enhance extraction efficiency. After centrifugation, the resulting extracts are transferred manually or by an automated liquid handler to new tubes or 96-well plates for further processing if needed, for example, drying and reconstitution, prior to LC-MS/MS analysis.

Common organic solvents or their mixtures with water, for example, methanol, methanol:water, or acetonitrile:water, at various ratios may be considered first as the extraction solvent. The water in the extraction solution is believed to be helpful for effective elution of the analytes off the DBS card. In contrast, use of pure acetonitrile may result in low extraction efficiency. Depending on the physicochemical properties of the analyte, addition of a pH modifier or buffer to the extraction solution may help improve extraction efficiency.

For a better extraction efficiency of DBS samples, addition of the organic and aqueous portions of an extraction solution in two steps is considered more useful compared to a single step approach. First, an aqueous solution (e.g., 5% methanol or 10% acetonitrile) is added to DBS discs to dissociate the blood matrix from the card materials, and then acetonitrile or methanol is added to precipitate the proteins of the blood matrix that are eluted from the DBS cards by the aqueous solvent in the first step. This approach is similar to protein precipitation method commonly used for aqueous samples. Surfactants, for example, Triton-X100, might be added to the aqueous solvent to ensure a complete dissociation of blood sample matrix from the card materials (Zheng et al., 2013). However, the presence of surfactants in the resulting sample matrix might lead to unwanted ion suppression in LC-MS/MS. A water-immiscible organic solvent, such as MTBE or ethyl acetate, can be added to the aqueous sample mixture containing the surfactants for further processing. This procedure is comparable to the conventional LLE for aqueous samples. Of the above, LLE appears to be an effective extraction method in removing the matrix background introduced by DBS cards (Liu et al., 2010; Li et al., 2011; Zheng et al., 2013).

What should be the most meaningful way of application of IS in a DBS-LC-MS/MS assay has been and is being debated and investigated both scientifically and logistically within the bioanalysis community (Zimmer et al., 2013). In a conventional DBS assay, sample disc is punched from a DBS card and mixed with the extraction solvent that contains appropriate IS. This is followed by necessary extraction procedure as described above prior to LC-MS/MS analysis. As the IS is in a solvent while the analyte of interest is in a "dried form" of blood on the disc, the elution of analyte from the DBS disc (solid phase) to the extraction solution (liquid phase) might not be properly tracked by the IS if the extraction method is not optimal due to whatever the causes. For the same reason, any change in extraction efficiency for the analyte of interest from the disc (dried form) due to changes in sample storage humidity and storage temperature and/or interaction of analyte with card materials may not be properly monitored by the IS either.

Other approaches are available for introducing IS in DBS assay. Each of these approaches is deemed to have some advantages/disadvantages over the other:

A. Treatment (spotting or spraying) of blank DBS cards with IS prior to blood (calibration standards, QCs, and study samples) spotting. This approach is ideal to ensure maximum compensation by the IS for the variation during DBS sample extraction and analysis. However, the process is apparently tedious and costly. A quality check has to be in place to ensure homogenous distribution of IS on DBS card prior to blood spotting. The possible uneven distribution of the IS due to chromatographic effect upon blood sample spotting also needs to be carefully assessed (Meesters et al., 2011; van Baar et al., 2013). Furthermore, it becomes necessary to track stability of IS on DBS cards, especially when it is spotted/sprayed periodically as the study moves forward, considering that late-stage studies may take a year or longer and may be conducted at multiple clinical sites.
B. Addition of IS to blood samples (calibration standards, QCs, and study samples) prior to spotting onto paper, by which the IS can be fully associated with blood components, especially the analyte of interest, for a maximum mimicking of the latter during sample extraction and analysis (van Baar et al., 2013). However, the procedure requires diligent and tedious mixing of accurate volumes of the blood (study samples, standards, and QCs) with accurate volumes of the IS working solution prior to spotting. This is not only impractical in many studies, but also requires extra safety measures when sampling from patients with infectious diseases.
C. Direct addition (spraying) of IS onto DBS samples (standards, QCs, and study samples) prior to extraction. Apparently, the key to success using this approach is the homogenous spraying of the IS onto the DBS and equally important is that addition of the IS to the surface of the DBS samples does not cause any chromatographic effects on the original spots (van Baar et al., 2013). Zimmer et al. (2013), via a high-precision microspray, applied a solution of a ^{14}C-labeled caffeine mimicking an IS onto DBS. Homogeneity of distribution and absorption of the "IS" across the major region of the DBS was checked with satisfaction. In addition, it was reported that the "IS" was able to penetrate into the dried blood (Zimmer et al., 2013).
D. Preparation of separate IS spots using the intended blood matrix. This can be carried out at a bioanalytical laboratory before and/or after study DBS samples are collected. As both the analyte and IS are in the dried blood, the approach is expected to minimize the variation during sample extraction and analysis. However, a recent report by van Baar et al. (2013) suggested that the use of two separate spots for analyte and IS might not fulfill the above expectation. Differences in blood spotting, extraction efficiency, and

others between the IS spot and analyte spot appear to be the areas that need to be thoroughly investigated during method development stage prior to assay validation and implementation.

It must be noted that different procedures of IS introduction to DBS assay can result in a marked difference in extraction recovery. Meesters et al. (2011) recently reported that the relative recovery of a model compound nevirapine ranged between 11.4% and 108%, highlighting the need for careful evaluation of IS procedures during method development and validation in order to ensure assay integrity.

14.2.4 Stability and Instability versus Nonspecific Interaction of Analyte with Card Materials

In general, DBS sampling is not recommended for compounds that are light sensitive and prone to oxidation upon exposure to air. Yellow light and opaque bags have to be employed for sample handling if there is a doubt that the analyte of interest has the potential of photooxidation (Li and Tse, 2010; Xu et al., 2013).

Unstable compounds, such as ester prodrugs, acylglucuronide metabolites, often require special attention in regulated LC-MS/MS bioanalysis (Li et al., 2011). Stabilizing those compounds on DBS is even more challenging due to the fact that the chemical stabilizers must be applied to the DBS card prior to blood spotting and potent enough to cease enzyme-catalyzed degradation of the compound of interest upon contact with blood sample. It is also equally important that the chemical treatment to the DBS card is stable and nontoxic (Li and Tse, 2010; Chen et al., 2011; Xu et al., 2013).

Chemically treated cards appear to offer some stabilizing effect for certain enzyme labile compounds, including ester prodrugs. A recent research by Heinig et al. for the stability assessment of oseltamivir (Tamiflu®) demonstrated the utility of collecting prodrug blood samples onto chemically treated DBS cards for stabilization (Heinig et al., 2011). Oseltamivir phosphate (OP) was unstable in rat EDTA blood with 35% conversion into oseltamivir carboxylate (OC) 15 minutes after sample collection. It was also noticed that drying the blood samples on untreated DBS cards under a stream of cold air within 30 minutes did not significantly reduce the conversion. Apparently, the observed instability of OP on untreated 226 cards is due to enzyme activity during the drying period of the blood spot. Unfortunately, it is not possible to mix a small volume (10–40 μL) of fresh blood sample with enzyme inhibitors and/or other stabilizers prior to blood spotting to prevent the enzymatic degradation. Although no significant further OP degradation occurred on untreated cards after 3 days of storage, the initial loss of approximately 50% presents a serious challenge for collecting study samples on untreated DBS cards. In contrast, an enhanced stability was shown for the DBS samples collected on chemically treated cards. During the 2-hour drying period of the DBS cards at ambient temperature, OP did not degrade in rat blood on DMPK-A and DMPK-B cards (conversion below 3%) (Heinig et al., 2011).

Similar concepts as for manufacturing chemically treated DMPK-A and DMPK-B cards have been applied for treating uncoated DBS cards with pH modifiers and/or enzyme inhibitors for the intended studies. For example, Liu demonstrated that citric acid-treated DBS cards could effectively stabilize enzyme labile compound in human blood at room temperature (Liu et al., 2011). Chen et al. combined tris-(2-carboxyethyl)-phosphine (TCEP), a disulfide bond reductant with good stability, and citric acid to chemically treat the uncoated DBS card. The TCEP–citric acid combination was proven to provide very good stabilization for the model compound tested (Chen et al., 2011).

To quickly explore the potential stability issue of analyte of interest on DBS cards, an accelerated stability test at 40°C/80% relative humidity (RH) was recommended by Xu et al. (2013) as part of a feasibility assessment of a DBS assay. A minimum of 1-week stability on DBS cards at ambient temperature as well as at 40°C/80% RH (to cover shipment from high-temperature/high-humidity area) was considered as the starting point. If instability is unveiled, the experiment could also serve as a stress test (forced-degradation study) to help identify degradation products as needed. In this case, further evaluation at lower temperature (4°C, ≤–20°C, or ≤–70°C) should be considered for improving storage stability for the analyte of interest in DBS samples.

In LC-MS/MS bioanalysis, a stable analyte is commonly defined as the one that is stable in aqueous matrices (blood, plasma, or serum) under various storage conditions, including after multiple freeze/thaw cycles, extended stay on laboratory bench top (room temperature stability), and extended storage at the defined temperature (≤−15°C or ≤−70°C). However, the same stable analyte might be marked "unstable" in DBS assay not due to its "true instability," but due to other reasons. Arguably, the reasons include, but are not limited to, altered extraction recovery of the stability DBS QCs as a result of sample aging and change in moisture of the DBS cards during the storage period compared to the corresponding standards and QCs freshly prepared. Unfortunately, the IS might "fail" to track the "altered" extraction efficiency for the analyte of interest in the stability DBS QCs. In a recent study, Li et al. (2012) described quantitative LC-MS/MS bioanalysis of acetaminophen (APAP) and its metabolites (acetaminophen-glucuronide and acetaminophen-sulfate) in human DBS samples. The overall extraction recovery was from 61.3% to 78.8% for the three analytes by direct extraction using methanol that contained the three respective stable isotope-labeled ISs. Interestingly, with increase (from 40% to 80%) in humidity and/or storage duration under high humidity (80%), a lower extraction recovery of APAP was observed.

In contrast, under a dry condition (i.e., ~0% humidity) or a higher storage temperature (e.g., 60°C), a higher extraction efficiency was seen for APAP QCs, especially those with high concentrations, when compared to regular QCs that were stored at room temperature under normal humidity (i.e., 40%). A possible interaction of APAP and 226 card materials (i.e., cellulose) was speculated as the cause for the above observations. With increase in humidity of storage environment, this "APAP–cellulose" interaction appears to have been elevated. In contrast, with decrease in humidity or at a higher storage temperature, this interaction might have been interrupted (Li, et al., 2012).

Again, every effort should be made to ensure a complete dissociation of the blood matrix, including the analyte of interest, from the card materials for a DBS assay. This is a key to understanding whether the observed "instability" for the analyte of interest is truly a stability issue.

14.3 DBS-LC-MS/MS METHOD VALIDATION

Although DBS samples are different from aqueous matrices (blood, plasma, serum, etc.), there should be no difference in performance requirement for a DBS-LC-MS/MS assay compared to a LC-MS/MS assay for aqueous samples. A DBS-LC-MS/MS assay method must be specific to the intended analyte and precise and reproducible so that it has the robustness needed to withstand repeated and long-term use in population or confirmatory newborn screening (Janzen et al., 2007), therapeutic drug monitoring (Li et al., 2011, 2012), preclinical and clinical study sample analysis (Barfield et al., 2008; Spooner et al., 2009), and other applications. The assay should be rugged to accommodate DBS samples with various blood properties and have a suitable dynamic range along with an optimal extraction recovery.

As per the most recently released US FDA draft guidance for industry on bioanalytical method validation (US Department of Health and Human Services, 2013), the FDA encourages the development and use of DBS sampling, but also points out that although the DBS sampling has been successful in individual cases, the method has not yet been widely accepted. A comprehensive validation will be essential prior to using DBS in regulated studies. This validation should address, at a minimum, the effects of the following issues: storage and handling temperature, homogeneity of sample spotting, HT, stability, carryover, and reproducibility including incurred sample reanalysis (ISR) (US Department of Health and Human Services, 2013).

14.3.1 Sensitivity and Selectivity

In general, the observed LC-MS/MS response for any analyte at the lower limit of quantification (LLOQ) level should be at least five times greater than any response detected from the blank DBS extract. The acceptable LLOQ is defined as the lowest analyte concentration, for which accuracy within ±20% of the nominal value and precision ≤20% from six replicates of measurement can be achieved for a predefined dynamic range (e.g., 50.0–50,000 ng/mL where 50.0 ng/mL is the intended LLOQ and 50,000 ng/mL is the intended upper limit of quantification (ULOQ)). Considering the limited blood volume from a disc punched from DBS samples, a larger disc or multiple discs would make the lower concentration of analyte measurable (Li and Tse, 2010). On the other hand, a larger disc from a small spot might result in a better reproducibility and a small disc from a large spot might be a source of poor assay precision due to the chromatographic effect (Hoogtanders et al., 2007).

The selectivity of a DBS-LC-MS/MS assay should be established by analyzing blank DBS samples prepared from six individual lots of blood with and without IS. Ideally, the six lots of blood should have different HTs for the intended study population. The mean LC-MS/MS response of the interfering peaks, if any, at the retention time region of the analyte or IS should be ≤20% of the mean response of the analyte at the LLOQ level or ≤5% of the mean response of the IS response at a working concentration. Five of the six lots should be free of interferences (Li and Tse, 2010).

14.3.2 Matrix Effect and Recovery

Matrix effect in MS/MS detection and assay recovery should be evaluated by preparing three extracts or solutions: (A) DBS QC (low and high or low, mid, and high) sample extracts at least in three replicates; (B) blank DBS sample extracts (at least in three replicates) post-fortified with both the analyte(s) and the IS with the concentrations the same as in A; and (C) neat solutions at least in three replicates with the concentrations of the analyte(s) and the IS the same as in A and B. All above extracts or solutions are assayed using LC-MS/MS with a set of calibration standards. The matrix effect and extraction recovery could be assessed, respectively, by comparing the response (peak area) of B and C, and A and B, with CV% of ≤15% from replicate measurements. A precise estimate of the amount (e.g., ng, μg) of the analyte from a single DBS punch (e.g., 3 mm) is necessary for the evaluation or accurately pipetting the prepared QC blood onto the card/paper and cutting the whole spot for the above assessment should be considered (Hoogtanders et al., 2007; ter Heine et al., 2008; Liang et al., 2009; van der Heijden et al., 2009).

14.3.3 Precision and Accuracy

The accuracy and precision of a DBS-LC-MS/MS method should be evaluated using QC samples prepared at least at three different concentration levels (low, middle, and high) and analyzed along with a set of non-zero calibration standards in three separate validation runs. The LLOQ samples must be assessed at least in one of the three validation runs. The intra- and inter-day accuracy, the bias (%) from the nominal concentration values, should be within ±15% for all QC levels except the LLOQs, for which a bias of ±20% is considered acceptable. The intra- and inter-day precision, assessed by the standard deviation divided by the mean (CV%) from the replicate analysis, should be ≤15% for the results of all QC levels except the LLOQs, for which a ≤20% CV is considered acceptable (Li and Tse, 2010).

To ensure that a method is really suitable for the intended population with various blood properties, the calibration standard samples should be prepared from a single lot of blood with HT of the medium value of the intended population, while the QC samples are recommended to be prepared from three different lots of blood with HT values that are low (e.g., 0.35), medium (e.g., 0.45), and high (e.g., 0.55), respectively. Accurate and precise QC results obtained from the above intra- and inter-runs should be expected if an assay method is rugged and robust (Li and Tse, 2010).

14.3.4 Dilution Integrity

To demonstrate a method is suitable for DBS samples with analyte concentration higher than the ULOQ, dilution integrity must be assessed. Due to physical differences of a DBS sample from conventional aqueous samples (e.g., blood, plasma, serum), it is not possible to dilute a DBS sample following the conventional way of diluting aqueous sample with blank matrix and then aliquoting a desired volume for analysis. A simple approach is to dilute the extracted dilution QC (DQC) sample using one or more of extracted DBS zero samples (containing IS only). The diluted DQC sample was then analyzed along with calibration standards and regular QCs. The obtained bias (%) and CV (%) from the DQCs in six replicate should be, respectively, within ±15% of the nominal values and ≤15% (Damen et al., 2009; Spooner et al., 2009) from the replicates of measurements.

Other approach has been reported. Reddy et al. assessed dilution integrity by analyzing a smaller disc of the DQCs using the regular disc of the calibration standards. The dilution factor was calculated based on the difference in the areas (πr^2) of the discs (3 mm for DQCs vs. 6 mm for standards) (Reddy et al., 2011).

14.3.5 Carryover

Carryover should be tested by injecting two extracted blank DBS samples sequentially immediately after an ULOQ sample injection. The response in the first blank matrix injection at the retention time of the analyte or IS should be less than 20% of the mean response of a LLOQ sample for the analyte and less than 5% of the mean response for the IS (Li and Tse, 2010).

14.3.6 Incurred Sample Reanalysis

A proper ISR should be performed for at least 20 study samples in single repeat to assess the performance of the assay method throughout the entire dynamic range. BLLOQ samples or samples with analyte concentrations less than three times of LLOQ should not be selected for ISR. The disc for ISR can be taken from a different spot of the same sample or from a different region of the same spot that was used for generating the initial result. The difference between the repeated data and the mean of two measurements (first and repeat) should be within ±20% for at least two-thirds of the selected samples.

The above practice is basically the same as for aqueous sample ISR. However, a better understanding of the difference between aqueous and DBS samples will definitely help trouble-shooting ISR issues related to DBS (Barfield et al., 2011).

- For aqueous sample, if treated correctly, it will always be a single homogenous sample. However, all (three or four) DBS spots on a single DBS card are basically discrete samples (spots). Depending on the sampling conduct, there might be a difference in analyte distribution across the entire area of a given spot and/or among the spots on the same DBS card. Although the differences might be within ±15% of each other region of the same spot or each other spot of the same DBS card, which does not trigger a concern as described previously, they definitely contribute additional layer of uncertainty to the bias (%) values of an ISR assessment.
- Contamination of a DBS card is more likely than a liquid sample as a card is only sealed from the atmosphere after a 2-hour drying period.

14.3.7 Stability

Autosampler stability of DBS sample extracts should be evaluated at QC low and high or QC low, middle and high levels. The data from the reinjections are processed against the original calibration curve. The obtained CV (%) and bias (%) values should be, respectively, ≤15% of the mean measurement and within ±15% of the nominal values.

In general, DBS samples can be stored in zip bags with or without desiccants. Depending on study needs, assessment of analyte stability in DBS samples can be carried out under conditions such as at 2–8°C, ≤–15°C, ≤–60°C, and/or ambient temperature (~22°C) to cover the intended storage period of study samples. The evaluation can be done at low and high (or low, middle, and high) QC concentration levels each in triplicate along with a set of calibration standards. In any case, the measured analyte concentrations in the stability QCs should be compared against the nominal values. The analyte of interest is considered stable if within ±15% bias (%) value is obtained.

14.3.8 Blood Volume Impact

The relationship between the area of dried blood spot and the volume of the blood spotted on the DBS card should be assessed by spotting increasing volumes (e.g., 10–40 μL blood volume with 5 or 10 μL increments) of DBS QC samples at two (low and high) or three (low, middle, and high) concentrations onto the intended cards. After drying, at least three replicates of discs are taken from the center of each DBS QC samples and analyzed along with calibration standards spotted using a middle volume (e.g., 20 μL) with middle HT (e.g., 0.4). The measured concentrations of the analyte of interest from the above DBS QC samples are compared against the nominal values. A bias within ±15% of the nominal values would suggest no apparent difference for the DBS samples made with blood volumes in the range tested.

14.3.9 Spreadability Impact

The possible effect of blood spreadability (or distribution) due to possible interaction of blood and/or the analyte with the materials of DBS card should be assessed during the validation. Discs (×3) are punched from the center and edge area of DBS QC samples at two (low and high) or three (low, middle, and high) concentration levels, followed by analysis along with calibration standards punched from the center area of the respective spots. The measured analyte concentrations from both the center and edge discs of the above QCs were compared against the nominal values and each other. Bias values within ±15% of the nominal concentrations and within ±15% of each other for the QC sample results from both the central and edge discs would suggest no apparent chromatographic effect (or distribution effect).

14.3.9.1 Hematocrit Impact The effect of HT on DBS analysis should be evaluated by preparing QC samples at two (low and high) or three (low, middle, and high) concentration levels using blood with HT values in the range of the intended study subjects (e.g., 0.2, 0.3, 0.4, 0.5, and 0.6). This is followed by analysis using calibration standards prepared using the blood with a middle HT value (e.g., 0.40). A difference in the measured QC sample concentrations within ±15% of the nominal values would demonstrate the HT effect, if any, is negligible.

14.3.9.2 Impact of Humidity and High Temperature Different from blood, plasma, or serum samples that have to be stored at very low temperature after collection prior to analysis, DBS samples can be stored in sealed plastic bags with or without desiccants after the blood on the cards become dried. Although there is, in general, less concern on DBS sample integrity than the regular aqueous samples when stored at ambient conditions, necessary evaluation on the possible impact of humidity and high temperature during sample collection and/or transfer prior to long-term storage at predefined conditions can be very useful in depicting the detailed procedure of collecting study DBS samples (Vu et al., 2011).

In this assessment, a set of DBS QC low and high (or low, middle, and high) samples, after complete dryness at ambient temperature in an open laboratory environment with humidity of ~40%, are placed in sealed containers with humidity inside manually adjusted at ~80% or ~0%. The high-humidity environment can also be created by storing DBS in a sealed plastic bag with wet tissues that are not in contact with DBS card. Humidity should be monitored with a hygrometer. Another set of these QCs can be placed in a HLPC column oven with temperature set to a predefined degree as per study needs, for example, ~40°C or 60°C. At 24-hour or a longer time (e.g., 1 week) interval, three replicates of discs were taken from the center of those QCs and analyzed along with calibration standards and regular QCs that were exposed to regular humidity (~40%) at ambient temperature (~22°C). The measured analyte concentrations were compared with the nominal values. A difference within ±15% of the nominal values for the measured analyte concentrations for the above QC samples would suggest the effect due to humidity and/or high temperature is negligible.

14.4 CONCLUSIONS

DBS are being explored as an important sampling tool in bioanalytics due to its many potential benefits. In order to gain widespread acceptance for use in regulated studies, however, DBS assay has yet to improve its reliability in delivering accurate and reproducible results over time. Continued efforts are needed to overcome technological hurdles especially in ensuring spot homogeneity, enhancing extraction recoveries of analytes from DBS substrates, and maximizing stability of unstable compounds.

REFERENCES

Adam BW, Alexander JR, Smith SJ, Chace DH, Loeber JG, Elvers LH, Hannon WH. Recoveries of phenylalanine from two sets of dried-blood-spot reference materials: prediction from hematocrit, spot volume, and paper matrix. *Clinical Chemistry* 2000;46(1):126–128.

Barfield M, Spooner N, Lad R, Parry S, Fowles S. Application of dried blood spots combined with HPLC-MS/MS for the quantification of acetaminophen in toxicokinetic studies. *Journal of Chromatography B, Analytical Technologies in the Biomedical and Life Sciences* 2008;870(1):32–37.

Barfield M, Ahmad S, Busz M. GlaxoSmithKline's experience of incurred sample reanalysis for dried blood spot samples. *Bioanalysis* 2011;3(9):1025–1030.

Chace DH, Millington DS, Terada N, Kahler SG, Roe CR, Hofman LF. Rapid diagnosis of phenylketonuria by quantitative analysis for phenylalanine and tyrosine in neonatal blood spots by tandem mass spectrometry. *Clinical Chemistry* 1993;39(1):66–71.

Chen B, Doherty J, Li W, Smith HT, Tse FLS. Evaluation of Chemically Treated DBS Cards for Stability Enhancement of Enzyme Labile Molecules in Quantitative LC-MS/MS Bioanalysis. Proceedings of the 59th ASMS Conference on Mass Spectrometry and Allied Topics, Denver, CO, USA, June 6–10, 2011.

Chen X, Zhao H, Hatsis P, Amin J. Investigation of dried blood spot card-induced interferences in liquid chromatography/mass spectrometry. *Journal of Pharmaceutical and Biomedical Analysis* 2012;61:30–37.

Cobb Z, de Vries R, Spooner N, Williams S, Staelens L, Doig M, Broadhurst R, Barfield M, van de Merbel N, Schmid B, Siethoff C, Ortiz J, Verheij E, van Baar B, White S, Timmerman P. In-depth study of homogeneity in DBS using two different techniques: results from the EBF DBS-microsampling consortium. *Bioanalysis* 2013;5(17):2161–2169.

Damen CW, Rosing H, Schellens JH, Beijnen JH. Application of dried blood spots combined with high-performance liquid chromatography coupled with electrospray ionisation tandem mass spectrometry for simultaneous quantification of vincristine and actinomycin-D. *Analytical and Bioanalytical Chemistry* 2009;394(4):1171–1182.

Guthrie R, Susi A. A simple phenylalanine method for detecting phenylketonuria in large populations of newborn infants. *Pediatrics* 1963;32:338–343.

Heinig K, Wirz T, Bucheli F, Gajate-Perez A. Determination of oseltamivir (Tamiflu®) and oseltamivir carboxylate in dried blood spots using offline or online extraction. *Bioanalysis* 2011;3(4):421–437.

Holub M, Tuschl K, Ratschmann R, Strnadová KA, Mühl A, Heinze G, Sperl W, Bodamer OA. Influence of hematocrit and localisation of punch in dried blood spots on levels of amino acids and acylcarnitines measured by tandem mass spectrometry. *Clinica Chimica Acta* 2006;373(1–2):27–31.

Hoogtanders K, van der Heijden J, Christiaans M, Edelbroek P, van Hooff JP, Stolk LM. Therapeutic drug monitoring of tacrolimus with the dried blood spot method. *Journal of Pharmaceutical and Biomedical Analysis* 2007;44(3):658–664.

Janzen N, Peter M, Sander S, Steuerwald U, Terhardt M, Holtkamp U, Sander J. Newborn screening for congenital adrenal hyperplasia: additional steroid profile using liquid chromatography-tandem mass spectrometry. *Journal of Clinical Endocrinology and Metabolism* 2007;92(7):2581–2589.

Kayiran SM, Ozbek N, Turan M, Gürakan B. Significant differences between capillary and venous complete blood counts in the neonatal period. *Clinical and Laboratory Haematology* 2003;25(1):9–16.

Koal T, Burhenne H, Römling R, Svoboda M, Resch K, Kaever V. Quantification of antiretroviral drugs in dried blood spot samples by means of liquid chromatography/tandem mass spectrometry. *Rapid Communications in Mass Spectrometry* 2005;19(21):2995–3001.

Li W, Tse FL. Dried blood spot sampling in combination with LC-MS/MS for quantitative analysis of small molecules. *Biomedical Chromatography* 2010;24(1):49–65.

Li W, Williams SM, Smith HT, Tse FL. Quantitative analysis of NIM811, a cyclophilin inhibitor, in human dried blood spots using liquid chromatography-tandem mass spectrometry. *Journal of Chromatography B, Analytical Technologies in the Biomedical and Life Sciences* 2011;879(24):2376–2382.

Li W, Doherty JP, Kulmatycki K, Smith HT, Tse FL. Simultaneous LC-MS/MS quantitation of acetaminophen and its glucuronide and sulfate metabolites in human dried blood spot samples collected by subjects in a pilot clinical study. *Bioanalysis* 2012;4(12):1429–1443.

Liang X, Li Y, Barfield M, Ji QC. Study of dried blood spots technique for the determination of dextromethorphan and its metabolite dextrorphan in human whole blood by LC-MS/MS. *Journal of Chromatography B, Analytical Technologies in the Biomedical and Life Sciences* 2009;877(8–9):799–806.

Liu G, Patrone L, Snapp HM, Batog A, Valentine J, Cosma G, Tymiak A, Ji QC, Arnold ME. Evaluating and defining sample preparation procedures for DBS LC-MS/MS assays. *Bioanalysis* 2010;2(8):1405–1414.

Liu G, Ji QC, Jemal M, Tymiak AA, Arnold ME. Approach to evaluating dried blood spot sample stability during drying process and discovery of a treated card to maintain analyte stability by rapid on-card pH modification. *Analytical Chemistry*. 2011;83(23):9033–9038.

Meesters R, Hooff G, van Huizen N, Gruters R, Luider T. Impact of internal standard addition on dried blood spot analysis in bioanalytical method development. *Bioanalysis* 2011;3(20):2357–2364.

Mei JV, Alexander JR, Adam BW, Hannon WH. Use of filter paper for the collection and analysis of human whole blood specimens. *Journal of Nutrition* 2001;131(5):1631S–1636S.

Millington DS, Kodo N, Norwood DL, Roe CR. Tandem mass spectrometry: a new method for acylcarnitine profiling with potential for neonatal screening for inborn errors of metabolism. *Journal of Inherited Metabolic Disease* 1990;13(3):321–324.

Newman MS, Brandon TR, Groves MN, Gregory WL, Kapur S, Zava DT. A liquid chromatography/tandem mass spectrometry method for determination of 25-hydroxy vitamin D_2 and 25-hydroxy vitamin D_3 in dried blood spots: a potential adjunct to diabetes and cardiometabolic risk screening. *Journal of Diabetes Science and Technology* 2009;3(1):156–162.

Parker SP, Cubit WD, The use of the dried blood spot samples in epidemiological studies. *Journal of Clinical Pathology* 1999;52:633–639.

Reddy TM, Tama CI, Hayes RN. A dried blood spots technique based LC-MS/MS method for the analysis of posaconazole in human whole blood samples. *Journal of Chromatography B,*

Analytical Technologies in the Biomedical and Life Sciences 2011;879(30):3626–3638.

Ren X, Paehler T, Zimmer M, Guo Z, Zane P, Emmons GT. Impact of various factors on radioactivity distribution in different DBS papers. *Bioanalysis* 2010;2(8):1469–1475.

Spooner N, Lad R, Barfield M. Dried blood spots as a sample collection technique for the determination of pharmacokinetics in clinical studies: considerations for the validation of a quantitative bioanalytical method. *Analytical Chemistry* 2009;81(4):1557–1563.

ter Heine R, Rosing H, van Gorp EC, Mulder JW, van der Steeg WA, Beijnen JH, Huitema AD. Quantification of protease inhibitors and non-nucleoside reverse transcriptase inhibitors in dried blood spots by liquid chromatography-triple quadrupole mass spectrometry. *Journal of Chromatography B, Analytical Technologies in the Biomedical and Life Sciences* 2008;867(2):205–212.

US Department of Health and Human Services, Food and Drug Administration, Center for Drug Evaluation and Research (CDER), Center for Veterinary Medicine (CVM), Guidance for Industry: Bioanalytical Method Validation (Draft Guidance), September 2013.

van Baar BL, Verhaeghe T, Heudi O, Rohde M, Wood S, Wieling J, de Vries R, White S, Cobb Z, Timmerman P. IS addition in bioanalysis of DBS: results from the EBF DBS-microsampling consortium. *Bioanalysis* 2013;5(17):2137–2145.

van der Heijden J, de Beer Y, Hoogtanders K, Christiaans M, de Jong GJ, Neef C, Stolk L. Therapeutic drug monitoring of everolimus using the dried blood spot method in combination with liquid chromatography-mass spectrometry. *Journal of Pharmaceutical and Biomedical Analysis* 2009;50(4):664–670.

Vu DH, Koster RA, Alffenaar JW, Brouwers JR, Uges DR. Determination of moxifloxacin in dried blood spots using LC-MS/MS and the impact of the hematocrit and blood volume. *Journal of Chromatography B, Analytical Technologies in the Biomedical and Life Sciences* 2011;879(15–16):1063–1070.

Wilhelm AJ, den Burger JC, Vos RM, Chahbouni A, Sinjewel A. Analysis of cyclosporin A in dried blood spots using liquid chromatography tandem mass spectrometry. *Journal of Chromatography B, Analytical Technologies in the Biomedical and Life Sciences* 2009;877(14–15):1595–1598.

Xu Y, Woolf EJ, Agrawal NG, Kothare P, Pucci V, Bateman KP. Merck's perspective on the implementation of dried blood spot technology in clinical drug development—why, when and how. *Bioanalysis* 2013;5(3):341–350.

Zheng N, Zeng J, Ji QC, Angeles A, Basdeo S, Aubry A –F, Ishani S, Jariwala N, Arnold ME. Detergent-Assisted Elution: Method Optimization to Improve Analyte Elution and Assay Performance for Dried Matrix Spots (DMS) by uHPLC-MS/MS. Proceedings of the 61st ASMS Conference on Mass Spectrometry and Allied Topics, Minneapolis, MN, USA, June 8–14, 2013.

Zimmer D, Hassler S, Betschart B, Sack S, Fankhauser C, Loppacher M. Internal standard application to dried blood spots by spraying: investigation of the internal standard distribution. *Bioanalysis* 2013;5(6):711–719.

15

CHALLENGES AND EXPERIENCES WITH DRIED BLOOD SPOT TECHNOLOGY FOR METHOD DEVELOPMENT AND VALIDATION

CHESTER L. BOWEN AND CHRISTOPHER A. EVANS

15.1 INTRODUCTION

The intent of bioanalytical method validation is to explore and investigate all possible parameters and scenarios that biological samples could be subjected to from the time of collection, through sample processing and transportation, in the course of sample storage, and finally to bioanalysis. This process ensures acceptable accuracy, precision, and ruggedness of an assay that is intended to be used for any regulatory or similar situations. These parameters are well understood for liquid samples, but not for novel sample formats, such as dried blood spots (DBS). In regard to DBS, several consortiums are actively meeting and discussing the scientific hurdles involved in developing and validating DBS and micro-sampling methods. Currently, one aspect of research by the Bioanalysis Working Group of the European Bioanalysis Forum (EBF) is focusing on four areas of DBS validation: sample dilution, stability, internal standard addition, and effects of hematocrit. The International Consortium for Innovation and Quality in Pharmaceutical Development (IQ) is another organization with a working group focusing on clinical study design, regulatory acceptance, and matrix bridging design that is actively looking to advance DBS and micro-sampling. These organizations have the advantage of drawing their membership from various companies from industry to academia. They are able to harness a wealth of scientific knowledge, share both the financial and scientific burden of the investigations, and collaborate to advance DBS and micro-sampling technology.

Many factors must be weighed when considering the use of a DBS assay. They include the following:

1. Sample collection;
2. Sample preparation and processing;
3. Sample transportation and storage variables; and
4. Bioanalysis options.

Nuances in the different phases from collection to bioanalysis must be investigated and the impact to the method must be assessed. Variables to consider during collection include subject's hematocrit level and collection procedure (finger prick (FP) or venous), while for sample preparation choice of spotting media (untreated, chemically treated, cellulose, noncellulose), spot size, drying temperature/conditions, and any additives that may need to be added to enhance analyte or metabolite stability should be considered. Sample transportation and storage variables include amount of desiccant added and type of storage bag, temperature and humidity extremes that are encountered during transit and storage, and length of storage time prior to analysis. Once samples are received for analysis a great number of bioanalysis options must be considered, including punching preference (manual or automated), punch size, extraction solvent, and instrumentation choice. Good method validation will stress the bioanalytical method, identify deficiencies, allow for improvement on those defunct areas for greater control, and build a rugged and reproducible method that will stand up to the scrutiny of any regulatory agency.

Dried Blood Spots: Applications and Techniques, First Edition. Edited by Wenkui Li and Mike S. Lee.

Each of these four discrete areas will be explored in greater detail within this chapter, along with a discussion on demonstrating assay control through the varying stages of a DBS assay.

15.2 SAMPLE COLLECTION

Factors to be considered at the time of sample collection include hematocrit levels of subjects and site of blood collection (venous or capillary). For adult humans, the normal definitive hematocrit levels are 34–54%, but they can change with age, sex and general health (Meditec, 2012). A range of 28–67% hematocrit should cover the majority of human blood samples likely to be received by an analytical laboratory. It should be noted that some compounds, along with associated disease state, are known to impact a subject's hematocrit level over time. Variability in viscosity associated with hematocrit could lead to differences in the diffusion of blood through and across the DBS media; hematocrit has also been noted to have an effect on analyte extraction recovery from media prior to analysis (Denniff and Spooner, 2010).

After the blood is spotted and dried, two standard approaches are employed to extract the analyte of interest from the DBS: (1) a disk is collected (punched) from within the spot or (2) the whole spot is eluted. Punching does not rely on the application of an accurate volume of blood onto the media substrate. Variability in assay performance due to spotting volume can be assessed during the validation. Spotting volume is generally found to have little quantitative effect if kept within reasonable limitations, for example, testing at 10 and 20 μL blood volumes for an assay based on a 15 μL blood volume (Spooner et al., 2009; Li and Tse, 2010). However, if different hematocrit values result in different blood spot sizes for a fixed volume of blood, then the volume of blood within this fixed diameter subsample of the DBS would also vary (Figure 15.1). An error may result when the analyte concentration is determined in the sample, if comparing to spiked calibration standards prepared at a different hematocrit level (Meditec, 2012). If entire spot elution is used, then deposition of a precise volume is essential for assay accuracy.

Regardless of which approach is used to obtain the sample (punch or whole spot), the impact of hematocrit should be assessed during method validation. Various methods include the addition of quality control samples at different hematocrit levels to show control of the assay, normalization of the sample concentration by the use of quality control samples with hematocrit levels similar to those of the samples, or preparation of calibration standards with a hematocrit level within the range of the samples. This evaluation is becoming an important parameter to assess during assay validation; quality control samples can be prepared with hematocrit levels that bracket the anticipated patient levels. Having data at

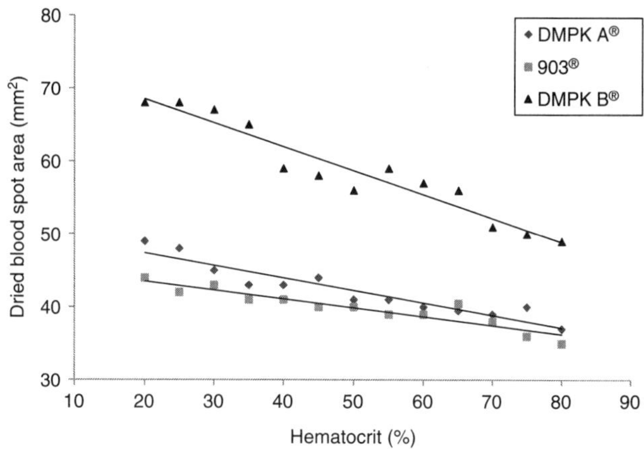

FIGURE 15.1 Relationship between human blood hematocrit and the area of the dried blood spot obtained on different paper substrates. Reprinted with permission from Denniff and Spooner (2010). Copyright 2010 Future Science Ltd.

these various levels of hematocrit within the control of the method validation will allow the researcher to understand what effect the extremes will have on assay bias.

Accurate spotting of a fixed volume is another option to eliminate the effect of hematocrit on spot size. Currently, at the writing of this book, it is difficult to spot accurate volumes in an efficient manner (Summer 2012). One vendor, Drummond Scientific Inc. (Broomall, PA, USA), is looking to optimize prototype units that can efficiently and accurately spot utilizing EDTA-coated capillary tubes (Bowen et al., 2012a). A capillary tube is inserted into a handheld dispenser and accurate, fixed volume spots are dispensed from a single capillary tube of blood.

Venous cannula and capillary FP collection are two common sites for blood collection prior to spotting on DBS media. In a recent comparison study, paracetamol was dosed to healthy human subjects and blood was sampled from both FP (three drops of 15 μL) and cannula (6 mL into an EDTA tube) (Spooner et al., 2010). The blood from the FP was spotted onto DBS media directly from the collection capillary (DBP-FP), while blood from the cannula was either spotted onto DBS media (DBS-Can) or hemolyzed with equal volume of water and frozen (WB-Can). Statistical analysis showed when collecting venous samples from a cannula, switching the sample type from whole blood to DBS can make a small difference in the measured concentrations (for paracetamol). Furthermore, when using DBS samples, switching the collection site from cannula to FP makes a difference on the low concentrations of paracetamol and the early post-dose time points (Figure 15.2). Differences in the early phase after administration may be due to the diffusion of the drug into the extravascular space through the capillary membrane, giving higher arterial than venous concentrations for early time points. The lack of interchangeability observed

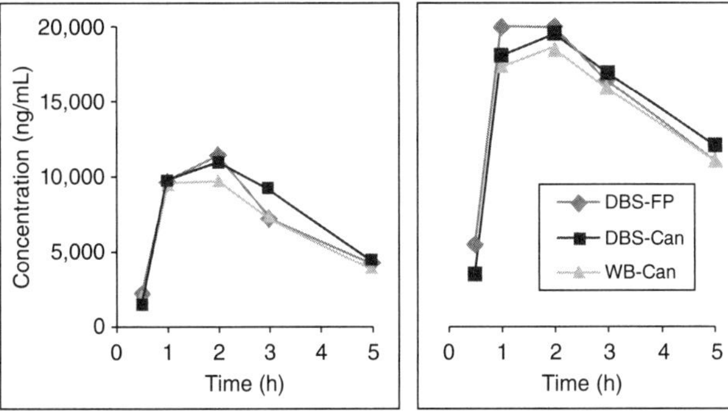

FIGURE 15.2 Comparison of blood collection for paracetamol. Reprinted with permission from Spooner et al. (2010). Copyright 2010 Future Science Ltd.

between both sampling site (FP and cannula) and sample type (DBS and WB) may not be limited to paracetamol. Proper sampling location should be taken into account when designing PK studies and proper training must be conducted with all staff involved on how to collect and prepare DBS samples.

15.3 SAMPLE PREPARATION AND PROCESSING

Variables that must be considered during sample processing include choice of spotting media (untreated vs. chemically treated media, cellulose vs. noncellulose media), spot size, drying temperature/conditions, and any additives that may need to be added to enhance the analyte or metabolite stability.

Choice of spotting media has a considerable impact on the method development and validation parameters, and therefore needs to be investigated and optimized. The variables include compound stability, extraction solvent, extraction efficiency and recovery, and any potential interaction between the compound(s) and the chemical treatment. In several experiments performed within GlaxoSmithKline (GSK) with several pharmaceutical entities, DBS has led to a pronounced increase in analyte stability when compared with the standard wet plasma assays (Bowen et al., 2012b). In these instances, the increase in compound stability can be attributed to removal of esterase activity (either by drying or by the effects of the card chemical treatment impregnated on the DBS cards) (D'Arienzo et al., 2010). Examples detailed in the literature have also shown that dehydration of the DBS samples minimizes hydrolysis of drugs that are predisposed to this type of degradation; thus DBS stability is enhanced in comparison to liquid samples (Wallmark and Lindberg, 1987).

Currently, as of the writing of this book, there are a variety of media choices for use in DBS assays. These include chemically treated cellulose media (GE DMPK A and DMPK B), untreated cellulose media (Ahlstrom 226 and 237, along with GE DMPK C and 903), and most recently released Agilent Bond Elut noncellulose-based media. The future holds great promise for a variety of media in formats that may reduce the diffusion characteristics of blood with various hematocrit levels and non-card-based spotting choices, but none of those are commercially available at the current time. Compound stability and extraction efficiency are also impacted by spotting media. In rare cases, the compound may interact with the additives on the chemically treated card. For instance, use of the DMPK B card with primary amines can lead to on-card derivatization of the amine. As shown in Figure 15.3, based on compound structure, a parent peak

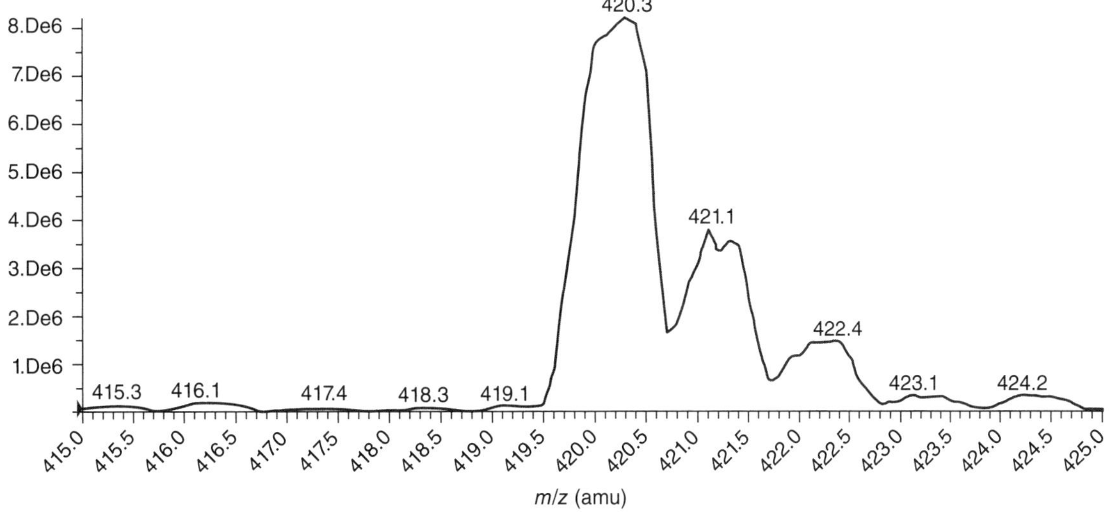

FIGURE 15.3 On-card derivatization of pharmaceutical compound by card additives on DMPK B media.

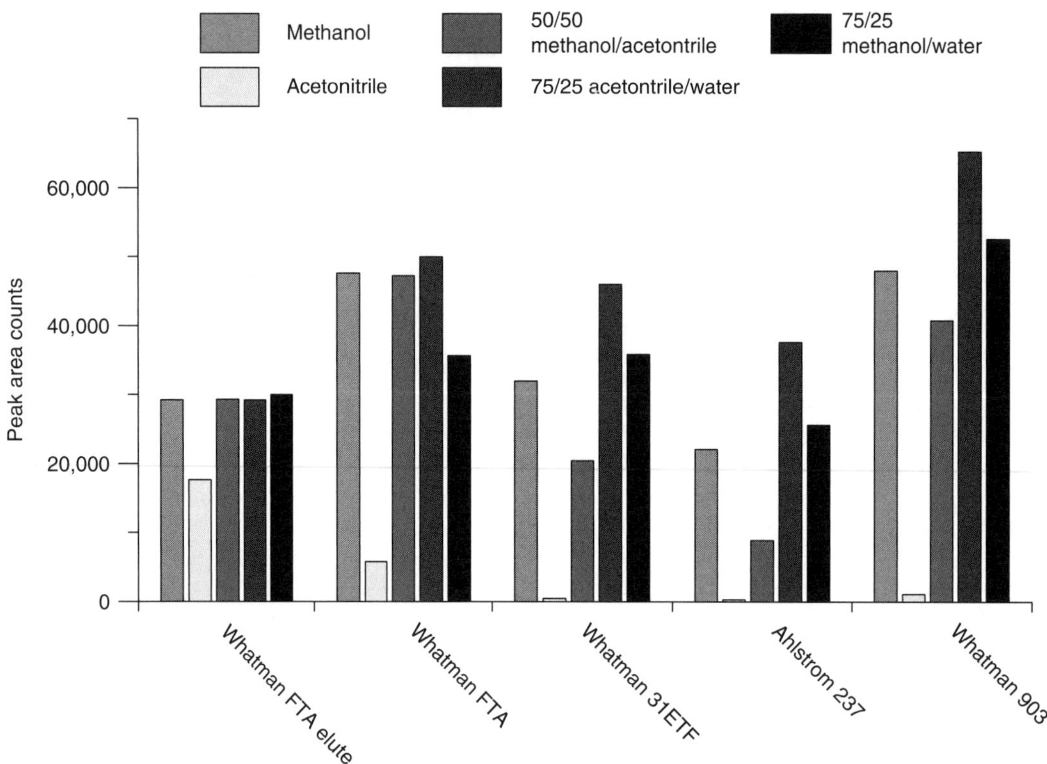

FIGURE 15.4 DBS extraction efficiency based on choice of optimal solvents.

at *m/z* 304 was expected; however due to on-card derivatization a peak was noted at *m/z* 420. Based on this mass shift, the on-card derivatization is consistent with two distinct modifications to the molecule (2 × 58 amu). Chemical derivatization is not unique to small-molecule applications of DBS; it was also noted during the method development of the peptide excendin-4 utilizing the DMPK B card (Kehler et al., 2010) (data not shown). Initially, instability of excendin-4 was noted on the DMPK A card, resulting in a loss of approximately 90% of parent over a 24-hour period. However, this stability issue was resolved through the addition of sucrose to the spotting media prior to the addition of the blood (Kehler et al., 2010). Sucrose is occasionally used to stabilize peptides and proteins in solution (Lee and Timashef, 1981), but to our knowledge this is the first application to the stabilization of a DBS assay. The chemical card treatments are not necessarily "inert" and should be adequately investigated, particularly as new chemical treatments are explored. Perhaps this reactivity can be exploited with future compounds and card additives.

Once spotting media has been selected, extraction solvent needs to be fully investigated and optimized. Mixtures of organic solvents (methanol or acetonitrile), addition of aqueous (water or low salt buffer), and varying pH levels should be explored. For example, sitamaquine was spotted on various media types and subjected to extraction with a diverse suite of solvents. As shown in Figure 15.4, the extraction efficiency can vary dramatically based on solvent choice. This optimization process is imperative to ensure a rugged and reproducible method capable of achieving the lower limits of quantification that many projects demand. Figure 15.5 illustrates, under optimal solvents, extraction efficiency can vary dramatically using different spotting media. Utilizing a spot presoak with water prior to solvent extraction has

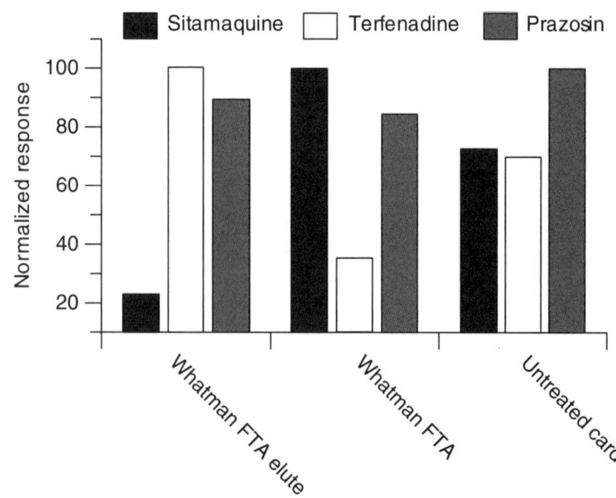

FIGURE 15.5 Variation of extraction efficiency based on spotting media.

also been shown to increase compound extraction recovery, especially in the case of polar analytes. Optimizing recovery during method development will mitigate other issues during the various stages of sample processing.

Furthermore, an assay ruggedness experiment to assess spot volume variation can be easily included in the validation parameters. This experiment will provide confidence that spot volume deviations will not impact assay accuracy. For example, if a method is validated using a 20 μL spot, quality control samples can be spotted at 10, 20, and 30 μL of blood. A representative punch can then be collected from each blood spot, subjected to solvent extraction followed by detection. If the variation in accuracy between the various spot sizes is within the acceptable, standard, assay bias ($\pm 15\%$), then the assumption can be made that "reasonable" variation in spot sizes will have minimal impact on the assay accuracy.

The normal procedure for card drying is approximately 2 hours on a drying rack, followed by placement into a zip-closure bag with a satchel containing 10 g of desiccant. These samples are then stored and shipped under ambient conditions until the time of analysis. Procedures of "speed" drying have been investigated that include blow down with nitrogen and heated drying. All of these procedures are acceptable assuming there is no detrimental effect on the compound stability and that all spotted cards (standards, quality control samples, and study samples) are capable of being prepared under the same conditions. These speed drying procedures require that the preclinical or clinical spotting location has the same drying device as the bioanalytical laboratory preparing the standards and quality control samples. Therefore, while it is a possibility, accelerated drying is not recommended.

Stabilization of phase II metabolites is an important consideration during bioanalytical method development and validation. Sample pretreatment is often used to prevent the conversion of any glucuronide metabolite back to parent under suboptimal conditions in liquid assays (i.e., plasma). If not controlled, this conversion can lead to an overestimation of the parent concentration. In recent experiments, four different glucuronide metabolites (minoxidil-NO, diclofenac-acyl, cotinine-*N*, and a GSK proprietary-*N*-glucuronide) were subjected to stability assessments in both dried and liquid forms under various stabilization options and storage conditions. Storage of both minoxidil-NO and cotinine-*N*-glucuronide in DBS at ambient conditions was equivalent to storage in liquid form at –20°C or –80°C. Interestingly, diclofenac-acyl-glucuronide required an acid stabilization in both liquid and dried forms, where a novel procedure was developed to pretreat nonchemically treated media with a mixture of acidic and ascorbic acid followed by overnight drying. Blood was then spotted onto the previously acidified card, stored under ambient conditions for up to 1 month, and compared to acidified blood stored at –80°C for an equivalent time span. Based on experimental data, storage of diclofenac-acyl-glucuronide on acidified media under ambient conditions was deemed

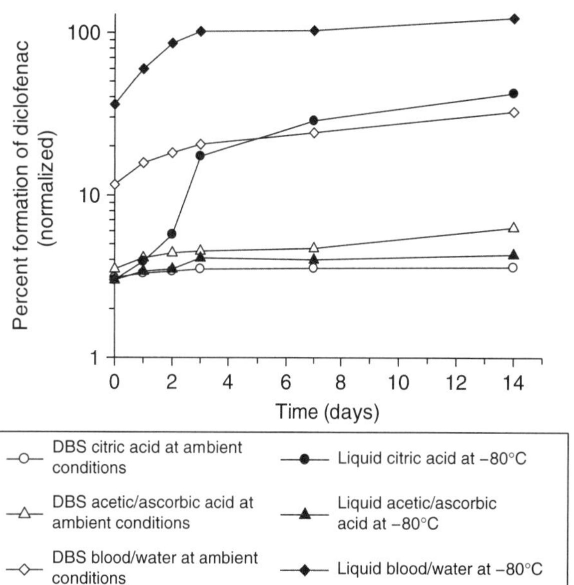

FIGURE 15.6 Comparison of diclofenac glucuronide storage conditions while monitoring diclofenac formation. With permission from Bowen et al. (2012b). Copyright 2012 Future Science Ltd.

equivalent to storage of acidified blood at –80°C for up to 1 month (Figure 15.6). However, when the on-card stability of the proprietary *N*-glucuronides was investigated it was determined to be acceptable following storage at –80°C, but extensive conversion back to parent was noted following storage in DBS under ambient conditions. This conversion resulted in doubling of parent concentrations after 2 weeks of ambient storage. The necessity for storage of the DBS media at –80°C negated one of the primary advantages of this technology; therefore progression of a suitable liquid bioanalytical method was prioritized and progressed (Bowen et al., 2012b).

15.4 SAMPLE TRANSPORTATION AND STORAGE VARIABLES

Prior to bioanalysis, sample transport and storage are critical considerations in any pharmacokinetic or toxicokinetic study design. Care must be taken to ensure the shipment is properly packaged and tracked to make certain it arrives at the desired, final destination in the appropriate timeframe, and that the integrity of the sample is not compromised. Plasma and other liquid matrices can be shipped under controlled conditions (i.e., frozen using dry ice); however, conditions must be established since DBS shipments are performed under ambient conditions. When dealing with biological specimens, environmental conditions may have a deleterious effect on the stability and conditions of the sample.

Laboratory environmental control and monitoring can normally be employed to ensure that controlled conditions are maintained during storage of DBS samples, both at the site of sample collection (animal facility or clinical setting) and at the bioanalytical site. However, environmental conditions (temperature, humidity, and UV light) may be difficult to control outside the laboratory environment. Monitoring of DBS sample preparation, shipment, and storage is often required.

Depending on the placement of in-life study portion and site responsible for bioanalysis, the distance between these locations could be as little as 0.5 miles or as great as 10,000 miles. In large-scale, late-stage clinical studies, samples may even be shipped to a central laboratory for cataloging and batching prior to bioanalysis. Based on the size and scope of the study, and the stability of the compound used in the study, several weeks or even months could pass before samples are analyzed, necessitating proper control conditions during storage and shipping. In a recent investigation, data loggers (portable environmental recording devices) were used to monitor the temperature, dew point, and humidity of a variety of DBS shipments. Data monitoring can occur as often as every 10 seconds (memory capability of 45 hours) to every 12 hours (memory capability of over 2 years). In this instance, temperature and humidity were monitored every 30 minutes, allowing a total of 330 days of ongoing data collection based on internal memory capability.

Supporting a recent DBS study, monitors and DBS samples were shipped in Jiffy bags, originating in Western Australia, transiting through California, with a final destination of King of Prussia, PA, USA, with a transit distance of approximately 11,600 miles. The average time for shipment of these samples encompassed 6 days and utilized two different aircrafts. This study was monitored for a period of 7 months (August 2010–March 2011), corresponding to 11 sample shipments and containers. Figure 15.7 illustrates a representative shipment of clinical DBS samples where the data logger recorded temperatures of less than 0°C, which correlated to in-flight storage of the airplane cargo hold. Monitoring many shipments around the world and over many seasons, these conditions were found to be consistent with numerous other monitored shipments (Bowen et al., 2011).

Based on these collective data and other experiences with DBS shipments, internal procedures for on-card validation

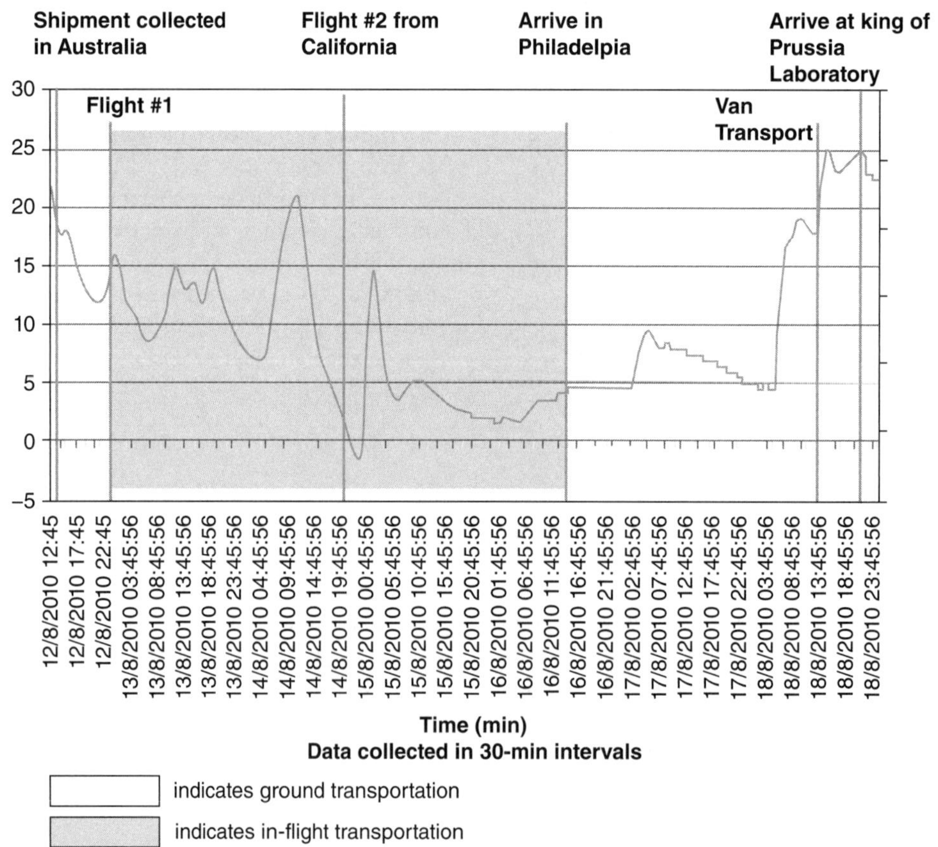

FIGURE 15.7 Temperature measurements for a package containing dried blood spot samples originating in Western Australia. Reprinted with permission from Bowen et al. (2011). Copyright 2011 Future Science Ltd.

TABLE 15.1 Assessment of the On-card Stability in a Human Dried Blood Spot Sample

	Low QC Level (30 ng/mL)			High QC Level (8000 ng/mL)		
	Ambient	−20°C	40°C	Ambient	−20°C	40°C
Mean	28.8	29.8	27.7	8234.7	7723.2	7497.5
%CV	3.9	8.0	2.5	3.5	4.5	5.0
Deviation from nominal (%)	−4.1	−0.6	−7.6	2.9	−3.5	−6.3
On-card stability	Storage assessment					
Ambient	4 months					
−20°C	48 h					
40°C	48 h					

temperature range criteria were modified to –20°C and 40°C in replicates of six at three times the LLOQ and 85% of the ULOQ. Table 15.1 illustrates the results of the on-card stability assessment performed at –20°C and 40°C for a representative compound. These data indicate that the temperature extremes encountered during sample transit had no deleterious effect on compound stability.

Furthermore, investigations have been completed exploring light sensitivity aspects of DBS sampling and storage, by investigating potential stabilization advantages for two light-sensitive compounds. Both nifedipine and omeprazole were separately spiked into rat blood and spotted on various media, allowed to dry, and stored in zip-closure bags with desiccant. Forced degradation was performed using an Atlas Suntest CPS + (Enhance) Tabletop Xenon Arc system with microprocessor control for up to 2 International Conference on Harmonization (ICH) light. Quantitatively, 1 ICH light is the equivalent of more than 1.2 million lux hours and an integrated near-ultraviolet energy of not less than 200 Wh/m^2 (Bowen et al., 2010a).

As shown in Figures 15.8 and 15.9, negligible photodegradation was observed from the DBS samples following either benchtop exposure or forced degradation scenarios up to 2 ICH. These findings contrast the liquid samples, where greater than 90% loss (independent of concentration level) was observed in both the water and plasma samples. The overall increase in light sensitivity on DBS cards can be attributed to the fact that solution-state reaction kinetics is faster than equivalent solid-state reaction kinetics (Thoma and Klimek, 1991).

The data demonstrated extended photostability utilizing DBS, at a variety of benchtop exposure or forced degradation scenarios up to 2 ICH. While the dataset only encompassed two compounds, omeprazole and nifedipine, the results were compelling. It is acknowledged that exposure for 2 ICH is severe and biological samples would never be exposed to those conditions; photo-protection was also noted at less severe conditions, particularly for nifedipine. Additional data

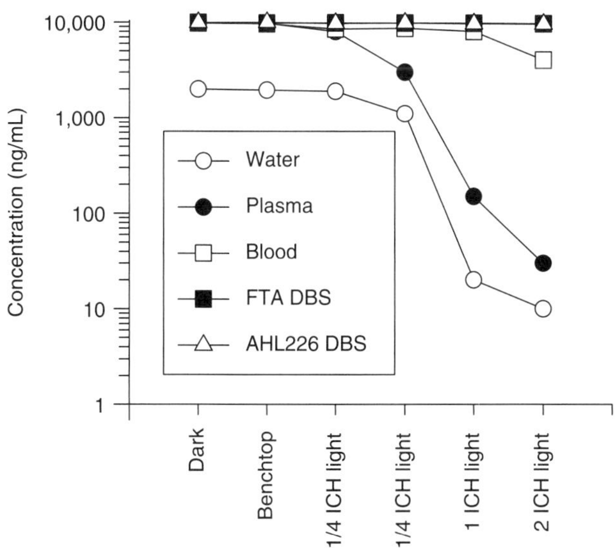

FIGURE 15.8 Photodegradation of omeprazole under various conditions. Reprinted with permission from Bowen et al. (2010a). Copyright 2010 Future Science Ltd.

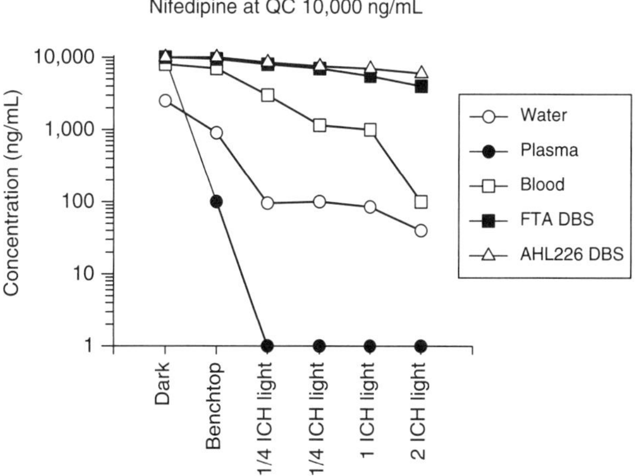

FIGURE 15.9 Photodegradation of nifedipine under various conditions. Reprinted with permission from Bowen et al. (2010a). Copyright 2010 Future Science Ltd.

should be generated; however, it appears that light-sensitive compounds stored in DBS are more stable than aqueous and biological matrices. Judicious care in sample handling must be taken once extraction from the DBS has been completed.

15.5 BIOANALYSIS OPTIONS AND CAPABILITY OF DIRECT ANALYSIS

Standard methods of bioanalysis for DBS assays include mass spectrometer detection coupled to a liquid chromatographic (LC) system for separation. The benefits of this system include orthogonal separation from the LC and selective detection due to proper tuning and MRM transition monitoring of the detector. While the more standard approach is spot punching, followed by extraction and detection, recent advances in instrumentation and automation have made direct elution and extraction of the DBS spots possible.

In the work published by Abu-Rabie and Spooner (2009), the CAMAG thin-layer chromatography mass spectrometer (TLC-MS) interface was used as a tool for the direct quantitative bioanalysis of drugs from DBS samples, using an MS detector, with or without high-performance liquid chromatography (HPLC) separation. A noted advantage to using direct extraction with the TLC-MS interface over manual extraction is the significant increase in response obtained for all test compounds. A relative increase in background noise was not observed, so increased assay sensitivity should be achievable. The use of this technique could easily enable analysis of analytes using DBS where lower limits of quantification are required.

In an attempt to combine LC separation with direct analysis, the feasibility of direct LC-SPE-MS^n from DBS punches has also been assessed. The TriVersa-NanoMate nano-electrospray system can be equipped with filled eTips (HILIC-eTips), allowing one to place a DBS punch into the back of the eTip and sequentially elute xenobiotics directly off the card and through the HILIC package into the ESI emitter. This configuration, when automated and coupled to a mass spectrometer, allows for sensitive and selective analysis of xenobiotics from DBS with a significantly reduced workload. The increase in sensitivity can be attributed to direct elution and "trapping" of all the analyte within the punched spot on the HILIC material. This contrasts the sensitivity obtained using the standard methodology of solvent extraction from the DBS disk (Bowen et al., 2010b).

Use of an orthogonal LC separation in both the direct elution and chip-based nanospray ionization has the advantage of providing a cleaner area of peak elution into the detector, therefore minimizing baseline noise and suppression. While the method for direct elution detailed by Abu-Rabie is fully automatable, the chip-based approach still requires punches to be placed and packed into the SPE-filled pipette tips prior to analysis.

An additional exploration into direct analysis by Wiseman et al. (2010) detailed experiments with DBS using desorption electrospray ionization (DESI). In the DESI process, ions are generated from the sample surface by way of bombardment with charged micro-droplets through the atmosphere. The spray impact causes the formation of microscopic liquid layers on the surface in which the condensed-phase analyte dissolves. This process is followed by desorption via momentum transfer when additional droplets collide with the liquid layer forcing the dissolved analyte into the gas (atmospheric air) phase in the form of micron-sized droplets (Costa and Cooks, 2007). Removal of the requirement to punch and extract the xenobiotics from the DBS would streamline the procedure and bring some advantage back to the bioanalytical laboratory. While detection limits using DESI were 10-fold lower than conventional punching method, this procedure is a nondestructive sampling with three-dimensional spots and one must consider the actual amount of DBS sampled by DESI.

Analysis of DBS utilizing paper spray ionization is another novel technique for DBS analysis. Paper spray uses a solvent electrospray to produce gas-phase ions from samples deposited on paper or other porous media by applying a high voltage to the wet substrate. Preliminary experiments indicate that paper spray has the potential for rapid quantitative analysis of drugs from whole blood. Through the use of a disposable cartridge to hold the DBS media, coupled to a fully automated mass spectrometer, this device could be used in a clinic or hospital for therapeutic drug monitoring (Manicke et al., 2011).

The biggest drawback to the direct analysis technology using DESI or paper spray, in its current state, is lack of chromatography to separate metabolites, conjugates, endogenous components, and card additive interferences. Most chemical entities will elute off the card together and potentially cause suppression of the intended *m/z* transition, decreasing sensitivity and selectivity. Application of the internal standard for direct analysis is also being actively investigated; however, at the current time there does not appear to be a standardized method.

15.6 CONCLUSIONS

The objective of any bioanalytical method is to generate consistent, reliable, accurate, and robust data. Validated analytical methods are the essence of achieving this goal. Method development and validation of analytical methods are essential components of the drug development process. The results from these validated methods are used to judge the quality, reliability, and consistency of analytical results. Researchers rely on accurate and precise bioanalytical results to make decisions on safety and efficacy. Validation of analytical methods is also required by most regulatory agencies, and adequate time and expertise invested in developing and validating robust analytical methods will ensure the method performs to the highest possible standards. New techniques for DBS analysis, the components to consider, and optimal sample volume will certainly evolve. Currently, more validation and vetting of the underlying science are warranted for this novel, but remarkable, technology. As companies embrace this technology and share experiences and regulatory agencies become accustomed to reviewing DBS data, validation criteria will become consistent and streamlined. While the implementation of DBS involves consideration of additional factors in comparison to traditional liquid-based assays, in certain instances the benefits can be quickly and effectively realized. In the various other chapters of this book, many of the factors and parameters that were mentioned here will be discussed in further detail.

REFERENCES

Abu-Rabie P, Spooner N. Direct quantitative bioanalysis of drugs in dried blood spot samples using a thin-layer chromatography mass spectrometer interface. *Analytical Chemistry* 2009;81:10275–10284.

Bowen CL, Hemberger MD, Kehler JR, Evans CA. Utility of dried blood spot sampling and storage for increased stability of photosensitive compounds. *Bioanalysis* 2010a;2:1823–1828.

Bowen CL, Kehler JR, Eikel D, Alpha C, Evans CA, Henion JD, Prosser SJ. Direct LC-SPE-MSn analysis using TriVersa-Nanomate and eTips for quantitation of xenobiotics from dried blood spots. Presented at the 2010 American Society for Mass Spectroscopy Conference, May 24–27, 2010; Salt Lake City, UT; 2010b.

Bowen CL, Dopson W, Kemp DC, Lewis M, Lad R, Overvold C. Investigations into the environmental conditions experienced during ambient sample transport: impact to the dried blood spot sample shipment. *Bioanalysis* 2011;3:1625–1633.

Bowen CL, Karlinsey M, Licea-Perez H, Jurusik K, Kenney J, Siple J. Proposed plasma micro-sampling device for implementation in pre-clinical studies. Presented at the 2012 American Society for Mass Spectroscopy Conference, May 20–24, 2012; Vancouver, Canada; 2012a.

Bowen CL,Volpatti J, Cades J, Licea-Perez H, Evans CA. Evaluation of glucuronide metabolite stability in dried blood spots. *Bioanalysis* 2012b;4:2823–2832.

Costa AB, Cooks RG. Simulation of atmospheric transport and droplet thin-film collisions in desorption electrospray ionization. *Chemical Communications (Cambridge)* 2007;3915–3917.

D'Arienzo C, Ji Q, Discenza L, Cornelius G, Hynes J, Cornelius L, Santella J, Olah T. DBS sampling can be used to stabilize prodrugs in drug discovery rodent studies without the addition of esterase inhibitors. *Bioanalysis* 2010;2:1415–1422.

Denniff P, Spooner N. The effect of hematocrit on assay bias when using DBS samples for the bioanalysis of drugs. *Bioanalysis* 2010;2:1385–1395.

Kehler J, Bowen C, Boram S, Evans C. Application of DBS for quantitative assessment of the peptide excendin-4; comparison of plasma and DBS method by UHPLC-MS/MS. *Bioanalysis* 2010;2:1461–1468.

Lee JC, Timashef SN. The stabilization of proteins by sucrose. *Biological Chemistry* 1981;256:7190–7201.

Li W, Tse FLS. Dried blood spot sampling in combination with LC–MS/MS for quantitative analysis of small molecules. *Biomedical Chromatography* 2010;24:49–65.

Manicke NE, Yang Q, Wang H, Oradu S, Ouyang Z, Cooks RG. Assessment of paper spray ionization for quantification of pharmaceuticals in blood spots. *International Journal of Mass Spectrometry* 2011;300:123–129.

Meditec. 2012. http://www.meditec.com/resourcestools/medical-reference-links/normal-lab-values

Spooner N, Lad R, Barfield M. Dried blood spots as a sample collection technique for the determination of pharmacokinetics in clinical studies: considerations for the validation of a quantitative bioanalytical method. *Analytical Chemistry* 2009;81:1557–1563.

Spooner N, Ramakrishnam Y, Barfield M, Dewit O, Miller S. Use of DBS sample collection to determine circulating drug concentrations in clinical trials: practicalities and considerations. *Bioanalysis* 2010;2:1515–1522.

Thoma K, Klimek R. Formulation and stability testing of photolabile drugs. *Pharmazeutische Industrie* 1991;53:504–507.

Wallmark B, Lindberg P. Mechanism of action of omeprazole. *Pharmacology* 1987;1:158–161.

Wiseman J, Evans CA, Bowen CL, Kennedy J. Direct analysis of dried blood spots utilizing desorption electrospray ionization (DESI) mass spectrometry. *Analyst* 2010;135:720–725.

16

CLINICAL IMPLICATIONS OF DRIED BLOOD SPOT ASSAYS FOR BIOTHERAPEUTICS

MATTHEW E. SZAPACS AND JONATHAN R. KEHLER

16.1 INTRODUCTION

The development and use of dried blood spot (DBS) assays have been carried out in newborn screening (Guthrie and Susi, 1963). Since that time, DBS has been used for the qualitative and quantitative assessment of both small molecules (Spooner et al., 2009; Barfield and Wheller, 2011; Li et al., 2011; Reddy et al., 2011) and protein biomarkers (Tsao et al., 1986; Young et al., 2003; Funk et al., 2008; Brindle et al., 2010; Bastani et al., 2012; Ia Marca et al., 2012). The interest in DBS methodologies stems from the advantages that have been attained and discussed in multiple publications and reviewed by Li and Tse (2010). Some advantages commonly cited are (1) less invasive sampling technique, (2) smaller sampling size, (3) no requirement for freezer storage, (4) greater stability of certain compound classes, and (5) reducing the risk for the spread of infectious disease. Within the pharmaceutical industry this technology has been primarily applied to small molecule quantitation; however, the use of DBS for support of pharmacokinetic (PK) or toxicokinetic (TK) studies in a GLP or GCP environment for the quantitation of biotherapeutics has generated increased attention (Kehler et al., 2010, 2011; Prince et al., 2010;Lin et al., 2011, 2012; Liu et al., 2011; Sleczka et al., 2012).

This chapter focuses on recent literature in the development and validation of DBS assays to support preclinical and clinical biotherapeutic studies. This focus includes a discussion of DBS card types that have been used for both small and large molecule quantitation as well as new card types that may be amenable to biotherapeutic DBS assays. We also review method development and validation work that has been done using both immunoassay and LC-MS/MS along with the advantages that have been recognized using DBS with biotherapeutics. Finally, we explore the use of DBS assays in the bioanalytical laboratory including some areas of the DBS methodology that needs to be more fully understood before this can become a routine technique in a clinical setting for accurate biotherapeutic quantitation.

16.2 REVIEW OF RECENT LITERATURE

16.2.1 Card Types

The selection and use of the proper card type are crucial for the success of biotherapeutic assays utilizing DBS techniques. Currently, there are a variety of card chemistries available from different vendors. GE Healthcare (Little Chalfont, UK) offers three card types (DMPK-A, DMPK-B, and DMPK-C) that can be used for both small molecule and peptide/protein quantitative DBS work. These cards are manufactured from cellulose material to ISO standards (ISO9001:2008) to ensure batch consistency. DMPK-A and DMPK-B cards are pretreated with a proprietary chemical mixture that can lyse cells and denature proteins which will inactivate endogenous enzymes. On the other hand, DMPK-C cards do not contain any chemicals. GE also has introduced color indicating cards that enable the collection and storage of clear biosamples such as CSF, saliva, and urine in the DMPK-A, -B, and -C format. Perkin Elmer (Greenville, SC, USA) offers a single card type for bioanalysis called Ahlstrom 226. The card is made of 100% pure cotton linter filter paper validated for even and uniform sample distribution and is not treated with any chemicals. Finally, Agilent

Dried Blood Spots: Applications and Techniques, First Edition. Edited by Wenkui Li and Mike S. Lee.

Technologies (Sweden) is selling the Bond Elut DMS (dried matrix spotting) card. This card differs from the others by being composed of a noncellulose material. Agilent claims that the noncellulose paper type reduces nonspecific binding of analytes of interest, improving MS analyte response and increasing signal-to-noise ratios. Also the spot size, homogeneity, and recovery are reproducible and independent of hematocrit levels which are a concern with other paper types (Denniff and Spooner, 2010; Ren et al., 2010). As can be seen, there is interest from various vendors on the production of DBS cards. This interest could lead to further research and manufacturing of new and improved paper types with various properties that may be amenable to either small molecule and/or large molecule quantitation.

16.2.2 Quantitation of Biopharmaceutical Compounds from DBS Cards (PK, TK, Biomarkers)

Table 16.1 contains a summary of recent literature references for the quantitation of peptide and protein biotherapeutics using DBS methodologies. These references feature quantitation studies performed with either immunoassay, LC-MS/MS, or direct MS detection. A few of these publications are highlighted below.

16.2.2.1 Immunoassay Currently, immunoassay is considered the "gold standard" for the detection and quantitation of biopharmaceuticals for the support of PK and TK studies as it provides an extremely sensitive platform to develop quantitative assays in biological matrices. The first publication to use a DBS assay for the quantitation of a therapeutic antibody was published in 2010 by Prince et al. (2010). In this publication, the authors developed and validated quantitative methods for two therapeutic monoclonal antibodies (AMG A and AMG B). During method development the DMPK-A, DMPK-B, and DMPK-C DBS cards were evaluated for recovery. It was determined that the highest signal was attained when extracting the monoclonal antibody therapeutic off the DMPK-C cards while the DMPK-A cards showed 56% less signal and the DMPK-B cards showed no signal. The authors suggest that this result was due to lower elution efficiency from the DMPK-A and DMPK-B cards. After fully validating a quantitative DBS immunoassay method the authors conclude that precise, accurate, and sensitive DBS methods can be developed for large molecules that can be used in drug discovery and early development programs. In addition, a publication by Lin et al. (2012) demonstrated that DBS could be used successfully for quantitation of an anti-CD20 monoclonal antibody therapeutic in human blood using immunoassay. The method was fully validated, including recovery of the anti-CD20 antibody at both high and low hematocrit levels as well as the evaluation of compound stability on DBS cards at various temperatures, exposure to sunlight, and storage at various humidity

TABLE 16.1 Summary of Recent Literature on the Quantitation of Biotherapeutics Using DBS Methods

Analyte	Methodology	Card Type	Assay Range	Stability	Citation
Exendin-4 peptide	LC-MS/MS	DMPK-C	10–2000 ng/mL	8 days @ room temp	Kehler et al. (2010)
Large protein therapeutic (>70 kDa)	LC-MS/MS and immunoassay	DMPK-A (LC-MS/MS) DMPK-C (immunoassay)	500–50,000 ng/mL (LC-MS/MS) 25–1000 ng/mL (immunoassay)	7 days @ room temp (LC-MS/MS and immunoassay)	Kehler et al. (2011)
Ceruloplasmin	LC-MS/MS and immunoassay	Whatman Protein Saver 903	200–950 μg/mL (LC-MS/MS)	None specified	deWilde et al. (2008)
Therapeutic mAb	Immunoassay	DMPK-C	100–50,000 ng/mL	At least 3 days	Prince et al. (2010)
Anti-CD20 mAb	Immunoassay	IDBS-1004	100–2500 ng/mL	7 days @ room temperature	Lin et al. (2012)
Hemoglobin	Direct surface sampling coupled with high-resolution mass spectrometry	Standard NHS blood spot cards	Qualitative assessment	None specified	Edwards et al. (2011)
KAI-9803 (therapeutic peptide)	LC-MS/MS	Citric acid treated DMPK-C card	Stability assessed at 1 and 10 μg/mL	At least 48 h at room temperature	Liu et al. (2011)
Exendin-4 peptide	Immunoassay	IDBS-1004	100–5000 pg/mL	7 days @ room temperature, 4°C, and −70°C	Lin et al. (2011)
Pegylated-Adnectin-1 and an Fc-fusion protein	LC-MS/MS	DMPK-C	Various ranges depending on species	14 days @ room temperature	Sleczka et al. (2012)

levels (discussed in greater detail in Section 16.2.3). Although the authors do not show results from a clinical trial, they do state that the method can be used for quantification of biomolecular drugs and biomarkers.

Finally, although not discussed in the publications above, one additional consideration to performing DBS assays in a preclinical or clinical setting is the tedious nature of performing dilutions for immunoassay quantitation. Since the range of an immunoassay is often only one or two orders of magnitude, several dilutions may need to be performed for the samples to fall within a given range. To allow matrix-matched dilutions to be made using a DBS method, blank DBS punches need to be extracted prior to performing dilutions. This adds to the already time-consuming nature of these assays and could preclude it from high-throughput analysis without the use of robotics.

16.2.2.2 LC-MS/MS The use of LC-MS/MS to quantitate peptides and proteins for support of PK and TK studies from different matrices has gained considerable attention as the result of its increased range and selectivity compared with immunoassay methods. In addition, there have been some notable publications on the detection and quantitation of peptides and proteins from DBS cards using LC-MS/MS. These large molecule DBS-LC-MS/MS methods have been developed for both PK/TK support or as a method for detection of biomarkers and other endogenous peptides and proteins.

Kehler et al. (2010) reported on a fully validated method for the determination of the 39-amino acid peptide Exendin-4 (MW = 4186 g/mol) using DBS-LC-MS/MS. In this method, the authors spotted monkey blood onto a DMPK-C DBS card, extracted a 4-mm disc from the center of the DBS card with methanol containing isotopically labeled internal standard followed by injection onto a LC-MS/MS. For quantitation, the $[M+H]^{5+}$ precursor ion of 838 was monitored with a product ion of 396. The fully validated method has a range of 10–2000 ng/mL and demonstrated that DBS can successfully be applied to the quantitation of intact peptide therapeutics using LC-MS/MS.

To extend this work, Kehler et al. published a paper (Kehler et al., 2011) on work using LC-MS/MS for the quantitation of a 70 kDa protein therapeutic. Since intact proteins often have poor ionization efficiencies compared with smaller peptides (Szapacs et al., 2010) and form charge envelopes, distributing the signal from the parent molecule over several charge states (Heck and Van Den Heuvel, 2004), the authors wanted to determine if a large molecule therapeutic could be quantitated by performing on-card digestion followed by LC-MS/MS quantitation. In this paper, DMPK-A cards were spotted with blood containing a protein therapeutic. A 4-mm punch was taken from the center of the blood spot and transferred to a tube containing the enzyme Lys-C. The on-card digestion was allowed to proceed overnight with subsequent peptide extraction and quantitation of a representative peptide using LC-MS/MS. The LC-MS/MS method allowed quantitation over the range of 500–50,000 ng/mL. In addition, a DBS immunoassay method for the same protein therapeutic was developed and validated over the range of 25–1000 ng/mL using the same size blood spot punch on a DMPK-C card. This work illustrates that both immunoassay and LC-MS/MS can be successfully used for the quantitation of a protein therapeutic but care must be given to choosing the card type prior to sampling as the two assays were optimal using different card types.

The first literature reference using enzymatic digestion of a biomarker protein directly off a DBS card was published in 2008 by deWilde et al. (2008). In this article, the authors wished to develop a DBS-LC-MS/MS method to screen patients who are at risk for Wilson disease. Wilson disease is characterized by the accumulation of copper in various organs and can be fatal if left untreated. The biomarker ceruloplasmin (120 kDa) is decreased in patients at risk for developing Wilson disease (Lutsenko et al., 2002; Das and Ray, 2006). The authors used tryptic digestion directly off the blood spot card followed by selection of several representative peptides for quantitation. The LC-MS/MS method showed good correlation with immunoassay method and showed that LC-MS/MS could be used for quantitation of large protein biomarkers and biotherapeutics.

16.2.2.3 Direct Analysis As can be seen from the above references, the workflow for the quantitation of a therapeutic peptide or protein from a DBS can be tedious (Figure 16.1). The steps normally include spotting the sample on the card, punching the spot, and digesting and/or extracting

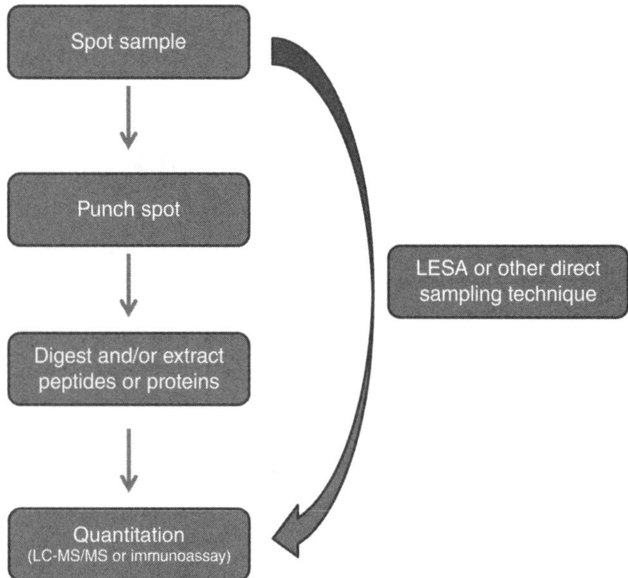

FIGURE 16.1 Dried blood spot workflow with and without direct sampling techniques.

the peptides or proteins of interest with subsequent quantitation using LC-MS/MS. However, the use of direct sampling would eliminate the multiple steps involved and allow the user to directly measure the therapeutic peptide or protein of interest directly from the solid support into the gas phase of the mass spectrometer. A recent publication by Edwards et al. (2011) demonstrated that hemoglobin variants could be identified by extracting the intact proteins from a standard NHS blood spot card utilizing liquid extraction surface analysis (LESA) on an Advion TriVersa Nanomate with detection of the intact hemoglobin variants using an Orbitrap Velos mass spectrometer. This system allowed the "unambiguous determination of hemoglobin variants" with no prior sample preparation. This work indicates that similar experiments could be carried out with therapeutic peptides and proteins, eliminating the multiple extraction and digestions steps currently needed for quantitation using DBS methodologies.

16.2.3 Stability of Peptides/Proteins on DBS Cards

One of the commonly cited advantages for using DBS for small molecules has been that the stability of certain compound classes can be extended when the blood is spotted onto a card compared to frozen matrix (Bowen et al., 2010). Can the same be said for biotherapeutics? Discussed below and summarized in Table 16.2 is review of recent literature on the study of the stability of biotherapeutic molecules as well as protein biomarkers when spotted onto DBS cards.

Kehler et al. (2010) reported that the 39-amino acid peptide, Exendin-4, was stable for at least 8 days at room temperature when spotted onto DMPK-C cards without the presence of protease inhibitors. In contrast, Exendin-4 was found to be stable in plasma at room temperature for only 3 hours following treatment with the protease inhibitor aprotonin. In addition, Liu et al. (2011) has experimented with the stability of KAI-9803, a compound made up of two peptides linked by a disulfide bond. KAI-9803 was found to be very unstable with a half-life of 46.5 minutes in human whole blood at room temperature. To prevent degradation, the sample collection procedure employed mixing fresh blood with a solution of 7.5% trichloroacetic acid at 4°C for stabilization. The addition of stabilizing reagents to a plasma or blood sample requires the reagent to be accurately pipetted in a timely manner and mixed thoroughly. These steps can be problematic in a clinical setting as the technicians are often dealing with multiple samples from multiple patients simultaneously. Since KAI-9803 is under evaluation for use in a critical care setting, new procedures were needed to collect PK samples in a worldwide, multisite clinical study. Therefore, the team wanted to determine if the compound could be stabilized on a blood spot card by pretreatment of the card. The authors found that pretreating the card with citric acid allowed the pH of a spotted blood sample to be lowered within 1 minute of spotting and prevented degradation of KAI-9803, allowing storage of the samples for at least 48 days at room temperature with no degradation.

Finally, Skogstrand et al. (2008) undertook studies to determine the stability of multiple inflammatory markers in plasma, serum, and DBS. The markers were quantitated using validated immunoassay procedures. Since cytokines are an unstable class of molecules (Flower et al., 2000), care must be taken to immediately centrifuge blood samples and freeze the plasma for long-term storage. In general, the authors found that the inflammatory markers measured in plasma and serum that were collected from blood stored at either 4°C, RT, or 35°C for 4–48 hours increased with longer incubation times and with increased temperatures. The authors found that cytokines were relatively stable when stored at

TABLE 16.2 Summary of Various DBS Card Types for Quantitation of Biotherapeutics

Card Type	LC-MS/MS Exendin Peptide[a]	LC-MS/MS of KAI-9803 Peptide[b]	LC-MS/MS 70 kDa Protein[c]	Immunoassay 70 kDa Protein[c]	Antibody[d]
DMPK-A	Extract was not stable, cards pretreated with sucrose allowed extract stability	Not tested	Used for validation	Low signal	Slightly lower signal
DMPK-B	No peak indicative of Exendin-4 was detected	Not stable	Same response as DMPK-A, extract was not stable	Not tested	No signal
DMPK-C	Used for validation	Stable on citric acid treated DMPK-C card	No peak indicative of representative peptide	Used for validation, equivalent signal to blood:buffer	Used for validation

Source: Data from Kehler et al. (2010), Liu et al. (2011), Kehler et al. (2011), Prince et al. (2010).
[a]LC-MS/MS detection of intact Exendin-4 peptide.
[b]LC-MS/MS detection of the intact KAI-9803 peptide therapeutic.
[c]LC-MS/MS detection of a representative peptide and immunoassay detection for >70 kDa therapeutic protein.
[d]Immunoassay detection of therapeutic antibody.

various temperatures on DBS cards for multiple days compared to a DBS sample that was frozen immediately after preparation for the same period of time. The authors concluded that most cytokines are more stable in DBS than in liquid blood samples.

These publications indicate that certain classes of peptides and proteins may be more stable when spotted onto DBS cards than in plasma or whole blood. Peptides and proteins can be enzymatically catabolized during whole blood incubation, even at 4°C, prior to plasma preparation that often requires the addition of a protease inhibitor cocktail for stability. Since the addition of protease inhibitors in a multisite clinical study becomes difficult to standardize across sites, the use of DBS could serve as a general method for collection that allows many classes of peptides and proteins to be stabilized without the need for protease inhibitor cocktail.

16.2.4 Hematocrit

The amount of drug quantitated from a blood spot must be accurate for DBS to be used for routine regulated preclinical and clinical bioanalysis. One of the original advantages of DBS was the idea that differences in spotting technique and spotting blood volume were corrected on the card (i.e., a slightly smaller blood volume will spread over a smaller area and a larger blood volume will spread over a larger area) (Spooner et al., 2009). In this case, if a fixed punch is taken from the DBS card the volume sampled will be constant even when slightly different blood volumes are spotted. This fixed punch approach was thought to correct for slight differences that are found when different people or different techniques are used for spotting the cards. However, it was soon found that different hematocrit levels can lead to variation in the dimensions of the blood spot as well as variation in extraction efficiency causing bias in quantitation of the therapeutic as a fixed punch size is used to sample the spot (Denniff and Spooner, 2010; Ren et al., 2010) (Figure 16.2). To remedy this problem, a recent paper (Youhnovski et al., 2011) described the use of a technique coined precut dried blood spot (PCDBS) that was shown to minimize the effect of hematocrit on DBS sampling. The authors precut 6.1 mm discs out of either DMPK-A or DMPK-C cards, adhered these punches to a carrier card with a piece of double-sided tape, and spotted blood spiked with naproxen at different hematocrit levels. The spots were dried and eluted using sonication in methanol which increased extraction efficiency with subsequent quantitation by LC-MS/MS. The PCDBS method showed no bias ($\pm 3\%$) with respect to hematocrit level (from 25% to 75%) while the fixed punch DBS method showed biases from –14.2% (25% hematocrit) up to 46.2% (75% hematocrit) compared to a sample at 45% hematocrit. This technique appears to eliminate the bias associated with DBS sampling at different hematocrit levels; however, the amount of blood spotted onto the cards needs to be accurate to give an accurate result. This eliminates the advantage of using DBS to correct for slight differences in spotting volume and technique when used on a multisite clinical study. Additional work will need to be done in this area to better understand the effect of various blood properties on the blood spot dimensions and extraction efficiency before DBS can be used for routine clinical support.

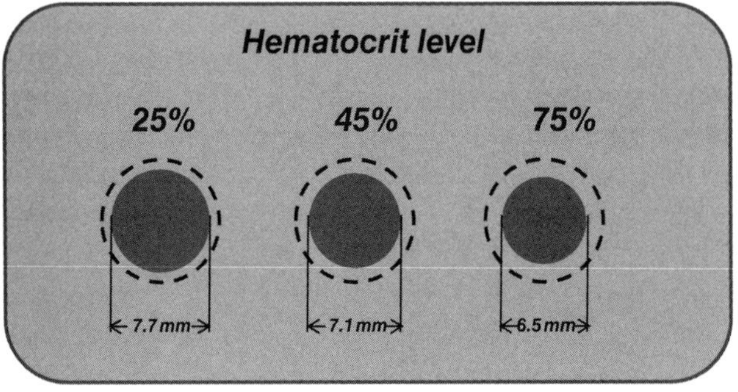

FIGURE 16.2 The effect of hematocrit on blood spot size.

16.3 KEY CONSIDERATIONS FOR USING DBS FOR PK/TK SUPPORT OF BIOTHERAPEUTICS

In this chapter, we have reviewed literature on the use of DBS for the quantitation of biotherapeutic molecules ranging in size from peptides (~4000 Da) to monoclonal antibodies (~150 kDa). Although there is not a vast amount of literature on this topic, a variety of methods (immunoassay, intact LC-MS/MS, enzymatic digestion-LC-MS/MS, and direct sampling) have already been used for quantitation of biotherapeutics from DBS cards. Throughout the studies reviewed it appears that accurate and precise methods can be developed for the quantitation of biotherapeutics using DBS. In addition certain peptides or proteins may be more stable when spotted onto DBS cards than they are in plasma or whole blood giving DBS sampling techniques an advantage over traditional plasma sampling. However, there are still questions about hematocrit that have to be resolved before DBS can offer a routine strategy for subject sampling in a regulated environment for support of biotherapeutic PK and TK studies. Taking all of these into account we believe that DBS assays can be used for a diverse set of biotherapeutics and offer an alternative sampling technique for both PK and TK study support when plasma sampling is not feasible. Considerations to using DBS include the proper selection of card type which may be dependent on assay type (immunoassay, LC-MS/MS, or other) as well as the stability of the molecule on the card. Finally, there is exciting new work being done with direct sampling that may provide a simplified extraction/quantitation workflow that can be used with no additional sample preparation.

REFERENCES

Barfield M, Wheller R. Use of dried plasma spots in the determination of pharmacokinetics in clinical studies: validation of a quantitative bioanalytical method. *Analytical Chemistry* 2011;83:118–124.

Bastani NE, Gundersen TE, Blomhoff R. Dried blood spot (DBS) sample collection for determination of the oxidative stress biomarker 8-epi-PGF (2α) in humans using liquid chromatography/tandem mass spectrometry. *Rapid Communications in Mass Spectrometry* 2012;26:645–652.

Bowen CL, Hemberger MD, Kehler JR, Evans CA. Utility of dried blood spot sampling and storage for increased stability of photosensitive compounds. *Bioanalysis* 2010;2:823–1828.

Brindle E, Fujita, M, Shofer J, O'Connor KA. Serum, plasma, and dried blood spot high-sensitivity C-reactive protein enzyme immunoassay for population research. *Journal of Immunological Methods* 2010;362:112–120.

Das S, Ray K. Wilson's disease: an update. *Nature Clinical Practice Neurology* 2006;2:482–493.

Denniff P, Spooner N. The effect of hematocrit on assay bias when using DBS samples for the quantitative bioanalysis of drugs. *Bioanalysis* 2010;8:1385–1395.

deWilde A, Sadilkova K, Sadilek M, Vasta V, Hahn SH. Tryptic peptide analysis of ceruloplasmin in dried blood spots using liquid chromatography-tandem mass spectrometry: application to newborn screening. *Clinical Chemistry*. 2008;54:1961–1968.

Edwards RL, Creese AJ, Baumert M, Griffiths P, Bunch J, Cooper HJ. Hemoglobin variant analysis via direct surface sampling of dried blood spots coupled with high-resolution mass spectrometry. *Analytical Chemistry* 2011;83:2265–2270.

Flower L, Ahuja RH, Humphries SE, Mohamed-Ali V. Effects of sample handling on the stability of interleukin 6, tumour necrosis factor-alpha and leptin. *Cytokine* 2000;12:1712–1716.

Funk WE, Waidyanatha S, Chaing SH, Rappaport SM. Hemoglobin adducts of benzene oxide in neonatal and adult dried blood spots. *Cancer Epidemiology, Biomarkers and Prevention* 2008;17:1896–1901.

Guthrie R, Susi A. A simple phenylalanine method for detecting phenylketonuria in large populations of newborn infants. *Pediatrics* 1963;32:338–343.

Heck AJ, Van Den Heuvel RH. Investigation of intact protein complexes by mass spectrometry. *Mass Spectrometry Reviews* 2004;23:368–389.

la Marca G, Malvagia S, Materazzi S, Della Bona ML, Boenzi S, Martinielli D, Dionisi-Vici C. LC-MS/MS method for simultaneous determination on a dried blood spot of multiple analytes relevant for treatment monitoring in patients with tyrosinemia type I. *Analytical Chemistry* 2012;84:1184–1188.

Kehler JR, Bowen CL, Boram SL, Evans CA. Application of DBS for quantitative assessment of the peptide exendin-4; comparison of plasma and DBS method by UHPLC-MS/MS. *Bioanalysis* 2010;2:1461–1468.

Kehler J, Akella N, Citerone D, Szapacs M. Application of DBS for the quantitative assessment of a protein biologic using on-card digestion LC-MS/MS or immunoassay. *Bioanalysis* 2011;3:2283–2290.

Li W, Tse F. Dried blood spot sampling in combination with LC-MS/MS for quantitative analysis of small molecules. *Biomedical Chromatography* 2010;24:49–65.

Li Y, Henion J, Abbott R, Want P. Dried blood spots as a sampling technique for the quantitative determination of guanfacine in clinical studies. *Bioanalysis* 2011;3:2501–2514.

Lin Y, Khetarpal R, Zhang Y, Song H, Li SS. Combination of ELISA and dried blood spot technique for the quantification of large molecules using exenatide as a model. *Journal of Pharmacological and Toxicological Methods* 2011;64: 124–128.

Lin Y, Zhang Y, Li C, Li L, Zhang K, Li S. Evaluation of dry blood spot technique for quantification of an anti-CD20 monoclonal antibody drug in human blood samples. *Journal of Pharmacological and Toxicological Methods* 2012;65:44–48.

Liu G, Ji QC, Jemal M, Tymiak AA, Arnold ME. Approach to evaluating dried blood spot sample stability during drying process and discovery of a treated card to maintain analyte stability by rapid on-card pH modification. *Analytical Chemistry* 2011;83:9033–9038.

Lutsenko S, Efremov R, Tsivkovskii R, Walker J. Human copper-transporting ATPase ATP7B (the Wilson's disease protein): biochemical properties and regulation. *Journal of Bioenergetics and Biomembranes* 2002;34:351–362.

Prince PJ, Matsuda KC, Retter MW, Scott G. Assessment of DBS technology for the detection of therapeutic antibodies. *Bioanalysis* 2010;2:1449–1460.

Reddy TM, Tama CL, Hayes RN. A dried blood spots technique based LC-MS/MS method for the analysis of posaconazole in human whole blood samples. *Journal of Chromatography B* 2011;879:3626–3638.

Ren X, Paehler T, Zimmer M, Guo Z, Zane P, Emmons GT. Impact of various factors on radioactivity distribution in different DBS papers. *Bioanalysis* 2010;8:1469–1475.

Skogstrand K, Ekelund CK, Thorsen P, Vogel I, Jacobsson B, Norgaard-Pedersen B, Hougaard DM. Effects of sample handling procedures on measurable inflammatory markers in plasma, serum and dried blood spot samples. *Journal of Immunological Methods* 2008;336:78–84.

Sleczka BG, D'Arienzo CJ, Tymiak AA, Olah TV. Quantitation of therapeutic proteins following direct trypsin digestion of dried blood spot samples and detection by LC-MS-based bioanalytical methods in drug discovery. *Bioanalysis* 2012;4:29–40.

Spooner N, Lad R, Barfield M. Dried blood spots as a sample collection technique for the determination of pharmacokinetics in clinical studies: considerations for the validation of a quantitative bioanalytical method. *Analytical Chemistry* 2009;81:1557–1563.

Szapacs ME, Urbanski JJ, Kehler JR, Wilson R, Boram SL, Hottenstein CS, Citerone DR. Absolute quantification of a therapeutic domain antibody using ultra-performance liquid chromatography-mass spectrometry and immunoassay. *Bioanalysis* 2010;9:1597–1608.

Tsao D, Hsiao KJ, Wu JC, Chou CK, Lee SD. Two-site enzyme immunoassay for alpha-fetoprotein in dried-blood samples collected on filter paper. *Clinical Chemistry* 1986;32:2079–2082.

Youhnovski N, Bergeron A, Furtado M, Garofolo F. Pre-cut dried blood spot (PCDBS): an alternative to dried blood spot (DBS) technique to overcome hematocrit impact. *Rapid Communications in Mass Spectrometry* 2011;25:2951–2958.

Young SP, Stevens RD, An Y, Chen YT, Millington DS. Analysis of a glucose tetrasaccharide elevated in Pompe disease by stable isotope dilution-electrospray ionization tandem mass spectrometry. *Analytical Biochemistry* 2003;316:175–180.

17

POTENTIAL ROLE FOR DRIED BLOOD SPOT SAMPLING AND BIOANALYSIS IN PRECLINICAL STUDIES

QIN C. JI AND LAURA PATRONE

17.1 INTRODUCTION

The fundamental goal of nonclinical studies is to limit the risk to patients as much as possible whenever a new active substance is used as a medicinal product in humans (Olejniczak et al., 2001). Nonclinical toxicology studies are designed to reduce patient risk by characterizing toxic effects relative to the target organ(s) of toxicity, dependence of dose on the toxicologic response, and its relationship to exposure and to assess the reversibility of the toxic response in nonhuman species prior to and concurrent with clinical studies. Therefore, it is essential to elicit a toxic response in our nonclinical studies to fully evaluate these endpoints. Doses must be sufficiently high to characterize the toxicity of a compound, using doses up to the maximum tolerated dose (MTD) or other appropriate limiting doses (including those that achieve large exposure multiples relative to the effective clinical dose or saturation of exposure) or use the maximum feasible dose (MFD), so that any potentially clinically relevant effects can be fully explored (International Conference on Harmonisation and European Medicines Agency, 2009).

17.2 IMPORTANCE OF EXPOSURE MEASUREMENTS

Measurement of exposure to xenobiotics to determine MTD, MFD, or exposure multiples relative to the efficacious clinical dose is accomplished through toxicokinetic evaluations. In nonclinical toxicity studies, toxicokinetics (TK) is the description of the rate of entry of a compound into the plasma (or serum or whole blood) and what happens with the compound once it is in the body with respect to time, based on a limited number of plasma (or serum or whole blood) samples within a study (Liu et al., 2010). The primary objective of TK is to describe the systemic exposure achieved in animals and its relationship to dose level and the time course of the toxicity study (International Conference on Harmonisation, 1995). Toxicokinetic evaluations analyze what the body does with a compound administered at a comparatively high dose relative to the therapeutic dose to focus primarily on the mean exposure data in relation to risk assessment (Ploemen et al., 2007). Correlation with exposure to determine the dose–response relationship is critical to determine the nature of the toxic response and the safety margins relative to the presumed efficacious exposure in the clinic (International Conference on Harmonisation, 1995). Secondary objectives relate the exposure achieved in toxicity studies to clinical signs of toxicity, as well as any associated clinical or anatomic pathology changes, and contribute to the assessment of the relevance of these findings to clinical safety. Additionally, this information is used to support the choice of species and design of subsequent toxicity studies and to know whether biological processes, such as absorption or metabolism, change as the administered dose changes. Toxicokinetic data often can be extrapolated from laboratory animals to humans through mathematical modeling.

The principal hypothesis underlying studies of the metabolism and TK of a chemical is that the adverse biological effects of the substance are correlated with its concentration in the tissues of the organism (i.e., microorganisms,

Dried Blood Spots: Applications and Techniques, First Edition. Edited by Wenkui Li and Mike S. Lee.

plants, or animals, including human beings) in which the toxic effect is observed. Consequently, it is always necessary to both identify and quantify the toxic chemicals that might be present in a sample (World Health Organization [WHO], 1986). The quantification of the chemical in the biological system is usually done by evaluating levels in the blood, currently most commonly performed in plasma, since quantification of substances in whole blood historically can be problematic due to high-performance liquid chromatography (HPLC) column blockage by red blood cells, high protein content, and the presence of many other high or low molecular weight substances over a wide dynamic range (Moore and Tebbett, 1987; Neu et al., 2012).

17.2.1 Limited Blood Volume

Key to the evaluation of not only TK, but also hematology, serum chemistry, and additional biomarkers in *in vivo* toxicology studies is the collection of blood. Blood-based (plasma or serum) media are preferred for analysis clinically and nonclinically. The challenge is not only to obtain this information without compromising the study, but also to enable collection for other critical blood-based endpoints. Current TK analysis requires large volumes of blood to generate the necessary plasma or serum, typically 300–600 μL per time point depending on the analyte(s) being evaluated. This volume is typically not limiting for larger animal species such as dog or nonhuman primates (although it can have an impact on immunogenicity evaluations requiring multiple bleeds in these larger species). However, the relatively large volume of blood required severely limits testing for rodents. Obtaining this volume for TK evaluations from rodents can be difficult, especially if there are many other blood-based endpoints. Additionally, if the animal's health is compromised, collecting a sufficient volume can be challenging. Due to the limited blood volume, TK analysis in rodents is performed as composite profiles. A typical TK period can cover five to six time points in a 24-hour period, and with a composite profile rats are typically bled two to three times during that period with a best practice maximum weekly blood collection of 1200 μL (Diehl et al., 2001). If more than one analyte is to be evaluated, or if other blood-based endpoints are evaluated such as for clinical pathology endpoints or immunogenicity or other special analyses, then more blood is often required and toxicokinetic satellite animals are frequently used. Blood volumes are more restrictive with studies that involve mice. A best practice maximum weekly blood collection of 100 μL is typical, and thus, most blood collections for TK are a terminal procedure (Morton et al., 1993; Diehl et al., 2001; Crawford et al., 2011) and require many satellite animals to be added to the study. The use of composite sampling in combination with TK satellite groups further distances the correlation between exposure and specific effects in individual animals.

17.2.2 Blood Collection Methods

The method of blood collection varies depending on the species or size of the animal. Blood is often obtained from rats by the lateral tail vein or by the jugular vein. Blood can be collected from mice by the tail tip, orbital venous sinus, saphenous, or submandibular vein (Diehl et al., 2001; Golde et al., 2005; Clark et al., 2010). In either species, if it is to be a terminal bleed, then blood can also be collected via cardiac puncture. Blood collection from the tail often requires incubation of the animals to dilate blood vessels of the tail, adding another source of stress to the animal. Bleeding from the orbital venous sinus in the mouse often requires anesthesia (Hoff, 2000), further complicating the procedure.

Dried blood spot (DBS) sample collection offers many advantages over standard plasma samples including a significant reduction in the volume of blood necessary for collection. Blood volumes can be decreased from 300–600 μL (for plasma needs) to 60 μL (four spots per card at 15 μL per spot) or less. Processing of blood samples at the time of collection is simplified because centrifugation to plasma or serum, pipetting of subaliquots, freezing and defrosting of samples, all of which can introduce errors in the analysis (Ferrari, 2011), are not required. Data quality can be improved by enabling serial sampling from the same animal for more accurate TK or pharmacokinetic (PK) profiling (Clark et al., 2010). DBS enables us to use fewer animals by eliminating the need for TK satellite animals in many cases (Barfield et al., 2008; Turpin et al. 2010; Wan et al., 2012) and to refine the way animals are used by reducing stress associated with withdrawing large blood volumes. For example, incubation of animals for dilation of blood vessels or larger bore needles would no longer be necessary, further addressing two of the three Rs of humane animal experimentation (Reduce, Refine, Replace) by reducing the number of animals used and refining our methods with which they are used (National Research Council, 2011). Additional advantages of DBS include enhanced stability of some compounds or pro-drugs that would normally require the addition of esterase inhibitors to stabilize the compound of interest once collected from the animal (D'Arienzo et al., 2010) or in other situations in which the compound is unstable in liquid format (Liu et al., 2011a).

Microsampling, defined as the removal of small quantities of a sample, less than 100 μL per time point (Wan et al., 2012), has enabled serial sampling from a single mouse to allow better characterization of PK/TK properties. Traditionally, only composite TK profiles can be obtained in small animal studies due to the physiological limitations, such as the total blood volume that is allowed to be drawn over a certain period of time (Morton et al., 1993; Diehl et al., 2001). Reduction in the volume of blood drawn per time point enables the utilization of serial rather than composite sampling for better correlation of exposure and effect

(Clark et al., 2010; Li et al., 2011). From a sample collection perspective, DBS offers improved safety with handling, shipping, and storage at room temperature. Finally, DBS has been shown to improve data quality, improve compound stability for many compounds and their metabolites, and afford considerable cost savings when one considers the decreased number of animal and associated husbandry.

With appropriate validation, DBS assays can be used in place of the commonly used plasma LC/MS/MS methods (Barfield et al., 2008). The validation procedures for DBS bioanalytical assays must also include parameters that are not typically assessed in plasma assays: the type of card used, the volume spotted, the detection limit, hematocrit range, extraction efficiency, internal standards (ISs), and stability on the card throughout the handling process (Meier-Davis et al., 2012).

17.3 PROGRESS AND CHALLENGES IN BIOANALYTICAL ASSAY DEVELOPMENT

In the drug development process, critical decisions are made based on the information generated through various data analyses. The concentration information of drugs, metabolites, and biomarkers in biological matrices is a critical part of these data. The quality of the bioanalysis process which defines, executes, and documents the entire workflow from sample collection to data reporting is essential for the reliability of the concentration information. With sample collection using DBS rather than a plasma format, the bioanalytical assay used for the analysis of these samples, especially assays with liquid chromatography-tandem mass spectrometry (LC-MS/MS) detection, has been investigated extensively in the past several years.

17.3.1 DBS Sample Preparation and Analyte Elution Efficiency

Method development of the LC-MS/MS assay consists of three main parts: sample preparation, chromatographic elution, and tandem mass spectrometry detection. Sample preparation has been a focus of discussions for method development for DBS LC-MS/MS assays since the sample format is changed from conventional liquid plasma to DBS. The main objective of sample preparation is removing the matrix background from biological matrix samples as much as possible while maintaining the analyte(s) integrity, as well as achieving an extensive and consistent recovery of the analyte(s). The first and one of the most critical steps in DBS sample preparation is to have an effective transfer of the analyte(s) from the DBS cards to the liquid extraction solution. The effectiveness of this step can be evaluated using a parameter that we called "elution efficiency" (EE) (Liu et al., 2010). EE can be calculated by comparing the extraction of a liquid blood sample to that of a DBS sample. EE = A/B, where A is the area ratio of analyte to the IS obtained from the DBS sample and B is the area ratio of analyte to IS obtained from the blood sample. There is one assumption for this calculation: that matrix effects from either the blood extract or DBS extract will be compensated for by the IS. In most cases, this calculation should hold true when a stable isotope-labeled IS is used. In most cases of DBS sample analysis, DBS samples are punched out prior to the addition of IS. Therefore, the IS does not compensate for the variation in the EE. High EE is necessary for a higher and, more importantly, a consistent analyte recovery for standards, quality controls, and study samples.

It has been observed that DBS cards will bring in extra matrix components that may not exist when plasma or blood samples are used. The extent of the matrix interference varies appreciably among different types of cards. Generally, DBS cards with a chemical additive, such as™ FTA™ and FTA Elute™ cards, could cause a significant matrix effect that needs to be addressed through sample preparation (Liu et al., 2010). Although the use of nontreated DBS cards results in less matrix interference, in all cases, optimizing the extraction procedure is critical for any card type in order to achieve high EE and reduce any matrix effects. When not addressed adequately, significant reduction of instrument sensitivity after consecutive injections of sample extracts has been observed, especially for DBS samples with chemically treated DBS cards (Liu et al., 2010). Although pure organic solvents, such as methanol or acetonitrile, with extensive sonication/mixing have been successfully used as extraction solvents for DBS assays, a significant percentage of aqueous solution in the extraction solvent will improve the EE with a shorter period of mixing or sonication (Liu et al., 2010).

17.3.2 DBS Sample Dilutions

Another important challenge of sample preparation for a DBS bioanalytical assay is related to the dilution of high-concentration samples. Dilution is critical for nonclinical studies since analysis of samples from the high-dose group is required in order to achieve sufficient sensitivity for samples from the low-dose group. For plasma samples, the sample dilution can be achieved simply by mixing a small, fixed volume of the plasma sample with a fixed, large volume of blank plasma prior to the sample extraction. The concentration of the analyte in the original sample can be backcalculated using the dilution factor. However, for DBS samples, due to the solid nature of the sample, it is almost impossible to dilute the sample prior to sample preparation. The most common approach for DBS sample dilution has been to dilute the DBS extracts with extracts of blank DBS samples. The IS can be added prior to or after sample dilution. Consistent "elution efficiency" of the analyte from sample to sample is essential because there is no IS compensation until the analyte is eluted into the liquid phase during the sample preparation step.

Another method of sample dilution, the so-called "donut dilution," is to cut a small diameter of the analyte DBS sample and combine it with a large diameter blank DBS sample punch out (the center piece with same diameter as the analyte DBS sample punched, hence, the shape of a donut is emulated) in the sample preparation. The dilution factor is based on the ratio of the area of the sample punch out and the area of blank punch out. Since the dilution step is performed prior to sample extraction, the variation of sample extraction can be largely compensated with the addition of IS. The major limitation of this approach is the limited range of the dilution factor.

More recently, a new approach, termed "internal standard tracking dilution," was proposed for DBS sample dilution (Liu et al., 2011b). In this approach, a concentrated IS working solution is used for the sample to be diluted before further sample processing instead of using the same concentration of IS for all standards, QCs, and analyte samples; subsequently, the processed sample was roughly diluted into the assay linear response range before LC-MS/MS analysis. The advantage of this strategy is that the dilution step is tracked by the IS and is no longer a volume-critical step. In addition, this "internal standard-tracked dilution" is compatible with automatic sample dilution using a liquid handler.

17.3.3 Impact of Hematocrit on Blood-to-Plasma Concentration Ratio

Although significant progress has been made in bioanalysis using DBS as a sample collection and storage format, there are still unresolved challenges that limit the extensive applications of this technology for nonclinical studies. One challenge is related to the fact that blood, rather than plasma, concentration is the direct result of measurement. Another challenge involves the variation of spot size for the same blood volume and nonuniform analyte distribution related to the variation of the blood matrix composition.

Although there have been discussions that blood could be a form of matrix used in drug development for PK and TK studies, plasma is still the predominantly used matrix. When a DBS sample is collected and analyzed, the blood concentration is obtained. Conversion of blood concentration to plasma concentration requires a constant conversion factor (blood-to-plasma concentration ratio for each sample). The conversion factor needs to be independent of the individual animal, sample collection time, and dose. However, the effect of these factors on the blood and plasma concentration ratio has not been extensively studied. In addition, as shown in the following equation (Hinderling, 1997),

$$R = 1 + (\mathrm{Cbc}/\mathrm{Cp} - 1) \times \mathrm{Hc}$$

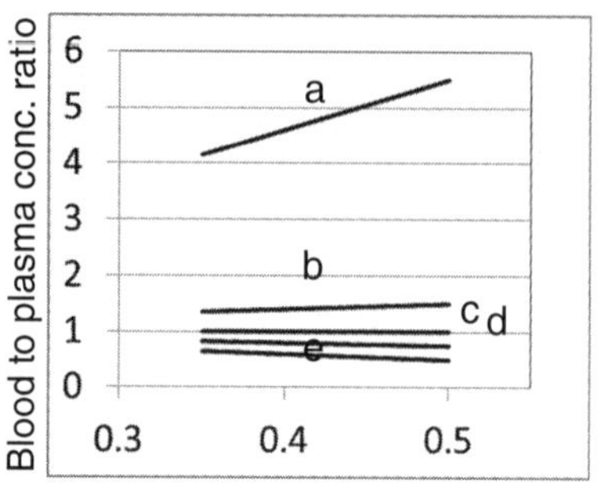

Line	Blood cell:plasma concentration ratio
a	10
b	2
c	1
d	0.5
e	0

FIGURE 17.1 Theoretical impact on the blood-to-plasma concentrations by hematocrit at various blood cell/plasma drug distributions. Reproduced with permission from Ji et al. (2012) Copyright 2012 Future Science Ltd.

where R is the ratio of blood-to-plasma concentration, Cbc is the red blood cell drug concentration, Cp is plasma drug concentration, and Hc is the hematocrit.

Assuming the Cbc/Cp ratio is a constant when plasma protein binding is not saturated, as demonstrated in Figure 17.1 (Ji et al., 2012), R is dependent on the hematocrit. The degree of impact of hematocrit on R is affected by the Cbc/Cp ratio as well. When Cbc/Cp is close to 1, the effect of hematocrit on R is greatly reduced. When Cbc/Cp is much less than 1 or larger than 1, the effect of hematocrit on R becomes significant. It is worth emphasizing that when Cbc/Cp is much larger than 1, the analyte mainly resides in the red blood cell and blood could therefore be the preferred matrix for analyte concentration measurement as pointed out by Emmons and Rowland (2010).

17.3.4 Impact of Hematocrit on Bioanalytical Measurement and Potential Solutions

The initial promise of DBS sample collection was that it was unnecessary to accurately measure the volume of blood collected on a DBS card because a fixed diameter punch size would be equivalent to same sample volume. This approach was based on early experimental observations that the analyte(s) is evenly and consistently distributed across the entire card spot, that the spot size for the same blood volume is identical from sample to sample, and that the spot size will proportionally increase (or decrease) with the increase (or decrease) of blood volume spotted. Therefore, if a 10, 15, or 20 μL blood volume was spotted, the concentration results should be the same as long as a fixed punch size is used, because of the consistent manner in which a spot spreads (Barfield et al., 2008; Liang et al., 2009). Later, it was recognized that this promise only holds true to a certain degree (Liang et al., 2009; Denniff and Spooner, 2010). The drug distribution on a DBS card can be affected by the properties of the blood (especially the hematocrit), the DBS card used, the drug analyte properties, and the environment at the time of spotting blood samples onto the card. Particularly, the impact of hematocrit has been examined (Liang et al., 2009; Denniff and Spooner, 2010). Generally, it has been found that higher hematocrit samples will generate a positive bias and lower hematocrit samples will tend to produce negative-biased results. The hematocrit of blood samples for different nonclinical species could be ranged from 38 to 52 (Dawes et al., 2010). The degree of this impact may not be a concern for discovery studies. However, it must be evaluated for GLP nonclinical studies, especially if the drug has an effect on blood chemistry. If the degree of impact is well beyond the normal acceptance level, then this issue must be resolved prior to the decision of using DBS sample collection for the drug development program.

Multiple approaches have been proposed to address the impact of hematocrit on bioanalytical assay performance (Fan and Lee, 2012; Capiau et al., 2013). Card manufacturers are working on developing new cards that are less susceptible to the impact of blood matrices on the DBS analyte distribution and spot size. A blood component, potassium, was investigated as a potential marker for blood hematocrit changes and therefore could be used as an internal calibrator to compensate the analytical bias due to hematocrit variation of individual sample (Capiau et al., 2013). Other approaches have also been developed in which an accurate blood volume is deposited for each spot and, subsequently, the whole spot is utilized for analysis. While it is a challenge to implement accurate blood volume measurement during sample collection in clinical studies, it is quite possible to implement this approach in nonclinical studies. Nonclinical studies are normally conducted in facilities with well-trained staff and advanced equipment. With further optimization of sample collection devices, handling processes, and proper staff training, the requirements of DBS sample collection with accurate blood sample volumes could be met for nonclinical studies.

DBS is well positioned as a technology that can provide assay performance in discovery nonclinical studies (Clark et al., 2010; D'Arienzo et al., 2010). When it is employed in drug development, especially for the GLP nonclinical studies, extensive validation must be performed to ensure that assay performance during sample analysis is well within study acceptance criteria (Timmerman et al., 2011). Validation criteria generally include a set of parameters that are commonly evaluated in both DBS and plasma assays and a set of parameters that are unique to DBS assays. Depending on how sample collection and sample analysis are defined, the assay validation criteria could be different as well. The following discussion is not intended to be comprehensive, but rather highlights some unique aspects of DBS LC-MS assays.

One of the important aspects that could affect validation parameters is related to how DBS samples are collected and how the DBS samples are used for sample analysis. In the situation where no accurate blood volume spotting is used for sample collection and sample analysis is performed with fixed area punching, one of the main features of validation is ensuring the consistency of the sampling using fixed area punching. Since the analyte distribution on a DBS card may depend on a number of factors, these factors need to be evaluated for each assay unless it is proved that their impacts are minimal and independent of the analyte of the measurement. Sample collection card lot-to-lot validation should be evaluated. However, this evaluation could be eliminated when a reliable manufacture quality control system is in place and is well demonstrated. Sample spotting processes should be well defined and a similar spotting process should be used at the study site and in the laboratory for the preparation of quality control sample and calibration standards. For example, if blood that is collected from the animal is chilled prior to spotting on the cards, it is important that the analytical laboratory chills the blood sample during validation, due to differences in the spreading behavior of blood at different temperatures. The effect of spotting blood volume variation should be validated to cover expected blood volume variation of real study DBS samples.

In the situation where accurate blood volume spotting is required for the sample collection and a whole blood spot is used for sample analysis, the method validation related to the sample spotting process can be eliminated. As the nonclinical study site staff becomes more skillful in collecting blood samples in accurate volumes, this approach reduces the bioanalytical work load significantly and eliminates the concerns of inaccurate sampling using fixed punch area.

17.3.5 Potential Differences between DBS Samples and *Ex Vivo* Calibration Standards/Quality Control Samples

As with LC-MS bioanalytical assays that involve tissue, there are concerns about the difference between *in vivo* samples and quality control samples prepared by spiking the analyte in blank blood. The difference of analyte distribution in and binding within matrix components between *in vivo* samples and *ex vivo* calibration standards/quality control samples could be more profound for blood samples than plasma samples. Because of this, true sample concentration may not be obtained even when excellent calibration standard and quality control sample results are obtained for the assay (Brockman et al., 2007). Knowledge about analyte red blood cell distribution could help decide how much this aspect should be investigated. It is important to keep this in consideration during the assay development and validation experimental design. For example, an evaluation of the minimum equilibrium time (after the analyte is spiked in the blank blood matrix) prior to spotting blood quality control samples to DBS cards after analyte(s) has been added to the blank blood may be necessary. Furthermore, extraction recovery is a more important validation parameter for the DBS method than for the plasma method. As discussed earlier in this section, a requirement (e.g., greater than 90%) of the EE should be established, in addition to the requirement of consistent extraction recovery and EE.

17.3.6 Sample Storage Stability

Stability evaluation should cover the entire process from sample collection to sample analysis with defined temperature and humanity range and other storage conditions. Of particular note, the time period of the sample drying process, especially for compounds with potential instability issues, needs to be investigated. The physical format of the DBS sample changes from wet to dry during the drying process; direct comparison of the cut sample spot at these two time points can still be affected by extraction recovery differences in addition to stability. This extraction recovery difference contribution can be eliminated by comparison of the homogenates of sample spots before drying and after drying (Liu et al., 2011a).

17.4 CONCLUSIONS

In conclusion, with the continuous effort of the scientific community to overcome a few obstacles (such as blood hematocrit effect on assay performance), DBS sample collection and bioanalysis is going to be well positioned for the support of discovery nonclinical studies and fits the need of preclinical development studies well when assay validation is successfully completed. The objectives of the nonclinical study need to be well aligned with the overall project development plan and design of future clinical studies. DBS sampling, as one of the microsampling techniques, offers the advantage of sample storage in a dried format, which could be critical for some analytes that might otherwise be unstable when stored in a liquid format. In addition, sample storage at room temperature offers the advantage of easy sample shipment and storage. With two fundamental characteristics of small sampling volume and dried sample format, DBS sampling has led to redefining the future direction of sample collection in drug development and the healthcare industry.

ACKNOWLEDGMENTS

The authors thank Drs Mark Arnold, Timothy Olah, Anne-F. Aubry and other BMS DBS initiative team colleagues for valuable discussions.

DISCLAIMER

The opinions in this article are from the authors and do not reflect the opinion and position of the organization with which the authors are employed.

REFERENCES

Barfield M, Spooner N, Lad R, Parry S, Fowles S. Application of dried blood spots combined with HPLC-MS/MS for the quantification of acetaminophen in toxicokinetic studies. *Journal of Chromatography B* 2008;870:32–37.

Brockman AH, Hatsis P, Paton M, Wu JT. Impact of differential recovery in bioanalysis: the example of bortezomib in whole blood. *Analytical Chemistry* 2007;79:1599–1603.

Capiau S, Stove VV, Lambert WE, Stove CP. Prediction of the hematocrit of dried blood spots via potassium measurement on a routine clinical chemistry analyzer. *Analytical Chemistry* 2013;85:404–410.

Clark GT, Haynes JJ, Bayliss MAJ, Burrows L. Utilization of DBS within drug discovery: development of a serial microsampling pharmacokinetic study in mice. *Bioanalysis* 2010;2(8):1477–1488.

Crawford E, Gordon J, Wu JT, Musselman B, Liu R, Yu S. Direct analysis in real time coupled with dried spot sampling for bioanalysis in a drug-discovery setting. *Bioanalysis* 2011;3:1217–1226.

D'Arienzo CJ, Ji QC, Discenza L, Cornelius G, Hynes J, Cornelius L, Santella JB, Olah T. DBS sampling can be used to stabilize prodrugs in drug discovery rodent studies without the addition of esterase inhibitors. *Bioanalysis* 2010;2:1415–1422.

Dawes ML, Liu G, Schuster AE, Shuster D, D'Arienzo C, Ji QC, Arnold ME. Investigation of Hematocrit Effect on Bioanalytical Assay Performance Using Fixed Volumes and Fixed Punch Size Dried Blood Spot Samples, DIA Workshop Dried Blood Spot Sampling in the Pharmaceutical Industry: Three Years of Experience and Implementation, Philadelphia, PA, 2010.

Denniff P, Spooner N. The effect of hematocrit on assay bias when using DBS samples for the quantitative bioanalysis of drugs. *Bioanalysis* 2010;2:1385–1395.

Diehl KH, Hull R, Morton D, Pfister R, Rabemampianina Y, Smith D, Vidal JM, van de Vorstenbosch C. A good practice guide to the administration of substances and removal of blood, including routes and volumes. *Journal of Applied Toxicology* 2001;21:15–23.

Emmons G, Rowland M. Pharmacokinetic considerations as to when to use dried blood spot sampling. *Bioanalysis* 2010; 2:1791–1796.

Fan L, Lee JA. Managing the effect of hematocrit on DBS analysis in a regulated environment. *Bioanalysis* 2012;4:345–347.

Ferrari L. Dried blood spot technology gaining favor. *Genetic Engineering and Biotechnology News* 2011;31(9):52–53.

Golde WT, Gollobin P, Rodriguez LL. A rapid, simple, and humane method for submandibular bleeding of mice using a lancet. *Laboratory Animals* 2005;34:39–43.

Hinderling P. Red blood cells: a neglected compartment in pharmacokinetics and pharmacodynamics. *Pharmacological Reviews* 1997;49:279–295.

Hoff J. Methods of blood collection in the mouse. *Laboratory Animals* 2000;29:47–53.

International Conference on Harmonisation. ICH Topic S3A—Toxicokinetics: The Assessment of Systemic Exposure in Toxicity Studies (I); 1995.

International Conference on Harmonisation and European Medicines Agency. ICH Topic M 3 (R2) Non-clinical Safety Studies for the Conduct of Human Clinical Trials and Marketing Authorization for Pharmaceuticals; 2009.

Ji QC, Liu G, D'Arienzo CJ, Olah TV, Arnold ME. What is next for dried blood spots? *Bioanalysis* 2012;4:2059–2065.

Li F, Ploch S, Fast D, Michael S. Perforated dried blood spot accurate microsampling: the concept and its applications in toxicokinetic sample collection. *Journal of Mass Spectrometry* 2011;47:655–667.

Liang X, Li Y, Barfield M, Ji QC. Study of dried blood spots technique for the determination of dextromethorphan and its metabolite dextrorphan in human whole blood by LC-MS/MS. *Journal of Chromatography B: Analytical Technologies in the Biomedical and Life Sciences* 2009;877:799–806.

Liu G, Patrone L, Snapp HM, Batog A, Valentine J, Cosma G, Tymiak A, Ji, QC, Arnold ME. Evaluating and defining sample preparation procedures for DBS LC/MS/MS assays. *Bioanalysis* 2010;2:1405–1414.

Liu G, Ji QC, Jemal M, Tymiak AA, Arnold ME. Approach to evaluating dried blood spot sample stability during drying process and discovery of a treated card to maintain analyte stability by rapid on-card pH modification. *Analytical Chemistry* 2011a;83:9033–9038.

Liu G, Snapp HM, Ji QC. Internal standard tracked dilution to overcome challenges in dried blood spots and robotic sample preparation for liquid chromatography/tandem mass spectrometry assays. *Rapid Communications in Mass Spectrometry* 2011b;25:1250–1256.

Meier-Davis SR, Meng M, Yuan W, Diehl L, Arjmand FM, Lucke RM, Huang B, Wen J, Shudo J, Nagata T. Dried blood spot analysis of donepezil in support of a GLP 3-month dose-range finding study in rats. *International Journal of Toxicology* 2012;31:337–347.

Moore CM, Tebbett IR. Rapid extraction of anti-inflammatory drugs in whole blood for HPLC analysis. *Forensic Science International* 1987;34:155–158.

Morton D, Abbot D, Barclay R, Close BS, Ewbank R, Gansk D, Heath M, Mattic S, Poole T, Seamer J, Southee J, Thompson A, Trussel B, West C, Jennings M. Removal of blood from laboratory mammals and birds: first report of the BVA/FRAME/RSPCS/UFAW Joint Working Group on Refinement. *Laboratory Animals* 1993;27:1–22.

National Research Council. *The Three Rs, in Guide for the Care and Use of Laboratory Animals*. Washington, DC: National Research Council; 2011.

Neu V, Delmotte N., Kobold U, Dallffer T, Herrmann R, Eltz H, Huber CG. On-line solid-phase extraction high-performance liquid chromatography-tandem mass spectrometry for the quantitative analysis of tacrolimus in whole blood hemolyzate. *Analytical and Bioanalytical Chemistry* 2012;404:863–874.

Olejniczak K, Günzel P, Bass R. Preclinical testing strategies. *Drug Information Journal* 2001;35:321–336.

Ploemen JPHTM, Kramer H, Krajnc EI, Martin I. The use of toxicokinetic data in preclinical safety assessment: a toxicologic pathologist perspective. *Toxicologic Pathology* 2007;35:834–837.

Timmerman P, White S, Globig S, Lüdtke S, Brunet L, Smeraglia J. Recommendation on the validation of bioanalytical methods for dried blood spots. *Bioanalysis* 2011;3:1567–1575.

Turpin PE, Burnett JEC, Goodwin L, Foster A, Barfield M. Application of the DBS methodology to a toxicokinetic study in rats and transferability of analysis between bioanalytical laboratories. *Bioanalysis* 2010;2:1489–1499.

Wan KX, Reimer MT, Metchkarova MP, Richter AJ, Paramadilok A, Kavetskaia O, El-Shourbagy T. Toxicokinetic evaluation of atrasentan in mice utilizing serial microsampling: validation and sample analysis in GLP study. *Bioanalysis* 2012;4:1351–1361.

WHO. Principles of Toxicokinetic Studies: International Programme on Chemical Safety Environmental Health Criteria. EHC 57; 1986.

18

CLINICAL AND BIOANALYTICAL EVALUATION OF DRIED BLOOD SPOT SAMPLING FOR GENOTYPING AND PHENOTYPING OF CYTOCHROME P450 ENZYMES IN HEALTHY VOLUNTEERS

THEO DE BOER, IZAAK DEN DAAS, JAAP WIELING, JOHAN WEMER, AND LINGSING CHEN

18.1 INTRODUCTION

The use of dried blood spot (DBS) sampling for newborns has been common practice since the early 1960s. Recently, the DBS method has also been used for therapeutic drug monitoring (Edelbroek et al., 2009) and pharmacokinetic (PK) studies (Spooner et al., 2009). More specific, the application of DBS for determination of the cytochrome P450 (CYP450) probes—dextromethorphan (CYP2D6), caffeine (CYP1A2), flurbiprofen (CYP2C9), midazolam (CYP3A4/5), omeprazole (CYP2C19), and rosiglitazone (CYP2C8)—using LC-MS/MS (Liang et al., 2009; Lad, 2010) and the application of DBS for genetic analysis (Wijnen et al., 2007; Hollegaard et al., 2009) were described as well.

In this chapter, we present the applicability of DBS method for phenotyping and genotyping of CYP450 enzymes in healthy volunteers as previously published (De Boer et al., 2011). It has been identified that CYP450 iso-enzymes play an important role in drug metabolism and genetic polymorphism. Poor metabolizers have low drug metabolizing enzyme activity compared to the majority of individuals defined as extensive metabolizers. This variability may result in a different PK, pharmacodynamic (PD), and adverse event profile. CYP450 enzyme activities can be evaluated by phenotyping and genotyping procedures. The CYP450 enzyme CYP3A4 accounts for about 25% of all liver CYP450s. The CYP3A family consists of several iso-enzymes whereof CYP3A4 and CYP3A5 are the most commonly expressed forms. Phenotyping of CYP3A4 has been done extensively with midazolam (MDZ) up to 7.5 mg as a single oral dose by determining the PK profile (AUC) of MDZ rather than the ratio of the main metabolite (1-OH MDZ) to MDZ based on extensive or limited sampling (Lee et al., 2006).

The results of the analysis of MDZ between the DBS method, plasma, and WB (drawn by finger puncture and venipuncture) are compared and discussed, whereby several methodological parameters were studied (i.e., punch width, amount of dots analyzed, storage time stability). In addition, genotyping of CYP3A4, CYP3A5, CYP2D6, and CYP2C19 is compared between DBS and venous WB samples and the subject and phlebotomist satisfaction of venous blood sampling versus DBS sampling was evaluated using a standardized questionnaire after completion of the clinical study.

For analysis of the DBS samples a method for the determination of MDZ and 1-OH MDZ in human DBS using LC-MS/MS was validated according to accepted validation criteria (Food and Drug Administration, 2001; Viswanathan et al., 2007). On-card stability of the analytes was also included (Edelbroek et al., 2009; Spooner et al., 2009; Li and Tse, 2010). The conventional methods for the determination of concentrations of MDZ and its metabolite 1-OH MDZ in human plasma and WB were fully validated as well, but are not described in detail.

Dried Blood Spots: Applications and Techniques, First Edition. Edited by Wenkui Li and Mike S. Lee.

18.2 EXPERIMENTAL

18.2.1 Chemicals, Solvents, and Blank Matrices

MDZ (listed in the Netherlands as a list II drug in the Opium Law) for preparation of calibration standards (CALs) and quality control samples (QCs) was obtained through the Pharmacy of QPS Netherlands from Spruyt Hillen (IJsselstein, The Netherlands); 1-OH MDZ was obtained from Sigma-Aldrich (Zwijndrecht, The Netherlands); MDZ-d5 was obtained from Syncom (Groningen, The Netherlands); and [$^{13}C_3$] 1-OH MDZ was obtained from Toronto Research Chemicals (North York, Canada). Acetic acid, acetonitrile (ACN), ammonium formate, diethylether, methanol (MeOH), sodium bicarbonate ACS reagent, sodium hydroxide, and formic acid were purchased from Merck (Darmstadt, Germany). Purified and deionized water was prepared by a Milli-Q system (Millipore, Billerica, MA, USA). Human EDTA WB and plasma for the preparation of calibration and quality control samples were obtained from healthy volunteers. All volunteers were in good physical health as determined by history taking, physical examination, and routine clinical tests.

18.2.2 Chromatographic System

The UHPLC used was an Infinity 1290 system (Agilent Technologies, Amstelveen, the Netherlands). The analytical column used was a reversed phase Luna PFP(2) (100 mm × 2.0 mm i.d., 3 μm; Phenomenex, Utrecht, the Netherlands). The following LC solvent program (400 μL/min) was used: 0.01 M ammonium formate solution pH 5.0 with (0–1.0 min): 30% ACN; (1.0–2.0 min): linear gradient from 30% to 60% ACN; (2.0–4.0 min): 60% ACN; (4.0–4.1 min): linear gradient from 60% to 30%; and finally (4.1–8 min): 30% ACN. The column temperature was 60°C. The LC eluate was introduced into an API-4000 triple quadrupole MS/MS system (Applied Biosystems, Foster City, CA, USA; OS: Analyst Software 1.4.1) equipped with an electrospray ionization source. The MS/MS system was operated in positive ion mode. The detailed mass spectrometer conditions were as follows: vaporizer temperature (TEM), 600°C; ionization spray voltage, 5500 V; collision gas (CAD, nitrogen), adjusted at 5 on a Sciex scale of 0–12; curtain gas (CUR, nitrogen), adjusted at 25 psig (10–50). The optimized MRM transitions as well as the chemical specifications of the analytes are presented in Figure 18.1. It was anticipated that small deviations in the masses (*m/z* ± 0.5) could occur because of tuning differences and used MS system.

18.2.3 Preparation of Stock Solutions, Calibration Standards, and Quality Control Samples

The analytes (including the internal standards (ISs)) were dissolved in MeOH to obtain stock solutions of 1.00 mg/mL, respectively. Stock solutions were prepared based on the active entity of the analytes, that is, concentrations were normalized with regard to impurities, water content, and counterions present in the reference materials. Aliquots of the stock solutions were further diluted with MeOH to obtain working

Chemical specification	MDZ	MDZ-d5	1-OH MDZ	[$^{13}C_3$] 1-OH MDZ
Structure	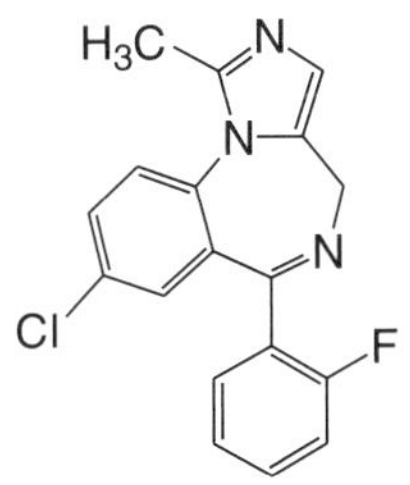	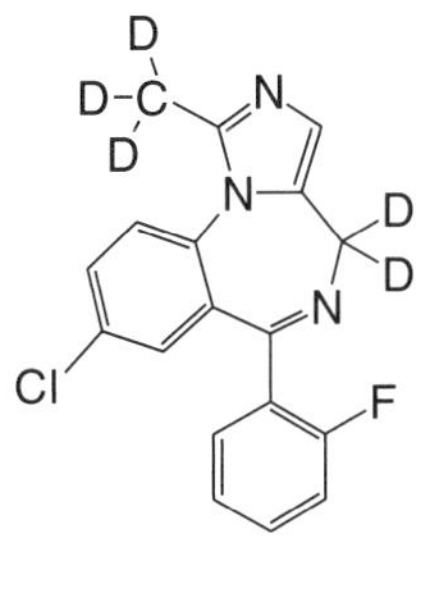	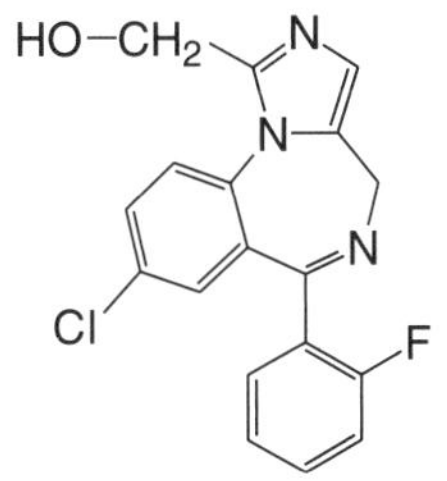	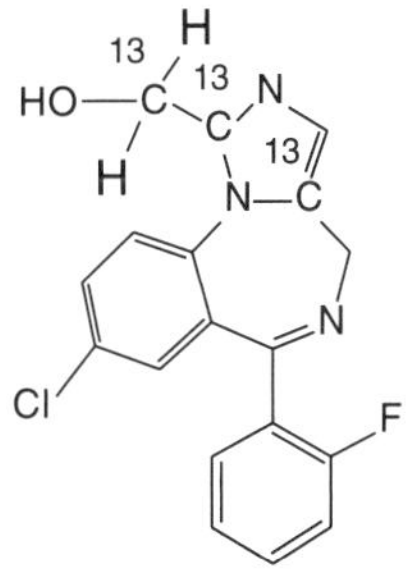
Emperical formula	$C_{18}H_{13}ClFN_3$	$C_{18}H_8D_5ClFN_3$	$C_{18}H_{13}ClFN_3O$	$^{13}C_3C_{15}H_{13}ClFN_3O$
Molar weight (g/mole)	325.8	330.8	341.8	344.8
MRM parameters				
Q1 mass (*m/z*)	326.1	331.1	342.1	345.1
Q3 mass (*m/z*)	291.2	296.2	324.1	171.3
Declustering potential (V)	101.0	101.0	91.0	91.0
Collision energy (eV)	37.0	37.0	31.0	53.0
Collision exit potential (V)	14.0	14.0	16.0	10.0

FIGURE 18.1 Chemical specifications and scan parameters of MDZ, 1-OH MDZ, and their corresponding ISs. Reproduced with permission from Wiley.

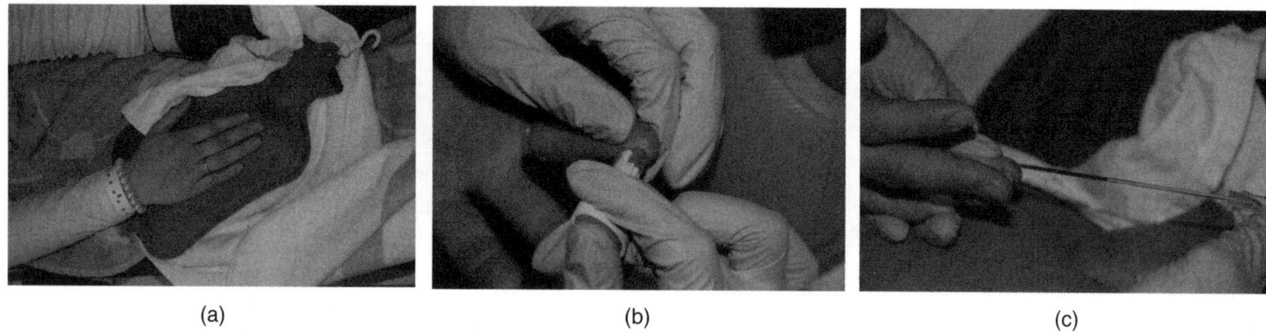

FIGURE 18.2 Blood sampling procedure by finger puncture. From top to bottom: (a) prewarming of the puncture side; (b) puncture using a single-use automatic lancing device; and (c) blood collection using an EDTA-filled capillary (100 μL). Reproduced with permission from Wiley.

standard solutions for the preparation of the calibration standards (CALs) and quality control samples (QCs). The spiked CALs and QCs were divided into individual tubes until use for analysis. For the analysis of the DBS samples, spiked WB was spotted onto Ahlstrom 226 IDBS Bioanalysis Cards (ID Biological Systems, Greenville, UK). The DBS cards were prepared freshly before use by spotting 20 μL of the CALs and QCs on the bioanalysis cards using a calibrated pipette. The samples were allowed to dry in a horizontal (flat) position for at least 2 hours at room temperature.

18.2.4 Blood Sampling Procedures

All healthy volunteers signed an informed consent. Two vials of 2 mL venous blood were collected from these subjects by a venous cannula or venipuncture, one for WB and one for plasma. In addition to the venous blood sampling, blood was also collected from finger puncture using a single-use, automatic lancing device (Accu check safe-T-pro at 2.3 mm depth, Roche Diagnostics, Woerden, the Netherlands). Before punching the site was prewarmed, cleansed with 70% alcohol, and air dried. The first drop was wiped after which blood was collected using EDTA-filled capillaries (Bilbate, Daventry, Northants, UK). The DBS sample was prepared on bedside by the previously trained phlebotomist by spotting approximately 20 μL WB from the capillaries on the bioanalysis cards. The sampling procedure is shown in Figure 18.2.

WB taken by venipuncture was collected in 2 mL EDTA-filled collection tubes after which the tubes were transported immediately to a dedicated sample preparation room, where 20 μL spots were pipetted by a trained technician on the bioanalysis cards. The cards were allowed to dry in a horizontal (flat) position (Figure 18.3a) for at least 2 hours before storage at room temperature in plastic bags containing a desiccant (Figure 18.3b). Plasma was prepared by a trained technician by centrifugation at 1500 × g for 10 minutes at room temperature. Plasma and remaining WB were stored under controlled conditions until analysis.

(a)

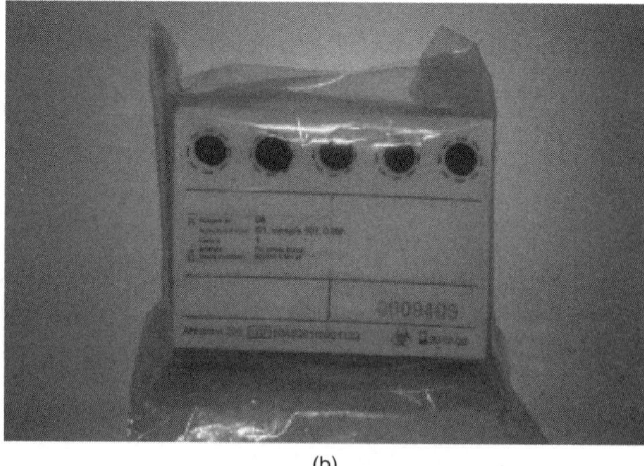

(b)

FIGURE 18.3 Preparation of the bioanalysis cards: (a) the cards were allowed to dry in a horizontal (flat) position for at least 2 hours. (b) After drying, the cards were stored in plastic bags containing a desiccant. Reproduced with permission from Wiley.

18.2.5 Sample Pretreatment and Sample Preparation

DBS. Depending on the used DBS method, 3 mm, 2 × 3 mm (two discs), or 6 mm discs were punched from the center of the bioanalysis cards ensuring that an area completely filled with blood was obtained. In case a fully filled disc could not be punched, the spot could not be used for analysis. The disc was transferred to a 1.5 mL micro test tube to which 50.0 μL Milli-Q water was added to pre-wet the cards in order to release the WB from the paper. Extraction of the analytes was performed by addition of 1000 μL MeOH and 10.0 μL IS solution. The samples were extracted for 45 minutes using a tumble mixer. After the extraction period, the samples were centrifuged for 5 minutes at 14,000 rpm after which the supernatant was transferred to a clean tube. The extract was evaporated until dryness under a gentle stream of nitrogen at 45°C. The dry residue was redissolved in 50.0 μL of Milli-Q water/ACN (2:3 v/v) and 10 μL was injected for LC-MS/MS analysis.

WB and plasma. A blood or plasma volume of 100 μL was pipetted to a 10 mL glass tube to which 10 μL IS solution and 200 μL of a sodium bicarbonate solution (0.7M; pH 9.7) was added. Extraction of the analytes was performed by addition of 3 mL diethylether. The samples were extracted for 10 minutes using a tumble mixer. After the extraction period, the samples were centrifuged for 5 minutes at 4000 rpm after which the supernatant was transferred to a clean tube. The extract was evaporated until dryness under a gentle stream of nitrogen at 45°C. The dry residue was redissolved in 100 μL of Milli-Q water/ACN (2:3 v/v) and 10 μL was injected for LC-MS/MS analysis.

18.2.6 Genomic DNA Isolation

DBS. DNA was isolated from DBSs using the Qiagen (Valencia, CA, USA) QIAamp DNA Mini Kit utilizing the manufacturer's protocol specific for purification of total genomic DNA from blood spotted and dried on filter paper. Genomic DNA samples were quantified using the NanoDrop (Wilmington, DE, USA) ND-1000 spectrophotometer to determine the DNA concentration and purity (A260/A280 ratio).

WB and plasma. Genomic DNA was isolated from human whole blood collected in K_2-EDTA using the Qiagen QIAamp DNA Mini Kit. Genomic DNA samples were quantified using the NanoDrop ND-1000 spectrophotometer to determine the DNA concentration and purity (A260/A280 ratio).

18.2.7 Genotyping Assays

The Pyrosequencing (Qiagen, Valencia, CA, USA) technology platform was used for genotyping the CYP2D6*3, CYP2D6*6, CYP3A4*1B, CYP3A4*2, and CYP3A5*3C polymorphisms. All assays were developed and validated or qualified at QPS. Analysis was performed using the Pyrosequencing PSQ HS96 and associated software.

The TaqMan® (Applied Biosystems, Foster City, CA, USA) technology platform was used for genotyping the CYP2D6*4, CYP2C19*2, and CYP2C19*3 polymorphisms. All assays were obtained from Applied Biosystems and listed as follows: CYP2D6*4 (#C__27102431_B0), CYP2C19*2 (#C__25986767_70), and CYP2C19*3 (#C_27861809_10). All assays were validated by the manufacturer as well as validated or qualified by QPS. The exact sequences of the primers and probes are the proprietary property of Applied Biosystems. PCR products were generated from genomic DNA and interrogated via TaqMan® allelic discrimination analysis using the ABI Prism 7900HT and SDS version 3.2 software.

18.2.8 Validation Procedures for the Quantitative Analysis of MDZ and 1-OH MDZ in Human DBS Samples by LC-MS/MS

The method for the determination of MDZ and 1-OH MDZ in human DBS samples was fully validated (Food and Drug Administration, 2001) for the use of 6 mm (diameter) discs taken from the center of a fully covered spot. A partial validation (accuracy and precision) was performed for the use of a 3 mm disc and for the use of two 3 mm discs for analysis. Development and validation of the three analytical methods for the determination of MDZ and 1-OH MDZ in plasma, WB, and DBS were carried out under controlled and monitored environmental conditions and conducted in accordance with good laboratory practice guidelines.

18.2.9 Clinical Study Design

The study was conducted as an open-label study to determine the usefulness of the DBS method for phenotyping and genotyping of CYP450 enzymes in healthy volunteers. Subjects were healthy male (n = 4) and female (n = 8) Caucasian volunteers aged 19–25 years with a body mass index (BMI) ≥18 kg/m^2 and ≤25 kg/m^2. All subjects were in good physical and mental health as determined by history taking, physical examination, routine clinical laboratory tests, electrocardiogram, and vital sign recordings. The clinical study protocol was approved by the independent Medical Ethics Review Committee "Stichting Beoordeling Ethiek Bio-Medisch Onderzoek" in Assen, the Netherlands. Written informed consent was obtained from each subject prior to any screening activity. Subjects received a single dose of MDZ (Dormicum® tablet 7.5 mg, Hoffman-LaRoche AG, Grenzach-Wyhlen, Germany). Venous blood samples for PK analysis were collected up to 12 hours after dosing (predose and at t = 0.5, 0.75, 1, 1.25, 1.5, 2, 2.5, 3, 4, 5, 6, 8, and 12 hours, n = 14). Blood taken from finger puncture was collected up to 8 hours after dosing (t = 0.5, 0.75, 1, 1.25, 1.5,

2, 2.5, 3, 5, and 8 hours, $n = 10$). As the terminal half-life of both MDZ and 1-OH MDZ is short (2 hours), collecting blood samples up to 12 hours after dosing is sufficient to obtain almost all area under the curve with little (if any) extrapolation.

The following primary, secondary, and explorative objectives were investigated:

1. *Primary objective.* Comparison of the MDZ concentrations in DBS, plasma, and WB, all drawn by venipuncture
2. *Secondary objective.* Comparison of the MDZ concentrations in DBS drawn by finger puncture and venipuncture
3. *Secondary objectives.* Comparison of the MDZ concentrations in DBS:
   a. using one and two discs of 3 mm (disc number)
   b. using 3 and 6 mm discs (disc size)
   c. after 0, 2, and 4 weeks storage at room temperature (on-card stability of study samples)
4. *Secondary objective.* Comparison of genotyping of CYP3A4, CYP3A5, CY2D6, and CYP2C19 between DBS and WB drawn by venipuncture
5. *Secondary objective.* Comparison of the AUC of MDZ between the use of limited sampling and extensive sampling
6. *Secondary objective.* Investigation of the feasibility of measuring 1-OH MDZ by DBS analysis
7. *Explorative objective.* Evaluation of subject and phlebotomist satisfaction of DBS method versus venous blood sampling using a standardized questionnaire

18.3 RESULTS AND DISCUSSION OF DBS METHOD VALIDATION

The results of the validation of the method for the determination of MDZ and 1-OH MDZ in human DBS samples by LC-MS/MS are summarized below.

18.3.1 Selectivity with Respect to Endogenous and Exogenous Substances

Milli-Q water was compared in each method to Std A to assess exogenous substances. No responses higher than 20% of the Std A response were observed. Blank human WB response (six different donors) was also compared in all methods to the Std A response for each analyte. No responses higher than 20% of the Std A response were observed. In addition, for all methods, and all blank samples, the response of interfering peaks at the retention time of the IS was found to be less than 5% of the response found for the IS at the concentration level of the IS used in the respective assay. These results indicate that the methods provide acceptable selectivity (i.e., there were no significant interfering substances for any of the analytes in the blank samples tested).

18.3.2 Assessment of Mutual Interference of MDZ, 1-OH MDZ, and ISs

The response of interfering peaks from MDZ and 1-OH MDZ at the retention time of the ISs was found to be less than 5% of the response found for the IS at the concentration level of the corresponding IS used in the assay. Vice versa, the responses of interfering peaks from the ISs at the retention time of MDZ and 1-OH MDZ were found to be less than 20% of the response found for the corresponding analyte in Std A. Also no interfering peaks were observed at the retention time of 1-OH MDZ when injecting 50.0 ng/mL MDZ. A small peak was observed at the retention time of MDZ when injecting 50.0 ng/mL 1-OH MDZ. The minor interference was related to the impurity of the reference material. As the response of the test solution of 1-OH MDZ (50.0 ng/mL) was three times higher than the response of the 1-OH MDZ peak in the highest calibration standard (100 ng/mL), the small peak was judged acceptable.

18.3.3 Linearity

The results of the back-calculated concentrations as well as the calibration curve parameters are presented in Table 18.1. The correlation coefficient (expressed as R) of all curves was >0.9950. Summary statistics covering all data of all accepted calibration curves illustrate that the accuracy and precision of the calibration samples were in compliance with generally accepted requirements (Shah et al., 2000) during the period in which the samples were analyzed.

18.3.4 Accuracy and Precision

The intra-run precision of the QC Low (0.300 ng/mL), QC Med (50.0 ng/mL), and QC High (100 ng/mL) samples was found to be lower than or equal to 8.5%, whereas the intra-run precision of QC LLOQ (0.100 ng/mL) was found to be lower than or equal to 9.1% for MDZ and 1-OH MDZ. The intra-run accuracy of QC Low, QC Med, and QC High samples was found to be between (and including the limits) 5.6% and 13.0% for MDZ and 1-OH MDZ, whereas the intra-run accuracy of QC LLOQ was found to be 6.4% (MDZ) and 18.0% (1-OH MDZ), respectively. The inter-run precision of the QC Low, QC Med, and QC High samples was found to be lower than or equal to 9.2% for MDZ and 1-OH MDZ, whereas the inter-run precision of QC LLOQ was found to be 9.1% (MDZ) and 15.2% (1-OH MDZ), respectively. The inter-run accuracy of QC Low, QC Med, and QC High

TABLE 18.1 Final Concentrations (ng/mL) of Spiked Whole Blood Calibration Standards (Std) and Mean Back-Calculated Concentrations (ng/mL) Including Mean Calibration Curve Parameters

Standard/ Control	Nominal Concentrations	MDZ	1-OH MDZ
Blank	0	0.0	0.0
Blank with IS	0	0.0	0.0
Std A	0.100	0.0996	0.0995
Std B	0.200	0.201	0.199
Std C	0.500	0.507	0.497
Std D	2.00	1.99	1.99
Std E	20.0	19.9	20.0
Std F	50.0	49.6	49.3
Std G	80.0	77.5	78.0
Std H	100	104	104
Bias%		≤|3.7|	≤|4.3|
CV%		≤8.8	≤9.2
Number of curves		Six in duplicate	Six in duplicate
Slope		0.131	0.0630
Intercept		0.00370	0.00289
R		0.9967	0.9974

IS, internal standard; R, correlation coefficient.
Blank sample consisted of blank human whole blood. Blank with IS sample consisted of blank human whole blood to which IS was added to determine selectivity toward the IS. The ISs used are presented in Figure 18.1.
Reproduced with permission from Wiley.

samples was found to be between (and including limits) 3.2% and 12.6% for MDZ and 1-OH MDZ, whereas the inter-run accuracy of QC LLOQ was found to be 5.3% (MDZ) and 9.5% (1-OH MDZ), respectively.

18.3.5 Dilution Integrity

Dilution experiments were performed with samples (QC OC: over-curve sample) spiked at 10 times the QC high concentrations. The samples were diluted 10-fold with extraction solvent.

As the inter-run accuracies of QC OC were –4.3% (MDZ) and –6.7% (1-OH MDZ), respectively, and the inter-run precision of QC OC was lower than or equal to 9.7% for both MDZ and 1-OH MDZ, dilutions with extraction solvent (MeOH) up to 10-fold were allowed.

18.3.6 Recovery

Recoveries were consistent over the entire concentration range (approximately 40%). As expected, the recoveries of the respective ISs were much higher in comparison to MDZ and 1-OH MDZ as they do not have to be extracted from the bioanalysis cards.

18.3.7 Matrix Effect (Ion Suppression or Enhancement)

No matrix effect was observed, as the coefficients of variation of the normalized matrix effect between four different sources of blank human WB were found to be 2.8% for MDZ and 4.3% for 1-OH MDZ.

18.3.8 Carryover

No carryover in the six blank human DBS samples that were analyzed after the Std H and QC high samples was observed.

18.3.9 Stability Experiments

The bias of the QC samples that were placed at RT for 24 hour did not deviate more than ±15% from the results of the originally injected QC samples (≤–5.7% for MDZ and ≤–5.9% for 1-OH MDZ). It is therefore not required to handle human WB samples that are used for preparation of CAL and QC samples under special conditions. The bias of the QC samples that were placed at RT for 4 weeks did not deviate more than ±15% from the results of the nominal value (≤4.2% for MDZ and ≤–5.3% for 1-OH MDZ). These long-term (on-card) stability data indicate that samples can be stored for 4 weeks at RT before analysis. The mean results of the processed samples after storage at 2–8°C for 3 days did not differ more than ±15% from the results of the original injected QC samples (0.0% for MDZ and 2.8% for 1-OH MDZ). These data indicate that samples can be processed and stored under cooled conditions for 3 days prior to analysis. Finally, the stock solutions for MDZ, 1-OH MDZ, and their ISs were stable for 15 months when stored at 2–8°C.

18.3.10 Disc Size and Disc Number Integrity

The partial validation of the method for the use of 3 mm discs instead of 6 mm discs (disc size) resulted in an increase of the LLOQ for 1-OH MDZ to 0.300 ng/mL. At all other concentrations, the partial validation for the determination of MDZ and 1-OH MDZ on 3 mm discs exhibited adequate precision and accuracy over a concentration range of 0.100–100 ng/mL. The partial validation for the determination of MDZ and 1-OH MDZ on 2 × 3 mm discs (disc number) exhibits adequate precision and accuracy over a concentration range of 0.100–100 ng/mL.

A sensitive and straightforward method was developed and validated for the quantification of MDZ and 1-OH MDZ in DBS samples using LC-MS/MS. No compromises were made regarding the LLOQ of the DBS method with regard to the complementary in-house developed methods for the determination of MDZ and 1-OH MDZ in plasma and WB, except for the partial validation using the 1 × 3 mm DBS disc. When using 3 mm discs instead of 6 mm discs, the

LLOQ of 1-OH MDZ was raised to 0.300 ng/mL in order to be able to fulfill the acceptance criteria. The fully validated and partially validated DBS methods adhered to the regulatory requirements for selectivity, sensitivity, precision, accuracy, recovery, carryover, matrix effect, and stability. Also the fully validated plasma and WB assays (data not shown) that were previously developed at our laboratory adhered to these criteria. The data indicates that the methods can be applied to the analysis of samples in early clinical development studies: PK, pharmacogenomics (PG), and safety assessments.

18.4 RESULTS AND DISCUSSION OF CLINICAL STUDY

18.4.1 Clinical Conduct

The study medication MDZ was well tolerated by all subjects. The observed adverse events were in line with the pharmacological profile of MDZ. Ten of the 12 volunteers experienced a total of 15 adverse events. Of these, 14 were considered to be possibly related to the administration of MDZ. These were sleepiness in 10 volunteers, headache in 3 volunteers, and tiredness in 1 volunteer. All adverse events were of a mild intensity and limited duration except tiredness which was of moderate intensity. Most of the related adverse events lasted up to a few hours post-dose and all were resolved within 10 hours post-dose.

18.4.2 Bioanalysis

The three methods described in this chapter have been successfully applied in our laboratory for the quantitation of MDZ and 1-OH MDZ in human DBS, plasma, and WB samples originating from the clinical trial conducted in our pharmacology unit. Bioanalysis was carried out under controlled and monitored environmental conditions and conducted in accordance with good laboratory practice guidelines. Calibration and quality control samples used for batch acceptance were in compliance with generally accepted requirements (Food and Drug Administration, 2001) during the period in which the samples were analyzed. The mean MDZ concentrations obtained from the applied sampling procedures are shown in Figure 18.4. The figure illustrates that the PK curves of MDZ in DBS taken by finger puncture and venipuncture are almost identical, whereas the plasma PK curve is higher (as expected) and the WB curve is lower than that of the mean DBS curve. In theory the mean WB curve should be in the same range as the mean DBS curve as the matrices are similar. This anomaly cannot be properly explained, but could be related to the difference between hematocrit values in CAL and QC samples and the study samples, as higher hematocrit in evaluation samples tends to give higher concentrations in DBS analyses (Fan et al., 2011). In Table 18.2, an overview of five major PK parameters, that is, the maximum concentration (C_{max}), the time at which C_{max} occurs (T_{max}), the area under the plasma concentration versus time curve up to infinite time (AUC_{inf}), the area under the plasma

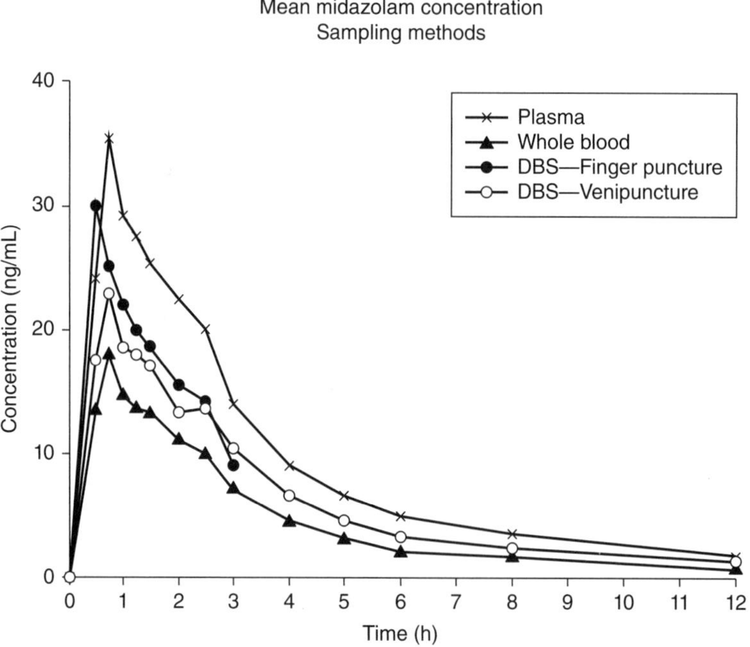

FIGURE 18.4 Mean plasma concentrations of MDZ (ng/mL) versus time (h) for the applied sampling methods (DBS from finger puncture, DBS from venipuncture, whole blood, and plasma). Reproduced with permission from Wiley.

TABLE 18.2 Summary Table PK Parameters Sampling Methods

Sampling Method	PK Variable	Unit	N	Mean
DBS—finger puncture	C_{max}	ng/mL	12	36.80
	T_{max}	h	12	1.139
	AUC_{inf}	h ng/mL	12	83.11
	$AUC_{(0-12)}$	h ng/mL	12	79.80
	$t_{1/2}$	h	12	2.586
DBS—venipuncture	C_{max}	ng/mL	12	29.00
	T_{max}	h	12	1.229
	AUC_{inf}	h ng/mL	12	80.95
	$AUC_{(0-12)}$	h ng/mL	12	73.15
	$t_{1/2}$	h	12	4.143
Plasma	C_{max}	ng/mL	12	43.43
	T_{max}	h	12	1.250
	AUC_{inf}	h ng/mL	12	117.57
	$AUC_{(0-12)}$	h ng/mL	12	108.24
	$t_{1/2}$	h	12	3.696
WB	C_{max}	ng/mL	12	23.37
	T_{max}	h	12	1.188
	AUC_{inf}	h ng/mL	12	57.64
	$AUC_{(0-12)}$	h ng/mL	12	53.90
	$t_{1/2}$	h	12	3.445

DBS, dried blood spot; WB, whole blood. Reproduced with permission from Wiley.

concentration versus time curve up to last sampling point, that is, 12 hours ($AUC_{(0-12)}$), and the terminal phase half-life ($t_{1/2}$), is given for the results of the analysis in DBS (finger puncture and venipuncture), plasma, and WB. As MDZ is usually analyzed in plasma, the relationship between the AUC obtained for the two matrices is of major importance. As can be seen from Table 18.2 the correlation between WB and plasma calculated by the ratio of $AUC_{(0-12)}$ for both matrices was 53.90/108.24 = 0.498. As can be seen from Table 18.2 the calculated WB/plasma ratio (0.498) deviates from the ratio that is obtained when $AUC_{(0-12)}$ is calculated for DBS_{veno}/plasma. In this case the ratio is 73.15/108.24 = 0.676. The correlation between DBS_{veno} and plasma as well as WB and the correlation between plasma and WB are depicted in Figures 18.5a, 18.5b, and 18.5c. A strong correlation is observed between the sampling techniques, expressed by their regression coefficients ($R^2 > 0.963$).

18.4.3 DBS Finger versus Venipuncture

Assessment of the correlation between the DBS concentrations of MDZ obtained by finger puncture (DBS_{finger}) and DBS_{veno} indicated a strong correlation between both sampling techniques which will offer opportunities to apply the less invasive finger prick for blood collection. The $AUC_{(0-12)}$ of the MDZ profile obtained from DBS_{finger} is comparable to the $AUC_{(0-12)}$ of the MDZ profile obtained from DBS_{veno} (Table 18.2). This result implies that blood collection by finger puncture is a less invasive alternative to venipuncture for phenotyping CYP3A4. This also relates to collected blood volumes (DBS_{veno} = 14 × 2 mL = 28 mL and DBS_{finger} = 10 × 0.1 = 1.0 mL) from each subject. However, closer analysis of the correlation between the two techniques as shown in Figure 18.6 indicates that the regression coefficient is much lower than those observed for the other techniques. The lower regression coefficient (0.754) is mainly caused by some extreme outliers. Comparison between DBS_{finger}, plasma, and WB exhibits similar results (R^2 = 0.784 and R^2 = 0.762, respectively).

18.4.4 Disc Numbers, Disc Size, Storage Stability

Secondary objectives which are related to disc number, disc size, and incurred sample reanalysis (ISR of 100% of study samples) after 2 and 4 weeks storage at RT are represented in Table 18.3. From the table it can be concluded that in comparison to the results of the 1 × 6 mm disc, the disc size has no significant influence on the $AUC_{(0-12)}$ whereas the number of discs shows an increase of 29.4% of the $AUC_{(0-12)}$. ISR of 100% of the study samples after storage for 2 weeks at RT had no significant effect (9.7%) on the $AUC_{(0-12)}$, whereas after storage for 4 weeks at RT a slight decrease (−19.7%) was observed. This decrease in AUC could be the consequence of a change in recovery (note that recovery was only ~40%) or a change in spot size due to extensive drying.

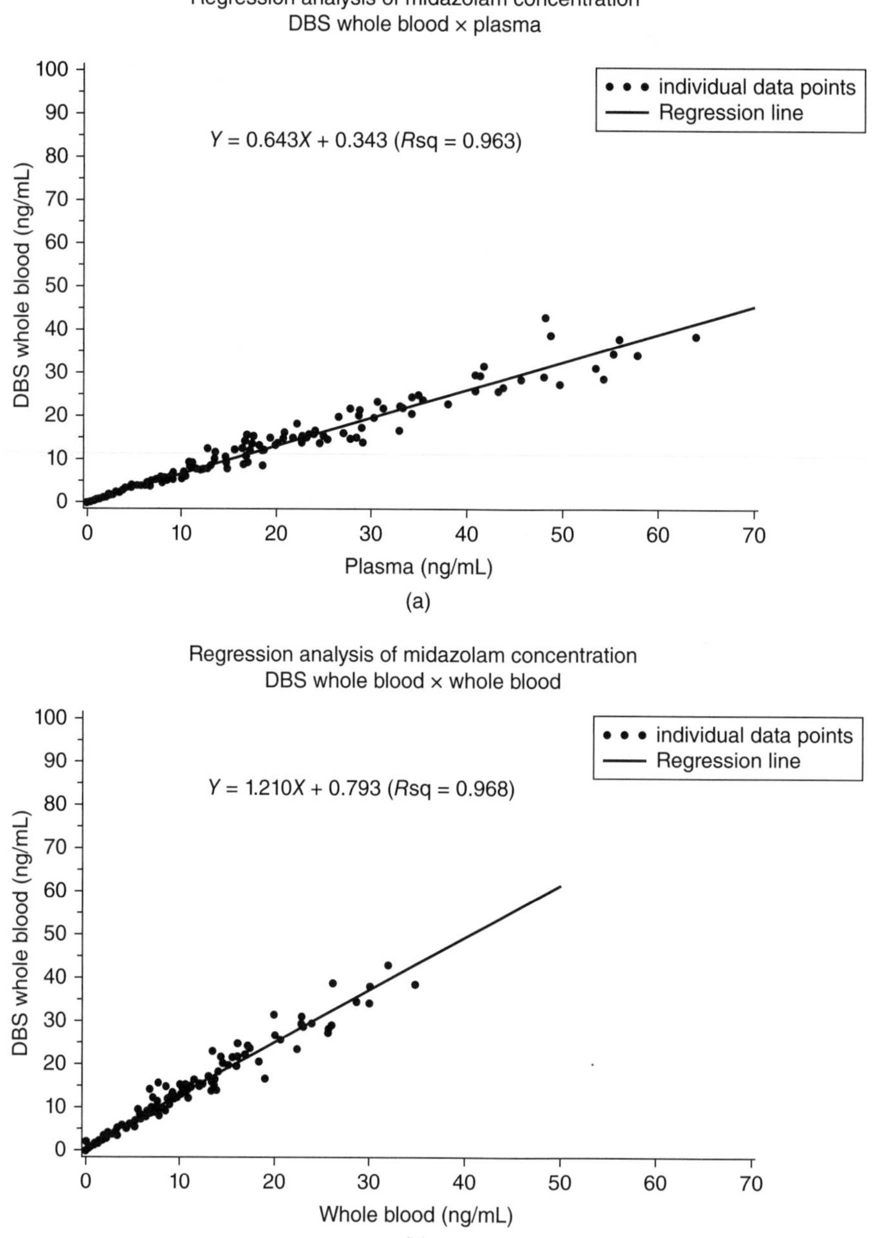

FIGURE 18.5 Regression analysis of MDZ concentrations (ng/mL) in DBS_{veno} versus plasma (a), DBS_{veno} versus whole blood (b), and plasma versus whole blood (c). Reproduced with permission from Wiley.

18.4.5 Limited versus Extensive Sampling

As described by Lee et al. (2006), for the exploration of the effect of limited and extensive sampling, $AUC_{(0-12)}$ values were compared descriptively across three models: a model estimated on all concentration time points (extensive model, i.e., 1 × 6 mm disc), a three-point limited-sampling model (t = 0.5, 2, and 6 hours), and a two-point limited-sampling model (t = 0.5 and 6 hours). The limited-sampling models indicate (depicted in Table 18.4) an increase in $AUC_{(0-12)}$ in comparison to the extensive modeling: 44.0% for three-point model and 59.6% for two-point model, respectively. Moreover, the linear regression plots for AUC presented in Figure 18.7 indicate that limited models can be predictive for MDZ AUC, although for both limited-sampling models an increase in AUC is observed, the correlation between the extensive model and the three-point limited model is good (R^2 = 0.918), whereas no correlation between the extensive

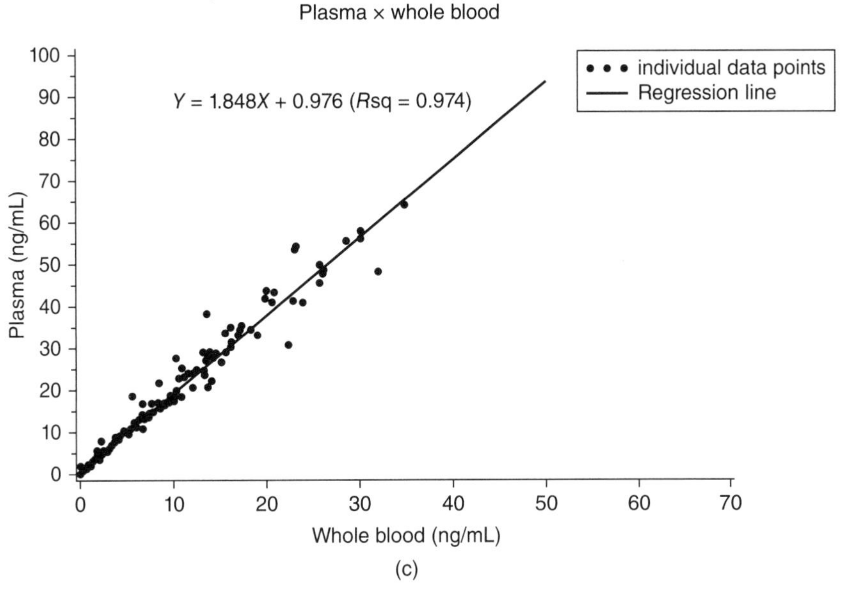

FIGURE 18.5 (*Continued*)

model and the two-point limited model is observed (R^2 = –0.368).

18.4.6 Analysis of 1-OH MDZ

As CYP3A4 activity is accurately predicted by the AUC of MDZ concentrations rather than the ratio of the main metabolite (1-OH MDZ) to MDZ, the analysis of 1-OH MDZ was not a primary objective. However, during this study, the feasibility of measuring 1-OH MDZ by DBS analysis was also investigated. The mean 1-OH MDZ concentrations obtained from the applied sampling procedures are shown in Figure 18.8. The results for the main metabolite are essentially similar to the results obtained for MDZ (Figure 18.4).

18.4.7 Genotyping

The descriptive statistics of the genotyping data are presented in Table 18.5. As can be seen from the table, a 100% match for each polymorphic gene allele examined was achieved, either by pyrosequencing or TaqMan technology platform. It can

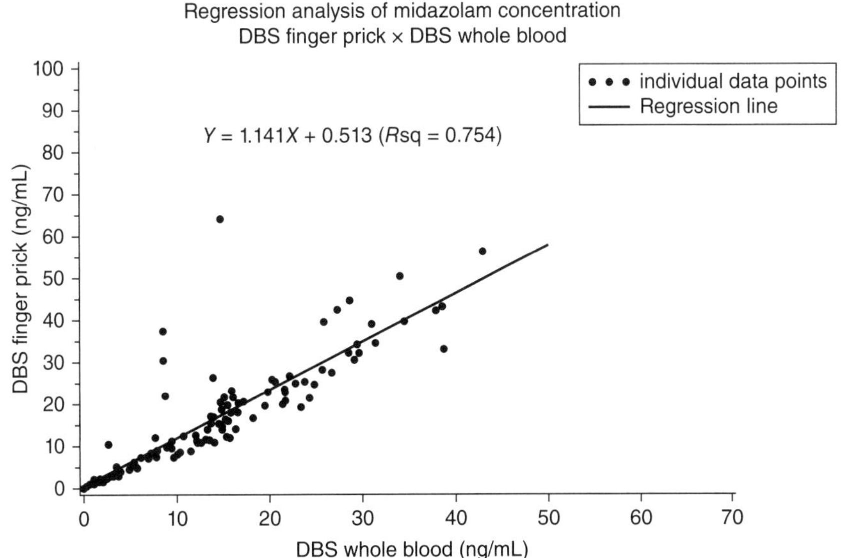

FIGURE 18.6 Regression analysis of MDZ concentrations (ng/mL) in DBS taken by finger puncture versus DBS taken by venipuncture. Reproduced with permission from Wiley.

TABLE 18.3 Results of Subject (Healthy Volunteers, $n = 12$) and Staff (Phlebotomists, $n = 8$) Evaluation of DBS Sampling

	Category	Selection	Number	Percentage
Subject	FP	None	2	16.7
		Moderate	1	8.3
		Mild	9	75.0
	FP tolerated per time point	1	2	16.7
		2	3	25.0
		3	7	58.3
	FP allowed per day	6–10	3	25.0
		11–15	9	75.0
	Participate again?	Yes	12	100
	Cannula (rating pain/discomfort)	None	7	58.3
		Mild	5	41.7
	Preferred sample type	Cannula	3	75.0
		FP	9	25.0
Staff	Spotting on paper	Extremely easy	2	25.0
		Easy	6	75.0
	Drying blood spot	Extremely easy	4	50.0
		Easy	3	37.5
		Neither easy or difficult	1	12.5
	Sample collection/processing	Extremely easy	1	12.5
		Easy	4	50.0
		Neither easy or difficult	1	12.5
		Not applicable	2	25.0
	FP: sufficient space for drying at bench	Yes	2	25.0
		No	4	50.0
		Not applicable	2	25.0
	Previous training/experience with DBS	No	7	87.5
		Yes	1	12.5
	Ease of blood collection by finger puncture	Easy	5	62.5
		Neither easy or difficult	3	37.5
	FP puncture needed for one time point	1	8	100

FP, finger puncture. Reproduced with permission from Wiley.

TABLE 18.4 Summary Table of PK Parameters and Secondary Objectives (Disc Size, Disc Number, and Storage Time of DBS Cards)

Secondary Objective	Measurement	PK Variable	Unit	N	Mean
Blood spot	1 × 6 mm disc	AUC_{inf}	h ng/mL	12	80.95
	(T = <1 week)	$AUC_{(0-12)}$	h ng/mL	12	73.15
Number of discs	1 × 3 mm disc	AUC_{inf}	h ng/mL	12	86.65
		$AUC_{(0-12)}$	h ng/mL	12	78.64
	2 × 3 mm disc	AUC_{inf}	h ng/mL	12	103.11
		$AUC_{(0-12)}$	h ng/mL	12	94.69
Storage time at RT	1 × 6 mm disc	AUC_{inf}	h ng/mL	12	88.19
(100% ISR)	(T = 2 weeks)	$AUC_{(0-12)}$	h ng/mL	12	80.26
	1 × 6 mm disc	AUC_{inf}	h ng/mL	12	63.91
	(T = 4 weeks)	$AUC_{(0-12)}$	h ng/mL	12	58.74
Limited sampling	Two-point model	$AUC_{(0-12)}$	h ng/mL	9	116.76
	Three-point model	$AUC_{(0-12)}$	h ng/mL	12	105.35

RT, room temperature; ISR, incurred sample reanalysis. Reproduced with permission from Wiley.

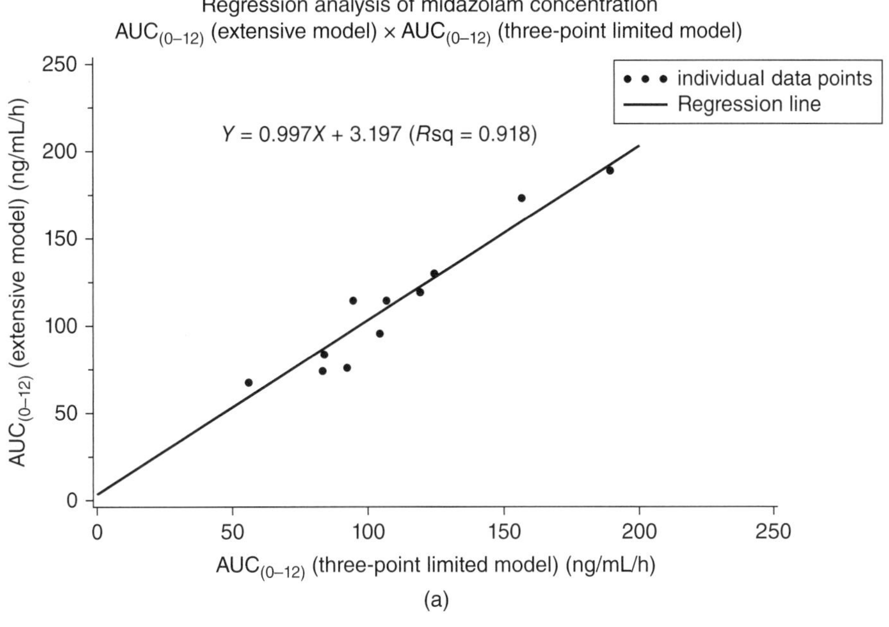

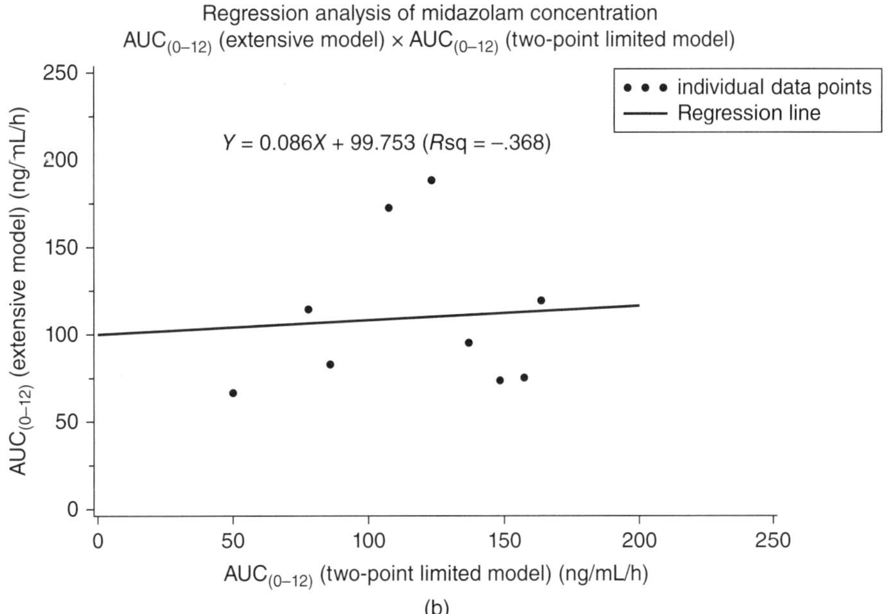

FIGURE 18.7 Regression analysis of MDZ concentrations (ng/mL) in DBS taken by venipuncture using a three-point limited model (a) and a two-point limited model (b). Reproduced with permission from Wiley.

be concluded from the table that DBS can be considered as a complementary method for genotyping studies although the DBS samples generally produced a slightly lower genomic DNA yield and slightly lower A260/A280 ratios.

18.4.8 Subject and Phlebotomist Satisfaction

To complete the clinical study, the phlebotomists (i.e., study nurses) and the subjects (healthy volunteers) were asked to complete a questionnaire regarding evaluation and comparison of the used sampling techniques. A summary of the results is given in Table 18.3. Generally, subjects preferred the cannula sampling method, but finger puncture for DBS sampling was also considered to be an acceptable sampling method for most subjects, the same when performed at multiple time points within 24 hours. The staff reported similar opinions on ease of use of DBS sampling versus venous sampling. The use of the DBS cards

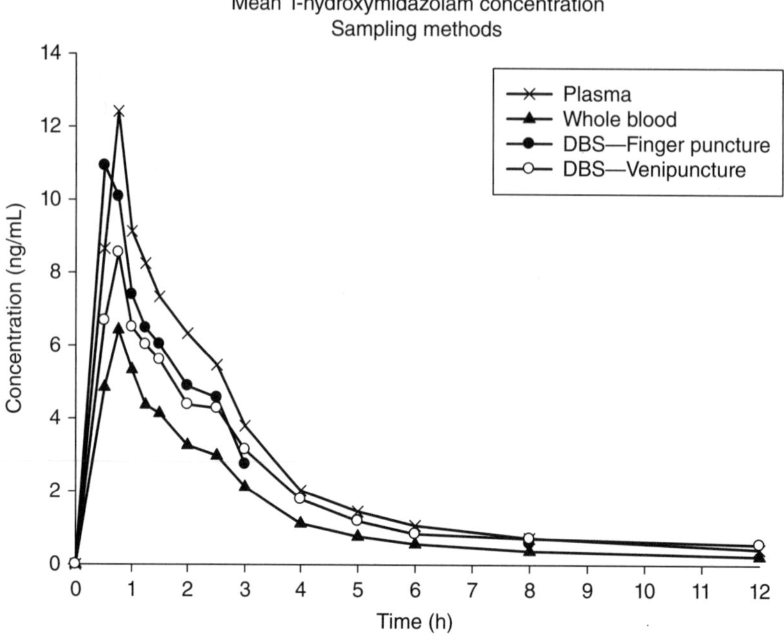

FIGURE 18.8 Mean plasma concentrations of 1-OH MDZ (ng/mL) versus time (h) for the applied sampling methods (DBS from finger puncture, DBS from venipuncture, whole blood, and plasma). Reproduced with permission from Wiley.

(identification, labeling) and spotting and treatment of bioanalysis (paper) cards were all considered (extremely) easy for all staff involved ($n = 8$) of which only one had little previous experience.

TABLE 18.5 Descriptive Statistics of the Genotyping Data for 12 Subjects

Allele	Genotype	DBS (*N*)	WB (*N*)
CYP2C19*2	*1/*1	7	7
	*1/*2	4	4
	*2/*2	1	1
CYP2C19*3	*1/*1	12	12
CYP2D6*3	*1/*1	11	11
	*1/*3	1	1
CYP2D6*4	*1/*1	6	6
	*1/*4	4	4
	*4/*4	1	1
CYP2D6*6	*1/*1	12	12
CYP3A4*1B	*1/*1	12	12
CYP3A4*2	*1/*1	12	12
CYP3A5*3C	*1/*1	10	10
	*1/*3C	2	2

DBS, dried blood spot; WB, whole blood. Reproduced with permission from Wiley.

18.5 SUMMARY

DBS sampling as an alternative to venous sampling was successfully implemented in our CPU and bioanalytical laboratory. Three fully validated LC-MS/MS assays for the determination of MDZ and 1-OH MDZ in plasma, WB, and DBS (6 mm discs) as well as two partially validated LC-MS/MS assays for the determination of MDZ and 1-OH MDZ in DBS (3 mm and 2 × 3 mm discs) were successfully applied to the analysis of the samples originating from this clinical trial. In this particular study design, a good correlation was observed in the PK profile between analysis in DBS_{finger}, DBS_{veno}, plasma, and WB for MDZ and 1-OH MDZ. It can therefore be concluded that DBS sampling proved to be suitable for PK analysis (including metabolites) and PG profiling. Finally, subjects and staff did not report critical objections.

ACKNOWLEDGMENTS

The authors thank Matt Barfield, Paul Abu-Rabie, and Neil Spooner of GSK (Ware, UK) for valuable discussions resulting in the setup of the clinical study. The authors also thank John Dinan of ID Biological Systems LLC (Greenville, UK) for kindly providing Ahlstrom 226 IDBS sheets for setting up the initial experiments.

REFERENCES

De Boer T, Wieling J, Meulman E, Reuvers M, Renkema G, den Daas I, van Iersel T, Wemer J, Chen L. Application of dried blood spot sampling combined with LC-MS/MS for genotyping and phenotyping of CYP450 enzymes in healthy volunteers. *Biomedical Chromatography* 2011;25:1112–1123.

Edelbroek PM, van der Heijden J, Stolk LML. Dried blood spot methods in therapeutic drug monitoring: methods, assays, and pitfalls. *Therapeutic Drug Monitoring* 2009;3:327–336.

Fan L, Lee J, Hall J, Tolentino EJ, Wu H, El-Shourbagy T. Implementing DBS methodology for the determination of Compound A in monkey blood: GLP method validation and investigation of the impact of blood spreading on performance. *Bioanalysis* 2011;3(11):1241–1252.

Food and Drug Administration. *Guidance for Industry: Bioanalytical Method Validation.* Rockville, MD: US Department of Health and Human Services, FDA, Center for Drug Evaluation and Research; 2001.

Hollegaard MV, Grove J, Thorsen P, Nørgaard-Pedersen B, Hougaard DM. High-throughput genotyping on archived dried blood spot samples. *Genetic Testing and Molecular Biomarkers* 2009;13(2):173–179.

Lad R. Validation of individual quantitative methods for determination of cytochrome P450 probe substrates in human dried blood spots with HPLC-MS/MS. *Bioanalysis* 2010;2(11):1849–1861.

Lee LL, Bertino JS, Nafziger AN. Limited sampling models for oral midazolam: midazolam plasma concentrations, not the ratio of 1-hydroxymidazolam to midazolam plasma concentrations, accurately predicts AUC as a biomarker of CYP3A activity. *Journal of Clinical Pharmacology* 2006;46:229–234.

Li W, Tse FLS. Dried blood spot sampling in combination with LC-MS/MS for quantitative analysis of small molecules. *Biomedical Chromatography* 2010;24:49–65.

Liang X, Li Y, Barfield M, Ji QC. Study of dried blood spots techniques for the determination of dextromethorphan and its metabolite dextrorphan in human whole blood by LC-MS/MS. *Journal of Chromatography B: Biomedical Sciences and Applications* 2009;877:799–806.

Shah VP, Midha KK, Findlay JWA, Hill HM, Hulse JD, McGilveray IJ, McKay G, Miller KJ, Patnaik RN, Powell ML, Tonelli A, Viswanathan CT, Yacobi A. Bioanalytical method validation—a revisit with a decade of progress. *Pharmaceutical Research* 2000;17(12):1551–1557.

Spooner N, Lad R, Barfield M. Dried blood spots as a sample collection technique for the determination of pharmacokinetics in clinical studies: considerations for the validation of a quantitative bioanalytical method. *Analytical Chemistry* 2009;81:1557–1563.

Viswanathan CT, Bansal S, Booth B, DeStefano AJ, Rose MJ, Sailstad J, Shah VP, Skelly JP, Swann PG, Weiner R. Workshop/Conference Report—quantitative bioanalytical methods validation and implementation: best practices for chromatographic and ligand binding assays. *Pharmaceutical Research* 2007;24(10):1962–1973.

Wijnen PA, Op den Buijsch RA, Cheung SC, van der Heijden J, Hoogtanders K, Stolk LM, van Dieijen-Visser MP, Neef C, Drent M, Bekers O. Genotyping with a dried blood spot method: a useful technique for application in pharmacogenetics. *Clinica Chimica Acta* 2007;388(1–2):189–191.

19

APPLICATION OF DRIED BLOOD SPOT SAMPLING IN CLINICAL PHARMACOLOGY TRIALS AND THERAPEUTIC DRUG MONITORING

Kenneth Kulmatycki, Wenkui Li, Xiaoying (Lucy) Xu, and Venkateswar Jarugula

19.1 INTRODUCTION

Use of dried blood spot (DBS) sampling in clinics in fact began in the early 1960s with the semiquantitative measurement of phenylalanine in newborns to screen for phenylketonuria, an autosomal recessive genetic disorder, in which a mutated phenylalanine hydroxylase prevents the breakdown of phenylalanine (Guthrie and Susi, 1963). Other disorders such as biotinidase deficiency, congenital hypothyroidism, cystic fibrosis, and tyrosinemia type 1 are examples in which blood spot sampling coupled with an analytical technique has been employed to screen newborns for disease (Mei et al., 2001; la Marcia et al., 2008). Collection of a few drops of blood from a heel or toe-stick has been a useful method in the clinic for screening of inheritable disorders in infants.

In recent years, analysis of DBSs to monitor disease has gained prominence for patients located in rural areas, especially in those underdeveloped and resource-limited areas. DBS sampling has become very important in providing patient care, as inadequate monitoring of human immunodeficiency virus (HIV)-1 viral load has been reported to result in treatment failure and disease progression (Johannessen et al., 2009a, 2009b). In addition, use of DBS sampling with a corresponding analytical method has been reported to aid in early infant HIV-1 diagnosis and reduce mother-to-child HIV-1 transmission (Sherman et al., 2005; Mirochnick et al., 2009; Cook et al., 2011). Methods for analyzing antimalarial drugs from DBSs to monitor blood drug concentration such as for mefloquine have also been developed for monitoring antimalarial drug levels in patients in developing countries (Bergqvist et al., 1988, Burhenne et al., 2008).

DBS sampling coupled with liquid chromatography-mass spectrometry/mass spectrometry (LC-MS/MS) or other analytical techniques is becoming an emerging method in support of clinical pharmacology studies and therapeutic drug monitoring due to the ease of sample collection, versatility, application to special populations, and convenience of sample shipping. In contrast to the many successes in using DBS samples for the analysis of markers of disease, there have been limited cases in the public domain for the evaluation of drug exposure using DBS sampling in clinical pharmacology trials and therapeutic drug monitoring. This chapter is to capture some technical considerations when implementing blood spot sampling for those studies.

19.2 ADVANTAGES AND DISADVANTAGES OF DRIED BLOOD SPOT SAMPLING IN CLINICAL PHARMACOLOGY TRIALS AND THERAPEUTIC DRUG MONITORING

In DBS sample collection, although the first two drops of capillary blood are generally discarded due to the possible contamination by interstitial fluid, the total blood volume required is low (~100 μL) for each collection. Thus, DBSs are ideal for pediatric studies and critically ill patients. The low blood volume requirement in combination with the possibility that blood sampling can be conducted in a nonclinic setting (e.g., at patient's residence) at low consumable costs with a minimum training for the patients or caregivers makes this method of sample collection very desirable (Williams and McDade, 2009; Spooner et al., 2010). Typical bioanalytical equipment such as centrifuges and freezers are

Dried Blood Spots: Applications and Techniques, First Edition. Edited by Wenkui Li and Mike S. Lee.

not required as the collection cards can be stored and shipped at room temperature. Accordingly, samples are, in general, not required to be packed with dry ice or stored in freezers (van Amsterdam and Waldrop, 2010). In particular, some commercial DBS sampling cards (e.g., DMPK A card and DMPK B card from GE Whatman) are coated with surfactants that inactivate pathogens and enzymes, denature proteins, and stabilize compounds providing samples that are not a biohazard.

Disadvantages of using this collection method are in general compound dependent. Apparently, volatile or air-sensitive drugs cannot be used for blood spot collection due to instability. Light-sensitive compounds may also not be appropriate for this methodology as special handling of the cards may be required. In addition, a lower limit of quantitation (LLOQ) in the low-to-mid picogram range presently is not attainable using DBS-LC-MS/MS approach considering the very limited sample volume. Depending on the blood-to-plasma distribution, pharmacokinetic (PK) exposure such as observed maximum concentration (C_{max}) and area under the curve (AUC) may not be similar between matrices (e.g., blood/DBS vs. plasma) due to other factors such as the hematocrit. Although drug metabolites can be measured using DBS sampling in drug development, typically plasma is the preferred matrix to conduct metabolite profiling. Presently, bioanalytical measurement of analytes from blood spot cards is time consuming as automation for sample analysis is limited. The same types of bioanalytical detection principles as for traditional plasma and aqueous whole blood can be applied for the bioanalysis of analytes in DBSs (EMA, 2012; CDER, 2013). A comprehensive validation is required prior to using DBS in regulated studies evaluating at a minimum sample storage and handling temperature, homogeneity of spotting, hematocrit, stability, carryover, and reproducibility including incurred sample repeats (CDER, 2013). During development correlative studies with traditional sampling should also be performed and sponsors are encouraged to obtain feedback early in development from health authority (CDER, 2013).

DBS sampling typically involves a simple finger, toe, or heel stick to collect a few drops of capillary blood onto filter paper imbedded on a labeled card (Spooner et al., 2010). However, as capillary blood is collected, any potential differences in drug concentration between capillary and venous blood should be considered. Differences in drug concentration comparing capillary to venous blood is generally minimal despite the fact that capillary blood may be more reflective of arterial blood as it has not drained through the tissue beds (Chiou, 1989; Emmons and Rowland, 2010). Any potential differences in drug concentration between capillary and venous blood may be due to, for example, physiochemical properties of the drug and would likely occur at early sampling time points taken to assess drug concentration prior to distribution equilibrium and may not be relevant when distribution equilibrium has been reached. Thus, collection of capillary blood can generally be considered in most cases to be similar to that of venous collection. However, exceptions exist as neonate capillary blood has a higher hematocrit compared to venous blood, which can result in a lower serum/plasma fraction for drugs that bind highly to red blood cells; also the sample site (e.g., finger tip vs. toe tip) may contribute to differences in measured drug concentration (Chiou, 1989; Kayiran et al., 2003). For successful collection of blood spots, various sampling details need to be considered in order to produce useable samples for quantitation of drug concentration in the clinic.

19.3 DRIED BLOOD SPOT COLLECTION IN CLINICAL PHARMACOLOGY TRIALS AND THERAPEUTIC DRUG MONITORING

Although DBS sample collection is a simple and convenient method, a few important points should be considered for safe collection and validity of the collected samples. All human blood specimens should be considered potentially infectious and handled as though transmission of disease may occur. Thus, precautions should be taken to protect all persons from accidental exposure to infectious agents when collecting blood drops. Once blood drops are applied to the card, the sample is generally not regarded as a biohazard risk, and thus, may be mailed to the bioanalytical laboratory for analysis (Edelbroek et al., 2009). As mentioned, the first two drops of blood are discarded due to the possible contamination from interstitial fluid. Typically, blood spot collection cards have four circles with only one drop of blood added from the subject's finger, heel, or toe onto each circle on the card without touching the filter paper. This procedure results in a total of four DBSs on each card per time point. Immediately after all four blood spots are collected, the card should be placed either in a drying rack or on a flat surface and allowed to dry for 2 hours at room temperature away from direct sunlight. After a drying period (e.g., 2 hours), the DBS cards are then put into a zip-lock bag that contains packets of desiccants and a humidity indicator. The cards can then be shipped to the laboratory for analysis either by a clinical site or mailed by the patient from their residence. Collection of blood spots is a very straightforward procedure with only a few important considerations and therefore minimal training is required for collection, shipping, and handling to provide samples suitable for analysis of drug levels.

19.4 PHARMACOKINETIC–PHARMACODYNAMIC CONSIDERATIONS OF DRIED BLOOD SPOT SAMPLING IN CLINICAL PHARMACOLOGY TRIALS AND THERAPEUTIC DRUG MONITORING

Accurate measurement of systemic drug and biomarker concentration is essential to determine pharmacokinetic and pharmacodynamic (PK/PD) relationships of drugs. When examining drug PK/PD relationships, generally total drug concentration is measured as a surrogate of unbound drug concentration (Benet and Zia-Amirhosseini, 1995). However, a fundamental premise is that pharmacological response to drug is determined by the unbound drug (free fraction) (Abdel-Rahman and Kauffman, 2004). In most cases, equilibrium exists between unbound circulating drug concentration and unbound tissue concentrations at steady state. Thus, measuring total circulating drug concentrations can be considered a surrogate to the free drug concentrations at the receptor site. The assumption for this to be valid is that drug plasma-free fraction, hematocrit (proportion of blood volume that is occupied by red blood cells), and blood cell partitioning remain constant. Understanding of this role of distribution to red blood cells is essential to assess the relevance of using DBSs as a matrix for measuring systemic drug levels. Historically, plasma has been the matrix of choice for the measurement of systemic drug concentrations in clinical PK/PD studies and therapeutic drug monitoring partially due to the difficulties in handling and analyzing aqueous blood samples although exceptions do exist such as for tacrolimus (Minematsu et al., 2004). As blood is collected for DBSs, distribution of drug into red blood cells should be considered as correction factors that may need to be employed when comparing drug levels measured from plasma to DBSs.

19.4.1 Blood Plasma Distribution

In general, understanding the significance of drug partitioning into red blood cells is critical in collecting PK samples. Distribution of drugs into red blood cells is usually quite rapid as shown for cyclosporine but may also be very slow, as for chlorthalidone which takes several hours (Fleuren et al., 1977; Atkinson et al., 1984; Kawai et al., 1994). For certain molecules (e.g., cyclosporine and tacrolimus), blood cell binding can be concentration and temperature dependent. This can make interpretation of data challenging (Akhlaghi and Trull, 2002; Staatz and Tett, 2004). When binding of compounds to red blood cells is high as in the case of cyclosporine, blood may be considered a more suitable matrix for assessment of systemic drug exposure. In fact, reports show that it may be preferable to measure systemic concentrations using blood spot collection combined with LC-MS/MS for monitoring of cyclosporine and tacrolimus (Keevil et al., 2002, 2005).

In general, hydrophilic, acidic drugs and therapeutic proteins have a high affinity to plasma proteins. As a consequence, partitioning of those molecules into red blood cells either does not occur or is minimal, and thus concerns of a possible inaccurate measurement of these drugs from DBSs would involve the probable large variations in plasma protein binding of those molecules. In contrast, basic drugs ($pK_a > 7$) are, in general, predominantly ionized at physiologic pH and bind well to acidic phospholipids that can bind to a greater extent to red blood cells (Rowland and Emmons, 2010). For these basic drugs, several factors such as free fraction of drug in the plasma, hematocrit, and red blood cell partitioning should be considered. The blood-to-plasma ratio should be evaluated *in vitro* to determine whether plasma protein binding or affinity to red blood cells is an important factor. The blood-to-plasma ratio generally ranges from a lower limit of around 0.55 (low blood cell uptake) to a higher limit of ≥2 (high blood cell uptake) (Emmons and Rowland, 2010). It should be noted that drugs may also bind to white blood cells and platelets in the blood but their fractional volume is low, thus playing a minor role in drug partitioning. However, fluctuations in the hematocrit can influence drug concentration measured from DBSs, which may need to be taken into account in order to better understand the concentration of drug measured using the DBSs versus conventional plasma.

19.4.2 Impact of Hematocrit

Hematocrit is the abundance of red blood cells in blood. In general hematocrit remains fairly constant in adults [male 47% (±6); female 40% (±6)] but can change with disease (<25% with severe anemia) and life style (e.g., mountaineers) and is higher in newborns (52%) (Cash and Sears, 1989; Adamson and Longo, 2012). Thus, changes in hematocrit may need to be taken into account for interpretation of drug concentrations when using DBS sampling versus plasma sample collection for drugs that are highly bound to red blood cells. In addition, coadministration of drugs that can change the expression of red blood cells needs to be considered (Ridley et al., 1994).

For drugs with low variability in free fraction in plasma and partitioning into erythrocytes, if hematocrit remains constant, either plasma or blood/DBS can be used as a matrix to quantify drug concentrations. To use DBS as a surrogate of unbound drug concentration it is essential to understand factors that may cause variability in plasma protein binding for drugs with a blood–plasma ratio near the lower limit of 0.55 (minimal erythrocyte uptake). The variability in erythrocyte affinity becomes very important for drugs that have a large blood-to-plasma ratio (≥2.0). Ignoring the impact of hematocrit in this case can lead to serious errors in measuring drug concentration. Thus, correction factors which would be determined from a bridging study may be required if comparing drug concentrations from DBSs to plasma samples for drugs that have large blood-to-plasma ratios.

19.4.3 Bridging between Dried Blood Spots and Plasma Samples

Various correction factors need to be considered to bridge drug concentrations measured from DBSs to that obtained from plasma samples if plasma has been employed as the matrix for the PK assessment of a drug of interest. Hematocrit is the most important parameter to be considered, especially when drugs bind extensively to red blood cells (i.e., red blood cell/plasma ratio ≥ 2) (O'Broin et al., 1995; Emmons and Rowland, 2010). For example, hematocrit was used to correct 25-hydroxy vitamin D_3 concentrations measured from DBSs using the following formula (Eyles et al., 2009).

$$\text{DBS}_{(25\text{OHD3})\text{nM}} \times 1/(1 - \text{Hematocrit fraction}) = \text{Plasma}_{(25\text{OHD3})\text{nM}}$$

Although this formula yielded an acceptable adjustment of 25-hydroxy vitamin D_3 concentrations measured from DBSs compared to plasma it does not include the factor of binding of the analyte of interest to red blood cells. Taking this into account, the f_{bc}, the fraction of drug bound to red blood cells, was included in the following formula proposed by Li and Tse (2010):

$$(\text{DBS}_{[\text{analyte}]}/[1 - \text{Hematocrit}]) \times (1 - f_{bc}) = \text{Plasma}_{(\text{analyte})}$$

The f_{bc} can be obtained from *in vitro* studies. If the fraction of drug bound to red blood cells is not constant, then additional factors should be included to adjust for concentration-dependent red blood cell binding.

Another example of using correction factors for comparing drug concentrations from DBSs to plasma samples is measurement of nevirapine and efavirenz systemic concentrations (Kromdijk et al., 2012). This study was conducted to compare concentrations of nevirapine and efavirenz from DBSs to plasma sampling in HIV-1-infected male subjects. Due to concerns that changes in hematocrit may alter interpretation of drug levels, both the mean hematocrit for men and patient-specific hematocrit were used to correct for any differences in drug concentration between the DBS and plasma analysis. The following formula was used to adjust for hematocrit as well as compound-specific protein binding:

$$[\text{DBS}_{[\text{analyte}]}/(1 - \text{Hematocrit})] \times f_{bpp} = \text{Plasma}_{[\text{analyte}]}$$

The f_{bpp} (fraction bound to plasma proteins) used for calculations was 0.6 and 0.995 for nevirapine and efavirenz, respectively, and the hematocrit value used was 0.45 as the outpatient population consisted of males enrolled in this study. The formula used in the study was adapted from Li and Tse (2010), which showed differences in DBS and plasma concentrations could be explained by applying compound-specific protein binding and patient-specific hematocrit.

Correction factors can be applied for measurement of drugs from DBSs allowing for comparison between the two matrices when required. However, dried plasma spots may be used to measure systemic drug concentration to avoid using correction factors.

19.4.4 Dried Plasma Spots as an Alternative to Dried Blood Spots

Dried plasma spots have also been suggested as an alternative matrix to aqueous plasma or blood samples. Presently centrifugation is required to isolate plasma to produce dried plasma spots. However, once plasma is applied onto the filter paper on the collection card, shipping and handling are similar to that for DBS samples. As plasma generally is light in color, one apparent drawback is that after application of the plasma onto the filter paper, the plasma spot can be difficult to visualize. Therefore, a color indicating technique has been suggested (Barfield and Wheller, 2011). Although centrifugation is a required step, by using dried plasma spot sampling, any potential issues related to red blood cell binding of the analyte of interest and changes in the hematocrit requiring correction factors for DBS can be avoided (Flowers and Cook, 1999; Knuchel et al., 2006; Brookes et al., 2010; Kolocouri et al., 2010; Barfield and Wheller, 2011).

19.5 DRIED BLOOD SPOT SAMPLING IN THERAPEUTIC DRUG MONITORING

19.5.1 Drug Monitoring and Dried Blood Spots

In therapeutic drug monitoring, doses of drugs are selected based on systemic concentration and the patient's clinical condition to provide efficacious drug therapy in the absence of toxicity (Burton et al., 2006). To adjust drug dosages based on systemic exposure, the assumption is that there is a strong relationship between drug concentration and the desired pharmacological effect (i.e., PK/PD relationship) and/or toxicity. In reality, only a small group of drugs are routinely monitored therapeutically due to their narrow therapeutic index (i.e., small difference between minimum effective and minimum toxic concentration) with examples shown in Table 19.1. As only a small blood volume is to be collected in DBS samples, therefore, the related assay should have sufficient sensitivity to measure the analytes of interest (e.g., digoxin and digitoxin) in the DBS samples.

Due to the difficulties in obtaining sufficient blood samples with venipuncture from special populations such as neonates and patients with venous damage or critically ill, DBS sampling is an appealing alternative for therapeutic drug monitoring (Neef et al., 2008). In addition, DBS sampling via a simple finger stick would be preferable when sampling needs to be conducted frequently to determine whether the drug concentrations are at steady state. This is especially necessary for patients with hepatic or renal impairment, for drugs posing concerns regarding potential drug–drug interactions, or when trough blood samples are difficult to be obtained due to dosing early in the morning or late in the evening. More interestingly, as collection of DBS samples via finger, toe, or heel pricks is convenient and requires much less training

TABLE 19.1 Concentration Ranges of Some Typical Drugs in Which Therapeutic Drug Monitoring Is Conducted

Class	Drug	Typical Plasma Concentration Ranges	
Antiepileptics	Ethosuximide	40–100 μg/mL	
	Phenobarbital	10–30 μg/mL	
	Phenytoin	10–20 μg/mL	
	Primidone	5–12 μg/mL	
Antiarrhythmics	Digoxin	0.8–2 ng/mL	
	Lidocaine	1–5 μg/mL	
	Procainamide	4–8 μg/mL	
	Quinidine	1–4 μg/mL	
Antibiotics	Amikacin	Peak	20–30 μg/mL
		Trough	<10 μg/mL
	Gentamicin/tobramycin	Peak	4–8 μg/mL
		Trough	<2 μg/mL
Antimanics	Lithium	0.6–0.8 mEq/L	
Antiretrovirals[a]	Atazanavir	Trough	100/150 ng/mL
	Efavirenz	Trough	1000 ng/mL
	Lopinavir/ritonavir	Trough	1000 ng/mL
	Nevirapine	Trough	3400 ng/mL
Immunosuppressants[b]	Cyclosporine	100–400 ng/mL	
	Everolimus	Trough	5–15 ng/mL
	Tacrolimus	Trough	5–20 ng/mL
	Sirolimus	Trough	12–20/16–24 ng/mL

[a]Patients with wild-type HIV.
[b]Whole blood monitored (summarized from Winters, 1994; Back et al., 2006; Burton et al., 2006; cyclosporine prescribing information, 2009; everolimus prescribing information, 2010; sirolimus prescribing information, 2011; tacrolimus prescribing information, 2012).

and/or practice than for the other matrices such as venous blood or plasma. DBS samples may be collected either in the clinic by clinic staff or at the patients' residence by patients or their guardians either at scheduled times or when a drug-related adverse effect occurs. Monitoring drug concentration in these DBS samples allows for a better understanding of the drug PK/PD or PK/safety relationship for disease management (Williams and McDade, 2009). Furthermore, by having the opportunity to take several samples at different time points rather than one sample at a predefined time point, intra-subject variability in systemic concentration in relation to drug response and toxicity can be accurately assessed. However, when patients collect blood spot samples, there should be an accurate method in place to document and communicate when the samples are collected. Bar code reading devices can be used for this purpose. Use of the correct DBS card for each time point and the exact date and time window of blood sample collection, for example, can be documented by taking a picture of the DBS card using a cell phone loaded with a bar code reader application before and after sampling.

Bioanalytical methods using DBS collection have been used for monitoring the concentration of antiretroviral drugs such as amprenavir, atazanavir, efavirenz, indinavir, lopinavir, nelfinavir, nevirapine, ritonavir, saquinavir, and sarunavir (ter Heine et al., 2009). Other examples of where blood spot collection has been used are when monitoring lamotrigine levels during pregnancy, metformin levels in diabetics, and antimalarial treatment in developing countries (Rowell et al., 1988; de Haan et al., 2004; AbuRuz et al., 2006).

19.5.2 Pediatric Drug Monitoring and Dried Blood Spots

It has been reported that up to 67% of drugs prescribed to children and 92% to critically ill neonates have been used in an unlicensed or off-label manner (t'Jong et al., 2000, 2002; Conroy et al., 2000). On the other hand, due to growth and maturation, marked changes may occur throughout the growth period, resulting in age-specific differences in PK that emphasize the need to understand pediatric dosing (Kearns et al., 2003). Some important technical aspects for therapeutic drug monitoring for pediatrics are number of samples (peak or trough), conditions of sampling, and volume of samples to quantify parent drug (Patel et al., 2010). With this regard, application of DBS sampling is of great value.

To note, of particular importance is the limitation that only a very small volume of blood can be collected from neonates, for example, 1.2 mL every 4 weeks from newborns weighing 500 g; thus, due to this limitation DBS may be the preferred method of blood collection (EMA, 2001, 2006, 2009). In addition, samples can be collected by the guardian with minimal training, and ease of storage and shipment makes this method ideal for monitoring pharmacotherapy in pediatric patients. Due to these advantages, blood spot collection has been implemented to monitor drug concentrations, for example, anti-HIV-1 (Koal et al., 2005), immunosuppressant (Hoogtanders et al., 2007; Wilhelm et al., 2009) and antiepileptic drugs (la Marcia et al., 2009), which potentially can be used for pediatric drug monitoring.

Correction factors for comparison of drug measurement from DBS to plasma may need to be employed due to the higher hematocrit in young infants. Hematocrit is one of the most important physiological parameters to consider when using blood spot collection for infants to measure systemic drug concentrations. As the hematocrit changes from birth to childhood (ranging from 36% to 52%), this parameter needs to be taken into account for drug binding to red blood cells. Since blood viscosity increases due to increased hematocrit, diffusion properties of the blood drop applied to the filter paper may be affected, affecting the drug measurement. However, hematocrit tends to be stable after 2 years of age and whether a correction factor would be required will need to be evaluated for the specific drug (Pandya et al., 2011). Similar to the pediatric patient population DBS collection can be used to monitor systemic drug levels in transplant patients for the prevention of organ rejection and drug toxicity.

19.5.3 Transplantation Drug Monitoring and Dried Blood Spots

For some immunosuppressants, such as everolimus, with narrow therapeutic index and high oral variability, minimum concentration (C_{min}) is commonly used as an indicator of the possible acute rejection episodes and toxicity. Therefore, therapeutic drug monitoring is desirable (Kirchner et al., 2004). In a study to determine whether blood spot sampling would be useful to monitor everolimus systemic levels, venous blood samples at various time points up to 5 hours were collected by laboratory staff via venipuncture to prepare the whole blood and DBS samples. In the meanwhile, DBS samples were collected via finger stick by the patient. The concentration–time profiles of everolimus for the three types of samples were compared. As shown in Figure 19.1, the profiles were all similar (van der Heijden et al., 2009).

DBS collection has been shown to be a promising alternative to venipuncture for monitoring systemic concentrations of tacrolimus (Hoogtanders et al., 2007). In renal transplant patients, the hematocrit and albumin concentrations in blood are generally low immediately after transplantation surgery and can increase as the patient recovers. These can impact the measured drug concentration in blood. For example, a

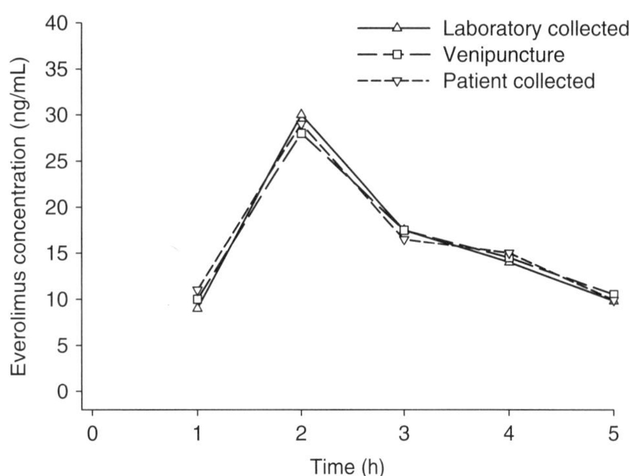

FIGURE 19.1 Everolimus concentration–time profiles from the analysis of laboratory-collected blood dried spot (DBS), venipuncture, and patient-collected DBS (Adapted from van der Heijden et al. (2009) with permission from Elsevier.

correlation has been reported between tacrolimus clearance and both hematocrit and albumin during the first 12 weeks posttransplantation. The low hematocrit values resulted in a decreased total drug concentration measured in whole blood (Undre and Schäfer, 1998). Thus, it would appear that total drug clearance has increased although, in fact, the clearance of unbound drug remains unchanged due to the low hematocrit (Staatz and Tett, 2004). Therefore, when using aqueous whole blood or DBSs to assess tacrolimus systemic concentrations within at least 12 weeks of transplantation, change in hematocrit and/or albumin blood concentrations must be included in the interpretation of the measured analyte concentrations (Undre and Schäfer, 1998; Mix et al., 2003).

Area under the curve may be used to estimate drug exposure for monitoring tacrolimus levels as reported by Cheung et al. (2008) when using DBS samples collected by patients. This limited sampling strategy of collecting six time points was believed to better predict drug exposure than a single collection at a designated time point and potentially performed by patients at their residence. The self-collection of six samples within a 24-hour time frame was deemed acceptable by patients in another recent study using acetaminophen as a model compound (Spooner et al., 2010).

19.5.4 DBS Biomarker Measurement in Drug Monitoring

The combination of systemic drug level and pharmacodynamic biomarker measurement using DBS will allow for a better understanding of the PK/PD relationship to tailor dosing regimens in therapeutic drug monitoring when needed. For example, inflammatory markers (e.g., interleukin-1β, tissue necrosis factor-α and -β) have been analyzed using blood spot sampling (Skogstrand et al., 2005). The main challenge, other than proper sample collection, to employing blood spot collection for measurement of biomarkers is assay development.

19.6 EXAMPLES OF DRIED BLOOD SPOT SAMPLING IN PEDIATRIC AND ADULT CLINICAL PHARMACOLOGY TRIALS

19.6.1 DBS Sampling for Pediatric Studies

DBS sampling can play a very important role in measuring drug levels in pediatric patients enrolled in clinical pharmacology studies especially for neonates. For example, a population PK study was conducted in which metronidazole dose recommendations for preterm neonates were determined (Suyagh et al., 2011). Prior to this study little was known regarding doses of metronidazole to use in preterm neonates. From the analysis of metronidazole from DBSs a population PK model was developed and dosing regimen recommended preventing administration of metronidazole doses higher and more often than needed to avoid adverse events. Collection of blood spots will be used to monitor drug formulation excipients kinetics in neonates to address concerns of toxicity (Yakkundi and Turner, 2011). Larger population-based studies are also reported for pediatric patients from whom DBS collection was used for biomarker quantification (McDade et al., 2007). Collection of DBS samples in these community-based settings was advantageous due to the ease of sample collection, storage, and transport to the laboratories for analysis.

19.6.2 DBS Sampling for Clinical Pharmacology Studies with Adults

DBS sampling is becoming an important tool in support of clinical pharmacology studies, in particular, when clinical settings are limited. For example, a drug–drug interaction study between chloroquine and chlorpheniramine was conducted in Nigeria using blood spot collection (Okonkwo et al., 1999). In the current practice, a thorough understanding of the feasibility of using DBSs for quantitation of drugs in clinical studies should be known prior to initiating this blood collection method in clinical pharmacology studies.

In a recent pilot clinical trial, we assessed the feasibility of collecting blood spots at selected time points via finger stick by healthy volunteers themselves after oral administration of 500 mg acetaminophen. Acetaminophen and its glucuronide and sulfate metabolites were quantified by LC-MS/MS. Prior to study initiation; a laboratory manual was supplied to clinical staff and the subjects. Procedures required for blood spot collection were provided to train site staff and subjects. To obtain results from at least 12 subjects, a total of 15 adult healthy volunteers (18–65 years) were enrolled in the study. The number of subjects enrolled was based on the bioequivalence guidance of requiring at least 12 subjects to assess PK (CDER, 2003; EMA, 2010), which is also in line with several publications with acetaminophen PK obtained from 8 to 12 subjects (Rawlins et al., 1977; Yin et al., 2001). The enrolled subjects signed the written informed consent. They confirmed to have no concurrent medical conditions or not taking any medications that would alter absorption, distribution, metabolism, or excretion of acetaminophen or pose an unacceptable risk. In addition, no other investigational drugs were to be taken within 30 days of dosing up to 5 half-lives at screening. Subjects were advised not to take any acetaminophen containing products or consume alcoholic beverages at least 48 hours prior to receiving a single 500 mg oral dose of acetaminophen.

To assess the utility of LC-MS/MS for quantitation of acetaminophen and its glucuronide and sulfate metabolites, blood spot samples were collected up to 6 hours post-dose. The subjects were fasted overnight (at least 10 hours) prior to receiving their acetaminophen dose and continued to fast for up to 3 hours post-dose. Blood spot samples were collected at pre-dose, 0.5, 1, 4, 6, and 8 hours post-dose by self-blood sampling onto 226 DBS cards that were bar code labeled with patient information and collection time point. After drying, the cards were packed for shipping to the bioanalytical laboratory. A total of 15 subjects completed the study.

The acetaminophen and metabolite concentration–time profiles are shown in Figure 19.2 and PK parameters are listed in Table 19.2. DBS concentrations of acetaminophen increased rapidly following oral administration with median peak concentrations at approximately 0.5 hours post-dose.

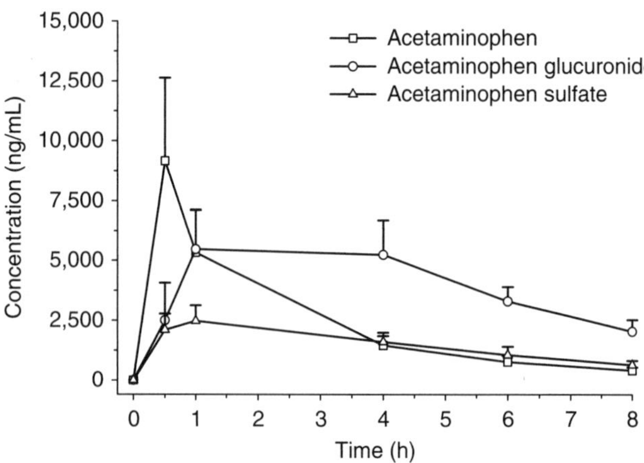

FIGURE 19.2 Concentration–time profiles of acetaminophen, acetaminophen glucuronide, and acetaminophen sulfate metabolites measured from dried blood spots following a single oral 500 mg dose of acetaminophen to healthy adult subjects (Adapted from Li et al. (2012) with permission from Future Science Ltd.

TABLE 19.2 Arithmetic and Mean (SD) of the Pharmacokinetic Parameters of Acetaminophen, Acetaminophen Glucuronide, and Acetaminophen Sulfate in Healthy Adult Subjects

Analyte[a,b]	AUC0–8h (ng·h/mL)	C_{max} (ng/mL)	T_{max} (h)	$T_{1/2}$ (h)
Acetaminophen	20,958 ± 5991	9236 ± 3337	0.58 (0.5;1.1)	2.11 ± 0.54
Acetaminophen glucuronide	32,590 ± 6809	5847 ± 1452	1.22 (1.4;12)	2.50 ± 0.31
Acetaminophen sulfate	12,431 ± 3052	2602 ± 593	1.13 (0.7;1.4)	2.94 ± 0.52

Source: Adapted from Li et al. (2012) with permission from Future Science Ltd.
[a]Values shown as mean (standard deviation), T_{max} shown as median and range.
[b]AUC0–8h, C_{max}, T_{max} ($n = 15$), $T_{1/2}$ acetaminophen ($n = 15$), acetaminophen glucuronide ($n = 6$), acetaminophen sulfate ($n = 9$).

The peak concentrations of the glucuronide metabolite were lower than the parent. However, the mean AUC0–8h was greater than acetaminophen as shown in Table 19.2. The sulfate metabolite was similar to the parent with median peak concentration reached around 1 hour and AUC substantially less than both the parent and the glucuronide metabolite. For all analytes, blood concentrations decreased rapidly with mean terminal elimination half-life of 2–3 hours. The PK parameters of acetaminophen measured in this study are similar to literature values and the ratios of parent-to-metabolite are also in line with those previously reported with the sulfate metabolite having lower exposure than acetaminophen and glucuronide metabolite having higher exposure than the parent (Rawlins et al., 1977; Forrest et al., 1982; Yin et al., 2001). This pilot study shows that collection of blood spots for drug and metabolite concentration assessment is feasible and can be performed either in the clinic or by patients themselves. Additional recent clinical trials using blood spot collection are shown in Table 19.3.

It is worth to note that DBS sampling is not necessarily suitable for all clinical pharmacology studies. From a bioanalytical perspective, in addition to method validation activities that are captured in the FDA (2013) and EMA (2011) guidance for conventional PK sample matrices (blood, plasma, serum, etc.), necessary evaluations need to be conducted to ensure the bioanalytical data quality and integrity. Those evaluations include, but are not limited to, impact of hematocrit, sample size, and punching location; extraction efficiency; internal standard addition; dilution integrity; and analyte–cellulose interaction plus other activities. From a PK perspective, only limited DBS samples (e.g., 6 time points) can be collected within 24 hours; according to a recent publication, this dramatically limits the number of samples that can be taken for blood spot collection (Spooner et al., 2010) for accurate assessment of PK parameters, especially the bioavailability and bioequivalence studies. Typically up to 16 blood samples may need to be collected within a 24-hour time period to characterize the concentration–time profile for assessment of bioequivalence; thus blood spot collection method may not be suitable for bioavailability and bioequivalence studies. However, a relative bioavailability study has been conducted in a rural area in a developing country in which blood spot samples were collected (Ogbonna et al., 2007). The number of samples collected each day was limited with DBS samples collected over many days due to the long half-lives of the drugs evaluated which made blood spot sampling practical for this bioavailability study.

DBS sampling can be very useful for assessing polymorphic metabolizing enzyme activity in clinical studies. Daali et al. (2012) used flurbiprofen, a substrate of cytochrome P450 (CYP)2C9, as a probe drug in a recent study to determine whether collection of blood spots from a single time point from healthy volunteers can be used to assess CYP2C9 activity by using flurbiprofen. This study was conducted in three sequential periods with subjects ($n = 10$) administered a single 50 mg oral dose of flurbiprofen alone; then after at least a 1-week washout period, coadministered a single 50 mg oral dose of flurbiprofen and 400 mg fluconazole (an inhibitor of CYP2C9); then after at least a 2-week washout period, first administered 600 mg rifampicin (an inducer of CYP2C9) orally once daily for 4 days, then coadministered a single 50 mg dose of flurbiprofen with 600 mg rifampicin on the next day. For each period, urine was collected for up to 8 hours post-dose, sparse blood collection was performed at 0 (pre-dose) and 0.5, 1, 2, 4, 6, and 8 hours post-dose for both capillary whole blood via finger stick on Whatman 903 filter paper cards and venous blood for plasma. Flurbiprofen and its metabolite 4′-hydroxyflurbiprofen were measured using a validated assay for DBSs, plasma, and urine (Zgheib et al., 2007; Déglon et al., 2011).

Comparing the measured analyte concentrations in plasma to DBSs, the C_{max} and AUC were approximately 50% lower in blood spot samples than in plasma. This result was thought to be due to the presence of red blood cells in the DBS diluting the sample, but analyte blood/plasma distribution ratio and hematocrit should have contributed to this significant difference (Emmons and Rowland, 2010). Metabolic ratios in urine showed a significant decrease, but metabolite/parent (4′′-hydroxyflurbiprofen:flurbiprofen) ratios increased when comparing flurbiprofen administration alone to coadministration with fluconazole or rifampicin, respectively. These results are in line with what was observed for the analysis of flurbiprofen and its metabolite in DBS and plasma samples. As shown in Figure 19.3, mean metabolic ratios against both the total and unconjugated parent in both plasma and blood spots are similar statistically at the 2-hour time

TABLE 19.3 Examples of Clinical Trials Completed or in Progress Using Blood Spot Collection

Brief Title[a]	Sponsor	Dried Blood Spot (DBS) Implementation	*N*	Phase	Status
Developing newborn screening for infants with primary immunodeficiency	National Human Genome Research Institute	Collection of DBS samples from infants with immunodeficiency diseases to determine what diseases can be detected by screening DBS for T-cell products	100	NA	Completed
The intensive pharmacokinetics sub-study of Encore1 (ENCORE1-PK)	Kirby Institute	To determine the utility of DBS collection for measuring efavirenz systemic concentration	40	3	Ongoing, not recruiting
HIV diagnosis in hospitalized Malawian infants	University of North Carolina, Chapel Hill	Comparison of standard of care (DBS–DNA/PCR analysis and presumptive diagnosis) to expedited PCR	300	NA	Ongoing, not recruiting
Pharmacokinetic study to characterize phenotyping metrics of the "Basel" cocktail after CYP induction or inhibition	University Hospital, Basel, Switzerland	Use of DBS to examine pharmacokinetics of the "Basal" cocktail after administration of cytochrome P450 inhibitors and inducer	16	1	Completed
Pharmacokinetics, safety and tolerability of escalating rifapentine doses in healthy volunteers	John Hopkins University	For method development to determine rifapentine concentrations from DBSs on sampling paper	37	1	Completed
Safety and effectiveness of lopinavir/ritonavir in individuals who have failed prior HIV therapy	AIDS Clinical Trials Group	Evaluation of DBS for HIV-1 RNA and drug resistance testing	123	NA	Ongoing, not recruiting
Cytochrome P450 3A4 and 1A2 phenotyping for the individualization of treatment with sunitinib or erlotinib in cancer patients	Cantonal Hospital of St. Gallen	Measurement of sunitinib, erlotinib, and probe drugs in DBS to assess whether DBS can be used as an alternative to venous blood sampling to assess drug exposure in cancer patients	60	4	Not yet recruiting
Immune response to an HIV DNA plasmid vaccine prime followed by adenovirus boost in HIV-uninfected individuals	National Institute of Allergy and Infectious Diseases	Evaluation of the use of DBSs for measurement of inflammatory mediators	45	1	Recruiting
Pompe prevalence study in patients with muscle weakness without diagnosis	Centre Hospitalier Universitaire de Nice	Confirmation of diagnosis	400	NA	Completed
Detection of pompe disease in adult patients with myopathies of uncertain origin or with asymptomatic hyper-CK-emia	Jordi Perez Lopez	Detection of disease	50	NA	Not yet recruiting
Clinical study of an Aluvia-based HAART regimen for prevention of mother-to-child HIV transmission in Africa	University of Zambia	Survival and negative DBS (PCR) for infants 3 months postweaning	280	4	Active, not recruiting
Community-based evaluation of a pilot PMTCT project in Kafue district	University of North Carolina, Chapel Hill	DBS to test for HIV in children and adults	4129	NA	Completed
Use of individual pharmacokinetically (PK)-guided sunitinib dosing: A feasibility study in patients with advanced solid tumors	The Netherlands Cancer Institute	Measurement of sunitinib systemic concentrations using DBS	30	1	Recruiting
Screening for early detection and prevention of pompe disease in Israel using tandem mass spectrometry	Rambam Health Care Campus	DBS sampling for determination of α-glucosidase activity measured by LC-MS/MS	10,000	NA	Not yet recruiting

NA, not applicable.

[a]Studies from http://clinicaltrials.gov/ (accessed April 2012).

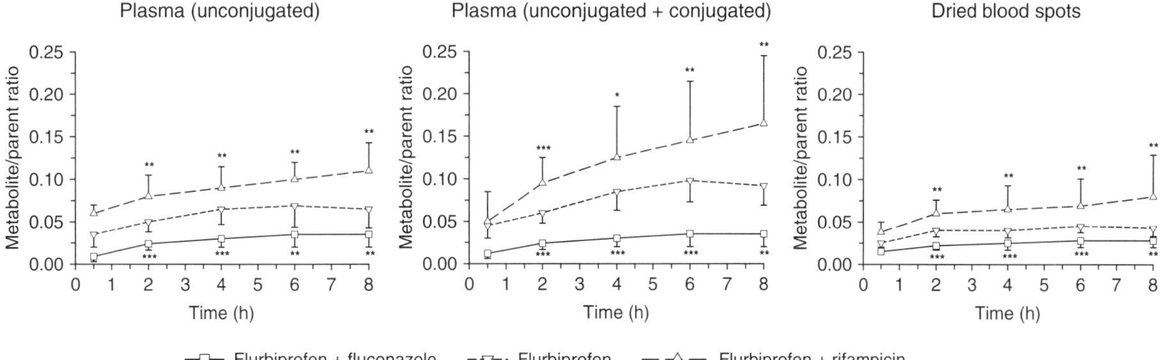

FIGURE 19.3 Metabolite/parent ratios (4′-hydroxyflurbiprofen/flurbiprofen) after administration of flurbiprofen to healthy subjects (n = 10) in plasma (conjugated, unconjugated + conjugated) and dried blood spots (*P <0.05, **P <0.01, ***P <0.001) (Adapted from Daali et al. (2012) with permission from Nature Publishing Group.

post-dose between the treatments. Although the differences in the measured parent and metabolite concentration in plasma versus blood spots were possibly due to the hematocrit and analyte blood/plasma distribution, those factors should have no impact on phenotyping in examining the ratio of measured parent-to-metabolite concentration in a given matrix (Streetman and Bertino, 2000; Daali et al., 2008).

19.7 CONCLUSIONS

DBSs are becoming an established sampling method in developing countries to monitor drug therapy in clinical trials. Use of DBS sample collection is gaining acceptance in developed countries due to the ease of sample collection and shipping. Of particular importance is the application of this sampling technique to special populations such as pediatrics and critically ill patients as blood volume for sampling is limited. In addition, incorporation of biomarker measurements in DBSs can enable a better understanding of the drug concentration–effect relationship. The use of biomarker measurements will allow for a better approach to a personalized exposure–response strategy for patients compared to conventional blood sampling as blood spot samples may be taken virtually anywhere at anytime and performed by either the clinician or patient.

REFERENCES

Abdel-Rahman SM, Kauffman RE. The integration of pharmacokinetics and pharmacodynamics: understanding dose-response. *Annual Review of Pharmacology and Toxicology* 2004;44:111–136.

AbuRuz S, Millership J, McElnay J. Dried blood spot liquid chromatography assay for therapeutic drug monitoring of metformin. *Journal of Chromatography B* 2006;832:202–207.

Adamson JW, Longo DL. Chapter 57: Anemia and polycythemia. In: *Harrison's Online*. Columbus, OH: McGraw-Hill Company; 2012.

Akhlaghi F, Trull AK. Distribution of cyclosporin in organ transplant recipients. *Clinical Pharmacokinetics* 2002;41:615–637.

Atkinson K, Britton K, Biggs J. Distribution and concentration of cyclosporin in human blood. *Journal of Clinical Pathology* 1984;37:1167–1171.

Back D, Gibbons S, Khoo S. An update on therapeutic drug monitoring for antiretroviral drugs. *Therapeutic Drug Monitoring* 2006;28:468–473.

Barfield M, Wheller R. Use of dried plasma spots in the determination of pharmacokinetics in clinical studies: validation of a quantitative bioanalytical method. *Analytical Chemistry* 2011;83:118–124.

Benet LZ, Zia-Amirhosseini P. Basic principles of pharmacokinetics. *Toxicologic Pathology* 1995;23:115–123.

Bergqvist Y, Churchill FC, Mount DL. Determination of mefloquine by electron-capture gas chromatography after phosgene derivatization in biological samples and in capillary blood collected on filter paper. *Journal of Chromatography* 1988;428:281–290.

Brookes S, Woodmansey K, Love I. The development of dry plasma spot analysis and a comparison with dry blood spots and conventional plasma bioanalysis using methylene blue. *Chromatography Today* 2010;3:36–39.

Burhenne J, Riedel KD, Rengelshausen J, Meissner P, Müller O, Mikus G, Haefeli WE, Walter-Sack I. Quantification of cationic anti-malaria agent methylene blue in different human biological matrices using cation exchange chromatography coupled to tandem mass spectrometry. *Journal of Chromatography B* 2008;863:273–282.

Burton ME, Shaw LM, Schentag JJ, Evans WE. *Applied Pharmacokinetics & Pharmacodynamics Principles of Therapeutic Drug Monitoring*. Philadelphia, PA: Lippincott Williams & Wilkins; 2006.

Cash JM, Sears DA. The anemia of chronic disease: spectrum of associated diseases in a series of unselected hospitalized patients. *American Journal of Medicine* 1989;87:638–644.

Cheung CY, van der Heijden J, Hoogtanders K, Christiaans M, Liu YL, Chan YH, Choi KS, van de Plas A, Shek CC, Chau KF, Li CS, van Hooff J, Stolk L. Dried blood spot measurement: application in tacrolimus monitoring using limited sampling strategy and abbreviated AUC estimation. *European Society for Organ Transplantation* 2008;21:140–145.

Chiou WL. The phenomenon and rationale of marked dependence of drug concentration on blood sample site: implications in pharmacokinetics, pharmacodynamics, toxicology and therapeutics (Part I). *Clinical Pharmacokinetics* 1989;17:175–199.

Conroy S, Choonara I, Impicciatore P, Mohn A, Arnell H, Rane A, Knoeppel C, Seyberth H, Pandolfini C, Raffaelli MP, Rocchi F, Bonati M, 't Jong G, de Hoog M, van den Anker J on behalf of the European Network for Drug Investigation in Children. Survey of unlicensed and off label drug use in paediatric wards in European countries. *British Medical Journal* 2000;320:79–82.

Cook RE, Ciampa PJ, Sidat M, Blevins M, Burlison J, Davidson MA, Arroz JA, Vergara AE, Vermund SH, Moon TD. Predictors of successful early infant diagnosis of HIV in a rural district hospital in Zambézia, Mozambique. *Journal of Acquired Immune Deficiency Syndrome* 2011;56:e104–e109.

Cyclosporine Prescribing Information; 2009. http://www.pharma.us.novartis.com/product/pi/pdf/neoral.pdf (accessed January 20, 2014).

Daali Y, Cherkaoui S, Doffey-Lazeyras F, Dayer P, Desmeules JA. Development and validation of a chemical hydrolysis method for dextromethorphan and dextrophan determination in urine samples: application to the assessment of CYP2D6 activity in fibromyalgia patients. *Journal of Chromatography B* 2008;861:56–63.

Daali Y, Samer C, Déglon J, Thomas A, Chabert J, Rebsamen M, Staub C, Dayer P, Desmeules J. Oral flurbiprofen metabolic ratio assessment using a single-point dried blood spot. *Clinical Pharmacology and Therapeutics* 2012;91:489–496.

de Haan GJ, Edelbroek P, Segers J, Engelsman M, Lindhout D, Dévilé-Notschaele M, Augustijn P. Gestation-induced changes in lamotrigine pharmacokinetics: a monotherapy study. *Neurology* 2004;63:571–573.

Déglon J, Thomas A, Daali Y, Lauer E, Samer C, Desmeules J, Dayer P, Mangin P, Staub C. Automated system for on-line desorption of dried blood spots applied to LC/MS/MS pharmacokinetic study of flurbiprofen and its metabolite. *Journal of Pharmaceutical and Biomedical Analysis* 2011;54:359–367.

Edelbroek PM, van der Heijden J, Stolk LML. Dried blood spot methods in therapeutic drug monitoring: methods, assays, and pitfalls. *Therapeutic Drug Monitoring* 2009;31:327–336.

Emmons G, Rowland M. Pharmacokinetic considerations as to when to use dried blood spot sampling. *Bioanalysis* 2010;2:1791–1796.

European Medicines Agency. *Clinical Investigation of Medicinal Products in the Paediatric Population*. London, UK: Committee for Medicinal Products for Human Use (CHMP); January 2001. http://www.emea.europa.eu/docs/en_GB/document_library/Scientific_guideline/2009/09/WC500002926.pdf (accessed January 20, 2014).

European Medicines Agency. *The Role of Pharmacokinetics in the Development of Medicinal Products in the Pediatric Population*. London, UK: Committee for Medicinal Products for Human Use (CHMP); June 2006. http://www.ema.europa.eu/docs/en_GB/document_library/Scientific_guideline/2009/09/WC500003066.pdf (accessed January 20, 2014).

European Medicines Agency. *Guidelines on the Investigation of Medicinal Products in the Term and Preterm Neonate*. London, UK: Committee for Medicinal Products for Human Use (CHMP); October 2009. http://www.ema.europa.eu/docs/en_GB/document_library/Scientific_guideline/2009/09/WC500003750.pdf (accessed January 20, 2014).

European Medicines Agency. *Guidelines on the Investigation of Bioequivalence*. London, UK: Committee for Medicinal Products for Human Use (CHMP); August 2010. http://www.emea.europa.eu/docs/en_GB/document_library/Scientific_guideline/2010/01/WC500070039.pdf (accessed January 20, 2014).

European Medicines Agency. *Guideline on bioanalytical method validation. Committee for Medicinal Products for Human Use*. London, UK: Committee for Medicinal Products for Human Use (CHMP); July 2011. http://www.ema.europa.eu/docs/en_GB/document_library/Scientific_guideline/2011/08/WC500109686.pdf (accessed January 20, 2014).

European Medicines Agency. *Guideline on Bioanalytical Method Validation*. London, UK: Committee for Medicinal Products for Human Use (CHMP); February 2012. http://www.ema.europa.eu/docs/en_GB/document_library/Scientific_guideline/2011/08/WC500109686.pdf (accessed January 20, 2014).

Everolimus Prescribing Information; 2010. http://www.pharma.us.novartis.com/product/pi/pdf/zortress.pdf (accessed January 20, 2014).

Eyles D, Anderson C, Ko P, Jones A, Thomas A, Burne T, Mortensen PB, Nørgaard-Pedersen B, Hougaard DM, McGrath J. A sensitive LC/MS/MS assay of 25OH vitamin D_3 and 25OH vitamin D_2 in dried bloodspots. *Clinical Chimica Acta* 2009;403:145–151.

Fleuren HLJ, van Rossum JM. Nonlinear relationship between plasma and red blood cell pharmacokinetics of chlorthalidone in man. *Journal of Pharmacokinetics and Biopharmaceutics* 1977;5:359–375.

Flowers CH, Cook JD. Dried plasma spot measurements of ferritin and transferrin receptor for assessing iron status. *Clinical Chemistry* 1999;45:1826–1832.

Forrest JAH, Clements JA, Prescott LF. Clinical pharmacokinetics of paracetamol. *Clinical Pharmacokinetics* 1982;7:93–107.

Guthrie R, Susi A. A simple phenylalanine method for detecting phenylketonuria in large populations of newborn infants. *Pediatrics* 1963;32:338–343.

Hoogtanders K, van der Heijden J, Christiaans M, van de Plas A, van Hooff J, Stolk L. Dried blood spot measurement of tacrolimus is promising for patient monitoring. *Transplantation* 2007;83:237–238.

Johannessen A, Trøseid M, Calmy A. Dried blood spots can expand access to virological monitoring of HIV treatment in resource-limited settings. *Journal of Antimicrobial Chemotherapy* 2009a;64:1126–1129.

Johannessen A, Garrido C, Zahonero N, Sandvik L, Naman E, Kivuyo SL, Kasubi MJ, Gundersen SG, Bruun JN, de Mendoza C. Dried blood spots perform well in viral load monitoring of patients who receive antiretroviral treatment in rural Tanzania. *Clinical Infectious Diseases* 2009b;49:976–981.

Kawai R, Lemaire M, Steimer JL, Bruelisauer A, Niederberger W, Rowland M. Physiological based pharmacokinetic study on a cyclosporin derivative, SDZ IMM 125. *Journal of Pharmacokinetics and Biopharmaceutics* 1994;5:327–365.

Kayiran SM, Özbek N, Turan M, Gürakan B. Significant differences between capillary and venous complete blood counts in the neonatal period. *Clinical and Laboratory Haematology* 2003;25:9–16.

Kearns GL, Abdel-Rahman SM, Alander SW, Blowey DL, Leeder JS, Kauffman RE. Developmental pharmacology—drug disposition, action, and therapy in infants and children. *The New England Journal of Medicine* 2003;349:1157–1167.

Keevil BG, Tierney DP, Cooper DP, Morris MR, Machaal A, Yonan N. Simultaneous and rapid analysis of cyclosporin A and creatinine in finger prick blood samples using liquid chromatography tandem mass spectrometry and its application in C2 monitoring. *Therapeutic Drug Monitoring* 2002;24:757–767.

Keevil BG, Roberts D, Preziosi R, Webb NJA. Fingerpick blood samples can be used to accurately measure tacrolimus levels by tandem mass spectrometry. *Therapeutic Drug Monitoring* 2005;27:229–230.

Kirchner GI, Meier-Wiedenbach I, Manns MP. Clinical pharmacokinetics of everolimus. *Clinical Pharmacokinetics* 2004;43:83–95.

Knuchel MC, Tomasik Z, Speck RF, Lüthy R, Schüpbach J. Ultrasensitive quantitative HIV-1 p24 antigen assay adapted to dried plasma spots to improve treatment monitoring in low-resource settings. *Journal of Clinical Virology* 2006;36: 64–67.

Koal T, Burhenne H, Römling R, Svoboda M, Resch K, Kaever V. Quantification of antiretroviral drugs in dried blood spot samples by means of liquid chromatography/tandem mass spectrometry. *Rapid Communications in Mass Spectrometry* 2005;19:2995–3001.

Kolocouri F, Dotsikas Y, Loukas YL. Dried plasma spots as an alternative sample collection technique for quantitative LC-MS/MS determination of gabapentin. *Analytical and Bioanalytical Chemistry* 2010;398:1339–1347.

Kromdijk W, Mulder JW, Rosing H, Smit PM, Beijnen JH, Huitema ADR. Use of dried blood spots for the determination of plasma concentrations of nevirapine and efavirenz. *Journal of Antimicrobial Chemotherapy* 2012;67:1211–1216.

la Marcia G, Malvagia S, Casetta B, Pasquini E, Donati MA, Zammarchi E. Progress in expanded newborn screening for metabolic conditions by LC-MS/MS in Tuscany: update on methods to reduce false positive tests. *Journal of Inherited Metabolic Disease Short Report* 2008;31(Suppl 2):S395–S404.

la Marcia G, Malvagia S, Filippi L, Luceri F, Moneti G, Guerrini R. A new rapid micromethod for the assay of phenobarbital from dried blood spots by LC-tandem mass spectrometry. *Epilepsia* 2009;50:2658–2662.

Li W, Tse FLS. Dried blood spot sampling in combination with LC-MS/MS for quantitative analysis of small molecules. *Biomedical Chromatography* 2010;24:49–65.

Li W, Doherty JP, Kulmatycki K, Smith HT, Tse FLS. Simultaneous LC-MS/MS quantitation of acetaminophen and its glucuronide and sulfate metabolites in human dried blood spot samples collected by subjects in a pilot clinical study. *Bioanalysis* 2012;4:1429–1443.

McDade TW, Williams S, Snodgrass JJ. What a drop can do: dried blood spots as a minimally invasive method for integrating biomarkers into population-based research. *Demography* 2007;44:899–925.

Mei JV, Alexander JR, Adam BW, Hannon WH. Use of filter paper for the collection and analysis of human whole blood specimens. *The Journal of Nutrition* 2001;313:1631S–1636S.

Minematsu T, Sugiyama E, Kusama M, Hori S, Yamada Y, Ohtani H, Sawada Y, Sato H, Takayama T, Sugawara Y, Makuuchi M, Iga T. Effect of hematocrit on pharmacokinetics of tacrolimus in adult living donor liver transplant recipients. *Transplantation Proceedings* 2004;36:1506–1511.

Mirochnick M, Thomas T, Capparelli E, Zeh C, Holland D, Masaba R, Odhiambo P, Fowler MG, Weidle PJ, Thigpen MC. Antiretroviral concentrations in breast-feeding infants of mothers receiving highly active antiretroviral therapy. *Antimicrobial Agents and Chemotherapy* 2009;53:1170–1176.

Mix TCH, Kazmi W, Khan S, Ruthazer R, Rohrer R, Pereira BJG, Kausz AT. Anemia: a continuing problem following kidney transplantation. *American Journal of Transplantation* 2003;3:1426–1433.

Neef C, Touw DJ, Stolk LM. Therapeutic drug monitoring in clinical research. *Pharmaceutical Medicine* 2008;22:235–244.

O'Broin SD, Kelleher BP, Gunter E. Evaluation of factors influencing precision in the analysis of samples taken from blood spots on filter paper. *Clinical and Laboratory Haematology* 1995;17:185–188.

Ogbonna A, Ogbonna CIO, Ogbonna BO, Enyi AAU, Uneke JC. Evaluation of pharmacokinetic and bioequivalence of brands of sulphadoxine-pyrimethamine tablets used in intermittent preventative therapy for pregnant women in Nigeria. *The Internet Journal of Tropical Medicine* 2007;4.

Okonkwo CA, Coker HAB, Agomo PU, Ogunbanwo JA, Mafe AG, Agomo CO, Afolabi BM. Effect of chlorpheniramine on the pharmacokinetics of and response to chloroquine of Nigerian children with falciparum malaria. *Transactions of the Royal Society of Tropical Medicine and Hygiene* 1999;93:306–311.

Pandya HC, Spooner N, Mulla H. Dried blood spots, pharmacokinetic studies and better medicines for children. *Bioanalysis* 2011;3:779–786.

Patel P, Mulla H, Tanna S, Pandya H. Facilitating pharmacokinetic studies in children: a new use of dried blood spots. *Archives of Disease in Childhood* 2010;95:484–487.

Rawlins MD, Henderson DB, Hijab AR. Pharmacokinetics of paracetamol (acetaminophen) after intravenous and oral administration. *European Journal of Clinical Pharmacology* 1977;11:283–286.

Ridley DM, Dawkins F, Perlin E. Erythropoietin: a review. *Journal of National Medical Association* 1994;86:129–135.

Rowell V, Rowell FJ, Baker A, Laurie D, Sidki AM. A specific ELISA method for determining chloroquine in urine or dried blood spots. *Bulletin of the World Health Organization* 1988;66:211–217.

Rowland M, Emmons GT. Use of dried blood spots in drug development: pharmacokinetic considerations. *The AAPS Journal* 2010;12:290–293.

Sherman GG, Stevens G, Jones SA, Horsfield P, Stevens WS. Dried blood spots improve access to HIV diagnosis and care for infants in low resource settings. *Journal of Acquired Immune Deficiency Syndrome* 2005;38:615–617.

Sirolimus Prescribing Information; 2011. http://labeling.pfizer.com/showlabeling.aspx?id=139 (accessed January 20, 2014).

Skogstrand K, Thorsen P, Nørgaard-Pedersen B, Schendel DE, Sørensen LC, Hougaard DM. Simultaneous measurement of 25 inflammatory markers and neurotrophins in neonatal dried blood spots by immunoassay with xMAP technology. *Clinical Chemistry* 2005;51:1854–1866.

Spooner N, Ramakrishnan Y, Barfield M, Dewit O, Miller S. Use of DBS sample collection to determine circulating drug concentrations in clinical trails: practicalities and considerations. *Bioanalysis* 2010;2:1515–1522.

Staatz CE, Tett SE. Clinical pharmacokinetics and pharmacodynamics of tacrolimus in solid organ transplantation. *Clinical Pharmacokinetics* 2004;43:623–653.

Streetman DS, Bertino JS Jr, Nafziger AN. Phenotyping of drug-metabolizing enzymes in adults: a review of in-vivo cytochrome P450 phenotyping probes. *Pharmacogenetics* 2000;10:187–216.

Suyagh M, Collier PS, Millership JS, Iheagwaram G, Millar M, Halliday HL, McElnay JC. Metronidazole population pharmacokinetics in preterm neonates using dried blood-spot sampling. *Pediatrics* 2011;127:e367–e374.

Tacrolimus Prescribing Information; 2012. http://www.astellas.us/docs/prograf.pdf (accessed January 20, 2014).

ter Heine R, Beijnen JH, Huitema ADR. Bioanalytical issues in patient-friendly sampling methods for therapeutic drug monitoring: focus on antiretroviral drugs. *Bioanalysis* 2009;1:1329–1338.

't Jong GW, Vulto AG, de Hoog M, Schimmel KJM, Tibboel D, van den Anker JN. Unapproved and off-label use of drugs in a children's hospital [Letter]. *The New England Journal of Medicine* 2000;343:1125.

't Jong GW, van der Linden PD, Bakker EM, van der Lely N, Eland IA, Stricker BHC, van den Anker JN. Unlicensed and off-label drug use in a paediatric ward of a general hospital in the Netherlands. *European Journal of Clinical Pharmacology* 2002;58:293–297.

Undre NA, Schäfer A, European Tacrolimus Multicentre Renal Study Group. Factors affecting the pharmacokinetics of tacrolimus in the first year after renal transplantation. *Transplantation Proceedings* 1998;30:1261–1263.

United States Food and Drug Administration. *Guidance for Industry. Bioavailability and Bioequivalence Studies for Orally Administered Drug Products—General Considerations.* Rockville, MD: Center of Drug Evaluation and Research (CDER); March 2003. http://www.fda.gov/downloads/Drugs/GuidanceComplianceRegulatoryInformation/Guidances/ucm070124.pdf (accessed January 20, 2014).

United States Food and Drug Administration. *Guidance for Industry: Bioanalytical Method Validation.* Rockville, MD: Center of Drug Evaluation and Research (CDER); September 2013. http://www.fda.gov/downloads/drugs/guidancecomplianceregulatoryinformation/guidances/ucm368107.pdf (accessed January 20, 2014).

van Amsterdam P, Waldrop C. The application of dried blood spot sampling in global clinical trials. *Bioanalysis* 2010;2:1783–1786.

van der Heijden J, de Beer Y, Hoogtanders K, Christiaans M, de Jong GJ, Neef C, Stolk L. Therapeutic drug monitoring of everolimus using the dried blood spot method in combination with liquid chromatography-mass spectrometry. *Journal of Pharmaceutical and Biomedical Analysis* 2009;50:664–670.

Wilhelm AJ, den Burger JCG, Vos RM, Chahbouni A, Sinjewel A. Analysis of cyclosporin A in dried blood spots using liquid chromatography tandem mass spectrometry. *Journal of Chromatography B* 2009;877:1595–1598.

Williams SR, McDade TW. The use of dried blood spot sampling in the national social life, health and aging project. *Journal of Gerontology: Social Sciences* 2009;64B(S1):i31–i36.

Winters ME. *Basic Clinical Pharmacokinetics*, 3rd edn. Vancouver, WA: Applied Therapeutics Inc.; 1994.

Yakkundi S, Turner M. Use of dried blood spots to study excipient kinetics in neonates. *Bioanalysis* 2011;3:2691–2693.

Yin OQP, Tomlinson B, Chow AHL, Chow MSS. Pharmacokinetics of acetaminophen in Hong Kong Chinese subjects. *International Journal of Pharmaceutics* 2001;222;305–308.

Zgheib NK, Frye RF, Tracy TS, Romkes M, Branch RA. Evaluation of flurbiprofen urinary ratios as in vivo indices for CYP2C9 activity. *British Journal of Clinical Pharmacology* 2007;63:477–487.

20

AUTOMATION IN DRIED BLOOD SPOT SAMPLE COLLECTION, PROCESSING, AND ANALYSIS FOR QUANTITATIVE BIOANALYSIS IN PHARMACEUTICAL INDUSTRY

LEIMIN FAN, KATTY WAN, OLGA KAVETSKAIA, AND HUAIQIN WU

20.1 INTRODUCTION

Since the development and implementation of dried blood spot (DBS) for newborn screening in early 1960s, this technique has experienced a rapid expansion and growth in many areas of applications. The current drivers for sample volume reduction such as Replacement, Reduction, and Refinement (3Rs), pediatric applications, therapeutic drug monitoring (TDM), generic error screening, and diagnostics are directing further improvement and the development of DBS. Small volume of the initial sample needed for spotting and easy and cost-effective storage and transportation of the paper-based dried matrix are some major benefits that make this technique a very attractive and valuable alternative to the traditional liquid biological sample collection and handling procedures. Furthermore, the tremendous improvement in the sensitivity of one of the major analytical techniques, liquid chromatography/mass spectrometry (LC/MS), utilized in the areas of life sciences and pharmaceutical industry allowed for the collection of small volumes of biological fluid for qualitative and quantitative determination of the analyte(s) of interest.

The challenges of DBS sampling methodology associated with the sample and carrier homogeneity, sample integrity, and selection of appropriate site of blood collection are better understood now due to extensive and in-depth research. However, DBS is known as a sampling format associated with very laborious analytical method development and sample preparation. The application of automation techniques in conjunction with DBS to increase the efficiency to be comparable to traditional liquid sample handling techniques is necessary. A typical DBS sample processing workflow involves method development, sample collection, sample preparation, and sample analysis. Opportunities to introduce automation into each of the above four steps exist and can lead to significant productivity gains.

20.2 AUTOMATED METHOD DEVELOPMENT

Method development for DBS samples is not very different from the traditional liquid samples in terms of optimizing LC separations and MS detections. Efforts that involve method development are more directed to the preparation (spotting) of standards (STDs) and quality controls (QCs), to maximize recovery, to minimize matrix effects, and to handle other DBS-specific issues such as hematocrit effects and spotting volume variations.

Standards and QCs can be prepared with automated liquid handlers. The volume of analyte spiking solution needs to be kept at a minimum (<5% of total volume) to avoid diluting blood and resulting in the alteration of the characteristics of blood. A compromise with respect to the solvent may have to be made to ensure sufficient solubility of the analyte and to minimize lysing effect on the red blood cells. Desired levels of STDs and QCs can be made by programming the liquid handler to conduct (1) individual spiking or (2) serial dilutions. Once the STDs and QCs are made and gently mixed, they can be spotted onto the DBS cards using repeated

Dried Blood Spots: Applications and Techniques, First Edition. Edited by Wenkui Li and Mike S. Lee.

pipettor or a potential automated spotter. This process has been automated to certain degree. For example, liquid handlers or MALDI spotter developed by Leap Technologies may be used to accomplish this task.

A generic workflow and decision tree can be developed in each lab to streamline method development. Typically, the workflow should encompass (1) card screen and (2) extraction solvent screen. Different card types should be screened for best analytical recovery and best tolerability to volume variations. If different lots of cards and different sources of blood (i.e., blood from more than one animal or one subject) can be tested early at the method development stage, then a more robust final analytical method can be ensured.

20.3 AUTOMATED BLOOD COLLECTION AND SPOTTING

Along with automated DBS spot punching, extraction, and analysis, automation for blood sampling and spotting is also very attractive to bioanalytical scientists to improve efficiency and reduce manual errors. A number of manufacturers including Instech, DiLab, and BASi have developed automated blood collection systems for both liquid and DBS sample collections.

Instech DBS2, a system for both automated blood sampling collection and DBS spotting, features the collection and deposition of cannulated animal blood samples directly onto DBS paper discs. Each 6 cm diameter disc can hold 12 spots of 10–25 μL of blood. Individual blood samples can be spotted singly, in duplicate, or in triplicate. Due to the spotting volume as low as 10 μL, the DBS2 can collect multiple blood samples from a single mouse within a dosing interval to provide full pharmacokinetic (PK) profile. Uyeda et al. (2011) at Amgen used the DBS2 device and liquid chromatography/tandem mass spectrometry (LC-MS/MS) in a PK study of AMG 517, a potent and selective vanilloid receptor antagonist, with a jugular vein catheter for dosing and a carotid artery catheter for automated blood collection. Blood samples from intravenously dosed rats ($n = 3$) were collected at nine time points over a 24-hour period and analyzed by LC-MS/MS. The concentration–time data and PK parameters generated from automatically spotted samples were very similar to those derived from manually collected blood samples. The difference of AUC_{0-24h} values was about 14%.

Aubert et al. 2011 at CitoxLAB had similar results from their investigation with 25 mg/kg dosing of ibuprofen in 24 rats using Instech blood collection system. Comparable exposure data were reported between manual DBS and automated DBS methods. Despite the additional work of surgery to place a catheter into an animal, using automated blood collection instrument provides extra benefit of reducing animal handling during the sampling time, decreasing the animal stress, and improving the quality and repeatability of the blood samples.

AccuSampler® from DiLab® offers another automated blood sampling with solutions for stress-free, unattended sampling in freely moving animals. The system is equipped with a DBS option for automated spotting. Clark et al. (2011) studied propranolol orally administered to Han-Wistar rats attached to either a standard DiLab® AccuSampler® for manual DBS spotting or a retrofitted unit designed to directly collect the DBS samples. A 50 μL volume of blood was collected via the standard DiLab® AccuSampler® and manually spotted onto DBS cards. A 20 μL volume of blood was collected by the retrofitted unit and spotted onto DBS cards automatically. A total of six time points were collected over a 7-hour period. The DBS samples were analyzed by an LC/MS/MS method. The authors concluded that the conversion of a standard DiLab® AccuSampler® to an automated DBS collection system was easy and the results of both methods were indeed comparable.

Wong et al. (2011) investigated the feasibility of serial sampling from individual mice using BASi Culex Automated Blood Sampling system. The blood samples obtained from the automated sampling system was mixed with heparinized saline solution during collection time. The resulting diluted blood saline mixtures were manually spotted onto DBS cards. Full PK profiles from individual animals ($n = 3$, nine time points) were obtained after AMG 517 was administered intravenously. In addition, diluted whole blood (WB) samples were also collected for comparison purpose. The authors concluded the PK parameters calculated using the diluted WB or the DBS concentration data were comparable.

Although automated blood sampling for DBS is a very attractive tool, the animals need to be catheterized and automated sampling devices are needed for each animal. Both catheterization and the devices are quite expensive. The cost and complexity associated with this technology may not be justified for studies with large number of animals. In summary, one needs to consider scientific, ethical, and practical factors when planning studies using automated blood sampling.

20.4 AUTOMATION IN ANALYSIS OF DBS SAMPLES

The efforts that involve DBS analysis with automation have been significant. There are many developments in the areas of automated blood spot punching for off-line extraction and online extraction or direct extraction/measurement. To ease the laborious step of puncturing paper disc out from a DBS, improved manual or power punching tools and punching stations, semiautomatic, and automatic punching machines have been developed. Some automated punching machines are already available on the market. Some are still remaining as prototypes in laboratories. At Abbott Laboratories, a manual punch is made by modifying a rotating-barrel Yankee style screw driver with a punch tip attached to the end. It converts the downward motion to rotations of the punch tip. The punching action is lighter and less stressful. A spring-loaded push pin helps to release the punched DBS sample disc to a 96-well extraction plate. By adding a small motor to a manual DBS punch, Horizon Specialty (Product, Horizon Specialty) manufactures a semiautomated DBS punch that cuts a DBS by spinning the punching tip. As the manufacturer suggested, this low-cost tool "increases DBS productivity while relieving the discomfort and fatigue associated with repetitive manual punching operations." A similar battery-powered punching tool was also developed by GSK-Research Triangle Park (Bowen, 2010). Some other punching tools to improve efficiency include DBS Pneumatic Card Punch made by Analytical Sales & Services (Product, Analytical Sales and Service) and manual DBS punching station made by Tomtec (Product, Tomtec). The DBS Pneumatic Card Punch is driven by foot pneumatics that punches individual blood spot sample to a 96-well plate. A guiding tube underneath the punch head directs the punched DBS sample into a specific well. DBS cards are fed to the DBS Pneumatic Card Punch manually. The Tomtec Manual DBS Punch uses a lever to push its punch head. The DBS sample drops down to a 96-well plate for further extraction. More advanced DBS punch systems added automation for DBS card feeding, spot area detection, receiving plate well positioning, barcode scanning for sample ID, and sample well position correlation. These systems range from smaller unit to large fully automated systems. The smaller systems include Dried Blood Spot Processor from Hudson Robotics (product, Hudson Robotics), AutoDBS-1^{TM} Dried Blood Spot Processing Workstation from Tomtec, BSD-300 to BSD-600 series punching systems from Life Technologies (Product, Life Technologies), and Wallac DBS Punch (product number 1296-071) and Panthera-Puncher 9 (product number 2081-0010) from PerkinElmer (Product, PerkinElmer). These small systems have the capability to select correct punching position by either visual spot guide with a light beam or automatic spot area scanning. A fully automated system, such as BSD-1000 GenePunch from Life Technologies (Product, Life Technologies), has the load and walk-away functionalities that are designed for high-throughput laboratories. All operations are computer controlled to provide accurate punch. Hundreds of DBS cards can be loaded into a card magazine for continued process. According to a worklist, the card gripper selects a DBS card and transports it to the scanning module. The DBS card image is scanned to determine the spot position and the best point for punching. The DBS card is then moved to the punch module for punching to 96- or 384-well plates placed on a plate carousel. A barcode on the DBS card is used to provide positive sample identification and tracking. Up to six plates can be placed in a BSD-1000 GenePunch.

Punching out a paper disc from a DBS card and mixing with extraction solvents in an extraction plate is the most common extraction method of DBS analysis. However, thanks to the strength of filter paper, it is possible to manipulate DBS sample directly to separate the analyte from the supporting substrate. A stream of extraction solvents is forced to pass through a DBS to resolubilize the analyte. The resulting solution is then analyzed by LC/MS/MS to obtain analyte concentration. Déglon et al. (2009) reported an early prototype of online DBS in 2009. A paper disc (12 mm diameter) containing a 10 μL DBS was punched out from a DBS card and placed in an extraction cell. Internal standard (5 μL) was added to the spot prior to the extraction. Acetonitrile was used to desorb and elute the analyte from filter paper. Water with 0.1% formic acid was added to the desorption solvent stream by a T-connector to lower the acetonitrile concentration. The resulting mixture flew through an extractive column to trap the analyte. Using a switching valve, the analyte was eluted to an analytical column and a mass spectrometer. When saquinavir, imipramine, and verapamil were used as model compounds, the accuracy of low and high QCs were between 90.5% and 116.1% for the three compounds and relative standard deviations (RSDs) were lower than 15%. A newer edition of the prototype made by a rotation plate and a clamp was reported by Déglon again in 2011 (Déglon et al., 2011). There were 30 wells positioned on the rotation plate surface to host DBS paper discs (10 mm diameter) with 5 μL of blood spot. The clamp affixed to a DBS paper disc allowed extraction solvent flow through the paper disc to elute the analyte. Using this automated online DBS extraction system to analyze flurbiprofen (FLB) and 4-hydroxyflurbiprofen (OH-FLB), the QC samples (100–10,000 ng/mL for FLB and 10–1000 ng/mL for OH-FLB) showed relative bias between −5.1% and +6.7% for both analytes. The precision for both analytes was less than 12%.

There are two other flow-through online DBS extraction systems that are commercially available, SCAP^{TM} DBS system made by Prolab and Dried Blood Spot Autosampler manufactured by Spark Holland. Both systems are modified from online solid phase extraction (SPE) systems by changing the autosampler to DBS extraction stations. DBS card is moved from card racks by a gripper to the DBS clamp module. A clamp with opening in the center is used to hold the DBS card. Extraction solvent flows through the clamped area onto the DBS card to elute the analyte. The area of DBS sample eluted equals the opening area of the clamp. After the analyte is eluted from the DBS card, the extraction solution is diluted with water to lower the organic content. The solution with analyte flows through a SPE cartridge first to trap the analyte. Typical online SPE and LC/MS/MS analysis process is followed with analyte being eluted from the SPE cartridge through an analytical column to a mass spectrometer. The setup of Spark Holland DBS autosampler featured the elution of the analyte from the trapping column to a mass spectrometer directly. Ganz et al. published their method validation work on using SCAP DBS system to analyze bosentan and its three primary metabolites (Ganz et al., 2012). Good linearity over a dynamic range of 2–1500 ng/mL was obtained with $R^2 \geq 0.9951$. The analytical method was sensitive and selective with inter-day accuracy of 91.6–108.0% and precision of 3.4–14.6%. Ooms et al. (2011) reported method validation of analyzing four drugs using the Spark Holland Dried Blood Spot Autosampler. A single high-performance SPE cartridge (HySphere C18 SPE cartridge) was used for enrichment, clean-up, and LC separation of analytes prior to MS/MS analysis. The internal standard was introduced to the extraction stream by using a 20 μL loop. The precision of all four compounds was between 1.5% and 5.2%. The accuracy of all four compounds over the range of 1–1000 ng/mL was between 89% and 110%.

Another commercially available online DBS extraction system is Camag's DBS MS-500. Instead of using a clamp to extract a DBS card, it uses a TLC-MS interface to extract a fixed area of DBS sample. An optical card recognition module is used to determine the DBS card property, spot number and position, and sample barcode. An internal standard application module adds internal standard to DBS cards by spraying internal standard onto a DBS card before extraction. A wash station cleans all parts that contact with a DBS card to eliminate possible carryover. The extracted analyte solution is directed to an analytical column for direct LC/MS/MS analysis. Alternatively, it can be sent to a trapping column prior to an analytical column. Abu-Rabie and Spooner reported an experiment of manual extraction and direct extraction with TLC-MS interface to analyze 576 DBS and dried plasma spot (DPS) samples spiked with acetaminophen and sitamaquine (Abu-Rabie and Spooner, 2011). The analytes versus internal standard peak area ratio was consistent throughout 576 extractions. The %CV of the peak area ratio for both compounds was in the range of 2.3–4.1%, within typical bioanalytical method precision requirement of 15%. Carryover from direct extraction method appeared to be less than that from manual extractions.

Beyond traditional extraction and LC/MS/MS analysis methods, several direct DBS surface analysis methods have been reported in literature. Edwards et al. (2011) analyzed hemoglobin variants in DBS using liquid microjunction (LMJ) surface sampling technique coupled with a high-resolution mass spectrometer. At the time of analysis, a conductive tip containing 7 μL of wetting/aspirating solution comprising water/methanol/formic acid (48.5:48.5:3, v/v/v) was placed above a DBS. About 6 μL of the solution was dispensed out of the tip to form a LMJ between the DBS and the conductive tip. As the solution contacts the DBS, the analyte dissolves into the wetting solution. After about 5 seconds, the solution was aspirated back to the conductive tip. The solution is further transmitted to a Triversa Nanomate from Advion for nanospray ESI mass spectrometric analysis using a Thermo Fisher Scientific Orbitrap Velos mass spectrometer. Although this format still requires a significant amount of work before the LMJ surface sampling technique can be utilized in regulated bioanalysis, the operation of the system demonstrated a high degree of automation. Thus, this platform may be a good alternative for those mixed qualitative and quantitative works that require less precision and accuracy.

20.5 GENERAL ISSUES ASSOCIATED WITH DBS AUTOMATION

The fundamental driving force for automation is the validity of the methodology—being able to measure the concentration of drugs or drug candidates accurately and precisely. Once the DBS methodology becomes accepted (by the industry and the regulatory agencies) as a valid alternative sampling technique, the sheer volume of samples that can be collected with DBS will drive the automation because of the advantages of the methodology as described throughout the book and shared within the whole bioanalysis community.

The DBS field has been very innovative and dynamic. New ideas and new approaches are being developed to address challenges associated with this particular sampling technique. Different manufactures are also researching on utilizing different media (non-paper) for sample collections. With the exception of BSD Punching System, there is no fixed format yet industry wise regarding automation.

The current automation can be separated into two categories. One category highlights an off-line approach that focuses on automating the card punching and solvent additions. The other approach is an online approach that focuses on direct "flow through" without the need for punching. Both efforts involve "visualizing" the position of the spot, and therefore, the format being explored is primarily DMPK cards from GE Healthcare. However, the ability to accommodate the different formats and material of the sample collection media should be relatively easy to accomplish with the current development of robotics.

There are many challenges yet to overcome before large scale of automation efforts can be implemented.

Under optimal conditions, liquid WB sample and dried state DBS sample from the same origin should produce the same analytical results. However, multiple reports in the literature reported non-matching results from WB and DBS when parallel samples were collected and analyzed. A significant effort has been made to understand the cause of these non-matching results, and therefore, several fundamental aspects of DBS were investigated to establish its validity to use as an alternative sampling technique in quantitative bioanalysis.

To date, equivalency of DBS to WB must be established on a compound-by-compound basis. For high log D compound, a general negative bias (DBS value < WB value) is observed (Kavetskaia et al., 2011). The hematocrit and spotting volume may also have an impact on the values obtained by DBS versus the values measured using WB.

The majority of current DBS assays involved the addition of internal standard after the DBS samples are collected, at the moment of extraction (online or off-line). Therefore, the internal standard is not compensating for the variation of sample spotting process. Improved analytical performances were obtained when the IS was applied to liquid sample prior to spotting or sprayed to the cards prior to spotting (Chad et al., 2011). Feasibility to automate IS application to either liquid sample or card prior to spotting has not yet been explored widely. Now that DBS is becoming a more acceptable alternative sampling methodology in the pharmaceutical industry, automation of IS application is inevitable. For example, IS addition to liquid sample can be incorporated into the current automated sampling and spotting devices, such as Instech's second-generation sampler built on five precision peristaltic pumps with in-line sensors. IS-treated cards can also be manufactured by vendor or by analytical lab utilizing a spray device that is similar to the ones that are commonly used to apply matrix to tissue slice prior to MALDI imaging.

Different spot sizes can contribute significantly to analytical output and depend on the property of compounds (Fan et al., 2011). Automated or semiautomated spotting device that is accurate in volume and easy to use will be widely appreciated. The prototype dispenser from Drummond Scientific Company utilizes a fixed volume glass capillary (i.e., 50 μL) to precisely draw and deliver multiple fixed small volumes (i.e., 15 μL) of blood samples that can be spotted onto DBS collection media. The use of glass capillary (either coated or not coated with anticoagulant) interfaces well with tail nick (for most rodent tox studies) and finger prick (for human clinical studies).

Another innovative approach for the application of blood to DBS paper was developed at Abbott by using a glass capillary tube to transfer blood samples to a precut disc (Fan et al., 2012). The sample volume was precisely controlled by saturating precut DBS disc. The volume of blood inside the capillary tube does not need to be controlled. Therefore, this methodology provides a simple and easy application to those who are not trained for accurate volume delivery. In addition, no instrument is needed for volume measurement.

The effect of hematocrits on the analytical output has been investigated by multiple reach groups. For drugs that change blood chemistry, such as oncology drugs, analytical results can be skewed significantly due to drastic hematocrit difference between STDs and incurred samples. The addition of the IS prior to spotting followed by extraction of the whole spot or using precut paper discs in a collection cartridge (Fan et al., 2012) has demonstrated success to a certain degree. When the elution efficiency (Liu et al., 2010) is not the same between STDs and incurred samples due to different blood chemistry/property, DBS might not be applicable in this situation regardless of extraction methods.

20.6 FUTURE OUTLOOK

The future needs of DBS automation in quantitative bioanalysis depend largely on the scale of DBS applications in the preclinical and clinical samplings.

The immediate need for an automated handheld puncher to reduce hand stress will encourage vendors to develop electronic, motorized tools. These tools will likely be light weight, be easy to use, and have "visualization" (spot position locator) functions built in. Besides punching tools, dispensers with better design are also needed. Handheld automated blood dispenser/spotter with accurate and precise volume control is much more preferred over traditional air displacement or positive displacement pipettors. In most preclinical testing facilities and clinical sites, the technicians and nurses are not trained and familiar with pipettors.

With a better understanding of the fundamental processes regarding the interaction between blood and the collection media, improved collection media that are more tolerate to variable spotting volume and blood chemistry should be developed. DBS as a sampling technique can also be extended to other type of matrices such as CSF and urine, for example. When DBS becomes a more "universal" sampling technique, the need of future automation will focus on developing a versatile and integrated sample preparation platform that can handle both punching and solution additions.

In summary, compared to the mature liquid sample handling and processing, the current degree of automation on DBS sample collection and processing is low. The future of automation largely depends on the success of DBS applications. With all the efforts put from the industry and the instrument manufacturers, continuous improvement of additional automation can be achieved in the foreseeable future.

REFERENCES

Abu-Rabie P, Spooner N. Dried matrix spot direct analysis: evaluating the robustness of a direct elution technique for use in quantitative bioanalysis. *Bioanalysis* 2011;24:2769–2781.

Aubert N, Ameller T, Lefebvre C, Laurent S, Leuillet S, Legrand J. Automated Blood Sampling Associated or Not with Dried Blood Spot Sample Processing Provides Similar Exposure Parameters than Classical Pharmacokinetic Procedures; 2011. Presented at Eurotox 2011—Automated Blood Sampling.

Bowen C. Automatable Processes and Future Directions in Dried Blood Spot Analysis; 2010. http://www.americanlaboratory.com/913-Technical-Articles/514-Automatable-Processes-and-Future-Directions-in-Dried-Blood-Spot-Analysis/?adpi=2

Chad DC, Casey JL, Johnson CN, Sheaff DF, Laine JS, Zimmer D, Shane RN. Overcoming the Obstacles of Performing Dilutions and Internal Standard Addition to DBS Analysis Using HPLC/MS/MS. Presented at ASMS, Denver, Colorado, USA, June 2011.

Clark GT, Giddens G, Burrows L, Strand C. Utilization of dried blood spots within drug discovery: modification of a standard DiLab® AccuSampler® to facilitate automatic dried blood spot sampling. *Laboratory Animals* 2011;45:124–126.

Déglon J, Thomas A, Cataldo A, Mangin P, Staub C. On-line desorption of dried blood spot: a novel approach for the direct LC/MS analysis of micro-whole blood samples. *Journal of Pharmaceutical and Biomedical Analysis* 2009;49:1034–1039.

Déglon J, Thomas A, Daali Y, Lauer E, Samer C, Desmeules J, Daycr P, Mangin P, Staub C. Automated system for on line desorption of dried blood spots applied to LC/MS/MS pharmacokinetic study of flurbiprofen and its metabolite. *Journal of Pharmaceutical and Biomedical Analysis* 2011;54:359–367.

Edwards R, Creese A, Baumert M, Griffiths P, Bunch J, Cooper H. Hemoglobin variant analysis via direct surface sampling of dried blood spots coupled with high-resolution mass spectrometry. *Analytical Chemistry* 2011;83:2265–2270.

Fan L, Lee J, Hall J, Tolentino E, Wu H, El-Shourbagy T. Implementing DBS methodology for the determination of compound A in monkey blood: GLP method validation and investigation of the impact of blood spreading on performance. *Bioanalysis* 2011;11:1241–1252.

Fan L, Hall J, Strasburg D, Wu H, El-Shourbagy T. DMPD (Dried Matrix on Paper Disc)—A Precut DBS Method for Mouse PK Studies with Serial Bleeding. Presented at 60th ASMS, Vancouver, Canada, May 24, 2012.

Ganz N, Singrasa M, Nicolas L, Gutierrez M, Dingemanse J, Döbelin W, Glinski M. Development and validation of a fully automated online human dried blood spot analysis of bosentan and its metabolites using the Sample Card And Prep DBS System. *Journal of Chromatography B, Analytical Technologies in Biomedical and Life Sciences* 2012;885–886:50–60.

Kavetskaia O, Wan K, Fan P, Xu R, El-Shourbagy T. DBS in Regulated Bioanalysis. Presented at CPSA, Langhorne, Pennsylvania, USA, October 2011.

Liu G, Patrone L, Snapp HM, Batog A, Valentine J, Cosma G, Tymiak A, Ji QC, Arnold ME. Evaluating and defining sample preparation procedures for DBS LC–MS/MS assays. *Bioanalysis* 2010;2:1405–1414.

Ooms JA, Knegt L, Koster EHM. Exploration of a new concept for automated dried blood spot analysis using flow-through desorption and online SPE-MS/MS. *Bioanalysis* 2011;20:2311–2320.

Product from Analytical Sales and Service. http://www.analytical-sales.com/DBS.html (accessed January 2014).

Product from Horizon Specialty. http://www.horizon-specialty.com/dried-blood-spot.htm#dbs-rack (accessed January 2014).

Product from Hudson Robotics. http://www.hudsonrobotics.com/products/biological-research/dried-blood-spot/ (accessed January 2014).

Product from Life Technologies. http://www.lifetechnologies.com/us/en/home/industrial/human-identification/sample-punching-systems.html. (accessed January 2014).

Product from PerkinElmer. http://www.perkinelmer.com/Catalog/Product/ID/2081-0010 (accessed January 2014).

Product from Tomtec. http://www.tomtec.com/dried-blood-spots.html (accessed January 2014).

Uyeda C, Pham R, Fide S, Henne K, Xu G, Soto M, James C, Wong P. Application of automated dried blood spot sampling and LC–MS/MS for pharmacokinetic studies of AMG 517 in rats. *Bioanalysis* 2011;20:2349–2356.

Wong P, Pham R, Whitely C, Soto M, Salyers K, James C, Bruenner BA. Application of automated serial blood sampling and dried blood spot technique with liquid chromatography–tandem mass spectrometry for pharmacokinetic studies in mice. *Journal of Pharmaceutical and Biomedical Analysis* 2011;56:604–608.

21

BEYOND DRIED BLOOD SPOTS—APPLICATION OF DRIED MATRIX SPOTS

Shane R. Needham

21.1 INTRODUCTION

The benefits of using dried blood spots (DBSs) for the determination of drug concentrations in whole blood have been widely reported (Barfield et al., 2008; Rowland and Emmons, 2010; Clark et al., 2011; Liang et al., 2011; O'Mara et al., 2011; la Marca et al., 2012). These benefits include less invasive sample collection, simpler sample storage (ambient conditions), reduced risk of infection when handling the dried samples, and smaller sample volume requirements (Li and Tse, 2009). Other researchers have demonstrated the utility of DBS for the analysis of unstable molecules such as prodrugs (D'Arienzo et al., 2010; Liu et al., 2011), peptides (deWilde et al., 2008), and metabolites (Liang et al., 2009). Researchers have also used DBS for analysis of biomarkers from whole blood (Miller et al., 2011). Guthrie et al. used blood and urine as the matrix for assays developed for PKU screening of newborn infants in 1963 (Guthrie and Susi, 1963). Details of the research such as how the authors visually detected the urine spot on the paper media were not revealed. With the exception of this early paper from Guthrie in 1963, the DBS technique has not been used extensively for matrices other than whole blood.

21.2 DRIED MATRIX SPOT ANALYSIS

Whole blood with its highly red color is easily detected by the human eye when present on a white background of media such as paper. This is one reason the DBS technique is so attractive for sample collection and preparation methods with whole blood. This is also a reason that the technique has limitations for collection and analysis of other colorless (cerebral spinal fluid (CSF), synovial fluid, etc.) or semi-colorless fluids (urine, serum, plasma, etc.). Without being able to visually detect the spot, proper collection and optimal analysis are a challenge, as only a portion of the spot may be inadvertently sampled.

At the Annual American Society for Mass Spectrometry (ASMS) Conference, Christianson et al. presented seminal work on the analysis of drugs from biological fluids other than whole blood using DBS cards (Christianson et al., 2010a). This work was pivotal in the pharmaceutical industry and showed that matrices such as urine, tears, synovial fluid, CSF, and saliva were viable fluids for DBS analysis. Subsequent work facilitated the use of colorless fluids for analysis on a DBS card by the introduction of a color-indicating dye to the card before the sample is spotted (Christianson et al., 2010c). At that time, the term dried matrix spot (DMS) analysis was introduced for the use of the DBS technique with fluids other than whole blood. Previous researchers (Guthrie and Susi, 1963) had used DBS for fluids other than whole blood for other screening procedures as far back as 1957 when Roberts et al. spotted serum on paper media (Roberts and Hanson, 1957). However, to our knowledge, the DMS technique introduced in 2010 was the first use of the technique using a color-indicating dye and the first non-blood matrix application published in support of drug development. An illustration of a typical DMS card for the analysis of synovial fluid is shown in Figure 21.1. Additional applications have since been published from Barfield et al. (Barfield and Wheller, 2011) and Huie et al. (Huie and Takahashi, 2011). In early 2011, Barfield's group published data on the use of a commercially available color-indicating paper for the

Dried Blood Spots: Applications and Techniques, First Edition. Edited by Wenkui Li and Mike S. Lee.

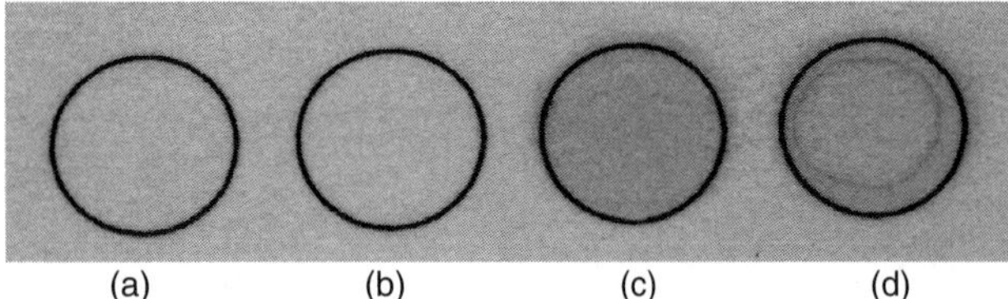

FIGURE 21.1 Ahlstrom 226 Specimen Collection Paper. (a) No color-indicating dye or sample spot; (b) no color-indicating dye with sample spot (synovial fluid); (c) Color-indicating dye with no sample spot; and (d) color-indicating dye with sample spot (synovial fluid). Reproduced from Christianson et al. (2010c) with permission from Future Science Ltd.

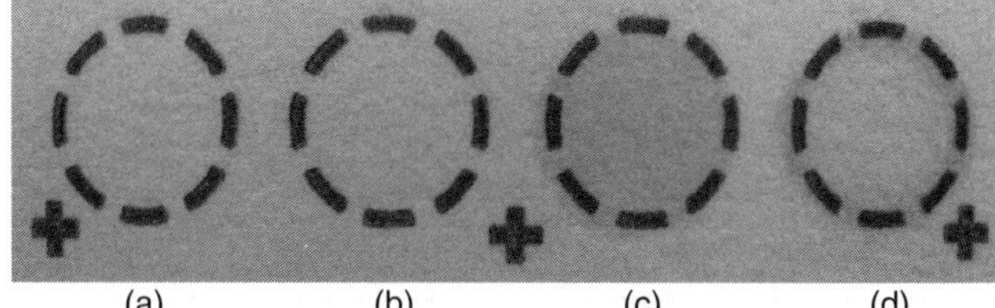

FIGURE 21.2 Whatman C Collection Card. (a) No color-indicating dye or sample spot; (b) no color-indicating dye with *in vitro* fluid sample spot; (c) color-indicating dye with no *in vitro* fluid sample spot; and (d) color-indicating dye with *in vitro* fluid sample spot.

analysis of paroxetine in dried plasma spots (DPS) in a clinical study with comparison to conventional plasma results. Huie's group presented data for the analysis of loxapine and phase II metabolites from rat urine. The data from Huie's group compared very well with traditional methods for urine analysis for loxapine and phase II metabolites. In 2011, Rago et al. employed the dried spot sampling technique to analyze serially collected *in vivo* rat CSF samples after a single dose of PHA-00543613 in CSF-cannulated rats (Rago et al., 2011). The DBS methods used by Rago's group enabled the collection of more time points and used fewer animals than traditional collection and analysis methods. In the research with Huie's group or Rago's group, neither researcher used a color-indicating dye for the DMS analysis.

Traditional analysis on fluids other than whole blood, such as synovial fluid, cerebrospinal fluid, and tears, can be difficult due to sample volume limitations (Christianson et al., 2010c). Additionally, as drugs are developed with more "drug-like" properties, they tend to be less water soluble making urine and *in vitro* assays (e.g., P450 microsomal assays, CACO-2 assays) and other highly aqueous matrices challenging, as drugs often precipitate out of the solution due to insolubility before analysis. An alternative to the traditional analysis of these fluids is to use the DMS technique where sample volumes less than 20 μL can be easily accommodated and the analyte of interest is dried on a card and not in solution. An example of utilizing DMS for *in vitro* analysis was presented where the lab compared results from the DMS and conventional analysis of dextromethorphan incubated in human liver microsomes (Christianson et al., 2012). The results compared well to conventional non-DMS analysis and sample storage and handling was facilitated as there was a minimized risk of the drug falling out of solution or losses due to nonspecific binding on the dried DMS card. An example of a typical application of analysis of dextromethorphan incubated with human liver microsomes and the reaction quenched using DMS is shown in Figure 21.2.

The desire for improved sampling of compartments other than blood mentioned above has led to the need for a technique such as DMS. One of the challenges of DBS is trying to compare historical plasma results with blood compartment results. Thus, researchers are developing dried plasma spot techniques (Kolocouri et al., 2010; Barfield and Wheller, 2011). DMS analysis will also facilitate the sampling of this blood compartment (i.e., plasma), as DPS are still difficult to detect visually on a card and a color-indicating dye may be needed. Barfield's group addressed this problem by using commercially available color-indicating cards. An illustration of the plasma spots dried on the color-indicating card is given in Figure 21.3.

Many of the applications we describe in this report are DMS techniques coupled with HPLC-MS. However, the DBS technique is also being used by non-chromatographic techniques (Kehler et al., 2011), direct desorption and direct spray methods (Wiseman et al., 2010; Manicke et al., 2011), and GC/MS techniques (Ingels et al., 2011). In all cases, applications are given relevant to the usefulness of DMS in the pharmaceutical industry for support of drug development.

A nice illustration of the versatility of the DMS technique for analysis of a variety of matrices is displayed in Table 21.1 (ASMS 2010 table). A picture of CSF spots using a novel color-indicating dye procedure is displayed in Figure 21.4.

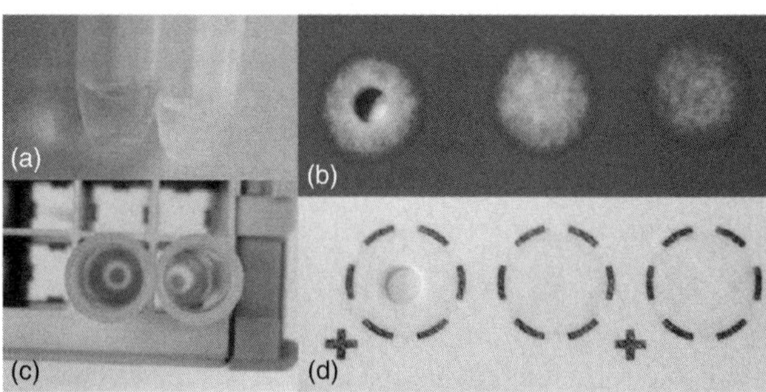

FIGURE 21.3 Illustrations of (a) a side on and (b) a top down view of two extracted DPS samples, indicating FTA DMPK paper on the left and 226 paper on the right using 100 μL of 70:30 MeOH–water (v/v); (c) three 20 μL dried plasma spots on indicating FTA; and (d) three 20 μL dried plasma spots on 226 paper. Reproduced from Barfield and Wheller (2011) with permission from American Chemical Society.

TABLE 21.1 Results for the HPLC/MS/MS Analysis of Indomethacin in Human Urine, Human Saliva, Human CSF, Human Tears, and Mini-pig Synovial Fluid

QC Level (ng/mL)	Intra-assay Accuracy and Precision	Matrix Factor	IS Normalized Matrix Factor
	Human urine		
400	99.3 ± 6.3	NA	NA
25	102 ± 2.0	NA	NA
3	105 ± 8.7	0.80	0.99
	Human saliva		
400	104 ± 6.4	NA	NA
25	107 ± 6.6	NA	NA
3	96.9 ± 9.3	0.71	1.0
	Human CSF		
400	104 ± 1.9	NA	NA
25	102 ± 3.6	NA	NA
3	98.7 ± 5.9	0.78	0.81
	Human tears		
400	114 ± 5.9	NA	NA
25	110 ± 2.5	NA	NA
3	109 ± 7.5	0.91	1.0
	Mini-pig synovial fluid		
400	90.5 ± 4.7	NA	NA
25	105 ± 4.4	NA	NA
3	94.1 ± 4.5	0.84	0.97

Numbers reported as accuracy ± precision (%CV) and represent the average of $n = 6$ replicates at each concentration level on 1 day of validation. Reproduced with permission from Future Science Ltd.

The method validation followed the 2001 FDA Guidance on Bioanalytical Method Validation and the Crystal City III White Papers (US FDA, 2001; Viswanathan et al., 2007).

21.2.1 Color-indicating Techniques

In order for the colorless fluid (CSF, synovial fluid, tears, etc.) to be easily visible on a DBS card, a color-indicating technique must be used. Although commercially available cards with many different properties (i.e., surfactants, substrate thickness, pore size) can be purchased from several manufacturers, a commercially available color-indicating fluid for use specifically for DMS is not available. Additionally, at this time, all the color-indicating techniques and fluids used have been proprietary thus investigation into the benefits and pitfalls of fluids with specific properties has been problematic. GE Healthcare has a DMS card with a specific dye imbedded on the card (Barfield and Wheller, 2011). With a commercially available color-indicating dyed card, there is a loss in the flexibility of application of the specific dye with a specific card and there is less control of where and how the dye is applied to the card. Additionally, as has been published at the Applied Pharmaceutical Analysis conference and the annual meeting of American Association of Pharmaceutical Scientists (AAPS) by Carpenter et al. (2010), one dye may not be optimal for all biological fluids or analyte properties (e.g., viscosity, pH) (Christianson et al., 2010c). As was shown by Barfield et al. with the commercially available color-indicating card from GE Healthcare for DPS analysis the method gave >37% matrix ion suppression (Barfield and Wheller, 2011). It is likely that the color-indicating fluid was a major contributor of these matrix effects as the authors demonstrated much less matrix effects from the conventional plasma assay.

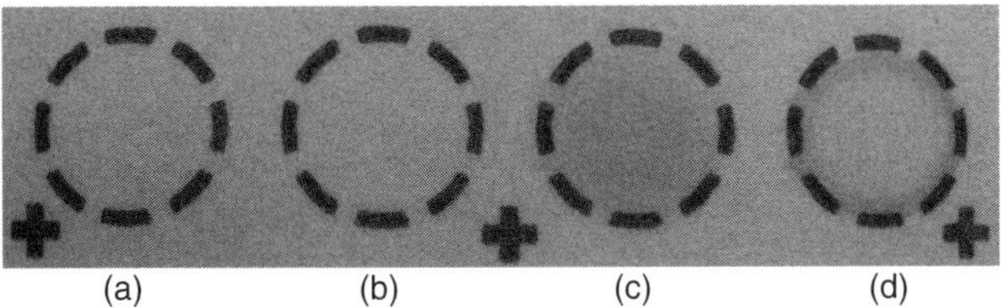

FIGURE 21.4 FTA DMPK-C Card with (a) no color-indicating dye or CSF sample spot; (b) no color-indicating dye with CSF sample spot; (c) color-indicating dye with no CSF sample spot; and (d) color-indicating dye with CSF sample spot.

The selection of a color-indicating dye is not trivial. A color-indicating fluid must have many properties for success with DMS. The properties include the following:

1. Show low background or interference for assay of the majority of molecules from biological fluids.
2. Potentially act as a solvent for simultaneous addition of internal standard (see discussion below).
3. Have minimal halo effects of the applied dye (Ren et al., 2010).
4. Be compatible with the fluid of interest and card of interest.
5. Ability to be applied uniformly and reproducibly with conventional laboratory tools (pipettes, etc.).
6. Not interact with additives embedded on card.

In showcasing the challenges of selecting a color-indicating dye for each analyte and matrix as needed, Christianson et al. analyzed tobramycin from tears (Christianson et al., 2010b – Applied Pharmaceutical Analysis). The method was accurate and precise and the color-indicating technique facilitated the visualization of the spot before analysis by HPLC/MS/MS. Refer to Figure 21.5 for visualization of the tear spots by the color-indicating technique used at the bioanalytical laboratory.

21.2.2 Addition of Internal Standard in Dried Matrix Spot Analysis

Since the use of DBS in drug development, one of the concerns raised by scientists and regulators is the nonoptimal

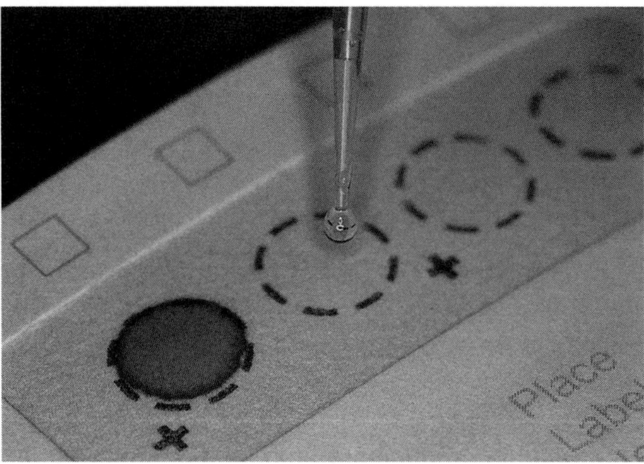

FIGURE 21.5 DMS analysis of tears on a FTA DMPK-C card with addition of a proprietary color-indicating technology.

addition of the internal standard (Vishwanathan, 2011). Typically, the internal standard is added (or is already contained) in the extraction fluid during sample preparation. Obviously, this is not optimal for the addition of the internal standard as the internal standard should be contained and mixed in the matrix as much as possible before further sample preparation. This approach allows the internal standard to best compensate for any variations during assay and should improve accuracy and precision.

As an example, researchers at Alturas Analytics, Inc. found that a DMS method developed for the analysis of dexamethasone from pig synovial fluid did not have sufficient precision to validate (precision greater than 15%); see Table 21.2 (Christianson et al., 2010c). With the addition of the color-indicating dye that contains an internal standard to the collection paper prior to sample spotting, the accuracy and precision were acceptable (Table 21.3). The data indicate that when color-indicating dye containing internal standard is added to the collection paper prior to spotting the sample, the method is approximately 10% more accurate and precise. This increase in the accuracy and precision was attributed to one or both of two factors: (1) the sample spot location was more evident due to the presence of the indicating dye; and (2) the internal standard was compensating for

TABLE 21.2 Intra-assay Accuracy and Precision for Dexamethasone in Pig Synovial Fluid without Color-indicating Dye or Internal Standard Addition Added Prior to Sample Spotting

QC Concentration (ng/mL)				
5.00	15.0	50.0	400	500
114.4 ± 13.4	106.8 ± 9.6	94.2 ± 19.1	93.0 ± 15.3	94.3 ± 10.4

Numbers reported as accuracy ± precision (%CV) and represent the average of 4 days of analysis. Reproduced with permission from Future Science Ltd.

TABLE 21.3 Intra-assay Accuracy and Precision for Dexamethasone in Pig Synovial Fluid

QC Concentration (ng/mL)				
5.00	15.0	50.0	400	500
101.4 ± 9.6	97.7 ± 6.2	104.3 ± 6.6	95.9 ± 5.8	102.1 ± 5.2

Numbers reported as accuracy ± precision (%CV) and represent the average of $n = 6$ replicates at each concentration level on 1 day of validation.

irregular spot removal from the collection paper during the punching process.

It is not uncommon to observe a spot that was not fully separated from the collection paper after the initial punching exercise is performed. When the spot is removed with a second punching action, the spot removed occasionally has some extra fiber attached to the spot. By adding the internal standard before the spot is removed, the extra fiber would contain both internal standard and the analyte of interest, thus the internal standard would compensate for any imprecise spots punched from the collection paper. See Christianson et al. at the annual ASMS conference and other references where the topic of addition of internal standard during DBS was thoroughly investigated (Christianson et al., 2011; Meesters et al., 2011). This method of internal standard addition is in theory superior to current methods for typical plasma, serum, and DBS bioanalysis, since the internal standard is added before the sample is placed into the collection device. Thus, the internal standard begins to compensate for any changes of the sample conditions that will occur during collection, shipment, and storage. The internal standard addition can be done in an automated fashion. See below for automation of DMS analyses.

21.2.3 Automation of Dried Matrix Spot Analysis

Since the implementation of DBS assays in the pharmaceutical industry, a concern for scientists has been the lack of automated tools for analysis. Typically, a manual punching tool is used like a Harris Uni-Core™ punch (http://www.tedpella.com/histo_html/unicore.htm). These tools require significant manual intervention and become dull after only several hundred punches. Ergonomically this is not optimal for a scientist and significantly limits throughput. Better punching devices have been manufactured including what we like to call a "semi-manual" punching device from TOMTEC (Figure 21.6). Although this device is still a manually operated instrument, the ergonomics of this device are much improved and throughput increases significantly.

Other fully automated DBS systems have been manufactured by several vendors. However, as with many fully automated devices, these tools are expensive and lack flexibility. Flexibility is important for many reasons including being

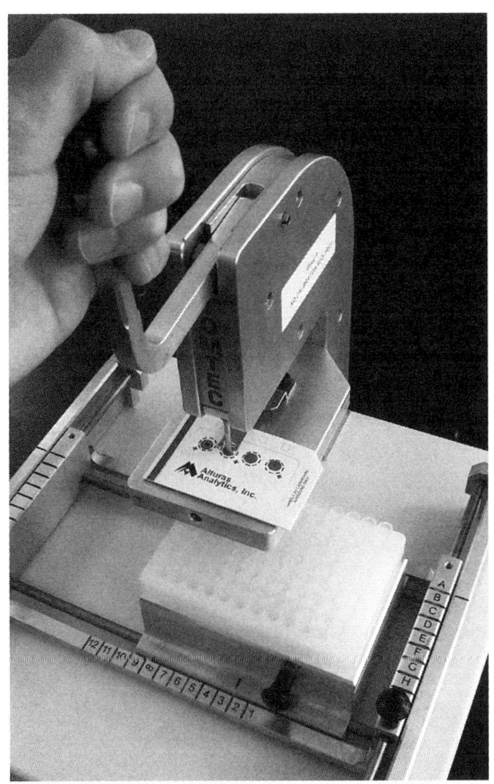

FIGURE 21.6 A "semi-manual" improved DBS/DMS punching device from TOMTEC.

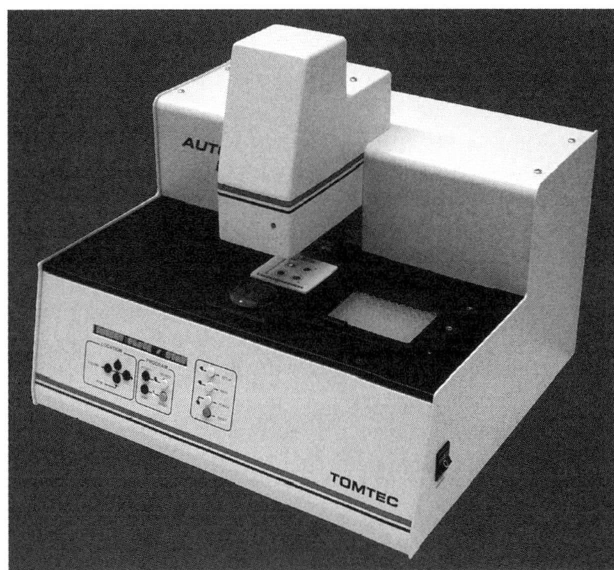

FIGURE 21.7 An automated DBS/DMS punching device from TOMTEC. Reproduced from Johnson et al. (2011) with permission from Future Science Ltd.

FIGURE 21.8 An illustration of an internal standard being added to DBS card using a liquid handling device. Reproduced from Johnson et al. (2011) with permission from Future Science Ltd.

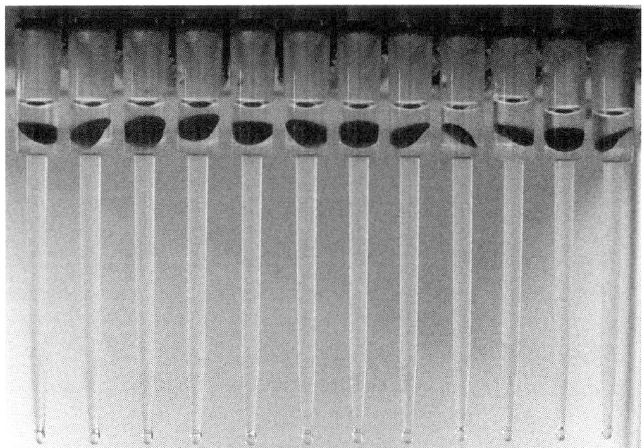

FIGURE 21.9 A DBS punch extraction in dispensing tips of liquid handling device. Reproduced from Johnson et al. (2011) with permission from Future Science Ltd.

able to spot and prepare cards in several manners without limitations to how they are loaded into the automated instrument. Additionally, if the automated system can perform more than just DBS assays, this further improves flexibility and allows for easier economic justification for purchase of the instrument.

As an example of automating DMS analyses with the use of conventional liquid handling devices, Johnson et al. employed improved punching devices along with TOMTEC Quadra and Apricot Designs Personal Pipettor devices to semiautomate DBS extraction of alprazolam, α-hydroxyalprazolam, and midazolam from human whole blood (Johnson et al., 2011a). The punching devices "punch" the spots directly into a rack of solid phase extraction tips. The instruments are then programmed to extract the punches using a variety of conditions and solvents then dispense the sample before injection or further processing. Later, the researchers presented progress at the AAPS conference with automation of the method by using a fully automated punching device from TOMTEC called the AUTO DBS 1 (Johnson et al., 2011b) (see Figure 21.7). For all the automated methods, the liquid handling devices were used to add internal standard to the DBS cards (Figure 21.8) and to extract the analytes from the DBSs (Figure 21.9) to be analyzed by HPLC/MS/MS. The automated and manual results showed statistically the same results.

21.3 CONCLUSIONS

Here we have shown that DMS analysis is a microsampling technique that is useful for sampling biological and *in vitro* fluids such as CSF, synovial fluid, tears, urine, and plasma. As the development of more efficacious drugs with fewer side effects continues, the need to sample these unconventional fluids for drug development will expand the applications of DMS. As scientists continue to investigate and improve the DMS technique, it is likely that the technique will extend into other analytical sampling and analysis fields that may not always involve biological fluids including GMP analytical testing, environmental matrices, forensics, botany, and clinical applications. For example, recent developments (Jokerst et al., 2012) where the researchers developed a paper-based analytical device for colorimetric detection of select foodborne pathogens show great promise.

REFERENCES

Barfield M, Wheller R. Use of dried plasma spots in the determination of pharmacokinetics in clinical studies: validation of a quantitative bioanalytical method. *Analytical Chemistry* 2011;83:118–124.

Barfield M, Spooner N, Lad R, Parry S, Fowles S. Application of dried blood spots combined with HPLC-MS/MS for the quantification of acetaminophen in toxicokinetic studies. *Journal of Chromatography B* 2008;870:32–37.

Carpenter A, Sheaff C, Christianson C, Needham S. Determination of Fluconazole in Human CSF Using Dried Blood Spot Technique and HPLC/MS/MS. Presented at the Annual Meeting of American Association of Pharmaceutical Scientists, New Orleans, LA, November 14–18, 2010.

Christianson C, Johnson C, Zimmer J, Needham S. Dried Blood Spot Analysis – Utilizing the Technique to Develop Assays in Rare or Limited Matrices. Presented at the 58th Annual American Society for Mass Spectrometry Conference, Salt Lake City, UT, May 2010a.

Christianson C, Johnson C, Sheaff C, Laine D, Zimmer J, Needham S. Alturas Analytics Dried Matrix Spot Tears. Presented at Applied Pharmaceutical Analysis 2010 Baltimore, MD, September 19–22, 2010b.

Christianson C, Zimmer J, Laine D, Needham S, Johnson C, Sheaff C, Carpenter A. Development and validation of an HPLC-MS/MS method for the analysis of dexamethasone from pig synovial fluid using dried matrix spotting. *Bioanalysis* 2010c;2:1829–1837.

Christianson C, Johnson C, Needham S. Overcoming the Obstacles of Performing Dilutions and Internal Standard Addition to DBS Analysis Using HPLC/MS/MS. Presented at the 59th Annual American Society for Mass Spectrometry Conference, Denver, CO, June 2011.

Christianson C, Johnson C, Zimmer J, Needham S. Dextromethorphan Metabolism Profiling by the Analysis of Dried Matrix Samples Collected from Liver Microsomes In-vitro Fluid and Analyzed by MFLC MS/MS. Presented at the 60th Annual American Society for Mass Spectrometry Conference, Vancouver BC, Canada, May 20–24, 2012.

Clark G, Giddens G, Burrows L, Strand C. Utilization of dried blood spots within drug discovery: modification of a standard DiLab® AccuSampler® to facilitate automatic dried blood spot sampling. *Laboratory Animals* 2011;45:124–126.

D'Arienzo CJ, Ji QC, Discenza L, Corneluis G, Hynes J, Cornelius L, Santella JB, Olah T. DBS sampling can be used to stabilize prodrugs in drug discovery rodent studies without the addition of esterase inhibitors. *Bioanalysis* 2010;2:1415–1422.

deWilde A, Sadilkova K, Sadilek M, Vasta V, Hahn SH. Tryptic peptide analysis of ceruloplasmin in dried blood spots using liquid chromatography-tandem mass spectrometry: application to newborn screening. *Clinical Chemistry* 2008;54:1961–1968.

Guthrie R, Susi A. A simple phenylalanine method for detecting phenylketonuria in large populations of newborn infants. *Pediatrics* 1963;32:338–343.

Huie K, Takahashi L. Dried Matrix Spot Analysis of Loxapine Phase II Metabolites in Rat Urine. Presented at the 17th North American Regional ISSX Meeting, Atlanta, GA, October 16–20, 2011.

Ingels AS, De Paepe P, Anseeuw K, Van Sassenbroeck D, Neels H, Lambert W, Stove C. Dried blood spot punches for confirmation of suspected γ-hydroxybutyric acid intoxications: validation of an optimized GC-MS procedure. *Bioanalysis* 2011;20:2271–2281.

Johnson C, Christianson C, Sheaff C, Laine D, Zimmer J, Needham S. Use of conventional bioanalytical devices to automate DBS extractions in liquid-handling dispensing tips. *Bioanalysis* 2011a;3(20):2303–2310.

Johnson C, Christianson C, Sheaff C, Laine D, Zimmer J, Needham S. Automated DBS Extractions in Liquid Handling Dispensing Tips and Internal Standard Addition Using Conventional Bioanalytical Instruments. Presented at the Annual Meeting of American Association of Pharmaceutical Scientists, Washington, DC, October 2011b.

Jokerst C, Adkins J, Bisha B, Mentele M, Goodridge L, Henry C. Development of a paper-based analytical device for colorimetric detection of select foodborne pathogens. *Analytical Chemistry* 2012;84(6):2900–2907.

Kehler J, Akella N, Citerone D, Szapacs M. Application of DBS for the quantitative assessment of a protein biologic using on-card digestion LC-MS/MS or immunoassay. *Bioanalysis* 2011;20:2283–2290.

Kolocouri F, Dotsikas Y, Loukas YL. Application of DBS for the quantitative assessment of a protein biologic using on-card digestion LC-MS/MS or immunoassay. *Analytical and Bioanalytical Chemistry* 2010;398:1339–1347.

la Marca G, Giocaliere E, Villanelli F, Malvagia S, Funghini S, Ombrone D, Filippi L, De Gaudio M, De Martino M, Galli L. Development of an UPLC-MS/MS method for the determination of antibiotic ertapenem on dried blood spots. *Journal of Pharmaceutical and Biomedical Analysis* 2012;61: 108–113.

Li W, Tse F. Dried blood spot sampling in combination with LC-MS/MS for quantitative analysis of small molecules. *Biomedical Chromatography* 2009;24:49–65.

Liang X, Li Y, Barfield M, Ji Q. Study of dried blood spots technique for the determination of dextromethorphan and its metabolite dextrorphan in human whole blood by LC-MS/MS. *Journal of Chromatography B* 2009;877:799–806.

Liang X, Yang L, Berezhkovskiy L, Liederer B, Nguyen T, Jia W, Schweiger M, Dean B, Deng Y. Evaluation of dried blood spot sampling following cassette dosing in drug discovery. *Bioanalysis* 2011;20:2291–2302.

Liu G, Ji Q, Jemal M, Tymiak A, Arnold M. Approach to evaluating dried blood spot sample stability during drying process and discovery of a treated card to maintain analyte stability by rapid on-card pH modification. *Analytical Chemistry* 2011;83:9033–9038.

Manicke NE, Yang Q, Wang H, Oradu S, Ouyang Z, Cooks RG. Assessment of paper spray ionization for quantitation of pharmaceuticals in blood spots. *International Journal of Mass Spectrometry* 2011;300:123–129.

Meesters R, Hooff G, van Huizen N, Gruters R, Luider T. Impact of internal standard addition on dried blood spot analysis in bioanalytical method development. *Bioanalysis* 2011;3(20):2357–2364.

Miller JH 4th, Poston PA, Karnes HT. Direct analysis of dried blood spots by in-line desorption combined with high-resolution chromatography and mass spectrometry for quantification of maple syrup urine disease biomarkers leucine and isoleucine. *Analytical and Bioanalytical Chemistry* 2011;400:237–244.

O'Mara M, Hudson-Curtis B, Olson K, Yueh Y, Dunn J, Spooner N. The effect of hematocrit and punch location on assay bias during quantitative bioanalysis of dried blood spot samples. *Bioanalysis* 2011;20:2335–2347.

Rago B, Lui J, Tan B, Holliman C. Application of the dried spot sampling technique for rat cerebrospinal fluid sample collection and analysis. *Journal of Pharmaceutical and Biomedical Analysis* 2011;55:1201–1207.

Ren Xiao, Paehler T, Zimmer M, Guo Z, Zane P, Emmons G. Impact of various factors on radioactivity distribution in different DBS papers. *Bioanalysis* 2010;2(8):1469–1475.

Roberts R, Hanson RP. Eastern equine encephalomyelitis virus isolated from three species of Diptera from Georgia. *Science* 1957;125:395–396.

Rowland M, Emmons G. Use of dried blood spots in drug development: pharmacokinetic considerations. *American Association of Pharmaceutical Scientists Journal* 2010;12:290–293.

US Food and Drug Administration [US FDA]. *Guidance for Industry: Bioanalytical Method Validation*. Rockville, MD: US Department of Health and Services, FDA, Center for Drug Evaluation and Research; 2001.

Viswanathan CT, Bansal S, Booth B, DeStefano AJ, Rose MJ, Sailstad J, Shah VP, Skelly JP, Swann PG, Weiner R. Workshop/conference report-quantitative bioanalytical methods validation and implementation: best practices for chromatographic and ligand binding assays. *American Association of Pharmaceutical Scientists Journal* 2007;9(1):E30–E42.

Vishwanathan C. DBS and Microsampling. Moving Past the Hype to Knowledge and Implementation. Presented at the Annual Meeting of American Association of Pharmaceutical Scientists, Washington, DC, October 2011.

Wiseman JM, Evans CA, Bowen CL, Kennedy JH. Direct analysis of dried blood spots utilizing desorption electrospray ionization (DESI) mass spectrometry. *Analyst* 2010;135:720–725.

WEBSITE REFERENCES

http://www.tedpella.com/histo_html/unicore.htm (accessed April 2012).

PART III

NEW TECHNOLOGIES AND EMERGING APPLICATIONS

22

DIRECT ANALYSIS OF DRIED BLOOD SPOT SAMPLES

Paul Abu-Rabie

22.1 INTRODUCTION

The preceding chapters in this book have discussed in great detail the recent interest in dried blood spots (DBS) for application in multiple fields, such as healthcare, environmental, forensic, and pharmaceutical studies. This interest has been sparked by the significant ethical, financial, and logistical advantages DBS sampling offers over traditionally used methods, which doubtless have also been thoroughly examined elsewhere in this book (Barfield et al., 2008; Edelbroek et al., 2009; Spooner et al., 2009; Al-Ghazawi et al., 2010; Barfield and Wheller, 2011; Suyagh et al., 2011). The third part of this book focuses on new technologies and emerging applications of DBS. The objective of this opening chapter is to provide an introduction and overview to a fascinating part of this topic, namely DBS direct analysis. This chapter aims to explain why DBS direct analysis is desirable, highlight the various options available, discuss the application and performance of these techniques (both from personal experience and from published accounts), and provide perspective on how direct analysis may be used in the short-term and long-term future.

In addition to the general aspiration to make analytical measurements as quick, simple, and cost effective as possible, the interest in using direct analysis techniques to analyze DBS samples has, at least in part, stemmed from the desire to counter some of the disadvantages of using DBS. What are these disadvantages? From an analytical perspective these are largely limited to increased complexity of sample extraction and decreased assay sensitivity compared to conventional plasma sampling techniques. The term "direct analysis" is used to describe techniques that eliminate the laborious manual extraction procedures traditionally used to analyze complex mixtures of samples. To understand the potential benefits on offer from a suitable DBS direct analysis technique, one must first take a close look at the conventional sampling and analysis procedure. Section 22.2 highlights pharmaceutical drug development as an example and compares wet plasma sampling and DBS sampling used in regulated quantitative bioanalysis. This application has been chosen as it involves more stringent criteria than that required in many other areas, and thus provides a full picture of what potentially can be involved in the bioanalytical process. Qualitative and screening applications can ignore many of the processes that are vital in regulated quantitative bioanalysis.

22.2 DBS IN CONTEXT

22.2.1 Sampling of Wet Plasma versus Dried Blood Spots in Pharmaceutical Drug Development

Conventionally, wet plasma sampling techniques have been used in quantitative and qualitative bioanalysis to support drug development studies due to plasma's ease of handling, shipping, and storage compared to whole blood. The conventional approach involves collecting whole blood in the clinic (or animal facility) from a volunteer (or animal) via venous cannula, which is then centrifuged under refrigeration to separate it into plasma and red blood cell fractions. Typically, the red blood cell fraction (which is typically around half of the original whole blood volume collected, depending on the hematocrit) will be discarded, and the plasma portion is transferred into a tube, placed on ice, and maintained frozen during transportation to, and storage at, the site of analysis. In order to derive the appropriate volume of plasma required for

Dried Blood Spots: Applications and Techniques, First Edition. Edited by Wenkui Li and Mike S. Lee.

quantitative bioanalysis, typically >500 μL of blood needs to be collected (Spooner et al., 2009).

It is at the sampling, transportation, and storage stages where the main advantages of DBS lie. In pharmaceutical drug development, following a finger prick to draw blood, typically three spots of blood (of approximately 15 μL) will be added to a DBS sample card using a pipette or glass capillary from the blood sample collected at each sample time point. The first spot is for primary analysis, with spots 2 and 3 being used for reanalysis or incurred sample reanalysis (ISR) when required. The sample cards used are typically composed of a cellulose or glass fiber substrate (which may or may not contain a chemical treatment) which is sandwiched between two layers of thin cardboard (Figure 22.1). Following the application of blood, DBS cards are allowed to dry for a minimum of 2 hours and then stored in plastic ziplock bags with sachets of desiccant, ready for transportation at ambient temperature to the site of analysis.

Sampling whole blood rather than plasma is always going to offer a volume advantage as a large proportion of the sample is not being separated and discarded. This volume reduction is the basis of the ethical advantages of DBS. The smaller volumes (<100 μL) of blood required for DBS samples enable pediatric and juvenile toxicology studies to be undertaken and reduce the need for satellite animals. There are physiological and ethical limitations of obtaining multiple serial plasma samples from individual animals, especially juvenile rodents, which means composite sampling is often required. This results in more animals having to be used and may result in lower quality toxicokinetic (TK) data (Barfield et al., 2008). The small blood volumes used in DBS enable serial bleeds to be taken from the same animal, increasing data quality. These significant reductions in the blood volume required allow for significant benefits in the 3Rs (reduction, refinement, and replacement) for animal use in drug development. DBS also offer the advantage of less invasive sampling (finger or heel prick, rather than venous cannula in human studies) which may aid recruitment of subjects for clinical studies. For rodent studies the lower volumes required also mean that animal warming can be eliminated or reduced, further enhancing the ethical advantages of the approach.

Further financial and logistic advantages come from eliminating the requirement for refrigerated centrifugation to produce plasma and the need to refrigerate samples during shipping and storage. Additionally, transportation and storage are further simplified due to the antimicrobial properties of the DBS sample, removing the requirements for special biohazard arrangements.

22.2.2 Conventional Manual Extraction of Wet Plasma versus Dried Blood Spot Samples in Pharmaceutical Drug Development

In terms of sampling, ethics, cost, transportation, and logistics, DBS have some clear advantages over conventional plasma analysis. Once at the site of analysis, however, some

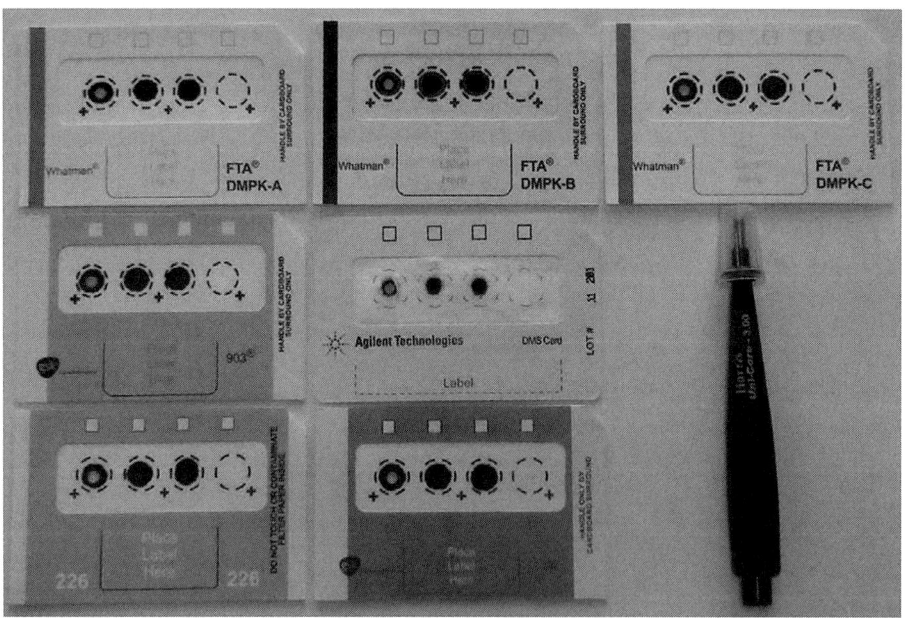

FIGURE 22.1 Dried blood spot 85 × 53 mm 4-spot cards typically used in sampling for pharmaceutical drug development studies. Three 15 μL spots of blood have been applied to each card. The first spot has had a 3 mm diameter disk punched out using Harris Uni-core punching device (also shown). Left to right: top row: Whatman FTA-DMPK-A, Whatman FTA-DMPK-B, Whatman FTA-DMPK-C; middle row: Whatman 903, Agilent DBS card; bottom row: Ahlstrom 226, Ahlstrom 237.

of the disadvantages become evident. The following section continues to use pharmaceutical drug development as an example. At the site of analysis a wet plasma sample is thawed and a volume (typically 20–150 μL) is accurately sub-aliquotted by pipette and taken through the manual extraction procedure. Sub-aliquotting has traditionally been carried out in the analytical laboratory, as an accurate sample volume is essential in generating good quality data. This step has always been deemed as difficult to manage when dealing with untrained personnel (in analytical methodology) across multiple study centers. Extraction or cleanup techniques prior to analysis range from very quick and simple protein precipitation methods, to more complex liquid–liquid extraction (LLE) and solid-phase extraction (SPE) methods. Concentration steps can also be utilized where low detection limits are required.

Elaborate laboratory automation is available that is compatible with all the above techniques and is capable of supporting high sample throughput (hundred of samples per analyst per day). Resulting extract volumes are typically in the region of 50–200 μL. From here, separation and detection of the compounds of interest is typically undertaken using liquid chromatographic (LC) separation coupled to triple quadrupole mass spectrometers. Typically 2–20 μL of extract will be introduced per injection. This technique offers proven selectivity and sensitivity and often high-performance liquid chromatography tandem mass spectroscopy (HPLC-MS/MS) cycle run times short enough to be compatible with high sample throughput requirements (~1.5–3 minutes per sample).

This wet plasma sampling and analysis technique, when validated and performed to internationally accepted guideline criteria (Shah et al., 2000), is acknowledged as being suitable for measuring drug exposures by regulatory authorities. Concentration of analyte in test samples is calculated by plotting the LC-MS/MS chromatographic peak area ratio (analyte/internal standard (IS)) response against a calibration line prepared at the site of analysis. Quality control (QC) samples, also prepared at the site of analysis, are used to ensure confidence in the data generated. As a guideline, calibration standard and QC samples need to be within $\leq \pm 15\%$ of their nominal value for study sample data to be deemed acceptable.

Current DBS manual extraction methods follow much the same procedure as above, but an additional step of punching out the DBS from the surrounding substrate is required prior to extraction. Typically a fixed diameter disc from the center of the DBS is punched out using a sharp cutting tool (Figure 22.1). In practice (within normal hematocrit ranges, and depending on species) an ~15 μL blood spot on cellulose substrate has a diameter of approximately 7–8 mm, and a 3 or 4 mm diameter disk is typically punched out (Figure 22.1). This step is necessary to ensure a fixed volume is sampled in each analysis (remember, only an approximate volume is spotted in the clinic) and is the equivalent of taking an accurate volume of wet plasma using a pipette.[1] Note that, to account for any inaccuracy in the volume of blood applied to the card in the clinic, it is proved during method validation that blood volume variance (e.g., of 15 ± 5 μL) does not produce a significant assay bias (Spooner et al., 2009).

The punched out disk is added to a sampling tube (typically 96-well format) and extracted using the same variety of techniques used for wet plasma analysis. DBS sample analysis is typically carried out by solvent extraction, which transfers analytes from the substrate paper into an injectable solution that is compatible with LC. The simplest form of extraction involves adding 100 μL of highly organic solvent (typically 70:30 (v:v) methanol:water), containing a suitable IS, to this tube and extracting the analyte of interest by agitating the sample for ~2 hours on an automated bench shaker. The sample is then centrifuged to move the disk to the bottom of the tube, and the supernatant is transferred to a fresh tube. The supernatant can then be analyzed using HPLC-MS/MS in exactly the same way as a wet plasma extract. LPE and SPE can also be used, usually in cases where a cleaner extract and/or lower limits of quantitation (LLQs) are required.

In addition to extra complexity of extraction, current DBS manual extraction methods also exhibit less sensitivity than the corresponding wet plasma extraction methods. The lower sensitivity can partially be due to extra ion suppression caused by chemically treated cards, but by far the biggest factor is the reduced quantity of sample being analyzed. A 3 mm diameter punch taken from the center of a 15 μL DBS sample typically corresponds to only around 2.5 μL of blood. Compare this to the 20–150 μL that is typically subsampled during wet plasma analysis, and we have at least a 10-fold decrease in the amount of material being extracted and eventually injected into the detection system. Thus, using manual extraction techniques it is a challenge to meet low LLQ requirements using DBS sampling techniques. To date generally DBS sampling has not been compatible with the development of respiratory drugs that typically need LLQs in the low picogram per milliliter region. The result of this is that not all drugs can be supported with DBS sampling and thus the important ethical, financial, and operational advantages on offer from the technique cannot be maximized throughout the pharmaceutical industry and beyond.

22.3 DIRECT ANALYSIS IN CONTEXT

22.3.1 DBS Direct Analysis—Why Is It Desirable?

The ways in which direct analysis can benefit DBS sampling depends on the application. For simplicity, the applications

[1] For the sake of simplicity, the variation in fixed volume spot size with hematocrit level has been ignored here (see Section 22.3.3.9 for more details).

and how they differentiate can be broadly categorized as follows:

1. *Regulated quantitative bioanalysis (e.g., drug development)*
   Drug concentration measurements must have the highest level of confidence (in both the identity of the compound and the quantity measured) and be determined according to internationally recognized guideline criteria. Throughput will often be an important factor. DBS samples will be collected by trained clinical personnel and shipped internationally from multiple study centers to a facility containing specialist preparation and instrument laboratories operated by trained personnel.
2. *Drug discovery and screening applications*
   This includes some drug discovery and qualitative and quantitative screening applications that do not require the same levels of acceptance criteria required for regulated quantitative bioanalysis. Potentially, this application opens up applicability to direct analysis techniques that do not provide the level of selectivity (or the level of selectivity is unproven) required for regulated quantitative bioanalysis. Throughput will often be an important factor. As above, DBS samples will have been collected by trained clinical personnel and analyzed in specialist facilities containing specialist preparation and instrument laboratories operated by trained personnel.
3. In situ *bioanalysis for therapeutic drug monitoring*
   "*In situ*" in this instance refers to analysis occurring at the site of sampling. DBS samples may be collected by trained (in sampling) personnel in medical clinics (which may be mobile) or possibly by patients themselves at home. In most cases high throughput will not be an important factor. Analysis will be carried out locally in the clinic by untrained (in MS) personnel or theoretically at the home of the patient with a suitable portable device.

Continuing the example of pharmaceutical drug development that was used above (Section 22.2), the extra effort required for DBS extraction may seem like a very minor increase in complexity compared to wet plasma analysis (and in isolation it certainly is). The problem is that in the context of high sample throughput quantitative bioanalysis, any additional analytical complexity, no matter how minor, is always highly undesirable. Indeed, some resistance to accepting the technique (from bioanalysts) has been observed for this reason alone. Anyone that has manually punched out 300 DBS samples in a single session will tell you that it is tedious and tiring! In addition there is also reluctance to change from established techniques that work and are accepted by regulatory authorities, which in the world of pharmaceutical development, for example, is vital. If one looks at "the bigger picture," DBS offer sufficient ethical, financial, and operational advantages to easily compensate for any minor additional analytical complexity. However, any reluctance to accepting DBS, regardless of how founded or unfounded it might be, is a barrier to maximizing the important ethical, financial, and operational advantages that DBS sampling undoubtedly offers.

A potential solution is the use of direct analysis techniques that eliminate the laborious manual extraction procedures traditionally used to analyze complex mixtures of samples. Utilizing direct analysis could not only compensate for the extra analytical complexity involved in DBS analysis, but also significantly simplify the entire bioanalytical procedure. Clearly, direct analysis would also potentially be of a similar benefit to conventional wet plasma analysis, wherever a cheaper, faster, and simpler alternative is desired. It so happens that dry samples, such as DBS, are particularly amenable to many existing direct analysis techniques, and the emergence of DBS as a "new" sampling technique has created an impetus to investigate different (hopefully improved) ways of performing bioanalysis.

Direct analysis is particularly desirable for DBS because, as discussed above, the manual extraction of these samples is more complex than for traditionally used wet plasma. Ideally, what is required is a technique where DBS samples, once shipped to the site of analysis, are loaded directly onto an automated direct analysis device. From this stage no further manual intervention is required. Conventional sample preparation, carried out in the bioanalytical "wet" laboratory, where extraction is carried out, is bypassed completely. The direct analysis device extracts the analyte of interest and is coupled to a suitable separation and detection technique (Figure 22.2). The concept sounds simple, but predictably, developing a technique and putting it into practice is anything but (see Section 22.3.3)!

Direct analysis is desirable for any DBS application where a quicker, simpler and possibly cheaper analysis alternative is advantageous. The majority of bioanalytical measurements require the use of complex HPLC-MS/MS methodology, and consequently the use of both specialist sample preparation and instrumentation laboratories, and trained personnel to perform the analysis. DBS have an obvious application in the area of therapeutic drug monitoring and other applications where it is desirable to run MS applications outside traditional laboratory settings, such as doctor's surgeries. For *in situ* chemical analysis applications such as point-of-care diagnostics the monitoring of drugs in whole blood is critical, as it is in therapeutic drug development, clinical disease treatment, and forensic applications. In this situation cheap and possibly easily portable devices are required that are simple enough for non-MS specialists to use and which would allow them to obtain immediate results. There are three areas

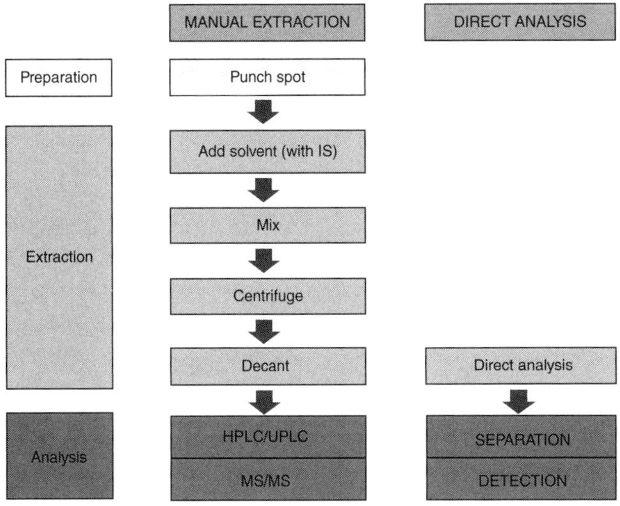

FIGURE 22.2 DBS manual extraction and direct analysis schematics.

that need to be developed to make this a reality (Cooks et al., 2006):

- A sampling technique that is simple enough for nonclinicians to perform with accuracy and produces samples that can be safely shipped, preferably without precautions such as freezing or refrigeration, and is free of contamination hazards
- A method of direct sample extraction and ionization
- MS instrument miniaturization

It has been established that DBS have the potential to be a suitable sampling technique for this purpose. In drug development and discovery environments MS miniaturization will usually not be relevant. However, for some *in situ* bioanalysis applications it could be vital. The concept of portable detectors is certainly not a new one and some interesting progress has recently been made in this area demonstrating that the practical application of this process is realistic (Van Berkel et al., 2008; Soparawalla et al., 2011).

22.3.2 DBS Direct Analysis—What Are the Options Available?

The methods of simplifying or eliminating conventional manual DBS sample extraction that are available largely depend on the intended application. *In situ* bioanalysis for therapeutic drug monitoring direct analysis applications will be largely limited to instrumentation designed specifically for this purpose. For example, Cooks et al. have recently proposed an automated paper spray (PS) technique coupled to a miniaturized MS for this purpose (Section 22.6.1.2) (Liu et al., 2010). For drug development, discovery, and screening applications, in fact anywhere where the use of elaborate preparation and instrument laboratories and trained personnel is available, more options are available. These options can be grouped into three categories:

- Automated DBS extraction
- Direct analysis, which comprises
  - Direct elution and
  - Direct desorption

Automated DBS analysis describes a process by which the manual burden of DBS extraction is relieved through automation. These processes are not direct analysis techniques, and thus will not be covered in great detail in this chapter. However, since for some applications they can be used to reach the same goal, it is important that they are considered as an alternative.

The term direct analysis covers a large and expanding number of techniques, many of them subtle variations on similar processes (see Sections 22.5 and 22.6 and Table 22.1, where the direct analysis techniques discussed in this chapter have been primarily categorized according to their primary mechanism of surface sampling). A number of groups have attempted to summarize and categorize these techniques by their extraction, or primary or secondary ionization mechanisms—a difficult task as in many cases these mechanisms are not fully understood (Cooks et al., 2006; Van Berkel et al., 2008; Venter et al., 2008; Weston, 2010; Harris et al., 2011; Corso et al., 2012). In this chapter, where the focus is on DBS direct analysis, a distinction has been made between direct elution and direct desorption techniques. Direct elution methods are similar to currently used manual extraction techniques in the sense that they use direct solvent interaction to extract compounds of interest from DBS samples and consequently produce a liquid extract that is potentially compatible with HPLC. Direct desorption MS techniques (also described as atmospheric pressure surface sampling/ionization MS, ambient ionization MS, ambient MS, ambient desorption ionization MS, etc.) use an alternative form of extraction (e.g., thermal, laser, or gas jet desorption), and thus do not produce a liquid extract and are not directly compatible with HPLC. The reason for highlighting differentiation on this parameter is due to the long-standing and widely used reliance on HPLC for separation and selectivity in many bioanalytical applications. Direct elution techniques are directly compatible with HPLC-MS/MS, and thus can utilize the selectivity on offer from the technique. This also means that the detectors currently commonly used (triple quadrupole MS) in these applications are still compatible with DBS direct elution.

In some applications (e.g., screening), noncompatibility with HPLC will not be an issue. However, where selectivity is of paramount importance, such as in regulated drug

TABLE 22.1 Summary of Direct Analysis Techniques Discussed in This Chapter

Primary Surface Sampling Process	Desorption/ Ionization Scheme	Desorption Method	Ionization Method	Dominant Ionization Process	Technique	Acronym	Notes	Reference	Commercial Availability	DBS Direct Analysis Evaluation (Published/ Unpublished)
Liquid and gas jet desorption	Direct ionization	Liquid extraction	Induce electrospray from solution on sample substrate	ESI	Paper spray	PS	Voltage applied to a wetted paper substrate containing DBS to create an ESI-like spray plume	Wang et al. (2010a)	Yes	Yes (published)
	Combined desorption/ ionization	Impact sample surface by charged solvent droplet		ESI-like	Desorption electrospray ionization	DESI	Charged droplet/gas jet surface impact	Takats et al. (2004)	Yes	Yes (published)
				SSI-like	Easy ambient sonic spray ionization	EASI	Neutral droplet, also known as desorption sonic spray ionization (DeSSI)	Haddad et al. (2008)	No	No
				ESI	Electrode-assisted desorption electrospray ionization	EADESI	Grounded conductive surface decoupled from nebulizing sprayer	Özdemir and Chen (2010)	No	No
	Multistep ionization	Neutral gas stream	Charged ESI species	Secondary (by ESI)	Neutral desorption extractive electrospray ionization	ND-EESI	No spray solvent or high voltage required	Chen et al. (2007)	No	No
		Charged gas stream	Charged ESI solvent species	Secondary (by ESI)	Extractive electrospray ionization	EESI	Uses charged solvent stream	Chen et al. (2006b)	No	No

Thermal desorption/ ionization	Combined desorption/ ionization	Impact sample surface by charged solvent ions or metastable atoms		APCI—corona discharge	Desorption atmospheric pressure chemical ionization	DAPCI	Chemical sputtering technique	Takats et al. (2005a)	No	No
				APCI-like	Plasma-assisted desorption/ ionization	PADI	Cold-plasma-assisted thermal desorption	Ratcliffe et al. (2007)	No	No
				APCI-like	Dielectric barrier discharge ionization	DBDI	Cold-plasma-assisted thermal desorption	Na et al. (2007a)	No	No
				APCI-like	Atmospheric pressure glow discharge desorption ionization	APGDDI	Also known as flowing pressure afterglow (FAPA)	Andrade et al. (2008a)	No	No
				APCI-like	Low-temperature probe	LTP	Cold-plasma-assisted thermal desorption	Andrade et al. (2008b)	No	No
				APCI—corona discharge	Desorption corona beam ionization	DCBI	Chemical sputtering technique	Wang et al. (2010b)	No	No
	Multistep ionization	Pyrolysis, thermal desorption, and heating	Charged species generated by APCI	APCI-like	Direct analysis in real time	DART	Multistep ionization TD/APCI variant	Cody et al. (2005)	Yes	Yes (published)
				APCI—corona discharge	Atmospheric pressure solids analysis probe	ASAP	Capillary containing analyte sample is inserted into source	McEwen et al. (2005)	Yes	No

(continued)

TABLE 22.1 (*Continued*)

Primary Surface Sampling Process	Desorption/ Ionization Scheme	Desorption Method	Ionization Method	Dominant Ionization Process	Technique	Acronym	Notes	Reference	Commercial Availability	DBS Direct Analysis Evaluation (Published/ Unpublished)
				Secondary (by APCI)	Thermal desorption/ atmospheric pressure chemical ionization	TD/APCI	First introduced in mid-1970s	Ebejer et al. (2005)	Yes	No
				Surface liberation of organic salts	Atmospheric pressure thermal desorption ionization	APTDI	TD/APCI variant	Chen et al. (2006c)	No	No
			Photons and charged species generated by APPI	DAPPI	Desorption atmospheric pressure photoionization	DAPPI	Couples thermal/chemical ionization to APPI reactive chemistry	Haapala et al. (2007)	No	No
			Charged ESI solvent species	ESI	Atmospheric pressure thermal desorption electrospray ionization	AP-TD-ESI	ESI variant	Basile et al. (2010)	No	No
		Laser desorption/ ablation and laser-induced shock-waves	Charged species generated by APCI	APCI—corona discharge	Laser diode thermal desorption	LDTD	Commercially available TD/APCI variant	Wu et al. (2007)	Yes	Yes

Laser desorption/ionization	Multistep ionization	Laser desorption/ablation and laser-induced shock-waves	Charged ESI solvent species	Secondary (by ESI)	Electrospray-assisted laser desorption/ionization	ELDI	First reported combination of laser desorption with ESI	Shiea et al. (2005)	No	No
				Secondary (by ESI)	Matrix-assisted laser desorption electrospray ionization	MALDESI	Addition of matrix required	Sampson et al. (2006)	Yes	No
				Secondary (by ESI)	Laser ablation with electrospray ionization	LAESI	Limited applicability to dry samples	Nemes and Vertes (2007)	Yes	Yes (unpub-lished)
				Secondary (by ESI)	Infrared laser-assisted desorption electrospray ionization	IR-LADESI	LAESI variant	Rezenom et al. (2008)	No	NO
				Secondary (by ESI)	Laser-induced acoustic desorption electrospray ionization	LIAD-ESI	Involves acoustic desorption	Cheng et al. (2009)	No	No
Direct elution (liquid extraction surface sampling/ionization)	Direct ionization	Confined liquid extraction	ESI or APCI	Secondary	Liquid microjunction surface-sampling probe	LMJ-SSP	Confined liquid stream with liquid microjunction surface contact	Van Berkel et al. (2009)	Yes (LESA)	Yes (published)
				Secondary	Sealing surface sampling probe	SSSP	Confined liquid stream with sealed surface contact	Luftmann (2004)	Yes	Yes (published)
				Secondary	Digital microfluidics	DMF	Miniaturization of solvent extraction procedure to “lab on a chip”-type device. Can be configured to spray extract directly to MS	Jebrail et al. (2011)	In develop-ment	Yes (published)

development, an alternative source of selectivity must be available. This may come from different types of detectors (e.g., those with accurate mass or ion mobility separation capabilities) not necessarily readily available in bioanalytical laboratories. Alternatively this could come from a separation process intertwined with the direct analysis technique itself.

In a regulated environment, confidence in selectivity must be proved and the process of getting a new technique approved can be long and intensive (which is a reflection of how important confidence in selectivity is for this application). Essentially, direct elution techniques are similar enough to currently used methodology that their use in a regulated environment can be considered to be achievable in a short time frame. The absence of HPLC compatibility, associated regulatory issues, and the potential extra costs involved in purchasing or developing different types of instrumentation in this environment mean that direct desorption techniques are likely to be a longer term goal for this application. In other applications, these barriers do not exist and DBS direct desorption is far easier to apply as a routinely used technique.

The remainder of Section 22.3 summarizes the additional functionality and other factors that need to be considered for direct analysis techniques. It is important that these functions are understood as they dictate the suitability of the techniques for different applications. Sections 22.5 and 22.6 describe direct analysis techniques in more detail and assess their suitability for DBS analysis.

22.3.3 Not Just Getting Ions into Gas Phase. . . .Additional Functionality and Other Considerations

No matter how well an analysis technique fundamentally performs, it is of limited use if it cannot be made to complement the requirements of the potential application. There is much more involved in creating a useable direct analysis technique than just getting analyte ions into gas phase for MS detection. This section details the additional functionality and other factors that must be taken into consideration in making the routine use of DBS direct analysis become a reality. Figure 22.3, which complements each part of this section, demonstrates a theoretical DBS direct analysis work flow for use in regulated drug development bioanalysis. Each of the following subsections discusses part of this procedure. Again, regulated drug development has been chosen as the working example as this constitutes the most complex workflow. For other applications some of these parameters may not be applicable, but there is considerable overlap.

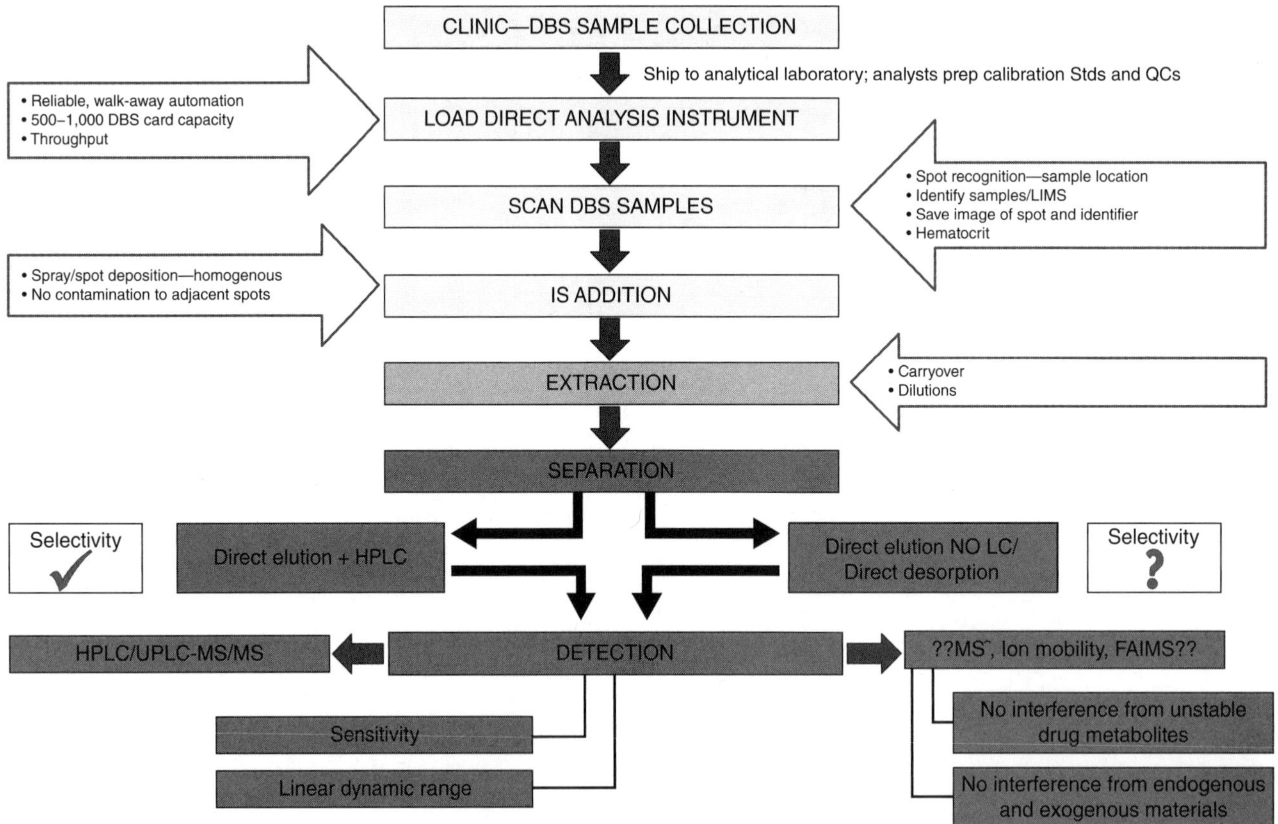

FIGURE 22.3 DBS direct analysis theoretical workflow.

*22.3.3.1 **Automation and Throughput*** The level of automation required for DBS direct analysis depends on the application. In its simplest form, a direct analysis setup consists of the extraction/ionization device coupled to an MS inlet with no additional functionality. In this arrangement each DBS sample would have to be manually loaded or inserted into the device for analysis. Many direct desorption techniques can be used in this manner, as can manually operated direct elution techniques such as the CAMAG thin layer chromatography (TLC)-MS interface (Figure 22.4). For low sample throughput requirements, such a simple setup may be entirely adequate. However, for other applications, such as routine drug development, discovery, and some screening applications, there may be a requirement to analyze hundreds of DBS samples per day, ideally without the need for any human interaction once the "run" or "batch" of multiple samples has been started. In these situations reliable "walk-away" automation, large sample capacities, and run times short enough to suit high-throughput applications are essential.

To offer a realistic alternative to conventional wet plasma sampling, DBS direct analysis needs to offer similar levels of throughput. Direct analysis sample cycle times need to be as close as possible to existing wet plasma analysis timescales, although there is some scope for compromise here due to the efficiency gains brought by removing manual extraction (manual extraction of several hundred DBS samples will take several hours). Sample capacity needs to be able to at least accommodate continuous overnight running (i.e., ~15 hours with no human interaction) with run times per sample as low as 1.5 minutes. This equates to a minimum sample capacity of ~600 samples. Typically, cards will be stacked in racks and mounted vertically or horizontally on an instrument deck or platform. Automated sample handling is then required to move DBS sample cards from these racks on the instrument platform to the area of extraction/ionization. For the widely used 85 × 53 mm 4-spot substrate cards currently favored by clinicians and bioanalysts (Figure 22.1), the sample handling will most likely be achieved by some sort of robotic arm that can pick out a sample card from the storage racks, transfer it around the deck so any additional functions can be performed (e.g., visual recognition, IS addition), and then move it into position so the sample can be extracted, before being returned.

Robust automation of this procedure is a challenge, due to the flexible nature of the 4-spot card format, where the paper substrate is sandwiched between two layers of thin card. The level of robustness desired is hard to quantify, but as an analogy, automation errors should not exceed the occurrence of automation errors or "bad injections" seen in conventional sample injection systems in HPLC instruments, if direct analysis techniques are to be deemed a suitable alternative. A number of commercially available DBS direct analysis systems have struggled to cope with this challenge, and not offered the automation reliability required from anything other than perfectly flat DBS cards (and they are often not perfectly flat, even straight from fresh packaging, before they have been shipped around the world). However, significant progress has been made in automation and the latest instruments utilize highly efficient/robust card handling with suitable error handling and generous card capacity.

(a)

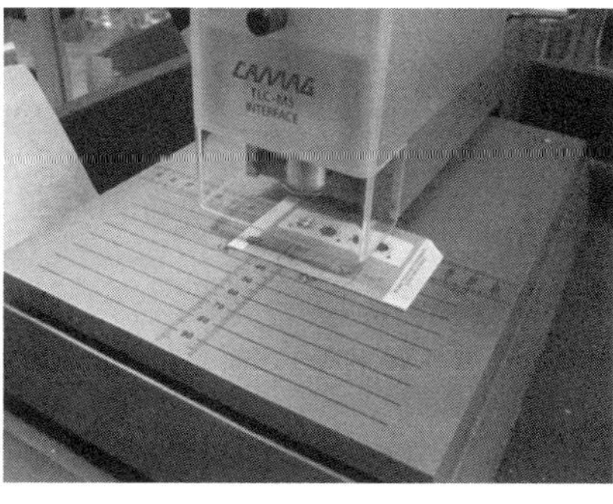

(b)

FIGURE 22.4 Examples of nonautomated direct analysis: (a) direct analysis in real time (DART) being used to analyze a tablet by manually inserting it between the heated gas stream of the DART probe and the MS inlet (photo from JEOL USA, Inc., used with permission); (b) CAMAG TLC-MS interface being used to directly elute a DBS sample. The DBS card is manually placed on the extraction platform, and the center targeted using a laser crosshair. A switch is pressed which lowers the extraction head onto the center of the spot using a pneumatic system. Extraction is then carried out by manually switching the flow of solvent "through" the DBS sample which then is directed into the HPLC mobile-phase flow and into the MS source.

Further challenges include making systems compatible with the variety of DBS substrate cards on the market, as they have different thicknesses, and different substrate material strengths (e.g., glass fiber-based substrates tend to be much easier to punch than cellulose). This latter parameter is an issue for automated punches and direct elution techniques where a sealed sampling area is created on the DBS sample (Section 22.5). A further challenge is to integrate compatibility with formats other than the 4-spot card that has been widely accepted in drug development, as used in some screening applications. There has also been some demand for a multipurpose integrated sample rack that combines the following functions:

- Drying, immediately after whole blood has been applied to the substrate card
- Shipping to the site of analysis
- Storage at the site of analysis
- Analysis of samples (i.e., rack could be loaded directly onto direct analysis instrument)

However, a solution that is cheap enough to produce (it would need to be sent out to multiple study centers) and compact enough not to be detrimental to shipping costs has yet to emerge. The majority of systems currently rely on the analyst to transfer sample cards to custom system storage racks.

Recent interest in DBS direct analysis has led to significant progression in the automation capabilities for a number of techniques. In the space of the last couple of years direct elution techniques have progressed from fully manual devices, to simple low-capacity automation, to large-capacity full functionality automation (Section 22.5). An example of direct desorption automation is demonstrated by IonSense, Inc., who have developed a "conveyer belt"-type approach to DART (Direct Analysis in Real Time) which can be used to raster a DBS sample between the gas jet and MS inlet and also alter the angle of sampling (see Section "Direct Analysis in Real Time"). This approach would be suitable for a number of direct desorption techniques. Cooks et al. have taken a slightly different approach with PS technology (Section 22.6.1.2). This technique requires the paper substrate containing the DBS to be cut to a sharp point. They have thus taken the logical step of using a custom-designed cartridge system that is used at the point of sample collection and is then shipped and loaded onto an automated PS direct analysis instrument at the site of analysis.

A number of existing automation platforms have been modified with DBS punch and visual recognition modules to enable the use of automated DBS extraction. This approach could have cost advantages over purchasing new and elaborate DBS direct analysis instrumentation (Section 22.4).

22.3.3.2 Visual Recognition and Sample Identification

Many automated direct analysis systems will require a visual recognition system to accurately locate the position of each spot on the sample card. Sample cards contain guide marks for applying blood (Figure 22.1), and experienced bioanalysts apply perfect blood spots within these guidelines with ease. However, clinical personnel who are new to the technique will often apply blood slightly outside these areas and very occasionally produce spoiled samples that are unsuitable for analysis (i.e., where the volume applied is obviously much smaller or larger than what it should be, the blood is smeared or spackled, the surface of the substrate is damaged, or a mistake is made) (Figure 22.5). In addition, there is some variation in the printing of substrate cards which means the position of the guide marks varies between manufacturers and also from one batch of the same card to another. Hence an integrated visual recognition system is absolutely vital for accurate sampling.

Current automated instruments use a camera or scanner for this purpose and need to be coupled to intelligent software that can identify a number of variations that would otherwise be identified by eye. These include the absence of DBS samples, "bad samples" (those that fall outside predetermined size and shape parameters), and whether a sample has already been sampled. Figure 22.5 highlights a number of examples of unacceptable blood spots that would be rejected for analysis. A visual recognition system also provides the means for a number of other potential secondary functions. These include the ability to identify each sample, perhaps through a bar code reader system or radio frequency (RF) emitter chip. This facility could be linked to laboratory information management systems (LIMS), ensuring that the

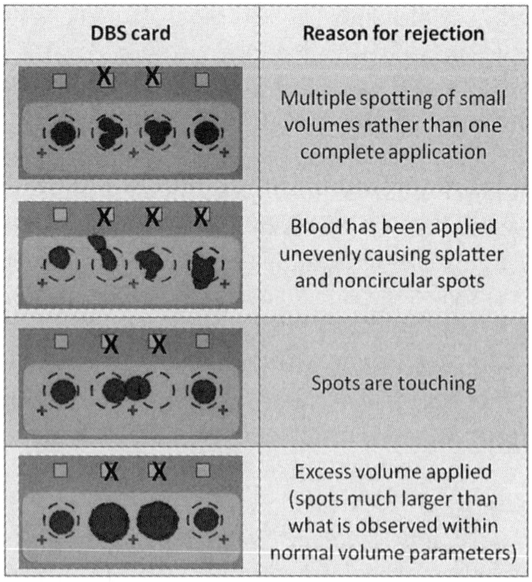

DBS card	Reason for rejection
	Multiple spotting of small volumes rather than one complete application
	Blood has been applied unevenly causing splatter and noncircular spots
	Spots are touching
	Excess volume applied (spots much larger than what is observed within normal volume parameters)

FIGURE 22.5 Examples of unacceptable DBS. "X" in check box above DBS denotes DBS samples that would typically be deemed to be unacceptable for analysis.

analysis and detector sequence lists are reconciled. An image of the DBS (before and after extraction) and identification may wish to be saved for regulatory reasons. Estimating the hematocrit level of a DBS sample has also been attempted, with some success, by measuring the intensity of color in the sample via the visual recognition system (see Section 22.3.3.9).

22.3.3.3 Internal Standard Addition For many quantitative analysis applications, such as drug development, IS is added to correct for variable MS performance (ionization and detection), for losses during preparation, and possibly for variability in sample injection volume. The ideal IS is a stable isotopically labeled analog of the drug of interest. A peak area ratio (the area under the analyte chromatographic peak divided by the IS area) is used for quantitation. The following summarizes the options available for DBS IS addition:

1. *IS incorporated into extraction/elution solvent*
   This method is widely used for manual extraction and some direct analysis techniques (e.g., direct elution and PS), but is obviously not an option for direct desorption techniques where a suitable extraction or elution medium (i.e., a solvent or liquid interaction) is not used. This method of adding IS is simple, widely used in drug development applications, and accepted by regulatory authorities. However, when applied in this way, the IS is not fully incorporated into the matrix components and sample paper prior to extraction and is therefore not being co-extracted with the analyte as part of the sample. Thus, only limited information on assay performance is offered by this approach.
2. *Substrate pretreated with IS*
   Incorporating blank substrate (such as the cellulose or glass fiber cards currently in use) with IS prior to applying the wet matrix would likely ensure the IS is integrated and extracted with the analyte. This is an ideal technique for large-scale applications such as certain therapeutic drug monitoring and screening applications, where a large number of samples will be analyzed for a small finite number of compounds of interest. However, the presence of an IS is likely to be less important in these applications. For other applications, such as drug development, this approach is unlikely to be logistically feasible when dealing with large numbers of studies, study centers, and compounds, if the cost and procedural simplifications on offer from DMS are to be kept intact.
3. *IS added to matrix before spotting onto paper at clinic*
   Adding IS to liquid matrix prior to spotting onto paper substrate would ensure the IS is fully associated with matrix components along with the compound of interest. This may be possible in practice for a small number of applications. However, in many cases this approach is even less logistically feasible than pretreating the substrate with IS, as it involves accurate volumes being dispensed in a clinical environment. This would overcomplicate the procedure in the clinic and move away from the simple process DBS sampling offers.
4. *IS applied to DBS prior to extraction*
   A technique that applies IS to DBS samples *prior* to manual or direct analysis would improve the integration of IS to the sample compared to currently used manual extraction techniques. However, this approach is only possible if the IS is given sufficient time prior to extraction for it to bind to the matrix and substrate components and does not adversely affect the distribution of the analyte. Such an IS application would preferably be carried out once samples have been shipped to the site of analysis (e.g., analytical laboratory), but before undertaking the extraction procedure. This procedure is potentially easily compatible with direct analysis techniques and could be easily configured into a fully automated procedure.

The only feasible methods for IS addition with drug development applications are options 1 or 4 above. As detailed above, adding IS at the time of extraction means that IS is not fully integrated into the matrix components and substrate of the sample prior to the extraction process and therefore does not correct for any variability during the extraction process. It should be noted that this method, when used in wet plasma analysis, is widely accepted by the appropriate regulatory authorities. However, this is not a reason not to try to improve bioanalytical techniques wherever possible, to ensure the best possible quality data are produced. Also, as previously mentioned, for many direct desorption techniques no suitable medium to add IS via an extraction solvent exists, and thus applying IS prior to analysis becomes the only (logistically feasible) option. Clearly, IS performance is an area in which the current manual extraction method can be improved on, and developing new DBS direct analysis techniques provides an opportunity to do this.

Alternatives have been investigated with the aim of finding a technique that could better integrate IS with DBS samples prior to extraction, while still offering acceptable reproducibility, maintaining the simplicity on offer from DMS sampling, and being compatible with manual extraction or direct analysis techniques. To date, research has focused on spray-type IS applications. The ideal technique would involve the homogenous application of IS solution across the width and depth of the DMS to ensure its full interaction with matrix and paper substrate components of the sample, in a manner similar to the analyte. The application of the IS would also not cause any significant chromatography effect that could significantly affect the concentration or distribution of analyte within the sampling area. Abu-Rabie et al. used a piezo electric spray technology (TouchSpray, from The Technology Partnership, Cambridgeshire, UK) to apply IS to DBS samples prior to manual extraction or direct elution and demonstrated that assay performance was at least as good as that offered by conventional IS application via the extraction solvent, or where IS is added to control blood prior to standard preparation (Abu-Rabie et al., 2011).

The most important parameter for applying IS to DBS using this approach is the mechanism of IS addition and the volume applied. If too large a volume of IS solution is applied in a small localized section of the sample (a static spray arc) it becomes "flooded" and wicks out to the surrounding area. Thus, a chromatographic effect that could significantly affect the concentration or distribution of analyte within the sampling area is risked (Figure 22.6a). To overcome this, a static spray arc with a larger diameter (covering the whole spot) can be applied (Figure 22.6b). The problem now becomes finding a balance between applying a small enough volume to avoid a significant mass-flow wicking effect and a large enough volume to ensure full depth penetrating of the DBS sample. Partial DBS depth penetration would still likely produce a performance improvement over conventional IS addition via an elution solvent, but it is clearly not optimal. In order to maximize the potential performance increases on offer, a suitable balance needs to be reached. It is likely that this could be achieved by optimizing the mechanism of application, the volume applied, and other parameters such as flow rates, dispensed particle size, solvent, and application distance.

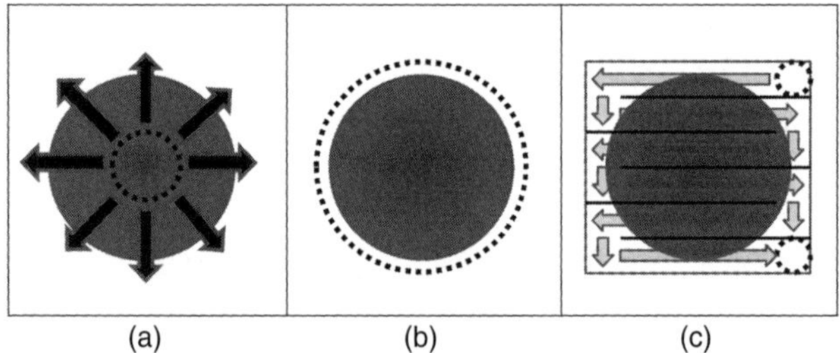

FIGURE 22.6 Methods of IS addition via static and dynamic spray-type technologies. Dashed line represents solvent spray arc diameter that initially impacts DBS surface. (a) To cover an entire DBS using a static spray arc smaller than the DBS diameter the DBS has to be flooded (~24 μL IS in methanol onto an 8 mm diameter 15 μL DBS) and relies on a wicking effect to reach the perimeter of the DBS. This causes a significant chromatographic effect which can effect analyte concentration for some compounds. (b) A larger diameter static spray arc can be used to cover the entire DBS. Lower application volumes (~8 μL) avoid the undesirable chromatographic effect but can limit IS penetration into the full depth of the DBS. It may be possible to reach an effective compromise using this technique. (c) Dynamic spray application can be used to apply larger IS volumes (up to ~40 μL, thus fully penetrating the depth of the DBS) without causing a significant chromatographic effect.

A subtle refinement of the technique has been used by CAMAG and integrated into their automated DBS-MS 500 direct elution platform (see Section 22.5.3.2). Rather than applying the IS solution as an arc covering the entire DBS, this system applies a fine spray covering only a very small section of the sample and uses automation to move the sample relative to the spraying device (a dynamic spray application). The result is a 10 × 10 mm grid of IS solution applied centered around the DBS (Figure 22.6c). This system has been shown to allow greater volumes to be applied (thus penetrating the full depth of the DBS sample) before the limit at which a significant chromatographic effect occurs is reached.

In this automated approach, after IS addition, the samples would then be left to dry for a suitable period of time before being analyzed using a direct analysis. Further understanding of the technique is required to identify the minimum drying times required prior to analysis. In practice, it is likely in an automated system that a full batch of DBS samples will have IS applied and be allowed to dry. The instrument will then automatically move on to the direct analysis phase. Another consideration of note is that contamination of IS to adjacent samples on a single card, or from one card to another, must be avoided.

22.3.3.4 Sensitivity Overall bioanalytical assay sensitivity is a combination of

- The quantity of sample available
- The quantity of compound recovered from the sample after extraction
- The sensitivity of the detector

For many applications, the therapeutic ranges or compound concentrations likely to be encountered are not likely to offer challenging levels of sensitivity for the extraction methods and detectors used. However, in drug development, the low levels of quantitation, especially in the analysis of respiratory drugs where low picogram per milliliter levels of detection are often required, push the limits of sensitivity on offer from even the latest, highly sensitive triple quadrupole mass spectrometers when analyzing wet plasma extracts. For this reason, DBS analysis (using manual extraction) is not currently compatible with respiratory compounds. A typical DBS-punched disk may contain around an order of magnitude less drug than in typical wet plasma samples. While the benefits of DBS sampling for this application have allowed the technique to progress despite this disadvantage, there is a desire to maximize the advantages of DBS by making it compatible with as many compounds as possible. Therefore, direct analysis techniques must at least match the sensitivity of manual DBS extraction to be a suitable alternative. Any extra sensitivity beyond this would be a considerable bonus. Potentially, increased assay sensitivity could come from

- Analyzing more sample than the small central sub-punch portion currently taken in manual extraction (e.g., 3 or 4 mm diameter disk punched from a 15 μL DBS)
- Maximizing assay recovery (thus producing a greater detector response for the same sample input)
- More sensitive detectors

Mass spectrometry (MS) instrumentation is continuously being refined, and more sensitive instrumentation is launched year on year. Typically, triple quadrupole MS is the most sensitive platform. However, even the latest triple quad MS does not have the sensitivity required to enable support for many respiratory-based compounds in drug development using manual DBS extraction. Other types of MS including accurate mass and high-resolution instruments and those utilizing other separation techniques (e.g., ion mobility) may be required for non-HPLC-compatible direct analysis techniques (see selectivity in Section 22.3.3.5). Huge improvements in sensitivity have been made in hybrid platforms in recent years and while they still trail the latest triple quads in most cases, the gap is closing.

The other routes for improving sensitivity are through sampling as much of the DBS as possible and maximizing recovery. A major limiting factor in manual extraction is the small sub-aliquot (3 or 4 mm punch) taken from a DBS sample (typically 15 μL). This may equate to only around 2.5 μL of blood, which is why sensitivity suffers compared to plasma analysis. The ability to sample more of or the entire sample (see Section 22.3.3.9) is an obvious route to improving sensitivity. Another limiting factor of current LC-MS/MS manual extraction techniques is that often only a small portion of the total extract can be injected onto a HPLC column if acceptable chromatography is to be maintained. For example, a 3 mm diameter DBS punch may be extracted in 100 μL methanol. Typically, only 2–20 μL (i.e., 2–20% of the compound extracted) of this extract can be injected on column. Trapping column approaches can be used to improve the amount of analyte that can be introduced to the detector, but this adds to cycle times and creates more complex methodology, which is highly undesirable in many high sample throughput applications.

Direct analysis techniques that bypass the assay-sensitivity limiting factor of on-column injection volumes being limited by the acceptability of resulting HPLC-MS/MS chromatography, or that otherwise increase recovery, could also increase sensitivity.

22.3.3.5 *Selectivity, Separation, and Types of Detectors* The selectivity required from DBS analytical techniques largely depends on the application being supported, and consequently how much confidence is required in the identity of the compound being measured. In regulated pharmaceutical drug analysis the utmost confidence in this factor is vital. HPLC coupled to highly sensitive triple quadrupole mass spectrometers is the current separation and detection method of choice in this field and is almost exclusively used. LC separation is used to separate analytes of interest from coexisting components and, in conjunction with tandem MS monitoring (detecting the parent ion of a molecule, and a specific product fragment ion following passage through a collision cell), provides the selectivity required.

Any extraction technique that produces a liquid extract (manual/automated DBS extraction, direct elution) is potentially compatible with HPLC-MS/MS. Many direct desorption techniques do not produce a liquid extract and/or are not compatible with HPLC-MS/MS. The extraction product is transferred directly to the MS without further separation (unless inherent in the extraction process). This results in a situation where a question mark is raised over the level of selectivity, and thus the level of confidence in analytical data, where direct desorption is used without LC (see Figure 22.3). In practice removing LC can result in poor sensitivity due to ion suppression and reduced selectivity, and risks assay interference via metabolite decomposition (e.g., N-oxides and glucuronides) into parent compounds during MS ionization. Potentially, circulating drug concentrations can be overestimated if fragile metabolites or pro-drugs, such as acyl glucuronides and N-oxide metabolites and esters, convert back to the parent drug. To overcome this issue, further understanding is required into the likelihood of these "risks" involved in analysis sans LC separation.

Eliminating LC would offer highly desirable and significant simplifications to the bioanalytical workflow, but cannot be at the expense of adequate selectivity. It is here that perhaps different types of detectors and separation techniques than those commonly found in drug development bioanalytical laboratories (where triple quad MS is prevalent) could be of assistance. It is possible that high mass accuracy and high-resolution MS, together with separation techniques such as ion mobility or high-field asymmetric waveform ion mobility MS (providing that fragmentation does not occur during the ionization process), or perhaps some other novel approach, could fill the selectivity gap vacated by LC and provide unambiguous assignment of the signals obtained.

For many other applications, HPLC compatibility, and the regulatory consequences of this, is not such a major issue. It is in these areas that direct analysis without LC has less barriers to overcome, so the benefits on offer can hopefully be reaped in a much shorter timescale.

22.3.3.6 *Cross-Contamination/Carryover* Contamination between samples must be controlled to ensure confidence in analytical results. As a target guideline, in quantitative pharmaceutical drug development ideally the level of contamination in blank samples following samples containing analyte should be ≤20% of the analyte response at the LLQ. In conventional wet plasma analysis, cross-contamination can be avoided by careful sample preparation. During subsequent LC-MS/MS detection, contamination is controlled through the use of integrated wash systems that clean the injection syringe between samples. A similar system is required for DBS direct analysis. Very large levels of carryover is often observed in direct elution systems, due to extract collecting in sealed sampling areas and the capillaries leading from them to the detection system. It has been demonstrated that suitable wash systems that rinse the sampling and extraction apparatus can reduce carryover levels to well within internationally accepted guideline criteria, and under what is observed in conventional wet plasma analysis (Abu-Rabie and Spooner, 2009, 2011). The integration of such systems is thus vital if DBS direct elution techniques are used in a regulated development environment (see Section 22.5.3.2).

Carryover in direct desorption systems has not yet been identified as a major issue and will obviously depend on the specific technique. Understanding of this will increase as direct desorption becomes more widely used in applications where guideline carryover level criteria must be adhered to.

22.3.3.7 ***Robustness and Reproducibility*** If direct analysis techniques are to be viewed as suitable candidates to replace existing manual extraction techniques, then similar levels of robustness and reproducibility must be provided. Experience with LC-MS/MS analysis of manually extracted DBS has demonstrated that typically many hundreds of samples can be analyzed before mass spectrometer performance is affected by buildup of matrix components (at which point the interface will require cleaning to remove the buildup of co-extracted matrix components which eventually cause a decrease in sensitivity). Similarly, HPLC columns (and pre-column filters) will typically last many hundreds of injections of manual DBS extracts before buildup of co-extracted components causes performance to decrease below acceptable limits.

One of the potential problems with direct analysis is that conventional cleanup steps are bypassed, often resulting in a "dirtier" extract or product being introduced into the MS. Clearly this has the potential to compromise the robustness of the analytical procedure. Large-scale robustness studies are required to identify the significance of this issue in high-throughput environments. For example, direct elution systems lacking trapping columns demonstrably produce a relatively dirty extract. However, chromatographic performance measured over many hundreds of consecutive DBS and dried plasma spot (DPS) samples has shown that direct elution can perform at least as well as the conventional manual extraction technique (see Section 22.5.3.3) (Abu-Rabie and Spooner, 2011).

Even where IS is added to correct for variation, there is a limit to how far the MS response can vary before it adversely effects the sensitivity and linear dynamic range of the assay. Reproducibility can be judged by the ability to meet guideline-acceptance criteria (where they exist). In the case of regulated drug development, the acceptance criteria applied to the accuracy and precision of multiple QC samples are a very direct way of identifying if a technique is suitable for the application and how well it compares to conventional manual DBS extraction (Shah et al., 2000). A number of DBS direct elution and direct desorption techniques have been shown to produce assay validation data that meet internationally recognized guideline criteria for regulated drug development (note that some of this was used without HPLC) (Abu-Rabie and Spooner, 2009; Wiseman et al., 2010; Manicke et al., 2011b). The acceptance criterion for this application states that the accuracy and precision of QC sample data should be ±15% ($n = 6$) across a range of concentrations covering the range of the assay.

22.3.3.8 ***Dilution and Linear Dynamic Range*** For many applications, a direct analysis solution also needs to be able to accommodate sample dilution requirements. Typically, in drug development applications, current LC-MS/MS methods supporting DBS manual extraction will have an assay range limited to 3 orders of magnitude (e.g., 1–1000 ng/mL) in line with the 3–4 orders of magnitude linear dynamic range typically on offer from triple quadrupole MS. If a DBS sample is found to contain a concentration beyond the higher limit of quantitation (HLQ), then a repeat analysis will be performed where the DBS extract is diluted by a known factor to bring it within range. This procedure is typically carried out by diluting a DBS extract with a matrix-matched diluent (blank control DBS extract containing IS) (Spooner et al., 2009). An alternative "doughnut punch" method has also been reported where a small-diameter punch (e.g., 1.1 mm diameter) is taken from both the sample to be diluted and a blank blood sample. The sample punch is added to the extraction tube, and the blank punch is discarded. A second 4 mm diameter punch is then taken from this blank sample, centered over the existing 1.1 mm diameter punch (hence the term "doughnut punch"). This second punch is then added to the extraction tube containing the 1.1 m diameter sample punch, resulting in an effective dilution of 11.1 in this example (Ewles et al., 2011).

How can DBS direct analysis techniques accommodate sample dilution? The ideal solution would be to use detectors that have a much larger linear dynamic range (a minimum of 5–6 orders of magnitude) than what we currently use (typically 3–4 orders of magnitude), so that sample dilution would no longer be necessary. Analytical methods could be developed with larger ranges that would encompass the higher concentration samples sometimes encountered in early stage drug development studies. Clearly, this approach is more of a work-around than a solution and unfortunately such detectors are not currently available and are not likely to be in the short-term future. A reasonably simple alternative is available for direct elution whereby the liquid eluate following extraction could be diverted and diluted with matrix-matched solvent before being redirected to the HPLC column and/or detector. This alternative approach could be reasonably easily integrated into an automated direct elution system. A dilution mechanism for DBS direct desorption techniques has yet to be established. It is hoped that as the sampling mechanisms occurring in DBS direct desorption become better understood, a reproducible way of subsampling (by a known factor) to bring detector responses within assay ranges can be incorporated into the techniques.

22.3.3.9 Hematocrit The relationship between DBS sample hematocrit and assay bias is likely to be one of the most important issues that must be overcome if DBS sampling is to be used to its full potential in the pharmaceutical industry. Current thinking is that this issue needs to be addressed before practical application of DBS analysis (in regulated bioanalysis) can progress to the next level, and any direct analysis technique needs to be compatible with this solution. A full discussion of the DBS hematocrit issue is beyond the scope of this chapter. However, this critical issue could have a major influence on the progression of DBS direct analysis and therefore the following section attempts to summarize the main issues.

It has been demonstrated that hematocrit values that deviate significantly from the "normal" expected range can significantly affect assay bias when using manual extraction of DBS samples for the quantitative bioanalysis of drugs (Denniff and Spooner, 2010). The overall assay bias caused by changes in hematocrit is made up of three components: area bias, recovery bias, and ion suppression bias (Figure 22.7) (Denniff and Spooner, 2011). Analyte recovery is dependent upon hematocrit level, with high hematocrit levels resulting in lower recoveries. There is evidence to suggest that as analyte recovery falls, the effect of hematocrit increases, resulting in a wider recovery bias range (Denniff et al., 2012). Thus, maximizing analyte recovery should reduce the range of the recovery bias as hematocrit changes. An improved method of IS addition (see Section 22.3.3.3) could assist here. In conventional manual extraction techniques (where the IS is not fully incorporated into the matrix components and sample paper prior to extraction) recovery bias occurs because the IS is not co-extracted with the analyte. If both the analyte and IS were extracted together, then variation in recovery would not be eliminated, but the bias would fall to zero. Where elevated levels of suppression or a correlation with hematocrit level is observed, the effects can be minimized by utilizing an additional sample cleanup step to remove interfering species and/or by modifying chromatography to move the analyte away from areas of suppression.

This leaves the area bias which can be visualized as being influenced by the viscosity of the whole blood sample, which is directly related to its hematocrit. For a fixed volume of blood on cellulose substrate, high hematocrit results in a relatively small spot area, low hematocrit a relatively large area. This means that the analyte density of each spot will be different (i.e., a fixed volume of blood will spread over different areas depending on the hematocrit value). To better understand this issue it may be helpful to consider the following exaggerated example: Two patients both have the same concentration of a drug circulating in their bodies. Patient A has a very high hematocrit while patient B has a very low hematocrit. Both patients have 15 μL of blood taken via finger prick. Both these samples contain the same amount of drug. However, patient A's DBS sample eventually covers 20 mm^2, while patient B's covers 40 mm^2. When a fixed diameter (say, 3 mm) sub-punch is taken from both DBS samples, a much larger proportion of the sample is taken from patient A. Thus, when the samples are subsequently analyzed, a much higher concentration of drug is found in patient A (as the blood, and hence the drug, has spread over a much smaller area) despite the drug concentrations circulating in the body actually being identical.

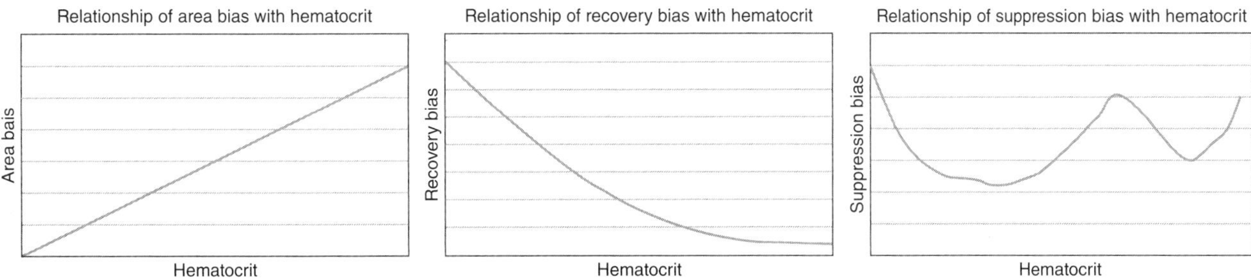

FIGURE 22.7 Components of hematocrit-based assay bias and how the bias varies with hematocrit level.

The ideal solution for controlling the area bias would be the use of a novel substrate that behaves independently of hematocrit, which would enable the current workflow to remain unchanged. This could be via a novel substrate material, substrate configuration, or the addition of a suitable modifying agent (that, e.g., perhaps eliminated differences in drying rates) to the current format. Likewise, a modifier could potentially be added to the sample prior to spotting. Unfortunately, such a substrate or modifier has yet to be reported. Currently, during method validation, the effect of hematocrit on assay bias is assessed over a range of normal hematocrit values for the matrix of interest. This ensures bias is within acceptable limits (typically $< \pm 15\%$). The problem is that hematocrit values of study samples may be unknown and certain disease states or exposure to certain drugs will take a patient's hematocrit level beyond normal ranges.

An alternative could be to extract the whole spot rather than a sub-punch, which would eliminate the effect of any DBS area variability. Unfortunately, this workflow would also require an accurate volume to be dispensed at the clinic. Of course, the reason for taking a sub-punch from the DBS in the first place is to avoid this! Additionally, it was also previously considered not possible for accurate volumes to be reliably dispensed in the clinic! To work around this problem, devices that could enable accurate volume dispensing of blood to be feasible in the clinic are being explored. This is no easy task as such a device would need to be cheap, disposable, and easy enough to use accurately by non-analysts in a clinical environment.

A number of groups have reported DBS analysis techniques related to the whole spot extraction concept. Li et al. (2011) reported the use of perforated dried blood spot (PDBS) which is essentially a simplified version of whole spot extraction. This approach partially precuts (or perforates) a small-diameter circular perimeter in paper substrate. An accurate amount of blood is deposited that largely fills the substrate area within the perforations and is allowed to dry. The entire DBS can then be easily pushed into an extraction tube using a single-use pipette tip. The technique is claimed to offer decreased wastage through complete sample utilization, no requirement for punching, ease of recovery assessments, and elimination of sampling influence due to hematocrit level (though crucially, this last advantage is no different to any other whole spot extraction technique). This technique could potentially be easily compatible with existing automation platforms with only minor modifications. Youhnovski et al. (2011) published a similar concept named precut dried blood spot (PCDBS) and reported good analytical performance and hematocrit independency. Likewise, Fan and Lee (2012) reported a similar commercially available approach named dried matrix in paper disc (DMPD).

Another option could be to apply a correction factor that modifies the analytical result based on the hematocrit value. This correction factor would rely on an accurate volume of blood being spotted and the hematocrit value of the samples and calibration standards being known. Such a correction factor could be derived from the knowledge of the difference between the hematocrit of standards and samples (Vu et al., 2011). It has been demonstrated that visual recognition systems can measure spot area (or even hematocrit level through sample color) with good accuracy, but calibration is required. Alternatively, automated analyzers are available that can measure hematocrit (from wet blood samples) in isolation. Using a correction factor is certainly worthy of consideration, as it could allow the current workflow to be maintained, albeit with some additional steps. The downside is the level of complexity this could add to the overall process as well as the additional time it could take to prove and implement such a technique in a regulated environment.

The hematocrit issue certainly complicates the ultra simple workflow on offer from DBS sampling, but the introduction of an elegant sampling device in the clinic should enable the benefits of DBS to be reaped without a significant compromise. What is the impact on direct analysis? For automated DBS analysis and direct elution the same techniques could be used with minor modifications (such as larger punches or sealed sampling areas to fully encompass the entire DBS). It was thought that a potential problem with whole spot direct elution could be that the extraction solvent would take the path of least resistance through the surrounding blank substrate rather than interact with the DBS. However, it has already been shown that this is not the case and it has been demonstrated that direct elution is compatible with whole spot elution without loss of sensitivity or chromatographic performance (Abu-Rabie, 2011). Another advantage of whole spot extraction/elution is that potentially sample volumes could be further reduced without effecting assay sensitivity, as the same amount of blood could be sampled in total as is currently sampled in the sub-punch (typically 3–4 mm diameter), eliminating the wastage that is currently encountered with manual extraction. Alternatively, current typical DBS volumes could be retained and many of the sensitivity issues with DBS would be eliminated.

It is not currently clear how hematocrit level effects direct desorption techniques, but it can be assumed that the area bias will be relevant in any analysis technique that samples a localized area rather than the whole sample. Further investigation is required into mechanisms of direct desorption sampling and how hematocrit variation will bias analytical results.

For many applications the hematocrit issue will not be relevant or significant, especially in situations where there is confidence that hematocrit levels are within "normal" levels. In the vast majority of cases, it has been shown that significant assay variation is only observed in extreme hematocrit levels only observed in critically ill patients (Denniff and Spooner, 2010).

22.3.3.10 Advances in Substrate The last few years have seen numerous innovations in DBS substrate material and card format from the likes of Agilent, Ahlstrom, and Whatman. Advancements to the substrate material and card format could assist many of the functions and considerations discussed in this section and influence the progression of DBS direct analysis. For example, the use of more robust, rigid cards, which would sandwich the substrate in a tough plastic material (or similar) rather than cardboard, would have numerous advantages. First, the cards would be more robust and offer more physical resistance to tearing and deformation during shipping and storage. Some form of flip-over or sliding cover could also offer further physical protection and some additional confidence against contamination during drying. Second, a rigid card would greatly assist ease of automation. The currently used cards bend and twist easily. The less variability there is in the size and shape of the sample format, the easier it is to reliably automate. A number of prototype plastic card variants have been produced by various card and substrate manufacturers, and it is possible that these superior performing cards may one day replace the current cardboard versions. Currently, the additional cost of these new formats and the popularity of the existing formats are inhibiting the development of these alternatives. Additionally, it has been proven that robust, reliable automation of the cardboard format cards is possible, and a new format could mean existing automation has to be modified. It would also be desirable not to significantly increase the depth of the cards so the ease of shipping and storage advantages can be maintained.

As previously discussed, minor modifications to the existing format have also been made in an attempt to assist with efficient sampling and the hematocrit issue such as PDBS, PCDBS, and dried matrix in paper disc (PMPD) (Section 22.3.3.9).

As discussed in Section 22.3.3.9 the most important substrate development would be the introduction of a substrate that behaves independently of hematocrit or at least produces a fixed spot area for a given volume, regardless of sample hematocrit (it is hoped that recovery and ion suppression variation with hematocrit can be controlled through optimized extraction and HPLC conditions). It is hoped that this can be achieved using the same paper-type substrate currently used. For some applications, the hematocrit issue could be deal breaker for the future of DBS usage. If a paper-type substrate that behaves independently of hematocrit, or a suitable method of applying accurate volumes of blood in the clinic (prior to whole spot extraction), does not emerge, then other microsampling variants may emerge to take the place of the current format. The most obvious idealized approach would be to use a substrate that can only physically absorb a fixed volume of blood, regardless of hematocrit, thus eliminating the issue of variable blood volumes being applied in the clinic and the effect of hematocrit on assay bias. If this could be achieved without resorting to anything that would compromise the cost advantages and simplicity of DBS, then it would be a very neat solution! It is likely that such a substrate, unlike the current flat paper format, will be precut or molded, or will incorporate some kind of physical or chemical barriers to limit the volume of blood that can be absorbed. Such a technique could also potentially significantly reduce wastage of both sample and substrate. To continue the goal of a simplified workflow, direct analysis instrumentation would need to modified and optimized to suit this new substrate, chiefly in the areas of automation and sample handling.

Another substrate innovation is the introduction of "indicating" cards that enable colorless fluids (such as plasma, urine, or cerebrospinal fluid) to be easily viewed (on "normal" substrate these fluids can be almost invisible). These substrates contain a dye that is displaced by the addition of a sample, leaving a lighter-colored area for easy identification (Spooner et al., 2009). Direct analysis visual recognition systems should ideally be compatible with this type of indicating substrate for full flexibility and versatility.

While the 85 × 53 mm 4-spot substrate card is particularly suitable for drug development applications, it is probable that alternative formats will be optimal for other applications. Ideally, for maximum flexibility, a direct analysis instrument should be compatible with these variants. Unfortunately, this seems unlikely given the difficulty in automating the current cards. A more feasible solution is likely to be a modular-type approach where different attachments can be fitted according to the format being used.

22.3.3.11 Communication, Compatibility with Other Systems, and Error Handling In this proposed workflow, the direct analysis system takes the place of an autosampler in a conventional HPLC-MS/MS system. Thus, the analysis system must be able to communicate in synchrony with the detector and HPLC system in the same way. It is also desirable for the visual recognition system to reconcile samples against a sample sequence list that may be submitted directly from a LIMS and in turn reconcile this against the detector sequence. Intelligent error handling is essential in any automated high-throughput "walk-away" system and becomes increasingly important in regulated environments where sample tracking is required. It is vital that any automation, visual recognition, or communication errors that occur are immediately corrected or, if a sample cannot be analyzed, this is reported at the end of the run.

22.4 AUTOMATED DBS EXTRACTION

Automated DBS extraction instrumentation uses a semi- or fully automated DBS "card-punching" device possibly in tandem with a liquid-handling robot. These techniques replicate the currently used DBS manual extraction technique described above, but automate one of more of the following steps to relieve some of the manual burden of DBS analysis:

1. Moving the cards from a position in a rack to the punch location
2. Locating the exact position of the DBS on the card
3. Cutting a disk from the center of the DBS sample
4. Placing the disk into a sample tube
5. Adding extraction solvent to the tube containing the disk
6. Shaking/mixing/agitating the tube + disk + extraction solvent
7. Centrifuging the tube
8. Transferring the supernatant to a fresh tube
9. Analyzing the supernatant using HPLC-MS/MS

The most basic example of DBS automation involves the use of powered handheld disc cutters which partially automate *step 3* above. These devices, akin to an electronic powered cordless screwdriver, simply use a rotary motor to revolve the cutting tool used to cut the disc from the center of the DBS sample. Clearly, this is a very minor improvement in automation over the conventional manual extraction method, but in its favor these devices are relatively cheap. In the absence of other alternatives, they have also proven to be popular with analysts sick of nursing aching wrists after punching hundreds of DBS samples by hand!

The next level of automation provides some additional functionality. Instruments such as the bench top BSD600 (http://www.luminexcorp.com) and Perkin Elmer Wallac DBS puncher (http://www.perkinelmer.com) punch out a spot at a time and deposit the disk into a 96-well plate (*steps 3* and *4*). However, these instruments still rely on human interaction (aided by a laser line targeting system) to locate the center of the DBS samples and load cards onto the system one at a time. More complex instrumentation such as the BSD1000 Genepunch (http://www.biospot-llc.com/BSD1000.htm) and Hudson Robotics dried blood spot processor (http://www.husdonrobotics.com) further automates the process, providing "walk-away automation." The BSD1000 incorporates a visual recognition system (essentially a camera and suitable recognition software to locate the center of each DBS sample), movement of both DBS cards and 96-well blocks, and an automated punch. This system can be programmed and left to punch out hundreds of DBS cards and the disks transferred to 96-well blocks or similar (*steps 1–4*), ready for the remainder of the extraction process.

The next step in functionality is achieved by utilizing liquid-handling robotics (to automate one or more of *steps 5–8*) coupled to (or integrated with) the DBS punching modules described above. There are a number of existing robotics platforms currently used in sample preparation laboratories that can carry out these functions and a number of manufacturers are developing DBS-specific instruments, such as the Hamilton easyPunch system (http://www.hamiltonrobotics). The advantage of some of these products is that they may already exist in analytical laboratories, and thus it may be possible to convert an existing model to DBS compatibility, thus offering a cost advantage. On a similar note, Johnson et al. (2011) recently reported on semiautomated DBS extraction methods using conventional liquid-handling devices and demonstrated that modest improvements to efficiency could be achieved with relatively little expense.

In theory, a fully automated device, located in an MS laboratory, could even transfer the final tubes into a HPLC autosampler ready for analysis. Systems that then automatically introduce the extracts into the HPLC-MS/MS system do not currently exist, but given enough time, space, and money, they are certainly possible to achieve. Such complex automation, however, would be a challenge to develop, probably impractically large, highly complex, and no doubt, very, very expensive. An integrated device that has a large enough capacity for high sample throughput usage (approximately 500 cards) and covers steps 1–8 does not currently exist (most likely due to the issues described above).

Automated DBS analysis instrumentation does offer some advantages over direct analysis techniques. Such instruments are commercially available and use exactly the same workflow that analysts are accustomed to. Modest levels of automation can also be relatively affordable, and in some cases it may be possible to modify existing robotic platforms, offering a cost advantage. The use of such instruments can clearly be of use in high sample throughput environments, but without investing prohibitive amounts of money to produce very complex integrated instruments, they do not provide the seamless, simplified workflows on offer from direct analysis techniques. In many situations these emerging direct analysis techniques offer significant advantages over automated DBS analysis that are worth the additional resource involved in their development.

22.5 DBS DIRECT ELUTION

DBS direct elution describes a number of similar techniques where the analyte of interest is extracted from a DBS sample through interaction with a suitable liquid solvent. These techniques can broadly be categorized as follows: online DBS; sealing surface sampling probe (SSSP); liquid microjunction surface-sampling probe (LMJ-SSP); a simplified version liquid extraction surface analysis (LESA); and DBS digital microfluidics (DMF). One of the attractions of direct elution is that a liquid eluate is produced which

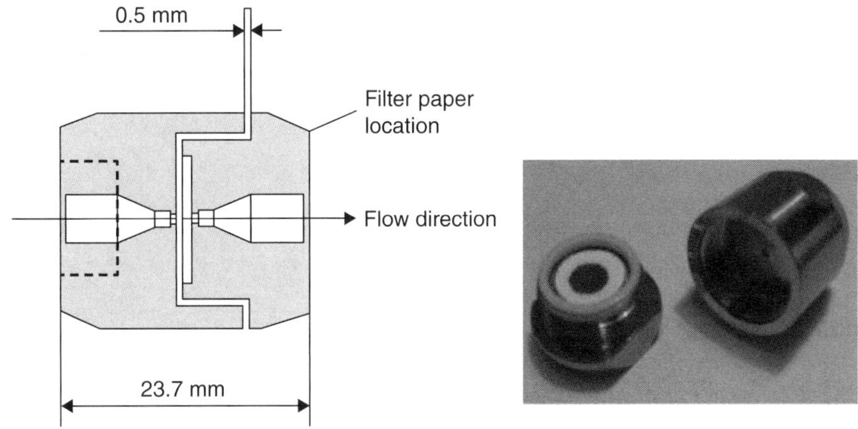

FIGURE 22.8 Desorption (inox) cell used in Déglon et al.'s original online DBS direct elution system. Reprinted with permission from Déglon et al. (2009). Copyright 2009 Elsevier.

can be separated and detected using LC-MS/MS, a technique that is readily available in the bioanalytical lab and is familiar to, and accepted by, bioanalytical scientists and, where applicable, regulatory authorities. For some applications direct elution therefore currently possesses a higher level of accessibility than direct desorption techniques that cannot be coupled to HPLC, or alternative separation techniques. Over the last few years commercially available and prototype technology has emerged from CAMAG (TLC-MS and DBS-MS 500) (http://www.camag.com), Spark Holland (http://www.sparkholland.com), Prolab (SCAP) (http://www.prolab.ch), and Advion (LESA and LMJ-SSP utilizing the Triversa Nanomate) (http://advion.com) and others that encompass a variety of approaches and varying levels of automation and additional functionality. This chapter attempts to summarize the various technologies that have emerged so far.

22.5.1 Online DBS Analysis

In 2009, Déglon et al. introduced a method of online DBS extraction into an LC-MS system without sample pretreatment. The procedure involved manually punching a disk from the center of the DBS on a filter paper sample (as performed for manual extraction) and placing the punch in a desorption (or inox) cell that was compatible with LC capillaries (Déglon et al., 2009) (Figure 22.8). HPLC users will recognize this device as being rather like an in-line HPLC precolumn filter cartridge, though the DBS punch is placed in the cell rather than a frit. An extraction solvent of choice is flowed through the cell using an LC pump which extracts (or desorbs) the compound of interest from the DBS and allows it to flow to a HPLC column for separation and then to the MS for detection. A trapping column setup was used to ensure the purification and separation of compounds before detection (Figure 22.9). Pump 1 used a high organic concentration

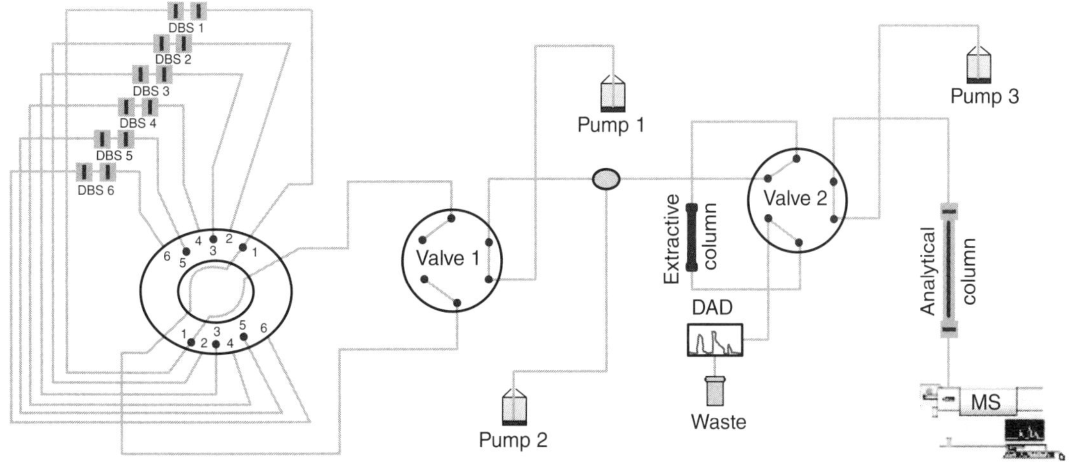

FIGURE 22.9 Global view of the online DBS procedure coupled to a column-switching LC/MS system. Reprinted with permission from Déglon et al. (2009). Copyright 2009 Elsevier.

solvent to desorb and elute the analytes from the DBS samples and transfer them toward the trapping column. Pump 2 used a high aqueous concentration solvent to adjust the chemical properties of the desorption mobile phase. Déglon et al. reported that while acetonitrile was optimum for solvent extraction, it did not effectively trap analytes in a reverse-phase trapping column. Thus, pump 2 was used to add water to reduce the organic content (to around 10% organic) to enable effective trapping. The extraction time was 5.5 minutes (which at a flow rate of 0.25 mL/min gave complete desorption), after which the trapping and analytical columns were connected. Trapped analytes were then transferred to the analytical column for 2 minutes, while the precolumn was regenerated, and finally, a generic gradient was used by pump 3 to deliver the analytical mobile phase with a constant flow rate. The total run time was 18 minutes, which, for many high sample throughput applications, would not be acceptable. There are advantages and disadvantages of using a trapping column coupled to an analytical column, over an analytical column-only approach. One potential advantage is that more analyte may be stored in a trapping column and subsequently transferred to the analytical column than may be possible with an analytical column-only system. Thus, sensitivity may be improved. Disadvantages include higher costs, complexity, and substantially increased run times.

The initial work performed by Déglon et al. demonstrated that the extraction step and analysis step could be integrated into a relatively simple process without sample pretreatment. This initial work demonstrated that this technique could offer reasonably good sensitivity (within therapeutic range of the drugs tested) using three small-molecule compounds and sufficient accuracy and precision to be suitable for regulated drug development DBS applications. A comparison of online DBS versus a validated LC-MS/MS procedure on patient samples was also undertaken and showed a close correlation. Interestingly, carryover observed was quoted as being (just) adequate for regulated development applications (carryover in a blank solution following an injection of double the HLQ was lower than 0.01%) despite the lack of a dedicated wash system.

In its initial form, online DBS extraction is of limited use in a high-throughput environment as not only do the DBS samples still need to be punched by hand, but the punches must then be manually inserted into the inox cells between extractions (using a 12-port valve it was shown that up to 6 DBS samples could be extracted per run). Additionally, very long run times were used. Clearly, this approach would benefit from automation, and the concept has been subsequently developed into commercially available and prototype online DBS instrumentation by various groups. These include the DBS-MS/MS (and DBS SPE-MS/MS) from Spark Holland, the commercially available SCAP DBS from Prolab Instruments, and prototype instrumentation reported by Miller et al. (2011) and Déglon et al. (2011b). These devices all work around the same principle. Instead of having to punch the DBS samples and fit the disk into an online cartridge for analysis, the samples are instead left intact and are clamped from either side (Figure 22.10) using automation. These clamps are essentially two halves of the desorption cell and house integrated capillaries that allow the extraction solvent to flow through the DBS sample and onto the HPLC column and detector.

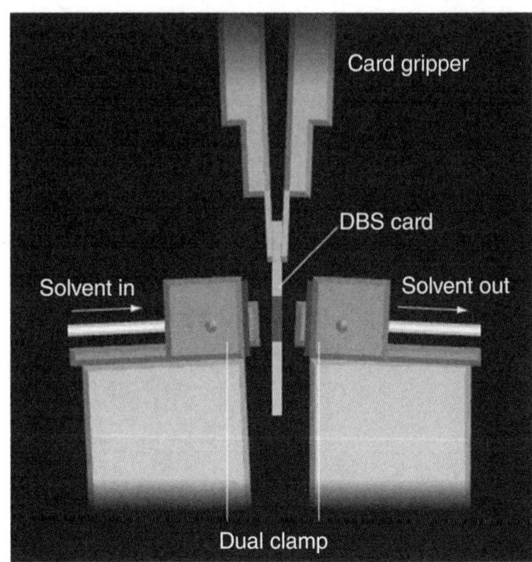

FIGURE 22.10 Automated online DBS desorption (Prolab SCAP schematic). The Spark Holland DBS SPE-MS/MS and automated prototypes from Déglon et al. use the same concept. Reproduced from Heinig et al. (2011) with permission from Future Science Ltd.

The Prolab SCAP (Sample card and prep) system is based around existing CTC Pal automation technology and has been designed to automate the 85 × 53 mm 4-spot substrate cards (Figure 22.11). This system is designed to work around a

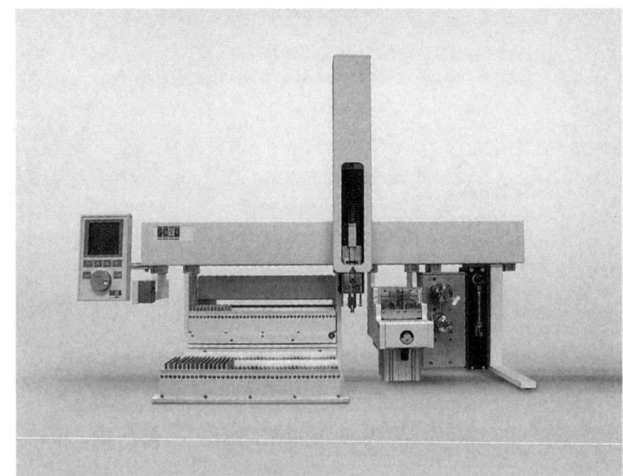

FIGURE 22.11 Prolab SCAP (Sample card and prep) system, commercially available online DBS instrument. Photo from Prolab, Inc., used with permission.

trapping column arrangement but can also be programmed to work with an analytical column only for shorter run times. The first version of this system had a small capacity (28 cards) and lacked both a wash system (other than rinsing the clamps) and a visual recognition system (relying on consistent printing alignment on the cards and accurate spotting). The latest versions have additional functionality, adding a camera for accurate spot location, larger trays (160 cards) for increased card capacity, more sophisticated error handling, and a new wash procedure that reported significant decreases in carry-over. The only provision for adding IS is via the extraction solvent, but its favor, the SCAP system, has a cost advantage over some of the commercially available alternatives. Recent experiences in our laboratory with the first version of this device was that the automation on offer worked well with perfectly straight DBS cards but struggled with bent or over/undersized cards. The simplistic error handling could not cope with such errors, which were catastrophic to the sample runs. However, there are a number of reports in the literature where the SCAP system has been evaluated and the same problems have not been highlighted, suggesting that it can be used reliably, perhaps as a result of the modifications made to later versions. Heinig et al. (2011) and Ganz et al. (2012) have both reported reliable automation performance and the ability to generate quantitative validation data within guideline-acceptance criteria.

The DBS SPE-MS/MS from Spark Holland has been in development for a number of years and was first formally reported on in the literature in 2011 (Ooms et al., 2011). The concept is the same as that used in the SCAP system: the DBS sample is clamped from either side and solvent is allowed to flow through to desorb the analyte. Similar levels of assay sensitivity, accuracy, and precision have been demonstrated using this prototype device. Recent developments with this system have focused on automation using either a CTC platform or a more elaborate robotic arm (http://www.sparkholland). Déglon et al. (2011a) developed a similar automated clamp, online DBS prototype based on their online DBS concept (Figure 22.12). The design is based on a rotating plate with multiple wells where the DBS punches are manually placed. While this system is likely to be perfect for some applications, the necessity to manually punch discs from DBS samples and transfer them to the device limits its use in high-throughput applications.

22.5.2 Liquid Microjunction Surface-Sampling Probe and Liquid Extraction Surface Analysis

The LMJ-SSP approach uses an unconfined liquid microjunction in contact with a sample surface to extract analytes of interest. A probe is positioned at an appropriate distance away from the surface and is configured so that an extraction liquid is brought to the surface and is then carried on to the ionization source through another probe that acts as a liquid conduit (Figure 22.13). This is achieved by reducing the self-aspiration flow rate of the probe to less than the flow rate

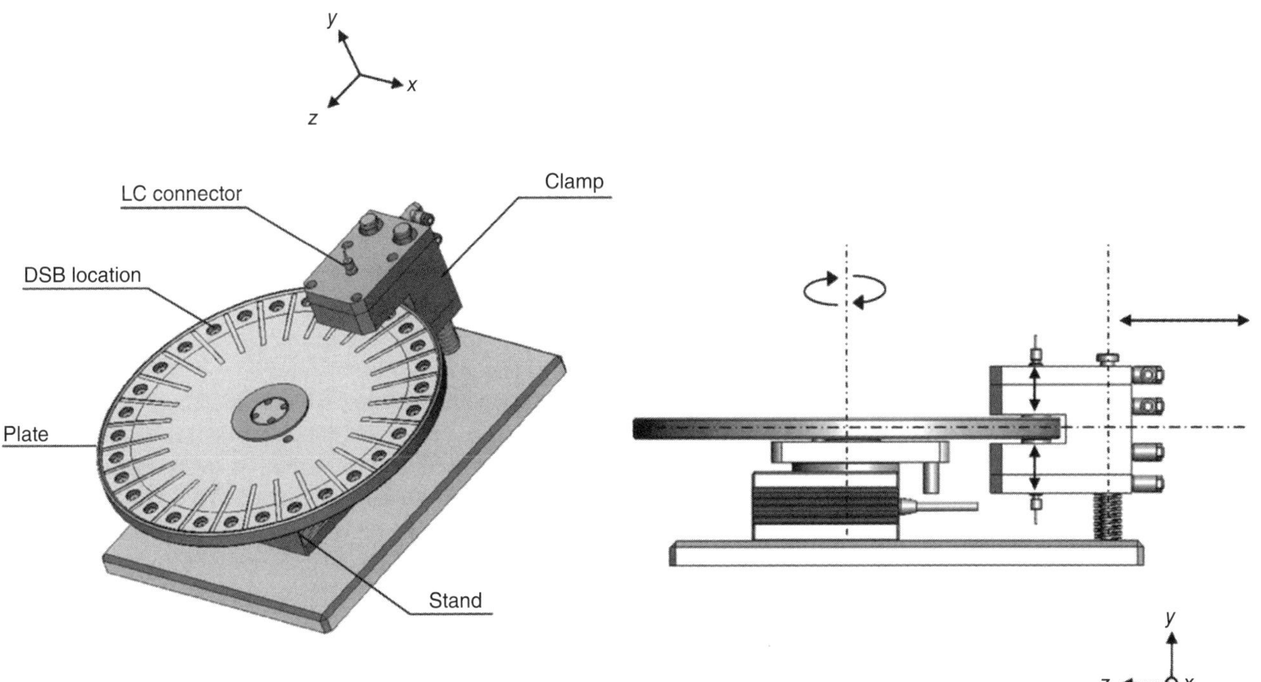

FIGURE 22.12 Automated online DBS from Déglon et al. Reprinted with permission from Déglon et al. (2011a). Copyright 2011 Elsevier.

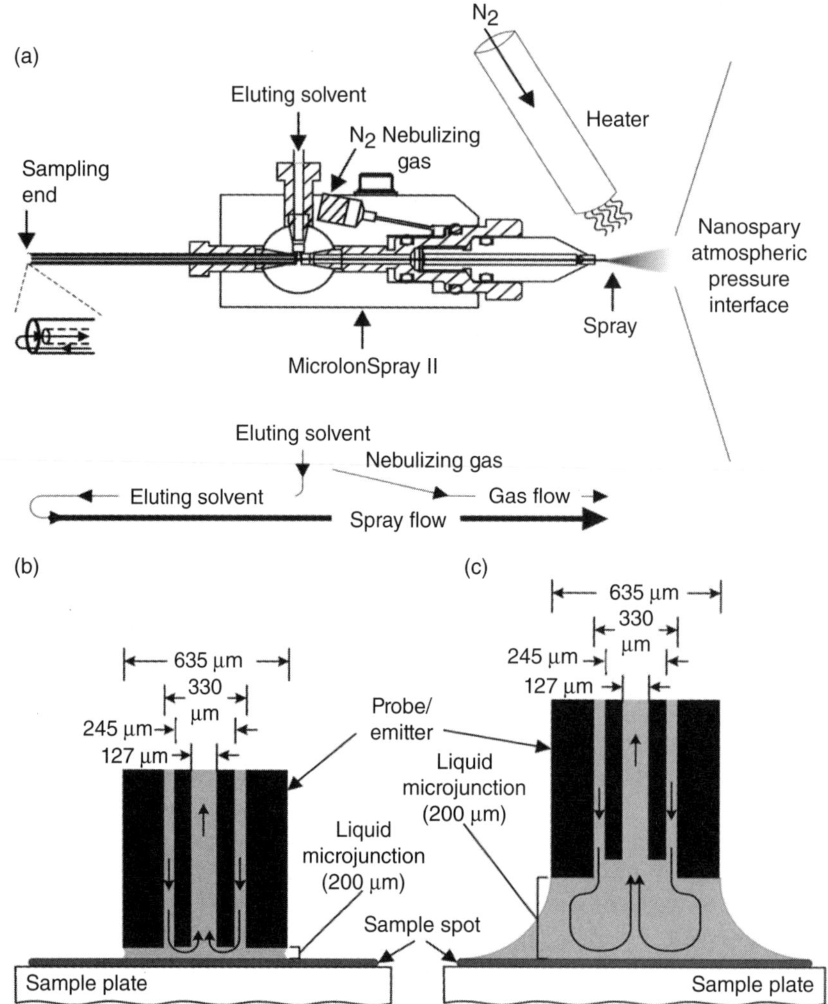

FIGURE 22.13 LMJ-SSP. (a) Schematic illustration of the liquid microjunction surface-sampling probe/ESI-MS experimental setup with details of the surface sampling probe/emitter. (b) Schematic illustration showing the close probe-to-surface spacing and narrow liquid microjunction used for spot sampling. (c) Schematic showing larger probe-to-surface spacing for spot sampling and the resulting liquid microjunction developed in an updated system. Note that the probe-to-surface liquid microjunction was actually formed in the horizontal position and is only shown in (b) and (c) in the vertical position for ease of viewing. Reprinted with permission from Van Berkel et al. (2009). Copyright 2009 American Chemical Society.

volume pumped into the probe. The self-aspiration rate of the probe is then increased (controlled by altering the nebulizing gas flow rate), allowing desorbed analytes on the sample surface to be aspirated back into the probe with the liquid that created the liquid microjunction and transferred to the MS. The technique has been reported as being fast enough to be compatible with high-throughput sample analysis (Kertesz and Van Berkel, 2010).

Unfortunately, this wall-less microjunction and dynamic flow approach is not suited to sampling porous surfaces, such as DBS on paper substrate, where the liquid microjunction cannot be maintained. However, it was recognized that the technique could be used without the continuous flow provided by the LMJ-SSP, potentially being compatible with any probe device capable of both dispensing and retrieving a solvent from the sample surface. Thus, in an attempt to overcome the LMJ-SSP issue with porous DBS samples, the technique has been coupled to the Advion Triversa Nanomate chip-based infusion nanoESI system, which is also capable of automating the process (Figure 22.14). The Nanomate system comprises a pipette-based liquid-handling robotic system coupled to chip-based electrospray ionization (ESI) technology. For DBS analysis the Nanomate is used in LESA mode and utilizes a static microjunction. Automated liquid

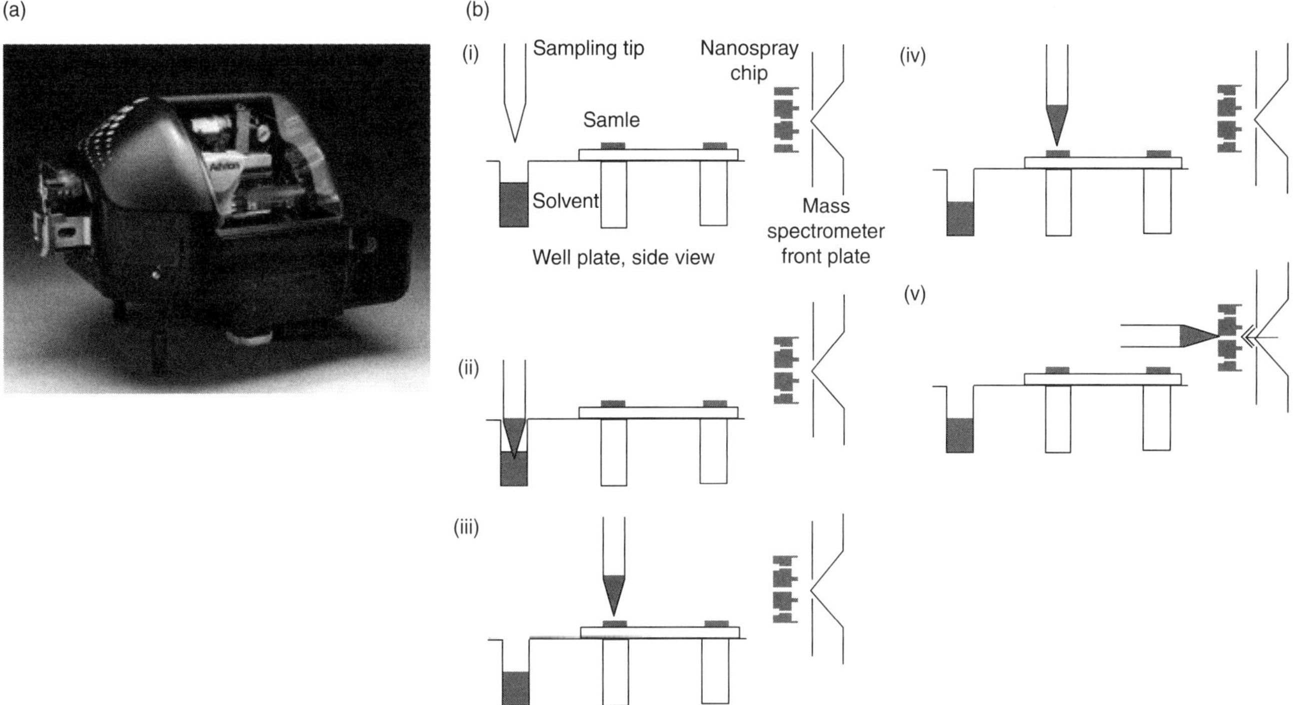

FIGURE 22.14 (a) Advion Triversa Nanomate chip-based infusion nanoESI system. Photo from Advion, used with permission. (b) For DBS analysis the Nanomate is used in LESA (liquid extraction surface analysis) mode and utilizes a static microjunction. Schematic shows the individual steps of the surface sampling process (Kertesz and Van Berkel, 2010).

handling is enabled by the use of a robotic arm that positions pipette tips around the Nanomate instrument deck. In LESA mode a pipette tip picks up a volume of extraction solvent, moves above a DBS sample, and is lowered to an appropriate distance above the sample. A portion of the extraction solvent is then dispensed onto the sample surface forming a static liquid microjunction, which is the mechanism of analyte desorption. The solution containing the extracted analyte is then aspirated back into the tip and transferred to a nanospray nozzle for MS analysis (Figure 22.14b). Kertesz and Van Berkel (2010) demonstrated performance of the technique using DBS samples containing sitamaquine using 4 mm diameter punches mounted onto plates using double-sided tape. Acceptable accuracy and precision was reported down to 100 ng/mL (LC separation was not used) using small volumes of highly organic extraction solvent (2 μL of MeCN/MeOH/H_2O/formic acid; 58/34/8/0.1 v/v/v/v).

Advantages of this approach include quick run times, the incorporation of an automation platform, and no carryover issues as pipette tips and nanospray nozzles are disposed off between samples. Disadvantages include relatively poor sensitivity compared to SSSP (see Section 22.5.3.1) and the requirement to manually punch disks from DBS samples. Fixed-area punched disks must be used as otherwise the extraction solvent will not form the microjunction and simply wick out to the surrounding areas on the porous substrate. However, the formation of the microjunction, even in this static incarnation on a punched disk, is tricky to perform reliably. A potential solution, reported by Henion (2011), reported the development of DBS substrate containing a non-porous silicon ring around the DBS area which would both aid the formation of the microjunction and enable analysis without the need to pre-punch the sample.

22.5.3 Sealing Surface Sampling Probe

22.5.3.1 Proof of Concept The SSSP concept is based on a device originally reported by Luftmann in 2004 which coupled TLC to MS detection (Luftmann, 2004). This device was later commercialized by CAMAG into the TLC-MS interface, a simple and relatively cheap, pneumatically driven device that is operated manually and used to directly elute samples from TLC plates and transfer the extract to mass spectrometer interfaces. A number of groups have recognized the potential of extending the use of this device to directly elute DBS samples (Figure 22.15). In a simple arrangement, the TLC-MS can be positioned between the HPLC pump and the HPLC column and MS, essentially replacing the autosampler in a typical LC-MS setup (Figure 22.15b). Prior to extraction, mobile phase from the HPLC pump bypasses

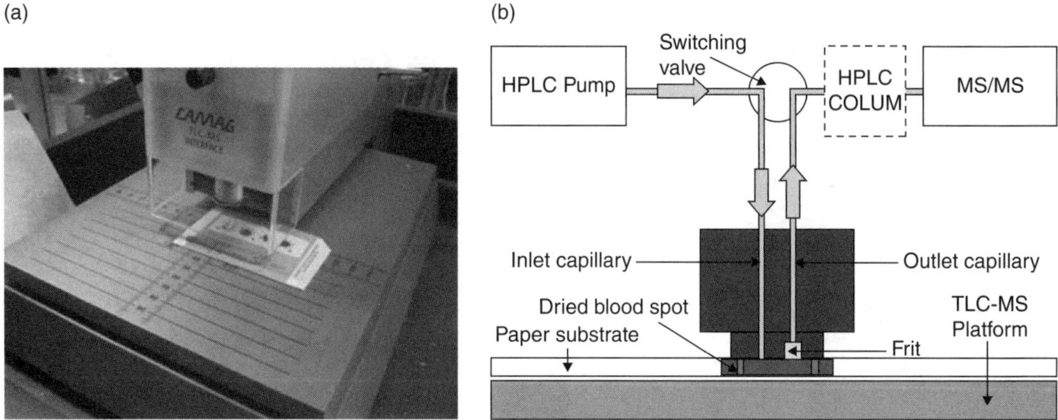

FIGURE 22.15 (a) CAMAG TLC-MS instrument which utilizes the SSSP concept. Photo shows the instrument being used to directly elute a DBS sample. (b) Schematic diagram of dried blood spot direct elution assembly using the CAMAG TLC-MS interface. Reprinted with permission from Abu-Rabie and Spooner (2009). Copyright 2009 American Chemical Society.

the TLC-MS and flows directly to the HPLC column (if used) and MS. DBS extraction is carried out by lowering the plunger (or extraction head) onto the center of a DBS sample typically forming a sealed sampling area of 4 mm diameter. The exact site of extraction is located using a laser crosshair (Figure 22.15). Solvent is then allowed to flow down the inlet capillary of the plunger, filling the sealed sampling area on the DBS. The continuous flow of the mobile phase forces the extract to flow up the outlet capillary and toward the HPLC column and MS. It is this action of solvent flow through the DBS that transfers the compound of interest from the DBS sample into the solvent flow.

In 2009, Abu-Rabie and Spooner (2009) and Van Berkel and Kertesz (2009) demonstrated that the TLC-MS interface was suitable for DBS analysis. Abu-Rabie et al. demonstrated that direct elution of DBS samples using the TLC-MS compared favorably with existing DBS manual extraction methods used for quantitative bioanalysis. In this initial evaluation, as many assay parameters as possible were unchanged from the manual extraction methods including chromatographic conditions such as mobile phases, flow rates, and HPLC columns. Hence, the mobile-phase conditions at the start of the extraction cycle were used as the "extraction solvent" and the flow rates were retained at 0.5–1 mL/min. It was found that short "extraction times" (the duration the extraction solvent was allowed to pass through the DBS sample), up to 5 seconds, maintained the chromatographic performance observed for manual extraction. Extending the extraction time further resulted in increasingly broad (and eventually unacceptable) chromatographic peak shape. It was immediately obvious from the initial evaluation that this direct elution technique warranted further investigation as not only was chromatographic performance maintained without method re-optimization, but assay sensitivity was considerably increased (see Section 22.5.3.4). Over a range of representative small-molecule test compounds an average 10-fold increase in assay sensitivity was observed. This work sparked interest in this technique as it appeared to offer multiple advantages: direct analysis with no sample pretreatment, compatibility with existing LC conditions, and significant increases in assay sensitivity. This last advantage is of course particularly of interest due to the sensitivity challenges faced when using DBS manual extraction.

In this study, Abu-Rabie and Spooner performed DBS bioanalytical method validations using the TLC-MS for test compounds sitamaquine and acetaminophen and were able to achieve linearity, accuracy, and precision within guideline-acceptance criteria. Carryover was found to be large enough to significantly bias subsequent samples. In order to generate acceptable data, carryover was controlled using a time-consuming procedure that involved back-flushing the outlet capillary column using a secondary pump. A rudimentary recovery evaluation was also carried out by performing successive multiple extractions from the same DBS sampling area (Figure 22.16). This study showed that despite direct elution producing a 15-fold increase in analytical response (for sitamaquine) compared to manual extraction, only around 17% of the total analyte extracted was eluted in the initial 2-second extraction (Table 22.2). This study also suggested that sensitivity could potentially be further increased with optimized extraction and chromatographic conditions. Abu-Rabie and Spooner also reported on DBS direct elution using this arrangement where the HPLC column was removed, and the elute was transferred directly to the MS. It was found that using the same mobile-phase conditions excellent accuracy and precision was achieved, but sensitivity was poor. Switching the extraction solvent to 70:30 (v/v)

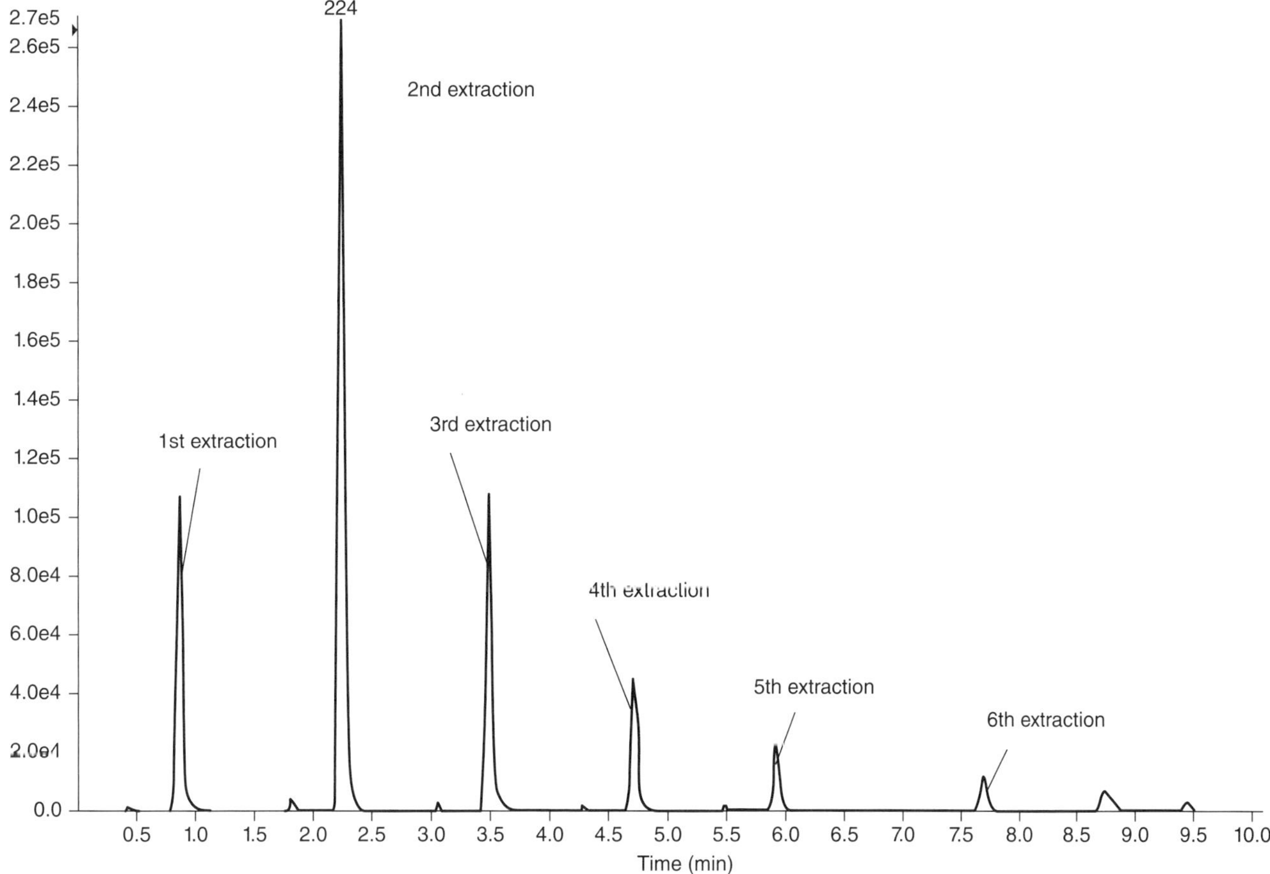

FIGURE 22.16 Mass chromatogram showing successive multiple extractions (2-second elution) from the same DBS sampling area using the TLC-MS interface. The sample was a 100 ng/mL sitamaquine human DBS. Reprinted with permission from Abu-Rabie and Spooner (2009). Copyright 2009 American Chemical Society.

TABLE 22.2 Peak Area of Successive Multiple Extractions Taken from the Same DBS Sampling Area Using the TLC-MS Interface with HPLC Separation (Sitamaquine 100 ng/mL DBS, Human Blood, Each Extraction Was for 2 s)

Extraction Number	Peak Area/Counts	Individual Peak Area Response Compared to Total (Six Extractions) (%)
1	325,718.4	17.2
2	895,814.6	47.3
3	385,444.5	20.4
4	165,614.5	8.7
5	80,534.4	4.3
6	39,781.3	2.1

Source: Reprinted with permission from Abu-Rabie and Spooner (2009). Copyright 2009 American Chemical Society.

methanol:water retained the accuracy and precision performance and greatly increased sensitivity, to the extent that it was comparable to what was achieved with chromatographic separation (Figure 22.17).

Around the same time Van Berkel et al. also reported good accuracy and precision using the TLC-MS for the quantitation of sitamaquine and acetaminophen in DBS samples. In this study, chromatographic separation was not used, and lower flow rates (0.2 mL/min) and longer extraction times (60 seconds) were utilized, with methanol (for sitamaquine) or methanol:formic acid 100/0.1 (v/v) (for acetaminophen) as the extraction solvent. A 60-second extraction of a blank sample (e.g., blank paper substrate) was used to wash the extraction head which reduced carryover to acceptable levels.

Following these initial reports, a number of other groups have published accounts of using the TLC-MS for direct elution of DBS samples in bioanalytical quantitation. For

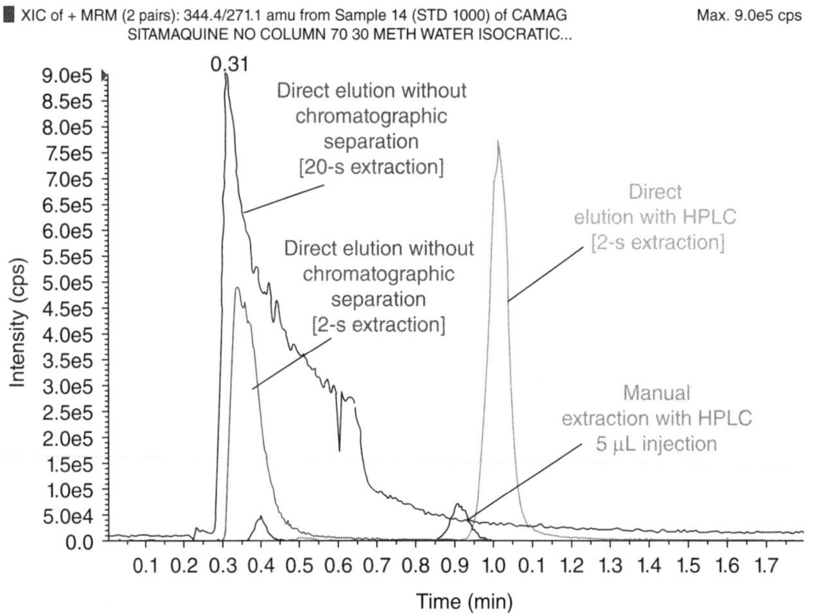

FIGURE 22.17 DBS direct elution with and without HPLC separation using the CAMAG TLC-MS interface. DBS samples containing 1000 ng/mL sitamaquine DBS were analyzed using manual extraction; direct elution with HPLC using mobile phase (isocratic) as the extraction solvent; and direct elution without HPLC using 70:30 (v/v) methanol:water as the extraction solvent. Reprinted with permission from Abu-Rabie and Spooner (2009). Copyright 2009 American Chemical Society.

example, Heinig et al. (2012) integrated the TLC-MS into a column-switching LC-MS/MS system that incorporated online SPE (a trapping column) for additional analyte collection and cleanup. Excellent sensitivity, linearity, accuracy, and precision data were reported. An extraction time of 45 seconds onto a trapping column was used before the TC and AC were connected.

22.5.3.2 SSSP Automation and Development The TLC-MS interface is a relatively cheap and simple device for performing DBS direct elution, but like many direct analysis devices at the proof of concept stage, its operation is entirely manual and requires constant user interaction. The SSSP mechanism is an example of a DBS direct analysis technique that has been developed from its original concept to meet high sample throughput bioanalytical requirements. The TLC-MS was not originally designed for DBS direct elution and thus required extensive modification to be suitable for high-throughput bioanalysis. To meet high-throughput direct analysis requirements the TLC-MS required card handling, extraction mechanism, and wash mechanism to be automated, in addition to the incorporation of the additional functionality outlined in Section 22.3.3.

The first modification was the design of a plunger head more suitable for paper substrate than the original version designed for TLC plates. The next major development was the introduction of the DBS-MS 16 prototype (Figure 22.18). This device automated the extraction of up to 4 DBS cards (a total of 16 DBS samples, hence the name). It lacks a true visual recognition system, instead relying on a laser crosshair to correctly position the cards on a moving platform which moves the spots in turn to under the extraction head for extraction. Therefore, accurate spotting within the marked region of the cards is essential if the center of the DBS is to be sampled. The most important progressive aspect of this device was the automation of the wash cycle post sample extraction. Not only does this function automate the lengthy and labor-intensive rinsing procedure that was required for the TLC-MS between samples, it also vastly improves the wash performance and achieves this well within the sample cycle time frame required for high-throughput bioanalysis. Carryover is reduced to levels below that observed for manual DBS extraction using a high-performance conventional autosampler (Figure 22.19) (Abu-Rabie and Spooner,

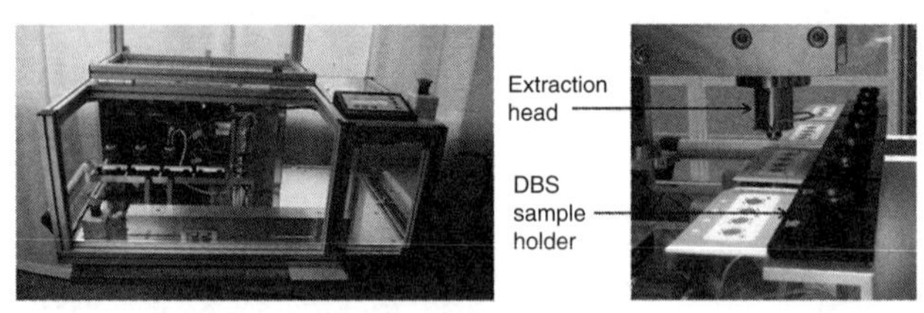

FIGURE 22.18 The CAMAG DBS-MS 16 SSSP prototype automated direct elution instrument. Copyright 2011 Future Science Ltd.

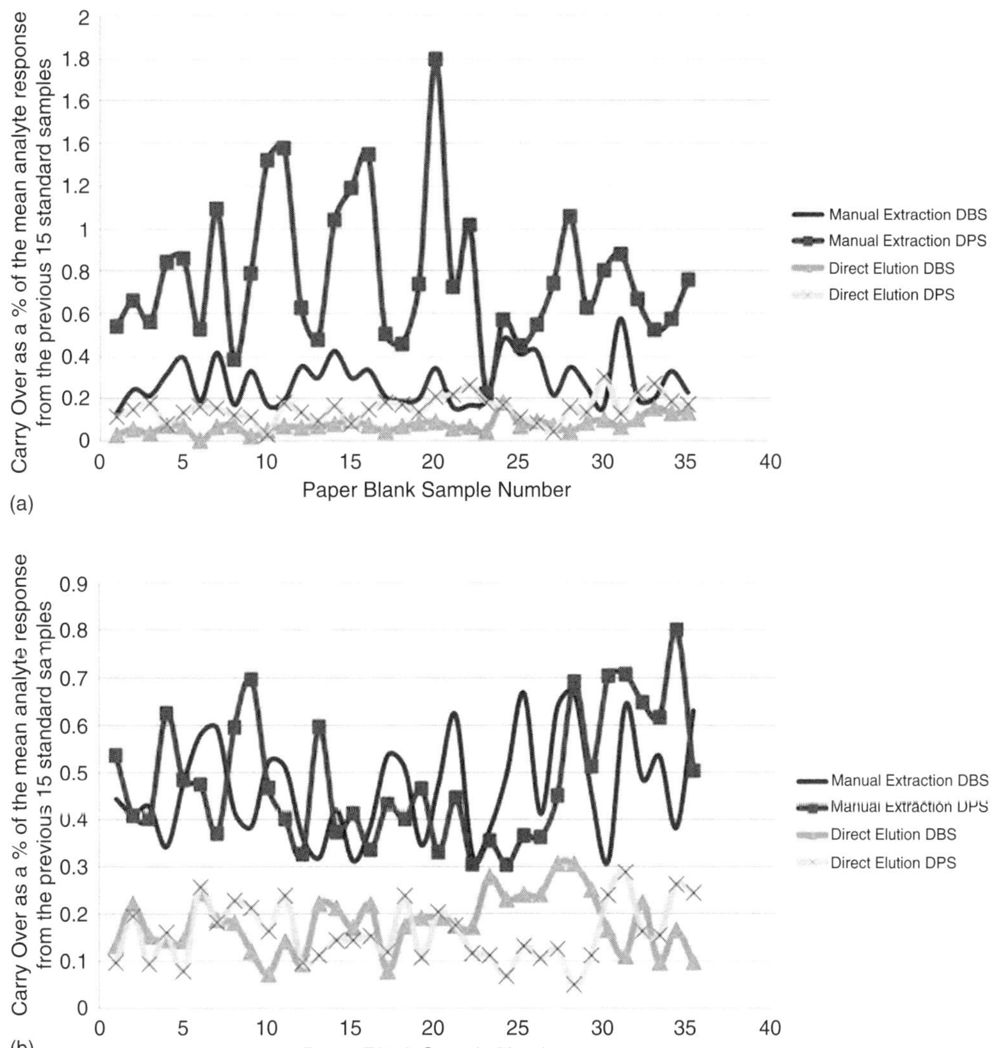

FIGURE 22.19 Sitamaquine and acetaminophen HPLC-MS/MS data over 576 consecutive dried blood spot and dried plasma spot samples using manual extraction (injecting extracts via a Waters Acquity autosampler) and direct elution (via the CAMAG DBS-MS 16). Peak area carryover for (a) sitamaquine and (b) acetaminophen was calculated using the analyte peak area response in a blank control sample expressed as a percentage of the mean response analyte peak area from the preceding 15 samples containing the drug (500 ng/mL sitamaquine, 10,000 ng/mL acetaminophen). Copyright 2011 Future Science Ltd.

2011). This is achieved by utilizing a "dual capillary" wash system that utilizes separate wash solvents that can be optimized for removing matrix contamination and compound contamination, respectively. The DBS-MS 16 also provides more options for extraction optimization. The only extraction parameter that could be easily adjusted on the TLC-MS was the "extraction time." It was complex in practice to configure the TLC-MS so that anything other than the mobile phase was used as the extraction solvent (which in most cases is unlikely to offer optimized extraction performance). The DBS-MS 16 incorporates additional pumps that allow an optimized extraction solvent to be used. The flow rate and volume of the extraction solvent used can also be optimized independently of the mobile-phase conditions.

The DBS-MS 16 prototype led to the creation of the commercially available CAMAG DBS-MS 500 (Figure 22.20). This instrument retains the SSSP extraction mechanism and wash system concept of the DBS and adds a large sample capacity (500 cards) and robust automated card handling via a robotic arm. It also incorporates an intelligent visual recognition system and has a pre-extraction integrated IS addition module. In fact, this direct analysis instrument currently incorporates all the primary additional functionality outlined in Section 22.3.3, bar the ability to handle DBS dilutions. The

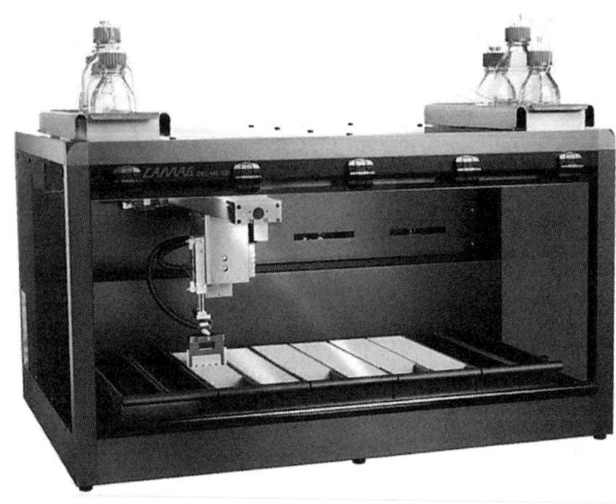

FIGURE 22.20 CAMAG DBS-MS 500, commercially available DBS direct elution system. Photo from CAMAG, used with permission.

downside to this robustness and functionality is the expense, which compared to the TLC-MS is now quite considerable, although still in the same range as existing liquid-handling robotics. Of course, this outlay could quickly be made back in resource savings if the sample throughput on offer is used to its potential.

22.5.3.3 SSSP Robustness One of the main concerns regarding direct elution of DBS samples is that it is likely to produce relatively dirty extracts due to the absence of sample cleanup. The problem is how this may affect the robustness of HPLC and MS instrumentation. If the performance of HPLC columns and MS instruments degrades rapidly due to the buildup of endogenous matrix material, then the technique is not robust enough for routine high-throughput usage. One potential solution is the use of an SPE or trapping column that cleans up the extract, possibly in tandem with an analytical column (focusing of compounds directly onto analytical columns is also possible due to the low desorption of contaminants). The disadvantages of this approach include increased costs and complexity, and increased sample run times.

In a high-throughput environment simplicity and fast run times are essential. For this reason the simplest method of direct elution should not be ruled out before deciding on a more complex, longer, and more expensive procedure. The availability of automated direct elution systems has made it much easier for the robustness of these systems to be evaluated over large number of samples. Abu-Rabie and Spooner (2011) demonstrated that direct elution using a simple analytical column-only LC-MS/MS arrangement offered similar robustness to conventional manual extraction of DBS samples over sample runs of up to 576 samples. This number of samples was deemed to be around the maximum used in a single unattended run and matched the capacity of a typical autosampler used in conventional manual DBS extraction analysis. Detector and chromatographic robustness was assessed, with the aim of providing an understanding of how DMS direct elution will perform in practice to support high sample throughput bioanalysis. To test this robustness, 576 consecutive samples were run using 2 test compounds for the following sample sets: manual extraction of DBS; manual extraction of DPS; direct elution of DBS; direct elution of DPS. Analyte and IS peak area, peak area ratio, and chromatographic parameters such as analyte peak asymmetry and number of theoretical plates were demonstrated to be as robust for direct elution as they were for conventional manual extraction of DBS. Of course, more work (and larger sample sets) is required to gain a better understanding of direct elution robustness, but initial studies have shown that simple analytical-only HPLC column setup may be feasible in practice.

22.5.3.4 Sensitivity and Extraction Solution Optimization Abu-Rabie and Spooner (2009) reported that DBS direct elution using the TLC-MS produced an average 10-fold increase in assay sensitivity compared to conventional manual DBS extraction using a set of representative small-molecule pharmaceutical compounds. The importance of this feature of DBS direct elution cannot be overstated. One of the main disadvantages of DBS manual extraction is the relatively low sensitivity compared to traditional wet plasma analysis. A direct analysis technique that could bridge this sensitivity gap could help maximize the advantages on offer from DBS by making this sampling technique feasible for compounds that need to be detected at lower concentrations. It is partly this reason that has seen the growth in interest in DBS direct elution in recent times. Online and SSSP-based DBS direct elution systems produce similar levels of sensitivity (if relative sampling areas and extraction times/recoveries are corrected for).

The limitation in assay sensitivity observed in manual DBS extraction is due to the small sample volume taken, typically a 3 mm diameter sub-punch from a 15 μL DBS, which corresponds to around only 2 μL of blood. In comparison, a typical plasma sample volume will be at least an order of magnitude larger. Far less compound is available to extract at the start of the assay; hence, DBS assay sensitivity is less than that for wet plasma analysis. Direct elution counters this disadvantage by enabling more analyte to be introduced to the MS for a given sample volume. A typical manual extraction involves extracting the DBS punch in 100 μL of methanol. A further reduction then occurs when using HPLC-MS/MS as typically only a small portion of this supernatant can be injected onto HPLC columns, if acceptable chromatographic peak shape is to be obtained. Typically, this is 2–20 μL of the supernatant (i.e., only 2–20% of the extracted analyte). Direct elution increases assay sensitivity

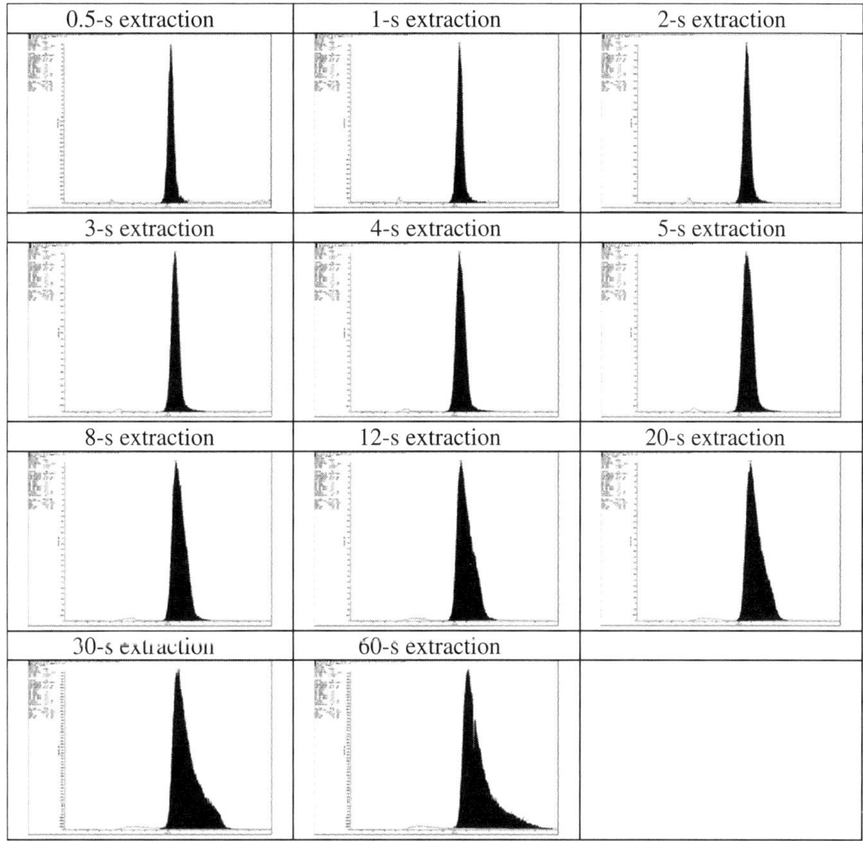

FIGURE 22.21 LC-MS/MS chromatograms showing the effect of extraction time on chromatography, when performing direct elution on human DBS samples with the TLC-MS interface using HPLC separation. Sitamaquine (100 ng/mL) DBS samples were used for each extraction. Reprinted with permission from Abu-Rabie and Spooner (2009). Copyright 2009 American Chemical Society.

by allowing larger quantities of analyte to be introduced to the HPLC column without ruining the chromatographic peak shape. Note that a minor part of the sensitivity increase also results from the 4 mm diameter sampling area used by the TLC-MS, which represents a 1.78-fold increase in spot area sampled compared to a typical 3 mm diameter DBS punch.

Abu-Rabie and Spooner (2009) demonstrated assay sensitivity increases with direct elution using very short (2-second) extraction times, with the mobile phase as the extraction solvent, at nonadjusted flow rates (0.5–1 mL/min), and without elaborate extraction optimization. At 1 mL/min, 16.6 μL of elute is transferred to the HPLC column per second. No corresponding increases in background MS response were observed. It is likely that exploring extraction optimization could further increase these sensitivity gains. Methods of achieving this include extending the extraction time to desorb more analyte. Simple recovery investigations using sitamaquine DBS samples indicated that the first 2-second elution only extracted approximately 17% of the total analyte that could be extracted (Figure 22.16 and Table 22.2). However, initial evaluations determined that the extraction time could only be extended so far before chromatographic peak shape becomes unacceptable (Figure 22.21) (Abu-Rabie and Spooner, 2009). It is likely that optimizing the extraction conditions (choice of extraction solvent, flow rate, etc.) will enable a larger quantity of compound to be extracted and transferred while retaining chromatographic performance. If the sample quantities are equalized, then the maximum increases in assay sensitivity on offer from DBS direct elution are in the region of 10- to 50-fold, depending on the maximum injection volume that can be used with manual extraction (Abu-Rabie and Spooner, 2009). While these absolute maximums are unlikely to be achieved in practice, it is hoped that extraction optimization will further increase assay sensitivity increases using direct elution.

If a larger quantity of analyte is being introduced via direct elution, then there is a risk that the analyte will be accompanied by a larger quantity of substrate and endogenous material (i.e., cellulose, blood, plasma, urine). As stated in Section 22.5.3.3, in the absence of a cleanup step clearly there is a concern that the introduction of relatively dirty extracts will negatively impact on the robustness of the technique. However, in the limited comparisons made to date direct elution has performed as robustly as conventional manual extraction (Abu-Rabie and Spooner, 2011).

The correct choice of extraction solvent can aid both assay sensitivity and robustness. Given sufficient extraction times, highly aqueous extraction solvents appear to more completely desorb the DBS sample, removing more endogenous matrix material from the paper substrate (the DBS sample is completely discolored back to the original color of the substrate). Conversely, organic extraction solvents do not discolor the samples (the color of the DBS after extraction is unchanged), which suggests that more matrix material is left intact. Aqueous extraction does not, however, produce an improvement in MS signal and suggests that an effective increase in analyte extraction does not occur. In fact, Déglon et al. (2009) reported that the use of water led to ionization source contamination resulting in rapid degradation of MS signal. Aqueous extraction appears to extract no more analyte from the DBS sample, but far more unwanted matrix components are extracted. Déglon et al. demonstrated that water-extracted DBS is characteristic of blood spectrum using UV–vis spectra with bands that correspond to the absorption of iron porphyrin of hemoglobin and protein. These bands are absent with organic solvent extraction of DBS. Thomas et al. discussed selective (i.e., analyte only) desorption using an organic extraction solvent as opposed to an aqueous solvent which also desorbs large amount of matrix material (i.e., hemoglobin and protein). It was shown that selective desorption was applicable to compounds with a wide range of polarities (including hydrophobic and hydrophilic drugs and their phase I and II metabolites). It was also reported that percolating organic solvent through DBS on cellulose substrate induced protein precipitation which enabled selective desorption of small molecules (Thomas et al., 2010). Highly organic extraction solvents appear to provide multiple benefits over highly aqueous solutions. Aqueous desorption can lead to rapid clogging of analytical systems, increased risk of matrix effects, and the need to change UHPLC columns and clean MS inlets more frequently.

22.5.4 Digital Microfluidics

In 2011, Jebrail et al. reported proof of concept for a fast and efficient DMF method for DBS analysis (Jebrail et al., 2011). DMF is a fluid-handling technique where discrete droplets of samples and regents are manipulated on an open surface by applying a series of electrical potentials to an array of electrodes. Droplet actuation is driven by electromechanical forces generated on free charges in the droplet meniscus (for conductive liquids) or on dipoles inside of the droplet (for dielectric liquids). Jebrail et al. demonstrated that sample analysis could be undertaken by spotting blood directly onto a "chip" device and allowing it to dry or by positioning a punched disk from a DBS (on paper substrate) onto the chip (Figure 22.22). Extraction solvent is dispensed onto the chip, driven onto the DBS, and actuated back-and-forth multiple times to facilitate extraction (Figure 22.23). Samples processed by this method can be collected and then analyzed off-line, or, to meet the automation requirements of high-throughput bioanalysis, the DMF platform can be directly coupled to a nanoelectrospray emitter for in-line MS analysis. To initiate analysis by MS, a droplet was driven to the entrance of a pulled-glass emitter on the chip, and after filling by capillary action, a voltage was applied to the top plate of

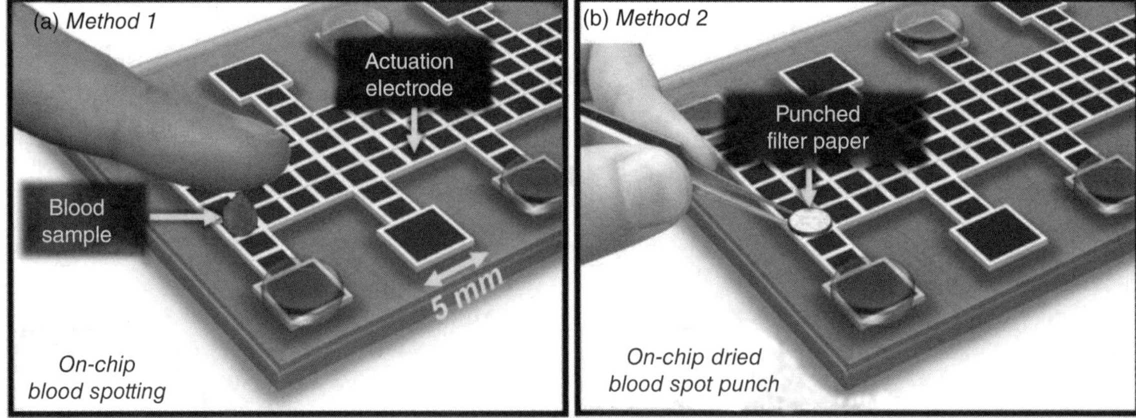

FIGURE 22.22 Two digital microfluidic methods designed to analyze dried blood spot samples. In method 1 (a), a 5 μL droplet of blood is spotted directly onto the device surface and allowed to dry. In method 2 (b), a 3.2 mm diameter punch from a filter paper bearing dried blood is positioned on the device surface. Reproduced from Jebrail et al. (2011) with permission from the Royal Society of Chemistry.

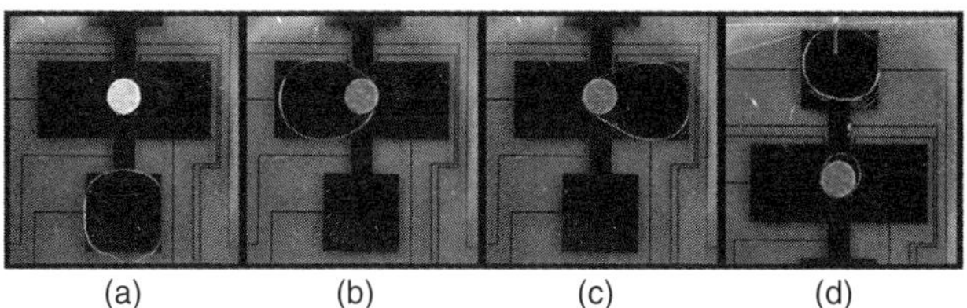

FIGURE 22.23 Frames from a video (left to right) illustrating a digital microfluidics (DMF) dried blood spot (DBS) extraction. (a) A reservoir is filled with extraction solvent (MeOH containing internal standard). (–b, c) The solvent is then driven onto the DBS, and actuated back-and-forth 10× before being allowed to incubate for 5 minutes. (d) After incubation, the droplet is moved to the final electrode, where it fills a pulled-glass emitter by capillary action. Photo courtesy of Aaron Wheeler, University of Toronto, Department of Chemistry, used with permission.

the DMF device to generate a nanoelectrospray into a mass spectrometer. Devices were translated horizontally in front of the mass spectrometer to switch between emitters. Potentially such emitters could be built into DMF DBS sampling devices. The mechanism for DBS DMF analyte extraction is automated solvent extraction; however, the execution is different enough from traditional large-scale wet lab extraction that it is worth recognizing it as a distinct technique in the context of this chapter.

In the initial publication, Jebrail et al. (2011) demonstrated that there was no significant difference in data generated using DMF or conventional DBS extraction techniques to quantify analytes that are commonly measured as biomarkers for amino acid metabolism disorders in newborn patients. Further work using sitamaquine as a test compound demonstrated that a linear MS response over 3 orders of magnitude with a limit of detection of 3 ng/mL (on a linear ion trap MS, see Section 22.7 for more details on comparative sensitivity) could be achieved using a 3 mm DBS punch (unpublished data). Initial testing also demonstrated that accuracy and precision data within guideline-acceptance criteria for regulated pharmaceutical development could be achieved for sitamaquine QC samples (unpublished data). Compatibility with HPLC methodology has yet to be reported, but potentially such technology could be integrated into a chip format for a seamless approach. Further testing is required to fully evaluate the performance on offer from this technique but DMF certainly offers considerable potential in terms of automation and miniaturization for DBS direct analysis. Further refinements to the technique, including simplified chip fabrication and improved liquid actuation, have recently been reported that are specifically aligned with optimizing DBS direct analysis. This work compared newborn screening of succinylacetone (a marker for hepatorenal tyrosinemia) in DBS samples using a conventional manual extraction technique and DMF and demonstrated no significant difference in the two sets of data (Shih et al., 2012).

22.6 DIRECT DESORPTION

For the purpose of this chapter the term direct desorption describes the host of direct analysis techniques that use a nonliquid elution basis for analyte extraction and do not produce a liquid extract. This term encompasses techniques that have also been referred to as atmospheric pressure surface sampling/ionization MS, ambient ionization MS, ambient MS, and ambient desorption ionization MS techniques, among others. In line with the scope of this chapter, direct desorption techniques that are only suitable for sampling liquids, and are thus not compatible with DBS direct desorption (without additional coupling to another process), have not been covered. To reflect current trends and short-term future prospects of DBS direct analysis, direct elution/liquid interface-based extraction techniques (Section 22.5) have been treated separately from direct desorption.

In the most general sense, direct desorption techniques sample analytes from surfaces, form ions from analytes, and then transport these ions into vacuum for detection by MS. Ionization occurs externally to the MS, with analyte ions, not the entire sample being introduced to the MS. These techniques typically use well-established ionization mechanisms such as ESI, chemical ionization (CI), or photoionization (PI). The advantage of these techniques is the configuration of the ion sources in an open air direct ionization format that enables direct sample analysis without extensive sample preparation. Direct desorption techniques allow the analysis of materials directly from their surfaces, without any preparation (i.e., extraction, preconcentration, dissolution, derivatization), hence, the interest from DBS users, where simplifying and speeding up the process of measuring these samples is highly desirable. The elimination of sample preparation results in rapid analysis, with the overall timescales being governed by the time required to present the sample to the MS. Most direct desorption techniques reported to date have not been coupled to additional pre-MS separation techniques, such as HPLC, and will often not be directly compatible due to the extraction medium. Alternative methods of separation (such as utilizing gas chromatography mass spectroscopy (GC/MS)) may be possible but to date relatively few examples have been reported. Elimination of LC (or any separation process that adds complexity to the analysis) would be a highly desirable simplification of the analytical process in high-throughput environments. However, as discussed in Section 22.3.3.5, for many applications this cannot be at the expense of confidence/selectivity in the resulting data.

The term direct desorption covers a large and expanding number of techniques, many of them subtle variations on similar processes (Table 22.1). A number of groups have attempted to summarize and group these techniques by their sampling, extraction, or primary or secondary ionization mechanisms. Categorizing these techniques is tricky as in many cases there is some debate over predominant mechanisms and there is considerable overlap between primary and secondary ionization (Cooks et al., 2006; Van Berkel et al., 2008; Venter et al., 2008; Weston, 2010; Harris et al., 2011; Huang et al., 2011; Corso et al., 2012). These techniques have been categorized (Table 22.1) according to their primary mechanism of surface sampling, namely liquid and gas jet desorption/ionization; thermal desorption/ionization; laser desorption/ionization; and liquid extraction (direct elution). A secondary categorization has then been assigned according to the entire desorption/ionization process (or processes) that occurs. Thus, each technique is categorized as either a direct ionization, combined desorption/ionization, or multistage ionization process, which are defined as follows:

- *Direct ionization* refers to techniques utilizing direct ionization of liquid analyte molecules without pretreatment. Generally these techniques are only suitable for liquid state sampling, so are not applicable for DBS direct desorption. Only processes that integrate a wetting step as part of the analysis (such as PS) fit into this category.
- *Combined desorption/ionization* processes involve the generation of metastable atoms or charged reactive species (e.g., charged solvent droplets, or ions) from an ambient ionization source, which are impacted with a sample surface for analyte desorption and ionization.
- In *multistep ionization* techniques, the initial step involves desorption of analyte molecules via thermal energy, laser irradiation, shockwaves, or nebulization. The second step involves these desorbed analyte species reacting with metastable atoms or charged solvent species (generated through ESI, atmospheric pressure chemical ionization (APCI), or PI) from an ambient ionization source to form analyte ions.

Each process leads to the formation of analyte ions which are then transferred to an MS inlet for detection. Table 22.1 includes separate headings for desorption method and ionization method. The desorption method is the predominant feature of interest for direct ionization techniques; combined desorption/ionization techniques share a single process across both categories while multistep ionization techniques have a discrete desorption and ionization process. An alternative classification method used by a number of groups have categorized direct desorption techniques into two main classes: those that utilize a solvent spray, such as desorption electrospray ionization (DESI), and incorporate ESI-like ionization; and others like DART, which rely on plasmas to create gas-phase ions, and exhibit APCI-like ionization characteristics. No attempt has been made here to include (and categorize) all of the published techniques that could be classed as "direct analysis," many of which are subtle variations of the same process. The scope of this chapter is to highlight the main types of direct desorption that have been reported, with an emphasis on applicability, or potential applicability, to DBS analysis. For this reason, some prominent techniques that have to date only shown applicability to liquid samples have not been included (e.g., droplet ESI, probe ESI, laser spray ionization).

In this chapter, the various categories of direct desorption will be introduced and discussed, focusing on those that have the most potential for DBS direct desorption compatibility, followed by a more in-depth evaluation of the techniques that have been directly applied to DBS analysis to date. While a detailed discussion of all the surface desorption and ionization mechanisms that occur is beyond the scope of this chapter (many of these mechanisms are not yet fully understood), some of the basics have been included to aid in understanding the different methods of desorption and ionization processes that occur. Interested readers are encouraged to review the references for further information.

22.6.1 Liquid and Gas Jet Desorption/Ionization

22.6.1.1 Desorption Electrospray Ionization The interest and rapid emergence of numerous ambient surface sampling and ionization combinations for use with MS appears to have been sparked from the introduction of DESI in 2004 (Takats et al., 2004). DESI is one of the most utilized direct desorption techniques in a wide range of applications such as the analysis of pharmaceuticals, chemical warfare agents, textiles, counterfeit and illicit drugs, forensics, peptides and proteins, metabolomics, and plants, among many others. Commercial DESI sources are available for most common mass spectrometers, thus assisting the incorporation of this technique in a wide range of industries. DESI combines features of electrospray and desorption ionization mechanisms. In this technique, a pneumatically assisted electrospray needle directs a high-velocity (90–200 m/s) spray of charged microdroplets toward a sample (Figure 22.24). A thin solvent film

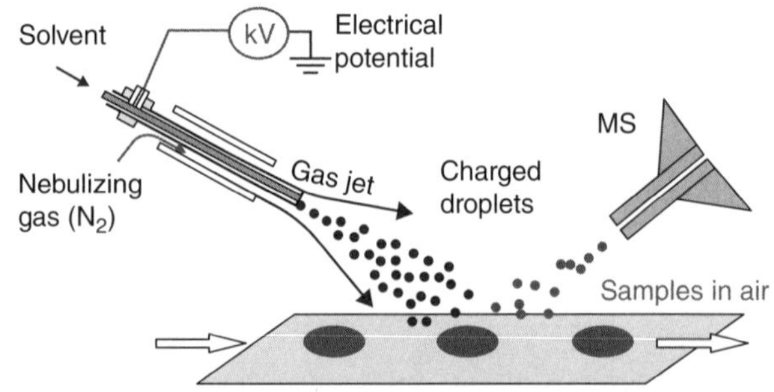

FIGURE 22.24 DESI schematic, demonstrating direct analysis of DBS samples. Reproduced from Ifa et al. (2010) with permission from the Royal Society of Chemistry.

is formed and rapid extraction and/or dissolution of analyte molecules occurs. The incoming jet of fresh droplets then dislodges liquid from this microfilm, resulting in an ejection cone of secondary droplets containing analyte, which is subsequently introduced into an MS inlet. Secondary droplet gas-phase ion formation occurs by ESI-like mechanisms. The complete desorption and ionization mechanisms are still not fully understood, but this account summarizes current views on the processes involved.

The angle at which the spray plume attacks the sample has been shown to affect analyte response and selectivity (often optimized at 45–55°). The dominant desorption/ionization process in DESI was thought to be "droplet pick-up," a liquid–solid microextraction/liquid film-droplet sputtering process (Takats et al., 2005a, 2005b). This theory was supported following further work that progressed the understanding of low kinetic energy solvent cluster ions impacting the sample surfaces (Gologan et al., 2005; Takats et al., 2005a, 2005b). The evaporation of desorbed droplets would then be expected to follow conventional ESI-like behavior (Kebarle, 2000). However, fluid dynamic simulations have since been used to investigate droplet scattering behavior which supports a multistage momentum transfer method, moving away from the single-stage droplet pick-up theory previously postulated and also ruling out a significant contribution by spray voltage or electrostatic forces toward formation and transportation of droplets (Costa and Cooks, 2008). This work suggested that a fine film appeared on the sample surface caused by the spray plume wetting the surface, which influenced condensed analyte extraction into the surface film. This process has been shown to be influenced by the duration taken to first wet and then extract from the sample surface, known as a solvation delay (Bereman and Muddiman, 2007). Progeny droplets are subsequently produced via momentum transfer from impacting droplets and nebulization gas, which are then transferred to the MS inlet.

The ability to utilize DESI for quantitative analysis of drugs has been demonstrated and shown to be particularly effective when an IS is present throughout the samples (Nyadong et al., 2008; Manicke et al., 2009). Any solvent suitable for ESI can be used in DESI, and additives or other chemical reagents can be used to perform "reactive DESI." Cooks et al. first reported that if the DESI spray is doped with selective chemical reagents, then it is possible to perform ion/molecule reactions between the charged microdroplets and an analyte exposed on a solid sample surface. Spray–surface and analyte–surface interactions influence the efficiency of desorption/ionization and analyte introduction into the MS (Cotte-Rodríguez et al., 2005; Chen et al., 2006a; Nyadong et al., 2007). DESI appears to offer a high level of applicability to bioanalytical challenges with wide mass ranges (Shin et al., 2007) and non-proximate analysis (Soparawalla et al., 2009) being reported. DESI has been demonstrated to work successfully when coupled to miniature MS. Improvements in vacuum and sensitivity of miniature MS instrumentation provide confidence that in the future a similar approach could be utilized for DBS direct analysis (Sanders et al., 2010). Successful experiments coupling DESI to ion mobility MS have also been reported providing some confidence that such a technique, which increases selectivity, could be utilized with DBS direct analysis (Harry et al., 2009; D'Agostino and Chenier, 2010; Galhena et al., 2010).

Desorption Electrospray Ionization and DBS Wiseman et al. demonstrated the potential to use DESI for DBS direct analysis in 2010, coupling the direct desorption technique to a triple quadrupole MS (Wiseman et al., 2010). In this proof of concept study the potential for eliminating sample preparation was not completely fulfilled, as each DBS was manually cut out (the whole spot, not a sub-punch) from the card and mounted on a microscope glass slide. DBS sample analysis involved scanning the DBS sample surface in one direction at a constant velocity (optimized at 200 μm/s), with the spray oriented toward the MS inlet. DESI-MS/MS chronograms depicted ion intensity across the DBS sample as the surface was moved horizontally beneath the DESI spray (Figure 22.25). Based on signal intensity, signal stability, and matrix effects, the optimal solvent composition was found to be 80:20 acetonitrile:water (v/v) containing 0.1% formic acid. Signal response was shown to increase with solvent flow rate over an order of magnitude from 3 to 30 μL/min, but was limited to 15 μL/min to avoid carryover. DBS calibration standards and QCs were scanned sequentially and peak areas were calculated to produce a standard curve for quantitation.

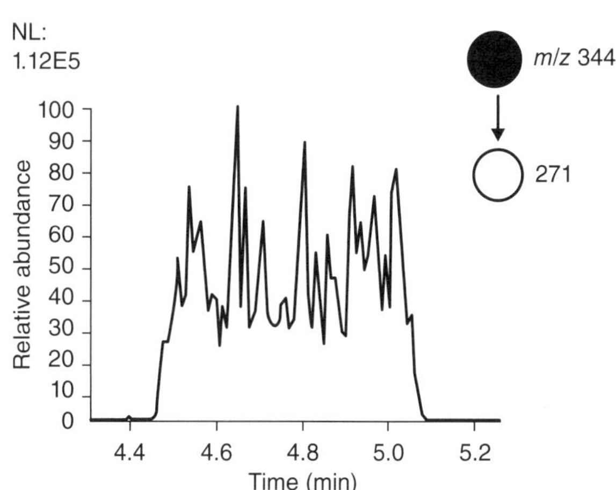

FIGURE 22.25 Representative DESI-MS /MS selected reaction monitoring (SRM) chronogram of 10,000 ng/mL sitamaquine in DBS. Reproduced from Wiseman et al. (2010) with permission from the Royal Society of Chemistry.

The technique demonstrated limits of detection down to 10 ng/mL for a small set of representative small-molecule pharmaceutical compounds (sitamaquine, terfenadine, and prazosin, with verapamil used as a common IS) and excellent linearity over 3 orders of magnitude. The authors proposed that the total volume of blood in a 3 mm DBS sub-punch analyzed by manual DBS extraction is likely to be greater than the total volume sampled by DESI, which will have influenced the detection limits. If variation in detector sensitivity is ignored, this is an order of magnitude less sensitive than what is typically achieved using manual extraction of DBS followed by LC-MS/MS. While good accuracy (%bias) was observed at mid to high QC concentrations, generally precision (%CV) and bias at lower concentrations were outside of the guideline-acceptance criteria for regulated bioanalysis. A number of strategies were suggested to improve performance. These include utilizing a larger DESI spray plume to increase the DBS area being sampled which should increase the signal response and limit of detection/quantitation. Possibilities include improving desolvation of secondary droplets by directly heating the nebulizing gas or using a heated auxiliary gas to sweep the surface. Selectivity was reported to be "sufficient given that no significant interfering signals were observed." This statement is unlikely to be unchallenged in all applications bar regulated pharmaceutical bioanalysis.

Wiseman et al. found that of the cellulose-based paper substrates evaluated, untreated cards gave the highest signal performance. It was noted that chemically treated cards showed significant ion suppression, similar to what is observed with ESI-MS/MS for the small-molecule compounds evaluated in this work. A relationship demonstrating decreasing analyte response with increasing IS concentration was also reported. This was also proposed to mirror typical ESI behavior, where this relationship is thought to be associated with analyte competition for charge in the secondary droplets emitted from the sample surface.

22.6.1.2 Paper Spray PS is a technique closely related to both DESI and nanoelectrospray that is particularly well suited to DBS analysis. Paper spray mass spectrometry (PSMS) was first reported in 2010 as a convenient way of directly analyzing molecules in bioanalytical samples for application in areas such as biomedical research, clinical diagnosis, drug development and therapeutic drug monitoring (Wang et al., 2010a). The technique involves cutting a porous substrate, typically chromatographic paper, into a triangular shape and applying a biological sample. A solvent is applied to the paper, followed by a high voltage, causing a spray of charged droplets to be induced at the tip of the paper triangle, which is positioned in front of a mass spectrometer inlet. For DBS analysis this has been shown to work either by directly applying blood to the paper substrate or by taking a sub-punch from a DBS sample and fixing it to the paper triangle (Figure 22.26).

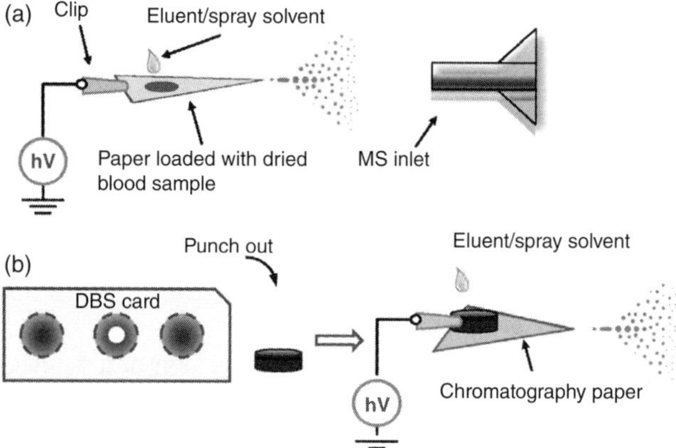

FIGURE 22.26 Schematic of PS. (a) Schematic of a typical paper spray experiment. (b) Schematic for quantitation from a DBS by taking a punch from the spot and adding an IS solution prior to ionization. Reprinted with permission from Manicke et al. (2011a). Copyright 2011 Elsevier.

Some work has been carried out to investigate the ionization mechanism employed by PS, though it is not yet fully understood. Liu et al. (2010) compared PS with nanospray ionization and found that PS required much higher spray voltage than nanospray. In positive ion mode PS showed an increase in both MS intensity and spray current with increasing spray voltage. In comparison, the nanospray signal intensity increased to a plateau over the voltage range (0.5–2 kV), then decreased to zero (2–4.5 kV), while the nanospray current increased with increasing spray voltage. It was concluded that capillary action was likely to be responsible for fluid transport in the paper. The network of cellulose in a typical paper substrate provides microchannels for liquid transport. It is thought that the high electric potential between the paper triangle and MS inlet generates an electric charge that accumulates at the apex of the triangle. As with ESI, the liquid is broken by Coulombic force to form charged droplets. These droplets then undergo desolvation to produce dry ions. It was shown that a Taylor cone was observed at the tip of the triangle (Tepper and Kessick, 2009). The ionization mechanism may be similar to ESI methods that use wicking (Liu et al., 2010). In these methods, when the maximum capillary flow rate through the substrate exceeds the electrospray flow rate, a single electrospray forms from a droplet at the end of the wick. Alternatively, when the maximum capillary flow rate is less than the electrospray flow rate, a multitude of microscopic nanoelectrospray sources are formed from within the surface of the wick tip. Interestingly, the macroscopically sharp tips may be needed for either effective fluid transport or possibly to provide high electric fields at the tip if it is the tip, and not the individual fluid channels that are responsible for ionization. The exact mechanism has yet to be determined.

In contrast to positive ion mode, significant differences have been observed in negative ion mode between PS and nanospray. Using methanol:water (1:1) as the solvent, a stable spray could not be observed even with negative voltages up to 6 kV. However, electrons were found to be emitted at >3.5 kV with the radical ion $[M]^{-\cdot}$ being generated. Electron emission also exists in nanospray in negative ion mode (2 kV) but it does not interfere with ion formation. At voltages above 2.5 kV, currents measured in PS are significantly higher than for nanospray. It is possible that this is due to the onset voltage of electrospray and that of electrical discharge. For PS, electrospray onset voltage appears to be lower in positive mode, allowing stable ESI to occur. However, in negative ion mode, the onset of electrical discharge is always lower than the ESI voltage onset, which significantly degrades electrospray performance. It was noted that using pure methanol as the solvent allowed stable electrospray in negative ion mode. The above explanation is supported by the fact that it is difficult to induce discharge in pure methanol. It is assumed that the charge accumulation required for ionization is much lower when using pure methanol as the surface tension of the solvent is much lower (Gabellca and De Pauw, 2005). A "survival yield" method, which is used to judge the internal energy deposition associated with soft ionization techniques, was used to measure internal energy distributions of typical ions generated by PS (Nefliu et al., 2008). PS and nanospray produced very similar internal energy distributions, suggesting that these two ionization methods may follow the same mechanism.

Paper Spray and DBS The fact that PS is based around analysis from a paper substrate clearly makes it a prime candidate for application with DBS. Wang et al. (2010a) demonstrated quantitation of imatinib (an FDA-approved drug for the treatment of chronic myelogenous leukemia) in whole blood utilizing MS/MS and produced a linear response well beyond the therapeutic range. Manicke et al. (2011b) measured therapeutic drugs quantitatively over several orders of magnitude demonstrating excellent bias and precision performance well within guideline-acceptance criteria. This study also demonstrated that PSMS is capable of high sensitivity with DBS samples containing pharmaceutical drugs compared to many other direct desorption techniques. Detection limits ranged from <100 pg/mL for proguanil and benzethonium chloride to >100 ng/mL for acetaminophen and ibuprofen, suggesting that PS will be applicable to cover the therapeutic range of many drugs, but not all. To put this in context, this places PSMS as approximately an order of magnitude more sensitive for some compounds than DBS manual extraction methods, while for others it is an order of magnitude less sensitive, highlighting the influence of compound class on the applicability of the technique (see Section 22.7). Overall, the sensitivity, linearity, accuracy, and precision data generated in this study suggest that PSMS has considerable potential as a DBS direct analysis technique.

The use of an IS has been shown to work when applied via the solvent or when preprinted onto the paper. Manicke et al. reported a PS evaluation where a labeled IS was dropped onto paper and allowed to dry, before a blood spot containing drug was added on top of the IS (Espy et al., 2011). Precision was acceptable (8%, compared to 3% where IS was added to blood before spotting) and interestingly recovery was lower for the drug than the IS. In this work, the authors suggest the possibility of the drug, but not the IS, being fully embedded in the dried blood matrix. It is possible that the drug may not be recovered from the matrix as efficiently due to the solvent not fully wetting and penetrating the dried blood, or because of hydrophobic interactions between the drug and dried blood components. Alternatively, the structure of the substrate may dictate local variations in how blood is absorbed by the paper, and there may be pathways through the blood that do not contain as much blood. It is possible that solvent could preferentially take these "paths of less resistance." This is a factor that could possibly be eliminated or reduced by using more homogeneous substrates. As part of the same study, where IS was added via the extraction solvent, recovery of drug, relative to IS, was very low compared to the other methods of IS addition. This is likely to be due to the relatively slow recovery of drug from the paper substrate, compared to IS already in solution. Precision was relatively poor using this method (16%). In a third method, a DBS sub-punch was taken, and an IS solution was added to the punch and allowed to dry. This punch is then placed on top of a paper triangle and PS is carried out as previously described. Variance was found to be 8%. Variations in blood hematocrit, the volume of blood applied, and inconsistencies in blood collection could lead to greater variability.

One drawback of PS is that for the process to work efficiently (with no sample preparation) the substrate format will most likely need to be specifically designed (to incorporate the macroscopic fine point on the substrate, and possibly the ability to easily apply a voltage), rather than using the popular 4-spot cards. This has been addressed by the introduction of an elaborate fully automated PS instrument (Espy et al., 2011; www.quantiontech.com). This device is based around handling specific cartridges that house the DBS sample on a triangular-shaped paper substrate, which takes the place of current DBS cards.

As for many direct desorption techniques, methods will need to be developed to avoid overestimation of circulating drug concentrations using PSMS for MS analysis of drugs known to metabolize into fragile compounds. One specific possibility for PSMS is the use of selective on-paper chemical reactions (Vaz et al., 2010) where potentially a metabolite could be converted into a stable form (Liu et al., 2010). PS has been shown to incorporate chemical separation by utilizing separation by paper chromatography (Wang et al., 2010a).

Dye solutions were shown to be separated, then detected by PSMS. PS using chromatographic paper thus integrates three analytical procedures: sample collection, analyte separation, and analyte ionization. It is suggested that this technique could be used to separate chemicals in biological fluids prior to *in situ* ionization. It would be quite a leap to apply this method of separation to components in DBS samples, but nonetheless it is an intriguing idea, especially given the current lack of compatibility of PS with LC-MS/MS.

As with DESI, online derivatization can be achieved by adding reagents into the spray solution. Improvements have been observed for the analysis of steroids in urine (Huang et al., 2007). Derivatization can be required where analytes of interest have low ionization efficiencies and low concentrations in mixtures. If the derivatization reagents are stable on paper, then they potentially could be added to the paper before they are used to load and analyze the samples. In a similar manner to IS addition prior to sample load, derivatization will only be suitable for certain applications. Another recent development is the use of silica-coated paper substrate which has been shown to provide increases in recovery and sensitivity (Zhang et al., 2012).

22.6.1.3 Other ESI-Related Techniques (EESI, ND-EESI, EASI) and DESI Variants A number of variations on ESI-based techniques have emerged that are subtly different to DESI and offer alternative analytical performance. Neutral desorption extractive electrospray ionization (ND-EESI) (Chen et al., 2007) is a derivative of extractive electrospray ionization (EESI) (Chen et al., 2006b) that utilizes a neutral gas stream (rather than a charged solvent stream in EESI) to initially desorb analyte molecules co-occurring with an ESI plume. In this arrangement both the liquid and high voltage are eliminated. Desorption is carried out using a moderate velocity room temperature gas jet. Ionization of desorbed analyte is then carried out via a secondary mechanism. The main advantages of ND-EESI are the non-proximate sample introduction, a selectivity gain compared to solution-based ESI, and the elimination of voltages from ambient sampling (Chen and Zenobi, 2008; Williams and Scrivens, 2008). ND-EESI has been reported as being tolerant of complex matrixes due to the specialist mechanism of the EESI process. Charge transfer collisions between the primary charged ESI solvent spray and a secondary neutral plume (delivered by the ND sampling transfer line) in close proximity to the source bring about analyte ionization. A major limitation, most likely a result of the absence of a heated gas stream, is a difficulty in desorbing larger molecules (above m/z 300).

Atmospheric pressure thermal desorption-secondary ionization (AP-TD/SI) (Basile et al., 2010) is a variant of ND-EESI, though here it is categorized as a thermal desorption/ionization technique (see Section "TD/APCI and APTDI"). In this technique thermal desorption is coupled to APCI or ESI secondary ionization. A similar advancement to ND-EESI/EESI was demonstrated by removing the spray voltages used in DESI which resulted in what was originally termed desorption sonic spray ionization (DeSSI) (Haddad et al., 2006), now known as easy ambient sonic spray ionization (EASI) (Haddad et al., 2008). EASI is essentially created from the DESI setup by turning off the variable high voltage and increasing the nebulization gas velocity. Here, only the force from high-velocity nebulizing gas is used to produce gas-phase ions. The majority of species detected by DeSSI are also detected with DESI, but the reverse is not always true.

A modification to DESI, termed "nano-DESI," has been reported that uses a primary capillary to deliver a low-flow solvent stream to the sample surface, while simultaneously a liquid bridge is formed with a nanospray capillary which is positioned close to the MS inlet (Roach et al., 2010). This technique in fact more closely resembles LMJ-SSP (see Section 22.5.2) than DESI. The potential benefit of this configuration is that it separates the desorption and ionization steps which can lead to improved sampling efficiency. Electrode-assisted DESI (EADESI) has demonstrated improvements in ion efficiency by using a high-voltage needle and grounded conductive surface decoupled from the nebulizing sprayer (Özdemir and Chen, 2010).

Spray-based photon/energy techniques such as electrospray-assisted laser desorption ionization (ELDI), desorption atmospheric pressure photoionization (DAPPI), and matrix-assisted laser desorption electrospray ionization (MALDESI), which could be classed in this category, are covered in laser desorption/ionization (Section 22.6.3).

22.6.2 Thermal Desorption/Ionization

The majority of thermal or plasma-based desorption/ionization direct analysis techniques rely on the principles of APCI. As in traditionally used LC-MS, APCI-based direct desorption is generally applicable to polar and nonpolar, low molecular weight analytes rather than larger biomolecules. In these techniques a direct current or RF electrical discharge is generated between electrode pairs which are in contact with the flowing gas (usually nitrogen or helium). This generates a stream of ionized molecules, excited-state neutrals, radicals, and electrons. These plasma species are directed toward a sample surface, and secondary heating of the plasma gas stream to enhance desorption may occur. Generally, applicability is with analytes below 1000 Da. The differences between the plasma-based direct desorption variants are based on the following parameters (Harris et al., 2011):

- Presence or removal of plasma species in the flowing gas prior to sample interaction

- Whether the plasma gas is stream heated (either via an additional heating element or induced by the discharge current) or used at ambient temperature?
- Operation of discharge current (ac pulsed, dc, etc.) and configuration of discharge (point-to-lane or annular)
- Whether the plasma created is within a larger (>several millimeters) or smaller (micrometer-sized) gap?

22.6.2.1 Thermal Multistep Ionization Techniques

TD/APCI and APTDI Thermal desorption atmospheric pressure chemical ionization (TD/APCI) was first introduced in the mid-1970s. Heated gas is passed over a sample surface to desorb the sample intact from the condensed phase to the vapor phase (Ebejer et al., 2005). The sample can be ionized by an array of ion/molecule chemistries once in the gas phase, typically APCI. Non-proximate detection (sampling remote from the mass spectrometer) has been utilized using a TD/APCI approach using heated sample lines for chemical contraband detection applications, and selectivity was gained by utilizing tandem mass spectrometry (MS/MS) (Van Berkel et al., 2008). Interestingly, the application of heat can result in the gas-phase ion liberation of some ionic species. Using atmospheric pressure thermal desorption ionization (APTDI), organic salts can be thermally desorbed as ions into the gas phase (Chen et al., 2006c).

Secondary ionization by APCI involves a reagent-ion population being generated through an ion/molecule cascade caused by a corona discharge. In the simplest example, in positive ion mode a high voltage is applied via a corona discharge needle ionizing nitrogen in air. A series of reactions then occur with nitrogen and water vapor, with hydronium ion-water clusters being produced as the major reagent-ion species. Molecules with a gas-phase proton affinity (or basicity) higher than that of water can be protonated by these hydronium ion-water clusters. When a negative high voltage is applied to the corona electrode (negative ion mode), electrons are emitted, typically producing OH^- as the reagent ion. All species in the gas phase with a higher gas-phase acidity than water will be ionized by the OH^- via proton transfer. However, the gas-phase ion chemistry can be complicated where charge transfer and cation attachment (in positive ion mode) or electron capture and anion attachment (in negative ion mode) cause ionization. The ions ultimately formed can be influenced by the gas-phase ion chemistry, which can be altered by the introduction of doping agents into the discharge area (Moini, 2007).

APTDCI AND DBS In 2011, Corso et al. reported an atmospheric pressure thermal desorption chemical ionization (APTDCI) tandem MS method using a heated gas flow and electrical discharge for the direct analysis of sterols and acylcarnitines in DBS and DPS samples for application in clinical disease diagnosis (Corso et al., 2011). In this report, APTDCI was compared to the current standard MS/MS analysis screening method for these compounds, which involves preparation and derivatization prior to analysis. It is notable for this application that Corso et al. state that chromatographic separation is not required for the existing analysis, as a preparative butyl esterification derivatization step is used to improve both the sensitivity and selectivity for acylcarnitine analysis. Here direct analysis without derivatization is desirable to simplify analytical operations. In cases where acylcarnitines cannot be differentiated on the basis of mass alone, Corso et al. suggest that additional investigations, such as GC-MS analysis or *in vitro* studies using stable isotope labels, are necessary for clarification. The APTDCI technique (Figure 22.27) employed by Corso et al. (2011) appears to be suited to DBS analysis as the diameter of the heated nitrogen gas stream is wide enough (15 mm) to completely cover the surface of spots up to 10 mm in diameter. The technique showed great promise but reservations were made over the lack of sensitivity on offer.

ASAP The atmospheric pressure solids analysis probe (ASAP) technique can be used for the direct analysis of liquids and semivolatile solids. ASAP was created by modifying a corona discharge APCI source so that a glass capillary could be inserted into the heated gas stream. Analyte present in the sample added to a capillary is thermally desorbed by the heated gas and subsequently ionized by APCI and analyzed by MS (McEwen et al., 2005). ASAP has proved to be applicable to the types of compounds typically amenable

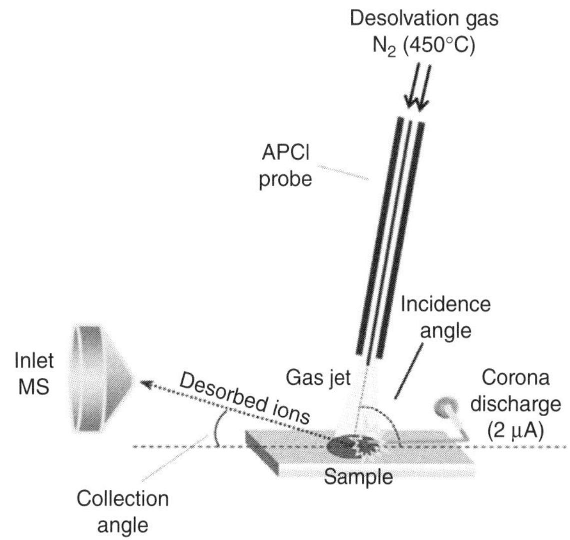

FIGURE 22.27 Schematic representation of the atmospheric pressure thermal desorption chemical ionization (APTDCI) mechanism of desorption/ionization. Reprinted with permission from Corso et al. (2011). Copyright 2011 Elsevier.

to APCI (e.g., fresh biological samples, lipids, fatty acids, drugs, capsaicins, and carotenoids).

LDTD Laser diode thermal desorption (LDTD) is a commercially available variant of TD/APCI that utilizes an IR laser to thermally desorb samples. As its name suggests, a thermal desorption process is dominant here, rather than laser beam surface impact. Typical applications include high-throughput cytochrome P450 inhibition analysis (Wu et al., 2007). In many applications, this technique cannot be classed as a true direct analysis technique as some sample preparation is required. Typically, an organic solvent, often the final product of an analytical extraction scheme, is evaporated in a specific type of well, resulting in the metallic surface being coated with the analyte. An infrared laser is then used to desorb the analytes which releases neutral gas-phase molecules which are subsequently ionized using APCI.

LDTD AND DBS Swales et al. (2011) reported the use of LDTD-APCI-MS/MS to quantitate metformin and sitagliptin in DBS samples in a pharmaceutical discovery environment. While this method was described as a simple, rapid, and robust high-throughput assay, it was not a true direct analysis technique as the DBS samples were punched out and extracted with solvent prior to LDTD analysis.

Direct Analysis in Real Time The use of DART has been widely reported in multiple fields of interest since its introduction by Cody et al. in 2005 including analysis of drugs, foods, plants, explosives, cells, forensics, and chemical warfare agents, among many others (Cody et al., 2005). In addition to DESI, DART has been largely responsible for the rapid growth in the interest in direct analysis. DART is essentially a multistep ionization TD/APCI variant. The sampling method involves thermal desorption, followed by APCI. In brief the technique is based around plasma discharge formation in a heated helium gas stream, common to many APCI methods. It is thought that the mechanism involves Penning ionization (reactions of electronic or vibronic excited-state species with reagent molecules and analytes). Using a CI process, water molecules react with metastable helium atoms, followed by sample ionization by thermal desorption via the heated gas stream (Harris and Fernández, 2005).

A schematic representation of the DART process is illustrated in Figure 22.28. Helium gas is flowed through a needle

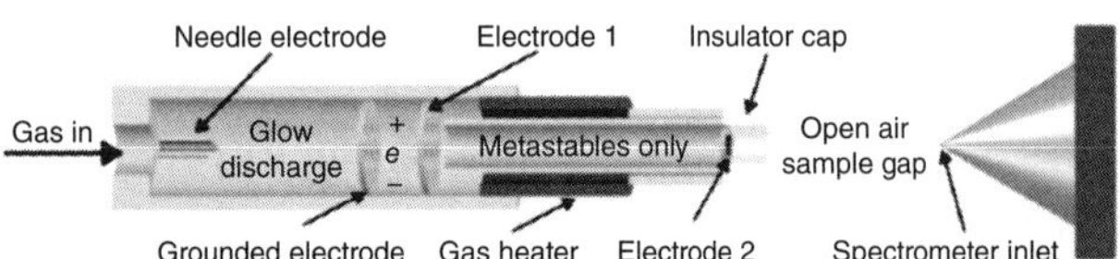

FIGURE 22.28 Direct analysis in real time (DART) schematic (JEOL USA, Inc., used with permission).

electrode with a potential, resulting in a glow discharge. This produces ions, electrons, and metastable (gaseous excited-state) species. Ions are removed from the gas stream by perforated electrodes downstream (electrode 1), leaving only metastables. Neutral metastable species carried by the gas through a heated chamber pass through a grid electrode (electrode 2) and enter the open atmosphere. Heating the gas controls both thermal desorption and sample pyrolysis. The grid electrode serves two purposes. It prevents ion–ion and ion–electron recombination and is also a source of electrons (through Penning ionization of neutral species or surface Penning ionization). The sample to be analyzed is then placed between the resulting gas flow and the MS inlet (open air sample gap), where ionization occurs. Ions are directed to the MS inlet by both the gas flow and a slight vacuum on the spectrometer inlet. DART uses a variant APCI mechanism. The reagent-ion population is created via the gas-phase reactions of metastable helium atoms produced in the discharge. The metastable helium atoms react rapidly with ambient species in the ambient atmosphere to create a reagent-ion population capable of ionizing gas-phase analytes. In ambient air, where water vapor traces are present, protonated water clusters are a dominant reagent and provide protons for gas-phase species protonation.

Cody et al. (2005) describe the ionization process as follows: During positive ion formation the metastable helium atoms generated by the source react with atmospheric water to produce clusters of ionized water:

$$\mathrm{He(2^3S) + H_2O \rightarrow H_2O^{+\bullet} + He(1^1S) + e^-}$$
$$\mathrm{H_2O^{+\bullet} + H_2O \rightarrow H_3O^+ + OH^\bullet}$$
$$\mathrm{H_3O^+ + \mathit{n}H_2O \rightarrow [(H_2O)_{\mathit{n}+1}H]^+}$$
$$\mathrm{[(H_2O)_\mathit{n}H]^+ + M \rightarrow MH^+ + \mathit{n}H_2O}$$

The $\mathrm{He(2^3S)}$ electronic excited state has an energy of 19.8 eV and a reaction cross section of 100 Å for water ionization. After reacting with the excited-state helium metastable the protonated water that is formed can then react with the analyte to form a protonated molecule.

In negative ion formation metastable helium atoms react with a neutral (N) (e.g., the grid electrode) to form electrons via Penning ionization:

$$\mathrm{M^* + N \rightarrow N^{+\bullet} + M + e^-}$$

Negative ions are formed when electrons undergo electron capture by O_2 which produces O_2^-, which reacts with desorbed analyte. As shown in the scheme below the generated electrons are thermalized with atmospheric gas (G) collisions and react with gaseous oxygen to form negative oxygen ions. These oxygen anions can then react with sample molecules (S) to produce analyte negative ions (S is assumed to be a sample that contains hydrogen in the scheme below):

$$e^{-*} + G \rightarrow e^{-} + G^{*}$$
$$e^{-} + O_2 \rightarrow O_2^{-\bullet}$$
$$O_2^{-\bullet} + S \rightarrow [S - H]^{-} + OOH^{\bullet}$$
$$O_2^{-\bullet} + S \rightarrow S^{-\bullet} + O_2$$
$$O_2^{-\bullet} + S \rightarrow [S + O_2]^{-\bullet *} + G \rightarrow [S + O_2]^{-\bullet} + G^{*}$$

Distribution of reagent ions formed has been demonstrated to be influenced by the ion polarity, the gas used (nitrogen or helium), and the presence of gas stream dopants.

Generally, sensitivity is maximized through heating the gas stream to 200°C and beyond. Reported data strongly suggest that TD is the dominant process in surface desorption. Contributions to desorption from sputtering or ionized water cluster/metastable species bombardment have been suggested (Cody et al., 2005). However, the observation that the majority of excitation energy in low kinetic energy metastable atom surface interaction is used to eject electrons from the surface, together with the very short atmospheric pressure metastable helium atom life span, brings this suggestion into question. Recent work backs up the theory that the same mechanisms occur regardless of the initial method of CI, resulting in the same ion chemistries prevailing in any APCI analysis (McEwen and Larsen, 2009). Recent investigations into DART ionization have suggested that gas velocity and temperature gradients play an important role in ion transmission within the DART ionization region to the MS inlet. Optimal sample placement is a balance of two competing factors, rapid heating to induce efficient thermal desorption and a sufficient neural population, and the losses resulting from blocking the gas trajectory lines when a sample is placed between the DART source outlet and the MS inlet (Harris and Fernández, 2005).

In-source fragmentation of labile molecules in DART is an established phenomenon (Fernández et al., 2006). It is also well established that the appearance of fragmentation ions is influenced by metastable gas temperature (Lapthorn and Pullen, 2009). Recent work has shown that DART is a "harder" ionization technique than ESI, but there will be an overlap between the two processes depending on the MS used. When the glow discharge was sustained using helium, it appeared that thermal ion activation and increased in-source activation within the first differentially pumped section of the MS was occurring, as indicated by an increase in DART internal energy with increasing gas flow rates and temperatures. Metastable-simulated desorption leading to internal energy deposition pathways, and energy release from proton affinity differentials did not appear to occur. However, at the upper limits of glow discharge gas flow rates and temperatures, high-energy helium metastable-induced fragmentation was observed (Harris et al., 2010). Ionization in atmospheric pressure ion sources does not follow single, clearly defined mechanisms. For example, it has been shown that while positive ions in DART are generally generated by proton transfer, certain experimental conditions can also produce molecular ions by charge exchange mechanisms (Cody, 2009). Molecular ions are initially produced which react further, creating charge or proton transfer products.

Not utilizing a solvent spray, one of the advantages of DART may also be a disadvantage, as the chemical sensitivity observed with reactive DESI may not be possible. While there have been doubts over DART's ability to produce reliable and reproducible quantitative data, good quantitative analysis performance has been reported on a number of occasions (Zhao et al., 2008; Lapthorn and Pullen, 2009; Nilles et al., 2009).

Investigations of coupling DART–MS to HPLC have been reported, albeit unfortunately not in a configuration that progresses DBS direct analysis. Eberherr et al. (2010) reported HPLC coupled to DART-TOF-MS that utilized previously validated HPLC methods to test an experimental setup that combined HPLC with DART using a time-of-flight MS for detection. The LC to MS coupling via DART was achieved using a simple capillary interface that directed the LC eluent close to the ionization region of the DART. The main aim of this work was to determine if DART-TOF-MS could be regarded as a suitable detector for HPLC even when MS-incompatible mobile phases (such as phosphate buffers) are used. It certainly provides an intriguing avenue of interest to explore, but from a DBS perspective DART is not being used as a direct sampling technique in this instance (i.e., the sample is in liquid form prior to detection). In a similar scenario, coupling of GC to DART has also been reported (Cody, 2009). Here the gas chromatograph was used only as a sample introduction method. Integrating a separation step in this way involved sample preparation, putting the procedure at odds with the main advantage of direct desorption. (It should be noted however that the aim of this work was to demonstrate the difference between mass spectra obtained using conditions that favor proton transfer and those that favor molecular ion formation.) To progress DBS direct desorption, the additional selectivity would need to be integrated with or directly coupled to the sampling mechanism, thereby retaining the inherent advantage of direct desorption techniques.

Edison et al. (2011) reported the use of DART coupled to a high-resolution orbitrap MS to aid efficiency in food produce pesticide screening. Interestingly, this group claimed that the selectivity traditionally offered via LC/MS or GC/MS separation was replaced through a combination of using a DART helium temperature gradient (100–350°C over 3 minutes) that allowed separation of analytes based on volatility differences and the high resolution on offer from the Thermo Exactive Orbitrap™ MS. This technique demonstrates a way in which additional selectivity may be coupled to direct desorption without traditional LC or GC methods. This work also demonstrated passing the DART helium stream through a porous substrate, rather than over or around it, as it typically reported. Such a technique could prove very useful for DBS analysis (see Section "DART and DBS"). The coupling of DART with Fourier transform ion cyclotron resonance (FT-ICR) MS has also been reported (Rummel et al., 2010), providing an example of where selectivity could be enhanced by the use of ultra-high-resolution MS to unambiguously detect analytes.

DART AND DBS To date, successful DBS analysis from typically used cellulose-based substrate using DART has not been reported, presumably due to the significant lack of sensitivity compared to what can be achieved using conventional manual DBS extraction methods. Experience with this technique in our laboratory over a representative set of pharmaceutical small molecules demonstrated a loss in sensitivity of around 3 orders of magnitude compared to manual DBS extraction. Strategies to improve quantitative analytical performance have focused on using a substrate material more applicable to DART analysis. Crawford et al. (2011) reported much improved sensitivity (similar to what would be expected from manual DBS extraction) and good quantitative performance using DART to analyze DBS on glass slides. Wider applicability is likely to be an issue, as in many situations using glass slides as a "substrate" is unlikely to be compatible with a robust workflow in practice. However, this work does demonstrate that an alternative substrate that allows the DART gas jet to permeate the full depth of the DMS (rather than just the surface) could enhance the sensitivity on offer. Gauze-type arrangements have been suggested which theoretically could significantly help.

To increase sample throughput and applicability, some simple automation had been incorporated with the DART apparatus. IonSense has developed a "conveyer belt"-type option to the DART source apparatus, which can be used to raster a DBS sample between the gas jet and MS inlet and also alter the angle of sampling (Figure 22.29). The current incarnation will not meet high-throughput, high capacity requirements, but demonstrates a possible direction for fully automated direct desorption instrumentation.

FIGURE 22.29 DART automation: schematic of a DART-MS setup. Whole blood samples directly spotted onto glass slides were automatically driven into the DART desorption ionization region. Reproduced from Crawford et al. (2011) with permission from Future Science Ltd.

Desorption Atmospheric Pressure Photoionization DAPPI couples thermal/chemical desorption and APPI reactive chemistry, producing a technique suitable for sampling a wide range of surface polarities. DAPPI is similar to desorption atmospheric pressure chemical ionization (DAPCI) (see Section "DAPCI and DCBI"), except that a PI process rather than a corona discharge is responsible for initiating the reagent-ion population. Haapala et al. (2007) utilized a microchip nebulizer to pass a heated solvent vapor jet over the sample surface and then used a UV lamp to photoinitiate ion–molecule reactions responsible for gas-phase ionization (Figure 22.30). Thermal desorption is carried out by a microfabricated nebulizer chip that localizes the gas jet onto the sample. A toluene or acetone solvent stream from the chip protonates or photoionizes analytes via gas-phase chemical reactions. Desorbed molecules may also be directly photoionized by unconfined high-energy UV photons that are generated from a PI krypton lamp positioned above the desorption region. Typically, protonated molecules of high proton affinity analytes and molecular ions are observed with this

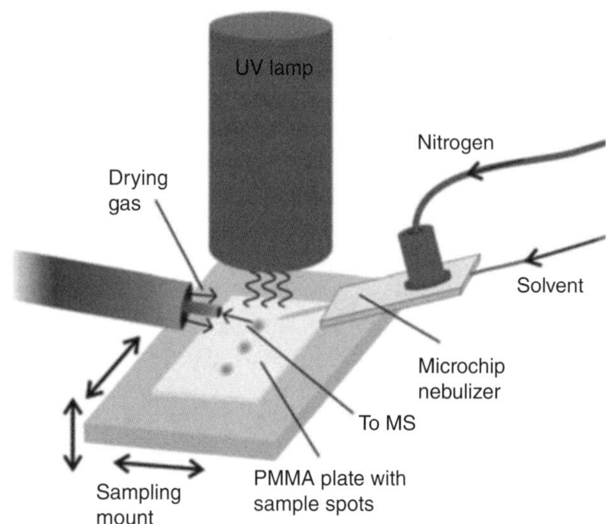

FIGURE 22.30 DAPPI: schematic view of the desorption atmospheric pressure photoionization setup. Reprinted with permission from Haapala et al. (2007). Copyright 2007 American Chemical Society.

technique. DAPPI has been used in the analysis of pesticides, pharmaceuticals, and illegal drugs.

Microplasmas The use of micro-fabricated microplasmas for direct desorption has been demonstrated and reported to offer numerous advantages including simplicity, low cost, low power consumption, and the possibility of batch producing the devices, seemingly making the technique attractive for miniaturized MS and field-based detection applications. Reportedly, small water cluster ions are likely to be largely responsible for proton transfer, producing mass spectral data similar to those generated using other plasma-based ambient ionization techniques (Symond et al., 2010).

22.6.2.2 Combined Desorption/Ionization Combined desorption/ionization techniques involve sample surface impact by charged solvent ions or metastable atoms. Many of these techniques (such as low-temperature plasma probe (LTP), plasma-assisted desorption/ionization (PADI), and flowing atmospheric pressure afterglow (FAPA)) rely on ionization at the sample surface using a locally positioned plasma source, as opposed to a non-proximate plasma used in DART. The advantage of such techniques is that they remove factors such as high temperatures and voltages that may be unsuitable for some sampling requirements.

Flowing Atmospheric Pressure Afterglow FAPA, also known as flowing afterglow-atmospheric pressure glow discharge desorption ionization (APGDDI), shares many similarities with DART (Andrade et al., 2008a, 2008b). The following parameters highlight the ways in which FAPA is different from DART:

- Prior to sample interaction, plasma species are not filtered by electrodes. This results in the potential for additional reactions.
- They operate in different systems (FAPA in the glow-to-arc, DART in the corona-to-discharge).
- The FAPA gas stream is heated through joule heating within the electrical discharge (rather than with a heating element in DART).

The use of FAPA coupled to GC/MS has been reported and was shown to offer reproducible and sensitive performance (Shelley and Hieftje, 2010). Like the LC and GC couplings to DART that have been reported (see Section "Direct Analysis in Real Time"), here GC has been used as an external sampling (and separation technique), which moves the process away from a true direct analysis technique, due to the preparation steps involved. While this approach will obviously add complexity and time to sample analysis compared to "pure" direct desorption, the benefits were demonstrated to be a significant reduction in ion suppression and improvements to analyte identification.

DAPCI and DCBI DAPCI and desorption corona beam ionization (DCBI) are related to both plasma and spray-based direct desorption techniques and can also be classed as chemical sputtering ionization techniques. DAPCI is another TD/APCI variant, this time based around charge exchange between an ionized vapor-phase compound (e.g., toluene) and a physisorbed analyte molecule (Takats et al., 2005a). The DAPCI ion source was created to investigate the DESI mechanism of ionization/desorption by charged solvent species in the source followed by analyte molecule surface ionization by charge transfer—a chemical sputtering process. An emitter (a capillary containing a coaxially aligned electrode) is aimed at the target surface, as in DESI. An inert sheath gas flows through the capillary at high velocity, and a voltage is applied to the electrode, creating a corona discharge at its tip, ionizing the solvent vapor. DAPCI mechanisms are similar to other APCI sources, but desorption mechanisms are not always well understood. With a heated sheath gas, TD is likely to be the most prominent process. However, in other situations, minute surface particles may be released by the high-velocity gas that can be ionized in the gas phase. Where desorption/ionization has been demonstrated without either high-velocity gas or heat, it has been suggested that desorption is enabled by buildup of static charge on the dielectric sample surface. DAPCI-MS has been used for analysis of drug compounds, agricultural chemicals, and explosives, among others.

DCBI (Wang et al., 2010b) is a technique that works in two ways, either akin to temperature-ramped DART (Nilles et al., 2010) or using the DAPCI mechanism. Here, plasma species are not filtered, which results in a thin visible corona beam being emitted from the ion source. This enables specific area of surfaces to be sampled. The primary desorption mechanism is likely to be thermal, with chemical sputtering by plasma species also possibly contributing. A DAPCI-like mechanism can be utilized when solvent is selectively added to the probe.

PADI, DBDI, and LTP PADI (Ratcliffe et al., 2007), dielectric barrier discharge ionization (DBDI) (Na et al., 2007a, 2007b) (Figure 22.31), and LTP are "plasma-assisted" direct analysis techniques that appear to utilize TD as part of the sampling mechanism, despite being referred to as "cold-plasma" techniques. Applying an alternating voltage between two electrodes (one covered by a dielectric layer) enables a plasma to be produced in a flowing helium stream. Analyte desorption and ionization is carried out by directly applying this plasma jet to the sample surface. It has been suggested that the combined effect of metastable helium energy transfer, radical–surface interactions, and ion impact is

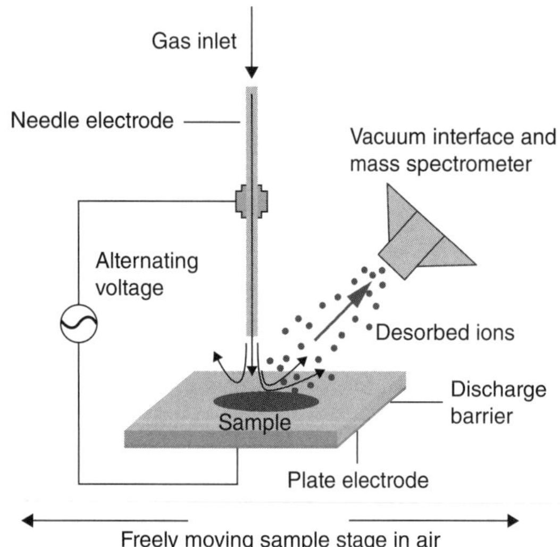

FIGURE 22.31 DBDI: direct detection of explosives on solid surfaces by mass spectrometry with an ambient ion source based on dielectric barrier discharge. Reprinted with permission from Na et al. (2007a). Copyright 2007 John Wiley & Sons.

responsible for analyte desorption and ionization, but the mechanism is not fully understood. One initial publication speculates that ion–molecule reactions, direct electron ionization, and metastable Penning ionization are responsible for positive ion mode ionization, while in negative ion mode, molecular ion species may be produced by ionization via direct and dissociative electron attachment to oxygen species (Ratcliffe et al., 2007).

DBDI uses an unheated plasma to interact with the sample surface. Dielectric barrier discharge generates this plasma, produced using millimeter-sized gaps in a number of possible configurations (Figure 22.31). Here, electrodes are positioned away from the discharge chamber and are not in contact with the plasma. Cold nonequilibrium plasma is generated by these discharges that are highly influenced by space charge effects causing local field distortions. Because one or more electrodes is coated with a dielectric material that cannot pass a direct current, the gap must contain an electric field that is high enough to enable gases to break down. LTP modified the DBDI configuration (where the sample acted as the dielectric) enabling plasma to be targeted toward specific regions (Harper et al., 2008). LTP-MS/MS has been shown to offer good quantitative performance and analyte sensitivity from complex mixtures (Huang et al., 2009).

22.6.3 Laser Desorption (Ablation)/Ionization Techniques

The basic principle behind laser desorption (ablation)/ionization direct analysis techniques is relatively simple. Material from the sample surface is desorbed/ionized by a focused pulsed laser beam, and the ions produced are subsequently introduced into an MS. These methods are therefore multistep ionization techniques, the initial step involving laser desorption/ionization or desorption via laser-induced shockwaves, followed by ESI or APCI. However, as with many direct analysis approaches, the true desorption and ionization mechanisms that occur are not always fully understood and may be a complex combination of processes. In general, as a solid surface absorbs a laser pulse it is excited and analytes are desorbed, forming a reactive area on the surface. This surface region is thought to participate in radical reactions, electron transfer, and cationization which contribute to producing the resulting ionic products. Typically, many more neutral species than ions are generated by laser desorption. The use of a secondary, independently optimized, ionization process is thus used to create optimal overall ionization efficiency and flexibility in the ions produced.

As shown in Table 22.1, many laser-based techniques couple laser desorption/ablation to ESI secondary ionization. These techniques utilize UV or IR laser desorption/ionization to produce plumes of neutrals. These plumes are transferred within either a stream of reactive species from a plasma source or a charged droplet cloud from an ESI source located above the site of desorption. The most commonly accepted term for laser sampling ESI techniques utilizing IR radiation is laser ablation electrospray ionization (LAESI). Where UV is utilized, ELDI is the most commonly used term. Where the sample is cocrystallized with matrix and either IR or UV is utilized, the term MALDESI is generally used.

Traditionally, laser desorption/ablation techniques struggle to produce multiply charged ions, which is a disadvantage as the production of multiply charged species is important as it allows molecules of a larger mass range to be detected in optimum m/z ranges.

22.6.3.1 LAESI/IR-LADESI Laser-assisted electrospray ionization or LAESI (Nemes and Vertes, 2007) and infrared laser-assisted desorption electrospray ionization (IR-LADESI) (Rezenom et al., 2008) utilize IR radiation to ablate analytes from a sample surface. Typically, laser wavelength is selected according to sample composition. For samples containing water, typically an IR laser at 2940 nm with 5-nanosecond pulses at 50 Hz (with ~100 μJ energy) is used to excite OH vibrations resulting in rapid microscale phase explosion from the sample surface (Nemes et al., 2010). The ablated material expands longitudinally upwards until it meets an electrospray plume. Here analytes either react with gas-phase species to produce charged ions which are sucked into the MS inlet or dissolve in the electrospray plume. Due to differential analyte enrichment during ablation, quantitation can be tricky with these techniques. However, reports of adequate quantitation have been published (Sripadi et al., 2009). LAESI was reported as capable of direct desorption of wet biological samples, but appears to be of limited applicability to dried samples, putting into question the suitability of these approaches for DBS analysis without some sample pretreatment.

Commercially available, automated high-throughput systems are available, such as the Protea Biosciences LAESI DP-1000 (https://proteabio.com). Protea Biosciences have demonstrated DBS direct analysis by spraying on a water repellent coating onto both sides of the cards after blood has been spotted. Once visibly dry, water is added to the DBS samples, prior to analysis via direct desorption LAESI-MS. While this incorporates more sample preparation than is ideal, it could be automated reasonably simply, making LAESI a viable DBS direct desorption candidate if the performance on offer shows enough potential.

22.6.3.2 ELDI Combining laser desorption with ESI was first reported as ELDI (Shiea et al., 2005). Here the desorbed analyte is ionized through reaction with charged solvent droplets, protonated solvent species, or gas-phase ions created via ESI. The secondary ionization process used is similar to EESI, a "fused droplet" ESI process (Chang et al., 2002; Chen et al., 2006b). Typical desorption processes with ELDI involve pulsed UV lasers with 5-nanosecond pulses operating at 10 Hz, with pulse energies in the region of 150–300 μJ. Cheng et al. (2010) reported the analysis of substances from paper substrate suggesting that ELDI-MS DBS analysis is worthy of investigation.

22.6.3.3 MALDESI MALDESI (Sampson et al., 2006) was created by combining spray-based ionization with laser ablation desorption. Compared to traditional laser desorption ionization (LDI) these methods offer additional scope and selectivity. An additional matrix preparation step is featured with MALDESI prior to sampling with the aim of improving both reproducibility and sensitivity. A voltage is also applied to the target plate as with traditional MALDI. It was demonstrated that the primary ionization pathway in MALDESI was via laser-desorbed neutral plume interaction with electrospray-charged droplets (Dixon and Muddiman, 2010). While minimal sample preparation will be acceptable in some situations, the requirement to add matrix in MALDESI will rule it out for many applications, especially if other, zero sample preparation analysis methods are available. The use of MALDI-MS/MS to analyze DBS samples has been reported, but only where significant sample preparation has been used. Thus, MALDI-MS/MS cannot be categorized as a true DBS direct analysis technique (Meesters et al., 2010).

22.6.3.4 LIAD Laser-induced acoustic desorption (LIAD), which can be classed as a laser-based or acoustic desorption technique, uses multiple laser pulses to generate acoustic shockwaves which propagate through a thin metal foil (with low thermal conductivity and large thermal expansion coefficient), to which the sample is attached, which desorbs sample analytes. This process will usually be coupled to a secondary ionization source. The ionization mechanism is thought to involve neutral molecules either fusing into charge solvent spray or reacting with the charge spray through ion–molecule reactions, as in fused droplet ESI (Cheng et al., 2009).

22.6.4 Multimode Techniques

Desorption electrospray/metastable-induced ionization (DEMI) is an interesting technique that combines the benefits of DESI and metastable-induced chemical ionization (MICI, a DART variant) and bypasses some of the limitations (limited polarity range in DESI and limited molecular weight in MICI) (Nyadong et al., 2009). The DEMI ion source utilizes three unique operation modes: spray (DESI-like)-only; MICI-only; and DEMI (multimode). These operation modes allow for the detection of a wider range of analytes. MICI-only mode is reportedly well suited for the analysis of low-polarity, low molecular weight compounds in powdered, solid, or dissolved samples, while spray-only mode is suitable for the analysis of high molecular weight, high-polarity compounds. The application of heat to the nebulizer gas in spray-only mode has been demonstrated to improve spray solvent analyte solubility, which can result in a significant (order of magnitude) increase in sensitivity. Interestingly, the DEMI mode allows the simultaneous detection of compounds within a much broader range of polarities and molecular weights than each of the individual modes.

22.7 CONCLUSIONS: DIRECT ANALYSIS COMPARISON, CURRENT TRENDS, AND SHORT-TERM AND LONG-TERM PROSPECTS

The use of DBS samples in bioanalysis is not new (e.g., it has been used in newborn screening for decades), but the recent surge in interest has involved areas new to the technique such as pharmaceutical discovery and development, and therapeutic drug monitoring. Thus, a number of different requirements exist, depending on the application being supported, and this reflects the levels of performance, compatibility, selectivity, sensitivity, and additional functionality required from a DBS direct analysis technique. In many of these new DBS application areas, the evaluation of the technique is still ongoing. Indeed, in many cases, the routine use of DBS for bioanalytical support is far from guaranteed.

Regardless of the application, the ideal DBS direct analysis technique for any application will be fast, simple to perform, robust, and reliable, and in many cases, will need to offer a high level of selectivity and sensitivity. However, depending on the application, other parameters and additional functionality may also be important, such as full automation, sample capacity, the ability to add IS prior to sampling/desorption, the ability to perform sample dilution, and a method of controlling sample carryover. The large number of different criteria to consider, and the way in which the relative importance of these varies for different DBS applications, makes comparing DBS direct analysis techniques a difficult task. Also, any current evaluation of direct analysis techniques must address the caveat that the parameters deemed to be important at present could well shift as DBS usage and development matures. In addition, the multitude of direct analysis techniques reported are at a wide range of stages of development. Some of these techniques have been tested using a wide range of representative compounds in the DBS format while other techniques have had very limited DBS testing. Further, many direct analysis techniques that are potentially applicable to DBS analysis have yet to be tested at all.

Nonetheless, to conclude this chapter, an attempt has been made to summarize the performance and compatibility of the most prominent direct analysis techniques that have been applied to DBS analysis as well as how they compare to automated manual extraction (Table 22.3). This table is an updated version of a similar attempt to summarize DBS direct analysis techniques by Déglon et al. in 2011, albeit from a slightly different perspective and some additional experiences (Déglon et al., 2011b).

It should be noted that Table 22.3 only includes DBS direct analysis techniques where results and evaluations have been adequately published to date or where the author has direct experience. A number of the direct desorption techniques described in Section 22.6 (but not present in Table 22.3) are being investigated for DBS direct analysis application but have yet to be adequately developed and/or the results published. It must also be noted that Table 22.3 should only be used as a current guide to DBS direct analysis performance and compatibility since the content is likely to change as these techniques are further evaluated and developed.

Summarizing this topic is not a straightforward task. In addition to some subjective criteria, the relative importance of many parameters will likely depend on the specific application. Even where absolute values are being compared, such as limits of quantitation, numerous variables will need to be considered. These variables include sensitivity of the detector coupled to the direct analysis technique, the quantity sampled, whether an additional separation technique is being utilized, and the criteria used to determine these values. For example, in terms of sensitivity, the inherent variability in MS performance from occasion to occasion, lab to lab, and between different models of detector makes published limits of detection data difficult to reconcile. Also, certain techniques have demonstrated excellent sensitivity with some test compounds, and poor sensitivity with others (sensitivity has been measured against what can be achieved with conventional manual extraction), which means that rating a technique as "good" or "bad" in this respect does not tell the whole story. In these cases, a compromise has been used. For example, DESI demonstrates sensitivity similar to what is achieved with conventional manual DBS extraction for some compounds, but is up to 2 orders of magnitude less sensitive for others. For this reason, the sensitivity ratings on the table are weighted primarily toward recent experiences in my laboratory with these techniques (where assessments have been made using a unique set of test compounds; a range of representative pharmaceutical small molecules), with a secondary weighting toward techniques that have been widely tested

TABLE 22.3 Summary and Comparison of the Current Methods Available to Analyze DBS Samples

		Direct Elution				Direct Desorption		
Parameter	Manual Extraction	Online DBS	SSSP	LMJ-SSP (LESA)	Digital Microfluidics	DESI	DART	Paper Spray
Automation/ throughput	***	***	****	**	**	*	**	***
Sensitivity	***	****	****	**	***	**	*	***
Quantitative performance	****	****	****	**	***	**	**	****
HPLC compatibility[a]	Yes	Yes	Yes	No	No	No	No	No
Commercial availability	Yes	Yes	Yes	Yes	No	Yes (not DBS specific)	Yes (not DBS specific)	Yes (proposed)
Cost range	* up to ****	** up to ***	** up to ****	***	tbd	**	**	tbd
Robustness	****	****	****	tbd	tbd	tbd	tbd	tbd

The table refers to maximum current performance, hence manual extraction is being considered as a technique coupled to instrumentation that provides the best level of automation and sample throughput available. tbd, to be determined. Parameter performance: low (*), medium (**), high (***), very high (****).
[a]Where the direct analysis technique is being used as the sampling mechanism.

by a range of representative molecules. In a similar fashion, a technique that has demonstrated quantitative performance over a range of compounds, and on multiple occasions, will rate higher than a technique that has shown the same performance level but for few applications. To put sensitivity comparisons in context, it should be noted that in most cases a typical conventional manual extraction DBS assay exhibits roughly an order of magnitude less sensitivity than a corresponding conventional plasma-based assay, due to much smaller amount being sampled (i.e., a 25 μL plasma sample vs. a 3 mm diameter DBS punch). The remainder of this section summarizes the current performance and future prospects of the three principle methods of improving the efficiency of DBS analysis.

22.7.1 Automated DBS Extraction

Manual DBS extraction offers a reasonably quick, simple, cheap, and reliable method of analysis for many applications. Automating the punching and liquid transfer steps can be used to relieve some of the manual burden. This automated technique has been used in high-throughput environments and demonstrated to work effectively. It also has the advantage of not significantly deviating from current methods of sample bioanalysis. However, the additional complexity of DBS bioanalysis over conventional plasma extraction techniques, particularly in high-throughput applications, and the inherent sensitivity decrease encountered with microsampling, means that this technique has sometimes proved unpopular with laboratory analysts, and this has proved to be a barrier to DBS acceptance. Without investing prohibitive amounts of money to produce very complex integrated instruments, automated DBS extraction cannot provide the seamless, simplified workflows on offer from direct analysis techniques. For this reason, if DBS sampling ever reaches its potential and replaces plasma analysis as a primary method of bioanalysis, then direct analysis will provide a more efficient method of analysis in applications where justified by sample throughput requirements. Thus, the direct analysis techniques that are emerging offer significant additional advantages that warrant the additional resource involved in their development.

22.7.2 Direct Elution

DBS direct elution describes a number of similar automated or semiautomated techniques where the analyte of interest is extracted through contact with a suitable extraction solvent. One of the advantages of DBS direct elution that differentiates it from direct desorption is that a liquid eluate is produced which means that the technique can easily be made compatible with LC-MS/MS separation and detection (Abu-Rabie and Spooner, 2009). The compatibility is a key feature, as LC-MS/MS is the predominantly used analysis technique in many high sample throughput quantitative bioanalysis applications. Furthermore, this area has generated considerable momentum in the recent interest in DBS sampling. LC-MS/MS is a technique that is readily available in many bioanalytical laboratories and is familiar to, and accepted by, bioanalysts and regulatory authorities. Thus, because of its compatibility with the existing bioanalytical LC-MS/MS workflow, using DBS direct elution in practice is a realistic short-term goal, even for regulated applications where any changes to existing workflows can be extremely difficult, and slow, to implement. In the longer term, it is possible that DBS direct elution could be coupled to different types of detector, or an alternative separation technique that replaces the selectivity lost if HPLC is not used. These potential applications would lead to an even more seamless bioanalytical workflow, akin to what is offered by direct desorption techniques. Initial work has also shown that there is some promise in utilizing direct elution without HPLC (Abu-Rabie and Spooner, 2009). This highlights the flexibility of the direct elution technique that should be compatible with both currently used detectors and those in the future that may make other direct desorption techniques a realistic option.

Essentially, one of the main advantages of direct elution is that the concept is not significantly different to currently used DBS (or wet plasma or blood) liquid manual extraction LC-MS/MS techniques, and yet it offers considerable advantages over automated DBS extraction. HPLC coupled to highly sensitive triple quadrupole mass spectrometers is the current separation and detection method of choice in many quantitative bioanalytical applications and is very widely used. Triple quadrupole mass spectrometers are utilized as they offer very high levels of sensitivity and are relatively cheap (compared to accurate mass MS). LC separation is used to separate analytes of interest from coexisting components and, in conjunction with tandem MS monitoring (detecting the parent ion of a molecule, and a specific product fragment ion following passage through a collision cell), provides the selectivity required. MS instrumentation is the most expensive component in the bioanalytical workflow, so compatibility with preexisting resource is a major advantage in the introduction of any new technique. DBS direct elution's compatibility with HPLC-MS/MS means that triple quadrupole mass spectrometers, which are prevalent in high sample throughput quantitative bioanalysis facilities, can be used and the level of selectivity required for regulated applications is retained. Therefore, DBS direct elution has a financial and practical advantage over direct desorption techniques for some applications.

Another important benefit on offer from DBS direct analysis is the increase in assay sensitivity that can be readily achieved. Compared to conventional manual DBS extraction, an order of magnitude increase in assay sensitivity is regularly observed using existing HPLC-MS/MS methodology (Abu-Rabie and Spooner, 2009). The importance of this feature cannot be overstated, as it breeches the sensitivity gap faced in switching from conventional plasma sample volumes (20–50 μL) to DBS microsampling (a 3 mm diameter punch contains ~2 μL blood). This increase in sensitivity, coupled with the most sensitive detectors, could make DBS sampling applicable to a section of compounds currently not compatible with the technique. For example, respiratory-based pharmaceutical compounds, which typically have LLQ requirements in the low picogram per milliliter region, could be supported using DBS. Clearly, the benefit here is that the wider the range of compounds that can be supported, the further the ethical and financial benefits on offer can be maximized.

DBS direct elution coupled to HPLC is currently the closest direct analysis technique available that is compatible with supporting regulated quantitative bioanalysis applications. Fully automated systems are now commercially available that enable significant efficiency gains in high sample throughput environments. Instruments such as the Prolab SCAP system and CAMAG DBS-MS 500 offer much of the additional functionality required for high-throughput compatibility. The only major gap in the proposed direct analysis bioanalytical workflow described in Figure 22.3 (Section 22.3.3) is the ability to dilute samples. However, it is envisioned that dilution steps could be addressed relatively easily (see Section 22.3.3.8).

22.7.3 Direct Desorption

DBS direct desorption is still at a very early stage of development, but hopefully this chapter has demonstrated that a number of techniques show massive potential for gains in efficiency over a range of applications. If the goal is to find the simplest and quickest DBS analysis technique, then direct desorption techniques are theoretically the most favorable option. The potential advantages on offer from direct desorption such as even simpler bioanalytical workflows are significant. The elimination of LC would transform the way bioanalysis is performed and the environments where analytical instrumentation could be used for non-regulated applications (e.g., surgeries). However, the challenges described above clearly show that the use of direct desorption for supporting some applications such as regulated drug development studies must overcome a series of barriers and, thus, currently can only be considered a longer term goal. For other applications, such as drug screening, drug discovery, and therapeutic drug monitoring, where the same level of confidence may not be required and the same regulatory restrictions do not exist, there are far fewer barriers to the use of direct desorption techniques.

One of the potential advantages of direct desorption is the elimination of LC separation. This elimination of LC is also one of the challenges that must be overcome if it is to be introduced to a regulated bioanalytical environment. The elimination of LC would be a welcome and significant simplification of the bioanalytical workflow. However, in practice removing LC can result in poor sensitivity due to ion suppression and reduced selectivity and risks assay interference via metabolite decomposition (e.g., N-oxides, glucuronides) into parent compounds during MS ionization. Perhaps more selective detectors and different types of separation techniques than those commonly found in the bioanalytical laboratory could assist in this regard. Cutting-edge, accurate mass, and high-resolution hybrid mass spectrometers with ion mobility capabilities are now much closer in sensitivity to high-performance triple quadrupole instruments, making their utilization a realistic possibility. However, it should be noted that these instruments are generally much more expensive than triple quadrupole MS.

It is clear that utilizing DBS direct elution will be a significant change to current bioanalytical workflows, and this means a significant input of resource (financial and time) is required to develop these techniques and underpin the underlying science behind them to ensure that they are acceptable for regulatory approval. However, this is not the only barrier to the practical application of direct desorption. Extensive investigation has yet to identify a direct desorption technique that offers adequate sensitivity across a range of representative pharmaceutical (small-molecule) compounds. DBS analysis using DART on the widely used paper substrate format has demonstrated sensitivity considerably lower than that currently achievable from manual DBS extraction. DESI and, in particular, PS have demonstrated sensitivity close to and, in some cases, exceeding the sensitivity obtained with manual DBS extraction for some compounds, but have also been equally disappointing for other compounds. No direct desorption technique has matched the sensitivity on offer from direct elution. For some applications the inherent differences in sensitivity will not be an issue as the desorption techniques will still offer enough sensitivity to put the technique within therapeutic range. However, for other techniques, sensitivity will be a limiting factor that could exclude the use of the techniques all together. The next few years promise to produce a number of significant challenges for both DBS as a sampling technique and direct analysis as a means to support it.

REFERENCES

Abu-Rabie P. Developing a Fully Automated Dried Blood Spot Direct Analysis Technique for High Sample Throughput Quantitative Bioanalysis. Presented at the 59th ASMS Conference on Mass Spectroscopy and Allied Topics, Denver, CO, USA, June 5–9, 2011.

Abu-Rabie P, Spooner N. Direct quantitative bioanalysis of drugs in dried blood spot samples using a thin-layer chromatography mass spectrometer interface. *Analytical Chemistry* 2009;81(24):10275–10284.

Abu-Rabie P, Spooner N. Dried matrix spot direct analysis: evaluating the robustness of a direct elution technique for use in quantitative bioanalysis. *Bioanalysis* 2011;3(24):2769–2781.

Abu-Rabie P, Denniff P, Spooner N, Brynjolffssen J, Galluzzo P, Sanders G. Method of applying internal standard to dried matrix spot samples for use in quantitative bioanalysis. *Analytical Chemistry* 2011;83(22):8779–8786.

Al-Ghazawi M, AbuRuz S, Al-Hiar Y. A simple dried blood spot assay for therapeutic drug monitoring of lamotrigine. *Chromatographia* 2010;71:1093–1099.

Andrade FJ, Shelley JT, Wetze WC, Webb MR, Gamez G, Ray SJ, Hieftje GM. Atmospheric pressure chemical ionization source. 1. Ionization of compounds in the gas phase. *Analytical Chemistry* 2008a;80(8):2646–2653.

Andrade FJ, Shelley JT, Wetze WC, Webb MR, Gamez G, Ray SJ, Hieftje GM. Atmospheric pressure chemical ionization source. 2. Desorption-ionization for the direct analysis of solid compounds. *Analytical Chemistry* 2008b;80(8):2654–2663.

Barfield M, Wheller R. Use of dried plasma spots in the determination of pharmacokinetics in clinical studies: validation of a quantitative bioanalytical method. *Analytical Chemistry* 2011;83(1):118–124.

Barfield M, Spooner N, Lad R, Parry S, Fowles S. Application of dried blood spots combined with HPLC-MS/MS for the quantification of acetaminophen in toxicokinetic studies. *Journal of Chromatography B* 2008;870:32–37.

Basile F, Zhang S, Shin YS, Drolet B. Atmospheric pressure-thermal desorption (AP-TD)/electrospray ionization-mass spectrometry for the rapid analysis of Bacillus spores. *Analyst* 2010;135(4):797–803.

Bereman MS, Muddiman DC. Detection of attomole amounts of analyte by desorption electrospray ionization mass spectrometry (DESI-MS) determined using fluorescence spectroscopy. *Journal of the American Society for Mass Spectrometry* 2007;18:1093–1096.

Chang DY, Lee CC, Shiea J. Detecting large biomolecules from high-salt solutions by fused-droplet electrospray ionization mass spectrometry. *Analytical Chemistry* 2002;74(11):2465–2469.

Chen H, Zenobi R. Neutral desorption sampling of biological surfaces for rapid chemical characterization by extractive electrospray ionization mass spectrometry. *Nature Protocols* 2008;3:1467–1475.

Chen H, Cotte-Rodríguez I, Cooks RG. cis-Diol functional group recognition by reactive desorption electrospray ionization (DESI). *Chemical Communications* 2006a;6(6):597–599.

Chen H, Venter A, Cooks RG. Extractive electrospray ionization for direct analysis of undiluted urine, milk and other complex mixtures without sample preparation. *Chemical Communications* 2006b;(19):2042–2044.

Chen H, Ouyang Z, Cooks RG. Thermal production and reactions of organic ions at atmospheric pressure. *Angewandte Chemie – International Edition* 2006c;45(22):3656–3660.

Chen H, Wortmann A, Zenobi R. Neutral desorption sampling coupled to extractive electrospray ionization mass spectrometry for rapid differentiation of biosamples by metabolomic fingerprinting. *Journal of Mass Spectrometry* 2007;42(9):1123–1135.

Cheng SC, Cheng TL, Chang HC, Shiea J. Using laser-induced acoustic desorption/electrospray ionization mass spectrometry to characterize small organic and large biological compounds in the solid state and in solution under ambient conditions. *Analytical Chemistry* 2009;81(3):868–874.

Cheng SC, Lin YS, Huang MZ, Shiea J. Applications of electrospray laser desorption ionization mass spectrometry for document examination. *Rapid Communications in Mass Spectrometry* 2010;24(2):203–208.

Cody RB. Observation of molecular ions and analysis of nonpolar compounds with the direct analysis in real time ion source. *Analytical Chemistry* 2009;81(3):1101–1107.

Cody RB, Laramée JA, Durst HD. Versatile new ion source for the analysis of materials in open air under ambient conditions. *Analytical Chemistry* 2005;77(8):2297–2302.

Cooks RG, Quyang Z, Takats Z, Wiseman JM. Ambient mass spectrometry. *Science* 2006;6(311):1566.

Corso G, D'Apolito O, Garofalo D, Paglia G, Dello Russo A. Profiling of acylcarnitines and sterols from dried blood or plasma spot by atmospheric pressure thermal desorption chemical ionization (APTDCI) tandem mass spectrometry. *Biochimica et Biophysica Acta – Molecular and Cell Biology of Lipids* 2011;1811(11):669–679.

Corso G, D'Apolito O, Gelzo M, Paglia G, Dello Russo A. A powerful couple in the future of clinical biochemistry: in situ analysis of dried blood spots by ambient mass spectrometry. *Bioanalysis* 2012;2(11):1883–1891.

Costa AB, Cooks RG. Simulation of atmospheric transport and droplet-thin film collisions in desorption electrospray ionization. *Chemical Physics Letters* 2008;464:1–8.

Cotte-Rodríguez I, Takáts Z, Talaty N, Chen H, Cooks RG. Desorption electrospray ionization of explosives on surfaces: sensitivity and selectivity enhancement by reactive desorption electrospray ionization. *Analytical Chemistry* 2005;77(21):6755–6764.

Crawford E, Gordon J, Wu JT, Musselman B, Liu R, Yu S. Direct analysis in real time coupled with dried spot sampling for bioanalysis in a drug-discovery setting. *Bioanalysis* 2011;3(11):1217–1226.

D'Agostino PA, Chenier CL. Desorption electrospray ionization mass spectrometric analysis of organophosphorus chemical warfare agents using ion mobility and tandem mass spectrometry. *Rapid Communications in Mass Spectrometry* 2010;24(11):1617–1624.

Déglon J, Thomas A, Cataldo A, Mangin P, Staub C. On-line desorption of dried blood spot: a novel approach for the direct LC/MS analysis of μ-whole blood samples. *Journal of Pharmaceutical and Biomedical Analysis* 2009;49(4):1034–1039.

Déglon J, Thomas A, Daali Y, Lauer E, Samer C, Desmeules J, Dayer P, Staub C. Automated system for on-line desorption of dried blood spots applied to LC/MS/MS pharmacokinetic study of flurbiprofen and its metabolite. *Journal of Pharmaceutical and Biomedical Analysis* 2011a;54(2):359–367.

Déglon J, Thomas A, Mangin P, Staub C. Direct analysis of dried blood spots coupled with mass spectrometry: concepts and biomedical applications. *Analytical and Bioanalytical Chemistry* 2011b;8:2485–2448.

Denniff P, Spooner N. The effect of hematocrit on assay bias when using DBS samples for the quantitative bioanalysis of drugs. *Bioanalysis* 2010;2(8):1385–1395.

Denniff P, Spooner N. Effect of Hematocrit on Quantitative Dried Blood Spots. Presented at the 19th International Reid Bioanalytical Forum, Guildford, UK, July 4–7, 2011.

Denniff P, Hughes A, Merchan A, Spooner N. Issues with DBS and HCT – Possible Solutions. Presented at the British Mass Spectrometry Society Annual Meeting, April 17–18, 2012.

Dixon RB, Muddiman DC. Study of the ionization mechanism in hybrid laser based desorption techniques. *Analyst* 2010;135(5):880–882.

Ebejer KA, Brereton RG, Carter JF, Ollerton SL, Sleeman R. Rapid comparison of diacetylmorphine on banknotes by tandem mass spectrometry. *Rapid Communications in Mass Spectrometry* 2005;19(15):2137–2143.

Eberherr W, Buchberger W, Hertsens R, Klampfl CW. Investigations on the coupling of high-performance liquid chromatography to direct analysis in real time mass spectrometry. *Analytical Chemistry* 2010;82(13):5792–5796.

Edelbroek PM, Van der Heijden J, Stolk LML. Dried blood spot methods in therapeutic drug monitoring: methods, assays, and pitfalls. *Therapeutic Drug Monitoring* 2009;31:327–336.

Edison SE, Lin LA, Gamble BM, Wong J, Zhang K. Surface swabbing technique for the rapid screening for pesticides using ambient pressure desorption ionization with high-resolution mass spectrometry. *Rapid Communications in Mass Spectrometry* 2011;25(1):127–139.

Espy R, Manicke N, Ouyang Z, Cooks RG. Paper Spray Mass Spectrometry as an Ambient, Quantitative Method for Point of Care Therapeutic Drug Monitoring from Blood. Presented at the 59th ASMS Conference on Mass Spectrometry and Allied Topics, Denver, Colorado, June 5–9, 2011.

Ewles MF, Turpin PE, Goodwin L, Bakes DM. Validation of a bioanalytical method for the quantification of a therapeutic peptide, ramoplanin, in human dried blood spots using LC-MS/MS. *Biomedical Chromatography* 2011;25(9):995–1002.

Fan L, Lee JA. Managing the effect of hematocrit on DBS analysis in a regulated environment. *Bioanalysis* 2012;4(4):345–347.

Fernández FM, Cody RB, Green MD, Hampton CY, McGready R, Sengaloundeth S, White NJ, Newton PN. Characterization of solid counterfeit drug samples by desorption electrospray ionization and direct-analysis-in-real-time coupled to time-of-flight mass spectrometry. *ChemMedChem* 2006;1(7):702–705.

Gabellca V, De Pauw E. Internal energy and fragmentation of ions produced in electrospray sources. *Mass Spectrometry Reviews* 2005;24(4):566–587.

Galhena AS, Harris GA, Kwasnik M, Fernández FM. Enhanced direct ambient analysis by differential mobility-filtered desorption electrospray ionization-mass spectrometry. *Analytical Chemistry* 2010;82(22):9159–9163.

Ganz N, Singrasa M, Nicolas L, Gutierrez M, Dingemanse J, Döbelin W, Glinski M. Development and validation of a fully automated online human dried blood spot analysis of bosentan and its metabolites using the Sample Card And Prep DBS System. *Journal of Chromatography B: Analytical Technologies in the Biomedical and Life Sciences* 2012;885–886:50–60.

Gologan B, Green J, Alverez J, Laskin J, Cooks R. Ion/surface reactions and ion soft-landing. *Physical Chemistry Chemical Physics* 2005;7(7):1490–1500.

Haapala M, Pól J, Saarela V, Arvola V, Kotiaho T, Ketola RA, Franssila S, Kostiainen R. Desorption atmospheric pressure photoionization. *Analytical Chemistry* 2007;79(20):7867–7872.

Haddad R, Sparrapan R, Eberlin MN. Desorption sonic spray ionization for (high) voltage-free ambient mass spectrometry. *Rapid Communications in Mass Spectrometry* 2006;20(19):2901–2905.

Haddad R, Sparrapan R, Kotiaho T, Eberlin MN. Easy ambient sonic-spray ionization-membrane interface mass spectrometry for direct analysis of solution constituents. *Analytical Chemistry* 2008;80(3):898–903.

Harper JD, Charipar NA, Mulligan CC, Zhang X, Cooks RG, Ouyang Z. Low-temperature plasma probe for ambient desorption ionization. *Analytical Chemistry* 2008;80(23):9097–9104.

Harris GA, Fernández FM. Simulations and experimental investigation of atmospheric transport in an ambient metastable-induced chemical ionization source. *Analytical Chemistry* 2005;81(1):322–329.

Harris GA, Hostetler DM, Hampton CY, Fernandez FM. Comparison of the internal energy deposition of direct analysis in real time and electrospray ionization time-of-flight mass spectrometry. *Journal of the American Society for Mass Spectrometry* 2010;21(5):855–863.

Harris GA, Galhena AS, Fernandez FM. Ambient sampling/ionization mass spectrometry: applications and current trends. *Analytical Chemistry* 2011;83:4508–4538.

Harry EL, Reynolds JC, Bristow AWT, Wilson ID, Creaser CS. Direct analysis of pharmaceutical formulations from non-bonded reversed-phase thin-layer chromatography plates by desorption electrospray ionisation ion mobility mass spectrometry. *Rapid Communications in Mass Spectrometry* 2009;23(17):2597–2604.

Heinig K, Wirz T, Bucheli F, Gajate-Perez A. Determination of oseltamivir (Tamiflu®) and oseltamivir carboxylate in dried blood spots using offline or online extraction. *Bioanalysis* 2011;3(4):421–437.

Heinig K, Wirz T, Gajate-Perez A. Sensitive determination of a drug candidate in dried blood spots using a TLC-MS interface integrated into a column-switching LC-MS/MS system. *Bioanalysis* 2012;2(11):1873–1882.

Henion JD. LESA of DBS Cards via Chip-Based Nanoelectrospray for Drug and Drug Metabolite Monitoring Studies. Presented at the 59th ASMS Conference on Mass Spectroscopy and Allied Topics, Denver, CO, USA, June 5–9, 2011.

http://www.advion.com/biosystems/triversa-nanomate/triversa-mass-spectrometry.php (accessed March 2012).

http://www.biospot-llc.com/BSD1000.htm (accessed March 2012).

http://www.camag.com (accessed March 2012).

http://www.hamiltonrobotics.com/ (accessed March 2012).

http://www.hudsonrobotics.com/products/biological-research/dried-blood-spot (accessed March 2012).

http://www.luminexcorp.com/Products/Instruments/DriedBioSampleAutomation/index.htm (accessed March 2012).

http://www.perkinelmer.com (accessed March 2012).

http://www.prolab.ch/scap.html (accessed March 2012).

http://www.sparkholland.com (accessed March 2012).

https://proteabio.com/LAESI (accessed April 2012).

Huang G, Chen H, Zhang X, Cooks RG, Ouyang Z. Rapid screening of anabolic steroids in urine by reactive desorption electrospray ionization. *Analytical Chemistry* 2007;79(21):8327–8332.

Huang G, Ouyang Z, Cooks RG. High-throughput trace melamine analysis in complex mixtures. *Chemical Communications* 2009;(5):556–558.

Huang MZ, Cheng SC, Cho YT, Shiea J. Ambient ionization mass spectrometry: a tutorial. *Analytica Chimica Acta* 2011;702(1):1–15.

Ifa DR, Wu C, Ouyang Z, Cooks RG. Desorption electrospray ionization and other ambient ionization methods: current progress and preview. *Analyst* 2010;135:669.

Jebrail MJ, Yang H, Mudrik JM, Lafrenière NM, McRoberts C, Al-Dirbashi OY, Fisher L, Wheeler AR. A digital microfluidic method for dried blood spot analysis. *Lab on a Chip - Miniaturisation for Chemistry and Biology* 2011;11(19):3218–3224.

Johnson CJL, Christianson CD, Sheaff CN, Laine DF, Zimmer JSD, Needham SR. Use of conventional bioanalytical devices to automate DBS extractions in liquid-handling dispensing tips. *Bioanalysis* 2011;3(20):2303–2310.

Kebarle P. A brief overview of the present status of the mechanisms involved in electrospray mass spectrometry. *Journal of Mass Spectrometry* 2000;35:804–817.

Kertesz V, Van Berkel GJ. Fully automated liquid extraction-based surface sampling and ionization using a chip-based robotic nanoelectrospray platform. *Journal of Mass Spectrometry* 2010;45(3):252–260.

Lapthorn C, Pullen F. "Soft" or "hard" ionisation? Investigation of metastable gas temperature effect on direct analysis in real-time analysis of voriconazole. *European Journal of Mass Spectrometry* 2009;15(5):587–593.

Li F, Zulkoski J, Fast D, Michael S. Perforated dried blood spots: a novel format for accurate microsampling. *Bioanalysis* 2011;3(20):2321–2333.

Liu J, Wang H, Manicke NE, Lin JM, Cooks RG, Ouyang Z. Development, characterization, and application of paper spray ionization. *Analytical Chemistry* 2010;82:2463–2471.

Luftmann H. A simple device for the extraction of TLC spots: direct coupling with an electrospray mass spectrometer. *Analytical and Bioanalytical Chemistry* 2004;378(4):964–968.

Manicke NE, Kistler DR, Ifa R, Cooks G, Quyang Z. High-throughput quantitative analysis by desorption electrospray ionization mass spectrometry. *Journal of the American Society for Mass Spectrometry* 2009;20:321–325.

Manicke NE, Yang Q, Wang H, Oradu S, Ouyang Z, Cooks RG. Assessment of paper spray ionization for quantitation of pharmaceuticals in blood spots. *International Journal of Mass Spectrometry* 2011a;300:123–129.

Manicke NE, Abu-Rabie P, Spooner N, Ouyang Z, Cooks RG. Quantitative analysis of therapeutic drugs in dried blood spot samples by paper spray mass spectrometry: an avenue to therapeutic drug monitoring. *Journal of the American Society for Mass Spectrometry* 2011b;22(9):1501–1507.

McEwen CN, Larsen BS. Ionization mechanisms related to negative ion APPI, APCI, and DART. *Journal of the American Society for Mass Spectrometry* 2009;20(8):1518–1521.

McEwen CN, McKay RG, Larsen BS. Analysis of solids, liquids, and biological tissues using solids probe introduction at atmospheric pressure on commercial LC/MS instruments. *Analytical Chemistry* 2005;77(23):7826–7831.

Meesters RJW, Van Kampen JJA, Reedijk ML, Scheuer RD, Dekker LJM, Burger DM, Hartwig NG, Gruters RA. Ultrafast and high-throughput mass spectrometric assay for therapeutic drug monitoring of antiretroviral drugs in pediatric HIV-1 infection applying dried blood spots. *Analytical and Bioanalytical Chemistry* 2010;398(1):319–328.

Miller JHIV, Poston PA, Karnes HT. Direct analysis of dried blood spots by in-line desorption combined with high-resolution chromatography and mass spectrometry for quantification of maple syrup urine disease biomarkers leucine and isoleucine. *Analytical and Bioanalytical Chemistry* 2011;400(1):237–244.

Moini M. Atmospheric pressure chemical ionization: principles, instrumentation, and applications. In: Gross ML, Caprioli RM (eds), *Encyclopedia of Mass Spectrometry*, Vol. 6. Amsterdam: Elsevier; 2007. pp. 344.

Na N, Zhang C, Zhao M, Zhang S, Yang C, Fang X, Zhang X. Direct detection of explosives on solid surfaces by mass spectrometry with an ambient ion source based on dielectric barrier discharge. *Journal of Mass Spectrometry* 2007a;42(8):1079–1085.

Na N, Zhao M, Zhang S, Yang C, Zhang X. Development of a dielectric barrier discharge ion source for ambient mass spectrometry. *Journal of the American Society for Mass Spectrometry* 2007b;18(10):1859–1862.

Nefliu M, Smith JN, Venter A, Cooks RG. Internal energy distributions in desorption electrospray ionization (DESI). *Journal of the American Society for Mass Spectrometry* 2008;19(3):420–427.

Nemes P, Vertes A. Laser ablation electrospray ionization for atmospheric pressure, in vivo, and imaging mass spectrometry. *Analytical Chemistry* 2007;79(21):8098–8106.

Nemes P, Woods AS, Vertes A. Simultaneous imaging of small metabolites and lipids in rat brain tissues at atmospheric pressure by laser ablation electrospray ionization mass spectrometry. *Analytical Chemistry* 2010;82(3):982–988.

Nilles JM, Connell TR, Durst HD. Quantitation of chemical warfare agents using the direct analysis in real time (DART) technique. *Analytical Chemistry* 2009;81(16):6744–6749.

Nilles JM, Connell TR, Durst HD. Thermal separation to facilitate direct analysis in real time (DART) of mixtures. *Analyst* 2010;135(5):883–886.

Nyadong L, Green MD, De Jesus VR, Newton PN, Fernandez FM. Reactive desorption electrospray ionization linear ion trap mass spectrometry of latest-generation counterfeit antimalarials via noncovalent complex formation. *Analytical Chemistry* 2007;79:2150–2157.

Nyadong L, Hohenstein EG, Johnson K, Sherrill CD, Green MD, Fernandez FM. Desorption electrospray ionization reactions between host crown ethers and the influenza neuraminidase inhibitor oseltamivir for the rapid screening of Tamiflu®. *Analyst* 2008;133:1513–1533.

Nyadong L, Galhena AS, Fernández FM. Desorption electrospray/metastable-induced ionization: a flexible multimode ambient ion generation technique. *Analytical Chemistry* 2009;81(18):7788–7794.

Ooms JA, Knegt L, Koster EHM. Exploration of a new concept for automated dried blood spot analysis using flow-through desorption and online SPE-MS/MS. *Bioanalysis* 2011;3(20):2311–2320.

Özdemir A, Chen CH. Electrode-assisted desorption electrospray ionization mass spectrometry. *Journal of Mass Spectrometry* 2010;45(10):1203–1211.

Ratcliffe LV, Rutten FJM, Barrett DA, Whitmore T, Seymour D, Greenwood C, Aranda-Gonzalvo Y, McCoustra M. Surface analysis under ambient conditions using plasma-assisted desorption/ionization mass spectrometry. *Analytical Chemistry* 2007;79(16):6094–6101.

Rezenom YH, Dong J, Murray KK. Infrared laser-assisted desorption electrospray ionization mass spectrometry. *Analyst* 2008;133(2):226–232.

Roach PJ, Laskin J, Laskin A. Nanospray desorption electrospray ionization: an ambient method for liquid-extraction surface sampling in mass spectrometry. *Analyst* 2010;135(9):2233–2236.

Rummel JL, McKenna AM, Marshall AG, Eyler JR, Powell DH. The coupling of direct analysis in real time ionization to Fourier transform ion cyclotron resonance mass spectrometry for ultrahigh-resolution mass analysis. *Rapid Communications in Mass Spectrometry* 2010;24(6):784–790.

Sampson JS, Hawkridge AM, Muddiman DC. Generation and detection of multiply-charged peptides and proteins by matrix-assisted laser desorption electrospray ionization (MALDESI) Fourier transform ion cyclotron resonance mass spectrometry. *Journal of the American Society for Mass Spectrometry* 2006;17(12):1712–1716.

Sanders NL, Kothari S, Huang G, Salazar G, Cooks RG. Detection of explosives as negative ions directly from surfaces using a miniature mass spectrometer. *Analytical Chemistry* 2010;82(12):5313–5316.

Shah VP, Midha KK, Findlay JWA, Hill H, Hulse JD, McGilveray IJ, McKay G, Miller KJ, Patnaik RN, Powell ML, Tonelli A, Viswanathan CT, Yacobi A. Bioanalytical method validation—a revisit with a decade of progress. *Pharmaceutical Research* 2000;17(12):1551–1557.

Shelley JT, Hieftje GM. Fast transient analysis and first-stage collision-induced dissociation with the flowing atmospheric-pressure afterglow ionization source to improve analyte detection and identification. *Analyst* 2010;135(4):682–687.

Shiea J, Huang MZ, Hsu HJ, Lee CY, Yuan CH, Beech I, Sunner J. Electrospray-assisted laser desorption/ionization mass spectrometry for direct ambient analysis of solids. *Rapid Communications in Mass Spectrometry* 2005;19(24):3701–3704.

Shih SCC, Yang H, Jebrail MJ, Fobel R, McIntosh N, Al-Dirbashi OY, Chakraborty P, Wheeler AR. Dried blood spot analysis by digital microfluidics coupled to nanoelectrospray ionization mass spectrometry. *Analytical Chemistry* 2012;84(8):3731–3738.

Shin YS, Drolet B, Mayer R, Dolence K, Basile F. Desorption electrospray ionization-mass spectrometry of proteins. *Analytical Chemistry* 2007;79(9):3514–3518.

Soparawalla S, Salazar GA, Perry RH, Nicholas M, Gooks RG. Pharmaceutical cleaning validation using non-proximate large-area desorption electrospray ionization mass spectrometry. *Rapid Communications in Mass Spectrometry* 2009;23:131–137.

Soparawalla S, Tadjimukhamedov FK, Wiley JS, Ouyang Z, Cooks RG. In situ analysis of agrochemical residues on fruit using ambient ionization on a handheld mass spectrometer. *Analyst* 2011;136(21):4392–4396.

Spooner N, Lad R, Barfield M. Dried blood spots as a sample collection technique for the determination of pharmacokinetics in clinical studies: considerations for the validation of a quantitative bioanalytical method. *Analytical Chemistry* 2009;81:1557–1563.

Sripadi P, Nazarian J, Hathout Y, Hoffman EP, Vertes A. In vitro analysis of metabolites from the untreated tissue of Torpedo californica electric organ by mid-infrared laser ablation electrospray ionization mass spectrometry. *Metabolomics* 2009;5(2):263–276.

Suyagh M, Collier P, Millership J, Iheagwaram G, Millar M, Halliday H, McElnay J. Metronidazole population pharmacokinetics in preterm neonates using dried blood-spot sampling. *Pediatrics* 2011;127:367–374.

Swales G, Gallagher RT, Denn M, Raimund MP. Simultaneous quantitation of metformin and sitagliptin from mouse and human dried blood spots using laser diode thermal desorption tandem mass spectrometry. *Journal of Pharmaceutical and Biomedical Analysis* 2011;55:544–551.

Symond JM, Galhena AS, Fernández FM, Orlando TM. Microplasma discharge ionization source for ambient mass spectrometry. *Analytical Chemistry* 2010;82(2):621–627.

Takats Z, Wiseman JM, Gologan B, Cooks RG. Mass spectrometry sampling under ambient conditions with desorption electrospray ionization). *Science* 2004;306:471.

Takats Z, Cotte-Rodriguez N, Talaty H, Chen H, Cooks G. Direct, trace level detection of explosives on ambient surfaces by desorption electrospray ionization mass spectrometry. *Chemical Communications* 2005a;15:1950–1952.

Takats Z, Wiseman JM, Cooks RG. Ambient mass spectrometry using desorption electrospray ionization (DESI): instrumentation, mechanisms and applications in forensics, chemistry, and biology. *Journal of Mass Spectrometry* 2005b;40(10):1261–1275.

Tepper G, Kessick R. Nanoelectrospray aerosols from microporous polymer wick sources. *Applied Physics Letters* 2009;94:084106.

Thomas A, Déglon J, Steimer T, Mangin P, Daali Y, Staub C. On-line desorption of dried blood spots coupled to hydrophilic interaction/reversed phase LC/MS/MS system for the simultaneous analysis of drugs and their polar metabolites. *Journal of Separation Science* 2010;33(6–7):873–879.

Van Berkel GJ, Kertesz V. Application of a liquid extraction based sealing surface sampling probe for mass spectrometric analysis of dried blood spots and mouse whole-body thin tissue sections. *Analytical Chemistry* 2009;81(21):9146–9152.

Van Berkel GJ, Pasilis SP, Ovchinnikova, O. Established and emerging atmospheric pressure surface sampling/ionization techniques for mass spectrometry. *Journal of Mass Spectrometry* 2008;43:1161–1180.

Van Berkel GJ, Kertesz V, King RC. High-throughput mode liquid microjunction surface sampling probe. *Analytical Chemistry* 2009;81(16):7096–7101.

Vaz ADN, Wang WW, Bessire AJ, Sharma R, Hagen AE. A rapid and specific derivatization procedure to identify acyl-glucuronides by mass spectrometry. *Rapid Communications in Mass Spectrometry* 2010;24(14).2109–2121.

Venter A, Nefliu M, Cooks RG. Ambient desorption ionization mass spectrometry. *Trends in Analytical Chemistry* 2008;27(4):284–290.

Vu DH, Koster RA, Alffenaar JWC, Brouwers JRBJ, Uges DRA. Determination of moxifloxacin in dried blood spots using LC-MS/MS and the impact of the hematocrit and blood volume. *Journal of Chromatography B: Analytical Technologies in the Biomedical and Life Sciences*, 2011;879(15–16):1063–1070.

Wang H, Lui J, Cooks GR, Ouyang Z. Paper spray for direct analysis of complex mixtures using mass spectrometry. *Angewandte Chemie—International Edition* 2010a;49:877–880.

Wang H, Sun W, Zhang J, Yang X, Lin T, Ding L. Desorption corona beam ionization source for mass spectrometry. *Analyst* 2010b;135(4):688–695.

Weston DJ. Ambient ionization mass spectrometry: current understanding of mechanistic theory; analytical performance and application areas. *Analyst* 2010;135:661–668.

Williams JP, Scrivens JH. Coupling desorption electrospray ionisation and neutral desorption/extractive electrospray ionisation with a travelling-wave based ion mobility mass spectrometer for the analysis of drugs. *Rapid Communications in Mass Spectrometry* 2008;22:187–196.

Wiseman JM, Evans CA, Bowen CL, Kennedy JH. Direct analysis of dried blood spots utilizing desorption electrospray ionization (DESI) mass spectrometry. *Analyst* 2010;135(4):720–725.

Wu J, Hughes CS, Picard P, Letarte S, Gaudreault M, Levesque J-F, Nicoll-Griffith DA, Bateman KP. High-throughput cytochrome P450 inhibition assays using laser diode thermal desorption-atmospheric pressure chemical ionization-tandem mass spectrometry. *Analytical Chemistry* 2007; 79: 4657.

www.quantiontech.com/Products_IS (accessed April 2012).

Youhnovski N, Bergeron A, Furtado M, Garofolo F. Pre-cut dried blood spot (PCDBS): an alternative to dried blood spot (DBS) technique to overcome hematocrit impact. *Rapid Communications in Mass Spectrometry* 2011;25(19):2951–2958.

Zhang Z, Xu W, Manicke NE, Cooks RG, Ouyang Z. Silica coated paper substrate for paper-spray analysis of therapeutic drugs in dried blood spots. *Analytical Chemistry* 2012;84(2):931–938.

Zhao Y, Lam M, Wu D, Mak R. Quantification of small molecules in plasma with direct analysis in real time tandem mass spectrometry, without sample preparation and liquid chromatographic separation. *Rapid Communications in Mass Spectrometry* 2008;22(20):3217–3224.

23

PAPER SPRAY IONIZATION FOR DIRECT ANALYSIS OF DRIED BLOOD SPOTS

JIANGJIANG LIU, NICHOLAS E. MANICKE, R. GRAHAM COOKS, AND ZHENG OUYANG

23.1 INTRODUCTION

Dried blood spots (DBS) have been demonstrated to be an effective means of storage of blood samples for neonatal screening of inborn diseases. Recently, DBS has been applied for therapeutic drug monitoring (TDM) (Barfield et al., 2008; Spooner et al., 2009). Blood samples are typically collected by finger stick or heel stick and deposited on paper cards to form the DBS (Patton et al., 2007; Li and Tse, 2010). A series of sample preparation steps is required for the analysis of DBS using high-performance liquid chromatography mass spectrometry (HPLC-MS), including punching discs from the blood card, extraction and spiking internal standards, filtration, and chemical derivatization. Highly quantitative and specific information is obtained for clinical diagnosis (Chace et al., 2003; Barfield et al., 2008; Spooner et al., 2009; Li and Tse, 2010; Abu-Rabie et al., 2011).

In recent years, ambient ionization has evolved as a new concept for direct analysis of complex samples using mass spectrometry to analyze samples in the ambient environment with minimum sample preparation or chromatographic separation. Starting with the first two methods in 2004, desorption electrospray ionization (DESI) (Takats et al., 2004) and direct analysis in real time (DART) (Cody et al., 2005), more than 30 ambient ionization methods have been developed. Among these methods, DESI (Wiseman et al. 2010), laser ablation electrospray ionization (LAESI) (Nemes and Vertes, 2007), and liquid extraction sampling nanoelectrospray (Kertesz and Van Berkel, 2010) have been applied for DBS analysis. While qualitative analysis using ambient ionization methods has been demonstrated repeatedly (Wiseman et al., 2008; Shrestha et al., 2010; Zhou et al., 2010), quantitation of the chemicals in blood samples is mandatory for pharmaceutical and clinical applications. Recently, direct analysis of DBS with a high quantitative precision has been achieved with paper spray ionization (Liu et al., 2010; Wang et al., 2010). Paper spray was first introduced in 2009 as a means of direct sampling and ionization technique for analysis of complex condensed phase mixtures using mass spectrometry. Multiple versions of paper spray ionization sources have been developed for commercial triple quadrupole and ion trap mass spectrometers. A full MS-based solution for point-of-care (POC) analysis is being explored using paper spray with miniature mass spectrometer. Incorporation of internal standards is important for the quantitation while simple operation procedure also needs to be established without requiring lab techniques such as pipetting or sampling extraction.

23.2 PAPER SPRAY IN PRACTICE

The laboratory setup of paper spray ionization is simple, as shown in Figure 23.1a. A paper triangle is cut out from blood card or other porous substrate, with a size of approximately 10 mm in height and 5 mm in the base. For DBS analysis, the blood sample (typically 1–20 μL) is deposited on the paper substrate to form a blood spot and allowed to dry. During the mass spectrometry analysis, a DC voltage of 3–5 kV and a solvent of 10–40 μL are applied on the paper triangle to form a spray. The solvent extracts the chemicals from the DBS and the charged droplets that carry these chemicals are generated at the tip of the paper triangle. Dry ions are subsequently generated and analyzed by a mass spectrometer equipped

Dried Blood Spots: Applications and Techniques, First Edition. Edited by Wenkui Li and Mike S. Lee.

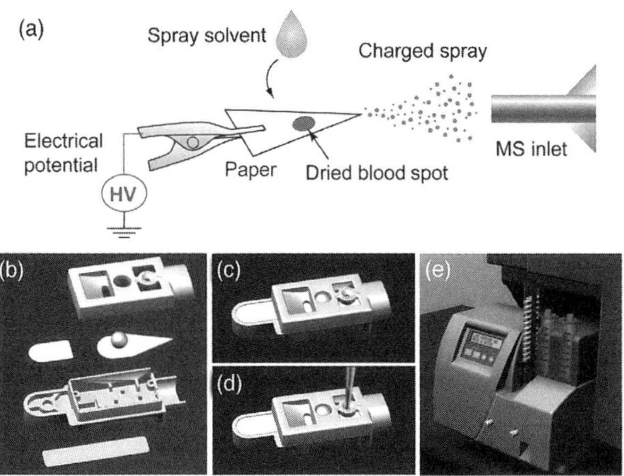

FIGURE 23.1 (a) Schematic setup of the paper spray ionization. Photos of the paper spray cartridge and the front-end source for mass spectrometer. (b) Configuration of the cartridge (c), assembled cartridge (d), blood sample loading on the cartridge (d), paper spray ion source mounted on a mass spectrometer. Adapted with permission from Liu et al. (2010) and AMIPurdue. Copyright 2010 American Chemical Society.

with an atmospheric pressure interface. The entire analysis process takes less than 60 seconds and does not involve separate steps for sample treatment or chromatographic separation. The low cost of the consumables, namely, paper substrate and tens of microliters of solvent, and the low volume of blood consumed make paper spray attractive as a potential method for DBS analysis in both analytical laboratories and physicians' offices.

Since the proof-of-concept demonstration of paper spray ionization, a serious effort has been put into transfer of the technology (Manicke et al., 2011a, 2011b; Yang et al., 2012a, 2012b) to deliver a relatively robust and user-friendly paper spray ion source for routine analysis using commercial mass spectrometers. As shown in Figures 23.1b–23.1e, a paper spray cartridge and an automated paper spray ion source have been developed in a joint effort between the academic labs at Purdue and the Alfred Mann Institute for Biomedical Development at Purdue (AMIPurdue). The paper spray cartridge is a low-cost disposable unit, which consists of a polymer holder, a paper substrate, and a metal ball as the electrode. Multiple cartridges with preloaded samples can be put into the paper spray source, where they are sequentially analyzed through paper spray ionization and tandem mass spectrometry. The analysis conditions can be preset according to the target chemicals, which include the polarity of ionization, the spray voltage, and the paper spray solvent.

23.3 FUNDAMENTALS OF PAPER SPRAY

Analysis of DBS or other dried samples on paper using paper spray involves two consecutive steps, the extraction of the analyte chemicals from the sample and spray ionization. The spray ionization process is like electrospray and as such is soft in that it generates intact molecular ions of the analytes. A comparison between paper spray and nanoESI is shown in Figures 23.2a and 23.2b (Wang et al., 2010). Similar spray plumes were observed in both methods. The "survival yield" method was used to characterize the internal energy of the ions generated by measuring the fragmentation fractions of a series of compounds, including *p*-chlorobenzylpyridinium chloride (*p*-Cl), *p*-methylbenzylpyridinium bromide (*p*-CH_3), *p*-methoxybenzylpyridinium tetrafluoroborate (*p*-OCH_3), *p*-nitrobenzylpyridinium bromide (*p*-NO_2), and *p*-cyanobenzylpyridinium chloride (*p*-CN). The breakdown curves (Figure 23.2c) and the corresponding internal energy distributions (Figure 23.2c) are similar with only a slightly higher (~0.4 eV) internal energy in the ions generated by nanoESI.

In spite of the similarity in the energy distributions, large differences were observed in the spray voltage and current between paper spray and nanoESI. In the positive ion mode, nanoESI is optimized at about 2 kV (Figure 23.3a), while a much higher voltage is required to initiate paper spray (Figure 23.3b). The spray currents in both paper spray and nanoESI increase with the spray voltage, while different trends were observed in the ion intensities. The spectra obtained with paper spray and nanoESI typically exhibit similar patterns and ion intensities (Figures 23.3c and 23.3d) (Manicke et al., 2011b). In the negative ion mode, significant differences have also been observed for the paper spray and nanoESI (Figure 23.3e) (Wang et al., 2010). When using methanol/water as spray solvent for paper spray, no spray plume was observed even at a high voltage of 6 kV. By applying 1,4-benzoquinone, a vapor-phase electron-capture agent, the emissions of electrons from the paper triangle were proved based on the formation of radical anion $[M]^{-\bullet}$ in the spectrum (Figure 23.3f). This phenomenon could be related to the competition of electrical discharge and the electrospray when the negative high voltage is applied on the paper triangle. In the negative ionization mode, spray solvents of high surface tension, such as water-based solvents, can make the onset voltage higher for electrospray than that for electrical discharge, depending on the configuration of the paper spray setup. In this case, the dominant ionization mode is based on electron emission and associated secondary ionization modes through subsequent reactions, similar to the processes that occur in atmospheric pressure chemical ionization (APCI). Pure methanol, acetonitrile, and some other polar organic solvents can work well to generate paper spray in the negative ionization mode and the spray plumes similar to those in positive mode have been observed.

All paper substrates have microscopically porous structures but a macroscopically sharp tip is required to generate a spray. The porous structure allows the spray solvent to wick through the substrate to the tip of the triangle, where

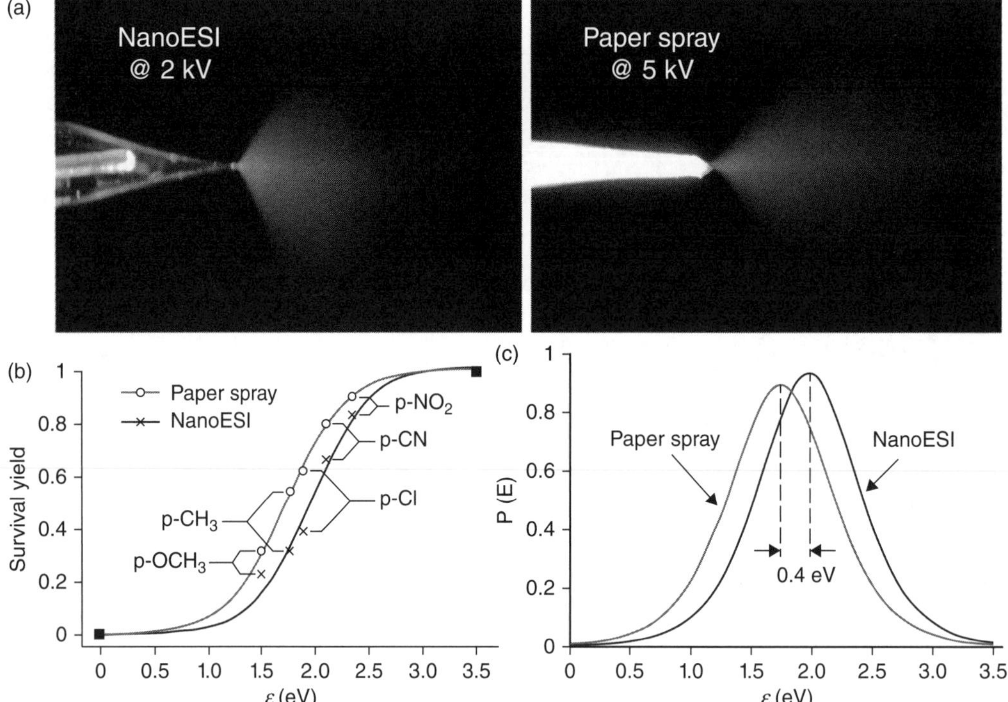

FIGURE 23.2 Comparison of nanoESI with paper spray ionization in the positive ion mode. (a) Optical images of the spray plume generated using nanoESI and paper spray, respectively. Spray voltage: 2 kV for nanoESI and 5 kV for paper spray. (b) Survival ion yields for paper spray and nanoESI. (c) Internal energy distributions for paper spray and nanoESI. Adapted with permission from Wang et al. (2010). Copyright 2010 John Wiley & Sons.

the macroscopic structure of the sharp tip provides a localized high electric field to generate the spray of charged droplets from which ions are generated (Figure 23.4a) (Yang et al., 2012b). The spray plumes occur only on the corners of the paper substrate and the expansion of the plume decreases as the tip angle increases (Figures 23.4b–23.4d). The total spray current increases with an increase in the spray voltage (Figure 23.4e) while the optimized voltage to obtain the highest ion intensity is dependent on the tip angle (Figure 23.4f).

The spray plumes generated on the paper tip are visible under strong illumination. The measurement of droplets' size and velocity could help to better understand the mechanism of paper spray ionization and quantitatively compare paper spray ionization with other spray ionization methods. A phase Doppler particle anemometer was used to measure the droplets' size and velocity during the paper spray ionization (Espy et al., 2012b). The preliminary results indicate that the size of droplets ranges from sub-micrometer to over 10 μm depending on the experimental conditions, including the spray solvent and the stage of solvent depletion during the ionization process. The velocity of the droplets varies from 3 to 10 m/s as the spray voltage increases from 3.5 to 5.5 kV. The experimental data point to two main ionization mechanisms: in the presence of excess solvent bulk flow occurs, multiple Taylor cone-jets are produced, and droplets have a range of sizes but a constant velocity. Under solvent-poor conditions, corona discharge contributes to ionization. The tolerance to the ionization source position is important when coupling to a mass spectrometer for analytical applications. The variation in ion intensity with position of the paper spray emitter was characterized by moving the paper triangle area while monitoring the ion intensity (Figure 23.5a) (Liu et al., 2010). A 2D contour map of the recorded ion intensity is shown in Figure 23.5b and a stable high-intensity ion current can be obtained within an area of about 5 × 10 mm (*x* by *y*, Figure 23.5b), which indicates that accurate positioning of the paper triangle is not required for the implementation of paper spray ionization.

23.4 CHEMICAL INTERACTIONS DURING PAPER SPRAY

The paper substrate and the spray solvent play important roles in the process of paper spray ionization. They interact with each other and both interact with the sample that contains the analytes of interest during the extraction and spray formation. During sample deposition, such as the preparation

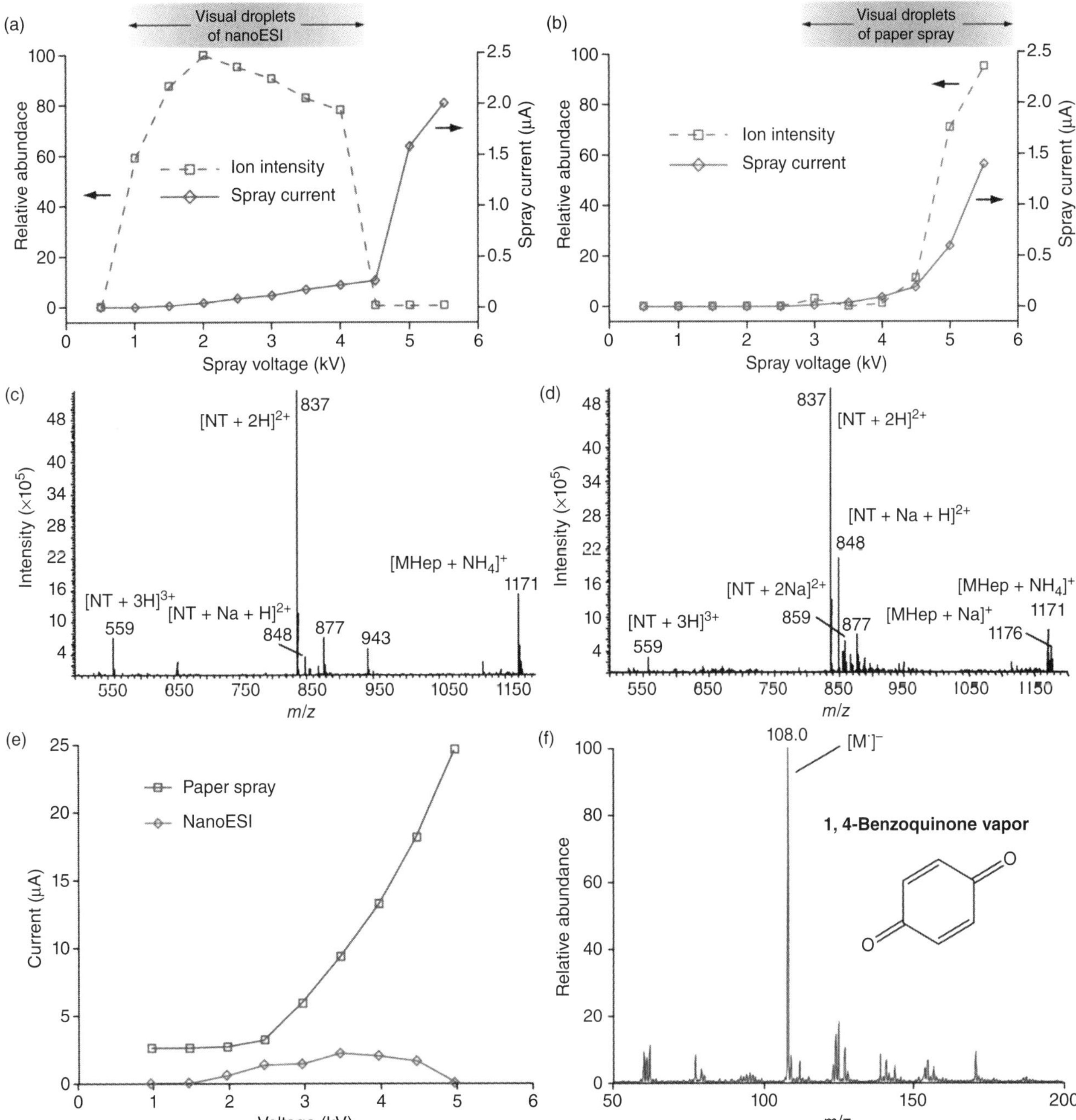

FIGURE 23.3 (a) NanoESI and (b) paper spray show different trends in ion intensities (cocaine in methanol, 200 ng/mL, *m/z* 304) and spray currents as the spray voltage is increased from 0.5 to 5.5 kV. Full-scan mass spectra obtained in the positive ion mode with (c) nanoESI and (d) paper spray, respectively. Tested sample: 10 μM neurotensin and maltoheptaose in methanol/water (1:1, v:v) with 10 mM ammonium acetate. (e) Measured current for nanoESI and paper spray in the negative ion mode. (f) Paper spray spectrum obtained from 1,4-benzoquinone vapor in the negative ion mode. High voltage, –4 kV; solvent, methanol/water (1:1, v:v). 1,4-Benzoquinone solid was placed close to the paper tip to produce 1,4-benzoquinone vapor. Adapted with permission from Wang et al. (2010) and Manicke et al. (2011b). Copyright 2011 John Wiley & Sons.

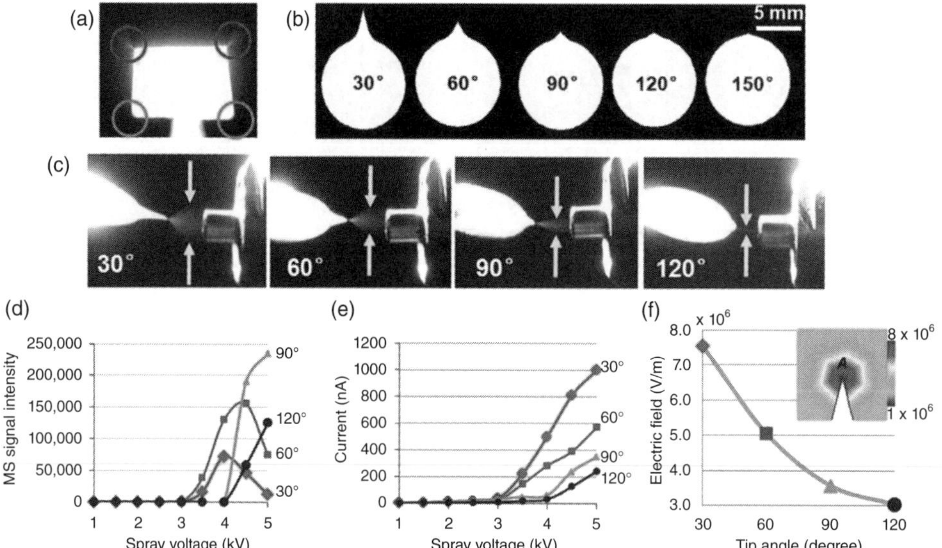

FIGURE 23.4 (a) Spray plumes on a rectangular paper substrate. The generation of spray only occurs at the sharp corners, rather than round corners. (b) Paper substrates with different tip angles of 30°, 60°, 90°, 120°, and 150°. (c) Spray plumes generated on different paper substrates with tip angles of 30°, 60°, 90°, and 120°. (d) Signal intensity of cocaine fragment ion *m/z* 182 and (e) total spray current as a function of the spray voltage as it is increased from 1 to 5 kV. (f) Electric field strength at the paper substrate tip as a function of tip angle from 30° to 120°. Inset: zoomed-in view of electric field distribution at a paper substrate tip of 30°. Adapted with permission from Yang et al. (2012b). Copyright 2012 Elsevier.

of a DBS, the interactions between the cellulose in the paper and analytes in the sample have an impact on the ease of the differential elution of the analytes from the sample. When the spray solution is applied to the paper substrate that contains the DBS and driven to the tip of the substrate by the high-voltage-induced solvent flow, a fast chemical extraction and separation occurs as is also the case in paper chromatography. As shown in Figure 23.6, two dyes, methylene blue and methyl violet, in a sample spot on a piece of chromatography paper can be well separated by methanol, which can also be used for the paper spray during the direct analysis of the separated spots on the paper (Wang et al., 2010). A mixture of methylene blue and methyl violet is spotted on the chromatography paper and allowed to separate by dipping one end of the paper into methanol. Colored blue and violet strips were formed on the paper after 90-second

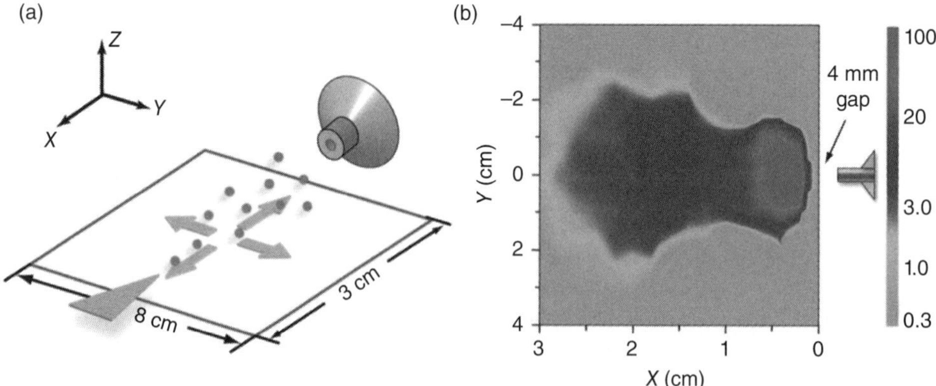

FIGURE 23.5 Characterization of the tolerance of paper spray ionization to the position of the paper tip. Cocaine solution (methanol/water, 1:1 v:v, 200 ng/mL, continuously applied on the paper) and chromatography paper were used for the test. (a) Illustration of the paper movement over the *x–y* plane. (b) 2D contour plot of the normalized intensity of *m/z* 304 as a function of the paper tip location on the *x–y* plane. Adapted with permission from Liu et al. (2010). Copyright 2010 American Chemical Society.

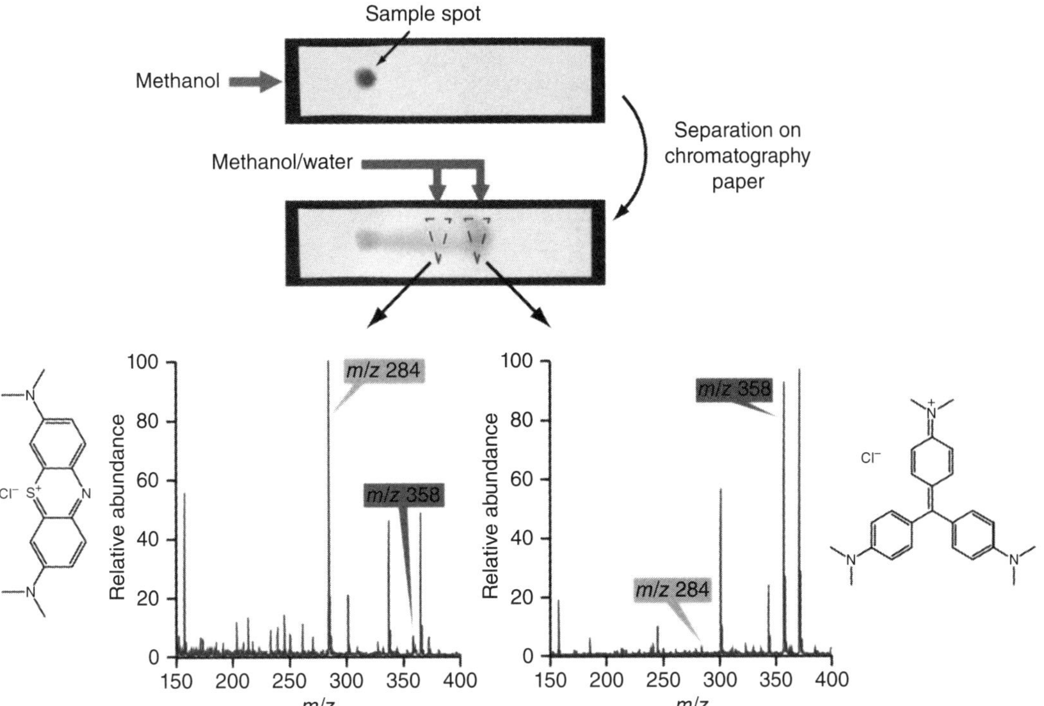

FIGURE 23.6 Separation of two dyes on chromatography paper, analyzed with paper spray ionization. Methylene blue (*m/z* 284) and methyl violet (*m/z* 358) in water at the concentration of 1 mg/mL were used; 100 μL of the mixture solution was spotted onto one end of the chromatography paper (4 cm × 0.5 cm) to form a spot. Two dyes were separated by dipping the end of the paper into methanol. After 90-second elution, the two triangles indicated by dashed lines were cut out from the paper and used for paper spray ionization test. The spectra obtained from the two paper triangles show significant difference on the distribution of the two dyes. Adapted with permission from Wang et al. (2010). Copyright 2010 John Wiley & Sons.

elution. Two triangles were subsequently cut off from the paper along the dashed lines and tested with paper spray ionization to evaluate the separation of the two chemicals. The ion intensities of the two chemicals are measured from the two cut paper triangles and their ratios are 7.1 and 0.063, respectively, indicating that the dyes are spatially separated on the chromatography paper. While this demonstration shows two separate steps of chemical separation and MS analysis with the same paper substrate and solvent for each step, direct analysis of samples can be done using paper spray with these two steps integrated. Note too, that chromatographic separation is neither sought nor even advantageous in the current implementations of paper spray ionization which depends on transport but not on separation of analytes.

Both the physical structure and chemical property of the paper substrate can affect the liquid transport on the paper as well as ionization of the analytes. A number of commercially available papers were tested for paper spray ionization. The paper spray spectra of a cocaine solution obtained with four filter papers of different pore sizes, glass fiber paper, and chromatography paper, respectively, are shown in Figure 23.7 (Liu et al., 2010). The spectrum of highest signal-to-noise ratio (S/N) for cocaine was obtained with the chromatography paper, which is typically used for blood spot analysis with paper spray ionization. In later studies, more types of papers and other porous media have been used. MS/MS is required for direct analysis of complex samples such as DBS and it also prevents deleterious chemical interferences due to the composition of the paper material.

Extraction of the analytes from the samples on the paper and the transport of the extracts to the tip of the paper are expected to be dependent on the chemical properties of the analytes and their competitive interactions with the paper substrate and the spray solvent. The solvent transport through the paper is also dependent on the relative ratio between the solvent amount and the substrate volume. In a preliminary investigation based on the analysis of 500 ng propranolol deposited on a paper substrate, it was found that analyte elution was influenced by the speed of feeding the spray solvent (Manicke et al., 2011b). When 10 μL of 75% methanol/water solvent was applied all at once onto the paper, a relatively constant ion intensity was observed for propranolol

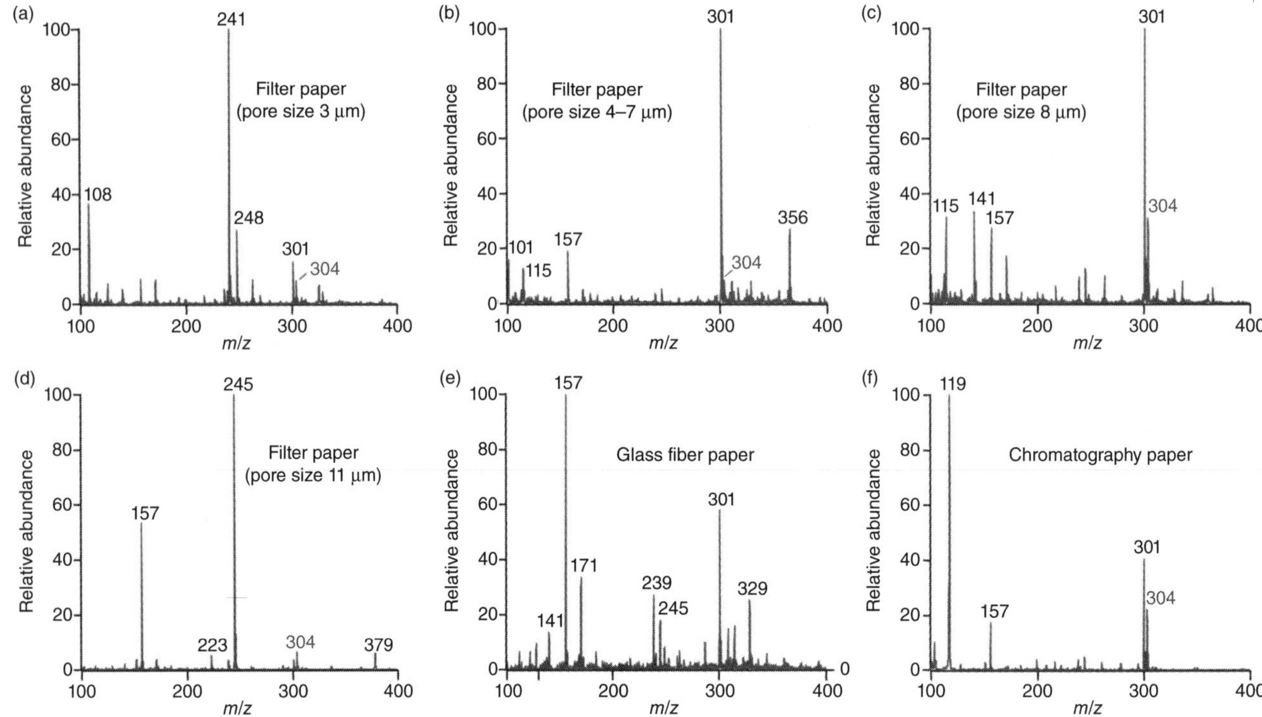

FIGURE 23.7 Paper spray spectra of cocaine (200 ng/mL in methanol/water, 1:1, v:v) recorded using different papers, including Whatman filter paper with different pore sizes: (a) 3 μm, (b) 4–7 μm, (c) 8 μm, and (d) 11 μm, (e) glass fiber paper, and (f) chromatography paper. Adapted with permission from Liu et al. (2010). Copyright 2010 American Chemical Society.

(Figure 23.8a); however, when the solvent was continuously fed to the paper via wicking action, the ion signal quickly reached a maximum and then decreased gradually and eventually stayed at a constant level for a long time (Figure 23.8b). For the quantitation of propranolol with propranolol-d7 mixed in the sample as the internal standard, the same elution patterns were found for the propranolol and propranolol-d7 (Figure 23.8c), the ratio of which is used for quantitation.

The elution of analytes during paper spray ionization is a process that involves the competition between retention by the paper substrate and extraction/transportation by the spray solvent. Variation in the properties of the paper substrate and the solvent is expected to provide optimized conditions for the elution of chemicals in the blood spots that differ in polarity or hydrophobicity. A silica-coated paper was tested for the analysis of therapeutic drugs in DBS using paper spray ionization (Zhang et al., 2012). In comparison with chromatography paper, the morphology of ordinary cellulosic paper was significantly modified by the silica coating (Figure 23.9) and this affected the diffusion of the blood during the DBS formation (Figure 23.9f). Isopropanol was found to be efficient for the extraction of verapamil from DBS on this paper, while addition of dichloromethane significantly improved desolvation during the spray and thereby the overall efficiency of the paper spray. In comparison with the use of chromatography paper and methanol/water solution, a 5- to 50-fold improvement in limit of quantification (LOQ) was achieved for the analysis of verapamil, citalopram, amitriptyline, lidocaine, and sunitinib in DBS using the silica-coated paper and isopropanol/dichloromethane solvent.

23.5 QUANTITATION OF DRUGS AND METABOLITES

Ambient ionization methods show great potential in direct analysis of complex biological samples. However, quantitative performance, in addition to qualitative confirmation and adequate limit of detection (LOD), is most important for clinical analysis. With appropriate introduction of internal standards, which will be discussed further later, paper spray has been shown to be a powerful tool for quantitation of target chemicals in the blood samples. The analysis of a series of DBS that contain sitamaquine and its internal standard, sitamaquine-d10, is shown in Figure 23.10. The concentration of sitamaquine varied from 1000 to 5 ng/mL and the concentration of sitamaquine-d10 was kept at 100 ng/mL (Manicke et al., 2011a). Although the total ion currents fluctuated, the ratio of sitamaquine to sitamaquine-d10

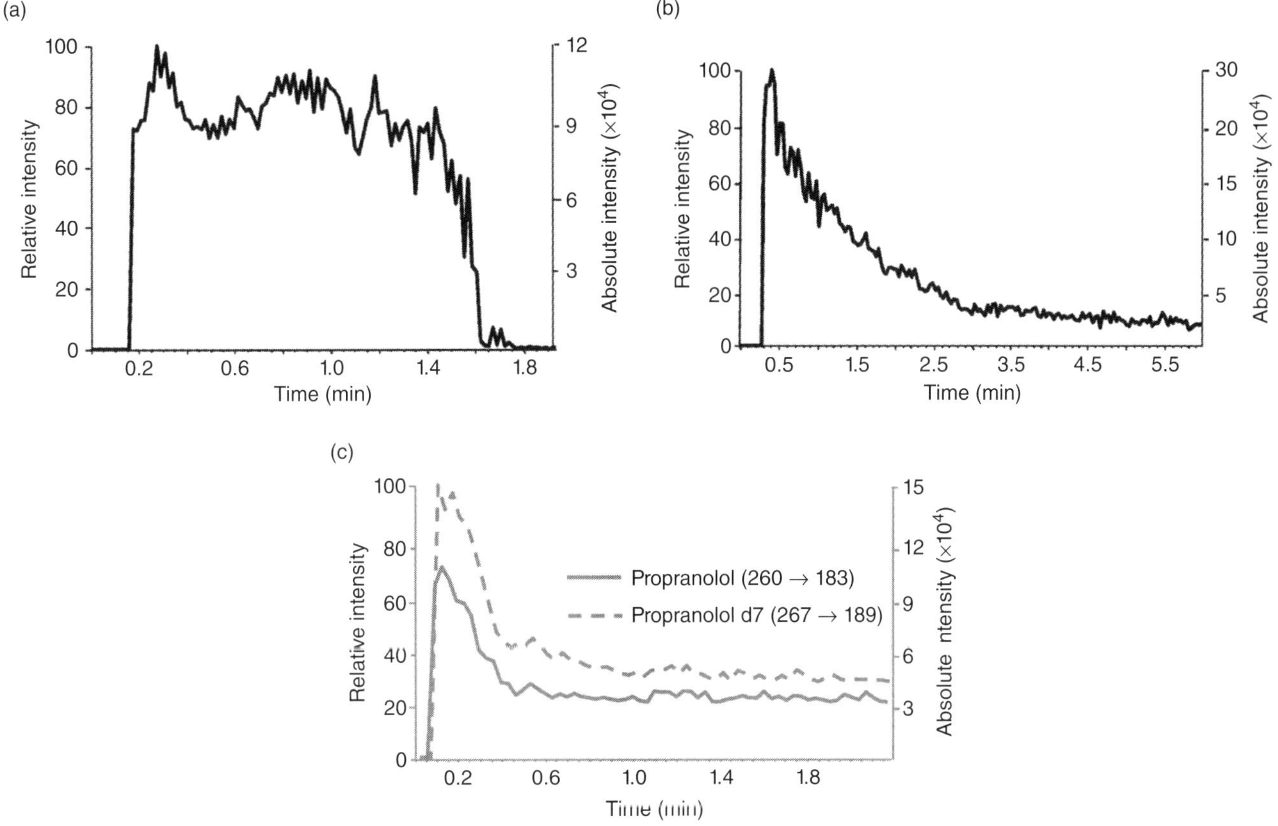

FIGURE 23.8 Ion chronograms (signal intensity vs. elution time) for propranolol (500 ng) tested with paper spray ionization. The deposited propranolol was eluted by either (a) applying 10 μL of solvent all at once or (b) allowing solvent to wick through the paper continually, (c) for propranolol and propranolol-d7 for the addition of the IS solution to the DBS punch, followed by the elution with solvent. Adapted with permission from Manicke et al. (2011b). Copyright 2011 Elsevier.

demonstrated good linearity over the range of concentrations of sitamaquine.

The quantitative analysis of amitriptyline in blood with deuterated forms of internal standards premixed into the sample using paper spray using selected reaction monitoring (SRM) is shown in Figure 23.11. [$^2H_6$]-Amitriptyline (2 μL, 1250 ng/mL) was used as internal standard and prespotted onto the paper before adding blood samples. A 10 μL of bovine blood spiked with amitriptyline was added onto the paper to fully cover the dried internal standard. Paper spray ionization was carried out with 25 μL methanol/water (95:5, v/v) after the blood spot had dried thoroughly. An LOQ of 0.9 ng/mL and good linearity over the therapeutic range were obtained. For biofluid analysis, relative matrix effects may alter the performance of the test when samples are collected from different individuals. The assessment of relative matrix effects by constructing standard curves is illustrated in Figure 23.11b. Five lots of blood samples obtained from five donors were tested, and the standard lines were generated by plotting the instrument responses versus drug concentrations. The variation in the five slopes is only 1.3% and the precision of each concentration measurements was less than 15% except at the lower limit of quantitation. These results indicate this particular assay can be considered as a reliable method which is free of relative matrix effects. In general, it is expected and has been observed that the presence of a stable isotopically labeled internal standard negates relative matrix effects. The reproducibility of the direct analysis of the drug compounds in blood using paper spray has been characterized with a set of experiments to repeatedly analyze imatinib in blood at different concentrations within a period of three successive days. Both precision and accuracy were acceptable for intraday and interday assessments over the entire range of tested concentrations from 4 to 8000 ng/mL (Table 23.1); this meets the requirement for TDM.

Paper spray has also been applied for quantitative analysis in neonatal screening. DBS analysis is of great importance for the neonatal screening to prevent irreversible damage to the neonate babies who have inborn metabolic disorders. There

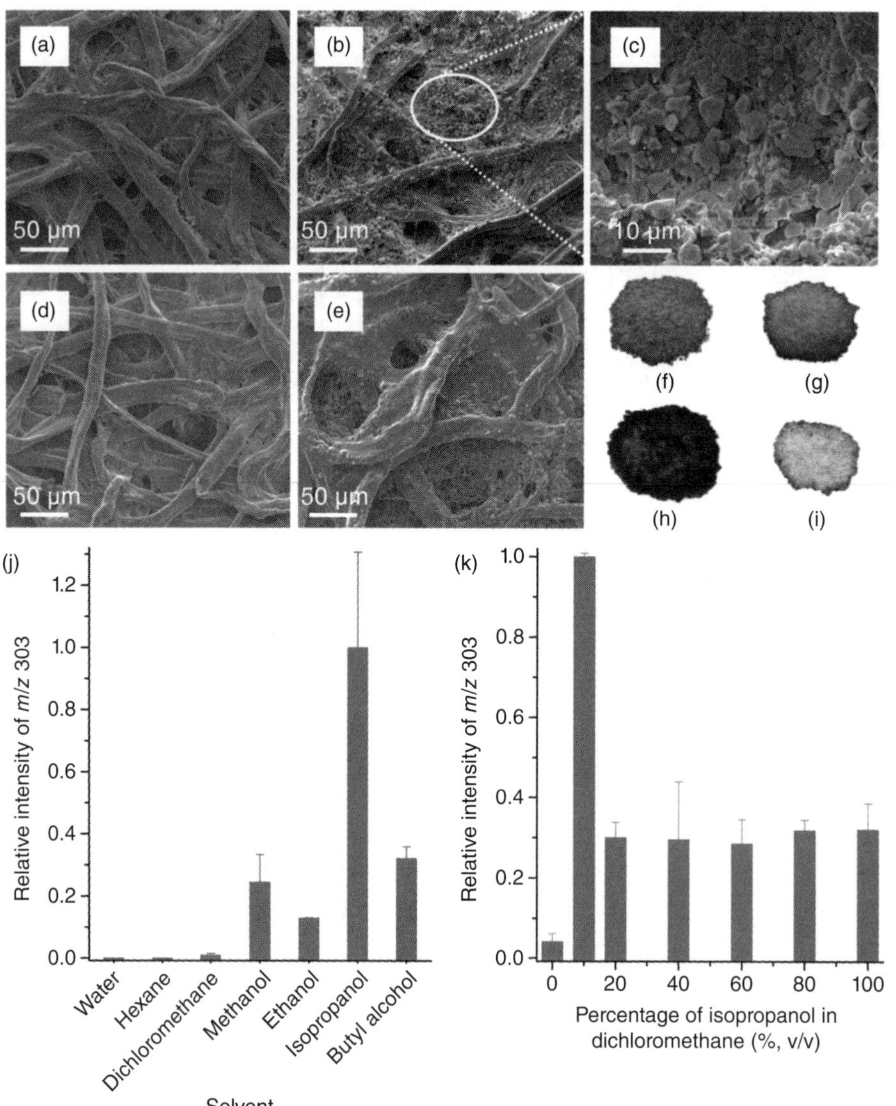

FIGURE 23.9 SEM images of (a) chromatography paper and (b) silica-coated paper without dried blood spots. (c) Close-up image of the selected area in (b). SEM images of (d) chromatography paper and (e) silica-coated paper with dried blood spots. Blood spots on chromatography paper showing (f) top view and (g) back view; silica-coated paper (h) top view and (i) back view. (j) Signal intensities obtained from verapamil in blood (500 ng/mL) with paper spray ionization eluted with different spray solvents. (k) Signal intensities obtained from verapamil in blood spots with paper spray ionization eluted with mixtures of isopropanol and dichloromethane. Silica-coated paper was used for all tests. Adapted with permission from Zhang et al. (2012). Copyright 2012 American Chemical Society.

are some 29 conditions which should be monitored during neonatal screening, which include organic acidurias, amino acidurias, disorders of fatty oxidation, hemoglobinopathies, and other conditions. The fatty acid oxidation disorder is an important metabolic disorder associated with fatty acid transport and mitochondrial oxidation. Diagnosis of fatty acid oxidation disorder could be done by evaluating the level of acylcarnitines in blood, which has the function of transporting fatty acids across the mitochondrial membrane for fatty acid >β-oxidation. A panel of acylcarnitines in a single dried blood or serum spot has been simultaneously quantified with paper spray, which include acetylcarnitine (C2), propionylcarnitine (C3), isovalerylcarnitine (C5), hexanoylcarnitine (C6), octanoylcarnitine (C8), decanoylcarnitine (C9), lauroylcarnitine (C12), myristoylcarnitine (C14), palmitoylcarnitine (C16), and stearoylcarnitine (C18). The quantitative measurement was achieved by introducing internal standards of C2-d3, C3-d3, C5-d9, C8-d3, and C16-d3.

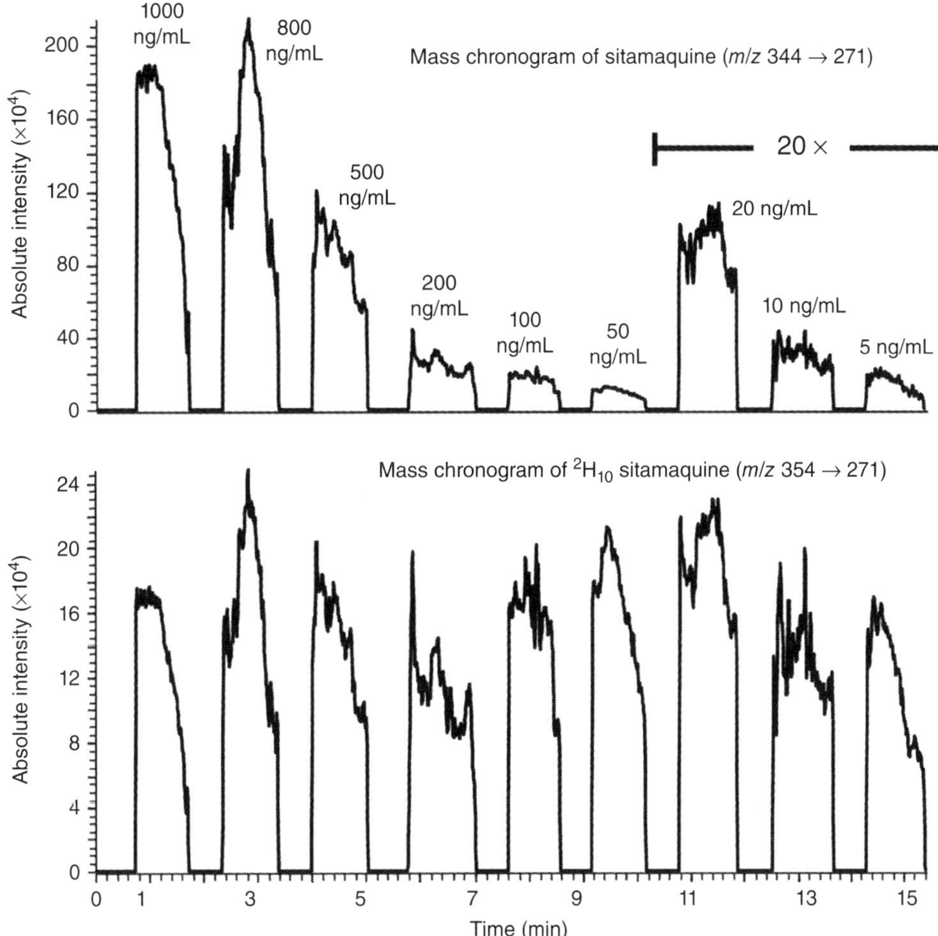

FIGURE 23.10 Chronograms that illustrate the total ion current obtained for the sequential analysis of sitamaquine using paper spray ionization in dried blood spots containing internal standard. Each peak in the chronogram represents the analysis of a different dried blood spot. (Upper panel) Total ion current chronograms for sitamaquine concentration 1000–5 ng/mL. (Bottom panel) Total ion current chronograms for the internal standard sitamaquine-d10 (100 ng/mL in all tests). Adapted with permission from Manicke et al. (2011b). Copyright 2011 Springer.

The experiments were carried out with the positively charged precursor ion scan set to record a common product ion *m/z* 85. Systemic optimization was performed on the spray solvents and the amount of sample loaded to enhance the performance of paper spray ionization. The spectra obtained from serum and blood in neutral loss scan mode with very high S/N are shown in Figure 23.12. Ten acetylcarnitine and five deuterated acetylcarnitine were identified in a single spectrum. The quantitative result shown in Figure 23.12 indicates that paper spray ionization could be utilized for quantitative analysis of acetylcarnitine with good detection limits. Although the low end of the normal range of acetylcarnitine in serum and blood is not fully covered by the dynamic range of this method, it is still a reliable method for clinical diagnosis since the level of acetylcarnitine in diseased individuals' blood is much higher than the normal range. The LODs for each acetylcarnitine in blood/serum are 3.5/2.7 μM for C2, 0.27/0.77 μM for C3, 0.26/0.2 μM for C5, 0.05/0.04 μM for C6, 0.06/0.04 μM for C8, 0.04/0.07 μM for C10, 0.01/0.06 μM for C12, 0.03/0.02 μM for C14, 0.24/0.31 μM for C16, and 0.35/0.10 μM for C18.

23.6 TOWARD POINT-OF-CARE ANALYSIS

Paper spray ionization has been demonstrated as a simple, fast, and low-cost method for the analysis of chemicals in blood samples. The possibility of routine application of paper spray for POC analysis can be well speculated. For a quantitative method, the incorporation of internal standards into blood samples is of critical importance to overcome matrix effects and other inevitable variation that arise from sample

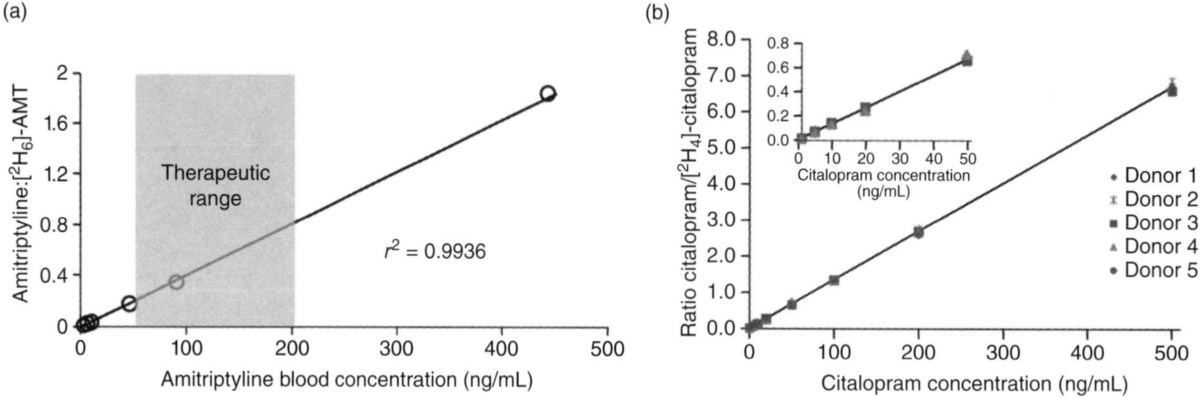

FIGURE 23.11 Quantitative analysis of amitriptyline and citalopram from dried blood spots with paper spray ionization. (a) Calibration curve for amitriptyline from dried bovine blood by pretreating the paper with the internal standard [$^2H_6$]-amitriptyline. (b) Paper spray analysis of blood from five different human donors spiked with citalopram and [$^2H_4$]-citalopram (internal standard). The slopes of the standard lines generated from the five different blood samples varied by 1.3%, indicating a lack of relative matrix effects that arise from the use of different blood lots. The inset plot shows the lower end of the sample concentration range. Adapted with permission from Manicke et al. (2011). Copyright 2011 Springer.

application, extraction, ionization, and ion transfer into and through the mass spectrometer. For paper spray ionization, spiking internal standard in the liquid blood before the deposition of blood on paper is the most accurate and precise way to achieve quantitative analysis. However, this approach is not feasible for DBS analysis in a POC manner because the blood sample is typically collected directly on the paper. The following methods were explored as alternatives: (i) prespotting the internal standard on the paper prior to spotting the blood, (ii) adding the internal standard solution to a punched out DBS on paper, and (iii) spiking the internal standard into the spray solvent. The results are shown in Table 23.2 (Manicke et al., 2011b).

For the first method, the internal standard solution (1.25 μL propranolol-d7, 2 μg/mL) is deposited on the paper and allowed to dry for immediate use or for storage. Then, prior to the paper spray measurement, a blood sample (2.5 μL, containing 1 μg/mL of propranolol) is spotted on the same paper to completely cover the area wetted by the internal standard solution. During the elution process, both

TABLE 23.1 Intraday and Interday Variations in Imatinib Analysis in Dried Blood Spots

	Intraday ($n = 6$ per day)			
Imatinib Concentration	Day 1	Day 2	Day 3	Interday ($n = 18$)
4 ng/mL (LOQ)				
Mean accuracy (%)	117	102	108	109
Standard deviation (%)	9.6	10.6	6.2	10.2
32 ng/mL				
Mean accuracy (%)	106	107	104	105
Standard deviation (%)	4.4	4.6	4.3	4.3
320 ng/mL				
Mean accuracy (%)	102	105	101	103
Standard deviation (%)	2.4	3.3	1.8	3.2
3200 ng/mL				
Mean accuracy (%)	104	106	101	104
Standard deviation (%)	1.8	1.8	2.8	2.8
8000 ng/mL				
Mean accuracy (%)	103	103	100	102
Standard deviation (%)	2.5	1.1	2.0	2.2

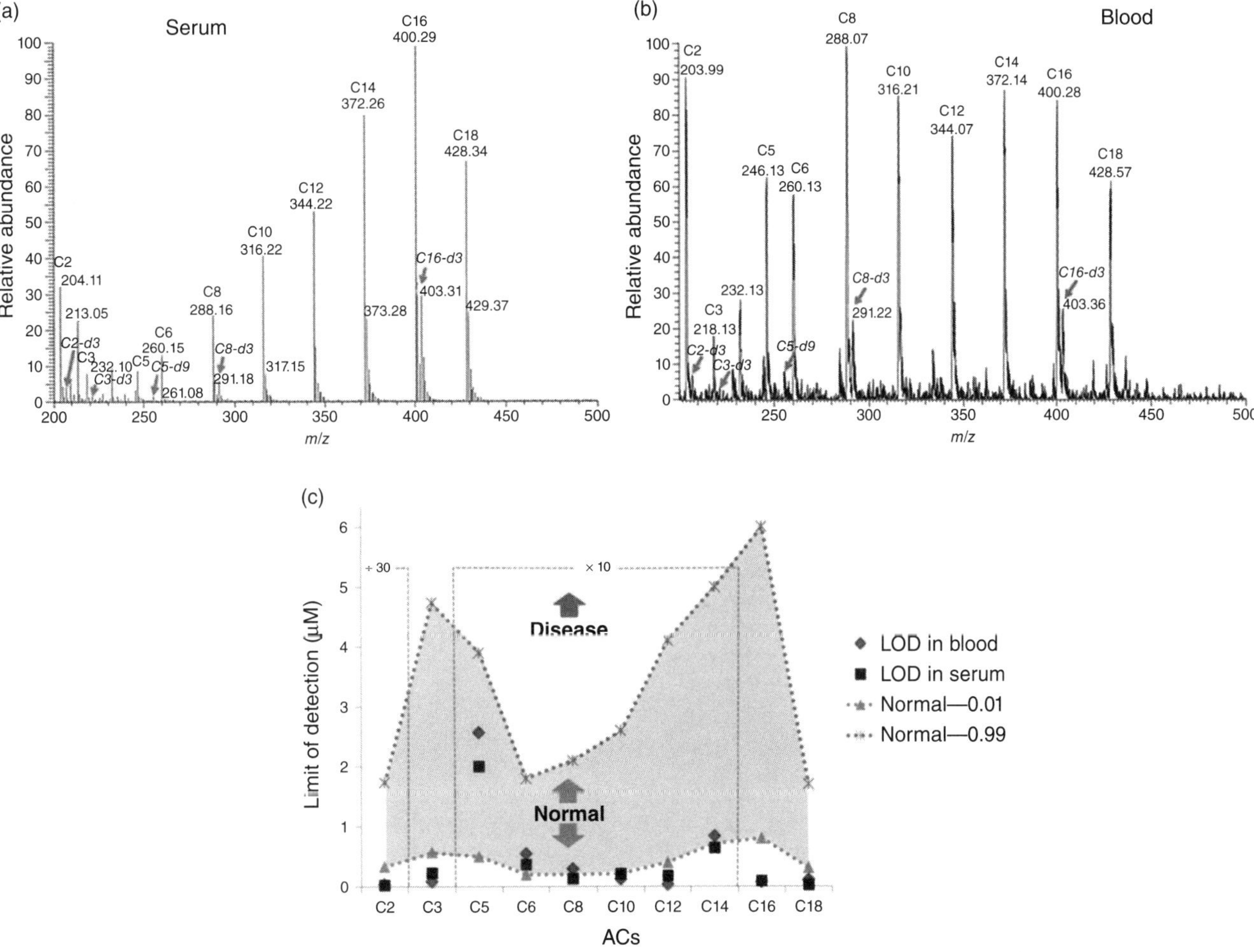

FIGURE 23.12 Spectra recorded in the precursor ion MS/MS mode for acylcarnitine in (a) dried serum spot and (b) dried blood spot, respectively. Acylcarnitine calibrators: C2, 5 μM; other acylcarnitines, 500 nM. Internal standards: C2-d3, 1 μM; C16-d3, 400 nM; others, 200 nM. (c) Limit of detection obtained with paper spray ionization for acylcarnitines in serum and in blood, respectively. The shaded area between the two dotted lines is the normal range of acylcarnitines. Adapted with permission from Yang et al. (2012a). Copyright 2012 Springer.

TABLE 23.2 Drug Recovery and Precision for Various Methods of Internal Standard Addition

Method of IS Addition	Mean Propranolol/Propranolol-d7 (AUC, $N = 7$)	Standard Deviation	Relative Standard Deviation (%)
Mixed in liquid blood	1.05	0.03	3
Dried on paper prior to blood spotting	0.73	0.06	8
Added to dried blood punch[a]	0.72	0.04	6
Present in spray solvent	0.26	0.04	16

Source: Taken with permission from Manicke et al. (2011b).
AUC, area under the curve.
[a]$N = 6$. One value rejected by Q-test at 95% confidence.

the internal standard and the analytes are extracted, ionized, and detected at the same time. Acceptable coefficients of variation (8%) were obtained with this method although the recovery of the analyte from the DBS (73%) is low. This low recovery may be due to the difficulty of solvent wetting and penetration into the blood spot.

The second method involves the direct application of an internal standard solution (17 μL propranolol-d7, 1 μg/mL) to the punched out DBS which covers a reproducible area from the paper. The punched disc was estimated to hold 17 μL blood that contains 1 μg/mL propranolol. The disc with DBS is then placed on another paper triangle to perform paper spray ionization. The additional spray solvent can flow through the punched disc and elute both the internal standard and analyte onto the paper triangle and allow for each to be ionized. The variance for this method is similar to that for the prespotted internal standard method (6% vs. 8%), indicating that the control on the amount of blood sample is as accurate as the pipetting method for blood spotting on paper. Again, the recovery of the analyte from the DBS is lower than theoretical level (72%).

Adding internal standard (133 ng/mL, propranolol-d7) into the spray solvent is the third and easiest way to introduce the internal standard for quantitative analysis of DBS (2 μL, containing 1 μg/mL propranolol). The elution of analyte and the mixing of analyte and internal standard occur at the same time. However, the precision of this method (RSD 16%) is worse than the previous two methods. Presumably the variability is due to the extraction of analyte from the DBS that is affected by the manual application and uneven distribution of the spray solvent onto the paper. Besides that, the ratio of analyte to internal standard is much lower than the theoretical value (0.26 vs. 1.0). A presumed explanation is that the recovery of analyte from blood is much more difficult than the recovery of internal standard from the spray solvent.

POC analysis of blood sample requires both rapid sampling and fast detection. Elimination of the blood drying step, which takes at least minutes, could benefit the POC analysis. A small amount of blood, for example, 1 μL, can be directly analyzed on paper immediately after spotting, but larger volumes of blood are desirable in order to improve the sensitivity and reproducibility in some cases. For 10–15 μL blood, the drying time of blood on paper could be as long as 2 hours. This problem related to a lengthy drying time could be simply solved by applying coagulants on the paper to clot the blood sample within seconds to realize the analysis of larger blood samples immediately. The blood spot analysis with and without adding alum to paper is shown in Figure 23.13 (Espy et al., 2012a). Oxygenated hemoglobin was transported to the tip of the paper, presumably indicating the movement of proteins and cellular components which could interfere with the formation of the Taylor cone at the tip. With prespotted alum on paper, the blood spot is quite stable during the spray process and the addition of alum did not introduce any contaminant ions.

A number of oncology drugs, including pazopanib, tamoxifen, imatinib, cyclophosphamide, paclitaxel, docetaxel, topotecan, and irinotecan, were tested with this method for comparison with the DBS method. The test results

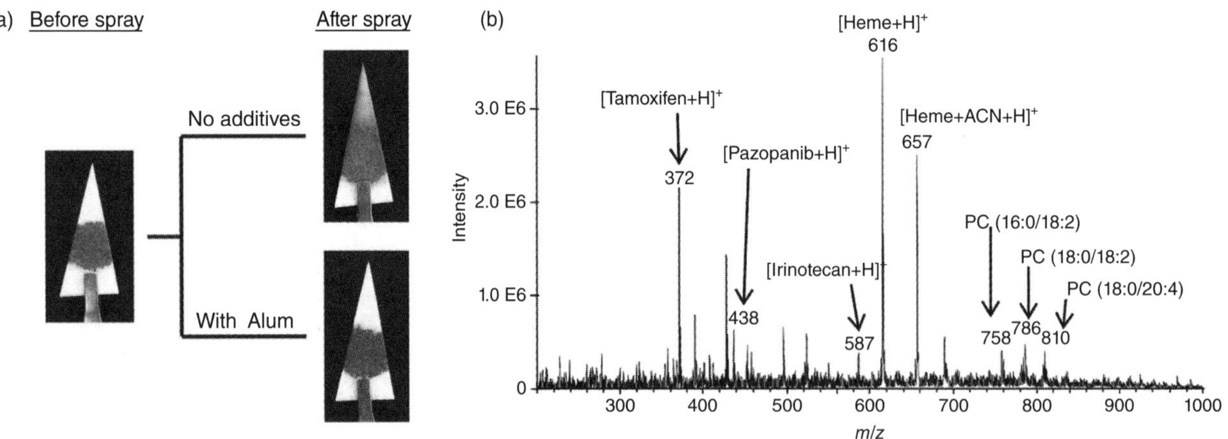

FIGURE 23.13 (a) Photos of paper triangles loaded with 10 μL whole blood with and without alum, a coagulant which was prespotted to clot the blood immediately before it was loaded. (b) Paper spray mass spectrum of 1 μg/mL tamoxifen, pazopanib, and irinotecan in 10 μL fresh whole blood with 0.45 mg of alum which was prespotted on paper before test. Spray solvent: 32 μL 50:50 methanol/acetonitrile; spray voltage: +3.0 kV. PC: phosphatidylcholine. Adapted with permission from Espy et al. (2012a). Copyright 2012 Royal Society of Chemistry.

TABLE 23.3 Limits of Detection for Oncology Chemotherapeutic Drugs Using Paper Substrate Prespotted with Alum

Drug	Therapeutic Range[a] (ng/mL)	Precursor Ion (*m/z*)	Fragment Ion (*m/z*)	Internal Standard (IS)	IS Precursor Ion	IS Fragment Ion	15 μL Dried Blood Spot LOD (ng/mL)	10 μL Fresh Blood w/Alum LOD (ng/mL)
Pazopanib	NA	438.1 $[M + H]^+$	357.1	$^2H_3{}^{13}C$-pazopanib	442.1 $[M + H]^+$	361.1	1	0.5
Tamoxifen	35–45	372.2 $[M + H]^+$	72.2	$^{13}C_2{}^{15}N$-tamoxifen	375.2 $[M + H]^+$	75.2	13	8
Imatinib	900–1800	494.2 $[M + H]^+$	394.0	Imatinib-d8	502.2 $[M + H]^+$	394.0	1	9
Cyclophosphamide	10,000–25,000	261.0 $[M + H]^+$	140.0	Ifosfamide	261.0 $[M + H]^+$	154.0	13	11
Paclitaxel	85–1000	876.5 $[M + Na]^+$	307.9	Docetaxel	830.3 $[M + Na]^+$	549.2	15	12
Irinotecan	100–10,000	587.7 $[M + H]^+$	124.5	Topotecan	422.2 $[M + H]^+$	377.0	5	13
Docetaxel	1000–4000	830.3 $[M + Na]^+$	549.2	Paclitaxel	876.5 $[M + Na]^+$	307.9	6	13
Topotecan	4–8	422.2 $[M + H]^+$	377.0	Irinotecan	587.7 $[M + H]^+$	124.5	13	17

Source: Taken with permission from Espy et al. (2012a).
[a]Therapeutic ranges for plasma concentrations.

are shown in Table 23.3. Acceptable LODs were obtained ranging from 0.5 ng/mL for pazopanib to 17 ng/mL for topotecan with the new method. These LODs are in the same range as those for the DBS tests obtained by paper spray, which indicates the applicability of paper spray with whole blood to POC TDM for clinical applications.

On-site detection methods are definitely necessary for POC analysis to realize immediate analysis after blood sampling. Handheld miniaturize mass spectrometers have been under development for several years (Ouyang and Cooks, 2009; Ouyang et al., 2009) and their capabilities have been demonstrated in POC analysis of DBS with paper spray ionization. The analysis of 4 ng atenolol in 0.4 μL blood on chromatography paper was achieved using a handheld miniature ion trap mass spectrometer as shown in Figure 23.14a. By using silica-coated paper, the performance of paper spray ionization with a Mini MS can be improved for the quantitative analysis (Zhang et al., 2012). As shown in Figure 23.14b, therapeutic drugs can be detected in blood with good LOQs about an order of magnitude better than that obtained by using chromatography paper for paper spray.

A proof-of-concept version of a paper spray MS system for POC analysis is shown in Figure 23.14c. This system can be used for TDM and many other applications. A discontinuous atmospheric pressure interface is used to allow a small pumping system to be used to transfer ions generated by paper spray into an ion trap mass analyzer for MS/MS analysis. An user-friendly interface is designed with simple instructions to lead the user to follow the analysis procedure and direct results are reported with the automated concentration calculation based on the calibrations saved in the database.

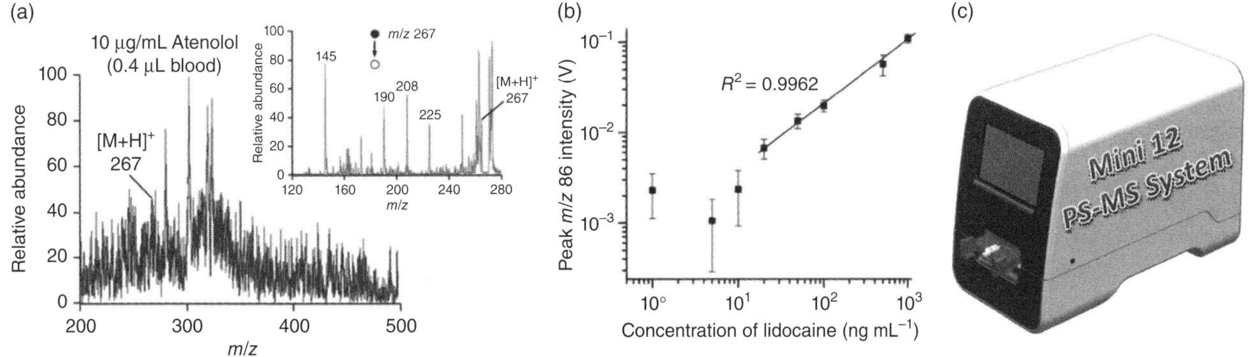

FIGURE 23.14 (a) Analysis of atenolol in blood (10 μg/mL in 0.4 μL blood) with paper spray ionization by using a handheld mass spectrometer (Mini 10). Spray voltage: 4.5 kV; spray solvent: methanol/water, 1:1, v/v, 10 μL. Inset: MS/MS spectrum of atenolol at *m/z* 267. (b) Linear dynamic range for lidocaine in blood spots with paper spray ionization by using a handheld mass spectrometer (Mini 11). Paper substrate: silica-coated paper; spray solvent: 9:1 dichloromethane/isopropanol. (c) Mini 12, a prototype miniature paper spray MS system for POC analysis of DBS. Adapted with permission from Wang et al. (2010), Zhang et al. (2012), and AMIPurdue. Copyright 2012 American Chemical Society.

23.7 OUTLOOK

Paper spray ionization is a very simple method to achieve ambient ionization of samples "as is." In the case of DBS the experiment is rapid and extremely sensitive. What is more surprising is that the data shows excellent reproducibility. Various methods of adding internal standards to the samples provide the most accurate quantitation results. The speed, specificity, and quantitative accuracy of paper spray ionization make direct POC analysis of whole blood a feasible experiment. The use of miniature mass spectrometers with paper spray should make the analysis attractive in cost and convenience.

ACKNOWLEDGMENT

This work was supported by National Science Foundation, National Institutes of Health, and the Alfred Mann Institute for Biomedical Development at Purdue University (AMIPurdue).

REFERENCES

Abu-Rabie P, Denniff P, Spooner N, Brynjolffssen J, Galluzzo P, Sanders G. Method of applying internal standard to dried matrix spot samples for use in quantitative bioanalysis. *Analytical Chemistry* 2011;83(22):8779–8786.

Barfield M, Spooner N, Lad R, Parry S, Fowles S. Application of dried blood spots combined with HPLC-MS/MS for the quantification of acetaminophen in toxicokinetic studies. *Journal of Chromatography B-Analytical Technologies in the Biomedical and Life Sciences* 2008;870(1):32–37.

Chace DH, Kalas TA, Naylor EW. Use of tandem mass spectrometry for multianalyte screening of dried blood specimens from newborns. *Clinical Chemistry* 2003;49(11):1797–1817.

Cody RB, Laramee JA, Durst HD. Versatile new ion source for the analysis of materials in open air under ambient conditions. *Analytical Chemistry* 2005;77(8):2297–2302.

Espy RD, Manicke NE, Ouyang Z, Cooks RG. Rapid analysis of whole blood by paper spray mass spectrometry for point-of-care therapeutic drug monitoring. *Analyst* 2012a;137:2344–2349.

Espy RD, Muliadi AR, Ouyang Z, Cooks RG. Spray mechanism in paperspray ionization. *International Journal of Mass Spectrometry* 2012b;325–327:167–171.

Kertesz V, Van Berkel GJ. Fully automated liquid extraction-based surface sampling and ionization using a chip-based robotic nanoelectrospray platform. *Journal of Mass Spectrometry* 2010;45(3):252–260.

Li WK, Tse FLS. Dried blood spot sampling in combination with LC-MS/MS for quantitative analysis of small molecules. *Biomedical Chromatography* 2010;24(1):49–65.

Liu J, Wang H, Manicke NE, Lin JM, Cooks RG, Ouyang Z. Development, characterization, and application of paper spray ionization. *Analytical Chemistry* 2010;82(6):2463–2471.

Manicke NE, Abu-Rabie P, Spooner N, Ouyang Z, Cooks RG. Quantitative analysis of therapeutic drugs in dried blood spot samples by paper spray mass spectrometry: an avenue to therapeutic drug monitoring. *Journal of the American Society for Mass Spectrometry* 2011a;22(9):1501–1507.

Manicke NE, Yang QA, Wang H, Oradu S, Ouyang Z, Cooks RG. Assessment of paper spray ionization for quantitation of pharmaceuticals in blood spots. *International Journal of Mass Spectrometry* 2011b;300(2–3):123–129.

Nemes P, Vertes A. Laser ablation electrospray ionization for atmospheric pressure, in vivo, and imaging mass spectrometry. *Analytical Chemistry* 2007;79(21):8098–8106.

Ouyang Z, Cooks RG. Miniature mass spectrometers. *Annual Review of Analytical Chemistry* 2009;2:187–214.

Ouyang Z, Noll RJ, Cooks RG. Handheld miniature ion trap mass spectrometers. *Analytical Chemistry* 2009;81(7):2421–2425.

Patton JC, Akkers E, Coovadia AH, Meyers TM, Stevens WS, Sherman GG. Evaluation of dried whole blood spots obtained by heel or finger stick as an alternative to venous blood for diagnosis of human immunodeficiency virus type 1 infection in vertically exposed infants in the routine diagnostic laboratory. *Clinical and Vaccine Immunology* 2007;14(2):201–203.

Shrestha B, Nemes P, Nazarian J, Hathout Y, Hoffman EP, Vertes A. Direct analysis of lipids and small metabolites in mouse brain tissue by AP IR-MALDI and reactive LAESI mass spectrometry. *Analyst* 2010;135(4):751–758.

Spooner N, Lad R, Barfield M. Dried blood spots as a sample collection technique for the determination of pharmacokinetics in clinical studies: considerations for the validation of a quantitative bioanalytical method. *Analytical Chemistry* 2009;81(4):1557–1563.

Takats Z, Wiseman JM, Gologan B, Cooks RG. Mass spectrometry sampling under ambient conditions with desorption electrospray ionization. *Science* 2004;306(5695):471–473.

Wang H, Liu JJ, Cooks RG, Ouyang Z. Paper spray for direct analysis of complex mixtures using mass spectrometry. *Angewandte Chemie International Edition* 2010; 49(5):877–880.

Wiseman JM, Ifa DR, Zhu Y, Kissinger CB, Manicke NE, Kissinger PT, Cooks RG. Desorption electrospray ionization mass spectrometry: imaging drugs and metabolites in tissues. *Proceedings of the National Academy of Sciences of the United States of America* 2008;105(47):18120–18125.

Wiseman JM, Evans CA, Bowen CL, Kennedy JH. Direct analysis of dried blood spots utilizing desorption electrospray ionization (DESI) mass spectrometry. *Analyst* 2010;135(4):720–725.

Yang Q, Manicke NE, Wang H, Petucci C, Cooks RG, Ouyang Z. Direct and quantitative analysis of underivatized acylcarnitines in serum and whole blood using paper spray mass spectrometry. *Analytical and Bioanalytical Chemistry* 2012a;404(5):1389–1397.

Yang Q, Wang H, Mass JD, Chappell WJ, Manicke NE, Cooks RG, Ouyang Z. Paper spray ionization devices for direct, biomedical analysis using mass spectrometry. *International Journal of Mass Spectrometry* 2012b;312:201–207.

Zhang Z, Xu W, Manicke NE, Cooks RG, Ouyang Z. Silica coated paper substrate for paper-spray analysis of therapeutic drugs in dried blood spots. *Analytical Chemistry* 2012;84:931–938.

Zhou M, McDonald JF, Fernández FM. Optimization of a direct analysis in real time/time-of-flight mass spectrometry method for rapid serum metabolomic fingerprinting. *Journal of the American Society for Mass Spectrometry* 2010;21(1):68–75.

24

DIRECT SOLVENT EXTRACTION AND ANALYSIS OF BIOMARKERS IN DRIED BLOOD SPOTS USING A FLOW-THROUGH AUTOSAMPLER

David S. Millington, Haoyue Zhang, M. Arthur Moseley, J. Will Thompson, and Peter Smith

24.1 INTRODUCTION

Dried blood spots (DBS) have been used for the past 50 years to collect specimens from newborns to screen for inherited metabolic diseases such as phenylketonuria (PKU) (Guthrie and Susi, 1963). Blood samples are collected at the point of birth by heel-stick, absorbed onto special filter paper, allowed to dry, and then mailed to a public health laboratory for analysis. The objective of this newborn screening (NBS) system is to identify conditions that pose a serious health risk if not recognized and diagnosed in the early neonatal period and initiate appropriate treatment. A method for the analysis of acylcarnitines in DBS using fast-atom bombardment ionization coupled to tandem mass spectrometry (MS/MS) was reported in 1990 that was later expanded to include analysis of several essential amino acids (Millington et al., 1990, 1991). This procedure was adapted for high-throughput screening by using 96-well plate technology with an autosampler and electrospray ionization tandem mass spectrometry (ESI-MS/MS) (Rashed et al., 1997). This method can be used to identify more than 30 metabolic disorders and has been adopted by NBS laboratories worldwide (Zytkovicz et al., 2001; Schulze et al., 2003; Frazier et al., 2006). The value of NBS programs has been significantly enhanced by the addition of MS/MS, and this subject is described in detail in Chapter 6 of this book. Acylcarnitines are now part of the standard diagnostic test panel in clinical biochemical genetics laboratories (Rinaldo et al., 2008; Millington, 2011) and targeted metabolomics platforms also generally include the analysis of acylcarnitines and amino acids (Kao et al., 2006; Newgard et al., 2009). The value of DBS as a noninvasive and convenient method of collecting, shipping, and storing blood samples for the analysis of drugs and other analytes by mass spectrometry is increasingly evident (Li and Tse, 2010; Keevil, 2011; Liu et al., 2011).

In the NBS or clinical diagnostic environment, the target analytes are solvent-extracted from a disc of 3–8 mm diameter punched out of the DBS and stable isotope-labeled internal standards (ISs) are added to facilitate their quantification by MS/MS. The extracts are usually dried, derivatized as either butyl or methyl esters, dried again, and reconstituted in the final matrix for analysis by ESI-MS/MS. Some NBS laboratories use a non-derivatized method (Nagy et al., 2003; Van Den Bulcke et al., 2011). Clinical laboratories analyze free carnitine and acylcarnitines by adding ISs to the plasma before deproteinization, extraction, and derivatization (Stevens et al., 2000; Rinaldo et al., 2008; Millington and Stevens, 2011). The sampling of analytes directly from DBS cards without punching and extraction has intuitive advantages primarily because of decreased sample handling, leading to lower cost, higher throughput, and potentially fewer laboratory errors. Several novel methods have recently been described for direct analysis of drugs and metabolites from blood spots, including desorption ionization (DESI) and paper spray (Wiseman et al., 2010; Manicke et al., 2011; Zhang et al., 2012). Prototype clamping devices that use flow-through solvent extraction coupled to LC-MS and LC-MS/MS for quantification of several therapeutic drug

Dried Blood Spots: Applications and Techniques, First Edition. Edited by Wenkui Li and Mike S. Lee.

molecules and metabolites have been reported (Deglon et al., 2009, 2011; Thomas et al., 2010; Miller et al., 2011). More recently, an article has been published that used the same DBS autosampler as the one used in the research described in this article to analyze the drug bosentan and its metabolites in DBS (Ganz et al., 2012). Here we review and describe in detail this new flow-through DBS autosampler and report several applications that demonstrate a range of analytical options with such a system.

24.2 MATERIALS AND METHODS

24.2.1 DBS Autosampler and MS/MS Instrument Configuration

The DBS system described here is designed such that it can be utilized for either flow-injection analysis or flow-through extraction with subsequent chromatographic separation online to the MS/MS detector. The introduction of ISs via a loop injector and the dilution and trapping of analytes on cartridges prior to analysis are additional options. The principle of operation is to pick up a single card of the drug metabolism/pharmacokinetic (DMPK) type, such as the Whatman™ FTA™-DMPK (GE Healthcare, Piscataway, NJ, USA), and hold it securely in a pneumatic clamp with the blood or plasma spot of interest directly in the path of a flowing stream of extraction solvent. The extracted analytes are carried directly to the mass spectrometer or in-line chromatographic column. The autosampler used in this study was an HTS PAL adapted with the SCAP DBS™ system for handling DBS cards, manufactured by Prolab GmbH (Reinach, Switzerland) and marketed by LEAP Technologies, Inc. (Carrboro, NC, USA). The autosampler consists of a rack for holding the sample cards, an arm for holding and moving cards from the rack to the clamp, a pneumatic clamping system for extracting the blood spots from the cards, two multiport valve systems for control of the extraction solvent from the gradient pump through the cards to the MS, and a syringe pump and sample loop for accurate introduction of IS. The device includes a programmable microcomputer to control the timing and sequence of events within a cycle and is recognized as a peripheral device by several leading mass spectrometry companies, including Waters Corporation (Milford, MA, USA).

In our applications, the LEAP-DBS autosampler was coupled to a Xevo TQ Mass Spectrometer (Waters) in a flow-injection configuration as shown in Figure 24.1. The mobile phase was typically provided at a flow rate of 50 μL/min by a standard Acquity ultra-performance liquid chromatography (UPLC) pump (Waters). All connections in this configuration were made using 0.005″ PEEK tubing and finger-tight fittings. During standby and manipulation of the DBS card into position for extraction, the flow continues directly from

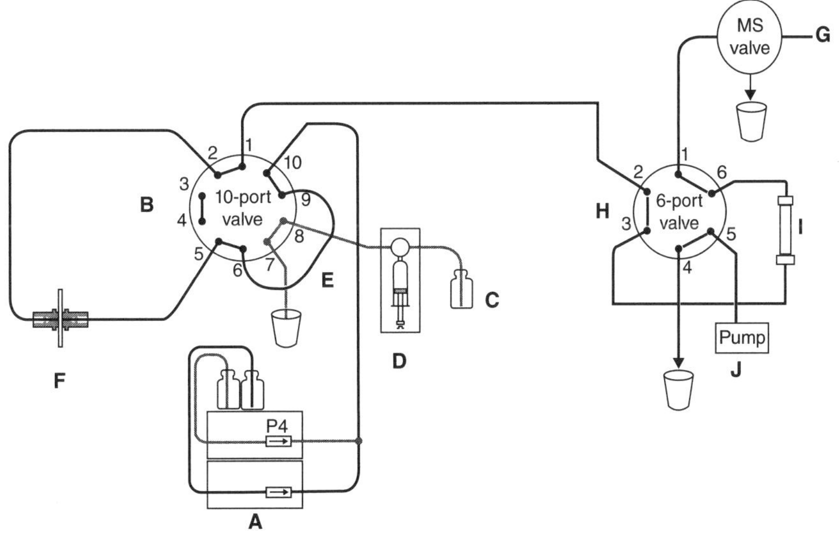

FIGURE 24.1 Configuration of the analytical system incorporating the DBS autosampler. A UPLC pump (A) delivers solvent via a 10-port valve (B) to a pneumatic clamp (F) into which a DBS card is inserted for flow-through analysis. Internal standards are carried through the DBS sample with the solvent from a loop (E) filled prior to analysis from a vial (C) using the syringe pump (D). The extracted analytes and their internal standards flow onto the mass spectrometer (G) via a 6-port valve (H). For flow-injection analysis, a stainless steel frit (I) filters particulates to prevent clogging of the electrospray inlet to the mass spectrometer. A UPLC column replaces the frit (I) for chromatographic separation. A second pump (J) is used to back-flush the frit or regenerate the column (I) between sample injections.

the pump (Figure 24.1, A) through the 10-port valve (B) via a 6-port valve (H), into the mass spectrometer (G). Prior to sample analysis, the injection syringe (D) was used to draw the IS solution from the vial (C) into the sample loop (E). The PAL arm then placed the sample card of interest in the clamp device (F) and simultaneously the clamp was activated (closed) and the 10-port valve switched to place the loop and DBS clamp device in line with the mass spectrometer, as shown in Figure 24.1. The device can accommodate clamps of various diameters. In the work described here, a 3 mm clamp was used exclusively. MS/MS acquisition was automatically initiated via 5 V contact closure detected by MassLynx v4.1 software. At the flow rate of 50 μL/min, a flow-through time of 120 seconds was adequate to extract most of the analytes from the clamped area of the DBS. To prevent contamination of the MS source and inlet with fibers from the DBS cards, a 2 μm in-line stainless frit (VICI, Houston, TX, USA) was utilized (I). To prevent fibers from accumulating and subsequently clogging the frit and to minimize sample carryover, the 6-port valve was switched between injections to enable a second HPLC pump (J) to back-flush the stainless steel frit.

In an alternative configuration, an HPLC or UPLC column was included (Figure 24.1, I) to permit the separation of eluted analytes. This configuration enabled a solvent gradient and, after the elution time, facilitated a washout cycle during solvent regeneration by closing the clamp without a DBS card in place and flushing the line to minimize carryover. For these studies, the flow rate was 400 μL/min and the extraction time was 40 seconds.

24.2.2 Direct Analysis of Acylcarnitines and Amino Acids from Dried Blood Spots

24.2.2.1 Preparation of Dried Blood Spots We have previously reported preliminary results using the PAL DBS autosampler to extract and analyze acylcarnitines from DBS as well as free and total carnitine in dried plasma spots (Thompson et al., 2012). In this work, a set of quality controls were analyzed that consisted of base, low, intermediate, and high pool samples (code numbers 122, 123, 124, and 125, respectively) and a set of proficiency test samples (code number 60) which were obtained from the Centers for Disease Control and Prevention (CDC, Atlanta, GA, USA). These DBS samples were provided as pre-spotted filter paper cards and it was necessary to mount them into blank DBS card holders (provided by LEAP Technologies) for analysis. For this particular analysis, kits that contain premixed isotope-labeled ISs for acylcarnitines and amino acids (NSK-B-1 and NSK-A-1 from Cambridge Isotope Laboratories, Andover, MA, USA) were also kindly supplied by CDC (courtesy of Dr Victor DeJesus, Atlanta, GA). The mixture of ISs was dissolved in 90:9.9:0.1 (v:v:v) MeOH:H_2O:HCOOH and placed in the IS vial (Figure 24.1, C) after dilution such that the 20 μL loop (Figure 24.1, E) should deliver approximately the same amounts of ISs to the extraction solvent for each analysis as used by the standard punch-and-extract method for a 3 mm diameter DBS punch.

24.2.2.2 Mass Spectrometry The solvent used for the flow-through analysis of DBS for acylcarnitines and amino acids was 90:9.9:0.1 (v:v:v) MeOH:H_2O:HCOOH. The mass spectrometer was operated in positive ion ESI mode with a capillary voltage of 3.0 kV. The source block and desolvation temperatures were 150°C and 350°C, respectively. The acylcarnitines and amino acids were quantified using three interlaced scan functions as follows: precursors of *m/z* 85 scan over the mass range *m/z* 150–500 for 1 second for acylcarnitines (cone voltage 20 V, collision energy 20 eV); neutral losses of 63 Da scan over the mass range *m/z* 110–200 for 0.4 seconds for basic amino acids (cone voltage 16 V, collision energy 15 eV); neutral losses of 46 Da scan over the mass range *m/z* 60–240 for 0.55 seconds for neutral amino acids (cone voltage16 V, collision energy 15 eV). The spectra acquired during the elution time of 120 seconds were combined into a single spectrum. After smoothing and baseline adjustment, the signal ratios for the analytes and their corresponding ISs were determined and reported using NeoLynx 4.1 software (Waters).

24.2.3 Analysis of Citrulline in DBS

The interest in a method for analysis of citrulline in DBS samples arises from the concept that this amino acid, which is produced almost exclusively in the gut, is a potential marker for intestinal damage resulting, for example, from exposure to ionizing radiation or chemotherapy (Lutgens et al., 2003; Blijlevens et al., 2004; Lutgens and Lambin, 2007). In certain field settings it may be practicable to collect DBS samples from individuals at risk and deliver them by courier to a laboratory equipped with a fast turnaround assay for this analyte. A modification of the amino acid analysis procedure described in Section 24.2.2 was employed to analyze citrulline in DBS. The objective of this research was to compare the DBS assay with a standard method for amino acid analysis in plasma from the same specimens.

24.2.3.1 Preparation of Calibrators and Samples A set of four calibrators was prepared from a sample of normal whole blood by standard addition of citrulline in concentrations of 0, 25, 50, and 100 μmol/L, respectively. Aliquots (35 μL) of each calibrator were spotted in quadruplicate onto DMPK cards (Perkin Elmer 226, supplied by ID Biologicals, Greenville, SC, USA) and allowed to dry at ambient temperature for at least 4 hours. A solution of 2,3,3,4,4,5,5-$^2H_7$-labeled citrulline (Sigma Chemical Co., St Louis, MO, USA) was prepared in 90:9.9:0.1 (v:v:v) MeOH:H_2O:formic acid (7.5 μmol/L) and placed in the IS vial (Figure 24.1, C).

The loop injector (Figure 24.1, E) provides approximately 0.15 nmol of this IS to each calibrator and sample during the flow-through analysis of DBS samples. Remnant, de-identified blood samples from human subjects were spotted in triplicate onto DMPK cards and allowed to dry. The use of such samples for research was approved by the Duke IRB. Some of these specimens were also analyzed by the standard clinical assay method in our laboratory that uses an amino acid analyzer (Model L-8800, Hitachi High Technologies America, Inc., Dallas, TX, USA).

*24.2.3.2 **Mass Spectrometry*** The solvent used for the flow-through analysis of citrulline and the mass spectrometer settings were essentially the same as those used for the analysis of amino acids described in Section 24.2.2.2. Data were acquired using a neutral loss of 63 Da scan function. The scan rate was 215 Da/s over the mass range from *m/z* 170 to 185, and the spectra acquired during the elution time of citrulline (120 seconds) were summed into a single spectrum. After smoothing and baseline adjustment, the signal ratio for citrulline (*m/z* 176) and the IS (*m/z* 183) was determined and reported using NeoLynx 4.1 software (Waters).

24.2.4 Analysis of Methylmalonic Acid in Plasma Spots

Methylmalonic acid (MMA) is of interest as a biomarker for two principal patient groups. One group is represented by inherited disorders of branched-chain amino acid catabolism at the level of the enzyme methylmalonyl-coenzyme A mutase or of the biosynthesis of its cofactor, vitamin B-12 (Matsui et al., 1983). Another group is represented by those with complications of vitamin B-12 deficiency caused or acquired by other factors, such as malnutrition or malabsorption (Oh and Brown, 2003). In either of these groups, the serum or blood MMA level is the most sensitive biomarker. Potential uses of the DBS autoanalyzer in this field include measurement of MMA in DBS as a second-tier test in NBS for samples with elevated C3-acylcarnitine (La Marca et al., 2007) and for the analysis of plasma or serum to assess vitamin B deficiency, as well as to monitor patients being treated for MMA due to an inherited metabolic disorder.

*24.2.4.1 **Preparation of Calibrators and Samples*** Calibrators were prepared by standard addition of 5, 10, 50, 100, 200, 500, and 1000 μmol/L of MMA (Sigma) to aliquots of a pooled human plasma sample. Quality control samples were leftover samples from the ERNDIM QAP program (www.erndimqa.nl/). To each sample aliquot (100 μL) was added a 5 μL aliquot of the IS solution, prepared by dissolution of $^2H_3$-MMA (Sigma) in MeOH:H_2O 80:20 (v:v). Aliquots (18 μL) of each calibrator and sample were spotted in quadruplicate onto the filter paper cards (Perkin Elmer 226, supplied by ID Biologicals). It should be noted that the volume of plasma required to fill the circles on these cards is approximately half that of the whole blood.

*24.2.4.2 **Liquid Chromatography and Mass Spectrometry*** For this project, an Acquity UPLC® system (Waters) was coupled with the LEAP-DBS autosampler. An autosampler method file was programmed into the PAL microprocessor controller that consisted of three phases: (i) sample extraction, (ii) data acquisition/ liquid chromatography, and (iii) back-flushing/column regeneration. A method file was also set up in the Waters MassLynx software to manage the solvent flow and acquisition of data from the mass spectrometer. This method file was coordinated with the PAL program to ensure timing consistency, and all the samples to be analyzed (calibrators, blanks, samples, quality controls) were incorporated into a single batch file. The entire batch process was initiated from the PAL and ran successfully without operator intervention. During phase (i), the DPS card was picked up from the tray and inserted into the clamp, which was then closed. The mass spectrometer scan function was initiated and the sample was exposed to the flowing mobile phase for 40 seconds, then the clamp was opened and the card moved back to the sample tray. In phase (ii), the 10-port valve was switched to bypass the clamp and allow the buffer to continue to flow through the column. The analytes were separated by liquid chromatography and data acquired by a Xevo-TQ™ mass spectrometer. During phase (iii), the clamp was closed again with no card in place and the 10-port valve switched to allow buffer to flow through the clamp. The flow was diverted to waste, not into the mass spectrometer. At the same time, the 6-port valve was switched, to back-flush and regenerate the column using the same starting buffer as used for the chromatography. An Acquity UPLC® BEH hydrophilic interaction column (HILIC) 1.7 μm 2.1 × 100 mm (Waters) was used for this assay under a flow rate of 400 μL/min with a programmed linear gradient from 15 mM NH_4Ac in MeCN:H_2O 86:14 (v:v) to15 mM NH_4Ac in MeCN:H_2O 78:22 (v:v) over 4.5 minutes. The mass spectrometer was employed in negative ion ESI mode with a capillary voltage of 2.9 kV and cone voltage of 16 V. The ion source and desolvation temperatures were 150°C and 350°C, respectively. The collision energy was 9 eV. Data were acquired by selected reaction monitoring (SRM) using the transition *m/z* 117 → 73 for MMA and the transition *m/z* 120 → 76 for the IS.

24.3 RESULTS AND DISCUSSION

24.3.1 Acylcarnitines and Amino Acids

We have reported preliminary results from the analysis of DBS samples for free carnitine in plasma and acylcarnitines in DBS using the LEAP-DBS autosampler elsewhere

(Thompson et al., 2012). We were able to demonstrate that the recovery of acylcarnitines from the DBS was generally similar (approximately 60–80%) to that observed using the standard punch-and-extract with derivatization methods, and that the reproducibility of the assay for the most abundant acylcarnitines in normal blood was within ±15%, which is also comparable with standard punch-and-extract methods. In this study, we analyzed acylcarnitines and amino acids in QA and PT samples provided by the CDC Newborn Screening Quality Assurance Program (courtesy of Dr Victor DeJesus). With isotope-labeled ISs added from the loop injector, a set of QA samples were analyzed in blinded fashion in a single batch and the results were compared with those provided by CDC. Examples of the raw data obtained from these analyses are shown in Figure 24.2. The signals that correspond to free carnitine and acylcarnitines from C2 to C18, plus seven ISs, marked by the * symbol, are shown from the analysis of a CDC QA sample (LP Lot #122). The spectrum was generated by accumulating all the individual parents of *m/z* 85 scans over the mass range from 150 to 500 during the elution of the DBS sample (2 minutes), after smoothing and baseline adjustment. The signal-to-noise ratios (*S/N*) for the target analytes and their ISs are excellent. The spectra for the analysis of neutral amino acids, generated by a neutral loss of 46 Da, and of basic amino acids, generated by a neutral loss of 63 Da scan function are shown in Figure 24.3. Again, the signals for the target analytes and their respective ISs, marked by the * symbol, are depicted.

Note that several large signals are observed in Figure 24.3 that are not derived from amino acids, and that there are marked differences in sensitivity between the different species. These differences are compensated for by the isotope-labeled ISs. The relative abundances of each pair of analyte : IS signals were determined and converted to approximate concentrations based on the known concentrations of the added IS and on the assumption of linear response (that a 1:1 molar ratio generates a 1:1 signal ratio). Each sample was analyzed in duplicate and the values obtained from each of the four QA lots (base, low, intermediate, and high) are compared with the CDC expected values, provided as ranges within the 95% confidence limits, in Tables 24.1 and 24.2. Overall, the reproducibility for the replicate analyses was within ±9% for the acylcarnitines and ±11% for the amino acids. The accuracy of the method was difficult to assess due to assumptions made regarding the IS concentrations; however, values for acylcarnitines nearly always fell within the 95% confidence interval from CDC. The measured values for acylcarnitines seemed to decrease toward the lower end of the confidence interval as the chain length increased. These results suggest a reduced extraction efficiency for the more lipophilic species. The values for most of the amino acids were at least 40% higher than the CDC certified

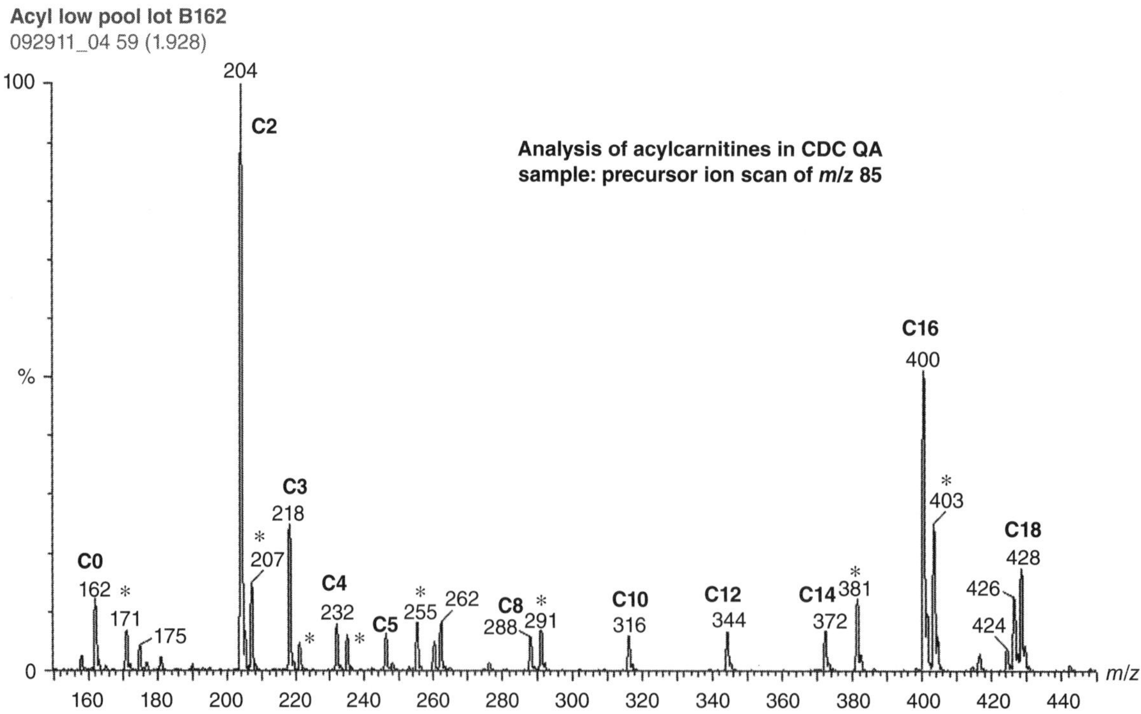

FIGURE 24.2 Mass spectrum of free carnitine (C0) and acylcarnitines (C2–C18) plus eight internal standards (marked *) generated by a precursor of *m/z* 85 scan function. The spectrum was acquired during flow-injection MS/MS analysis of a CDC QA sample.

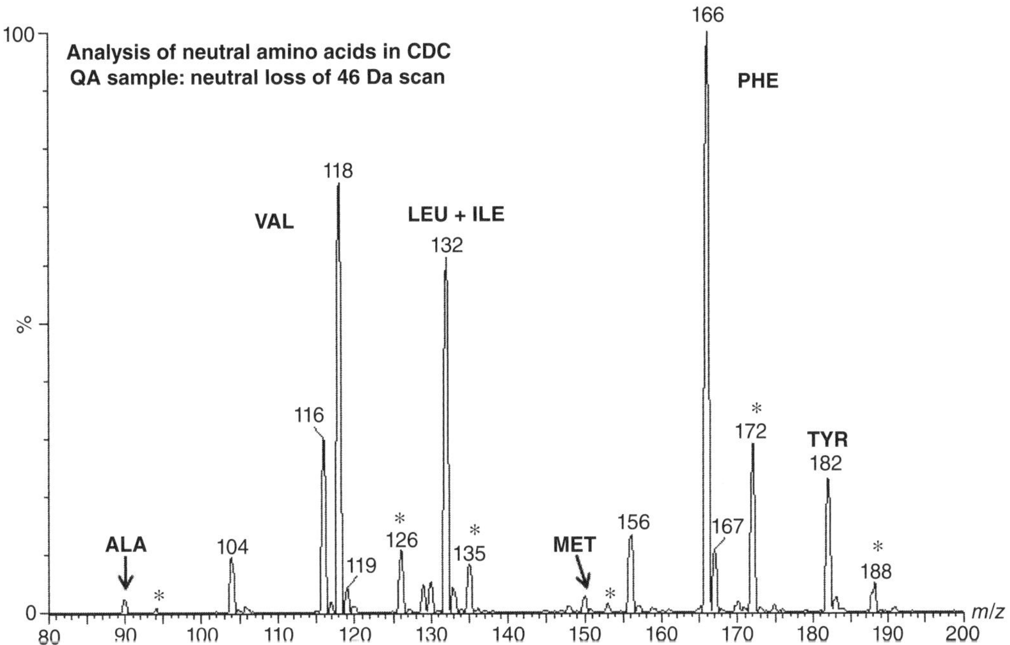

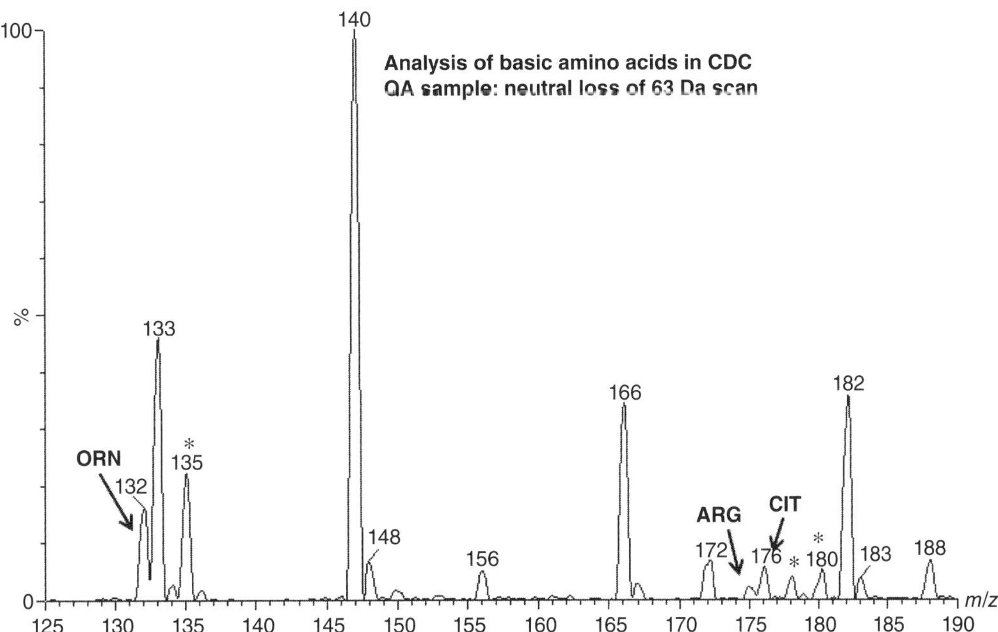

FIGURE 24.3 A neutral loss of 43 Da scan detects neutral amino acids including valine and phenylalanine, and a neutral loss of 63 Da scan detects basic amino acids including arginine and citrulline. These spectra were acquired during flow-injection MS/MS analysis of a CDC QA sample. The isotope-labeled internal standards are marked with the symbol*.

values, most likely because the required concentrations of ISs were underestimated. In practice, such discrepancies would be compensated for by using the CDC QC samples as calibrators, thereby generating more accurate values based on the slopes of the standard curves for each analyte. Even without such enhancements, the method as it stands is remarkably reproducible and could be used in a high-throughput setting to analyze samples of blood and/or plasma for multiple analytes without the need for conventional extraction and derivatization methods. In addition to the CDC QA samples,

TABLE 24.1 Duplicate Analysis of Acylcarnitines in CDC QA Samples Using DBS Autosampler Compared with CDC 95% Confidence Limits for Each Analyte (μmol/L)

Analyte	C2		C3		C4		C5		C5OH	
CDC Lot	BGL	CDC	BGL	CDC	BGL	CDC	BGL	CDC	BGL	CDC
BP (#121)	12.6	10.5–17.5	1.21	0.71–1.31	0.12	0.07–0.16	0.07	0.04–0.10	0.54	0.34–0.65
BP (#121)	12.5		1.22		0.18		0.12		0.56	
LP (#122)	28.5	19.6–25.1	4.83	3.27–4.65	1.24	0.66–0.94	0.71	0.38–0.56	1.03	0.73–1.05
LP (#122)	26.4		4.32		1.27		0.68		0.89	
IP (#123)	41.7	27.8–35.3	10.3	6.69–9.66	2.21	1.46–2.29	1.72	1.07–1.46	1.69	1.43–1.97
IP (#123)	40.6		10.6		3.12		1.92		1.89	
HP (#124)	43.9	32.8–43.5	13.8	8.88–13.8	4.32	2.79–3.92	3.48	2.20–2.68	2.55	1.91–2.85
HP (#124)	45.9		15.0		4.96		3.13		2.23	

Analyte	C8		C12		C16		C16OH		C18	
CDC Lot	BGL	CDC	BGL	CDC	BGL	CDC	BGL	CDC	BGL	CDC
BP (#121)	0.03	0.00–0.05	0.01	0.00–0.07	1.02	0.74–1.24	0.00	0.00–0.03	0.66	0.63–0.90
BP (#121)	0.05		0.02		1.00		0.01		0.70	
LP (#122)	0.68	0.40–0.58	0.57	0.33–0.70	4.33	3.51–4.61	0.24	0.27–0.44	1.51	1.34–1.92
LP (#122)	0.71		0.61		4.42		0.25		1.50	
IP (#123)	1.05	0.80–1.14	1.12	0.60–1.33	6.53	6.06–8.86	0.43	0.53–0.87	1.70	2.18–2.82
IP (#123)	1.16		1.04		7.30		0.40		2.00	
HP (#124)	2.47	1.73–2.97	2.27	1.52–2.89	9.15	8.39–11.7	1.02	1.38–1.86	3.68	4.14–5.28
HP (#124)	2.66		2.46		9.63		0.95		3.73	

a set of CDC proficiency test samples were also analyzed in blinded fashion and all expected out-of-range values were correctly identified (data not shown).

Several NBS laboratories are already using a non-derivatized method for screening newborns. However, in order to take full advantage of this type of DBS analyzer, the samples would have to be collected on compatible sample cards at the point of birth. By having the demographic information bar-coded on the collection card, and an optional camera mounted on the DBS autosampler, the patient's unique bar code could be embedded into the analysis report to minimize sampling and reporting errors.

24.3.2 Citrulline

A feasibility study was performed, to assess the reproducibility and accuracy using a set of calibrators and patient samples. The principle of the assay was similar to that of the basic amino acid analysis of DBS described in the previous section. In this case, the IS was $^2H_7$-labeled citrulline, made up in the eluting solvent and used to charge the loop injector with the IS solution (20 μL) for each DBS sample. A sample analysis is shown in Figure 24.4, where the signals for citrulline, arginine, and the citrulline ISs are marked on the spectrum. All blood samples analyzed by this method

TABLE 24.2 Duplicate Analysis of Amino Acids in CDC QA Samples Using DBS Autosampler Compared with CDC 95% Confidence Limits for Each Analyte (μmol/L)

Analyte	VAL		LEU/ILE		MET		PHE		TYR		CIT	
CDC Lot	CDC	BGL	CDC	BGL	CDC	BGL	CDC	BGL	CDC	BGL	CDC	BGL
BP (#121)	92–165	191	99–159	253	12–22	24	40–65	93	32–47	75	12–19	38
BP (#121)		195		216		30		84		69		53
LP (#122)	249–348	452	248–327	490	72–100	137	136–188	229	162–237	302	34–50	84
LP (#122)		457		404		115		254		346		76
IP (#123)	460–601	674	430–598	835	160–223	191	293–392	446	305–427	607	75–114	218
IP (#123)		670		750		203		472		614		199
HP (#124)	648–979	1088	623–898	1022	301–427	485	400–574	661	428–658	773	129–217	293
HP (#124)		1161		1087		380		706		937		263

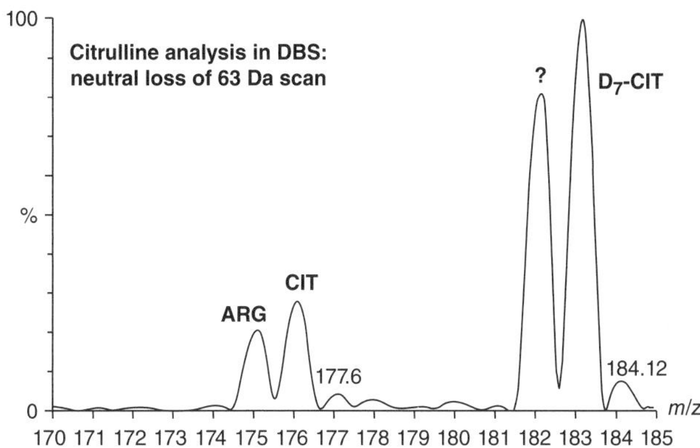

FIGURE 24.4 The analysis of citrulline in a dried blood spot specimen is accomplished by a neutral loss scan of 63 Da scan function. In this case, the internal standard is $^2H_7$-labeled citrulline. The example shown is from one of the calibrators used to generate a calibration curve.

showed similar concentrations of arginine and an unidentified signal at *m/z* 182. The theoretical contribution of isotope signals from these analytes to those of citrulline and its IS can readily be calculated by the software and accounted for if necessary.

We reasoned that such additional steps would not normally be necessary to generate a value within the experimental error of the method. The calibration curve for this assay is shown in Figure 24.5, which depicts the linear regression of the analyte:IS signal ratio versus added analyte concentration, performed for each of the calibrators in quadruplicate. The linearity is excellent, and the endogenous concentration of citrulline in the base blood pool was calculated from the intercept to be 32 μmol/L, which was the same value reported by the amino acid analyzer. A small set of blood samples that included four samples from normal adults and one sample

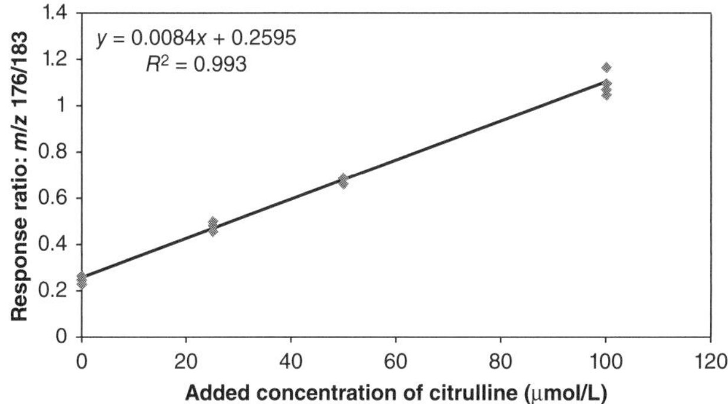

FIGURE 24.5 A calibration curve for citrulline in DBS. The plot shown is a linear regression of the signal ratio for citrulline to its internal standard (measured in quadruplicate) versus the concentration of citrulline added to the blood sample before spotting. This curve was used to quantify citrulline in DBS from patients and quality controls.

TABLE 24.3 Results of Citrulline Assays in DBS Samples

Analyte	Citrulline		
Method	DBS		AAA
Concn	μmol/L	(mean)	μmol/L
Subject A #1	30.7	30.1	
Subject A #2	29.4		
Subject B1 #1	33.0	34.2	
Subject B1 #2	35.4		
Subject B2 #1	30.6	30.8	
Subject B2 #2	31.0		
Control 1 #1	32.9	30.9	30.0
Control 1 #2	28.0		
Control 1 #3	31.7		
Subject C #1	18.3	18.9	16.7
Subject C #2	19.5		

from a patient known to have a rare metabolic disorder resulting in a reduced citrulline level was analyzed in duplicate for citrulline and the results are reported in Table 24.3. According to the duplicate values, the reproducibility of the assay is better than 10% and the accuracy is within 15% based on the results available from the amino acid analyzer for comparison (Table 24.3). Subsequent to this feasibility study, we have initiated a study to measure citrulline levels in mice exposed to radiation and compare them with a reference method. The results of this study will be published elsewhere.

24.3.3 Methylmalonic Acid

We describe here the preliminary results of a novel plasma assay for MMA using the DBS autosampler in conjunction with UPLC-MS/MS. The principle of this assay is to first add a fixed amount of the IS $^2H_3$-MMA to a fixed volume of each plasma sample, then spot the samples onto a DMPK card. In this case, it is necessary to introduce a chromatographic separation of MMA from its normally much more abundant biological isomer succinic acid. This was accomplished by placing a HILIC UPLC column between the DBS autosampler and the MS/MS. A linear gradient was used to effect the separation and analysis of MMA without any additional cleanup or derivatization steps such as those used in other published methods (Magera et al., 2000). There was some loss of column resolution when compared with the analysis of protein-precipitated plasma samples into the UPLC-MS/MS system (data not shown). However, the resolution was still adequate to resolve the MMA signal from that of its naturally more abundant biological isomer succinic acid, as shown in Figure 24.6. This shows an SRM chromatogram from one of the calibrators, from which it is evident that the calculation of the ratio of signals for MMA to its IS is straightforward. The total analysis time was

TABLE 24.4 Accuracy and Precision Data for DBS-UPLC-MS/MS Assay for MMA Based on the Replicate Analysis of Calibrators (n = 4)

Sample ID	Std Conc. (μmol/L)	Measured Conc. (μmol/L)	Measured Conc. (μmol/L)	Measured Conc. (μmol/L)	Measured Conc. (μmol/L)	Mean (μmol/L)	%CV	%Dev
Cal 1	5	3.9	4.7	4.8	4.9	4.6	9.9	–8.3
Cal 2	10	9	10.3	10.6	10.2	10	7.0	0.4
Cal 3	50	46.6	49.3	50.7	48.4	48.7	3.5	–2.5
Cal 4	100	91.3	100.7	103.8	99.2	98.7	5.4	–1.3
Cal 5	200	176.4	204.5	205.6	195	195.4	6.9	–2.3
Cal 6	500	563.1	567.8	565.9	567.6	566.1	0.4	13.2
Cal 7	1000	974	973.4	979.9	987.1	978.6	0.7	–2.1

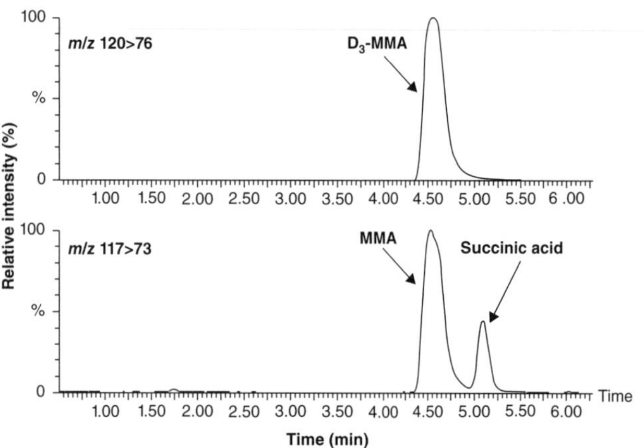

FIGURE 24.6 A chromatogram showing the analysis of methylmalonic acid (MMA) and its isotope-labeled internal standard in one of the calibrators used to generate a calibration curve. The signal from MMA is resolved from that of its biological isomer, succinic acid, by means of a UPLC column connected between the DBS autosampler and the tandem mass spectrometer.

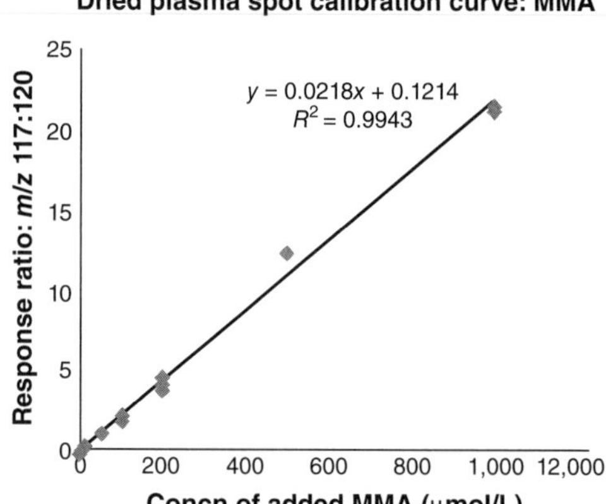

FIGURE 24.7 A calibration curve for MMA in DPS. The plot shown is a linear regression of the signal ratio for MMA to its internal standard versus the concentration of MMA added to the plasma before spotting and analysis. The curve was used to quantify MMA in patient and quality control samples.

7 minutes per cycle, which includes the time to regenerate the column and wash the injection system to eliminate carryover (see the experimental section 24.2.4.2). The calibration curve for MMA, generated by analysis of the calibrators in quadruplicate, is shown in Figure 24.7. The assay is linear from 10 to 1000 μmol/L. Reproducibility and accuracy, based on the replicate analysis of the calibrators, are excellent, as shown in Table 24.4. In Table 24.5, we show the results of analysis of some quality control samples from previous studies. Based on these results, we have full confidence in developing this assay as well as other plasma biomarker assays for application in a clinical laboratory setting.

The advantage of using DPS for such assays is that it avoids using the additional steps of protein precipitation, evaporation, and filtration that are otherwise necessary. A further application would be to analyze DBS samples for MMA as a second-tier newborn test for those specimens that screen positive for the propionylcarnitine (C3-carnitine) biomarker. The ability to rapidly test for MMA in such specimens would facilitate a distinction between a presumptive diagnosis of propionic or methylmalonic acidemia (La Marca et al., 2007).

TABLE 24.5 Comparison of Results for DBS-UPLC-MS/MS with Mean Values from All Laboratories Contributing Data for MMA Assay of Quality Assurance Samples (ERNDIM)

UPLC-MS/MS DPS (μmol/L)	Target Value (μmol/L)	Difference (%)
815	665	23
1186	1062	12
758	677	12
371	341	9
1125	1007	12
775	692	12

24.4 SUMMARY AND CONCLUSIONS

In practice, we have found the DBS PAL autosampler to be a robust and reliable device. Several hundred samples have been analyzed without any downstream clogging or maintenance. The ability to control the device using software from the mass spectrometer manufacturer facilitates batched analysis in a fully automated manner. The data obtained here demonstrate that the DBS PAL autosampler can be used to efficiently extract important endogenous metabolites such as acylcarnitines for direct analysis by MS/MS both with and without liquid chromatography. Additionally, from the data generated we may conclude that the results from the DBS autosampler with MS/MS are comparable with those obtained using the standard clinical assay methods for these analytes, which require several additional steps that typically include protein precipitation, SPE column chromatography, solvent evaporation, and reconstitution. Precision and accuracy are optimal when ISs are added to the sample before spotting onto the filter paper, but utilizing ISs in the loop injector and flowing these standards through the card has also performed very well. The recovery of analytes is surprisingly good, though not as efficient as with the more traditional punch-and-extract methods. However, the assays are not compromised provided that calibration and analysis are performed in the same manner. The results we have obtained thus far are highly encouraging, and we intend to develop and fully validate several new assays for routine application in this laboratory. The most significant advantages of this autosampler are its time and labor-saving features, which are attractive in a clinical diagnostic laboratory and in any high-throughput laboratory setting. By eliminating several steps that require sample transfers, potential laboratory errors are also minimized.

ACKNOWLEDGMENT

The authors are indebted to LEAP Technologies, Inc., for the extended loan of a DBS autosampler and for valuable technical assistance during this project.

REFERENCES

Blijlevens NM, Lutgens LC, Schattenberg AV, Donnelly JP. Citrulline: a potentially simple quantitative marker of intestinal epithelial damage following myeloablative therapy. *Bone Marrow Transplantation* 2004;34:193–196.

Deglon J, Thomas A, Cataldo A, Mangin P, Staub C. On-line desorption of dried blood spot: a novel approach for the direct LC/MS analysis of micro-whole blood samples. *Journal of Pharmaceutical and Biomedical Analysis* 2009;49:1034–1039.

Deglon J, Thomas A, Daali Y, Lauer E, Samer C, Desmeules J, Dayer P, Mangin P, Staub C. Automated system for on-line desorption of dried blood spots applied to LC/MS/MS pharmacokinetic study of flurbiprofen and its metabolite. *Journal of Pharmaceutical and Biomedical Analysis* 2011;54:359–367.

Frazier DM, Millington DS, McCandless SE, Koeberl DD, Weavil SD, Chaing SH, Muenzer J. The tandem mass spectrometry newborn screening experience in North Carolina: 1997–2005. *Journal of Inherited Metabolic Disease* 2006;29:76–85.

Ganz N, Singrasa M, Nicolas L, Gutierrez M, Dingemanse J, Dobelin W, Glinski M. Development and validation of a fully automated online human dried blood spot analysis of bosentan and its metabolites using the Sample Card And Prep DBS System. *Journal of Chromatography B-Analytical Technologies in the Biomedical and Life Sciences* 2012;885–886:50–60.

Guthrie R, Susi A. A simple phenylalanine method for detecting phenylketonuria in large populations of newborn infants. *Pediatrics* 1963;32:338–343.

Kao HJ, Cheng CF, Chen YH, Hung SI, Huang CC, Millington D, Kikuchi T, Wu JY, Chen YT. ENU mutagenesis identifies mice with cardiac fibrosis and hepatic steatosis caused by a mutation in the mitochondrial trifunctional protein beta-subunit. *Human Molecular Genetics* 2006;15:3569–3577.

Keevil BG. The analysis of dried blood spot samples using liquid chromatography tandem mass spectrometry. *Clinical Biochemistry* 2011;44:110--118.

La Marca G, Malvagia S, Pasquini E, Innocenti M, Donati MA, Zammarchi E. Rapid 2nd-tier test for measurement of 3-OH-propionic and methylmalonic acids on dried blood spots: reducing the false-positive rate for propionylcarnitine during expanded newborn screening by liquid chromatography-tandem mass spectrometry. *Clinical Chemistry* 2007;53:1364–1369.

Li W, Tse FL. Dried blood spot sampling in combination with LC-MS/MS for quantitative analysis of small molecules. *Biomedical Chromatography* 2010;24:49–65.

Liu G, Snapp HM, Ji QC. Internal standard tracked dilution to overcome challenges in dried blood spots and robotic sample preparation for liquid chromatography/tandem mass spectrometry assays. *Rapid Communications in Mass Spectrometry* 2011;25:1250–1256.

Lutgens L, Lambin P. Biomarkers for radiation-induced small bowel epithelial damage: an emerging role for plasma citrulline. *World Journal of Gastroenterology* 2007;13:3033–3042.

Lutgens LC, Deutz NE, Gueulette J, Cleutjens JP, Berger MP, Wouters BG, von Meyenfeldt MF, Lambin P. Citrulline: a physiologic marker enabling quantitation and monitoring of epithelial radiation-induced small bowel damage. *International Journal of Radiation Oncology, Biology, Physics* 2003;57:1067–1074.

Magera MJ, Helgeson JK, Matern D, Rinaldo P. Methylmalonic acid measured in plasma and urine by stable isotope dilution and electrospray tandem mass spectrometry. *Clinical Chemistry* 2000;46:1804–1810.

Manicke NE, Abu-Rabie P, Spooner N, Ouyang Z, Cooks RG. Quantitative analysis of therapeutic drugs in dried blood spot samples by paper spray mass spectrometry: an avenue to therapeutic drug monitoring. *Journal of the American Society for Mass Spectrometry* 2011;22:1501–1507.

Matsui SM, Mahoney MJ, Rosenberg LE. The natural history of the inherited methylmalonic acidemias. *New England Journal of Medicine* 1983;308:857–861.

Miller JH IV, Poston PA, Karnes TH. Direct analysis of dried blood spots by in-line desorption combined with high-resolution chromatography and mass spectrometry for quantification of maple syrup urine disease biomarkers leucine and isoleucine. *Analytical and Bioanalytical Chemistry* 2011;400:237–244.

Millington DS, Stevens RD. Acylcarnitines: analysis in plasma and whole blood using tandem mass spectrometry. In: Metz TO (ed.), *Methods in Molecular Biology: Metabolic Profiling,* Chapter 3. New York: Humana Press (Springer); 2011. pp. 55–72.

Millington DS, Kodo N, Norwood DL, Roe CR. Tandem mass spectrometry: a new method for acylcarnitine profiling with potential for neonatal screening for inborn errors of metabolism. *Journal of Inherited Metabolic Disease* 1990;13:321–324.

Millington D, Kodo N, Terada N, Chace DH, Gale D. The analysis of diagnostic markers of genetic disorders in human blood and urine using tandem mass spectrometry with liquid SIMS. *International Journal of Mass Spectrometry and Ion Physics* 1991;111:211–228.

Nagy K, Tákats Z, Pollreisz F, Szabó T, Vékey K. Direct tandem mass spectrometric analysis of amino acids in dried blood spots without chemical derivatization for neonatal screening. *Rapid Communications in Mass Spectrometry* 2003;17:983–990.

Newgard CB, An J, Bain JR, Muehlbauer MJ, Stevens RD, Lien LF, Haqq AM, Shah SH, Arlotto M, Slentz CA, Rochon J, Gallup D, Ilkayeva O, Wenner BR, Yancy WSJr, Eisenson H, Musante G, Surwit RS, Millington DS, Butler MD, Svetkey LP. A branched-chain amino acid-related metabolic signature that differentiates obese and lean humans and contributes to insulin resistance. *Cell Metabolism* 2009;9:311–326.

Oh RC, Brown DL. Vitamin B12 deficiency. *American Family Physician* 2003;67:979–986.

Rashed MS, Bucknall MP, Little D, Awad A, Jacob M, Alamoudi M, Alwattar M, Ozand PT. Screening blood spots for inborn errors of metabolism by electrospray tandem mass spectrometry with a microplate batch process and a computer algorithm for automated flagging of abnormal profiles. *Clinical Chemistry* 1997;43:1129–1141.

Rinaldo P, Cowan TM, Matern D. Acylcarnitine profile analysis. *Genetics in Medicine* 2008:10:151–156.

Schulze A, Lindner M, Kohlmuller D, Olgemoller K, Mayatepek E, Hoffmann GF. Expanded newborn screening for inborn errors of metabolism by electrospray ionization-tandem mass spectrometry: results, outcome, and implications. *Pediatrics* 2003;111:1399–1406.

Stevens RD, Hillman SL, Worthy S, Sanders D, Millington DS. Assay for free and total carnitine in human plasma using tandem mass spectrometry. *Clinical Chemistry* 2000;46:727–729.

Thomas A, Deglon J, Steimer T, Mangin P, Daali Y, Staub C. On-line desorption of dried blood spots coupled to hydrophilic interaction/reversed-phase LC/MS/MS system for the simultaneous analysis of drugs and their polar metabolites. *Journal of Separation Science* 2010,33:873–879.

Thompson JW, Zhang H, Smith P, Hillman S, Moseley AM, Millington DS. Extraction and analysis of carnitine and acylcarnitines by ESI-MS/MS directly from dried blood and plasma spots using a novel autosampler. *Rapid Communications in Mass Spectrometry* 2012;26:2548–2554.

Van den Bulcke T, Vanden Broucke P, Van Hoof V, Wouters K, Vanden Broucke S, Smits G, Smits E, Proesmans S, Van Genechten T, Eyskens F. Data mining methods for classification of medium-chain acyl-coA dehydrogenase deficiency (MCADD) using non-derivatized tandem MS neonatal screening data. *Journal of Biomedical Informatics* 2011;44:319–325.

Wiseman JM, Evans CA, Bowen CL, Kennedy JH. Direct analysis of dried blood spots utilizing desorption electrospray ionization (DESI) mass spectrometry. *Analyst* 2010;135:720–725.

Zhang Z, Xu W, Manicke NE, Cooks RG, Ouyang Z. Silica coated paper substrate for paper-spray analysis of therapeutic drugs in dried blood spots. *Analytical Chemistry* 2012;84:931–938.

Zytkovicz TH, Fitzgerald EF, Marsden D, Larson CA, Shih VE, Johnson DM, Strauss AW, Comeau AM, Eaton RB, Grady GF. Tandem mass spectrometric analysis for amino, organic, and fatty acid disorders in newborn dried blood spots: a two-year summary from the New England Newborn Screening Program. *Clinical Chemistry* 2001;47:1945–1955.

25

DEVELOPMENT OF BIOMARKER ASSAYS FOR CLINICAL DIAGNOSTICS USING A DIGITAL MICROFLUIDICS PLATFORM

David S. Millington, Ramakrishna Sista, Deeksha Bali, Allen E. Eckhardt, and Vamsee Pamula

25.1 INTRODUCTION

The increasing use of dried blood spots (DBS) as a method of sampling has prompted new and innovative analytical methods to extract and analyze DBS for biomarkers. The challenges presented by DBS include small blood sample volumes, the need to extract analytes prior to analysis, and the need to develop and validate quantitative assays. One of the most compatible technologies, at least in principle, for DBS samples is the use of microfluidics technology, such as that offered by the "lab-on-a-chip" (Manz et al., 1990). There are several versions of this type of technology, designed to perform liquid handling functions on a micro- or nano-scale (Whitesides, 2006). Arguably, the most versatile of these technologies is "digital microfluidics,"(DMF) in which fluids are manipulated in the form of discrete droplets rather than in fixed channels. The flexibility and versatility of DMF technology have allowed its application to fluorometric enzymatic assays, immunoassays, sample preparation for polymerase chain reaction (PCR), and metabolite analysis.

25.1.1 Digital Microfluidic Technology

DMF technology is based on the principle of the electrowetting effect (Pollack et al., 2002). When a voltage is applied across a liquid–solid interface, the interfacial tension is reduced, leading to a change in contact angle. This effect, shown in Figure 25.1a, can be used to manipulate droplets by applying voltages to an array of independently controllable surface electrodes. Individual droplets can be transported, merged, split, or dispensed entirely under software program control. All liquid operations are performed electrically without the need for any pumps, valves, or predefined channels. DMF cartridges are manufactured using a printed circuit board (PCB) to form the electrode array and an injection-molded plastic top-plate to form an operating chamber and to house samples and reagents. The cartridges are designed for single use and are more inexpensive and disposable compared with other microfluidic systems. The space between the top-plate and the PCB is filled with a hydrophobic fluid that prevents evaporation of the droplets, as indicated in Figure 25.1b. The versatility of this design facilitates scalability without compromising the basic functions of DMF. The volume of the individual droplets is scalable from 10 nL to several microliters, with typical volumes being in the range of 100–300 nL. Thus, it is practical to design small PCBs to perform a single, simple assay with a hand-held device and much larger arrays that could multiplex several assays on hundreds of samples simultaneously. When the cartridge is inserted into the analyzer (Figure 25.2), electrodes can be independently heated or activated magnetically, facilitating the development of different types of assays including immunoassays that use reagents immobilized on magnetic beads, DNA assays that include PCR amplification, and enzymatic assays that are typically conducted at 37°C. Incorporated into the analyzer is a built-in detector, such as a microfluorometer or microphotometer, depending on the type of assay.

Dried Blood Spots: Applications and Techniques, First Edition. Edited by Wenkui Li and Mike S. Lee.

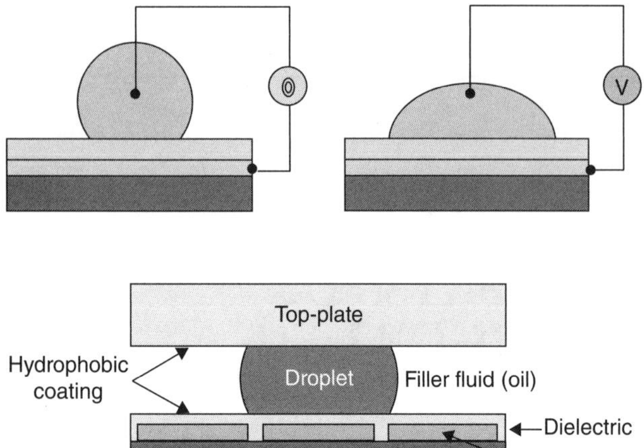

FIGURE 25.1 The principle of the electro-wetting effect. (Top) A droplet rests on the surface (at left) on an electrode coated with a dielectric. When a voltage is applied to the electrode, the droplet is attracted to and "flattens" on the surface (at right). Activation of an adjacent electrode will attract the droplet and thus, droplets can be moved independently on the cartridge entirely under software control and without leaving a residue. (Bottom) The droplets are sandwiched between the PCB and cover plate of a cartridge. The entire space is filled with an oil to prevent evaporation of the droplets.

25.1.2 Digital Microfluidic Assays on DBS and Whole Blood

Applications of a practical DMF system in the field of clinical diagnostics (Srinivasan et al., 2004; Sista et al., 2008b) and high-throughput neonatal screening (Millington et al., 2010) have been described previously. DMF is particularly well suited to neonatal screening, because specimens are typically collected in birthing centers as DBS and mailed to state public health laboratories for analysis. A multiplex diagnostic system for analysis of DBS in newborn screening laboratories could dramatically increase the number of tests that can be performed on a single platform. DMF has been employed to perform different types of assays using whole blood or plasma rather than DBS (Pollack et al., 2011). Several of these assays are adaptable to the DBS format and will be briefly described. The primary focus of this chapter is on results and discussion of fluorometric assays from DBS samples for newborn screening of lysosomal storage disorders (LSDs).

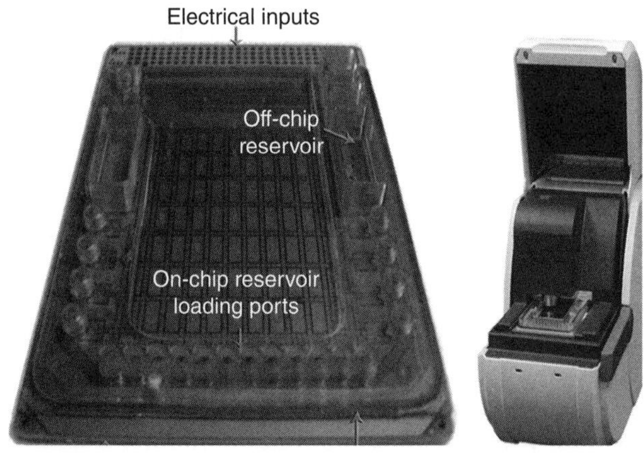

FIGURE 25.2 Digital microfluidic cartridge (at left) designed to perform up to 4 assays on 12 specimens. After loading sample and reagent reservoirs, the cartridge is inserted into the R-100 Analyzer (at right, not shown to scale) and all remaining steps are fully automated by digital control. The cartridges and analyzer are manufactured by Advanced Liquid Logic, Inc. Reprinted with permission from Sista et al. (2008b). Copyright 2008 Royal Society of Chemistry.

25.1.2.1 Fluorometric Assays LSDs are a group of over 40 rare inherited metabolic disorders that are characterized by lysosomal enzyme dysfunction, usually resulting from the deficiency of a single enzyme required for the metabolism of lipids, glycoproteins or mucopolysaccharides. Individually, LSDs occur with an incidence of less than 1:50,000. As a group, however, the incidence is thought to be about 1:5000–1:10,000. With the development of new therapies such as enzyme replacement and bone marrow transplant for some LSDs, the general interest in screening for a panel of treatable LSDs has acquired impetus from the fact that methods to screen for these conditions in DBS are also available (Zhou et al., 2011). The disorders of particular interest include Pompe disease (acid-α-glucosidase deficiency), Fabry disease (acid-α-galactosidase deficiency), Hurler syndrome (α-iduronidase deficiency), Hunter disease (α-iduronidate-2-sulfatase deficiency) and Gaucher disease (acid-α-glucocerebrosidase deficiency). Infants with these conditions are expected to benefit the most from NBS because they may present with irreversible organ damage if not diagnosed and treated early in life (Marsden and Levy, 2010). Several newborn screening programs have planned or have already started to screen newborns for one or more LSDs (Chien et al., 2009; Duffner et al., 2009; Orsini et al., 2012). The currently available methods, including microtiter plate fluorometry and tandem mass spectrometry (MS/MS) (Li et al., 2004; Zhang et al., 2008), target the residual activities of lysosomal enzymes in DBS samples. There are, however, significant challenges to the expansion from single to multiplex LSD enzymatic assays using these technologies including cost, time to result and limited availability of sample. These practical challenges present much less of an obstacle for the DMF platform, because of the greatly reduced scale in the amounts of sample and reagent requirement and additional advantages outlined in sections 25.3.1.1 and 25.3.1.2. We have developed fluorometric assays for several lysosomal enzymes on the DMF platform that are described in Section 25.2.2. Preliminary data have also been generated from

DBS using DMF to detect biotinidase deficiency and galactosemia by fluorometric assay (data not shown). Although newborn screening for these conditions is already routinely performed, the screening method is a laborious process in some programs. We are confident that this single, easy to use, inexpensive, and automated platform will facilitate the consolidation of several assay modalities to provide a variety of fluorometric newborn screening assays.

25.1.2.2 Immunoassays Current immunoassays used in newborn screening laboratories include the 17-OH progesterone assay for congenital adrenal hyperplasia, thyroxine and thyroid-stimulating hormone (TSH) for congenital hypothyroidism, and immunoreactive trypsinogen assay for cystic fibrosis. Advanced Liquid Logic Inc. (ALL) has developed several immunoassays on the DMF platform by using antibodies that are immobilized on magnetic beads (Sista et al., 2008a; Millington et al., 2010). Briefly, a droplet containing magnetic beads with bound captured antibodies is mixed with a sample droplet to capture the antigens. Then, it is mixed with a droplet containing enzyme-labeled secondary antibodies to result in a "sandwich' complex. The droplet is migrated to a magnetized electrode that holds the beads in place while excess supernatant is washed away on the cartridge with wash droplets and finally mixed with a fluorescent substrate droplet. The immunoassay is quantified from the measured fluorescence. These immunoassays can be performed in less than 5 minutes per assay. These same principles were applied to develop assays for TSH for congenital hypothyroidism and to develop a hypercoagulability screening panel including antithrombin III, protein C, protein S, and factor VIII antigens (Emani et al., 2012). Although these immunoassays were developed using whole blood samples, they are readily translatable to DBS and demonstrate the breadth of versatility of the DMF platform for newborn screening using either DBS or whole blood samples collected at the point of birth. This versatility could be an important future development in newborn screening, especially for programs that have no centralized public health laboratory.

25.1.2.3 Polymerase Chain Reaction and Detection of PCR Products DMF can be used for rapid (< 6 minutes) PCR thermocycling by shuttling a droplet through distinct temperature zones on a cartridge (Hua et al., 2010; Wulff-Burchfield et al., 2010; Schell et al., 2012). The amplified PCR product is labeled with a fluorescent indicator to facilitate detection. Of particular interest to newborn screening is the successful manipulation of nanoliter-sized droplets reconstituted from DBS for DNA extraction and real time PCR to quantify T-cell receptor excision circles (TRECs), the target for severe combined immunodeficiency syndrome (SCID) (Millington et al., 2010).

25.1.2.4 Metabolite Analysis Another potential application of DMF for DBS analysis has been reported recently (Jebrail et al., 2011). These authors have demonstrated the ability to extract and derivatize amino acids and acylcarnitines from DBS on a cartridge. The metabolites were analyzed off-line using tandem mass spectrometry. A prototype device that included built-in nano-electrospray was also described. This research indicates the potential of DMF to automate the procedure for biomarker analysis from DBS using tandem mass spectrometry.

25.2 MATERIALS AND METHODS

25.2.1 Digital Microfluidic Equipment

Most of the preliminary work described in this article was performed on a mid-sized cartridge, similar in area to a 96-well microtiter plate with inputs for up to 12 samples and 8 reagents, shown in Figure 25.2, left. From this design, a new cartridge was developed capable of accepting up to 44 samples and 5 reagents (Figure 25.3). Both of these cartridges have been used for multiplexed enzyme assays for LSDs. After the samples and reagents were loaded into the cartridge

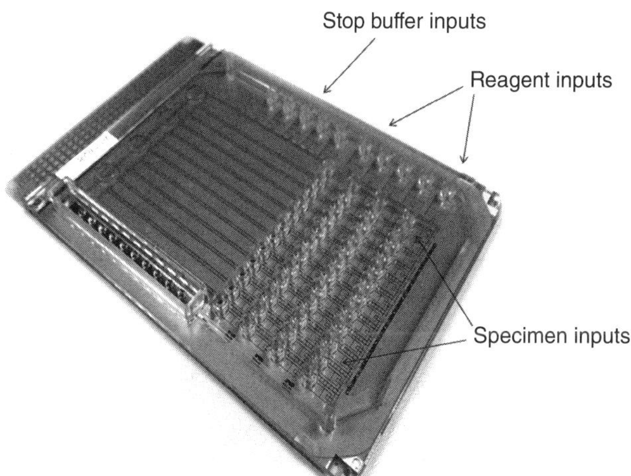

FIGURE 25.3 Digital microfluidic cartridge for multiplexing enzymatic assays for up to five lysosomal storage disorders on up to 44 dried blood spot extracts. Assays are performed by mixing independent droplets of 100 nL volume from the reagent and sample wells, enabling optimization of the conditions for each assay. The enzyme activity in each droplet releases the fluorophore 4-methylumbelliferone from the reagent for that particular assay. After quenching using the stop buffer, a built-in microfluorometer reads the fluorescence of each droplet in approximately 6 seconds. Four of the specimen input wells are used to calibrate the microfluorometer response for 4-methylumbelliferone. The overall time required to load reagents and samples, insert the cartridge into the analyzer, incubate, and generate the results is approximately 3.5 hours.

manually using standard Eppendorf pipettes, the cartridge was inserted into the instrument (Figure 25.2, right) and the selected program performed all subsequent steps in an automated fashion according to the requirements of the assay. The equipment and cartridges are components of the R-100 system manufactured by Advanced Liquid Logic (ALL), Inc. (Morrisville, NC, USA).

25.2.2 Enzymatic assays for Lysosomal Storage Disease

Full details of the enzymatic assays for individual LSDs have been published (Sista et al., 2011a, 2011b). Briefly, DBS punches of 3 mm diameter, corresponding to 3.1 μL of whole blood, were extracted in 96-well microtiter plates using 100 μL of extraction buffer (water with 0.1% Tween20) for 30 minutes on an orbital plate shaker. About 50 μL of each extract was pipetted into a sample reservoir on the DMF cartridge (Figure 25.2) using a multichannel Eppendorf pipettor (12 channels). Individual LSD reagents were prepared and loaded into the reagent reservoir wells on the cartridge; for example, for the Gaucher assay, 16 mM 4-methylumbelliferyl-β-D-glucopyranoside solution was prepared in 0.2 M/0.1 M citrate phosphate buffer, pH 5.2, with 1.5% sodium taurocholate, and for the Hurler assay, 2 mM 4-methylumbelliferyl-α-L-iduronide was prepared in 0.05 M sodium acetate, pH 3.5, with 0.01% Tween20, 75 mM NaCl, and 100 μM D-saccharic acid 1,4 lactone. The ability to perform these assays in independent droplets derived from a single DBS punch enables them to be fully optimized on the cartridge, while making the best use of the limited amount of sample available for additional assays.

25.2.2.1 Multiplex Assay Protocol for Five LSDs DBS samples were extracted according to the protocol described in Section 25.2.2. Specific reagent kits for each enzyme assay were prepared under GLP conditions, packaged in small lot sizes and stored at –80°C prior to use. The reagents (n = 5), calibrators (n = 4), samples (n = 44), and stop buffer were loaded into appropriate reservoirs on a cartridge designed to perform 5 LSD multiplexed enzyme assays simultaneously for up to 48 samples (Figure 25.3). The DMF protocol included mixing one droplet (300 nL) of the DBS extract with one droplet of the reagent solution. The double-sized reaction droplet was split into two unit-sized droplets. One droplet was discarded and the second droplet was incubated for 1 hour at 37°C. After incubation, one droplet of stop buffer (0.2 M sodium bicarbonate pH 10.0) was added to the reaction droplet and end-point fluorescence was measured at 370 nm excitation and 460 nm emission. The raw fluorescence values were converted to micromole of 4-methylumbelliferone (4-MU) produced per liter of blood per hour of incubation using a 4-MU calibration curve performed using the calibrators on each cartridge.

25.3 RESULTS AND DISCUSSION

25.3.1 Enzymatic Assays for Lysosomal Storage Disorders

25.3.1.1 Development of Fluorometric Enzyme Assays Using Digital Microfluidics Fluorometric enzyme activity assays for Pompe, Hunter, Fabry, Gaucher, and Hurler diseases were translated successfully from the microtiter plate to the DMF cartridge. In Figure 25.4, the results of samples from individuals affected with Pompe and Fabry disease (n = 6 and n = 7, respectively) are compared with the results from normal control DBS (n = 105). Incubation time for these assays was only 1 hour at 37°C, compared with the overnight (20 hours) incubations required by standard bench microtiter plate methods. Nevertheless, the DMF method was able to discriminate affected from control samples at least as effectively as the standard methods (Sista et al., 2011a).

Similarly, Figure 25.5 compares the results of assays for the Gaucher and Hurler enzyme activities, performed on the same groups of affected and normal control DBS samples, using DMF and bench microfluorometry. In this example, both DMF assays were performed from the same sample extract obtained from a single DBS punch, whereas the standard bench fluorometric assay method required a separate punch for assaying each LSD enzyme separately. The assay was able to discriminate affected patients (n = 6) from control newborn DBS samples (n = 105). A novel assay for Hunter disease using DMF has also been reported recently (Sista et al., 2011b) in which a serial enzyme cascade reaction was achieved in a single step with an incubation time of only 1 hour.

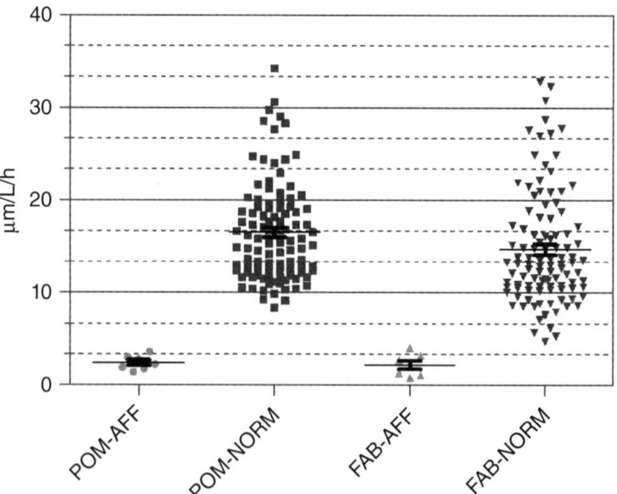

FIGURE 25.4 Enzyme activity determinations in DBS samples from a control group (n = 105) and patient groups for Pompe (n = 6) and Fabry (n = 7) disease. The assays were performed by digital microfluidics with an incubation time of 1 hour. There is clear discrimination between the affected and unaffected samples.

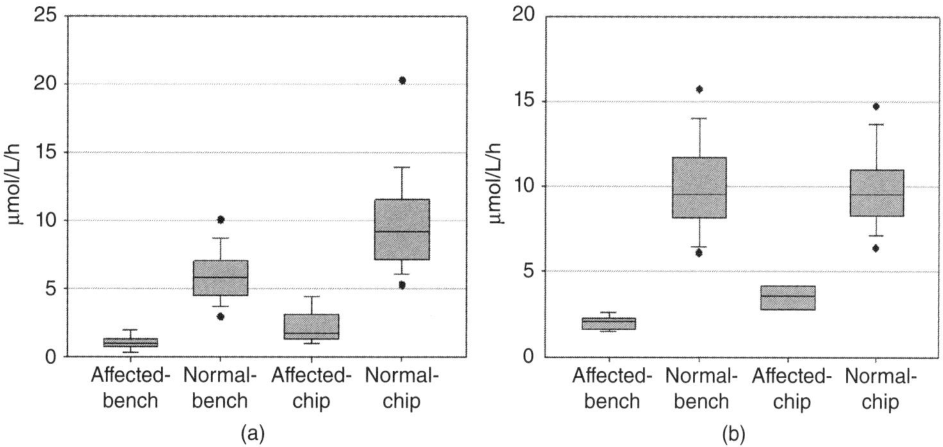

FIGURE 25.5 Comparison of benchtop microplate and digital microfluidic enzyme assays for Gaucher (a) and Hurler (b) disease using the same sets of DBS samples from normal controls (n = 105) and known affected individuals, showing that discrimination between affected and unaffected samples is accomplished using either method.

25.3.1.2 Multiplexing Up to Five Enzymatic Assays Using Digital Microfluidics A crucial step in the development of the DMF platform for newborn screening applications was to demonstrate that as many as five different assays could be performed from the extract of a single DBS punch. In Section 25.3.1.1 assays for two enzyme activities were multiplexed on the DMF platform. A 3-plex assay for Pompe, Fabry, and Gaucher was recently used in a pilot NBS program in the state of Illinois to screen approximately 8000 newborns from two participating hospitals (Burton et al., 2012). The results of this study led to the molecularly confirmed diagnosis of 7 cases of Fabry disease (out of 11 presumptive positive screens) and 2 with Gaucher disease (out of 22 presumptive positive screens), from a total of 8012 samples. There were only two positive screens for Pompe disease, neither of which was affected. Many of the Gaucher presumptive positive screens were from babies in the neonatal intensive care units of specific hospitals. The surprisingly high incidence of Fabry disease has been reported by other NBS programs using different technologies (Spada et al., 2006; Hwu et al., 2009). These early results warrant further investigation into the incidence of the LSDs. In fact, another pilot study using DMF to screen for LSDs is due to start in January 2013 in the state of Missouri using a new cartridge that has 44 sample inputs, 4 calibrant inputs, and up to 5 reagent inputs (Figure 25.3). The cartridge has been extensively validated for use in LSD screening and the detailed results will be reported elsewhere. One essential component of the validation process was to analyze quality control (QC) DBS specimens such as those circulated by the Centers for Disease Control (CDC) to all NBS programs engaged in LSD screening.

Enzymatic activities measured for all five enzymes tested using four different QC spots were stable and reproducible over a period of over 12 months. An example of the stability testing for ready to use reagent kits prepared for multiplexed testing is shown for Hunter disease in Figure 25.6. Coefficients of variation (CV) for various enzyme activity levels measured at different time points (day 0 through month 12) ranged from 2% to 16% (Pompe), 2% to 27% (Fabry), 2% to 25% (Hunter), 1% to 16% (Gaucher), and 1% to 19% (Hurler). This study also demonstrated that the stability of the prepared reagent kits, in storage at –80°C, was good for at least 6 months. Similar sets of QC samples were also analyzed at three different sites and at multiple times within the same 15-day time period by DMF, microtiter plate fluorometry, and tandem mass spectrometry (Wang et al., 2011). An example of the results is shown in Figure 25.7. Overall,

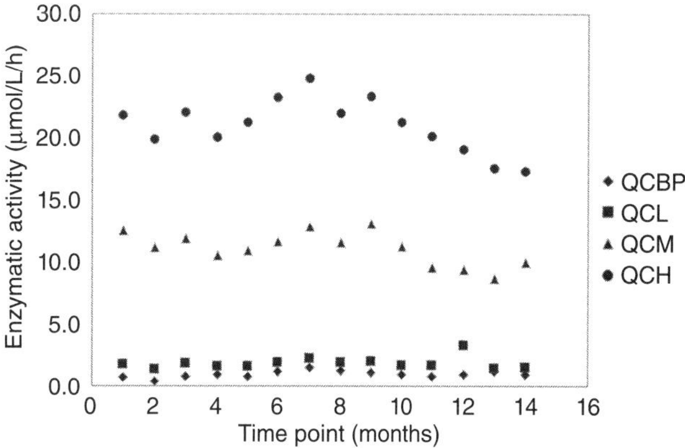

FIGURE 25.6 Reagent stability test for Hunter disease enzymatic assay using quality control DBS specimens derived from base (QCBP), low (QCL), medium (QCM), and high (QCH) activity pools on a digital microfluidic cartridge. The assay was performed at regular intervals over a 12-month period using fresh aliquots of reagent stored at –80°C.

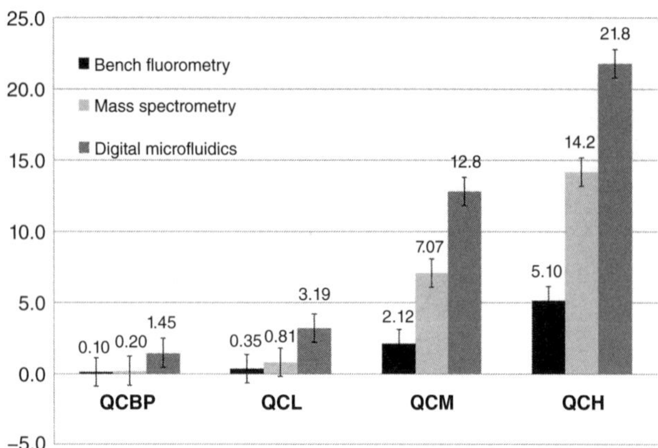

FIGURE 25.7 Interlaboratory comparison of enzymatic assay results for Pompe disease performed by different methods using the same sets of quality control (QC) dried blood spot specimens consisting of base pool (BP), low (L), medium (H), and high (H) activity QC samples. The scale on the *X*-axis is activity in μmol/L/h, and the error bars represent coefficients of variation for 15 determinations.

the performance of all three methods was similar in terms both of discriminating between the low, medium, and high QCs and of their reproducibility, expressed as CV. Differences in absolute activities are the result of methodological differences.

25.4 SUMMARY AND CONCLUSIONS

The DMF platform described in this chapter has evolved to become an interesting alternative to traditional testing methodologies. The platform is currently being validated and may be accepted by newborn screening programs as a viable alternative to tandem mass spectrometry for multiplexed LSD enzymatic assays in DBS samples. The performance metrics for both of these methods thus far appear to be comparable, but DMF uses portable equipment (Figure 25.2) that is much simpler and less expensive to purchase, operate, and maintain than MS/MS. Another attractive feature of DMF for high-throughput screening is the short turnaround time. The time required to load samples and reagents and to complete an analytical run (consisting of 40 samples plus 4 controls for 5 enzymes and 4 calibrators) takes approximately 3.5 hours. A workstation consists of four instruments managed by a single computer, and two such workstations can therefore assay over 600 DBS samples for all five enzymes in a single shift. The amount of sample consumed in each assay is miniscule compared with other methods, and reagent volumes are similarly very low. Furthermore, other assays currently performed in newborn screening programs, especially those dependent on fluorescent reagents, are amenable to DMF. In developing countries that have no centralized healthcare system for collection and analysis of DBS for newborn screening, point-of-care birth testing may be the only practical method to screen newborns for inherited metabolic conditions and indeed for more common conditions faced by developing countries, such as infectious diseases. The future prospects for applications of DMF both in newborn screening and point-of-care testing seem to be excellent, although more field studies are required to explore the full capabilities of this new technology.

REFERENCES

Burton B, Charrow J, Angle B, Widera S, Waggoner D. A pilot newborn screening program for lysosomal storage disorders (LSD) in Illinois. *Molecular Genetics and Metabolism* 2012;105:S23–S24.

Chien YH, Lee NC, Thurberg BL, Chiang SC, Zhang XK, Keutzer J, Huang AC, Wu MH, Huang PS, Tsai FJ, Chen YT, Hwu WH. Pompe disease in infants: improving the prognosis by newborn screening and early treatment. *Pediatrics* 2009;124:e1116–e1125.

Duffner PK, Caggana M, Orsini JJ, Wenger DA, Patterson MC, Crosley CJ, Kurtzberg J, Arnold GL, Escolar ML, Adams DJ, Andriola MR, Aron AM, Ciafaloni E, Djukic A, Erbe RW, Galvin-Parton P, Helton LE, Kolodny EH, Kosofsky BE, Kronn DF, Kwon JM, Levy PA, Miller-Horn J, Naidich TP, Pellegrino JE, Provenzale JM, Rothman SJ, Wasserstein MP. Newborn screening for Krabbe disease: the New York State model. *Pediatric Neurology* 2009;40:245–252.

Emani S, Sista R, Loyola H, Trenor CC, Pamula VK, Emani SM. Novel microfluidic platform for automated lab-on-chip testing of hypercoagulability panel. *Blood Coagulation and Fibrinolysis* 2012;23:760–768.

Hua Z, Rouse JL, Eckhardt AE, Srinivasan V, Pamula VK, Schell WA, Benton JL, Mitchell TG, Pollack MG. Multiplexed real-time polymerase chain reaction on a digital microfluidic platform. *Analytical Chemistry* 2010;82(6):2310–2316.

Hwu WL, Chien YH, Lee NC, Chiang SC, Dobrovolny R, Huang AC, Yeh HY, Chao MC, Lin SJ, Kitagawa T, Desnick RJ, Hsu LW. Newborn screening for Fabry disease in Taiwan reveals a high incidence of the later-onset GLA mutation c. 936+919G>A (IVS4+919G>A). *Human Mutation* 2009;30:1397–1405.

Jebrail MJ, Yang H, Mudrik JM, Lafreniere NM, McRoberts C, Al-Dirbashi OY, Fisher L, Chakraborty P, Wheeler AR. A digital microfluidic method for dried blood spot analysis. *Lab on a Chip* 2011;11:3218.

Li Y, Scott RC, Chamoles NA, Ghavami A, Pinto BM, Turecek F, Gelb MH. Direct multiplex assay of lysosomal enzymes in dried blood spots for newborn screening. *Clinical Chemistry* 2004;50:1785–1796.

Manz A, Graber N, Widmer HM. Miniaturized total chemical analysis systems: a novel concept for chemical sensing. *Sensors and Actuators B: Chemical* 1990;1:244–248.

Marsden D, Levy H. Newborn screening of lysosomal storage disorders. *Clinical Chemistry* 2010;56:1071–1079.

Millington DS, Sista R, Eckhardt A, Rouse J, Bali D, Goldberg R, Cotten M, Buckley R, Pamula V. Digital microfluidics: a future technology in the newborn screening laboratory? *Seminars in Perinatology* 2010;34:163–169.

Orsini JJ, Martin MM, Showers AL, Bodamer OA, Zhang XK, Gelb MH, Caggana M. Lysosomal storage disorder 4 + 1 multiplex assay for newborn screening using tandem mass spectrometry: application to a small-scale population study for five lysosomal storage disorders. *Clinica Chimica Acta* 2012;413:1270–1273

Pollack MG, Shenderov AD, Fair RB. Electrowetting-based actuation of droplets for integrated microfluidics. *Lab on a Chip* 2002;2:96-101.

Pollack MG, Pamula VK, Srinivasan V, Eckhardt AE. Applications of electrowetting-based digital microfluidics in clinical diagnostics. *Expert Review of Molecular Diagnostics* 2011;11(4):393-407.

Schell WA, Benton JL, Smith PB, Poore M, Rouse JL, Boles DJ, Johnson MD, Alexander BD, Pamula VK, Eckhardt AE, Pollack MG, Benjamin DK Jr, Perfect JR, Mitchell TG. Evaluation of a digital microfluidic real-time PCR platform to detect DNA of Candida albicans in blood. *European Journal of Clinical Microbiology and Infectious Diseases* 2012;31:2237–2245.

Sista RS, Eckhardt AE, Srinivasan V, Pollack MG, Palaki S, Pamula VK. Heterogeneous immunoassays using magnetic beads on a digital microfluidic platform. *Lab on a Chip* 2008a;8:2188–2196.

Sista R, Hua Z, Thwar P, Sudarasan A, Srinivasan V, Eckhardt A, Pollack M, Pamula V. Development of a digital microfluidic platform for point of care testing. *Lab on a Chip* 2008b;8:2091–2104.

Sista RS, Eckhardt AE, Wang T, Graham C, Rouse JL, Norton SM, Srinivasan V, Pollack MG, Tolun AA, Bali D, Millington DS, Pamula VK. Digital microfluidic platform for multiplexing enzyme assays: implications for lysosomal storage disease screening in newborns. *Clinical Chemistry* 2011a;57:1444–1451.

Sista R, Eckhardt AE, Wang T, Séllos-Moura M, Pamula V. Rapid, single-step assay for Hunter syndrome in dried blood spots using digital microfluidics. *Clinica Chimica Acta* 2011b;412:1895–1897.

Spada M, Pagliardini S, Yasuda M, Tukel T, Thiagarajan G, Sakuraba H, Ponzone A, Desnick R. High incidence of later-onset Fabry disease revealed by newborn screening. *American Journal of Human Genetics* 2006;79:31–40.

Srinivasan V, Pamula VK, Fair RB. An integrated digital microfluidic lab-on-a-chip for clinical diagnostics on human physiological fluids. *Lab on a Chip* 2004;4:310–315.

Wang T, Sista R, Graham C, Eckhardt A, Winger T, Xiong Y, Wu N, Shi Q, Dickerson G, Tolun A, Zhou H, Vogt R, Millington D, Pamula V, Bali D. Comparison of Enzyme Activities for Pompe, Fabry, and Gaucher Diseases on CDC's Quality Control Spots between Microplate Fluorometry, Mass Spectrometry, and Digital Microfluidic Fluorometry. Presented at the Association of Public Health Laboratories Newborn Screening and Genetic Testing Symposium, San Diego, CA, November, 2011.

Whitesides GM. The origins and the future of microfluidics. *Nature* 2006;442:368–373.

Wulff-Burchfield E, Schell WA, Eckhardt AE, Pollack MG, Hua Z, Rouse JL, Pamula VK, Srinivasan V, Benton JL, Alexander BD, Wilfret DA, Kraft M, Cairns C, Perfect JR, Mitchell TG. Microfluidic platform versus conventional real-time PCR for the detection of Mycoplasma pneumoniae in respiratory specimens. *Diagnostic Microbiology and Infectious Disease* 2010;67(1):22–29.

Zhang XK, Elbin CS, Chuang WL, Cooper SK, Marashio CA, Beauregard C, Keutzer JM. Multiplex enzyme assay screening of dried blood spots for lysosomal storage disorders by using tandem mass spectrometry. *Clinical Chemistry* 2008;54:1725–1728.

Zhou H, Fernhoff P, Vogt RF. Newborn bloodspot screening for lysosomal storage disorders. *Journal of Pediatrics* 2011;159:7–13e1.

26

APPLICATIONS AND CHEMISTRY OF CELLULOSE PAPERS FOR DRIED BLOOD SPOTS

Jacquelynn Luckwell, Åke Danielsson, Barry Johnson, Sarah Clegg, Mark Green, and Alan Pierce

26.1 INTRODUCTION

Drug metabolism and pharmacokinetic (DMPK) studies provide the insight into how potential new drug candidates are absorbed, distributed, metabolized, and excreted from the species of interest. The determination of these steps is imperative in the drug development process. These analyses have traditionally required large volumes of blood (generally 100–500 μL per subject per time point) in order to provide sufficient plasma volume for quantitative bioanalytical studies. As only a limited number of serial samples can be taken from each animal (e.g., mouse), composite sampling is often used, resulting in lower quality pharmacokinetic (PK) data due to greater experimental variability and an increase in the number of animals required. Larger volume requirements also make it difficult to conduct studies in juvenile subjects.

The use of dried blood spot (DBS) analysis has increased significantly in DMPK laboratories (Mei et al., 2001; Oliveira et al., 2002; Beaudette and Bateman, 2004; Barfield et al., 2008; Spooner et al., 2009; Barfield and Wheller, 2011; Tanna and Lawson, 2011; Todime et al., 2011; Wickremsinhe et al., 2011) and has made a notable contribution to the reduction, refinement, and replacement (3Rs) of animals (Festing and Wilkinson, 2007; Spooner, 2010a) in discovery stage research. The miniaturization of blood sampling (typically 15–20 μL per sample) has avoided the need for composite studies to be carried out in mice (Clark et al., 2010); therefore, the experimental variability between animals is reduced. The reduced sampling volume required for DBS proved to be advantageous to Guthrie and Suzi in establishing a rapid and economical screening phenylalanine method (Guthrie and Suzi, 1963) for detecting phenylketonuria in young children. The spotting of whole blood onto cellulose-based papers (Hannon et al., 2007; NSQAP, 2008; Pitt, 2010) has allowed samples to be safely stored and shipped at room temperature following air drying time for approximately 2 hours (Abu-Rabie and Spooner, 2010; Denniff and Spooner, 2010). Rather than generating plasma samples whereby the blood is centrifuged and separated prior to analysis, the use of whole blood DBS has resulted in a more streamlined work flow.

DBS offers the ability to extract compounds using organic solvent (e.g., methanol or acetonitrile) by direct elution of the compound(s) of interest from the cellulose-based cards. This solvent extraction approach has proved highly advantageous in high-throughput DMPK bioanalytical environments where a large number of samples can be processed in a short period of time.

However, the reduction in blood volume has posed analytical challenges where lower limits of assay quantification (Li and Tse, 2010) are required (e.g., inhalation compound development and micro-dose studies). This challenge has been somewhat helped by the introduction of highly sensitive mass spectrometers and the utilization of ultra high pressure liquid chromatography (UHPLC) with column switching (Mather et al., 2011; Clark and Haynes, 2011) in an attempt to remove matrix effects from phospholipids and other endogenous blood components (Larger et al., 2005; Lesnefsky et al., 2000; Kim et al., 1994). Traditionally, plasma samples have been extracted using a solid-phase extraction, liquid–liquid or protein precipitation extraction where a sample clean-up stage is carried out followed by a sample concentration stage. These

Dried Blood Spots: Applications and Techniques, First Edition. Edited by Wenkui Li and Mike S. Lee.

same sample preparation techniques can also be applied to the analysis of DBSs thus facilitating lower assay limits of quantification.

The physicochemical properties of the molecule should be considered for both the extraction and chromatographic detection of small molecules. Methanol or acetonitrile can be used or in combination with a volume of deionized water. Water can be used to extract very polar analytes. The physicochemical properties of the molecule can be manipulated by pH adjustment and allow for the extraction of ionizable analytes. Increasing the charge to improve solubility in water or reducing the charge to promote solubility in organic solvents is also used during extraction procedures. The extraction of moderately polar analytes may increase by adding 10–15% water to methanol, sometimes by adding water to the dry spot first, then allowing to soak for a few minutes before adding organic solvent. An extremely hydrophobic analyte can be extracted with a nonpolar solvent such as hexane, which also provides sample clean-up by leaving polar contaminants undissolved in the punch.

26.2 DBS CARDS

During method development, some labs screen different card types and solvent mixtures. This approach allows for the selection of the best combination to facilitate further method optimization.

It is important when spotting blood onto cellulose-based papers to ensure that the blood is uniformly spotted. Manufacturing control of cellulose-based papers ensures that inter- and intra-variabilities between batches of cellulose cards are within specification.

Any deviations in the parameters of card thickness, mass, and chemical content could potentially cause erroneous results which could be potentially detrimental to DMPK compound selection and development programs.

Whatman™ FTA™ DMPK cards from GE Healthcare Life Sciences have been tested to determine the variability of mass, thickness, chemical concentration, and dried blood area. The data tabulated were derived from validation work. Ongoing QC inspection for intra-batch variability is monitored to ensure results remain <10%.

Method.

Property	Analysis Method
Physical testing	Weight and thickness of 3 mm punch for all substrates
Chemical testing	*FTA DMPK-A* Chemical concentration by absorbance of eluate from 3 mm punch *FTA DMPK-B* Chemical concentration by conductivity of eluate from 3 mm punch
Dried blood spot area	Blood spot area for all substrates

The consistency of performance in PK applications is related to the reproducibility of the physical properties of the products.

Intra-batch variation.

Property	FTA DMPK-A (%)	FTA DMPK-B (%)	FTA DMPK-C (%)
Mass	<5	<6	<6
Thickness	<3	<3	<2
Dried blood spot area	<7	<5	<7
Chemical concentration	<5	<5	NA

Inter-batch variation.

Property	FTA DMPK-A (%)	FTA DMPK-B (%)	FTA DMPK-C (%)
Mass	<3	<4	<2
Thickness	<2	<4	<2
Dried blood spot area	<2	<4	<4
Chemical concentration	<4	<7	NA

The target specification for total variation in a PK assay is ±15%. Precision and accuracy are therefore extremely important and require a homogenous and high-quality sample. FTA™ DMPK cards are manufactured and tested to meet these criteria.

DBS microvolume sampling using specialized Whatman media has been shown to be both precise and accurate for a variety of compounds from different structural classes with acceptable inter- and intra-batch variability and is now being routinely employed in PK/toxicokinetic (TK) studies. The FTA™ DMPK-A and DMPK-B cards lyse cells and denature proteins on contact. Samples can be shipped and stored at ambient temperature and long-term stability has been demonstrated for analytes and metabolites sensitive to plasma enzymes.

26.3 QUALITY MANAGEMENT

The drive to ensure consistency of product quality and performance is promoted through a quality management system (QMS) which defines and dictates the systemic approach for product development ensuring that performance requirements are met and that robust manufacturing systems are established.

QMS provides the working environment for all activities in GEHC. Compliance dictates that a formal process is followed, documented, reviewed, and approved to ensure delivery on all essential product and manufacturing elements which include:

- Quality objectives and requirements for the product with criteria for design performance, product acceptance, intended use, and user needs.
- Required specifications, verification, validation, monitoring, measurement, inspection, and test activities specific to the product.
- Defined processes, documents, and resources specific to the product.
- Records needed to provide evidence that the realization processes and resulting product meet requirements.
- Process validation to demonstrate manufacturing reproducibility.
- Design validations performed on production or production equivalent units, lots, or batches of the finished product.
- Identification of the characteristics of the product that are essential for safe and proper use.
- Risk management to ensure mitigation of risk both for product use and equally as importantly during manufacture.

The manufacturing site is ISO 9001-2008 accredited, thus, ensuring that adherence to QMS requirements is maintained. QMS demands that testing during development and routine manufacture is performed to ensure that the highest standards of quality and reproducibility are established and maintained (refer to Figure 26.1).

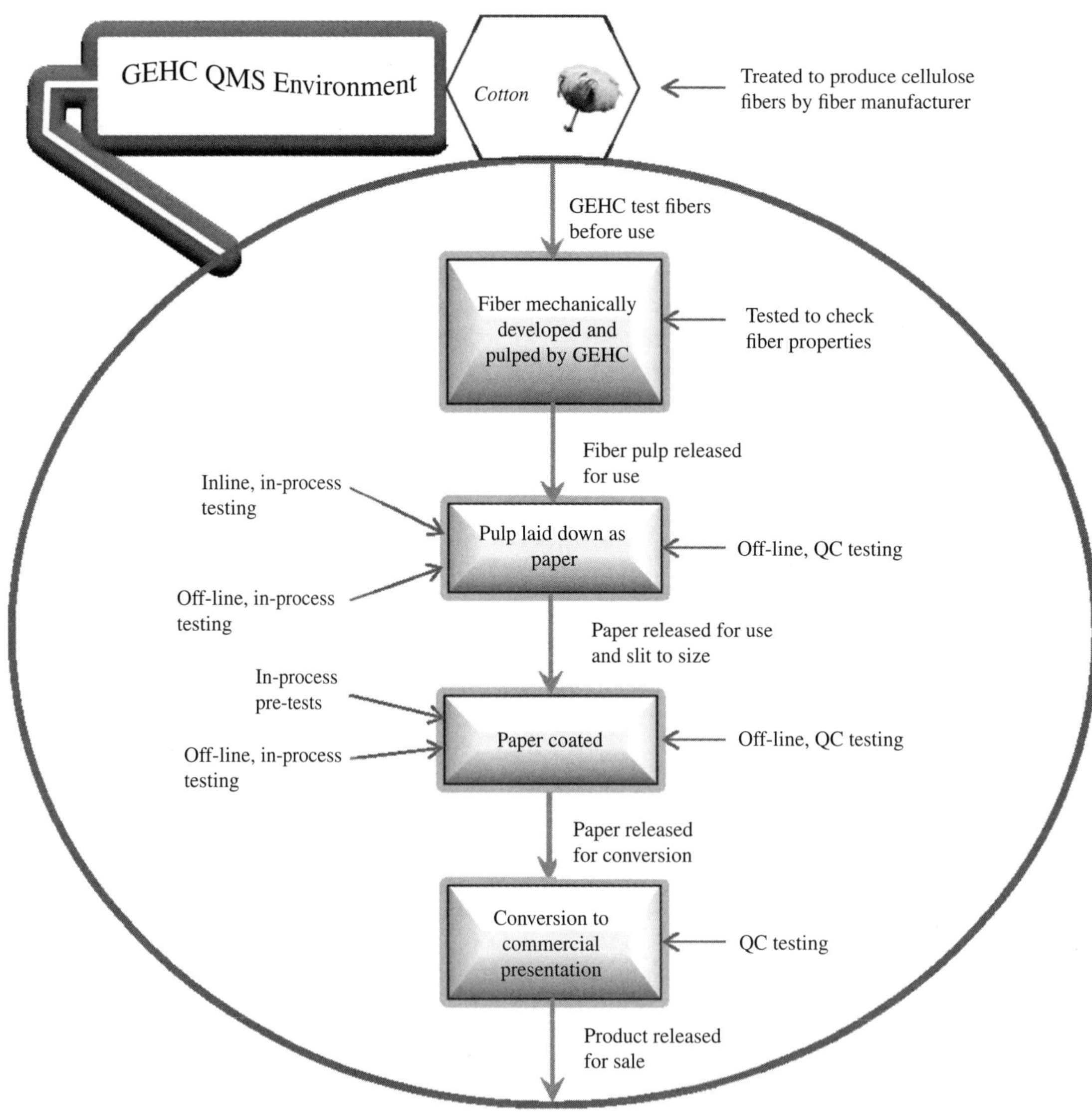

FIGURE 26.1 Process flow for the manufacture of Whatman FTA™ DMPK media.

26.4 PAPER MANUFACTURING

Whatman FTA^{TM} DMPK media are prepared using a proprietary process involving impregnation of pure cotton cellulose linter papers with a variety of chemical species to create the desired properties.

The basis for all Whatman FTA^{TM} DMPK media is cotton cellulose linter paper manufactured under strict controls to produce a highly reproducible material. The manifestation of QMS in the production setting is a theme of constant and intensive testing to provide the assurance of quality required by our customers.

Papermaking, regardless of the scale on which it is done, involves making a dilute suspension of fibers in water and allowing this suspension to drain through a screen so that a mat of randomly interwoven fibers is laid down. Water is removed from this mat of fibers by pressing and drying to make paper.

A modern paper mill is divided into several sections, roughly corresponding to the processes involved in making hand made paper. The industrial process is complex and requires highly trained and skilled operatives in order that paper of the required quality is reproducibly manufactured.

High-purity cotton cellulose is carefully mechanically refined to deliver fiber with the physical properties desired in the papers. The physical aspects of the developed fibers are examined to ensure that the paper made from them will have the desired properties. The fibers are then dispersed in water to make a slurry (the stock) before continuous laydown onto a porous screen.

Once the fiber is laid on the screen, the final stages of processing entail water drainage from the suspension: the wet paper sheet goes through presses and is dried before finally rolled into large reels of filter material. This seemingly simple process is in reality complex and challenging and requires tightly controlled manufacturing systems and operatives who are highly skilled, experienced, and well trained.

In-process tests during the paper manufacturing process may include continuous online monitoring of basis weight and thickness of the material combined with periodic manual assessments to ensure uniformity. The in-process tests are then confirmed by off-line QC testing with further specific functional testing to confirm that quality has been maintained.

The paper manufacturing process is allowed to proceed to the next stage only when QC is satisfied that the base paper meets quality requirements. The paper is then cut (slit) to a size appropriate for the coating process.

Manufacture of specialized coated papers is a time-consuming process. All stages of the manufacturing process from fiber development to coating are validated to ensure reproducibility and precision across and within multiple batches/lots of material. Ongoing quality of the product is maintained by rigorous checks, biological, chemical, and physical, both during the manufacturing process and post production.

The process known as coating involves the impregnation of paper with chemical systems. There are a number of processes that may be involved with coating (e.g., dipping, spraying, transfer printing). All of these processes are capable of producing a paper with the desired properties.

The GEHC proprietary process of necessity takes place in a clean room facility to prevent contamination of the product that could lead to confusing or even false data. High-quality cotton linter papers are used for the coating process to further ensure that final material is free from contamination.

Rigorous site processes are in place as required by QMS to ensure that the clean room and the manufacturing equipment are maintained to the appropriate standard. This process ensures that the sophisticated equipment does not impact on the high-quality requirements.

A mixture of traditional paper tests are used to determine paper quality by thickness, basis weight (also known as grammage), and air porosity. Specially designed tests are used to confirm suitability for use in the specialty markets. Further testing, in-process and off-line, occurs throughout the coating process and with adherence to QMS procedures this ensures that the product reaching the customer is of the highest quality and uniformity. The result of this stage of the process is a reel of coated paper that is now subjected to a barrage of off-line QC testing by specially trained analysts before release to the final stage of manufacture.

Presentation of the product for sale is the final stage of manufacture and is known as conversion. Conversion can be one of a multiple of processes; for a basic laboratory filter it can be as simple as cutting into an appropriately sized disc and packaging or as complex as Whatman FTA^{TM} DMPK media that need to be cut to size, have circles printed on one surface, and be mounted into a printed card or plastic mount, before packaging.

Whatever conversion process is required, QMS adherence remains paramount.

The adherence to QMS and rigorous testing is a constant theme of manufacture at GEHC. Included in the process is a commitment to constant improvement, during manufacture and testing, and listening to GEHC customers to develop new products that meet their needs.

26.5 FTA^{TM} DMPK CARD SELECTION

The choice of card is dictated by a combination of handling and performance criteria. Handling requirements may be influenced by operational or safety factors while performance depends on many factors such as the analyte(s) physicochemical properties, extraction solvent, and bioanalytical work flow. The identification of the most appropriate card chemistry can be determined by investigating analyte recovery from the three different card types. Manufacturers provide generic protocols with the cards which can be used to select the best card type.

FTA DMPK cards are divided into three and the capability of each is outlined below:

FTA DMPK-A.

- Blood spots dry within 2 hours.
- Blood spot area is approximately 20% smaller than DMPK-B or DMPK-C cards.
- Protein denaturing activity will inactivate endogenous enzymes.
- Cell lysis releases endogenous cellular materials onto the card.
- Stabilization of deoxyribonucleic acid (DNA) allows resampling of the blood spot for pharmacogenomics.
- Impregnated chemicals may interfere with analyte(s) mass spectrometry detection due to ion suppression.

FTA DMPK-B.

- Blood spots dry within 2 hours.
- Protein denaturing activity will inactivate endogenous enzymes.
- Cell lysis releases endogenous cellular materials onto the card.
- Stabilization of DNA allows resampling of blood spot for pharmacogenomics.
- Impregnated chemicals may interfere with analyte(s) mass spectrometry detection due to ion suppression.

FTA DMPK-C.

- Blood spots dry within 2 hours.
- No impregnated chemicals to interfere with bioanalysis.
- Proteins are not denatured so the cards may be better suited for protein-based biomolecules.

One pharmaceutical lab tested a panel of small molecule analytes of varying physicochemical properties using the three card types with a range of solvents. Based on the overall results, the FTA DMPK-B cards gave the best overall response to the greatest range of the compounds tested and helped develop a generic method to use these cards in early stage research. The FTA DMPK-C untreated card chemistry has demonstrated the best recovery for protein analysis.

26.6 APPLICATIONS USING DRIED BLOOD SPOT ANALYSIS

26.6.1 Whatman™ Indicating FTA™ DMPK Cards

Traditional (non-indicating) Whatman FTA DMPK cards are routinely used in DMPK laboratories for blood collection and storage. However, drug studies may involve the analysis of additional biological fluids in order to support toxicological and pharmacokinetic/pharmacodynamic (PK/PD) studies. The white color of non-indicating FTA DMPK cards makes it difficult to visualize colorless biological fluids (biofluids) such as plasma, urine, and cerebrospinal fluid (CSF), especially after the samples are dried. Whatman indicating FTA DMPK cards combine the advantages of the traditional FTA cards with the ability to clearly visualize colorless samples both immediately after application and after drying. The use of indicating cards for the analysis of dried plasma spots in the determination of PK in clinical studies is described by Barfield and Wheller (2011).

Whatman indicating FTA DMPK cards are available as FTA DMPK-A, -B, and -C IND cards (refer to Figures 26.2 and 26.3). When a sample is applied to an indicating FTA DMPK card, the blue dye is displaced. The process leaves a lighter-colored area for easy identification of colorless samples such as plasma, urine, and CSF both upon collection and after drying. Figure 26.4 illustrates the difference in appearance after spotting blood and plasma onto indicating cards. In actual practice, blood would be spotted onto non-indicating cards.

26.6.2 Recovery Testing Using Indicating versus Non-indicating FTA DMPK Cards

An experiment was designed to determine

1. the variation in mass spectrometric (MS) response of samples applied to each indicating FTA DMPK card (-A, -B, and -C IND) compared with its corresponding non-indicating card and
2. the intra-card variation in MS response of samples applied to the different types of indicating and non-indicating FTA DMPK cards.

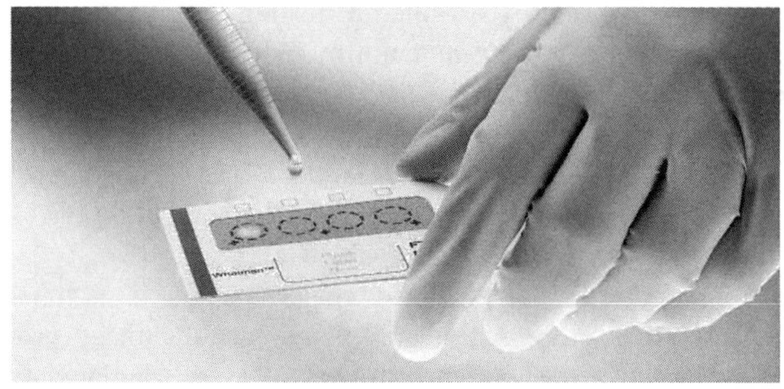

FIGURE 26.2 Collection of colorless biological samples is quick and easy with indicating FTA DMPK cards.

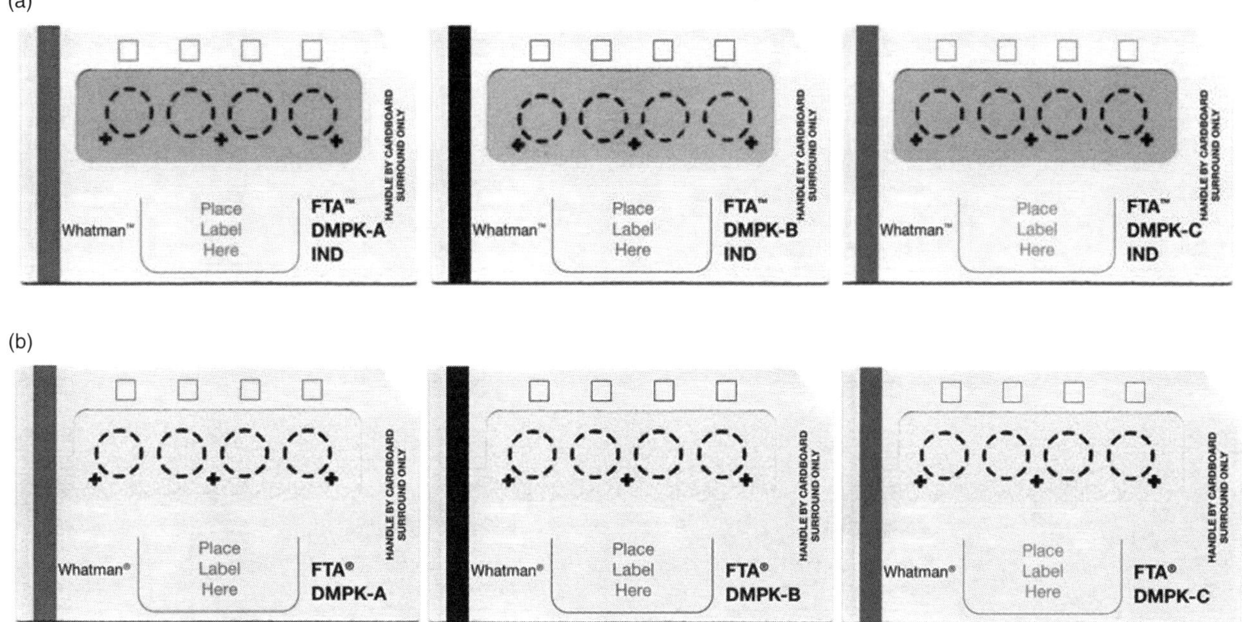

FIGURE 26.3 Whatman FTA DMPK indicating and non-indicating cards: (a) indicating FTA DMPK cards and (b) non-indicating FTA DMPK cards.

Three samples were prepared by adding chloroguanide, ibuprofen, or acetaminophen to human plasma at a concentration of 300 ng/mL. Each spiked plasma sample (15 μL; $n = 6$) was added to a Whatman indicating FTA DMPK-A, -B, and -C card and to the corresponding non-indicating card. The plasma spots were easily visualized on the indicating cards. After the cards were allowed to dry for 2 hours at room temperature, the plasma spots on the IND cards were still clearly visible. A 3 mm punch was removed from each spot and extracted with methanol containing each compound's stable isotope-labeled internal standard. Analytes were detected using an API3000 LC-MS/MS system (Applied Biosystems).

26.6.3 Variation between Indicating and Corresponding Non-indicating Card

For each card type, the mean peak area ratio response of test compound to internal standard was calculated using the following formula:

$$\frac{\text{Mean response of the spiked compound}}{\text{Mean response of the corresponding internal standard}}.$$

The peak area ratio values for indicating and non-indicating FTA DMPK-A, -B, and -C cards are graphically represented in Figure 26.5. Overall, no significant difference in signal was observed between the use of indicating and non-indicating cards.

The coefficient of variation (CV) for each compound/card combination was calculated. Table 26.1 includes a representative set of values for one compound on two card types. The CVs for all compound/card combinations are listed in Table 26.2. The data indicate that the CV for all compound/card combinations is below 10%.

Whatman indicating FTA DMPK cards enable easy visualization of colorless biofluids even after the cards are dried. Both the indicating and non-indicating versions of FTA DMPK cards have less than 10% CV within each card based on MS analysis of three test compounds. In addition, there is no significant difference in results obtained with indicating versus corresponding non-indicating cards for all card types.

26.6.4 FTA Cards for the Storage of DNA in Blood

FTA cards provide a solid medium support which enables DNA in blood samples to be stored and easily transported, allowing the recovery of the DNA for analysis. The analysis of DNA from blood is used in the diagnosis of genetic disease, in the determination of paternity, in forensics, and in animal breeding studies. Blood is applied to a cellulose-based paper which is coated with compounds which protect against degradation of DNA.

Cellulose papers may be comprised of a weak base, a chelating agent, and an anionic surfactant or detergent and optionally uric acid or a urate salt. Preferably, the composition results in an alkaline pH, between 8.0 and 9.5. The surfactant denatures proteins and the majority of any pathogenic organisms in the blood sample. The blood DNA, however, is protected from degradation due to the alkaline pH.

(a)

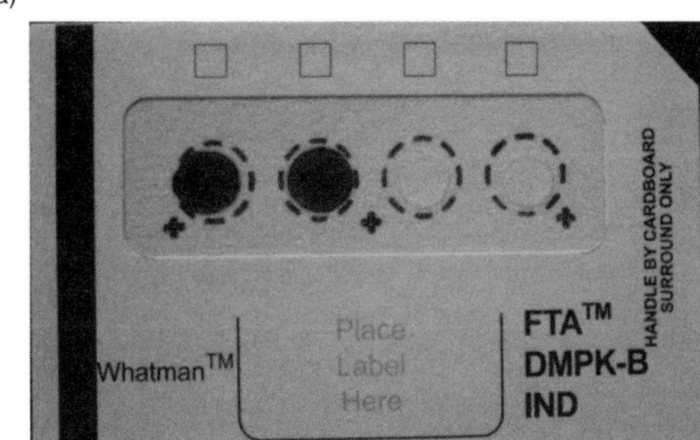

(b)

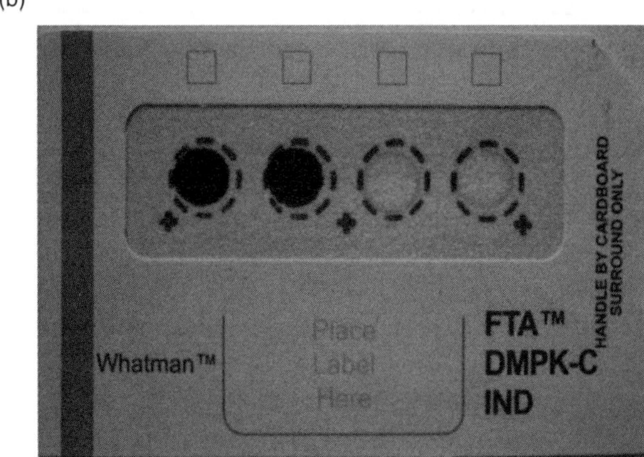

(c)

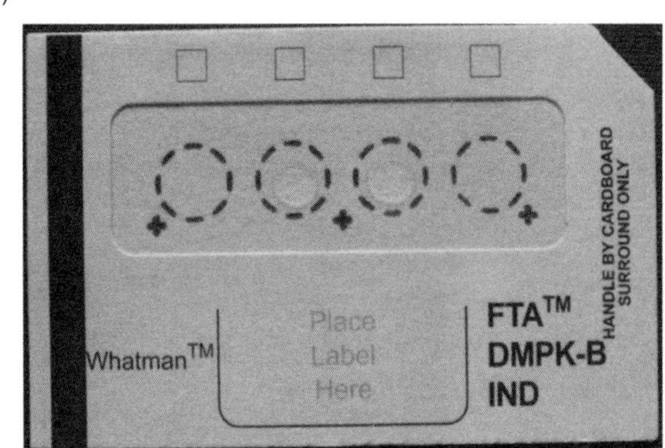

FIGURE 26.4 Indicating FTA DMPK cards. (a) 15 μL blood and 10 μL plasma spotted on FTA DMPK-B IND; (b) 15 μL blood and 10 μL plasma spotted on FTA DMPK-C IND; and (c) 5 μL urine spotted on FTA DMPK-B IND.

(a)

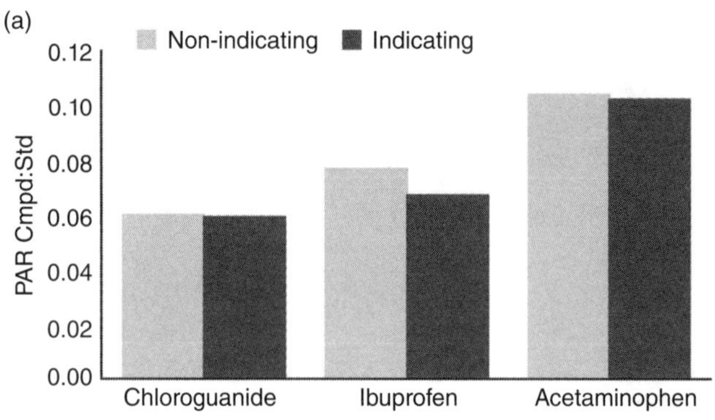

(b)

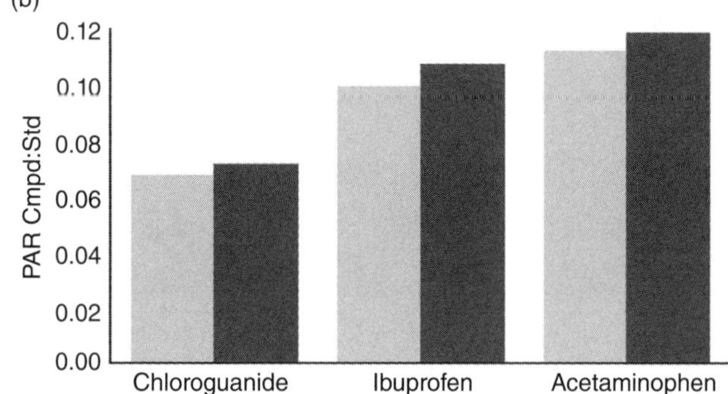

(c)

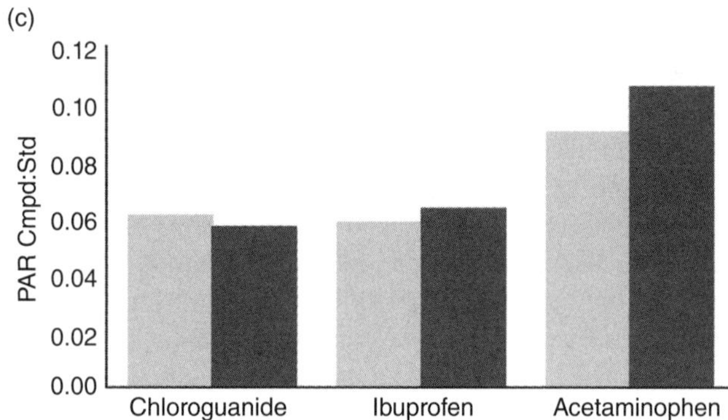

FIGURE 26.5 Mean peak area ratio mass spectrometric (MS) response for compounds relative to their isotope-labeled internal standard. Indicating and corresponding non-indicating cards were analyzed for (a) FTA DMPK-A; (b) FTA DMPK-B; and (c) FTA DMPK-C. PAR = peak area ratio response; Cmpd = compound; Std = internal standard.

As detailed in United States Patent 5496562, "solid medium and method for DNA storage" cellulose-based papers are coated with

i. a monovalent weak base (such as "Tris", tris-hydroxymethyl methane, either as the free base or as the carbonate;

TABLE 26.1 Mean, SD, and CV of Peak Area Ratio for Chloroguanide on Indicating and Non-indicating FTA DMPK-C Cards

Chloroguanide	Mean	SD	CV%
FTA DMPK-C NON	0.0497	0.00117	2.36
FTA DMPK-C IND	0.0469	0.00175	3.74

IND, indicating; NON, non-indicating.

TABLE 26.2 CV(%) of Peak Area Ratio for All Compound/Card Combinations

	Chloroguanide	Ibuprofen	Acetominophen
FTA DMPK-A			
NON	4.60	3.68	2.32
IND	5.88	2.27	2.67
FTA DMPK-B			
NON	7.43	6.17	4.99
IND	9.10	9.52	6.67
FTA DMPK-C			
NON	2.36	8.13	3.49
IND	3.74	7.47	6.88

ii. a chelating agent (such as EDTA, ethylene diamine tetracetic acid); and

iii. an anionic detergent (such as SDS, sodium dodecyl sulfate); and optionally

iv. uric acid or a urate salt

Uric acid or a urate salt has been found to be important in the long-term storage of DNA as it is converted into allantoin in acting as a "free-radical" trap that preferentially accepts free radicals, which would otherwise damage the base guanine in the DNA. Free radicals are generated by the spontaneous oxidation of thio groups in the denatured serum protein and may also be generated by iron in blood. The uric acid also acts as a component of the buffering system in that it is a weak acid. Uric acid also acts as an erodible surface in that it is sparingly soluble so that DNA-containing material dried onto its crystals will be released as the urate beneath them erodes.

A monovalent weak base, to cause an alkaline pH between 8.0 and 9.5, is used to allow the action of the chelating agent to bind divalent metal ions, calcium and magnesium, and metal ions such as iron. Both calcium and magnesium can promote DNA degradation by acting as cofactors for enzymes. Iron can readily undergo oxidation and reduction which can also damage nucleic acids by the production of free radicals. The anionic detergent causes most pathogens to be inactivated due to the nonspecific destruction of the secondary structure of the coat proteins, their internal proteins, and any membranes that they may be dependent upon.

DNA from blood samples applied to FTA has been successfully analyzed after 22 years. Specific loci on DNA can be amplified to discover the lengths of short tandem repeats (STRs), generating a DNA profile.

Figure 26.6 illustrates the STR analysis of 22-year-old blood stains on FTA using the ABI Profiler amplification system. Panel A (FAM dye) represents the loci D3S1358, vWA, and FGA. Panel B (JOE dye) represents amelgenin, D8S1179, D21S11, and D18S51. Panel C (NED dye) represents D5S818, D13S317, and D7S820.

Figure 26.7 illustrates the STR analysis of the same sample but using the ABI AmpFℓSTR Identifiler Direct amplification system. The Identifiler Direct method builds upon the markers used within the Profiler/COfiler systems by incorporating further loci and thus providing higher levels of discrimination between profiles.

In Figure 26.7, Panel A (FAM dye) represents loci D8S1179, D21S11, D7S820, and CSF1PO. Panel B (VIC dye) represents D3S1358, TH01, D13S317, D16S539, and D2S1338. Panel C (NED dye) represents D19S433, vWA, TPOX, and D18S51. Panel D (PET dye) represents amelogenin, D5S818, and FGA.

Prior to amplification with the Profiler/Cofiler reagents the blood spot samples were prepared by taking manually 1.2 mm diameter punches followed by washing according to a modified wash protocol, utilizing FTA purification reagent containing 60 μg/mL Proteinase K.

Processed punches from each sample were amplified using ABI AmpflSTR® Profiler™ and Cofiler™ Systems. The reaction volume used was 10 μL and the cycle number was reduced to 24 for all reactions. A 10 μL sample preparation (1 μL amplified sample, 0.5 μL GS ROX 500, 8.5 μL deionized formamide) was detected using an ABI 3130xl Genetic Analyzer. The default injection parameters were modified: 3 kV injection voltage and 15 seconds injection time. GeneMapper® Analysis Software version 3.2 was used to analyze the raw data.

A 1.2 mm punch was taken from the sample and placed into a 25 μL reaction for the Identifiler Direct amplification. Amplification was performed using the standard Identifiler Direct protocol for 27 cycles.

For analysis 10 μL of sample preparation (1 μL PCR product, 8.7 μL deionized formamide, and 0.3 μL GS500 LIZ ILS) was analyzed using 3130xl and standard Identifiler Direct run protocols.

Full concordance was observed between both testing kits, at the common loci.

The STR profiles obtained from 22-year-old blood demonstrate that FTA both protects the DNA during storage and provides a rapid, effective amplification method.

26.6.5 FTA Cards for the Storage of RNA

FTA cards can be used to prepare total RNA from a variety of samples for molecular analysis. Nucleic acids are instantly captured and stabilized, while pathogens and nucleases are inactivated. RNA can be easily purified using reverse transcriptase (RT)-PCR and northern blot analysis. RNA purified using FTA can serve as a template for RT-PCR analysis directly in solution or can be precipitated and loaded onto an agarose gel for northern blotting. RNA is eluted from the FTA card using an RNA processing buffer. Figure 26.8 illustrates a RT-PCR from FTA; RNA from HL60 cells are extracted with FTA or Competitor T.

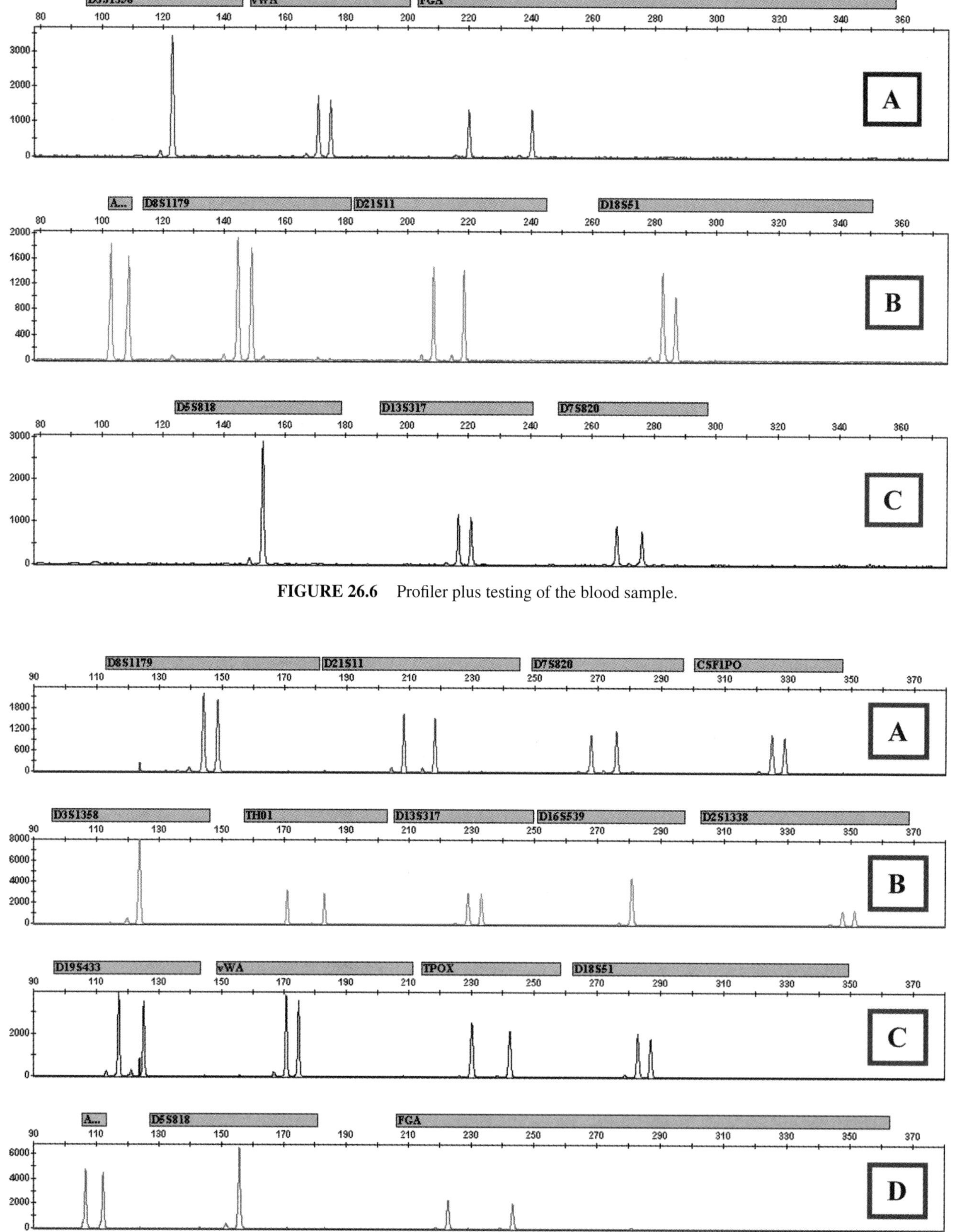

FIGURE 26.6 Profiler plus testing of the blood sample.

FIGURE 26.7 Identifiler Direct testing of the blood sample.

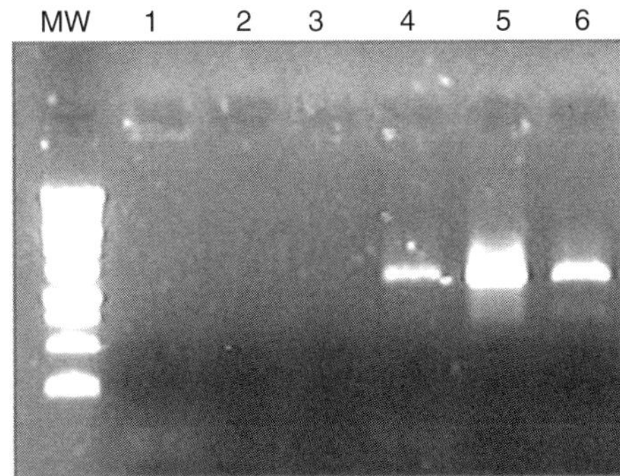

FIGURE 26.8 A RT-PCR from FTA; RNA from HL60 cells are extracted with FTA or Competitor T. Lanes 1 and 2: PCR using RNA from FTA (no DNA contamination). Lane 3: negative control (no template). Lane 4: PCR using RNA from Competitor T (DNA contamination). Lane 5: PCR using positive control K562 DNA. Lane 6: RT-PCR using RNA from FTA.

26.6.6 Detection of Protein Analytes Using FTA Cards

In a similar manner to small molecule analysis, the use of DBS on paper offers several benefits for protein analysis compared with conventional serum and plasma sampling (e.g., smaller volume sample collection, simplified storage and shipment). However, there are still relatively few reports on the analysis of proteins from DBS (Corran et al., 2008; Langkamp et al., 2008; Prince and Matsuda, 2008).

One likely reason for the slow adaption of DBS technology in protein analysis is that the dogma in experimental protein chemistry is to make every effort to avoid dramatic changes in a protein's physical or chemical environment, therefore, minimizing the risk of protein denaturation. Proteins that have lost their native conformation are known to be more prone to aggregation and to adsorption irreversibly to surfaces.

Also, commonly used protein analysis methods (e.g., ELISA, enzyme assays, and surface plasmon resonance) require native protein structure for reliable results.

There has been a concern that DBS technology is too harsh on proteins but recent studies have proven that is not the case. Indeed, it is now evident that DBS sampling for quantitative bioanalysis will be an accepted part of the future biopharmaceutical development landscape (Prince and Matsuda, 2008; Spooner 2010b).

Untreated cellulose cards (e.g., DMPK-C) demonstrated good recovery of therapeutic monoclonal antibodies (Prince and Matsuda, 2008) while treated paper optimized for analyses of small molecules and DNA did not. Protein recovery also depends on the elution conditions, and denaturing buffers should be avoided. Mild standard buffers like PBS, which are compatible with standard protein analysis techniques, have proven to work well. Using untreated cellulose cards and PBS buffer elution, GEHC have consistently achieved total protein recoveries around 60–80% of various protein analytes as monitored by immunological (ELISA) or functional (interaction analysis with Biacore; enzyme assay) methods (unpublished data).

26.7 HINTS AND TIPS IN USING DRIED BLOOD SPOTS

- It is important to have the sample spread evenly on the card surface as it quickly soaks in, giving a spot area directly proportional in area to the sample volume. The cards are designed and carefully manufactured for reproducible spread.
- For the application of blood to the card, a single large drop hanging from the pipette tip (or capillary tube) should touch the card. The pipette tip itself should not come in contact with the card. This way the sample spreads quickly and symmetrically on the card surface while it also soaks into the card, producing a reproducible and uniform spot. Spot formation is controlled by physical properties of the sample and the card, with minimal influence of application speed or direction and with minimal chromatographic separation in the card. Spot area will be proportional to sample volume and the sample amount per 3 mm punch will be highly reproducible. It is relatively easy to achieve this ideal application with small samples, up to about 25 μL as long as the pipette tip is not too fine. Larger volumes tend to dispense as multiple drops, increasing the risk of nonuniformity although some labs are able to obtain satisfactory spots with larger volumes.
- If the pipette or capillary tip were held in contact with the card during dispensing, then the liquid flow would tend to be more through the card and encourage local depletion of the chemical components in the card. The flow would probably be more directional and result in uneven sample distribution. These results would be more likely to vary between samples and between operators.
- Plasma and blood samples often contain an anticoagulant such as EDTA. The cards do not require anticoagulant (as long as the sample has not already begun to clot) but are compatible with them. EDTA- and heparin-coated capillaries and tubes are available for sample collection.
- Carryover of analytes from punch to punch is not a problem as long as the cards are thoroughly dry, preventing carryover of fibers. If required, then clean the punching tool by cutting a waste disc between samples from an unused part of the card.
- Experiments have shown very little variation between the drug concentration and where in the spot the sample

is punched although it is not recommended to sample from the very edge of the sample spot.

- Elution of analytes with organic solvents will leave protein precipitated in the cards.
- Coated DMPK-A and DMPK-B cards (both standard and indicating) contain chemicals to lyse cells and denature proteins. Some of these components will be eluted with the analyte of interest. If necessary, ion suppression effects can be eliminated by adjustment of the chromatographic conditions. DMPK-C cards contain pure cellulose without any chemical coating.
- In more qualitative assays such as newborn screening, the sample is sometimes applied as multiple separate spots to fill the sample area. This approach is not recommended for quantitative assays such as PK and TK.
- Frozen blood is not recommended because of cell lysis and protein aggregation during freezing and upon thawing. Anticoagulated blood can be stored for a week or longer at 4°C and should be gently mixed before use. Blood should be used from the same species as the analytical samples for the preparation of standards and quality controls. The appropriate control matrix should be used for samples other than blood.
- Aqueous buffers either with or without non-ionic detergent (e.g., 0.1% Tween™-20 or Triton™ X-100) are usually used as elution solvents for protein analytes. In general, an aqueous buffer with pH and salt concentration is selected to promote protein stability. Hydrophobic peptides will probably extract better with some methanol or acetonitrile added.
- Good air circulation and low ambient humidity are most important in the drying of DBS cards. Moderate heating will not damage the cards but may reduce analyte stability. Fans or vacuum desiccators may speed up the process.

REFERENCES

Abu-Rabie P, Spooner N. Study to assess the effect of age of control human and animal blood on its suitability for use in quantitative bioanalytical DBS methods. *Bioanalysis* 2010;2(8):1373–1384.

Barfield M, Wheller R. Use of dried plasma spots in the determination of pharmacokinetics in clinical studies: validation of a quantitative bioanalytical method. *Analytical Chemistry* 2011;83:118–124.

Barfield M, Spooner N, Lad R, Parry S, Fowles S. Application of dried blood spots combined with HPLC-MS/MS for the quantification of acetaminophen in toxicokinetic studies. *Journal of Chromatography B* 2008;870:32–37.

Beaudette P, Bateman KP. Discovery stage pharmacokinetics using dried blood spots. *Journal of Chromatography B* 2004;809:153–158.

Clark GT, Haynes JJ. Utilization of DBS within drug discovery: a simple 2D-LC-MS/MS system to minimize blood and paper-based matrix effects from FTA elute™ DBS. *Bioanalysis* 2011;3(11):1253–1270.

Clark GT, Haynes JJ, Bayliss MJ, Burrows L. Utilization of DBS within drug discovery: development of a serial microsampling pharmacokinetic study in mice. *Bioanalysis* 2010;2(8):1477–1488.

Corran PH, Cook J, Lynch C, Leendertse H, Manjurano A, Griffin J, Cox J, Abeku T, Bousema T, Ghani AC, Drakeley C, Riley E. Dried blood spots as a source of anti-malarial antibodies for epidemiological studies. *Malaria Journal* 2008;7:195.

Denniff P, Spooner N. Effect of storage conditions on the weight and appearance of dried blood spot samples on various cellulose-based substrates. *Bioanalysis* 2010;2(11):1817–1822.

Festing S, Wilkinson R. The ethics of animal research. *EMBO Reports* 2007;8(6):526–530.

Guthrie R, Suzi A. A simple phenylalanine method for detecting phenylketonuria in large populations of newborn infants. *Pediatrics* 1963;32:338–343.

Hannon HW, Whitley RJ, Davin B, Fernhoff P, Halonen T, Lavochkin M. *Blood Collection on Filter Paper for Newborn Screening Programs; Approved Standard – Fifth Edition*. CSLI document LA4-A5 (ISBN 1-56238-644-1).Wayne, PA: Clinical and Laboratory Standards Institute; 2007.

Kim H, Wang TL, Ma Y. Liquid chromatography/mass spectrometry of phospholipids using electrospray ionization. *Analytical Chemistry* 1994;66:3977–3982.

Langkamp M, Weber K, Ranke MB. Human growth hormone measurement by means of a sensitive ELISA of whole blood spots on filter paper. *Growth Hormone & Igf Research* 2008;18:526–532.

Larger PJ, Breda M, Fraier D, Hughes H, James CA. Ion-suppression effects in liquid chromatography-tandem mass spectrometry due to a formulation agent, a case study in drug discovery bioanalysis. *Journal of Pharmaceutical and Biomedical Analysis* 2005;39:206–216.

Lesnefsky EJ, Stoll MSK, Minkler PE, Hoppel CL. Separation and quantitation of phospholipids and lysophospholipids by high-performance liquid chromatography. *Analytical Biochemistry* 2000;285:246–254.

Li W, Tse FL. Dried blood spot sampling in combination with LC-MS/MS for quantitative analysis of small molecules. *Biomedical Chromatography* 2010;24:49–65.

Mather J, Ranville PD, Spooner, N, Evans CA, Smith NW, Plumb RS. Rapid analysis of dried blood spot samples with sub-2-μm LC-MS/MS. *Bioanalysis* 2011;3(4):411–420.

Mei JV, Alexander JR, Adam BW, Hannon H. Use of filter paper for the collection and analysis of human whole blood specimens. *Journal of Nutrition* 2001;131:1631S–1636S.

NSQAP. *NSQAP Monitor Neonatal Paper Quality and Performance*. Atlanta, GA: U.S. Department of Health and Human Services, Centres for Disease Control and Prevention, National Centre for Environmental Health; 2008.

Oliveira EJ, Watson DG, Morton NS. A simple microanalytical technique for the determination of paracetamol and its main metabolites in blood spots. *Journal of Pharmaceutical and Biomedical Analysis* 2002;29:803–809.

Pitt JJ. Newborn screening. *Clinical Biochemist. Reviews* 2010; 31:57–68.

Prince, PJ, Matsuda KC, Retter MW, Scott G. Assessment of DBS technology for the detection of therapeutic antibodies. *Bioanalysis* 2008;2(8):1449–1460.

Spooner N. A glowing future for dried blood spot sampling. *Bioanalysis* 2010a;2(8):1343–1344.

Spooner, N. Dried blood spot sampling for quantitative bioanalysis: time for a revolution? *Bioanalysis* 2010b;2(11):1781.

Spooner N, Lad R, Barfield M. Dried blood spots as a sample collection technique for the determination of pharmacokinetics in clinical studies: considerations for the validation of a quantitative bioanalytical method. *Analytical Chemistry* 2009;81:1557–1563.

Tanna S, Lawson G. Analytical methods used in conjunction with dried blood spots. *Analytical Methods* 2011;3:1709–1718.

Todime MR, Tama CI, Hayes RN. A dried blood spots technique based LC-MS/MS method for the analysis of posaconazole in human whole blood samples. *Journal of Chromatography B* 2011;879:3626–3638.

Wickremsinhe ER, Abdul BG, Huang NH, Richard JW, Hanes JL, Ruterbories KJ, Perkins EJ, Chaudhary AK. Dried blood spot sampling: coupling bioanalytical feasibility, blood-plasma partitioning and transferability to in vivo preclinical samples. *Bioanalysis* 2011;3(14):1635–1646.

27

DERIVATIZATION TECHNIQUES IN DRIED BLOOD SPOT ANALYSIS

Ann-Sofie M.E. Ingels, Nele Sadones, Pieter M.M. De Kesel, Willy E. Lambert, and Christophe P. Stove

27.1 INTRODUCTION

The dried blood spot (DBS) sampling technique has several advantages over a venipuncture, making it a cost-effective choice for the collection, transport, and storage of blood samples. Inherent to the sampling technique is the small sample volume available, ranging from 5 to 100 μL, compared with 1 mL or more obtained by venipuncture. Although this may represent an advantage in case of sampling patients with restricted or limited venous access, such as neonates and children, these small amounts may impose an analytical challenge and require efficient sample treatment and a sensitive detection (McDade et al., 2007; Garcia Boy et al., 2008; Edelbroek et al., 2009; Li and Tse, 2010). To achieve adequate method sensitivity for the analysis of different pharmaceutical compounds or biomarkers in DBS, even at lower concentration levels, the majority of DBS applications use tandem mass spectrometry (MS/MS) coupled to liquid chromatography (LC). Other analytical techniques such as direct MS/MS, LC coupled to fluorescence (LC-FLUO) or ultraviolet detection (LC-UV), or gas chromatography coupled to MS (GC-MS) or tandem MS (GC-MS/MS) have been demonstrated to be suitable alternatives (Déglon et al., 2010; Tanna and Lawson, 2011; Stove et al., 2012). Additionally, to achieve the required method sensitivity, DBS analysis may involve derivatization. This may lead to an improvement of the chromatographic properties of the analytes of interest, which consequently may also influence method sensitivity by enhancing volatility, separation efficiency, and/or selectivity (Lin et al., 2008). Derivatization is primarily known as a technique extending the molecular application range of GC (Kaal and Janssen, 2008). During sample workup of DBS GC-MS (/MS) applications, derivatization reactions such as silylation, alkylation, and/or acetylation have been performed (Shen et al., 2006; Déglon et al., 2010; Ingels et al., 2010). Also LC-UV or LC-FLUO applications may integrate a derivatization step during DBS analysis to improve detection sensitivity and selectivity by enhancing the UV properties or the fluorescence yield of the target analytes, respectively (Dale et al., 2003; Al-Dirbashi et al., 2005; Tanna and Lawson, 2011).

In contrast, derivatization is less commonly used for LC-MS/MS analysis, especially because omission of derivatization in an analysis is recognized as a major advantage, but also because variation and artifacts can be introduced (Wuyts et al., 2007; Xu et al., 2011). However, the integration of derivatization techniques could enhance the capabilities of certain MS/MS-based applications and may give rise to several advantages such as improved chromatography and improved mass spectrometric properties (e.g., ionization efficiency, *m/z*) of the target compounds (Xu et al., 2011). For example, the electrospray ionization (ESI) yield of neutral ketosteroids may be low as they lack a functional group that is easily ionized under normal conditions. This limitation can be overcome by introducing a chargeable moiety. In this way, derivatization of 17-hydroxyprogesterone to its positively charged hydrazone before LC-ESI-MS/MS analysis, resulted in a 10-fold gain in sensitivity (Lai et al., 2001; Santa, 2011).

Although, in theory, higher ionization yield and better selectivity are expected for small molecules or neutral compounds when using derivatization, it may be important to

Dried Blood Spots: Applications and Techniques, First Edition. Edited by Wenkui Li and Mike S. Lee.

evaluate the differences between derivatization and non-derivatization procedures on an analyte-per-analyte basis prior to selecting the most suitable sample workup protocol. This point was demonstrated by De Jesus et al., comparing the ionization efficiency for acylcarnitines and amino acids (AA) in derivatized (butyl ester) and non-derivatized forms (free acid) (De Jesus et al., 2010). The authors found that for the majority of the selected compounds, minor differences in quantitative results were observed between both methods, while mass spectrometric responses varied from more intense without derivatization, over being similar, to less than 66% of the mass spectrometric ion counts obtained by derivatization. Moreover, without derivatization the method may be less selective, not capable to differentiate isobaric acylcarnitines.

Different traditional derivatizing reagents are available for GC-MS, LC-FLUO and LC-UV based procedures, and some of these have been applied in (LC-) MS/MS applications. However, for the latter, limitations may be encountered such as suboptimal ionization efficiency and product ion yield. Furthermore, different LC behaviors are expected for the derivatized compounds. Therefore, efforts have recently been made to design derivatives specifically for (LC-) MS/MS based approaches (Santa, 2011; Xu et al., 2011; Niwa, 2012). Examples that illustrate the advantages gained as a result of derivatizing the target compounds in DBS are given below. However, as it is beyond the scope of this chapter to provide an exhaustive overview of current derivatization techniques and reagents utilized in GC, LC, and (LC-) MS/MS methods, we would like to refer to comprehensive reviews on this subject (Knapp, 1979; Segura et al., 1998; Kaal and Janssen, 2008; Lin et al., 2008; Santa, 2011; Todoroki et al., 2011; Xu et al., 2011; Niwa, 2012).

27.2 OVERVIEW OF DERIVATIZATION TECHNIQUES IN DBS ANALYSIS

The aim of this chapter is to present an overview of DBS methods utilizing derivatization published since 1990 up till now. Generally, formation of derivatives can be carried out during sample workup (pre-column) or post-column before the column eluate enters the detection system (Moffat et al., 2004). A few DBS methods reported in the 1980s opted for such a post-column derivatization technique in combination with LC–FLUO, mainly to avoid instability of the derivatives during sample workup and separation (Hayashi et al., 1983; Roesel et al., 1986). By mixing column eluates online with the derivatizing reagents, the derivatized target compounds are directly detected, but as these are also diluted this resulted in sensitivity loss that needed to be compensated for by the gain in sensitivity due to derivatization (Carducci et al., 2001). Although considerable improvements in method sensitivity and analysis time have been made in comparison with the original procedures, only a minority of recently reported DBS methods uses this derivatization technique (Abdulrazzaq and Ibrahim, 2001). Hence, contribution of the post-column derivatization technique in DBS analysis has been considered as too limited, and consequently, beyond the scope of this chapter.

The focus of this overview lies on derivatization techniques utilized during sample preparation. Therefore, we made a classification based upon the DBS sample workup procedure. The first group of methods has in common that the conducted DBS sample treatment is considered to be the "general" procedure (Table 27.1; Figure 27.1, general procedure). Tables 27.2, 27.3, and 27.4 summarize selected methods with modifications to this "general" procedure (Figure 27.1, modified sample workup procedures 1–3). The latter include direct derivatization, a procedure in which (extracting and) derivatizing solutions are applied in one single step with the DBS.

Important factors that contribute to the choice of a certain derivatization DBS sample workup procedure are the choice of derivatizing reagent and the circumstances required to form a stable derivative in a quantitative way. The choice of derivatization reagent depends on its turn on the physical and chemical properties of the target analytes and the instrument characteristics. The reaction yield can be influenced by the type and amount of derivatizing reagent, the pH, temperature, and time needed to complete the reaction. In addition, some reagents require an aprotic environment for the reaction to occur while others react well in aqueous media. Furthermore, as not all reagents can be injected directly into an analytical system, excess derivatization reagent may need to be removed prior to analysis (Rood, 1999; Lin et al., 2008; Xu et al., 2011).

27.3 GENERAL DBS SAMPLE WORKUP PROCEDURE INCLUDING DERIVATIZATION

An overview of selected procedures that apply the "general" procedure is shown in Table 27.1. These procedures have been widely applied in metabolic screening of newborn DBS and in follow-up monitoring of symptomatic patients (Carducci et al., 2006). Thereby, as shown in Table 27.1, various assay methods such as (LC-) MS/MS, and GC-MS and derivatization procedures have been utilized to achieve required method sensitivity and/or selectivity. All selected methods follow a similar sample workup (Figure 27.1: general procedure) and start with elution of the analytes of interest from the DBS, sometimes followed by an extra purification step such as solid-phase extraction (SPE) in order to increase analytical column lifetime and reduce MS cleaning (Higashi et al., 2008, 2011). Subsequently, (an aliquot of) the extract is transferred and evaporated under a stream of nitrogen before adding the derivatization reagent(s). Then,

TABLE 27.1 Selected Examples of DBS Methods Using the "General" Sample Workup Procedure

Assay Method	Type of Derivatization	Analyte(s) of Interest	Application	Selected References
LC-MS/MS	Hydrazone complex formation	17-OH-progesterone 17-OH-pregnolone	NBS	Lai et al., 2001; Higashi et al., 2008
	Alkylation: butylesterification	Guanidinoacetate Creatine	NBS	Bodamer et al., 2001
	Diels–Alder	25-OH-vitamin D_3 25-OH-vitamin D_2	NBS	Eyles et al., 2009
	Diels–Alder acetylation (2-step)	25-OH-vitamin D_3 3-epi-25-OH-vitamin D_3	NBS	Higashi et al., 2011
MS/MS	Alkylation: butylesterification	AA (Acyl)carnitine(s) GAA Creatine	NBS	Chace et al., 1993, 1995, 1996; Naylor and Chace, 1999; Fingerhut et al., 2001; Carducci et al., 2006; Chace et al., 2009; Jebrail et al., 2012
GC-MS	Alkylation: butylesterification acetylation (2-step)	AA	NBS	Deng et al., 2002b; Deng and Deng, 2003
	Silylation	AA	NBS	Deng et al., 2002a, 2005b, 2005c, Shen et al., 2006

AA, amino acids; GAA, guanidinoacetic acid; GC, gas chromatography; LC, liquid chromatography; MS/MS, tandem mass spectrometry; MS, mass spectrometry; NBS, newborn screening.

after completion of the derivatization, the excess reagent is removed by evaporation under a stream of nitrogen, followed by reconstitution of the derivatized extract prior to injection.

In an LC-MS/MS method developed to determine 17-hydroxyprogesterone and 17-hydroxypregnolone in DBS, the target compounds were derivatized according to this general procedure to enhance ionization during ESI-MS/MS (Higashi et al., 2008). In addition, direct MS/MS, without chromatographic separation, has become a well-established technique for the quantitative determination of several biomarkers in DBS after derivatization. Corresponding butylesters are prepared prior to analysis, in order to enhance sensitivity and to reduce potential background interferences by increasing *m/z* values as a result of mass gain. This procedure, first reported in the 1990s by Chace et al. (1993; 1995; 1996), has replaced historically used newborn screening (NBS) tests and has evolved to a single-run analysis using fully automated ESI-MS/MS, detecting over 65 metabolites and/or specific markers for disorders in amino acid, fatty acid, or organic acid metabolism (Naylor and Chace, 1999; Chace, 2009).

Under some conditions, a two-step derivatization reaction that involves the sequential application of non-compatible derivatization reagents may be required to obtain suitable derivatives. The usefulness of including a two-step reaction is illustrated by the following example. Eyles et al. (2009) and Higashi et al. (2011) both developed an LC-MS/MS method for the determination of 25-OH-vitamin D_3

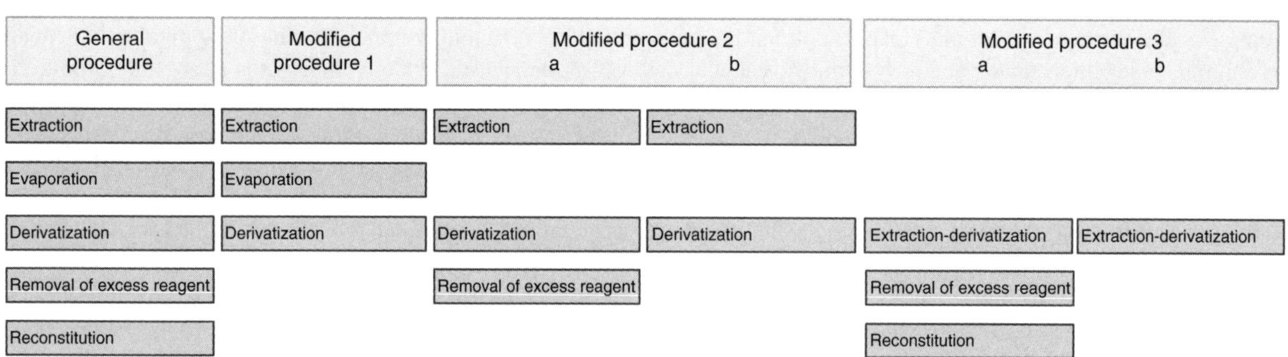

FIGURE 27.1 Schematic overview of the general and modified sample workup procedures in DBS analyses including a derivatization step.

in DBS to diagnose vitamin D deficiency. In the first method, 25-OH-vitamin D_2 and D_3 were derivatized using a single derivatization reaction to increase their ESI-MS/MS response. In addition, to separate the target analyte from a potentially interfering epimer (3-epi-25-OH-vitamin D_3) during LC, Higashi et al. (2011) performed an additional derivatization reaction, thereby enhancing method selectivity and avoiding overestimation of the vitamin D_3 level.

Although (LC-) MS/MS has been generally recognized as a powerful tool for NBS, it can be too expensive in some circumstances. Then, a possible and suitable alternative is GC-MS, known to be a relatively inexpensive, simple, but sensitive technique. For the determination of different AA in newborn DBS using GC-MS, previously reported two-step derivatization sample workup procedures using alkylation and acetylation have been improved by the use of single-step silylation, hence contributing to simplified sample workup procedures (Deng et al., 2002a, 2002b; Deng and Deng, 2003; Shen et al., 2006). Moreover, instead of classical silylation, microwave-assisted-silylation has been successfully applied to determine AA in DBS. Heating the sample mixture in a sealed vessel by microwave energy resulted in a strong reduction of the time required for energy transfer and, consequently, in faster completion of derivatization processes (Deng et al., 2005b, 2005c; Soderholm et al., 2010).

27.4 MODIFICATIONS

Several modifications have been made to the above mentioned "general" procedure, mainly to minimize and simplify the DBS sample preparation. Reported modified sample workup procedures can be classified into three groups. Procedures in the first group only differ from the "general" procedure in that the derivatized extract is not dried and reconstituted, but is directly injected into the analytical system (Table 27.2; Figure 27.1, modified sample workup procedure 1). In the second modified sample workup, the evaporation step between the extraction and derivatization is omitted (Figure 27.1: modified sample workup procedure 2). Furthermore, once derivatized, an aliquot of the sample mixture can be injected into the analytical system either after removal of the excess reagent (Table 27.3a) or directly (Table 27.3b). The third group achieves a simplified sample workup by using a direct derivatization technique, meaning that DBS extraction and derivatization of the extracted compounds are performed in one single step (Figure 27.1: modified sample workup procedure 3). To this end, DBS can be exposed to extraction and derivatization reagents simultaneously, or only to the latter, leading to "on spot" derivatization without the use of any extraction solvent. Similar to modified sample workup procedure 2, the final sample mixture can be injected either after removal of the excess reagent (Table 27.4a) or directly after derivatization (Table 27.4b). In the following subsections, these modified sample workup procedures are further explained and illustrated by several examples.

27.4.1 Modified Sample Workup Procedure 1: Extraction–Evaporation–Derivatization

A first modification of the "general" sample treatment consists of extraction of a DBS, derivatization of the dried extraction residue, and direct injection of an aliquot of the derivatized extract (Figure 27.1: modified sample workup procedure 1). Removal of the excess derivatizing reagent is no longer required if it can be injected directly without causing contamination or chemical damage of the column or detection system. In Table 27.2, selected examples of DBS methods using this sample workup procedure are shown.

In this context, a suitable strategy for GC-MS applications is silylation. Using silylating reagents N,O bis(trimethylsilyl)trifluoroacetamide (BSTFA), N-methyl-N-(trimethylsilyl)-trifluoroacetamide (MSTFA), and N-(tert-butyldimethylsilyl)-N-methyl-trifluoroacetamide (MTBSTFA), specific metabolites (for metabolomic profiling), 23 drugs of abuse, a free fatty acid (cis-4-decenoic acid), and 17 hydroxyprogesterone were successfully determined in DBS, the latter following microwave-assisted-silylation (Heales and Leonard, 1992; Deng et al., 2005a; Kong et al., 2011; Langel et al., 2011).

Several reagents for pre-column derivatization of non-fluorescent compounds to their fluorescent derivatives prior to reversed-phase high performance liquid chromatography (RP-HPLC) separation are also suitable for direct injection. In addition, those reagents can mostly be applied directly to the extraction mixture (see Section 27.4.2), but in some circumstances the extraction solvent needs to be evaporated as it may influence fluorescence yield and stability. For example, for the determination of methylmalonic acid in DBS, the DBS methanolic extract was dried and reconstituted in water, followed by addition of the pyrene reagent and catalysts in dimethyl sulfoxide. This procedure resulted in intense fluorescence of the dipyrene derivative of methylmalonic acid, which, in addition, could clearly be discriminated from that of the monomeric fluorescing compounds, thereby eliminating these interferences (Nohta et al., 2000, 2003; Al-Dirbashi et al., 2005).

Dansylhydrazine, originally used as derivatization reagent in LC-FLUO applications, has been successfully applied in DBS LC-MS/MS methods determining succinylacetone (SA), which shows very poor ionization efficiency in ESI-MS (Santa et al., 2011). Al-Dirbashi et al. developed a two-step derivatization reaction, as they concluded that applying dansylhydrazine alone to the DBS residue resulted in a mono-dansylhydrazone derivative with unfavorable

TABLE 27.2 Selected Examples of DBS Methods Using the Modified Sample Workup Procedure 1: Extraction–Evaporation–Derivatization

Assay Method	Type of Derivatization	Analyte(s) of Interest	Application	Selected References
GC-MS	Silylation	Metabolites (for metabolomic profiling)	MP	Kong et al., 2011;
		23 drugs of abuse	TOX	Langel et al., 2011;
		Free fatty acid (cis-4-decenoic acid)	NBS	Heales and Leonard, 1992;
		17-OH-progesterone	NBS	Deng et al., 2005a
	Silylation–Acetylation (2-step): formation of TMS-TFA-derivatives	AA Organic acids Glycines	NBS (DPS)	Yoon, 2007
LC-FLUO	Intramolecular-excimer fluorescence derivatization using pyrene reagent	Methylmalonic acid	NBS	Al-Dirbashi et al., 2005
LC-UV	Formation of AQC-derivatives	AA	NBS	Swenson and McWhinney, 2009
LC-MS/MS	Alkylation: butylesterification–hydrazone formation (2-step)	Succinylacetone	NBS	Al-Dirbashi et al., 2006, 2008

AA, amino acids; AQC, 6-aminoquinolyl-N-hydroxysuccinimidyl carbamate; DPS, dried plasma spots; FLUO, fluorescence; GC, gas chromatography; LC, liquid chromatography; MP, metabolomic profiling; MS/MS, tandem mass spectrometry; MS, mass spectrometry; NBS, newborn screening; TFA-, trifluoroacyl-; TMS-, trimethylsilyl-; TOX, toxicology; UV, ultraviolet.

chromatographic properties. Thus, to improve the chromatographic properties such as acceptable retention and less peak tailing and to increase ionization efficiency, SA was butylated prior to dansylation. Between the derivatization processes, the butylated extract was dried and after allowing the second reaction to occur, an aliquot of the resulting mixture was subjected to high performance LC (HPLC)- or ultrahigh performance LC (UHPLC)-MS/MS analysis (Al-Dirbashi et al., 2006, 2008).

27.4.2 Modified Sample Workup Procedure 2: Two-Step Extraction–Derivatization

GC and LC applications have been reported where the reagents are applied to the (aqueous) extraction mixture without prior evaporation (Figure 27.1: modified sample workup procedure 2). Derivatization reagents that react under the conditions of the extraction solvent are suitable candidates for this purpose. Moreover, some reagents have the additional advantage of reacting fast without heating. Table 27.3 shows selected examples of DBS methods using this modified sample workup procedure, divided into procedures where the excess reagent is removed prior to injection (Table 27.3a) and procedures where the final mixture is injected directly into the analytical system (Table 27.3b).

When using GC, the injection of not only excess reagent but also of residual water into the GC system is preferentially avoided (Table 27.3a). This can be achieved by extracting the derivatized target analytes from the (aqueous) reaction mixture. A GC-MS method developed for the determination of several AA in DBS after "*in situ*" derivatization with propyl chloroformate illustrates this principle. As water does not interfere with alkyl chloroformation, the chloroformate can be added directly to the eluate of an SPE step, isolating AA from a DBS extract. After completion of the reaction within 1 minute at room temperature, the resulting derivatives are isolated by liquid–liquid extraction (LLE) using a water-immiscible organic solvent, of which an aliquot was injected (Kawana et al., 2010). In another method, isobutyl chloroformate derivatized AA were extracted from an aqueous reaction mixture by solid-phase micro extraction (SPME), followed by desorption of the fiber into the GC-MS injector (Deng et al., 2004).

Likewise, excess derivatizing reagents enabling LC-UV detection may be removed by additional purification steps to avoid interference during analysis (Table 27.3a). Dale et al. (2003) used this approach for the determination of phenylalanine and tyrosine in DBS. Following extraction and the generation of phenylthiocarbamyl derivatives, the extracts were dried and reconstituted in 0.1 M sodium acetate buffer. The excess reagent still present in the final aqueous mixture was trapped in methylene dichloride along with other possibly interfering compounds prior to analysis of the clear aqueous phase by RP-HPLC with UV detection.

TABLE 27.3 Selected Examples of DBS Methods Using the Modified Sample Workup Procedure 2: Two-Step Extraction and Derivatization, Followed by Removal of the Excess Reagent (a) or by Direct Injection (b)

Assay Method	Type of Derivatization	Analyte(s) of Interest	Application	Selected References
		(a)		
GC-MS	Acetylation: alkyl chloroformation	AA	NBS	Deng et al., 2004; Kawana et al., 2010
LC-UV	Formation of phenylthiocarbamyl-derivatives	AA	NBS	Dale et al., 2003
		(b)		
LC-FLUO	Formation of isoindolic derivatives	AA	NBS	Lundsjo et al., 1990; Moretti et al., 1990; Vollmer et al., 1990; Kand'ar et al., 2009
	Formation of 2-substituted amino-4,5-diphenylimidazoles	GAA	NBS	Carducci et al., 2001
	Formation of 2-quinoxalinol derivatives	α-keto acids	NBS	Kand'ar et al., 2009
LC-MS/MS	Formation of PFB-derivatives	Seven-carbon sugars	NBS	Wamelink et al., 2011

AA, amino acids; GC, gas chromatography; LC, liquid chromatography; MS, mass spectrometry; NBS, newborn screening; UV, ultra-violet; FLUO, fluorescence; GAA, guanidinoacetic acid; MS/MS, tandem mass spectrometry; PFB-derivatives, O-2,3,4,5,6-pentafluorobenzyl-derivatives.

Similar procedures have been reported in combination with LC-FLUO or LC-MS/MS methods, having the advantage that an aliquot of the final reaction mixture can be injected directly (Table 27.3b). For example, pre-column derivatization of several AA, guanidinoacetic acid (GAA) and α-keto acids has been performed in order to achieve highly sensitive and selective LC-FLUO methods (Moretti et al., 1990; Carducci et al.; 2001; Kand'ar et al., 2009). A way to achieve reproducible results and avoid errors from instability of the derivative when applying certain derivatizing reagents pre- rather than post-column is controlling the time between derivatization and injection by an online derivatization technique (Lundsjo et al., 1990; Ziegler et al., 1992; Kand'ar et al., 2009). Online derivatization can be achieved by derivatizing in the autosampler, as has been demonstrated for several DBS-extracted AA, derivatized with o-phthaldialdehyde immediately prior to separation (Lundsjo et al., 1990; Kand'ar et al., 2009). An LC-MS/MS example is the determination of the derivatized seven-carbon sugar sedoheptulose, which proved to be substantially more sensitive, with no interferences of other sugars possibly present in the DBS. Additionally, as retention was increased due to derivatization, simple reversed-phase chromatography using a common mobile phase composition could be performed (Wamelink et al., 2011).

27.4.3 Modified Sample Workup Procedure 3: Direct Derivatization Techniques

The most convenient approach for derivatizing analytes from DBS is direct derivatization. This approach has been reported for the analysis of conventional matrices such as blood, plasma, and serum and of aqueous matrices such as drinks. In those cases, an aliquot of matrix (20–50 μL) is directly derivatized by adding an excess of derivatization reagent (up to 1 mL), thereby omitting the extraction step (Sabucedo and Furton, 2004; Van hee et al., 2004; Meyer et al., 2011). Implementing this principle in DBS analysis, extracting and derivatizing solutions could be applied simultaneously to the DBS or even only the derivatizing reagent(s) could be added without use of any extraction solvent (Figure 27.1: modified sample workup procedure 3). We coined the latter approach "on spot" derivatization. Table 27.4 shows selected examples of DBS methods using direct derivatization techniques, followed by removal of the excess reagent before injection (Table 27.4a) or by direct injection (Table 27.4b).

Depending on the type of derivatization reagents used in the first case, a proper extraction solvent should be selected, which still allows the reaction to occur and results in acceptable extraction recovery of the (derivatized) target analyte. For example, for the determination of antidepressants in

TABLE 27.4 Selected Examples of DBS Methods Using the Modified Sample Workup Procedure 3: Direct Derivatization, Followed by Removal of the Excess Reagent (a) or by Direct Injection (b)

Assay Method	Type of Derivatization	Analyte(s) of Interest	Application	Selected References
		(a)		
GC-MS/MS	Acetylation	Antidepressants	TDM	Déglon et al., 2010
GC-MS	On spot alkylation and acetylation	GHB	TOX	Ingels et al., 2010, 2011
	Transmethylation	Free fatty acids	NBS	Morton and Kelley, 1990; Kimura et al., 2002
MS/MS	Formation of 1-phenyl-3-methyl pyrazolone derivatives	Oligosaccharides	MD	Rozaklis et al., 2002; Ramsay et al., 2003
		(b)		
LC-MS/MS	Formation of DAABD-AE-derivatives	3-OH-glutaric acid Glutaric acid	NBS (DUS)	Al-Dirbashi et al., 2011
MS/MS	Hydrazone formation	Succinylacetone	NBS	Allard et al., 2004

GC, gas chromatography; GHB, gamma-hydroxybutyric acid; MD, metabolic disorder; MS/MS, tandem mass spectrometry; MS, mass spectrometry; NBS, newborn screening; TDM, therapeutic drug monitoring; TOX, toxicology; DAABD-AE, 4-2-(N,N-dimethylamino)ethylaminosulfonyl-7-(2-aminoethylamino)-2,1,3-benzoxadiazole; DUS, dried urine spots; LC, liquid chromatography.

DBS, a method was developed combining a fast and sensitive GC-MS/MS technique with direct derivatization of the DBS. Fluorinated agents were chosen to derivatize the analytes, increasing electro-affinity, which was important for MS/MS detection operating in negative chemical ionization (NICI) mode (this mode being highly selective and sensitive for compounds with high electron affinity). Although the use of methanol and water as extraction solvent resulted in higher extraction recovery of fluoxetine, norfluoxetine, reboxetine, and paroxetine, the authors selected an aprotic organic solvent (butyl chloride), as this could be applied simultaneously with the fluorinating agent, simplifying sample workup and still resulting in the required method sensitivity (Déglon et al., 2010).

In some circumstances, adequate sensitivity and selectivity can be achieved by applying the derivatization reagent (mixture) directly "on spot," without the use of additional (co-)extracting solvent. In this scenario, it is assumed that the reagent functions both as extracting and derivatizing agent, as no other solutions or solvents are added to the DBS. To our knowledge, up-to-date this approach of "on spot" derivatization has only been applied for the GC-MS-based determination of gamma-hydroxybutyric acid (GHB), a notorious club and date-rape drug, in DBS (Ingels et al., 2010, 2011). More specifically, GHB was derivatized by adding a mixture of trifluoroacetic acid anhydride (TFAA) and heptafluorobutanol (HFB-OH) directly "on spot." Although only a small volume of whole blood was analyzed (a 6-mm diameter punch from a DBS, corresponding to approximately 10 μL), adequate sensitivity was achieved (Ingels et al., 2011).

Although in the above-mentioned GC-methods, removal of the excess reagent was achieved by simple evaporation, in some circumstances, additional LLE and/or SPE is required in order to isolate the derivatized analyte and/or remove the excess reagent before injection (Table 27.4a). For example, after a single step transmethylation of free fatty acids present in DBS, the reaction was halted by adding 6% potassium carbonate and the derivatives were extracted in hexane, of which an aliquot was injected into the GC-MS (Kimura et al., 2002). In another example, the excess reagent used to derivatize oligosaccharides in DBS with a single step extraction/derivatization procedure was trapped in chloroform. The remaining aqueous layer was subjected to SPE, including wash steps to further remove any unincorporated reagent, and finally, the eluate containing the derivatives was dried and reconstituted before injection into the ESI-MS/MS system (Rozaklis et al., 2002; Ramsay et al., 2003).

As complicated procedures to remove excess reagent extend the sample workup associated with a simplified extraction/derivatization procedure, direct derivatization techniques in which an aliquot of the supernatant can be analyzed directly are even more practical (Table 27.4b). Since in this scenario the complete DBS sample preparation has been reduced to a single step, sample throughput is consequently only limited by the time required for the derivatization reaction to occur. To illustrate this, sample workup of an LC-MS/MS method determining 3-hydroxy-glutaric acid and glutaric acid in dried urine spots (DUS) was completed within 45 minutes since the direct derivatization step required 45 minutes at 60°C to obtain maximum derivatization yield. Here, internal standard solution and derivatizing reagents were added successively to the dried spot, hence eliminating a separate extraction step, and an aliquot of the final derivatized solution was analyzed directly after the reaction had been stopped. Derivatization was required to enhance ionization and fragmentation of the target compounds (Al-Dirbashi et al., 2011).

Special considerations will need to be taken into account when using direct derivatization techniques. This is because the derivatization yield reflects both elution of the target compounds from the DBS filter paper and the formation of the derivatives. An example is the simultaneous extraction of SA from DBS and production of SA derivatives by the use of a single step derivatization using a hydrazine-containing solution. The simultaneous extraction is assumed to occur via cleavage of covalently linked SA-protein adducts by hydrazine and results in extraction of SA as a hydrazone derivative. Moreover, although it might be possible to obtain a lower limit of quantification by including a purification step before MS/MS analysis and an additional derivatization step (see Section 27.4.1), the main advantages of the method are simplicity, the requirement of only minimal technical time, and the provision of enough sensitivity to detect positive cases (Allard et al., 2004).

Another point to consider with direct derivatization techniques is that the filter paper is derivatized as well. Normally after extraction, an aliquot of the solvent is transferred in order to evaporate, filter, or derivatize the extraction solvent containing the extracted compounds. Since the reagents are now applied directly to the filter paper present in the test tube, it might be relevant to check for interferences as a result of derivatizing the blank filter paper. In the case of derivatizing the blank filter paper used for applying whole blood from possible GHB-intoxicated patients, a small interference was seen at the retention time of GHB (<20% of the lower limit of quantification) (Ingels et al., 2010).

27.5 CONCLUSION AND FUTURE PERSPECTIVES

To conclude, DBS analysis has proven useful in a wide range of applications, because of the typical benefits associated with this sampling technique. Although LC-MS/MS is commonly used for DBS analysis, other techniques such as GC-MS (/MS) may also play a role in the determination of drugs or biomarkers in DBS. To determine trace levels of low molecular weight compounds in DBS, a derivatization step during DBS sample workup may be necessary or may help to achieve adequate method sensitivity. Dependent on the target compounds and the derivatizing reagents, different procedures are possible as shown in Figure 27.1. Moreover, a single DBS can be subjected to a combination of sample workup procedures including different derivatization techniques to extend the range of target compounds of a DBS method. For example, to include SA, most favorably extracted as hydrazone-derivative (see Section 27.4.3), in a method determining amino acids (AA) and acylcarnitines (AC) in newborn DBS, combinations of the "general" procedure (for AA and AC) and the direct derivatization technique (for SA) have been described (la Marca et al., 2008; Turgeon et al., 2008, Chace et al., 2009).

Although incorporating a derivatization procedure in DBS sample workup may significantly increase method sensitivity, this additional step may be experienced as laborious and tedious (e.g., requiring more technical time). Hence, replacement of manual preparation by automated techniques but also implementation of procedures using simpler sample workup such as direct derivatization may be less time-consuming and lead to increased sample throughput.

Thus, the introduction of automated techniques may result in high throughput DBS methods including derivatization, as illustrated by a recently developed fully automated sample preparation technique using a digital microfluidic method including extraction and derivatization of several AA in DBS by in-line or off-line nano-ESI-MS/MS (Jebrail et al., 2012). In addition, modified sample workup procedures such as direct derivatization techniques, using a single-step extraction and derivatization or "on spot" derivatization, may lead to a less time-consuming and less laborious sample workup. These procedures may be considered in cases where a derivatization step is needed, to ensure a minimal, economic, and fast sample treatment.

Consequently, a similar workload may be achieved as with classical DBS analyses, which only use extraction and purification procedures, but with additional gain in method sensitivity and selectivity due to derivatization. Hence, the application range of DBS analysis may be extended. Automated techniques are being developed such as paper-spray ionization MS/MS analysis with online derivatization, implemented by preloading reagents onto the paper or using solutions containing the derivatizing reagents. This approach was already successfully applied to detect cholesterol in dried serum spots via formation of a characteristic fragment ion, with enhanced MS/MS sensitivity and selectivity (Liu et al., 2011). Furthermore, we envisage that direct derivatization approaches may gain importance in the future, not only in GC, but also in LC-MS/MS and other MS(/MS)-based approaches, where they could be combined with online elution or could even precede direct desorption from the DBS (e.g., following application of a derivatizing reagent sprayed on the DBS).

REFERENCES

Abdulrazzaq YM, Ibrahim A. Determination of amino acids by ion-exchange chromatography on filter paper spotted blood samples stored at different temperatures and for different periods: comparison with capillary and venous blood. *Clinical Biochemistry* 2001;34:399–406.

Al-Dirbashi OY, Jacob M, Al-Hassnan Z, El-Badaoui F, Rashed MS. Diagnosis of methylmalonic acidemia from dried blood spots by HPLC and intramolecular-excimer fluorescence derivatization. *Clinical Chemistry* 2005;51:235–237.

Al-Dirbashi OY, Rashed MS, Brink HJ, Jakobs C, Filimban N, Al-Ahaidib LY, Jacob M, Al-Sayed MM, Al-Hassnan Z, Faqeih E. Determination of succinylacetone in dried blood spots and liquid urine as a dansylhydrazone by liquid chromatography tandem mass spectrometry. *Journal of Chromatography: B Analytical Technologies in the Biomedical and Life Sciences* 2006;831:274–280.

Al-Dirbashi OY, Rashed MS, Jacob M, Al-Ahaideb LY, Al-Amoudi M, Rahbeeni Z, Al-Sayed MM, Al-Hassnan Z, Al-Owain M, Al-Zeidan H. Improved method to determine succinylacetone in dried blood spots for diagnosis of tyrosinemia type 1 using UPLC-MS/MS. *Biomedical Chromatography* 2008;22:1181–1185.

Al-Dirbashi OY, Kolker S, Ng D, Fisher L, Rupar T, Lepage N, Rashed MS, Santa T, Goodman SI, Geraghty MT, Zschocke J, Christensen E, Hoffmann GF, Chakraborty P. Diagnosis of glutaric aciduria type 1 by measuring 3-hydroxyglutaric acid in dried urine spots by liquid chromatography tandem mass spectrometry. *Journal of Inherited Metabolic Disease* 2011;34:173–180.

Allard P, Grenier A, Korson MS, Zytkovicz TH. Newborn screening for hepatorenal tyrosinemia by tandem mass spectrometry: analysis of succinylacetone extracted from dried blood spots. *Clinical Biochemistry* 2004;37:1010–1015.

Bodamer OA, Bloesch SM, Gregg AR, Stockler-Ipsiroglu S, O'Brien WE. Analysis of guanidinoacetate and creatine by isotope dilution electrospray tandem mass spectrometry. *Clinica Chimica Acta* 2001;308:173–178.

Carducci C, Birarelli M, Santagata P, Leuzzi V, Antonozzi I. Automated high-performance liquid chromatographic method for the determination of guanidinoacetic acid in dried blood spots: a tool for early diagnosis of guanidinoacetate methyltransferase deficiency. *Journal of Chromatography. B: Biomedical Sciences and Applications* 2001;755:343–348.

Carducci C, Santagata S, Leuzzi V, Artiola C, Giovanniello T, Battini R, Antonozzi I. Quantitative determination of guanidinoacetate and creatine in dried blood spot by flow injection analysis-electrospray tandem mass spectrometry. *Clinica Chimica Acta* 2006;364:180–187.

Chace DH. Mass spectrometry in newborn and metabolic screening: historical perspective and future directions. *Journal of Mass Spectrometry* 2009;44:163–170.

Chace DH, Millington DS, Terada N, Kahler SG, Roe CR, Hofman LF. Rapid diagnosis of phenylketonuria by quantitative analysis for phenylalanine and tyrosine in neonatal blood spots by tandem mass spectrometry. *Clinical Chemistry* 1993;39:66–71.

Chace DH, Hillman SL, Millington DS, Kahler SG, Roe CR, Naylor EW. Rapid diagnosis of maple syrup urine disease in blood spots from newborns by tandem mass spectrometry. *Clinical Chemistry* 1995;41:62–68.

Chace DH, Hillman SL, Millington DS, Kahler SG, Adam BW, Levy HL. Rapid diagnosis of homocystinuria and other hypermethioninemias from newborns' blood spots by tandem mass spectrometry. *Clinical Chemistry* 1996;42:349–355.

Chace DH, Lim T, Hansen CR, De Jesus VR, Hannon WH. Improved MS/MS analysis of succinylacetone extracted from dried blood spots when combined with amino acids and acylcarnitine butyl esters. *Clinica Chimica Acta* 2009;407:6–9.

Dale Y, Mackey V, Mushi R, Nyanda A, Maleque M, Ike J. Simultaneous measurement of phenylalanine and tyrosine in phenylketonuric plasma and dried blood by high-performance liquid chromatography. *Journal of Chromatography B: Analytical Technologies in the Biomedical and Life Sciences* 2003;788:1–8.

De Jesus VR, Chace DH, Lim TH, Mei JV, Hannon WH. Comparison of amino acids and acylcarnitines assay methods used in newborn screening assays by tandem mass spectrometry. *Clinica Chimica Acta* 2010;411:684–689.

Déglon J, Lauer E, Thomas A, Mangin P, Staub C. Use of the dried blood spot sampling process coupled with fast gas chromatography and negative-ion chemical ionization tandem mass spectrometry: application to fluoxetine, norfluoxetine, reboxetine, and paroxetine analysis. *Analytical and Bioanalytical Chemistry* 2010;396:2523–2532.

Deng C, Deng Y, Wang B, Yang X. Gas chromatography-mass spectrometry method for determination of phenylalanine and tyrosine in neonatal blood spots. *Journal of Chromatography B: Analytical Technologies in the Biomedical and Life Sciences* 2002a; 780: 407–413.

Deng C, Shang C, Hu Y, Zhang X. Rapid diagnosis of phenylketonuria and other aminoacidemias by quantitative analysis of amino acids in neonatal blood spots by gas chromatography-mass spectrometry. *Journal of Chromatography B: Analytical Technologies in the Biomedical and Life Sciences* 2002b;775:115–120.

Deng C, Deng Y. Diagnosis of maple syrup urine disease by determination of l-valine, l-isoleucine, l-leucine and l-phenylalanine in neonatal blood spots by gas chromatography-mass spectrometry. *Journal of Chromatography B: Analytical Technologies in the Biomedical and Life Sciences* 2003;792:261–268.

Deng C, Li N, Zhang X. Rapid determination of amino acids in neonatal blood samples based on derivatization with isobutyl chloroformate followed by solid-phase microextraction and gas chromatography/mass spectrometry. *Rapid Communications in Mass Spectrometry* 2004;18:2558–2564.

Deng C, Ji J, Zhang L, Zhang X. Diagnosis of congenital adrenal hyperplasia by rapid determination of 17alpha-hydroxyprogesterone in dried blood spots by gas chromatography/mass spectrometry following microwave-assisted silylation. *Rapid Communications in Mass Spectrometry* 2005a;19:2974–2978.

Deng C, Yin X, Zhang L, Zhang X. Development of microwave-assisted derivatization followed by gas chromatography/mass spectrometry for fast determination of amino acids in neonatal blood samples. *Rapid Communications in Mass Spectrometry* 2005b;19:2227–2234.

Deng CH, Wang B, Liu LF. Fast diagnosis of neonatal phenylketonuria by gas chromatography-mass spectrometry following microwave-assisted silylation. *Chromatographia* 2005c;62:617–621.

Edelbroek PM, van der Heijden J, Stolk LM. Dried blood spot methods in therapeutic drug monitoring: methods, assays, and pitfalls. *Therapeutic Drug Monitoring* 2009;31:327–336.

Eyles D, Anderson C, Ko P, Jones A, Thomas A, Burne T, Mortensen PB, Norgaard-Pedersen B, Hougaard DM, McGrath J. A sensitive LC/MS/MS assay of 25OH vitamin D3 and 25OH vitamin D2 in dried blood spots. *Clinica Chimica Acta* 2009;403:145–151.

Fingerhut R, Roschinger W, Muntau AC, Dame T, Kreischer J, Arnecke R, Superti-Furga A, Troxler H, Liebl B, Olgemoller B, Roscher AA. Hepatic carnitine palmitoyltransferase I deficiency: acylcarnitine profiles in blood spots are highly specific. *Clinical Chemistry* 2001;47:1763–1768.

Garcia Boy R, Henseler J, Mattern R, Skopp G. Determination of morphine and 6-acetylmorphine in blood with use of dried blood spots. *Therapeutic Drug Monitoring* 2008;30:733–739.

Hayashi T, Tsuchiya H, Naruse H. Reversed-phase ion-pair chromatography of amino acids. Application to the determination of amino acids in plasma samples and dried blood on filter papers. *Journal of Chromatography* 1983;274:318–324.

Heales SJ, Leonard JV. Diagnosis of medium chain acyl coa dehydrogenase deficiency by measurement of cis-4-decenoic acid in dried blood spots. *Clinica Chimica Acta* 1992;209:61–66.

Higashi T, Nishio T, Uchida S, Shimada K, Fukushi M, Maeda M. Simultaneous determination of 17alpha-hydroxypregnenolone and 17alpha-hydroxyprogesterone in dried blood spots from low birth weight infants using LC-MS/MS. *Journal of Pharmaceutical and Biomedical Analysis* 2008;48:177–182.

Higashi T, Suzuki M, Hanai J, Inagaki S, Min JZ, Shimada K, Toyo'oka T. A specific LC/ESI-MS/MS method for determination of 25-hydroxyvitamin D3 in neonatal dried blood spots containing a potential interfering metabolite, 3-epi-25-hydroxyvitamin D3. *Journal of Separation Science* 2011;34:725–732.

Ingels ASME, Lambert WE, Stove CP. Determination of gamma-hydroxybutyric acid in dried blood spots using a simple GC-MS method with direct "on spot" derivatization. *Analytical and Bioanalytical Chemistry* 2010;398:2173–2182.

Ingels AS, De Paepe P, Anseeuw K, Van Sassenbroeck D, Neels H, Lambert W, Stove CP. Dried blood spot punches for confirmation of gamma-hydroxybutyric acid intoxications: validation of an optimized GC-MS procedure. *Bioanalysis* 2011;3:2271–2281.

Jebrail MJ, Yang H, Mudrik JM, Lafreniere NM, McRoberts C, Al-Dirbashi OY, Fisher L, Chakraborty P, Wheeler AR. A digital microfluidic method for dried blood spot analysis. *Lab on a Chip* 2012;11:3218–3224.

Kaal E, Janssen HG. Extending the molecular application range of gas chromatography. *Journal of Chromatography A* 2008;1184:43–60.

Kand'ar R, Zakova P, Jirosova J, Sladka M. Determination of branched chain amino acids, methionine, phenylalanine, tyrosine and alpha-keto acids in plasma and dried blood samples using HPLC with fluorescence detection. *Clinical Chemistry and Laboratory Medicine* 2009;47:565–572.

Kawana S, Nakagawa K, Hasegawa Y, Yamaguchi S. Simple and rapid analytical method for detection of amino acids in blood using blood spot on filter paper, fast-GC/MS and isotope dilution technique. *Journal of Chromatography B: Analytical Technologies in the Biomedical and Life Sciences* 2010; 878:3113–3118.

Kimura M, Yoon HR, Wasant P, Takahashi Y, Yamaguchi S. A sensitive and simplified method to analyze free fatty acids in children with mitochondrial beta oxidation disorders using gas chromatography mass spectrometry and dried blood spots. *Clinica Chimica Acta* 2002;316:117–121.

Knapp DR. *Handbook of Analytical Derivatization Reactions*. New York: John Wiley & Sons; 1979.

Kong ST, Lin HS, Ching J, Ho PC. Evaluation of dried blood spots as sample matrix for gas chromatography/mass spectrometry based metabolomic profiling. *Analytical Chemistry* 2011;83:4314–4318.

la Marca G, Malvagia S, Pasquini E, Innocenti M, Fernandez MR, Donati MA, Zammarchi E. The inclusion of succinylacetone as marker for tyrosinemia type I in expanded newborn screening programs. *Rapid Communications in Mass Spectrometry* 2008;22:812–818.

Lai CC, Tsai CH, Tsai FJ, Lee CC, Lin WD. Rapid monitoring assay of congenital adrenal hyperplasia with microbore high-performance liquid chromatography/electrospray ionization tandem mass spectrometry from dried blood spots. *Rapid Communications in Mass Spectrometry* 2001;15:2145–2151.

Langel K, Uusivirta H, Ariniemi K, Lillsunde P. Analysis of drugs of abuse by GC MS in dried blood spot sample matrix. Presented at the 49th TIAFT meeting; 2011 Sept 25—30; San Francisco, CA.

Li W, Tse FLS. Dried blood spot sampling in combination with LC-MS/MS for quantitative analysis of small molecules. *Biomedical Chromatography* 2010;24:49–65.

Lin DL, Wang SM, Wu CH, Chen BG, Lu RH. Chemical derivatization for the analysis of drugs by GC-MS—a conceptual review. *Journal of Food and Drug Analysis* 2008;16:1–10.

Liu J, Wang H, Manicke NE, Lin JM, Cooks RG, Ouyang Z. Development, characterization, and application of paper spray ionization. *Analytical Chemistry* 2011;82:2463–2471.

Lundsjo A, Hagelberg S, Palmer K, Lindblad BS. Amino acid profiles by HPLC after filter paper sampling: 'appropriate technology' for monitoring of nutritional status. *Clinica Chimica Acta* 1990;191:201–209.

McDade TW, Williams S, Snodgrass JJ. What a drop can do: dried blood spots as a minimally invasive method for integrating biomarkers into population-based research. *Demography* 2007;44:899–925.

Meyer MR, Weber AA, Maurer HH. A validated GC-MS procedure for fast, simple, and cost-effective quantification of glycols and GHB in human plasma and their identification in urine and plasma developed for emergency toxicology. *Analytical and Bioanalytical Chemistry* 2011;400:411–414.

Moffat AC, Osselton MD, Widdop B. *Clarke's Analysis of Drugs and Poisons*, 3rd edn. London: Pharmaceutical Press; 2004.

Moretti F, Birarelli M, Carducci C, Pontecorvi A, Antonozzi I. Simultaneous high-performance liquid chromatographic determination of amino acids in a dried blood spot as a neonatal screening test. *Journal of Chromatography* 1990;511: 131–136.

Morton DH, Kelley RI. Diagnosis of medium-chain acyl-coenzyme a dehydrogenase deficiency in the neonatal period by measurement of medium-chain fatty acids in plasma and filter paper blood samples. *Journal of Pediatrics* 1990;117:439–442.

Naylor EW, Chace DH. Automated tandem mass spectrometry for mass newborn screening for disorders in fatty acid, organic acid, and amino acid metabolism. *Journal of Child Neurology* 1999;14(Suppl 1):S4–S8.

Niwa M. Chemical derivatization as a tool for optimizing MS response in sensitive LC-MS/MS bioanalysis and its role in pharmacokinetic studies. *Bioanalysis* 2012;4:213–220.

Nohta H, Satozono H, Koiso K, Yoshida H, Ishida J, Yamaguchi M. Highly selective fluorometric determination of polyamines based on intramolecular excimer-forming derivatization with a pyrene-labeling reagent. *Analytical Chemistry* 2000;72:4199–4204.

Nohta H, Sonoda J, Yoshida H, Satozono H, Ishida J, Yamaguchi M. Liquid chromatographic determination of dicarboxylic acids based on intramolecular excimer-forming fluorescence derivatization. *Journal of Chromatography A* 2003;1010: 37–44.

Ramsay SL, Meikle PJ, Hopwood JJ. Determination of monosaccharides and disaccharides in mucopolysaccharidoses patients by electrospray ionisation mass spectrometry. *Molecular Genetics and Metabolism* 2003;78:193–204.

Roesel RA, Blankenship PR, Hommes FA. HPLC assay of phenylalanine and tyrosine in blood spots on filter paper. *Clinica Chimica Acta* 1986;156:91–96.

Rood D *A Practical Guide to the Care, Maintenance, and Troubleshooting of Capillary Gas Chromatographic Systems*. Weinheim: Wiley-VCH; 1999.

Rozaklis T, Ramsay SL, Whitfield PD, Ranieri E, Hopwood JJ, Meikle PJ. Determination of oligosaccharides in pompe disease by electrospray ionization tandem mass spectrometry. *Clinical Chemistry* 2002;48:131–139.

Sabucedo AJ, Furton KG. Extractionless GC-MS analysis of gamma-hydroxybutyrate and gamma-butyrolactone with trifluoroacetic anhydride and heptafluoro-1-butanol from aqueous samples. *Journal of Separation Science* 2004;27:703–709.

Santa T. Derivatization reagents in liquid chromatography/electrospray ionization tandem mass spectrometry. *Biomedical Chromatography* 2011;25:1–10.

Segura J, Ventura R, Jurado C. Derivatization procedures for gas chromatographic-mass spectrometric determination of xenobiotics in biological samples, with special attention to drugs of abuse and doping agents. *Journal of Chromatography B: Biomedical Sciences and Applications* 1998;713:61–90.

Shen X, Deng C, Wang B, Dong L. Quantification of trimethylsilyl derivatives of amino acid disease biomarkers in neonatal blood samples by gas chromatography-mass spectrometry. *Analytical and Bioanalytical Chemistry* 2006;384:931–938.

Soderholm SL, Damm M, Kappe CO. Microwave-assisted derivatization procedures for gas chromatography/mass spectrometry analysis. *Molecular Diversity* 2010;14:869–888.

Stove CP, Ingels ASME, De Kesel PMM, Lambert WE. Dried blood spots in toxicology: from the cradle to the grave? *Critical Reviews in Toxicology* 2012;42:230–243.

Swenson RJ, McWhinney BC. Utilising UPLC to develop a less invasive method for monitoring maple syrup urine disease patients. Presented at the AACB 47th Annual Scientific Conference; 2009 Sept 14–17; Brisbane, Australia.

Tanna S, Lawson G. Analytical methods used in conjunction with dried blood spots. *Analytical Methods* 2011;3:1709–1718.

Todoroki K, Yoshida H, Hayama T, Itoyama M, Nohta H, Yamaguchi M. Highly sensitive and selective derivatization-LC method for biomolecules based on fluorescence interactions and fluorous separations. *Journal of Chromatography B: Analytical Technologies in the Biomedical and Life Sciences* 2011;879:1325–1337.

Turgeon C, Magera MJ, Allard P, Tortorelli S, Gavrilov D, Oglesbee D, Raymond K, Rinaldo P, Matern D. Combined newborn screening for succinylacetone, amino acids, and acylcarnitines in dried blood spots. *Clinical Chemistry* 2008;54:657–664.

Van hee P, Neels H, De Doncker M, Vrydags N, Schatteman K, Uyttenbroeck K, Hamers N, Himpe D, Lambert W. Analysis of gamma-hydroxybutyric acid, DL-lactic acid, glycolic acid, ethylene glycol and other glycols in body fluids by a direct injection gas chromatography-mass spectrometry assay for wide use. *Clinical Chemistry and Laboratory Medicine* 2004;42:1341–1345.

Vollmer DW, Jinks DC, Guthrie R. Isocratic reverse-phase liquid-chromatography assay for amino-acid metabolic disorders using eluates of dried blood spots. *Analytical Biochemistry* 1990;189:115–121.

Wamelink MM, Struys EA, Jansen EE, Blom HJ, Vilboux T, Gahl WA, Komhoff M, Jakobs C, Levtchenko EN. Elevated concentrations of sedoheptulose in bloodspots of patients with cystinosis caused by the 57-Kb deletion: implications for diagnostics and neonatal screening. *Molecular Genetics and Metabolism* 2011;102:339–342.

Wuyts B, Stove V, Goossens L. Critical sample pretreatment in monitoring dried blood spot citrulline. *Clinica Chimica Acta* 2007;386:105–109.

Xu F, Zou L, Liu Y, Zhang Z, Ong CN. Enhancement of the capabilities of liquid chromatography-mass spectrometry with derivatization: general principles and applications. *Mass Spectrometry Reviews* 2011;30:1143–1172.

Yoon HR. Two step derivatization for the analyses of organic, amino acids and glycines on filter paper plasma by GC-MS/SIM. *Archives of Pharmacal Research* 2007;30:387–395.

Ziegler F, Le Boucher J, Coudray-Lucas C, Cynober L. Plasma amino-acid determinations by reversed-phase HPLC: improvement of the orthophthalaldehyde method and comparison with ion exchange chromatography. *Journal of Automatic Chemistry* 1992;14:145–149.

INDEX

NOTE: Locators followed by letter '*f*' and '*t*' refer to figures and tables respectively.

Acetaminophen, 163, 173
 glucuronide, 173, 222
 sulfate, 173, 222
Acid citrate dextrose, 11
Acylcarnitines, 62, 142, 314
AGP, *see* Alpha-1-acid glycoprotein
Ahlstrom 226, 188
 specimen collection paper, 236*f*
Alpha-1-acid glycoprotein (AGP), 123
Alpha-fetoprotein, 108
Ahlstrom 226, 181
Ahlstrom 237, 181
Amino acids, 61–62
Amitriptyline, quantitative analysis of, 305, 308*f*
Amphetamines, 147
ANC, *see* Antenatal clinics
ANC HIV surveillance, 80
Antenatal clinics (ANC), 76
Anticoagulants, 67–69
 acid citrate dextrose, 11
 EDTA, 11
 potassium oxalate, 11
 sodium heparin, 11
Antiretroviral therapy (ART), 80
 drugs, 220
APGDDI, *see* Flowing afterglow-atmospheric pressure glow discharge desorption ionization
APTDCI, *see* Atmospheric pressure thermal desorption chemical ionization
APTDI, *see* Atmospheric pressure thermal desorption ionization
AP-TD/SI, *see* Atmospheric pressure thermal desorption-secondary ionization
ART, *see* Antiretroviral therapy
ASAP, *see* Atmospheric pressure solids analysis probe
Atenolol, analysis of, 311, 311*f*
Atmospheric pressure solids analysis probe (ASAP), 283
Atmospheric pressure thermal desorption chemical ionization (APTDCI), 283, 283*f*
Atmospheric pressure thermal desorption ionization (APTDI), 283
Atmospheric pressure thermal desorption-secondary ionization (AP-TD/SI), 282
Avian respiratory viruses, 110

Bacterial inhibition assay (BIA), 5, 53
 in NBS, 59–60
Bang's method, 3
BDZ, *see* Benzodiazepines
Benzodiazepines (BDZ), 147
BIA, *see* Bacterial inhibition assay
Biomonitoring of environmental chemicals, application of DBS for, 135*t*
 analytical technologies, 136–137
 DDT, 134–136
 dichlorodiphenyldichloroethylene, 134
 hexachlorohexane, 134–136
 metals, 134
 nonhuman species, 137
 perchlorate, 136
 PFCs, 136
 protein adducts measurement, 136
Biotherapeutics (clinical implications of DBS assays)
 PK/TK support, 192
 recent literature review
  card types, 188–189
  hematocrit, 192

Dried Blood Spots: Applications and Techniques, First Edition. Edited by Wenkui Li and Mike S. Lee.

Biotherapeutics (clinical implications of DBS assays) (*Continued*)
peptides/proteins on DBS cards, 191–192
quantitation of biopharmaceutical compounds from DBS cards, 189–191, 191*t*
Bland–Altman comparisons, 118
Blood plasma distribution, 218

Caffeine (CYP1A2), 202
CAH, *see* Congenital adrenal hyperplasia
CAMAG DBS-MS 500, 292
Cannabinoids, 148
Canson 435 blotter paper, 4
Case studies
DBS in demographic and health surveys, 119–122
DBS in health initiatives for men study, 122–124
Cassette, 19
CDC, *see* Centers for Disease Control
Centers for Disease Control (CDC), 16
CH, *see* Congenital hypothyroidism
CHD, *see* Congenital heart disease
Chromatography, 60
CIE, *see* Counterimmunoelectrophoresis
Clinical and Laboratory Standards Institute (CLSI), 5
standard for blood collection
CLSI H04-A6, 2008, 21, 22
CLSI NBS-01-A6, 2013, 21, 22
Clinical pharmacology trials
advantages of DBS sampling, 216–217
DBS examples in adults, 222–225
DBS examples in pediatric, 222
disadvantages of DBS sampling, 217
PK/PD considerations of DBS sampling in, 217
blood plasma distribution, 218
DBS and plasma samples, 218–219
dried plasma spots, 219
hematocrit impact, 218
CLSI, *see* Clinical and Laboratory Standards Institute
CMV, *see* Cytomegalovirus
Coating, 335
Cocaethylene, 147
Cocaine, 147
Codeine, 148
Cold plasma techniques, 287
Competitive ELISA, 42–43, 43*f*
Congenital adrenal hyperplasia (CAH), 54
Congenital heart disease (CHD), 67
Congenital hypothyroidism (CH), 7, 54
Cotinine-*N*-glucuronide, 183
Counterimmunoelectrophoresis (CIE), 108
Cyclosporine, 218
Cystic fibrosis, 7, 53
Cytochrome P450 (CYP450) enzyme, 202
clinical study
bioanalysis, 208–209
clinical conduct, 208
DBS finger *versus* venipuncture, 209
disc numbers, 209
disc size, 209
genotyping, 211–213
limited *versus* extensive sampling, 210–211
1-OH MDZ, analysis of, 211
storage stability, 209
subject and phlebotomist satisfaction, 213–214
DBS method validation
accuracy and precision, 206–207
assessment of 1-OH MDZ, 206
carryover, 207
dilution integrity, 207
disc number, 207–208
disc size, 207
endogenous and exogenous substances, 206
linearity, 206
matrix effect, 207
recoveries, 207
stability experiments, 207
experimental
blood sampling procedures, 204
calibration standards preparation, 203–204
chemicals, solvents, and blank matrices, 203
chromatographic system, 203
clinical study design, 205–206
genomic DNA isolation, 205
genotyping assays, 205
quality control samples preparation, 203–204
sample preparation, 205
sample pretreatment, 205
stock solutions preparation, 203–204
validation procedures, 205
phenotyping of, 202
Cytomegalovirus (CMV), 32

Dansylhydrazine, 347
DAPPI, *see* Desorption atmospheric pressure photoionization
DART, *see* Direct analysis in real time
DBDI, *see* Dielectric barrier discharge ionization
DBS, *see* Dried blood spots
DBS assay
bioanalysis methods, 186
and capability of direct analysis, 186
sample collection, 180–181, 181*f*
sample preparation and processing, 181–183
sample transport and storage, 183–186
DBS autosampler
analysis of MMA in plasma spots
calibrators and samples preparation, 317
liquid chromatography, 317
mass spectrometry, 317
citrulline analysis
calibrators and samples preparation, 316–317
mass spectrometry, 317
direct analysis of acylcarnitines and amino acids
DBS preparation, 316
mass spectrometry, 316
and MS/MS instrument configuration, 315–316, 315*f*
results and discussion
acylcarnitines and amino acids, 317–320
citrulline, 320–321, 321*f*
MMA, 321–322

DBS-based ELISA methods, 47*f*
applications
anti-HCV antibody detection, 44
to detect hepatitis B virus, 43–44
to detect measles IgG and IgM, 45
diagnosis of infectious diseases, 43
EBV screening, 45
screening for blood-borne viral infections, 43
screening for DENV infection, 44
screening for hepatitis C virus, 44
for surveillance and to monitor measles, 45
DBS biomarker assays, 118–119
DBSC, *see* Dried blood spot card
DBS cards, 169–170, 333
DBS elution protocol
elution buffers, 46
filter paper, type of, 47
size, 46
DBS internal standard addition, 257–258
DBS-LC-MS/MS method
of acetaminophen, 173
card selection and assay sensitivity assessment, 169–170, 333
challenges, 169
hematocrit effect, 170
instable analyte, 173–174
sample extraction, 171–173
sample inhomogeneity, impact of, 171
stable analyte, 173–174
validation
accuracy and precision, 174–175
blood volume impact, 176
carryover, 175
dilution integrity, 175
incurred sample reanalysis, 175
matrix effect and recovery, 174
sensitivity and selectivity, 174
spreadability, 176
stability, 175–176
volume influence, 170–171
DBS-LC-MS/MS methods, 190
DBS-PCR technologies
for cystic fibrosis genotyping, 33, 33*t*
for detection of infectious diseases, 34, 35*t*
expansion of, 33–34
HIVDR monitoring in resource-limited settings, 34–35
HIV drug resistance surveillance, 34–35, 35*t*
nucleic acid extraction methods, 33
quantitation of infectious agent particles, 35–36
in sequencing-based molecular epidemiology analyses, 34
DDT, *see* Dichlorodiphenyltrichloroethane
Delta No. 310, 4
DEMI, *see* Desorption electrospray/metastable-induced ionization
Dengue virus (DENV), 44
DENV, *see* Dengue virus
Derivatization techniques, 351
in DBS analysis, 345
DBS sample workup procedure, 345–347, 346*t*
modifications
direct derivatization techniques, 349–351
extraction–evaporation–derivatization, 347–348
two-step extraction–derivatization, 348–349
DESI, *see* Desorption electrospray ionization
Desorption atmospheric pressure photoionization (DAPPI), 282, 286–287
Desorption corona beam ionization (DCBI), 287
Desorption electrospray ionization (DESI), 278–280
Desorption electrospray/metastable-induced ionization (DEMI), 289
Desorption sonic spray ionization (DeSSI), 282
DeSSI, *see* Desorption sonic spray ionization
Dexamethasone, 238
Dextromethorphan (CYP2D6), 202
Dichlorodiphenyldichloroethylene, 134
Dichlorodiphenyltrichloroethane (DDT), 132
Diclofenacacyl-glucuronide, 183
Dielectric barrier discharge ionization (DBDI), 287–288, 288*f*
Digital microfluidics (DMF)
assays on DBS and whole blood
fluorometric assays, 326–327
immunoassays, 327
metabolite analysis, 327
PCR, 327
cartridge, 327*f*
enzymatic assays for LSDs, 328
multiplex assay protocol for five LSDS, 328
results and discussion
enzymatic assays for LSDs, 328
multiplex assay protocol for five LSDS, 329–330
technology, 325
Direct analysis in real time (DART), 284–286, 284*f*
Direct analysis technique, 250*t*–253*t*, 290*t*
and automated DBS extraction, 263–264, 291
automation and throughput, 255–256
communication, 263
compatibility, 263
cross-contamination, 259–260
for DBS, 247–263
and DBS direct elution, 264–265, 291–292
digital microfluidics, 276–277
LMJ-SSP approach, 267–269
online DBS analysis, 265–267
sealing surface sampling probe, 269–276
DBS internal standard addition, 257–258
and DBS manual extraction, 249*f*
dilution and linear dynamic range, 260–261
direct desorption, 277–278, 292
hematocrit, 261
laser desorption (ablation)/ionization techniques
ELDI, 289
LAESI/IR-LADESI, 288–289
LIAD, 289
MALDESI, 289
LC separation, 259
liquid and gas jet desorption/ionization
DAPPI, 282
desorption electrospray ionization, 278–280

Direct analysis technique (*Continued*)
EASI, 282
EESI, 282
ELDI, 282
MALDESI, 282
ND-EESI, 282
paper spray, 280–282
manual extraction of wet plasma *versus* DBS in pharmaceutical drug development, 246–247
reproducibility, 260
robustness, 260
sampling in wet plasma *versus* DBS in pharmaceutical drug development, 245–246
selectivity, 259
sensitivity, 258–259
theoretical workflow, 254*f*
thermal desorption/ionization, 282–283
combined desorption/ionization techniques, 287–288
thermal multistep ionization techniques, 283–287
visual recognition and sample identification, 256–257
Direct desorption, 277–278
Direct ELISA, 42, 42*f*
DMF, *see* Digital microfluidics
DMS, *see* Dried matrix spot
DNA microextraction method, 33
from DBS specimens on filter paper, 33*t*
and Roche Amplicor HIV-1 DNA-PCR kit, 34*t*
Donut dilution, 198
DPS, *see* Dried plasma spots
Dried biological samples (DXS), 142
Dried blood spot card (DBSC)
cassette construction, 19
and DBS transportation, 19
and filter paper, 16–17
filter paper cards construction, 18–19
filter paper pre-treatment, 19–20
guidelines and printing, 17–18
flexographic method, 17
printing adaptations, 18
"tip-on" method, 18–19
wraparound cover, 19
Dried blood spots (DBS)
advantages, 8–9
class II medical device, 55–56
detection of metabolic disease in, 66–70
disadvantages, 9–12
in environmental population studies, 130–137
and filter paper, 3–7
historical applications, 6–8
history, 3–6
for HIVDR genotyping, 34
for HIV-related epidemiological studies, 77, 78*t*
in human health studies, 114–126
liquid *versus* dried blood specimens, 56–57
for molecular diagnostics, 32
monitoring, PCR technologies, 32
and packaging, 9
paper with biological matrix and substrates, 57–59, 58*f*
random sampling errors, 9
analyte extraction, 12
anticoagulants, 11
blood source, 12
chromatographic effects, 11
filter paper effects, 10
hematocrit effects, 11
humidity effects, 11
sample volume, 10–11
sample collection, 59
biosafety and infection control, 22
capillary tube, 25–26, 26*f*
common errors in, 29–30
drying, 27
labeling, 27
reagents and materials, 21–22
specimen collection, 22–27
umbilical cord blood, 27
venous blood, 27
sample storage, 27, 29*f*
common errors in, 29–30
sample transportation, 27–29
common errors in, 29–30
transportation, 19
use in forensic science, 140–149
use in pharmaceutical, 153–157
Dried matrix spot (DMS) analysis, 235–240
automation, 238–239
color-indicating techniques, 237
internal standard, 237–238, 239*f*
Dried plasma spots, 219
Dried plasma spots (DPS), 86
Drug resistance testing, 82–85
Duchenne muscular dystrophy, 7
DXS, *see* Dried biological samples

EADESI, *see* Electrodeassisted DESI
EASI, *see* Easy ambient sonic spray ionization
Easy ambient sonic spray ionization (EASI), 282
Eaton and Dikeman No. 613, 4
EBV, *see* Epstein–Barr virus
EDTA, *see* Ethylene diamine tetraacetic acid
EE, *see* Elution efficiency
EESI, *see* Extractive electrospray ionization
EIA, *see* Enzyme immunoassays
ELDI, *see* Electrospray-assisted laser desorption ionization
Electrodeassisted DESI (EADESI), 282
Electrophoresis, 60–61
Electrospray-assisted laser desorption ionization (ELDI), 282, 289
Electrospray ionization tandem mass spectrometry (ESI-MS/MS), 314, 346
ELISA, *see* Enzyme-linked immune-sorbent assays
Elution, 100, 106–107
Elution efficiency (EE), 197
Environmental population studies
advantages of using DBS for biomonitoring, 133
biomarkers of environmental chemicals, 132–133

biomonitoring using DBS, 131–132
disadvantages of using DBS, 133–134
introduction, 130–131
residual DBS storage, 133
Enzyme immunoassays (EIAs), 40
for HCV diagnosis, 40
Enzyme-linked immune-sorbent assays (ELISAs) technique, 40
antibody detection, 41–42, 42*f*
antigen detection, 42–43
applications (using DBS), 43–46
competitive ELISA, 42–43
direct ELISA, 42, 42*f*
optimization and validation (using DBS specimens)
DBS elution protocol, 46–47
defining cutoff values, 47–48
performance, influencing factors
anticoagulant, 48
cross-contamination, 48
DBS preparation, 48
punching, 48
sandwich ELISA, 41–42, 42*f*
Epstein–Barr virus (EBV), 44–45, 123
Erythrocyte acetylcholinesterase, 6
Erythrocyte protoporphyrin, 7
Erythrocyte volume fraction (EVF), 170
ESI-MS/MS, *see* Electrospray ionization tandem mass spectrometry
Ethylene glycol, 143
Everolimus, 221
EVF, *see* Erythrocyte volume fraction
E viruses, 32
Extraction techniques, for DBS analysis
liquid–liquid extraction, 165
protein precipitation, 164–165
solid-phase extraction, 165–166
Extractive electrospray ionization (EESI), 282

FAPA, *see* Flowing atmospheric pressure afterglow
Filter paper, 3, 16
advantages, 4–5
for biomonitoring, 131–132
and DBSC, 16–20
for lysozyme estimation, 7
pre-treatment, 19–20
for sample collection, 6*t*
for screening children, 4
sources, 16–17
urine samples, 6
for whole blood and serum, 4
FLB, *see* Flurbiprofen
Flexographic method, 17
Flowing afterglow-atmospheric pressure glow discharge desorption ionization (APGDDI), 287
Flowing atmospheric pressure afterglow (FAPA), 287
Fluorometry, 60
Flurbiprofen (CYP2C9), 202
Flurbiprofen (FLB), 231
Forensic science, application of DBSs
drugs of abuse detected, 145*t*–146*t*
amphetamines, 147
benzodiazepines, 147
cannabinoids, 148
cocaine and its metabolites, 147
opioids, 148
forensic metabolic pathology, 144*f*
metabolic autopsy, 141–143
forensic toxicology, 141–146
of drug detection in DBS, 148
FTA DMPK-A, 336
FTA DMPK-B, 336
FTA DMPK-C, 336
FTA DMPK cards, 335–336
and DBS, 341–342
indicating *versus* non-indicating, 336–337
protein analytes detection, 341
for storage of DNA in blood, 337–339
for storage of RNA in blood, 339–341

GA-I, *see* Glutaric acidemia type I
GAL, *see* Galactosemia
Galactosemia (GAL), 54
GE DMPK A, 181, 182, 188–189
GE DMPK B, 181, 182, 188–189
GE DMPK C, 181, 182, 188–189
Glucuronide metabolites, 183
Glutaric acidemia type I (GA-I), 143
Grade 903, 5

HBV, *see* Hepatitis B virus
HCC, *see* Hepatocellular carcinoma
HCV, *see* Hepatitis C virus
Hematocrit (HT), 170, 218
Hemoglobinopathies, 53
Hepacivirus, 95
Hepatitis B virus (HBV), 43–44
DBS for diagnosis and management
alpha-fetoprotein, 108
elution, 106–107
epidemiologic studies, 108–109
serological assays, 107
specimen collection, 104
storage and stability, 104–106
viral load assay, 107–108
diagnosis, 103
epidemiology, 102–103
transmission, 103
Hepatitis C virus (HCV), 44
diagnosis
nucleic acid tests, 96
serological tests, 95–96
dried blood use, 96–102, 97*t*
antibody detection assays, 100–101
elution, 100
sample collection and drying, 99
storage, 99–100

Hepatitis C virus (HCV) (*Continued*)
  epidemiology, 95
  genotype determination, 96
  introduction, 95
Hepatocellular carcinoma (HCC), 101
Herpes simplex virus (HSV), 32
HHV-6, *see* Human herpesvirus 6
Highly pathogenic avian influenza virus (HPAIV), 110
High-performance liquid chromatography (HPLC), 8, 60
HIV, *see* Human immunodeficiency virus
HIVDR, *see* HIV drug resistance
HIV drug resistance (HIVDR), 34
  genotyping analysis, 34, 35*t*
Homocystinuria (HCys), 54
HPAIV, *see* Highly pathogenic avian influenza virus
HPLC, *see* High-performance liquid chromatography
HSV, *see* Herpes simplex virus
HT, *see* Hematocrit
HTLV, *see* Human T-lymphotropic virus
HTS PAL, 315
Human African trypanosomiasis (HAT), 45
Human health studies, application of DBS in
  case studies
    DBS in demographic and health surveys, 119–122
    DBS in health initiatives for men study, 122–124
  clinical and non-clinical health research, 114–115
  DBS use in public and population health settings, 115–116
  methodological considerations
    DBS biomarker assays, 118–119
    DBS collection strategies, 116–117
    DBS storage, 117–118
    DBS *vs.* venous blood assay results, 118
    limitations of DBS, 119
    specimen treatment, biohazard, and safety issues, 117
Human herpesvirus 6 (HHV-6), 32
Human immunodeficiency virus (HIV), 32
  in Africa, 76–87, 77*f*
  DBS-based ELISA testing for, 43
  detection in DBS
    drug resistance testing, 82–85
    surveillance, 77–80, 81*t*
    viral load monitoring, 80–82
  epidemiological studies in resource-limited settings, 76–87
    emerging strategies, 86–87
  infection among infants, diagnosis of, 85–86
  and PMTCT, 85–86
  storage and transport of samples, 77
Human T-lymphotropic virus (HTLV), 32
4-Hydroxyflurbiprofen (OH-FLB), 231

IC-MS/MS, *see* Ion chromatography-MS/MS
IDMS, *see* Isotope dilution MS
IDSR, *see* Integrated Disease Surveillance
IEF, *see* Isoelectric focusing
Immunoassays, 8, 60
Inflammatory markers, 191, 221
Influenza
  clinical features, 109
  diagnosis, 109
  dry respiratory sample, 109–110
  introduction, 109
Integrated Disease Surveillance (IDSR), 45
Intravenous drug users (IDU), 101
Ion chromatography-MS/MS (IC-MS/MS), 136
Isoelectric focusing (IEF), 61
Isoleucine, 7
Isotope dilution MS (IDMS), 63

KAI-9803, 191

Laboratory information management systems (LIMS), 161, 256
Lactobacillus arabinosus, 7
LAESI, *see* Laser ablation electrospray ionization
Laser ablation electrospray ionization (LAESI), 288–289
Laser diode thermal desorption (LDTD), 284
Laser-induced acoustic desorption (LIAD), 289
LC coupled to fluorescence (LC-FLUO), 344, 345
LC-MS/MS assay, 197
LC-MS/MS bioanalysis, 173
LDTD, *see* Laser diode thermal desorption
LESA, *see* Liquid extraction surface analysis
Leu, *see* Leucine
Leucine (Leu), 7, 53
LIAD, *see* Laser-induced acoustic desorption
LIMS, *see* Laboratory information management systems
Liquid extraction surface analysis (LESA), 191
Liquid–liquid extraction (LLE), 165
Liquid microjunction (LMJ), 232
LLE, *see* Liquid-liquid extraction
LLOQ, *see* Lower limit of quantitation
LMJ, *see* Liquid microjunction
Lower limit of quantitation (LLOQ), 170, 174, 175, 217
Low-temperature plasma probe (LTP), 287
LSD, *see* Lysosomal storage diseases
LTP, *see* Low-temperature plasma probe
Lysosomal storage diseases (LSDs), 8, 54

Malaria, 45
MALDESI, *see* Matrix-assisted laser desorption electrospray ionization
Maple syrup urine disease (MSUD), 53
Matrix-assisted laser desorption electrospray ionization (MALDESI), 282, 289
Maximum feasible dose (MFD), 195
Maximum tolerated dose (MTD), 195
MDA, *see* 3,4-Methylenedioxyamphetamine
MDMA, *see* 3,4-Methylenedioxymethamphetamine
MDZ, *see* Midazolam
Measles, 32, 45
Medium-chain acyl CoA dehydrogenase (MCAD) deficiency, 142
Metabolic disease detection, in DBS, 66–67
  chemical interferences in NBS
    anticoagulants, 67–69
    premature or sick infants, 69
    total parenteral nutrition, 69
  lab test to diagnosis, 69–70
  and NBS, 67
  second-tier tests (accuracy improvement), 70

Metastable-induced chemical ionization (MICI), 289
Methionine (Met), 54
3,4-Methylenedioxyamphetamine (MDA), 147
3,4-Methylenedioxymethamphetamine (MDMA), 147
Methylmalonic acid (MMA), 317
  disorders, 143
MFD, *see* Maximum feasible dose
MICI, *see* Metastable-induced chemical ionization
Microsampling, 196
Microspheres, 60
Midazolam (MDZ), 202
Minoxidil-NO, 183
MMA, *see* Methylmalonic acid
Molar ratios, 66
Morphine, 148
MSUD, *see* Maple syrup urine disease
MTD, *see* Maximum tolerated dose

Nano-DESI, 282
Nasopharyngeal carcinoma (NPC), 44–45
Nasopharyngeal (NP), 109
NAT, *see* Nucleic acid testing
National Health and Nutrition Examination Survey (NHANES), 130
NBS, *see* Newborn screening
ND-EESI, *see* Neutral desorption extractive electrospray ionization
Neutral desorption extractive electrospray ionization (ND-EESI), 282
Neutral loss experiment, 61
Newborn screening (NBS), 3
  analytical methods in
    BIA, 59–60
    chromatography, 60
    electrophoresis, 60–61
    fluorometry, 60
    fundamental premise, 59
    immunoassays, 60
    molecular techniques, 66
    tandem mass spectrometry, 61–66
  applications of DBS, *see* Metabolic disease detection, in DBS
    DBS fundamentals, 55–59. *See also* Dried blood spots
    history and chronology, 53–54
    prognostications and predictions, 71
  for congenital hypothyroidism, 7
  expansion of, 53–54
  and metabolic screening, 70–71
Newborn Screening Quality Assurance Program (NSQAP), 55
NHANES, *see* National Health and Nutrition Examination Survey
Nifedipine, 185
nonhuman species, 137
NP, *see* Nasopharyngeal
NPC, *see* Nasopharyngeal carcinoma
NSQAP, *see* Newborn Screening Quality Assurance Program
Nucleic acid testing (NAT), 86, 96

Octanoic acid, 142
OH-FLB, *see* 4-Hydroxyflurbiprofen
Omeprazole (CYP2C19), 185, 202
Orbitrap Velos mass spectrometer, 191

Packed cell volume (PCV), 170
Packed red blood cells (PRBC), 99
PADI, *see* Plasma-assisted desorption/ionization
Paper manufacturing, 335
Paper spray ionization
  chemical interactions, 300–304
  chronograms illustration, 307
  fundamentals, 299–300
  ion chronograms, 305*f*
  nanoESI with, 299–300, 300*f*, 301*f*
  point-of-care analysis, 307–311
  quantitation of drugs and metabolites, 304–307
  spectra of cocaine, 304*f*
  tolerance, 300, 302*f*
Paper spray (PS), 280–282
Parenteral nutrition, 69
PBMC, *see* Peripheral blood mononuclear cells
PCB, *see* Printed circuit board
PCDBS, *see* Precut dried blood spot
PCR, *see* Polymerase chain reaction
PCV, *see* Packed cell volume
PD, *see* Pharmacodynamic
PEPFAR, *see* Emergency Plan for AIDS Relief
Perchlorate, 136
Perfluorohexane sulfonate (PFHxS), 136
Perfluorononanoate (PFNA), 136
Perfluorooctane sulfonamide (PFOSA), 136
Perfluorooctane sulfonate (PFOS), 136
Perfluorooctanoate (PFOA), 136
Peripheral blood mononuclear cells (PBMCs), 101
PFHxS, *see* Perfluorohexane sulfonate
PFNA, *see* Perfluorononanoate
PFOA, *see* Perfluorooctanoate
PFOS, *see* Perfluorooctane sulfonate
PFOSA, *see* Perfluorooctane sulfonamide
Pharmaceuticals, DBS in, 156
  early developments, 153–154
  hurdles, 155–156
  investigations, 154
  quantitative bioanalysis, 233
    automated blood collection and spotting, 230
    automated method development, 229–230
    DBS analysis, automation in, 230–232
    DBS automation issues, 232–233
  validation requirements, 154
Pharmacodynamic (PD), 202
Pharmacokinetic (PK), 188, 196, 217
Phe, *see* Phenylalanine
Phenylalanine, 53, 216
Phenylketonuria (PKU), 5
  blood spot screening for, 7
  DBS for, 53
Phthalates, 132
Pig synovial fluid, 238
Pivalic acid, 69
PK, *see* Pharmacokinetic

PKU, *see* Phenylketonuria
Plasma, 56–57, 82
Plasma-assisted desorption/ionization (PADI), 287
Plasma butyrylcholinesterase, 6
PMTCT, *see* Prevention of mother-to-child HIV transmission
Polymerase chain reaction (PCR), 8
Postmortem DBS, 142, 143
Postmortem metabolic profile, 142
Potassium oxalate, 11
PPT, *see* Protein precipitation
PRBC, *see* Packed red blood cells
Preclinical studies
  bioanalytical assay development
    DBS sample dilutions, 197–198
    DBS sample preparation and analyte elution efficiency, 197
    DBS samples *vs. ex vivo* calibration standards/quality control samples, 199–200
    impact of hematocrit on bioanalytical measurement and potential solutions, 198–199
    impact of hematocrit on blood-to-plasma concentration ratio, 198
    sample storage stability, 200
  exposure measurement
    blood collection methods, 196–197
    limited blood volume, 196
Precut dried blood spot (PCDBS), 192
Premature or sick infants, 69
Prevention of mother-to-child HIV transmission (PMTCT), 85–86
Printed circuit board (PCB), 325, 326*f*
Prolab SCAP system, 292
protein adducts measurement, 136
Protein precipitation (PPT), 164–165
PS, *see* Paper spray
Pseudo isotope dilution MS, 62–64, 65*f*
Punching
  devices, 160–162, 161*f*
  positions effects, 162–163
  volumetric issue, 162

Quality management system (QMS), 333–334
  process flow for manufacture of Whatman FTA™ DMPK media, 334*f*
Quasispecies, 95

Radial immunodiffusion (RID), 7
Radio frequency identification technology (RFID), 18
RBP, *see* Retinol-binding protein
Recommended uniform screening panel (RUSP), 8
Retinol-binding protein (RBP), 117
RFID, *see* Radio frequency identification technology
RID, *see* Radial immunodiffusion
Roche Amplicor HIV-1 DNA-PCR kit, 34
Rosiglitazone (CYP2C8), 202
Rubella, 32
RUSP, *see* Recommended uniform screening panel

Sandwich ELISA, 41–42, 42*f*
Schleicher and Schuell No. 589 paper, 4
Schleicher and Schuell No. 595 paper, 4
Schleicher and Schuell No. 903 paper, 16
SCID, *see* Severe combined immunodeficiency syndrome
SDS, *see* Sodium dodecyl sulfate
Sealing surface sampling probe (SSSP)
  automation and development, 272–274
  concept, 269–272
  robustness, 274
  sensitivity and extraction solution optimization, 274–276
Serological tests, 95–96
Severe combined immunodeficiency syndrome (SCID), 8, 53, 327
Sex hormone-binding globulin (SHBG), 123
SHBG, *see* Sex hormone-binding globulin
Short-chain acyl CoA dehydrogenase (SCHAD) deficiency, 142
Sickle cell disease, 54
Simian T-lymphotropic virus (STLV), 32
Sodium dodecyl sulfate (SDS), 170
Sodium heparin, 11
Solid-phase extraction (SPE), 165–166
Solid-phase micro extraction (SPME), 348
Solvation delay, 279
SPE, *see* Solid-phase extraction
Specimen collection, for DBS
  newborn heelsticks, 22–23, 23*f*
  older children and adults, 23–25
  for PCR, 32–36
  procedures, 22
Split barcode, 19*f*
SPME, *see* Solid-phase micro extraction
STLV, *see* Simian T-lymphotropic virus
SUAC, *see* Succinylacetone
Succinylacetone (SUAC), 54, 62
Sustained virologic response (SVR), 96
SVR, *see* Sustained virologic response
Syphilis test, 4

Tacrolimus, 218
Tandem mass spectrometry (MS/MS), 8, 53, 61, 142
  of acylcarnitines, 62, 142
  analysis of amino acids, 61–62
  DBS samples analysis, 168–169
  metabolic profiles *versus* quantification, 64–66
  precision and quantification improvements, 66
  pseudo isotope dilution MS, 62–64
  succinylacetone, 62
  thyroxine, 62
TBG, *see* Thyroidbinding globulin
T-cell receptor excision circles (TRECs), 8, 327
TCEP, *see* tris-(2-carboxyethyl)-phosphine
TD/APCI, *see* Thermal desorption atmospheric pressure chemical ionization
TDM, *see* Therapeutic drug monitoring
Tetrahydrocannabinol (THC), 148
THC, *see* Tetrahydrocannabinol
Therapeutic drug monitoring (TDM), 143, 220*t*
  advantages of DBS sampling, 216–217
  DBS sample collection in, 217

DBS sampling
DBS biomarker measurement, 221–222
drug monitoring and DBS, 219–220
pediatric drug monitoring and DBS, 220–221
transplantation drug monitoring and DBS, 221
disadvantages of DBS sampling, 217
PK/PD considerations of DBS sampling in, 217
blood plasma distribution, 218
DBS and plasma samples, 218–219
dried plasma spots, 219
Hematocrit impact, 218
Thermal desorption atmospheric pressure chemical ionization (TD/APCI), 283
Thyroidbinding globulin (TBG), 62
Thyroid-stimulating hormone (TSH), 54, 327
Thyroxine, 62
"Tip-on" method, 18–19
TK, *see* Toxicokinetics
Total parenteral nutrition (TPN), 69
Toxicokinetics (TK), 195
TPN, *see* Total parenteral nutrition
TRECs, *see* T-cell receptor excision circles
tris-(2-carboxyethyl)-phosphine (TCEP), 173
TSH, *see* Thyroid-stimulating hormone
Tyrosinemia type I, 54
Tyrosinemia type II, 54

Ultra-performance liquid chromatography (UPLC), 315
UPLC, *see* Ultra-performance liquid chromatography
US President's Emergency Plan for AIDS Relief (PEPFAR), 34

Valine, 7
Valproic acid, 69
Viral load monitoring, 80–82
Viral transport medium (VTM), 109–110
ViveST (sample collection system), 99
VTM, *see* Viral transport medium

WB samples, 207
Whatman C collection card, 236*f*
Whatman FTA DMPK cards, 336
Whatman No. 1 paper, 4
Whatman No. 3 paper, 4
WHO, *see* World Health Organization
World Health Organization (WHO), 19

Xenobiotics, 195
Xevo TQ mass spectrometer, 315